Multivariate Analysis

Multivariate Analysis

Second Edition

Kanti V. Mardia, John T. Kent, and Charles C. Taylor

WILEY

This second edition first published 2024

Edition History
© 1st Edition 1979 Academic Press

Registered Offices
John Wiley & Sons, Inc., 111 River Street, Hoboken, NJ 07030, USA
John Wiley & Sons Ltd, The Atrium, Southern Gate, Chichester, West Sussex, PO19 8SQ, UK

For details of our global editorial offices, customer services, and more information about Wiley products visit us at www.wiley.com.

Wiley also publishes its books in a variety of electronic formats and by print-on-demand. Some content that appears in standard print versions of this book may not be available in other formats.

Library of Congress Cataloging-in-Publication Data

Names: Mardia, K. V., author. | Kent, J. T. (John T.), joint author. | Taylor, C. C (Charles C.), joint author.
Title: Multivariate analysis / Kanti V. Mardia, John T. Kent, and Charles C. Taylor.
Description: Second edition. | Hoboken, NJ : Wiley, 2024. | Series: Wiley series in probability and statistics | Includes bibliographical references and index.
Identifiers: LCCN 2023002460 (print) | LCCN 2023002461 (ebook) | ISBN 9781118738023 (cloth) | ISBN 9781118892527 (adobe pdf) | ISBN 9781118892510 (epub)
Subjects: LCSH: Multivariate analysis.
Classification: LCC QA278 .M36 2023 (print) | LCC QA278 (ebook) | DDC 519.5/35–dc23/eng/20230531
LC record available at https://lccn.loc.gov/2023002460
LC ebook record available at https://lccn.loc.gov/2023002461

Cover Design: Wiley
Cover Images: © Liyao Xie/Getty Images, Courtesy of Kanti Mardia

Set in 9.5/12.5pt STIXTwoText by Straive, Chennai, India
Printed and bound by CPI Group (UK) Ltd, Croydon, CR0 4YY

C9781118738023_030624

To my daughters ***Bela*** *and* ***Neeta***

— *with Jainness* (Kanti V. Mardia)

To my son ***Edward*** *and daughter* ***Natalie***

(John T. Kent)

To my wife ***Fiona*** *and my children* ***Mike, Anna, Ruth, and Kathryn***

(Charles C. Taylor)

Contents

Epigraph

> *Everything is related with every other thing, and this relation involves the emergence of a relational quality. The qualities cannot be known* a priori, *though a good number of them can be deduced from certain fundamental characteristics.*
>
> – Jaina philosophy

The *Jaina Philosophy of Non-Absolutism* by S. Mookerjee, q.v.
Mahalanobis (1957).

Preface to the Second Edition

For over 40 years the first edition of this book (which was also translated into Persian) has been used by students to acquire a basic knowledge of the theory and methods of multivariate statistical analysis. The book has also served the wider statistical community to further their understanding of this field. Plans for the second edition started almost 20 years ago, and we have struggled with questions about which topics to add – something of a moving target in a field that has continued to evolve in this new era of artificial intelligence (AI) and "big data". Since the first edition was published, multivariate analysis has been developed and extended in many directions. This new edition aims to bring the first edition up to date by substantial revision, rewriting, and additions, while seeking to maintain the overall length of the book. The basic approach has been maintained, namely a mathematical treatment of statistical methods for observations consisting of several measurements or characteristics of each subject and a study of their properties. The core topics, and the structure many of the chapters, have been retained.

Briefly, for those familiar with the first edition, the main changes (in addition to updating material in several places) are:

- a new section giving Notation, Abbreviations, and Key Ideas used through the book;
- a new chapter introducing some nonnormal distributions. This includes new sections on elliptical distributions and copulas;
- a new chapter covering an introduction to *graphical models*;
- a completely rewritten chapter that begins from discriminant analysis and extends to nonparametric methods, classification and regression trees, logistic discrimination, and multilayer perceptrons. These topics are commonly grouped into the heading of *supervised learning*;
- the above chapter focuses on data in which group memberships are known, whereas "unsupervised learning" has more traditionally been known as cluster analysis, for which the current Chapter 14 has also been substantially updated to reflect recent developments;
- a new (final) chapter introduces some approaches to *high-dimensional data* in which the number of variables may exceed the number of observations. This includes shrinkage methods in regression, principal components regression, partial least squares regression, and functional data analysis;
- further development of discrete aspects, including log-linear models, the EM algorithm for mixture models, and correspondence analysis for contingency tables.

As a consequence of the above new and extended/revised chapters and in order to save space, we have omitted some material from this edition:

- the chapter on econometrics, since there are now dedicated books with an emphasis on statistical aspects (Maddala and Lahiri, 2009); (Wooldridge, 2019);
- the chapter on directional statistics, since there are now related dedicated books by one of the authors (Dryden and Mardia, 2016); (Mardia and Jupp, 2000).

Further changes to this Edition, bringing many subjects up to date, include new graphical representations (Chapter 1), an introduction to the matrix normal distribution (Chapters 2 and 5), elliptical distributions and copulas (Chapter 3), robust estimators for location and dispersion (Chapter 5), a revision of correspondence analysis and biplots (Chapter 9), and projection pursuit and independent component analysis (Chapter 9).

The figures in the first edition have been redrawn in their original style, using the statistical package R. A new Appendix C contains some specific R (R Core Team, 2020) commands applicable to most of the matrix algebra used in the book. In addition, an online addendum to Appendix C contains the data files used in this book as well as the R commands used to obtain the calculations for the examples and figures. This public repository is at github.com/charlesctaylor/MVAdata-rcode. In many cases, we have chosen to use base R functions to mimic the equations used in the text in preference to more "black-box" R functions. Note that intermediate steps in the calculation are generally rounded *only* for display purposes.

Multivariate analysis continues to be a research area of active development. We note that the Journal of Multivariate Analysis, in its 50th Anniversary Jubilee Edition (von Rosen and Kollo, 2022), has published a volume that describes the current state of the art and contains review papers. Beyond mainstream multivariate statistics, there have been developments in the applied sciences; one example in morphometrics is Bookstein (2018).

The first edition was published by Academic Press, and we are grateful to John Bibby for his contributions to that edition. For this edition, we thank the many readers who have offered their advice and suggestions. In particular, we would like to acknowledge the help of Susan Holmes for extensive discussions about a new structure as well as a draft of correspondence analysis material for Chapter 9.

We are extremely grateful to Wiley for their patience and help during the writing of the book, especially Helen Ramsey, Sharon Clutton, Richard Davies, Kathryn Sharples, Liz Wingett, Kelvin Matthews, Alison Oliver, Viktoria Hartl-Vida, Ashley Alliano, Kimberly Monroe-Hill, and Paul Sayer. Secretarial help at Leeds during the initial development was given by Christine Rutherford and Catherine Dobson.

Kanti would like to thank the Leverhulme Trust for an Emeritus Fellowship and Anna Grundy of the Trust for simplifying the administration process. Finally, he would like to express his sincere gratitude to his family for their continuous love, support, and tolerance.

We would be pleased to hear about any typographical or other errors in the text.

May 2022

Kanti V. Mardia
University of Leeds, Leeds, UK and
University of Oxford, Oxford, UK

John T. Kent
University of Leeds, Leeds, UK

Charles C. Taylor
University of Leeds, Leeds, UK

Preface to the First Edition

Multivariate Analysis deals with observations on more than one variable where there is some inherent interdependence between the variables. With several texts already available in this area, one may very well enquire of the authors as to the need for yet another book. Most of the available books fall into two categories, either theoretical or data analytic. The present book not only combines the two approaches but also emphasizes modern developments. The choice of material for the book has been guided by the need to give suitable matter for the beginner as well as illustrating some deeper aspects of the subject for the research worker. Practical examples are kept to the forefront, and, wherever feasible, each technique is motivated by such an example.

The book is aimed at final year undergraduates and postgraduate students in Mathematics/Statistics with sections suitable for practitioners and research workers. The book assumes a basic knowledge of Mathematical Statistics at undergraduate level. An elementary course on Linear Algebra is also assumed. In particular, we assume an exposure to Matrix Algebra to the level required to read Appendix A.

Broadly speaking, Chapters 1–6 and 12 can be described as containing direct extensions of univariate ideas and techniques. The remaining chapters concentrate on specifically multivariate problems that have no meaningful analogs in the univariate case. Chapter 1 is primarily concerned with giving exploratory analyses for multivariate data and briefly introduces some of the important techniques, tools, and diagrammatic representations. Chapter 2 introduces various distributions together with some fundamental results, whereas Chapter 3 concentrates exclusively on normal distribution theory. Chapters 4–6 deal with problems in inference. Chapter 7 [no longer included] gives an overview of Econometrics, while Principal Component Analysis, Factor Analysis, Canonical Correlation Analysis, and Discriminant Analysis are discussed from both theoretical and practical points of view in Chapters 8–11. Chapter 12 is on Multivariate Analysis of Variance, which can be better understood in terms of the techniques of previous chapters. The later chapters look into the presently developing techniques of Cluster Analysis, Multidimensional Scaling, and Directional Data [no longer included].

Each chapter concludes with a set of exercises. Solving these will not only enable the reader to understand the material better but will also serve to complement the chapter itself. In general, the questions have in-built answers, but, where desirable, hints for the solution of theoretical problems are provided. Some of the numerical exercises are designed to be run on a computer, but as the main aim is on interpretation, the answers are provided. We found NAG routines and GLIM most useful, but nowadays any computer center will have some suitable statistics and matrix algebra routines.

There are three Appendices A, B, and C, which, respectively, provide a sufficient background of matrix algebra, a summary of univariate statistics, and some tables of critical values. The aim of Appendix A on Matrix Algebra is not only to provide a summary of results but also to give sufficient guidance to master these for students having little previous knowledge. Equations from Appendix A are referred to as (A.x.x) to distinguish them from (l.x.x), etc. Appendix A also includes a summary of results in n-dimensional geometry that are used liberally in the book. Appendix B gives a summary of important univariate distributions.

The reference list is by no means exhaustive. Only directly relevant articles are quoted, and for a fuller bibliography, we refer the reader to Anderson et al. (1972) and Subrahmaniam and Subrahmaniam (1973). The reference list also serves as an author index. A subject index is provided.

The material in the book can be used in several different ways. For example, a one-semester elementary course of 40 lectures could cover the following topics. Appendix A; Chapter 1 (Sections 1.1–1.7); Chapter 2 (Sections 2.1–2.5); Chapter 3 (Sections 3.4.1, 3.5, and 3.6.1, assuming results from previous sections, Definitions 3.7.1 and 3.7.2); Chapter 4 (Section 4.2.2); Chapter 5 (Sections 5.1, 5.2.1a, 5.2.1b, 5.2.2a, 5.2.2b, 5.3.2b, and 5.5); Chapter 8 (Sections 8.1, 8.2.1, 8.2.2, 8.2.5, 8.2.6, 8.4.3, and 8.7); Chapter 9 (Sections 9.1–9.3, 9.4 (without details), 9.5, 9.6, and 9.8); Chapter 10 (Sections 10.1 and 10.2); Chapter 11 (Sections 11.1, 11.2.1–11.2.3, 11.3.1, and 11.6.1). Further material that can be introduced is Chapter 12 (Sections 12.1–12.3 and 12.6); Chapter 13 (Sections 13.1 and 13.3.1); Chapter 14 (Sections 14.1 and 14.2). This material has been covered in 40 lectures spread over two terms in different British universities. Alternatively, a one-semester course with more emphasis on foundation rather than applications could be based on Appendix A and Chapters 1–5. Two-semester courses could include all the chapters, excluding Chapters 7 and 15 on Econometrics and Directional Data, as well as the sections with asterisks. Mathematically orientated students may like to proceed to Chapter 2, omitting the data analytic ideas of Chapter 1.

Various new methods of presentation are utilized in the book. For instance, the data matrix is emphasized throughout, a density-free approach is given for normal theory, the union intersection principle is used in testing as well as the likelihood ratio principle, and graphical methods are used in explanation. In view of the computer packages generally available, most of the numerical work is taken for granted, and therefore, except for a few particular cases, emphasis is not placed on numerical calculations. The style of presentation is generally kept descriptive except where rigor is found to be necessary for theoretical results, which are then put in the form of theorems. If any details of the proof of a theorem are felt tedious but simple, they are then relegated to the exercises.

Several important topics not usually found in multivariate texts are discussed in detail. Examples of such material include the complete chapters on Econometrics, Cluster Analysis, Multidimensional Scaling, and Directional Data. Further material is also included in parts of other chapters: methods of graphical presentation, measures of multivariate skewness and kurtosis, the singular multinormal distribution, various nonnormal distributions and families of distributions, a density-free approach to normal distribution theory,

Bayesian and robust estimators, a recent solution to the Fisher–Behrens problem, a test of multinormality, a nonparametric test, discarding of variables in regression, principal component analysis and discrimination analysis, correspondence analysis, allometry, the jack-knifing method in discrimination, canonical analysis of qualitative and quantitative variables, and a test of dimensionality in MANOVA. It is hoped that coverage of these developments will be helpful for students as well as research workers.

There are various other topics that have not been touched upon partly because of lack of space as well as our own preferences, such as Control Theory, Multivariate Time Series, Latent Variable Models, Path Analysis, Growth Curves, Portfolio Analysis, and various Multivariate Designs.

In addition to various research papers, we have been influenced by particular texts in this area, especially Anderson (1958), Kendall (1975), Kshirsagar (1972), Morrison (1976), Press (1972), and Rao (1973). All these are recommended to the reader.

The authors would be most grateful to readers who draw their attention to any errors or obscurities in the book, or suggest other improvements.

January 1979

Kanti V. Mardia
John T. Kent
John C. Bibby

Acknowledgments from First Edition

First of all, we wish to express our gratitude to the pioneers in this field. In particular, we should mention M. S. Bartlett, R. A. Fisher, H. Hotelling, D. G. Kendall, M. G. Kendall, P. C. Mahalanobis, C. R. Rao, S. N. Roy, W. S. Torgeson, and S. S. Wilks.

We are grateful to the authors and editors who have generously granted us permission to reproduce figures and tables.

We are also grateful to many of our colleagues for their valuable help and comments, in particular Martin Beale, Christopher Bingham, Lesley Butler, Richard Cormack, David Cox, Ian Curry, Peter Fisk, Allan Gordon, John Gower, Peter Harris, Chunni Khatri, Conrad Leser, Eric Okell, Ross Renner, David Salmond, Cyril Smith, and Peter Zemroch. We are also indebted to Joyce Snell for making various comments on an earlier draft of the book, which have led to considerable improvements. We should also express our gratitude to Rob Edwards for his help in various facets of the book, for calculations, proofreading, diagrams, etc.

Some of the questions are taken from examination papers in British universities, and we are grateful to various unnamed colleagues. Since the original sources of questions are difficult to trace, we apologize to any colleague who recognizes a question of their own.

The authors would like to thank their wives, Pavan Mardia, Susan Kent, and Zorina Bibby.

Finally, our thanks go to Barbara Forsyth and Margaret Richardson for typing a difficult manuscript with great skill.

KVM
JTK
JMB

Notation, Abbreviations, and Key Ideas

Matrices and Vectors

- **Vectors** are viewed as column vectors and are represented using bold lower case letters. Round brackets are generally used when a vector is expressed in terms of its elements. For example,

$$\boldsymbol{x} = \begin{pmatrix} x_1 \\ \vdots \\ x_n \end{pmatrix}$$

in which the rth element or component is denoted x_r. The transpose of $\boldsymbol{x}$ is denoted $\boldsymbol{x}'$, so $\boldsymbol{x}' = (x_1, \ldots, x_n)$ is a row vector.
- **Matrices** are written using bold upper case letters, e.g. $\boldsymbol{A}$ and $\boldsymbol{\Gamma}$. The matrix $\boldsymbol{A}$ may be written as (a_{ij}) in which a_{ij} is the element of the matrix in row i and column j. If $\boldsymbol{A}$ has n rows and p columns, then the ith row of $\boldsymbol{A}$, written as a column vector, is

$$\boldsymbol{a}_i = \begin{pmatrix} a_{i1} \\ \vdots \\ a_{ip} \end{pmatrix}, \qquad i = 1, \ldots, n,$$

and the jth column is written as

$$\boldsymbol{a}_{(j)} = \begin{pmatrix} a_{1j} \\ \vdots \\ a_{nj} \end{pmatrix}, \qquad j = 1, \ldots, p.$$

Hence, A can be expressed in various forms,

$$\boldsymbol{A} = \begin{bmatrix} a_{11} & \cdots & a_{1p} \\ \vdots & \ddots & \vdots \\ a_{n1} & \cdots & a_{np} \end{bmatrix} = \begin{bmatrix} \boldsymbol{a}'_1 \\ \vdots \\ \boldsymbol{a}'_n \end{bmatrix} = \begin{bmatrix} \boldsymbol{a}_{(1)} & \cdots & \boldsymbol{a}_{(p)} \end{bmatrix}.$$

We generally use square brackets when a matrix is expanded in terms of its elements. Operations on a matrix $\boldsymbol{A}$ include
  - **transpose:** $\boldsymbol{A}'$
  - **determinant:** $|\boldsymbol{A}|$
  - **inverse:** $\boldsymbol{A}^{-1}$
  - **generalized inverse:** $\boldsymbol{A}^{-}$

where for the final three operations, $\boldsymbol{A}$ is assumed to be square, and for the inverse operation, $\boldsymbol{A}$ is additionally assumed to be nonsingular. Different types of matrices are given in Tables A.1 and A.3. Table A.2 lists some further matrix operations.

Random Variables and Data

- In general, a random vector and a nonrandom vector are both indicated using a bold lower case letter, e.g. $\boldsymbol{x} = (x_1, \dots, x_p)'$. Thus, the distinction between the two must be determined from the context. This convention is in contrast to the standard convention in statistics where upper case letters are used to denote random quantities, and lower case letters their observed values.
- The reason for our convention is that bold upper case letters are generally used for a data matrix $\boldsymbol{X}(n \times p)$, both random and fixed.
- In spite of the above convention, we very occasionally (e.g. parts of Chapters 2 and 10) use bold upper case letters $\boldsymbol{X} = (X_1, \dots, X_p)'$ for a random vector when it is important to distinguish between the random vector $\boldsymbol{X}$ and a possible value $\boldsymbol{x}$.
- The phrase "high-dimensional data" often implies $p > n$, whereas the phrase "big data" often just indicates that n or p is large.

Parameters and Statistics

Elements of an $n \times p$ data matrix $\boldsymbol{X}$ are generally written x_{ri}, where indices $r, s, \dots$, are used to label the *observations*, and indices $i, j, k, \dots$, are used to label the *variables*.

If the rows of a data matrix $\boldsymbol{X}$ are normally distributed with mean $\boldsymbol{\mu}$ and covariance matrix $\boldsymbol{\Sigma} = (\sigma_{ij})$, and $\boldsymbol{\Delta} = \text{Diag}\,(\boldsymbol{\Sigma})^{1/2}$, the following notation is used to distinguish various population and sample quantities:

Parameter	Sample	
$\boldsymbol{\mu}$	$\overline{\boldsymbol{x}}$	Mean vector
$\boldsymbol{\Sigma}$	$\boldsymbol{S} = \frac{1}{n}\boldsymbol{X}'\boldsymbol{X} - \overline{\boldsymbol{x}}\,\overline{\boldsymbol{x}}'$	Covariance matrix
	$\boldsymbol{S}_u = \frac{n}{n-1}\boldsymbol{S}$	Unbiased covariance matrix
$\boldsymbol{K} = \boldsymbol{\Sigma}^{-1}$	$\boldsymbol{K}^{\text{sample}} = \boldsymbol{S}^{-1}$	Concentration matrix
$\boldsymbol{P} = (\rho_{ij}) = \boldsymbol{\Delta}^{-1}\boldsymbol{\Sigma}\boldsymbol{\Delta}^{-1}$	$\boldsymbol{R} = (r_{ij})$	Correlation matrix

Distributions

The following notation is used for univariate and multivariate distributions. Appendix B summarizes the univariate distributions used in the Book.

$F(\cdot)$	cumulative distribution function/distribution function (d.f.)
$f(\cdot)$	probability density function (p.d.f.)
E	expectation
c.f.	characteristic function
d.f.	distribution function
T^2	Hotelling T^2
$N_p(\boldsymbol{\mu}, \boldsymbol{\Sigma})$	multivariate normal distribution in p-dimensions with mean $\boldsymbol{\mu}$ (column vector of length p) and covariance matrix $\boldsymbol{\Sigma}$ $(p \times p)$
$\text{var}(\cdot)$	variance–covariance matrix
$\text{cor}(\cdot)$	correlation matrix
$W_p(\boldsymbol{\Sigma}, m)$	Wishart distribution

The terms variance matrix, covariance matrix, and variance–covariance matrix are synonymous.

Matrix Decompositions

- Any symmetric matrix $\boldsymbol{A}(p \times p)$ can (by the spectral decomposition theorem) be written as

$$\boldsymbol{A} = \boldsymbol{\Gamma\Lambda\Gamma}' = \sum_i \lambda_i \boldsymbol{\gamma}_{(i)} \boldsymbol{\gamma}'_{(i)},$$

  where $\boldsymbol{\Lambda}$ is a diagonal matrix of eigenvalues of $\boldsymbol{A}$ (which are real-valued), i.e. $\boldsymbol{\Lambda} = \text{diag}(\lambda_1, \dots, \lambda_p)$, and $\boldsymbol{\Gamma}$ is an orthogonal matrix whose columns are standardized eigenvectors, i.e. $\boldsymbol{\Gamma} = [\boldsymbol{\gamma}_{(1)} \ \cdots \ \boldsymbol{\gamma}_{(p)}]$ and $\boldsymbol{\Gamma}'\boldsymbol{\Gamma} = \boldsymbol{I}$. See Theorem A.6.8.
- Using the above, we define the symmetric square root of a positive definite matrix $\boldsymbol{A}$ by

$$\boldsymbol{A}^{1/2} = \boldsymbol{\Gamma\Lambda}^{1/2}\boldsymbol{\Gamma}', \qquad \text{where} \qquad \boldsymbol{\Lambda}^{1/2} = \text{diag}(\lambda_1^{1/2}, \dots, \lambda_p^{1/2}).$$

- If $\boldsymbol{A}$ is an $n \times p$ matrix of rank r, then by the singular value decomposition, it can be written as

$$\boldsymbol{A} = \boldsymbol{ULV}',$$

  where $\boldsymbol{U}(n \times r)$ and $\boldsymbol{V}(p \times r)$ are column orthonormal matrices, and $\boldsymbol{L}(r \times r)$ is a diagonal matrix with positive elements. See Theorem A.6.8.

Geometry

Table A.5 sets out the basic concepts in n-dimensional geometry. In particular,

$\lVert\boldsymbol{x}\rVert = \sqrt{\boldsymbol{x}'\boldsymbol{x}}$	Length of a vector $\boldsymbol{x}$
$\lVert\boldsymbol{x} - \boldsymbol{y}\rVert$	Euclidean distance between $\boldsymbol{x}$ and $\boldsymbol{y}$
D^2	Squared Mahalanobis distance – one of the most important distances in multivariate analysis, since it takes account of a covariance, i.e. $D^2 = (\boldsymbol{x} - \boldsymbol{y})'\boldsymbol{\Sigma}^{-1}(\boldsymbol{x} - \boldsymbol{y})$

Table 14.6 gives a list of various distances.

Main Abbreviations and Commonly Used Notation

$\approx$	approximately equal to
$\perp\!\!\!\perp$	(conditionally) independent of
$\sim$	is distributed as
$A \backslash B$	the set of elements that are members of A but not B
$\lVert \boldsymbol{x} - \boldsymbol{y} \rVert$	Euclidean distance between $\boldsymbol{x}$ and $\boldsymbol{y}$
$\boldsymbol{A}'$	transpose of matrix $\boldsymbol{A}$
$\lvert \boldsymbol{A} \rvert$	determinant of matrix $\boldsymbol{A}$
$\boldsymbol{A}^{-1}$	inverse of matrix $\boldsymbol{A}$
$\boldsymbol{A}^{-}$	g-inverse (generalized inverse)
$\mathbf{1}$	column vector of 1s
$\mathbf{0}$	column vector or matrix of 0s
$\boldsymbol{B}$	between-groups sum of squares and products (SSP) matrix
$B(\cdot,\cdot)$	beta variable
$\mathrm{B}(\cdot,\cdot)$	normalizing constant for beta distribution (note nonitalic font to distinguish from the above)
BLUE	best linear unbiased estimate
$\mathrm{cov}(\boldsymbol{x},\boldsymbol{y})$	covariance between $\boldsymbol{x}$ and $\boldsymbol{y}$
χ^2_p	chi-squared distribution with p degrees of freedom
$\chi^2_p(1-\alpha)$	upper α critical value of chi-squared distribution with p degrees of freedom
c.f.	characteristic function
∂	partial derivative – multivariate examples in Appendix A.9
$\boldsymbol{D}$	distance matrix
D^2	squared Mahalanobis distance
d.f.	distribution function
δ_{ij}	Kronecker delta
diag	diagonal elements of a square matrix (as column vector) or diagonal matrix created from a vector
Diag	diag(diag($\boldsymbol{A}$)) (see above)
E	expectation
$F_{p,q}$	F distribution with degrees of freedom p and q
$F_{p,q}(1-\alpha)$	upper α critical value of F distribution with degrees of freedom p and q
$F(\cdot)$	cumulative distribution function
$f(\cdot)$	probability density function
$\Gamma(\cdot)$	gamma function
GLS	generalized least squares
$\boldsymbol{H}$	centering matrix
$\boldsymbol{I}$	identity matrix
ICA	independent component analysis
i.i.d.	independent and identically distributed
J	Jacobian of transformation (see Table 2.1)
$\boldsymbol{J}$	$\mathbf{1}\mathbf{1}'$
$\boldsymbol{K}$	concentration matrix ($= \boldsymbol{\Sigma}^{-1}$)

L	likelihood
l	log likelihood
LDA	linear discriminant analysis
log	logarithm to the base e (natural logarithm)
LRT	likelihood ratio test
MANOVA	multivariate analysis of variance
MDS	multidimensional scaling
ML	maximum likelihood
m.l.e.	maximum likelihood estimate
$\boldsymbol{\mu}$	mean (population) vector
$N_p(\boldsymbol{\mu}, \boldsymbol{\Sigma})$	multivariate normal distribution for p-dimensions (usually p omitted when $p = 1$)
OLS	ordinary least squares
$\boldsymbol{P}$	(population) correlation matrix (sometimes a matrix of counts, e.g. Section 9.5)
$\mathsf{P}(\cdot)$	probability
PCA	principal component analysis
p.d.	positive definite
p.d.f.	probability density function
p.s.d.	positive semi definite
$\mathbb{R}$	real numbers
ρ	correlation coefficient
$\boldsymbol{R}$	sample correlation matrix
$\boldsymbol{S}$	sample covariance matrix
$\boldsymbol{S}_u$	unbiased sample covariance matrix
$\boldsymbol{\Sigma}$	(population) covariance matrix
SLC	standardized linear combination
SSP	sums of squares and products
t_p	t distribution with p degrees of freedom
$\boldsymbol{T}$	total SSP matrix
T^2	Hotelling T^2 statistic
tr	trace
UIT	union intersection test
$V(\cdot)$	variance
$\text{var}(\boldsymbol{x})$	variance-covariance matrix of $\boldsymbol{x}$
$\boldsymbol{W}$	within-groups SSP matrix
$W_p(\boldsymbol{\Sigma}, m)$	Wishart distribution
$\boldsymbol{X}$	data matrix
$\overline{\boldsymbol{x}}$	sample mean vector
Λ	Wilks' Λ statistic
θ	greatest root statistic

1 Introduction

1.1 Objects and Variables

Multivariate analysis deals with data containing observations on two or more variables, each measured on a set of objects. For example, we may have the set of examination marks achieved by certain students, or the cork deposit in various directions of a set of trees, or flower measurements for different species of iris (see Tables 1.2, 1.4, and 1.3, respectively). Each of these data has a set of "variables" (the examination marks, trunk thickness, and flower measurements) and a set of "objects" (the students, trees, and flowers). In general, if there are n objects, $o_1, \ldots, o_n$ and p variables, $x_1, \ldots, x_p$, the data contains np pieces of information. These may be conveniently arranged using an $(n \times p)$ "data matrix", in which each row corresponds to an object, and each column corresponds to a variable. For instance, three variables on five "objects" (students) are shown as a (5×3) data matrix in Table 1.1.

Note that all the variables need not be of the same type: in Table 1.1, x_1 is a "continuous" variable, x_2 is a discrete variable, and x_3 is a binary variable. Note also that attribute, characteristic, description, measurement, and response are synonyms for "variable", whereas individual, observation, plot, reading, item, and unit can be used in place of "object".

1.2 Some Multivariate Problems and Techniques

We may now illustrate various categories of multivariate technique.

1.2.1 Generalizations of Univariate Techniques

Most univariate questions are capable of at least one multivariate generalization. For instance, using Table 1.2, we may ask, as an example, "What is the appropriate underlying parent distribution of examination marks on various papers of a set of students?" "What are the summary statistics?" "Are the differences between average marks on different papers significant?", etc. These problems are direct generalizations of univariate problems, and their motivation is easy to grasp. See, for example, Chapters 2–7 and 13.

Multivariate Analysis, Second Edition. Kanti V. Mardia, John T. Kent, and Charles C. Taylor.

Table 1.1 Data matrix with five students as objects, where x_1 is age in years at entry to university, x_2 is marks out of 100 in an examination at the end of the first year, and x_3 is sex.

	Variables		
Objects	x_1	x_2	x_3
1	18.45	70	1
2	18.41	65	0
3	18.39	71	0
4	18.70	72	0
5	18.34	94	1

1 indicates male; 0 indicates female.

1.2.2 Dependence and Regression

The data in Table 1.2, which were collected at the University of Hull in the early 1970s, formed part of an investigation into the merits of open-book vs. closed-book examinations. Marks (out of 100) were given for 88 students on each of five subjects; these observations were sorted (almost) according to the average. Initially, we may enquire as to the degree of dependence between performance on different papers taken by the same students. It may be useful, for counseling or other purposes, to have some idea of how final degree marks ("dependent" variables) are affected by previous examination results or by other variables such as age and sex ("explanatory" variables). This presents the so-called regression problem, which is examined in Chapter 7.

1.2.3 Linear Combinations

Given examination marks on different topics (as in Table 1.2), the question arises of how to combine or average these marks in a suitable way. A straightforward method would use the simple arithmetic mean, but this procedure may not always be suitable. For instance, if the marks on some papers vary more than others, we may wish to weight them differently. This leads us to search for a linear combination (weighted sum) which is "optimal" in some sense. If all the examination papers fall in one group, then *principal component analysis* and *factor analysis* are two techniques that can help to answer such questions (see Chapters 9 and 10). In some situations, the papers may fall into more than one group – for instance, in Table 1.2, some examinations were "open book", while others were "closed book". In such situations, we may wish to investigate the use of linear combinations within each group separately. This leads to the method known as *canonical correlation analysis*, which is discussed in Chapter 11.

The idea of taking linear combinations is an important one in multivariate analysis, and we will return to it in Section 1.5.

Table 1.2 Marks in open- and closed-book examination out of 100.

Mechanics (C)	Vectors (C)	Algebra (O)	Analysis (O)	Statistics (O)
77	82	67	67	81
63	78	80	70	81
75	73	71	66	81
55	72	63	70	68
63	63	65	70	63
53	61	72	64	73
51	67	65	65	68
59	70	68	62	56
62	60	58	62	70
64	72	60	62	45
52	64	60	63	54
55	67	59	62	44
50	50	64	55	63
65	63	58	56	37
31	55	60	57	73
60	64	56	54	40
44	69	53	53	53
42	69	61	55	45
62	46	61	57	45
31	49	62	63	62
44	61	52	62	46
49	41	61	49	64
12	58	61	63	67
49	53	49	62	47
54	49	56	47	53
54	53	46	59	44
44	56	55	61	36
18	44	50	57	81
46	52	65	50	35
32	45	49	57	64
30	69	50	52	45
46	49	53	59	37
40	27	54	61	61
31	42	48	54	68
36	59	51	45	51
56	40	56	54	35

(Continued)

Table 1.2 (Continued)

Mechanics (C)	Vectors (C)	Algebra (O)	Analysis (O)	Statistics (O)
46	56	57	49	32
45	42	55	56	40
42	60	54	49	33
40	63	53	54	25
23	55	59	53	44
48	48	49	51	37
41	63	49	46	34
46	52	53	41	40
46	61	46	38	41
40	57	51	52	31
49	49	45	48	39
22	58	53	56	41
35	60	47	54	33
48	56	49	42	32
31	57	50	54	34
17	53	57	43	51
49	57	47	39	26
59	50	47	15	46
37	56	49	28	45
40	43	48	21	61
35	35	41	51	50
38	44	54	47	24
43	43	38	34	49
39	46	46	32	43
62	44	36	22	42
48	38	41	44	33
34	42	50	47	29
18	51	40	56	30
35	36	46	48	29
59	53	37	22	19
41	41	43	30	33
31	52	37	27	40
17	51	52	35	31
34	30	50	47	36
46	40	47	29	17
10	46	36	47	39

Table 1.2 (Continued)

Mechanics (C)	Vectors (C)	Algebra (O)	Analysis (O)	Statistics (O)
46	37	45	15	30
30	34	43	46	18
13	51	50	25	31
49	50	38	23	09
18	32	31	45	40
08	42	48	26	40
23	38	36	48	15
30	24	43	33	25
03	09	51	47	40
07	51	43	17	22
15	40	43	23	18
15	38	39	28	17
05	30	44	36	18
12	30	32	35	21
05	26	15	20	20
00	40	21	09	14

O indicates open book, and C indicates closed book.

1.2.4 Assignment and Dissection

Table 1.3 gives three (50×4) data matrices (or one (150×5) data matrix if the species is coded as a variable). In each matrix, the "objects" are 50 irises of species *Iris setosa, Iris versicolor*, and *Iris virginica*, respectively. The "variables" are

$$x_1 = \text{sepal length}, \qquad x_2 = \text{sepal width},$$
$$x_3 = \text{petal length}, \qquad x_4 = \text{petal width}.$$

The flowers of the first two iris species (*I. setosa* and *I. versicolor*) were taken from the same natural colony but the sample of the third iris species (*I. virginica*) is from a different colony; for more general details on the data, see Mardia (2023). If a new iris of unknown species has measurements $x_1 = 5.1,\ x_2 = 3.2,\ x_3 = 2.7$, and $x_4 = 0.7$, we may ask to which species it belongs. This presents the problem of *discriminant analysis*, which is discussed in Chapter 12. However, if we were presented with the 150 observations of Table 1.3 in an unclassified manner (say, before the three species were established), then the aim could have been to dissect the population into homogeneous groups. This problem is handled by *cluster analysis* (see Chapter 14).

1.2.5 Building Configurations

In some cases, the data consists not of an $(n \times p)$ data matrix but of $n(n-1)/2$ "distances" between all pairs of points. To get an intuitive feel for the structure of such data, a

Table 1.3 Measurements (in cm) on three types of irises.

Iris setosa				*Iris versicolor*				*Iris virginica*			
Sepal length	**Sepal width**	**Petal length**	**Petal width**	**Sepal length**	**Sepal width**	**Petal length**	**Petal width**	**Sepal length**	**Sepal width**	**Petal length**	**Petal width**
5.1	3.5	1.4	0.2	7.0	3.2	4.7	1.4	6.3	3.3	6.0	2.5
4.9	3.0	1.4	0.2	6.4	3.2	4.5	1.5	5.8	2.7	5.1	1.9
4.7	3.2	1.3	0.2	6.9	3.1	4.9	1.5	7.1	3.0	5.9	2.1
4.6	3.1	1.5	0.2	5.5	2.3	4.0	1.3	6.3	2.9	5.6	1.8
5.0	3.6	1.4	0.2	6.5	2.8	4.6	1.5	6.5	3.0	5.8	2.2
5.4	3.9	1.7	0.4	5.7	2.8	4.5	1.3	7.6	3.0	6.6	2.1
4.6	3.4	1.4	0.3	6.3	3.3	4.7	1.6	4.9	2.5	4.5	1.7
5.0	3.4	1.5	0.2	4.9	2.4	3.3	1.0	7.3	2.9	6.3	1.8
4.4	2.9	1.4	0.2	6.6	2.9	4.6	1.3	6.7	2.5	5.8	1.8
4.9	3.1	1.5	0.1	5.2	2.7	3.9	1.4	7.2	3.6	6.1	2.5
5.4	3.7	1.5	0.2	5.0	2.0	3.5	1.0	6.5	3.2	5.1	2.0
4.8	3.4	1.6	0.2	5.9	3.0	4.2	1.5	6.4	2.7	5.3	1.9
4.8	3.0	1.4	0.1	6.0	2.2	4.0	1.0	6.8	3.0	5.5	2.1
4.3	3.0	1.1	0.1	6.1	2.9	4.7	1.4	5.7	2.5	5.0	2.0
5.8	4.0	1.2	0.2	5.6	2.9	3.6	1.3	5.8	2.8	5.1	2.4
5.7	4.4	1.5	0.4	6.7	3.1	4.4	1.4	6.4	3.2	5.3	2.3
5.4	3.9	1.3	0.4	5.6	3.0	4.5	1.5	6.5	3.0	5.5	1.8
5.1	3.5	1.4	0.3	5.8	2.7	4.1	1.0	7.7	3.8	6.7	2.2
5.7	3.8	1.7	0.3	6.2	2.2	4.5	1.5	7.7	2.6	6.9	2.3
5.1	3.8	1.5	0.3	5.6	2.5	3.9	1.1	6.0	2.2	5.0	1.5
5.4	3.4	1.7	0.2	5.9	3.2	4.8	1.8	6.9	3.2	5.7	2.3
5.1	3.7	1.5	0.4	6.1	2.8	4.0	1.3	5.6	2.8	4.9	2.0
4.6	3.6	1.0	0.2	6.3	2.5	4.9	1.5	7.7	2.8	6.7	2.0
5.1	3.3	1.7	0.5	6.1	2.8	4.7	1.2	6.3	2.7	4.9	1.8
4.8	3.4	1.9	0.2	6.4	2.9	4.3	1.3	6.7	3.3	5.7	2.1
5.0	3.0	1.6	0.2	6.6	3.0	4.4	1.4	7.2	3.2	6.0	1.8
5.0	3.4	1.6	0.4	6.8	2.8	4.8	1.4	6.2	2.8	4.8	1.8
5.2	3.5	1.5	0.2	6.7	3.0	5.0	1.7	6.1	3.0	4.9	1.8
5.2	3.4	1.4	0.2	6.0	2.9	4.5	1.5	6.4	2.8	5.6	2.1
4.7	3.2	1.6	0.2	5.7	2.6	3.5	1.0	7.2	3.0	5.8	1.6
4.8	3.1	1.6	0.2	5.5	2.4	3.8	1.1	7.4	2.8	6.1	1.9
5.4	3.4	1.5	0.4	5.5	2.4	3.7	1.0	7.9	3.8	6.4	2.0
5.2	4.1	1.5	0.1	5.8	2.7	3.9	1.2	6.4	2.8	5.6	2.2
5.5	4.2	1.4	0.2	6.0	2.7	5.1	1.6	6.3	2.8	5.1	1.5

Table 1.3 (Continued)

Iris setosa				*Iris versicolor*				*Iris virginica*			
Sepal length	**Sepal width**	**Petal length**	**Petal width**	**Sepal length**	**Sepal width**	**Petal length**	**Petal width**	**Sepal length**	**Sepal width**	**Petal length**	**Petal width**
4.9	3.1	1.5	0.2	5.4	3.0	4.5	1.5	6.1	2.6	5.6	1.4
5.0	3.2	1.2	0.2	6.0	3.4	4.5	1.6	7.7	3.0	6.1	2.3
5.5	3.5	1.3	0.2	6.7	3.1	4.7	1.5	6.3	3.4	5.6	2.4
4.9	3.6	1.4	0.1	6.3	2.3	4.4	1.3	6.4	3.1	5.5	1.8
4.4	3.0	1.3	0.2	5.6	3.0	4.1	1.3	6.0	3.0	4.8	1.8
5.1	3.4	1.5	0.2	5.5	2.5	4.0	1.3	6.9	3.1	5.4	2.1
5.0	3.5	1.3	0.3	5.5	2.6	4.4	1.2	6.7	3.1	5.6	2.4
4.5	2.3	1.3	0.3	6.1	3.0	4.6	1.4	6.9	3.1	5.1	2.3
4.4	3.2	1.3	0.2	5.8	2.6	4.0	1.2	5.8	2.7	5.1	1.9
5.0	3.5	1.6	0.6	5.0	2.3	3.3	1.0	6.8	3.2	5.9	2.3
5.1	3.8	1.9	0.4	5.6	2.7	4.2	1.3	6.7	3.3	5.7	2.5
4.8	3.0	1.4	0.3	5.7	3.0	4.2	1.2	6.7	3.0	5.2	2.3
5.1	3.8	1.6	0.2	5.7	2.9	4.2	1.3	6.3	2.5	5.0	1.9
4.6	3.2	1.4	0.2	6.2	2.9	4.3	1.3	6.5	3.0	5.2	2.0
5.3	3.7	1.5	0.2	5.1	2.5	3.0	1.1	6.2	3.4	5.4	2.3
5.0	3.3	1.4	0.2	5.7	2.8	4.1	1.3	5.9	3.0	5.1	1.8

Source: Fisher (1936) / John Wiley & Sons.

configuration can be constructed of n points in a Euclidean space of low dimension (e.g. $p = 2$ or 3). Hopefully, the distances between the n points of this configuration will closely match the original distances. The problems of building and interpreting such configurations are studied in Chapter 15, on *multidimensional scaling*.

1.3 The Data Matrix

The general $(n \times p)$ data matrix with n objects and p variables can be written as follows:

$$
\begin{array}{cc}
 & \overbrace{\begin{array}{ccccc} x_1 & \cdots & x_j & \cdots & x_p, \end{array}}^{\text{Variables}} \\
\text{Objects} \left\{ \begin{array}{c} o_1 \\ \vdots \\ o_r \\ \vdots \\ o_n \end{array} \right. &
\begin{array}{ccccc}
x_{11} & \cdots & x_{1j} & \cdots & x_{1p} \\
\vdots & & \vdots & & \vdots \\
x_{r1} & \cdots & x_{rj} & \cdots & x_{rp}. \\
\vdots & & \vdots & & \vdots \\
x_{n1} & \cdots & x_{nj} & \cdots & x_{np}
\end{array}
\end{array}
$$

Table 1.1 shows one such data matrix (5×3) with five objects and three variables. In Table 1.3, there are three data matrices, each having 50 rows (objects) and 4 columns (variables). Note that these data matrices can be considered from two alternative points of view. If we compare two columns, then we are examining the relationship between variables. On the other hand, a comparison of two rows involves examining the relationship between different objects. For example, in Table 1.1, we may compare the first two columns to investigate whether there is a relationship between age at entry and marks obtained. Alternatively, looking at the first two rows will give a comparison between two students ("objects"), one male and one female.

The general $(n \times p)$ data matrix will be denoted $\boldsymbol{X}$ or $\boldsymbol{X}(n \times p)$. The element in row r and column j is x_{rj}. This denotes the observation of the jth variable on the rth object. We may write the matrix $\boldsymbol{X} = (x_{rj})$. The rows of $\boldsymbol{X}$ will be written $\boldsymbol{x}_1', \boldsymbol{x}_2', \dots, \boldsymbol{x}_n'$. Note that $\boldsymbol{x}_r$ denotes the rth row of $\boldsymbol{X}$ *written as a column*. The columns of $\boldsymbol{X}$ will be written with subscripts in parentheses as $\boldsymbol{x}_{(1)}, \boldsymbol{x}_{(2)}, \dots, \boldsymbol{x}_{(p)}$; that is, we may write

$$\boldsymbol{X} = \begin{bmatrix} \boldsymbol{x}_1' \\ \vdots \\ \boldsymbol{x}_n' \end{bmatrix} = [\boldsymbol{x}_{(1)} \ \cdots \ \boldsymbol{x}_{(p)}],$$

where

$$\boldsymbol{x}_r = \begin{pmatrix} x_{r1} \\ \vdots \\ x_{rp} \end{pmatrix} \quad r = 1, \dots, n \qquad \text{and} \qquad \boldsymbol{x}_{(j)} = \begin{pmatrix} x_{1j} \\ \vdots \\ x_{nj} \end{pmatrix} \quad j = 1, \dots, p.$$

In general, we use square brackets for matrices and round brackets for vectors.

Note that on the one hand $\boldsymbol{x}_1$ is the p-vector denoting the p observations on the first *object*, while on the other hand $\boldsymbol{x}_{(1)}$ is the n-vector whose elements denote the observations on the first *variable*. In multivariate analysis, the rows $\boldsymbol{x}_1, \dots, \boldsymbol{x}_n$ usually form a random sample, whereas the columns $\boldsymbol{x}_{(1)}, \dots, \boldsymbol{x}_{(p)}$ do not. This point is emphasized in the notation by the use of parentheses.

Clearly when n and p are even moderately large, the resulting np pieces of information may prove too numerous to handle individually. Various way of summarizing multivariate data are discussed in Sections 1.4, 1.7, and 1.8.

1.4 Summary Statistics

We give here the basic summary statistics and some standard notation.

1.4.1 The Mean Vector and Covariance Matrix

An obvious extension of the univariate notion of mean and variance leads to the following definitions. The sample mean of the jth variable is

$$\bar{x}_j = \frac{1}{n} \sum_{r=1}^{n} x_{rj}, \tag{1.1}$$

and the sample variance of the jth variable is

$$s_{jj} = \frac{1}{n} \sum_{r=1}^{n} (x_{rj} - \bar{x}_j)^2 = s_j^2, \quad \text{say,} \quad j = 1, \dots, p. \tag{1.2}$$

The sample covariance between the ith and jth variables is

$$s_{ij} = \frac{1}{n}\sum_{r=1}^{n}(x_{ri} - \bar{x}_i)(x_{rj} - \bar{x}_j). \tag{1.3}$$

The vector of means,

$$\bar{\boldsymbol{x}} = \begin{pmatrix} \bar{x}_1 \\ \vdots \\ \bar{x}_p \end{pmatrix}, \tag{1.4}$$

is called the *sample mean vector*, or simply the "mean vector". It represents the center of gravity of the sample points $\boldsymbol{x}_r$, $r = 1, \ldots, n$. The $p \times p$ matrix

$$\boldsymbol{S} = (s_{ij}), \tag{1.5}$$

with elements given by (1.2) and (1.3), is called the *sample covariance matrix*, or simply the "covariance matrix".

The above statistics may also be expressed in matrix notation. Corresponding to (1.1) and (1.4), we have

$$\bar{\boldsymbol{x}} = \frac{1}{n}\sum_{r=1}^{n}\boldsymbol{x}_r = \frac{1}{n}\boldsymbol{X}'\boldsymbol{1}, \tag{1.6}$$

where $\boldsymbol{1}$ is a column vector of n ones. Also,

$$s_{ij} = \frac{1}{n}\sum_{r=1}^{n}x_{ri}x_{rj} - \bar{x}_i\,\bar{x}_j,$$

so that

$$\boldsymbol{S} = \frac{1}{n}\sum_{r=1}^{n}(\boldsymbol{x}_r - \bar{\boldsymbol{x}})(\boldsymbol{x}_r - \bar{\boldsymbol{x}})' = \frac{1}{n}\sum_{r=1}^{n}\boldsymbol{x}_r\boldsymbol{x}_r' - \bar{\boldsymbol{x}}\,\bar{\boldsymbol{x}}'. \tag{1.7}$$

This may also be written as

$$\boldsymbol{S} = \frac{1}{n}\boldsymbol{X}'\boldsymbol{X} - \bar{\boldsymbol{x}}\,\bar{\boldsymbol{x}}' = \frac{1}{n}(\boldsymbol{X}'\boldsymbol{X} - \frac{1}{n}\boldsymbol{X}'\boldsymbol{1}\boldsymbol{1}'\boldsymbol{X}),$$

using (1.6). Writing

$$\boldsymbol{H} = \boldsymbol{I} - \frac{1}{n}\boldsymbol{1}\boldsymbol{1}',$$

where $\boldsymbol{H}$ denotes the *centering matrix*, we find that

$$\boldsymbol{S} = \frac{1}{n}\,\boldsymbol{X}'\boldsymbol{H}\boldsymbol{X}, \tag{1.8}$$

which is a convenient matrix representation of the sample covariance matrix.

Since $\boldsymbol{H}$ is a symmetric idempotent matrix ($\boldsymbol{H} = \boldsymbol{H}', \boldsymbol{H} = \boldsymbol{H}^2$), it follows that for any p-vector $\boldsymbol{a}$,

$$\boldsymbol{a}'\boldsymbol{S}\boldsymbol{a} = \frac{1}{n}\,\boldsymbol{a}'\boldsymbol{X}\boldsymbol{H}'\boldsymbol{H}\boldsymbol{X}\boldsymbol{a} = \frac{1}{n}\,\boldsymbol{y}'\boldsymbol{y} \geq 0,$$

where $\boldsymbol{y} = \boldsymbol{H}\boldsymbol{X}\boldsymbol{a}$. Hence, the covariance matrix $\boldsymbol{S}$ is positive semidefinite ($\boldsymbol{S} \geq 0$). For continuous data, we usually expect that $\boldsymbol{S}$ is not only positive semidefinite but positive definite if $n \geq p + 1$; see Table A.7 for more specific definitions.

As in one-dimensional statistics, it is often convenient to define the covariance matrix with a divisor of $n-1$ instead of n. Set

$$\boldsymbol{S}_u = \frac{1}{n-1}\boldsymbol{X}'\boldsymbol{H}\boldsymbol{X} = \frac{n}{n-1}\boldsymbol{S}. \tag{1.9}$$

If the data forms a random sample from a multivariate distribution, with finite second moments, then $\boldsymbol{S}_u$ is an *unbiased* estimate of the true covariance matrix (see Theorem 2.11).

The matrix

$$\boldsymbol{M} = \sum_{r=1}^{n} \boldsymbol{x}_r\boldsymbol{x}_r' = \boldsymbol{X}'\boldsymbol{X} \tag{1.10}$$

is called the *matrix of sums of squares and products* for obvious reasons. The matrix $n\boldsymbol{S}$ can appropriately be labeled as the matrix of *corrected* sums of squares and products.

The *sample correlation coefficient* between the ith and the jth variables is

$$r_{ij} = s_{ij}/(s_i s_j). \tag{1.11}$$

Unlike s_{ij}, the correlation coefficient is invariant under both changes of scale and origin of the ith and jth variables. Clearly, $|r_{ij}| \leq 1$. The matrix

$$\boldsymbol{R} = (r_{ij}) \tag{1.12}$$

with $r_{ii} = 1$ is called the *sample correlation matrix*. Note that $\boldsymbol{R} \geq 0$ (i.e. $\boldsymbol{R}$ is positive semidefinite). If $\boldsymbol{R} = \boldsymbol{I}$, we say that the variables are uncorrelated. If $\boldsymbol{D} = \mathrm{diag}(s_i)$, then

$$\boldsymbol{R} = \boldsymbol{D}^{-1}\boldsymbol{S}\boldsymbol{D}^{-1}, \quad \boldsymbol{S} = \boldsymbol{D}\boldsymbol{R}\boldsymbol{D}. \tag{1.13}$$

Example 1.4.1 Table 1.4 gives a (28×4) data matrix (Rao, 1948) related to weights of bark deposits of 28 trees in the four directions north (N), east (E), south (S), and west (W).

It is found that

$$\bar{x}_1 = 50.536, \qquad \bar{x}_2 = 46.179, \qquad \bar{x}_3 = 49.679, \qquad \bar{x}_4 = 45.179.$$

These means suggest *prima-facie* differences with more deposits in the N–S directions than in the E–W directions. The covariance and correlation matrices are

$$\boldsymbol{S} = \begin{array}{c} \\ \mathrm{N} \\ \mathrm{E} \\ \mathrm{S} \\ \mathrm{W} \end{array} \begin{array}{c} \begin{array}{cccc} \mathrm{N} & \mathrm{E} & \mathrm{S} & \mathrm{W} \end{array} \\ \begin{bmatrix} 280.034 & 215.761 & 278.136 & 218.190 \\ & 212.075 & 220.879 & 165.254 \\ & & 337.504 & 250.272 \\ & & & 217.932 \end{bmatrix} \end{array},$$

$$\boldsymbol{R} = \begin{array}{c} \\ \mathrm{N} \\ \mathrm{E} \\ \mathrm{S} \\ \mathrm{W} \end{array} \begin{array}{c} \begin{array}{cccc} \mathrm{N} & \mathrm{E} & \mathrm{S} & \mathrm{W} \end{array} \\ \begin{bmatrix} 1 & 0.885 & 0.905 & 0.883 \\ & 1 & 0.826 & 0.769 \\ & & 1 & 0.923 \\ & & & 1 \end{bmatrix} \end{array}.$$

Since these matrices are symmetric, only the upper half needs to be shown.

Table 1.4 Weights of cork deposits (in centigrams) for 28 trees in the four directions.

N	E	S	W	N	E	S	W
72	66	76	77	91	79	100	75
60	53	66	63	56	68	47	50
56	57	64	58	79	65	70	61
41	29	36	38	81	80	68	58
32	32	35	36	78	55	67	60
30	35	34	26	46	38	37	38
39	39	31	27	39	35	34	37
42	43	31	25	32	30	30	32
37	40	31	25	60	50	67	54
33	29	27	36	35	37	48	39
32	30	34	28	39	36	39	31
63	45	74	63	50	34	37	40
54	46	60	52	43	37	39	50
47	51	52	43	48	54	57	43

Source: Rao (1948) / Oxford University Press.

Comparing the diagonal terms of $\boldsymbol{S}$, we note that the sample variance is largest in the south direction. Furthermore, the matrix $\boldsymbol{R}$ does not seem to have a "circular" pattern; e.g. the correlation between N and S is relatively high, while the correlation between the other pair of opposite cardinal points W and E is lowest. □

1.4.2 Measures of Multivariate Scatter

The matrix $\boldsymbol{S}$ is one possible multivariate generalization of the univariate notion of variance, measuring scatter about the mean. However, sometimes it is convenient to have a *single* number to measure multivariate scatter. Two common such measures are

1. the *generalized variance*, $|\boldsymbol{S}|$ (the determinant of $\boldsymbol{S}$) and
2. the *total variation*, $\text{tr}\,\boldsymbol{S}$ (the sum of the diagonal entries of $\boldsymbol{S}$).

A motivation for these measures is given in Section 1.5.3. For both measures, large values indicate a high degree of scatter about $\bar{\boldsymbol{x}}$, and low values represent concentration about $\bar{\boldsymbol{x}}$. However, each measure reflects different aspects of the variability in the data. The generalized variance plays an important role in maximum-likelihood estimation (Chapter 6), and the total variation is a useful concept in principal component analysis (Chapter 9).

Example 1.4.2 (Gnanadesikan and Gupta, 1970) An experimental subject spoke 10 different words seven times each, and five speech measurements were taken on each utterance.

For each word, we have five variables and seven observations. The (5×5) covariance matrix was calculated for each word, and the 10 generalized variances were as follows:

$$2.9, \quad 1.3, \quad 641.6, \quad 26\,828.8, \quad 262\,404.3,$$

$$169.2, \quad 3106.8, \quad 617\,671.2 \quad 6.7 \quad 3.0.$$

Ordering these generalized variances, we find that for this speaker the second word has the least variation and the eighth word has most variation. A general point of interest for identification is to find which word has least variation for a particular speaker. □

1.5 Linear Combinations

Taking linear combinations of the variables is one of the most important tools of multivariate analysis. A few suitably chosen combinations may provide more information than a multiplicity of the original variables, often because the dimension of the data is reduced. Linear transformations can also simplify the structure of the covariance matrix, making interpretation of the data more straightforward.

Consider a linear combination

$$y_r = a_1 x_{r1} + \cdots + a_p x_{rp}, \quad r = 1, \ldots, n, \tag{1.14}$$

where $a_1, \ldots, a_p$ are given. From (1.6), the mean $\bar{y}$ of the y_r is given by

$$\bar{y} = \frac{1}{n}\boldsymbol{a}' \sum_{r=1}^{n} \boldsymbol{x}_r = \boldsymbol{a}'\bar{\boldsymbol{x}}, \tag{1.15}$$

and the variance is given by

$$s_y^2 = \frac{1}{n}\sum_{r=1}^{n}(y_r - \bar{y})^2 = \frac{1}{n}\sum_{r=1}^{n}\boldsymbol{a}'(\boldsymbol{x}_r - \bar{\boldsymbol{x}})(\boldsymbol{x}_r - \bar{\boldsymbol{x}})'\boldsymbol{a} = \boldsymbol{a}'\boldsymbol{S}\boldsymbol{a}, \tag{1.16}$$

where we have used (1.7).

In general, we may be interested in a q-dimensional linear transformation,

$$\boldsymbol{y}_r = \boldsymbol{A}\boldsymbol{x}_r + \boldsymbol{b}, \quad r = 1, \ldots, n, \tag{1.17}$$

which may be written as $\boldsymbol{Y} = \boldsymbol{X}\boldsymbol{A}' + \boldsymbol{1}\boldsymbol{b}'$, where $\boldsymbol{A}$ is a $(q \times p)$ matrix, and $\boldsymbol{b}$ is a q-vector. Usually, $q \leq p$.

The mean vector and covariance matrix of the new objects $\boldsymbol{y}_r$ are given by

$$\bar{\boldsymbol{y}} = \boldsymbol{A}\bar{\boldsymbol{x}} + \boldsymbol{b}, \tag{1.18}$$

$$\boldsymbol{S}_y = \frac{1}{n}\sum_{r=1}^{n}(\boldsymbol{y}_r - \bar{\boldsymbol{y}})(\boldsymbol{y}_r - \bar{\boldsymbol{y}})' = \boldsymbol{A}\boldsymbol{S}\boldsymbol{A}'. \tag{1.19}$$

If $\boldsymbol{A}$ is nonsingular (so, in particular, $q = p$), then

$$\boldsymbol{S} = \boldsymbol{A}^{-1}\boldsymbol{S}_y(\boldsymbol{A}')^{-1}. \tag{1.20}$$

Here are several important examples of linear transformations that are used later in the book. For simplicity, all of the transformations are centered to have mean $\boldsymbol{0}$.

1.5.1 The Scaling Transformation

Let $\boldsymbol{y}_r = \boldsymbol{D}^{-1}(\boldsymbol{x}_r - \overline{\boldsymbol{x}})$, $r = 1, \dots, n$, where $\boldsymbol{D} = \text{diag}(s_i)$. This transformation scales each variable to have unit variance and thus eliminates the arbitrariness in the choice of scale. For example, if $\boldsymbol{x}_{(1)}$ measures lengths, then $\boldsymbol{y}_{(1)}$ will be the same whether $\boldsymbol{x}_{(1)}$ is measured in inches or meters. Note that $\boldsymbol{S}_y = \boldsymbol{R}$.

1.5.2 Mahalanobis Transformation

If $\boldsymbol{S} > 0$, then $\boldsymbol{S}^{-1}$ has a unique symmetric positive-definite square root $\boldsymbol{S}^{-1/2}$ (see (A.66)). The Mahalanobis transformation is defined by

$$\boldsymbol{z}_r = \boldsymbol{S}^{-1/2}(\boldsymbol{x}_r - \overline{\boldsymbol{x}}), \quad r = 1, \dots, n. \tag{1.21}$$

Then, $\boldsymbol{S}_z = \boldsymbol{I}$, so that this transformation eliminates the correlation between the variables and standardizes the variance of each variable.

1.5.3 Principal Component Transformation

By the spectral decomposition theorem, the covariance matrix $\boldsymbol{S}$ may be written in the form

$$\boldsymbol{S} = \boldsymbol{G}\boldsymbol{L}\boldsymbol{G}', \tag{1.22}$$

where $\boldsymbol{G}$ is an orthogonal matrix, and $\boldsymbol{L}$ is a diagonal matrix of the eigenvalues of $\boldsymbol{S}$, $l_1 \geq l_2 \geq \cdots \geq l_p \geq 0$. The principal component transformation is defined by the *rotation*

$$\boldsymbol{w}_r = \boldsymbol{G}'(\boldsymbol{x}_r - \overline{\boldsymbol{x}}), \quad r = 1, \dots, n. \tag{1.23}$$

Since $\boldsymbol{S}_w = \boldsymbol{G}'\boldsymbol{S}\boldsymbol{G} = \boldsymbol{L}$ is diagonal, the columns of $\boldsymbol{W}$, called principal components, represent *uncorrelated* linear combinations of the variables. In practice, one hopes to summarize most of the variability in the data using only the principal components with the highest variances and then reducing the dimension.

Since the principal components are uncorrelated with variances $l_1, \dots, l_p$, it seems natural to define the "overall" spread of the data by some symmetric monotonically increasing function of $l_1, \dots, l_p$, such as $\prod l_i$ and $\sum l_i$. From Section A.6, $|\boldsymbol{S}| = |\boldsymbol{L}| = \prod l_i$ and $\text{tr}\,\boldsymbol{S} = \text{tr}\,\boldsymbol{L} = \sum l_i$. Thus, the rotation to principal components provides a motivation for the measures of multivariate scatter discussed in Section 1.4.2.

Note that alternative versions of the transformations in Sections 1.5.1–1.5.3 can be defined using $\boldsymbol{S}_u$ instead of $\boldsymbol{S}$.

Example 1.5.1 *A transformation of the cork data* If in the cork data of Table 1.4, the aim is to investigate whether the bark deposits are uniformly spread, our interest would be in linear combinations such as

$$y_1 = \text{N} + \text{S} - \text{E} - \text{W}, \qquad y_2 = \text{N} - \text{S}, \qquad y_3 = \text{E} - \text{W}.$$

Here,

$$\boldsymbol{A} = \begin{bmatrix} 1 & -1 & 1 & -1 \\ 1 & 0 & -1 & 0 \\ 0 & 1 & 0 & -1 \end{bmatrix}.$$

From Example 1.4.1 and Equation (1.18), we find that the mean vector of the transformed variables is

$$\bar{\boldsymbol{y}}' = (8.857,\ 0.857,\ 1.000),$$

and the covariance matrix is

$$\begin{bmatrix} 124.12 & -20.27 & -25.96 \\ & 61.27 & 26.96 \\ & & 99.50 \end{bmatrix}.$$

Obviously, the mean of y_1 is higher than that of y_2 and y_3, indicating possibly more cork deposit along the north–south axis than along the east–west axis. However, $\text{var}(y_1)$ is larger. If we standardize the variables so that the sum of squares of the coefficients is unity, by letting

$$z_1 = (\text{N} + \text{S} - \text{E} - \text{W})/2, \qquad z_2 = (\text{N} - \text{S})/\sqrt{2}, \qquad z_3 = (\text{E} - \text{W})/\sqrt{2},$$

the variances are more similar:

$$\text{var}(z_1) = 31.03, \qquad \text{var}(z_2) = 30.63, \qquad \text{var}(z_3) = 49.75.$$

□

1.6 Geometrical Ideas

In Section 1.3, we mentioned the two alternative perspectives that can be used to view the data matrix. On the one hand, we may be interested in comparing the *columns* of the data matrix, that is the *variables*. This leads to the techniques known as *R-techniques*, so-called because the correlation matrix $\boldsymbol{R}$ plays an important role in this approach. R-techniques are important in principal component analysis, factor analysis, and canonical correlation analysis.

Alternatively, we may compare the *rows* of the data matrix, that is the different *objects*. This leads to techniques such as discriminant analysis, cluster analysis, and multidimensional scaling, which are known as *Q-techniques*.

These two approaches correspond to different geometric ways of representing the $(n \times p)$ data matrix. First, the columns can be viewed as p points in an n-dimensional space, which we call *R-space* or *object space*. For example, the correlation matrix has a natural interpretation in the object space of the centered matrix $\boldsymbol{Y} = \boldsymbol{HX}$. The correlation r_{ij} is just the cosine of the angle θ_{ij} subtended at the origin between the two corresponding columns,

$$\cos\theta_{ij} = \frac{\|\boldsymbol{y}'_{(i)}\|\ \|\boldsymbol{y}_{(j)}\|}{\|\boldsymbol{y}_{(i)}\|\ \|\boldsymbol{y}_{(j)}\|} = \frac{s_{ij}}{s_i s_j} = \boldsymbol{r}_{ij}.$$

Note that the correlation coefficients are a measure of similarity because their values are large when variables are close to one another.

Second, the n rows may be taken as n points in p-dimensional *Q-space* or *variable space*. A natural way to compare two rows $\boldsymbol{x}_r$ and $\boldsymbol{x}_s$ is to look at the Euclidean distance between them:

$$\|\boldsymbol{x}_r - \boldsymbol{x}_s\|^2 = (\boldsymbol{x}_r - \boldsymbol{x}_s)'(\boldsymbol{x}_r - \boldsymbol{x}_s).$$

An alternative procedure is to transform the data by one of the transformations of Sections 1.5.1 or 1.5.2 and then look at the Euclidean distance between the transformed rows. Such distances play a role in cluster analysis. The most important of these distances is the *Mahalanobis distance* D_{rs}, given by

$$D_{rs}^2 = \|\boldsymbol{z}_r - \boldsymbol{z}_s\|^2 = (\boldsymbol{x}_r - \boldsymbol{x}_s)'\boldsymbol{S}^{-1}(\boldsymbol{x}_r - \boldsymbol{x}_s). \tag{1.24}$$

The Mahalanobis distance underlies Hotelling's T^2 test and the theory of discriminant analysis. Note that the Mahalanobis distance can alternatively be defined using $\boldsymbol{S}_u$ instead of $\boldsymbol{S}$.

1.7 Graphical Representation

1.7.1 Univariate Scatters

For $p = 1$ and $p = 2$, we can draw and interpret scatter plots for the data, but for $p = 3$ the difficulties of drawing such plots can be appreciated. For $p > 3$, the task becomes hopeless, although computer facilities exist, which allow one to examine the projection of a multivariate scatter onto any given plane; see Wickham et al. (2015) and the references therein. However, the need for graphical representations when $p > 3$ is greater than for the univariate case since the relationships cannot be understood by looking at a data matrix. A simple starting point is to look into univariate plots for the p variables side by side. For the cork data of Table 1.4, such a representation is given in Figure 1.1, which indicates that the distributions are somewhat skew. In this case, the variables are measured in the same units, and therefore, a direct comparison of the plots is possible. In general, we should standardize the variables before using such a plot.

Figure 1.1 does not give any idea of relationships between the variables. However, one way to exhibit the interrelationships is to plot all the observations consecutively along the x-axis, representing different variables by different symbols. For the cork data, a plot of this type is given in Figure 1.2. It shows the very noticeable tree differences as well as differences in pattern that are associated with the given ordering of the trees. (That is, the experimenters appear to have chosen groups of small- and large-trunked trees alternately.) We also observe that the 15th tree may be an outlier. These features due to ordering

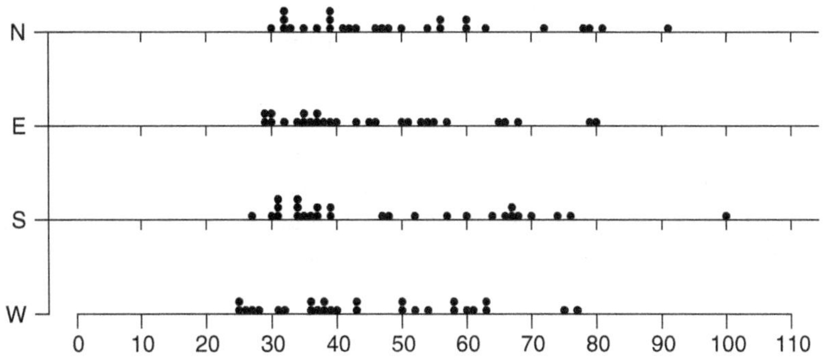

Figure 1.1 Univariate representation of the cork data of Table 1.4.

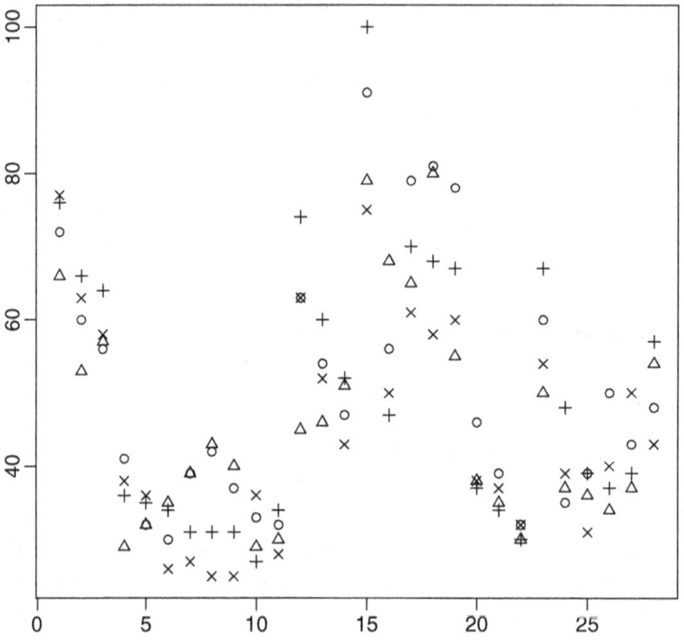

Figure 1.2 Consecutive univariate representation. Source: Adapted from Pearson (1956). ○ = N, △ = E, + = S, × = W.

would remain if the Mahalanobis residuals defined by (1.21) were plotted. Of course, the Mahalanobis transformation removes the main effect and adjusts every variance to unity and every covariance to zero.

1.7.2 Bivariate Scatters

Another way of understanding the interrelationships is to look into all $p(p-1)/2$ bivariate scatter diagrams of the data. For the cork data, with four variables, there are six such diagrams, and these could be presented as a *matrix* of simple scatterplots. Since the information is repeated in the upper and lower diagonals, there are various possibilities for these panels. One such example is shown in Figure 1.3 for the cork data.

Another method of graphing four variables in two dimensions is as follows. First, draw a scatter plot for two variables, say (N,E). Then, indicate the values of S and W of a point by plotting the values in two directions from the point (N,E). Figure 1.4 gives such a plot where W is plotted along the negative x-axis, and S is plotted along the negative y-axis using each point (N,E) as the origin. The markedly linear scatter of the points shows a strong dependence between N and E. A similar dependence between S and W is shown by the similar lengths of the two "legs" of each point. Dependence between N and S is reflected by the S legs being longer for large N, and a similar pattern is observed between N and W, E and

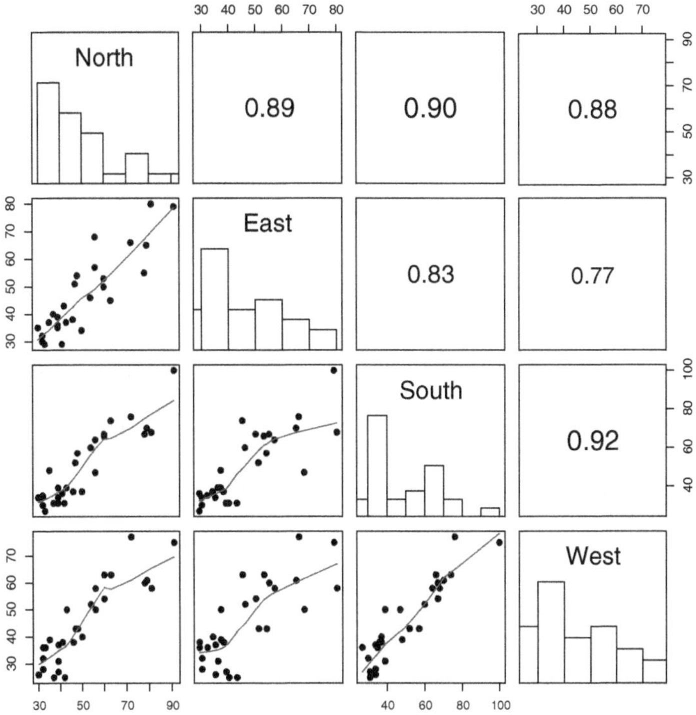

Figure 1.3 Matrix of scatterplots for the cork data. The upper diagonal shows the correlations (with text size proportional to the correlation), the lower diagonal shows a scatterplot (with a fitted smooth line), and the diagonal panels show the histograms.

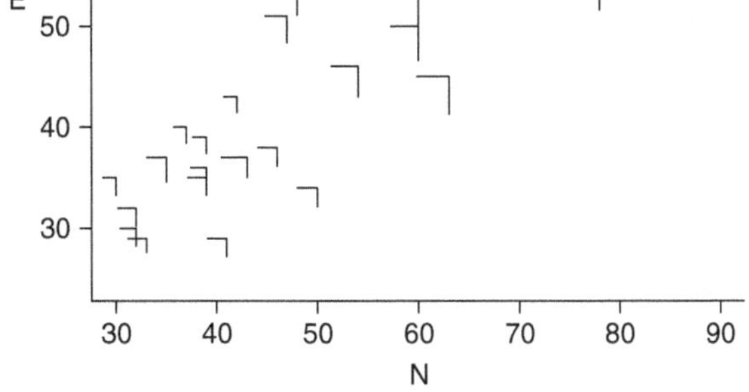

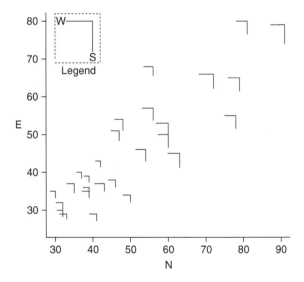

Figure 1.4 A glyph representation of the cork data of Table 1.4.

W, and E and S. This method of *glyphs* (Anderson, 1960) can be extended to several variables, but it does not help in understanding the multivariate complex as a whole. A related method due to Chernoff (1973) uses similar ideas to represent the observations on each object by a human face; see Mardia (2023) for an example using the iris data.

1.7.3 Harmonic Curves

Consider the rth data point $\boldsymbol{x}_r' = (x_{r1}, \dots, x_{rp})$, $r = 1, \dots, n$. An interesting method (Andrews, 1972; Ball and Hall, 1970) involves plotting the curve

$$f_{\boldsymbol{x}_r}(t) = \frac{x_{r1}}{\sqrt{2}} + x_{r2}\sin t + x_{r3}\cos t + \cdots + \begin{cases} x_{rp}\sin[pt/2] & p \text{ even} \\ x_{rp}\cos[(p-1)t/2] & p \text{ odd} \end{cases} \tag{1.25}$$

for each data point $\boldsymbol{x}_r$, $r = 1, \dots, n$, over the interval $-\pi \le t \le \pi$. Thus, there will be n harmonic curves drawn in two dimensions. Two data points are compared by visually studying the curves over $[-\pi, \pi]$. Note that the square of the L_2 distance

$$\int_{-\pi}^{\pi} [f_{\boldsymbol{x}}(t) - f_{\boldsymbol{y}}(t)]^2 \, dt$$

between two curves $f_{\boldsymbol{x}}(t)$ and $f_{\boldsymbol{y}}(t)$ simplifies to

$$\pi \|\boldsymbol{x} - \boldsymbol{y}\|^2,$$

which is proportional to the square of the Euclidean distance between $\boldsymbol{x}$ and $\boldsymbol{y}$.

Some practical hints for the plotting of harmonic curves with a large number of objects are given in Gnanadesikan (1977). Here, we use a small simulated data set to illustrate their use. In this data set, there are $n = 20$ observations with $p = 11$ variables. The means and standard deviations of the first 15 (group 1) observations and the remaining 5 (group 2) observations are given in Table 1.5. It can be noted that these differ mainly in the mean values of the last three variables.

The harmonic curves are plotted in Figure 1.5, for the original data, the data with each value increased by 10, the standardization $(\boldsymbol{x}_{(j)} - \overline{x}_j)/s_j$, $j = 1, \dots, p$, where $\overline{x}_j$ is the mean of the jth column $\boldsymbol{x}_{(j)}$, and a random permutation of the columns. Given that harmonic curves depend on the order in which the variables are written down, as well as the location and

Table 1.5 Sample means $\overline{\boldsymbol{x}}$ and standard deviations (s_j) for the two groups of data, with $n_1 = 15$ and $n_2 = 5$ observations.

	vars	1	2	3	4	5	6	7	8	9	10	11
Group 1	Mean	0.4	0.4	−0.1	−0.3	0.2	−0.4	0.0	0.2	0.1	0.0	0.1
	SD	0.9	1.2	1.0	0.6	1.2	1.3	1.1	0.8	0.8	1.1	1.3
Group 2	Mean	0.7	0.0	0.1	0.2	0.1	−0.4	−0.2	0.5	5.9	5.0	5.1
	SD	1.2	0.6	0.9	1.3	0.8	0.5	1.2	1.4	1.0	1.0	0.6

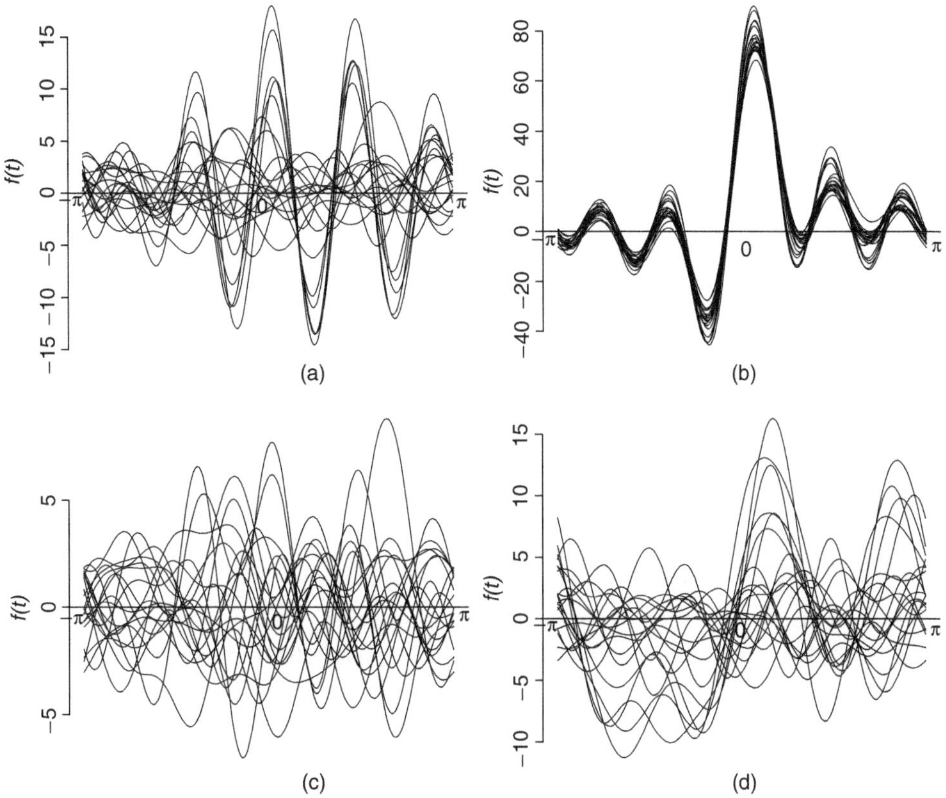

Figure 1.5 Harmonic curves for simulated data with $n = 10$ observations and $p = 11$ variables. (A summary of the two groups is given in Table 1.5.) (a) Original data; (b) data translated in all variables by 10; (c) each variable is standardized (by subtracting the mean and dividing by the standard deviation for that variable); (d) a random permutation $(7, 9, 3, 1, 10, 11, 6, 8, 5, 2, 4)$ of the variables.

scaling of the data, their interpretation requires some care. In the first plot, the second group of five observations gives curves that are away from the main group, but this is much less clear in the other plots.

One of the main uses of harmonic plots is to identify clusters in high-dimensional data. Further methods for cluster analysis are described in Chapter 14.

1.7.4 Parallel Coordinates Plot

In Figure 1.1, each observation is plotted once on each row of the plot. A parallel coordinates plot simply connects the points that belong to the same observation. For the cork data, this gives the plot shown in Figure 1.6. For these data, all the variables are measured in the same units, and so, it is appropriate to retain the scale for each axis. However, in general, each variable should be rescaled so that (for example) the range occupies $[0, 1]$. Such plots can then reveal clusters of observations and correlations between variables, provided that

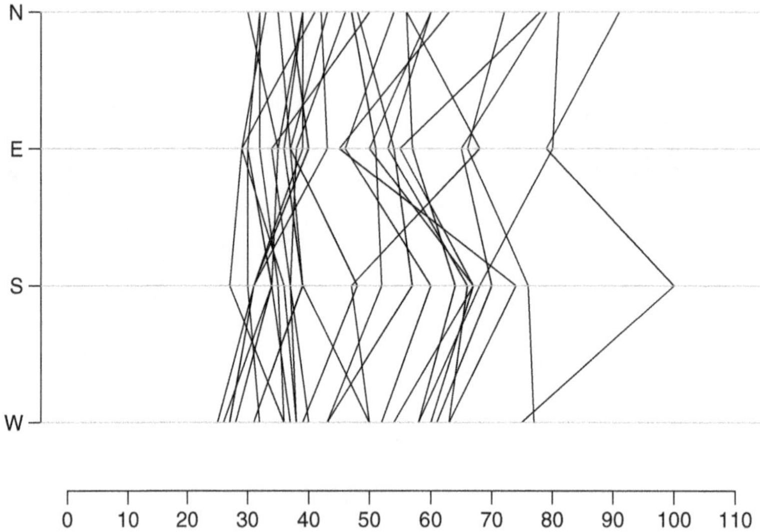

Figure 1.6 Parallel coordinates plot for the cork data.

the variables are appropriately (re-)ordered; see Wegman (1990) for some examples and extensions to this graphical tool.

1.8 Measures of Multivariate Skewness and Kurtosis

In Section 1.4, we have given a few basic summary statistics based on the first- and the second-order moments. In general, a kth-order central moment for the variables $i_1, \cdots, i_s$ is

$$M_{i_1,\ldots,i_s}^{(j_1,\ldots,j_s)} = \frac{1}{n}\sum_{r=1}^{n}\prod_{t=1}^{s}(x_{ri_t} - \overline{x}_{i_t})^{j_t},$$

where $j_1 + \cdots + j_s = k,\ j_t \neq 0,\ t = 1, \ldots, s$. As with mean and variance, we would like extensions to the multivariate case of summary measures such as $b_1 = m_3^2/s^6$ and $b_2 = m_4/s^4$, the univariate measures of skewness-squared, and kurtosis. Here,

$$m_k = \frac{1}{n}\sum_{r=1}^{n}(x_r - \overline{x})^k$$

is the (univariate) kth sample moment about the mean.

Using the invariant functions

$$g_{rs} = (\boldsymbol{x}_r - \overline{\boldsymbol{x}})'\boldsymbol{S}^{-1}(\boldsymbol{x}_s - \overline{\boldsymbol{x}}),$$

Mardia (1970b) has defined multivariate measures of skewness and kurtosis by

$$b_{1,p} = \frac{1}{n^2}\sum_{r,s=1}^{n} g_{rs}^3 \tag{1.26}$$

and

$$b_{2,p} = \frac{1}{n}\sum_{r=1}^{n} g_{rr}^2. \tag{1.27}$$

The following properties are worth noting:

(1) $b_{1,p}$ depends only on the moments up to third order, whereas $b_{2,p}$ depends only on the moments up to fourth order excluding the third-order moments.
(2) These measures are invariant under *affine* transformations

$$\boldsymbol{y} = \boldsymbol{A}\boldsymbol{x} + \boldsymbol{b}.$$

(Similar properties hold for b_1 and b_2 under changes of scale and origin.)
(3) For $p = 1$, $b_{1,p} = b_1$ and $b_{2,p} = b_2$.
(4) Let D_r be the Mahalanobis distance between $\boldsymbol{x}_r$ and $\overline{\boldsymbol{x}}$, and let

$$\cos\alpha_{rs} = (\boldsymbol{x}_r - \overline{\boldsymbol{x}})'\boldsymbol{S}^{-1}(\boldsymbol{x}_s - \overline{\boldsymbol{x}})/D_r D_s$$

denote the cosine of the Mahalanobis angle between $(\boldsymbol{x}_r - \overline{\boldsymbol{x}})$ and $(\boldsymbol{x}_s - \overline{\boldsymbol{x}})$. Then, Equations (1.26) and (1.27) reduce to

$$b_{1,p} = \frac{1}{n^2}\sum_{r=1}^{n}\sum_{s=1}^{n}(D_r D_s \cos\alpha_{rs})^3 \tag{1.28}$$

and

$$b_{2,p} = \frac{1}{n}\sum_{r=1}^{n} D_r^4. \tag{1.29}$$

Example 1.8.1 For the iris data shown in Table 1.2, the dimension p is equal to 4, and we compute $b_{1,4}$ and $b_{2,4}$ for each species and the pooled data. The results are shown in Table 1.6. It can be seen that the species have similar skewness and kurtosis values. □

Although $b_{1,p}$ cannot be negative, we note from (1.28) that if the data points are uniformly distributed on a p-dimensional hypersphere, then $b_{1,p}$ will be small (for example, typically less than 1 for $2 \leq p \leq 10$) since $\sum \cos\alpha_{rs} \simeq 0$ and D_r and D_s are approximately equal. We should expect $b_{1,p} \gg 0$ if there is a departure from spherical symmetry.

The statistic $b_{2,p}$ will pick up extreme behavior in the Mahalanobis distances of objects from the sample mean. The use of $b_{1,p}$ and $b_{2,p}$ to detect departure from multivariate normality is described in Chapter 6.

Table 1.6 Measures of multivariate skewness $b_{1,p}$ and kurtosis $b_{2,p}$ for each of the iris species ($n = 50, p = 4$) and the pooled data.

	I. setosa	*I. versicolor*	*I. virginica*	Pooled
Skewness	3.079	3.022	3.152	2.697
Kurtosis	26.54	22.88	24.30	23.74

Exercises and Complements

1.4.1 Under the transformation

$$\boldsymbol{y}_r = \boldsymbol{A}\boldsymbol{x}_r + \boldsymbol{b}, \quad r = 1, \ldots, n$$

show that

$$\text{(i) } \overline{\boldsymbol{y}} = \boldsymbol{A}\overline{\boldsymbol{x}} + \boldsymbol{b}, \qquad \text{(ii) } \boldsymbol{S}_y = \boldsymbol{A}\boldsymbol{S}_x\boldsymbol{A}',$$

where $\overline{\boldsymbol{x}}$ and $\boldsymbol{S}_x$ are the mean vector and the covariance matrix for the $\boldsymbol{x}_r$.

1.4.2 Let

$$\boldsymbol{S}(\boldsymbol{a}) = \frac{1}{n}\sum_{r=1}^{n}(\boldsymbol{x}_r - \boldsymbol{a})(\boldsymbol{x}_r - \boldsymbol{a})'$$

be the covariance matrix about $\boldsymbol{x} = \boldsymbol{a}$. Show that

$$\boldsymbol{S}(\boldsymbol{a}) = \boldsymbol{S} + (\overline{\boldsymbol{x}} - \boldsymbol{a})(\overline{\boldsymbol{x}} - \boldsymbol{a})'.$$

(i) Using (A.20) or otherwise, show that

$$|\boldsymbol{S}(\boldsymbol{a})| = |\boldsymbol{S}|\{1 + (\overline{\boldsymbol{x}} - \boldsymbol{a})'\boldsymbol{S}^{-1}(\overline{\boldsymbol{x}} - \boldsymbol{a})\},$$

and consequently,

$$\min_{\boldsymbol{a}} |\boldsymbol{S}(\boldsymbol{a})| = |\boldsymbol{S}|.$$

(ii) Show that $\min_{\boldsymbol{a}} \operatorname{tr} \boldsymbol{S}(\boldsymbol{a}) = \operatorname{tr} \boldsymbol{S}$.

(iii) Note that $|\boldsymbol{S}| = \prod l_i$ and $\operatorname{tr} \boldsymbol{S} = \sum l_i$ are monotonically increasing symmetric functions of the eigenvalues $l_1, \ldots l_p$ of $\boldsymbol{S}$. Use this observation in constructing other measures of multivariate scatter.

(iv) Using the inequality $g \le a$, where a and g are the arithmetic and geometric means of a set of positive numbers, show that $|\boldsymbol{S}| \le (n^{-1}\operatorname{tr} \boldsymbol{S})^n$.

1.4.3 (a) Fisher (1947) gives data relating to the body weight in kilograms (x_1) and the heart weight in grams (x_2) of 144 cats. For the 47 female cats, the sums and sums of squares and products are given by

$$\boldsymbol{X}_1'\boldsymbol{1} = \begin{pmatrix} 110.9 \\ 432.5 \end{pmatrix} \text{ and } \boldsymbol{X}_1'\boldsymbol{X}_1 = \begin{bmatrix} 265.13 & 1029.62 \\ 1029.62 & 4064.71 \end{bmatrix}.$$

Show that the mean vector and covariance matrix are $\overline{\boldsymbol{x}}$ and $\boldsymbol{S}$ given in Example 5.2.3.

(b) For the 97 male cats from Fisher's data, the statistics are

$$\boldsymbol{X}_2'\boldsymbol{1} = \begin{pmatrix} 281.3 \\ 1098.3 \end{pmatrix} \text{ and } \boldsymbol{X}_2'\boldsymbol{X}_2 = \begin{bmatrix} 836.75 & 3275.55 \\ 3275.55 & 13056.17 \end{bmatrix}.$$

Find the mean vector and covariance matrix for the sample of male cats.

(c) Regarding the 144 male and female cats as a single sample, calculate the mean vector and covariance matrix.

(d) Calculate the correlation coefficient for each of (a), (b), and (c).

1.5.1 Let $\boldsymbol{M} = \boldsymbol{X}'\boldsymbol{X}$, where $\boldsymbol{X}$ is a data matrix. Show that

$$m_{ii} = \boldsymbol{x}'_{(i)}\boldsymbol{x}_{(i)} = n(s_{ii} + \bar{x}_i^2), \qquad m_{ij} = \boldsymbol{x}'_{(i)}\boldsymbol{x}_{(j)} = n(s_{ij} + \bar{x}_i\bar{x}_j).$$

1.5.2 Show that the scaling transformation in Section 1.5.1 can be written as

$$\boldsymbol{Y} = \boldsymbol{H}\boldsymbol{X}\boldsymbol{D}^{-1}, \qquad \boldsymbol{Y}' = [\boldsymbol{y}_1, \ldots, \boldsymbol{y}_n].$$

Use the fact that

$$\boldsymbol{Y}'\boldsymbol{1} = \boldsymbol{0}, \qquad \boldsymbol{Y}'\boldsymbol{H}\boldsymbol{Y} = \boldsymbol{D}^{-1}\boldsymbol{X}'\boldsymbol{H}\boldsymbol{X}\boldsymbol{D}^{-1}$$

to show that

$$\bar{\boldsymbol{y}} = \boldsymbol{0}, \qquad \boldsymbol{S}_y = \boldsymbol{R}.$$

1.5.3 Show that the Mahalanobis transformation in Section 1.5.2 can be written as

$$\boldsymbol{Z} = \boldsymbol{H}\boldsymbol{X}\boldsymbol{S}^{-1/2}, \qquad \boldsymbol{Z}' = [\boldsymbol{z}_1, \ldots, \boldsymbol{z}_n].$$

Hence, following the method of Exercise 1.5.2, show that $\bar{\boldsymbol{z}} = \boldsymbol{0}$, and $\boldsymbol{S}_z = \boldsymbol{I}$.

1.5.4 For the mark data in Table 1.1, show that

$$\bar{\boldsymbol{x}}' = (18.458, 74.400, 0.400),$$

$$\boldsymbol{S} = \begin{bmatrix} 0.0159 & -0.4352 & -0.0252 \\ & 101.8400 & 3.0400 \\ & & 0.2400 \end{bmatrix},$$

$$\boldsymbol{S}^{-1} = \begin{bmatrix} 76.7016 & 0.1405 & 6.2742 \\ & 0.0160 & -0.1885 \\ & & 7.2132 \end{bmatrix},$$

$$\boldsymbol{S}^{-1/2} = \begin{bmatrix} 8.7405 & 0.0205 & 0.5521 \\ & 0.1013 & -0.0732 \\ & & 2.6274 \end{bmatrix}.$$

1.6.1 If

$$D_{rs}^2 = (\boldsymbol{x}_r - \boldsymbol{x}_s)'\boldsymbol{S}^{-1}(\boldsymbol{x}_r - \boldsymbol{x}_s),$$

we may write

$$D_{rs}^2 = q_{rr} + q_{ss} - 2q_{rs},$$

where

$$q_{rs} = \boldsymbol{x}'_r\boldsymbol{S}^{-1}\boldsymbol{x}_s.$$

Writing

$$g_{rs} = (\boldsymbol{x}_r - \bar{\boldsymbol{x}})'\boldsymbol{S}^{-1}(\boldsymbol{x}_s - \bar{\boldsymbol{x}}),$$

we see that

$$g_{rs} = q_{rs} + \bar{\boldsymbol{x}}'\boldsymbol{S}^{-1}\bar{\boldsymbol{x}} - \boldsymbol{x}'_r\boldsymbol{S}^{-1}\bar{\boldsymbol{x}} - \boldsymbol{x}'_s\boldsymbol{S}^{-1}\bar{\boldsymbol{x}}.$$

Therefore,

$$D_{rs}^2 = g_{rr} + g_{ss} - 2g_{rs}.$$

(i) Show that

$$\sum_r g_{rs} = 0 \quad \text{and} \quad \sum_r g_{rr} = \sum_r \operatorname{tr} \boldsymbol{S}^{-1}(\boldsymbol{x}_r - \overline{\boldsymbol{x}})(\boldsymbol{x}_r - \overline{\boldsymbol{x}})' = n \operatorname{tr} \boldsymbol{I}_p = np.$$

(ii) Therefore, show that

$$\sum_r D_{rs}^2 = np + ng_{ss} \qquad \text{or} \qquad g_{ss} = \frac{1}{n}\sum_r D_{rs}^2 - p$$

and

$$\sum_{r,s=1}^{n} D_{rs}^2 = 2n^2 p.$$

1.8.1 (Mardia, 1970b) Let $u_{pq} = M_{pq}/s_1^p s_2^q$, where

$$M_{pq} = \frac{1}{n}\sum_{r=1}^{n} (x_{r1} - \overline{x}_1)^p (x_{r2} - \overline{x}_2)^q.$$

Show that

$$\begin{aligned} b_{1,2} &= (1 - r^2)^{-3}[u_{30}^2 + u_{03}^2 + 3(1 + 2r^2)(u_{12}^2 + u_{21}^2) - 2r^3 u_{30} u_{03} \\ &\quad + 6r\{u_{30}(ru_{12} - u_{21}) + u_{03}(ru_{21} - u_{12}) - (2 + r^2)u_{12}u_{21}\}] \end{aligned}$$

and

$$b_{2,2} = (1 - r^2)^{-2}[u_{40} + u_{04} + 2u_{22} + 4r(ru_{22} - u_{13} - u_{31})].$$

Hence, for $s_1 = s_2 = 1,\ r = 0$, show that

$$b_{1,2} = M_{30}^2 + M_{03}^2 + 3M_{12}^2 + 3M_{21}^2$$

and

$$b_{2,2} = M_{40} + M_{04} + 2M_{22}.$$

Thus, $b_{1,2}$ accumulates the effects of M_{21}, M_{12}, M_{03}, and M_{30}, while $b_{2,2}$ accumulates the effects of M_{22}, M_{04}, and M_{40}.

1.8.2 Carry out a simulation experiment to see if you can determine something about the distribution of $b_{1,p}$ and $b_{2,p}$ for the case that the data come from a normal distribution with mean $\boldsymbol{0}$ and covariance matrix $\boldsymbol{I}$. Specifically, for a variety of n (with fixed $p = 2$) and a variety of p with fixed n generate, say, 100 samples of size n from $N_p(\boldsymbol{0}, \boldsymbol{I})$, and for each sample compute $b_{1,p}$ and $b_{2,p}$. Compute the sample mean (and variance) of these quantities and plot them against n and p as appropriate. You may also consider a histogram of a larger sample (maybe 10 000) for $n = 100$ (say) and $p = 5$ (say) to investigate the shape of the two distributions.

2

Basic Properties of Random Vectors

Introduction

The study of multivariate analysis requires an extension of the concepts used in univariate statistics. The chapter starts with basic notation for probability distributions and densities, together with marginal and conditional densities. Later sections discuss moments, including the mean vector and covariance matrix, characteristic functions, and nonlinear transformations. The most important multivariate distribution is the multivariate normal distribution, which is investigated in depth starting with the bivariate case.

2.1 Cumulative Distribution Functions and Probability Density Functions

Let $\boldsymbol{X} = (X_1, \ldots, X_p)'$ be a random vector. In this book, we usually do not distinguish notationally between a random vector and its realization. However, in this subsection we use upper case symbols for random variables. By analogy with univariate theory, the *cumulative distribution function* (c.d.f.) associated with $\boldsymbol{x}$ is the function F defined by

$$F(\boldsymbol{x}) = \mathsf{P}(\boldsymbol{X} \leq \boldsymbol{x}) = \mathsf{P}(X_1 \leq x_1, \ldots, X_p \leq x_p). \tag{2.1}$$

Two important cases are absolutely continuous and discrete distributions.

A random vector $\boldsymbol{x}$ is *absolutely continuous* if there exists a *probability density function* (p.d.f.), $f(\boldsymbol{x})$, such that

$$F(\boldsymbol{x}) = \int_{-\infty}^{\boldsymbol{x}} f(\boldsymbol{u})\, \mathrm{d}\boldsymbol{u}.$$

Here, $\mathrm{d}\boldsymbol{u} = \mathrm{d}u_1 \cdots \mathrm{d}u_p$ represents the product of p differential elements, and the integral sign denotes p-fold integration. Note that for any measurable set $D \subseteq \mathbb{R}^p$,

$$\mathsf{P}(\boldsymbol{X} \in D) = \int_D f(\boldsymbol{u})\, \mathrm{d}\boldsymbol{u} \tag{2.2}$$

and

$$\int_{-\infty}^{\infty} f(\boldsymbol{u})\, \mathrm{d}\boldsymbol{u} = 1. \tag{2.3}$$

Multivariate Analysis, Second Edition. Kanti V. Mardia, John T. Kent, and Charles C. Taylor.

For a *discrete* random vector $\boldsymbol{X}$, the total probability is concentrated on a countable (or finite) set of points $\{\boldsymbol{x}_j;\ j = 1, 2, \dots\}$. Discrete random vectors can be handled in the above framework if we define the probability function (p.f.) of $\boldsymbol{X}$

$$\begin{aligned} f(\boldsymbol{x}_j) &= \mathsf{P}(\boldsymbol{X} = \boldsymbol{x}_j), \quad && j = 1, 2, \dots, \\ f(\boldsymbol{x}) &= 0, && \text{otherwise}, \end{aligned} \tag{2.4}$$

and replace the integration in (2.2) by the summation

$$\mathsf{P}(\boldsymbol{X} \in D) = \sum_{j:\ \boldsymbol{x}_j \in D} f(\boldsymbol{x}_j). \tag{2.5}$$

However, most of the emphasis in this book is directed toward absolutely continuous random vectors.

The support S of $\boldsymbol{x}$ is defined as the set

$$S = \{\boldsymbol{x} \in \mathbb{R}^p : f(\boldsymbol{x}) > 0\}. \tag{2.6}$$

In examples, the p.d.f. is usually defined only on S, with the value zero elsewhere being assumed.

Marginal and conditional distributions Consider the partitioned vector $\boldsymbol{x}' = (\boldsymbol{x}_1', \boldsymbol{x}_2')$, where $\boldsymbol{x}_1$ and $\boldsymbol{x}_2$ have k and $(p - k)$ elements, respectively $(k < p)$. The function

$$\mathsf{P}(\boldsymbol{X}_1 \le \boldsymbol{x}_1) = F(x_1, \dots, x_k, \infty, \dots, \infty)$$

is called the *marginal cumulative distribution function* (marginal c.d.f.) of $\boldsymbol{X}_1$. In contrast $F(\boldsymbol{x})$ may then be described as the *joint* c.d.f., and $f(\boldsymbol{x})$ may be called the *joint* p.d.f.

Let $\boldsymbol{X}$ have joint p.d.f. $f(\boldsymbol{x})$. Then, the *marginal* p.d.f. of $\boldsymbol{X}_1$ is given by the integral of $f(\boldsymbol{x})$ over $\boldsymbol{x}_2$; that is,

$$f_1(\boldsymbol{x}_1) = \int_{-\infty}^{\infty} f(\boldsymbol{x}_1, \boldsymbol{x}_2)\, \mathrm{d}\boldsymbol{x}_2. \tag{2.7}$$

The marginal p.d.f. of $\boldsymbol{X}_2$, $f_2(\boldsymbol{x}_2)$, is defined similarly.

For a given value of $\boldsymbol{X}_1$, say, $\boldsymbol{X}_1 = \boldsymbol{x}_1$, the *conditional* p.d.f. of $\boldsymbol{X}_2$ is proportional to $f(\boldsymbol{x}_1, \boldsymbol{x}_2)$, where the constant of proportionality can be calculated from the fact that this p.d.f. must integrate to one. In other words, the conditional p.d.f. of $\boldsymbol{X}_2$, given $\boldsymbol{X}_1 = \boldsymbol{x}_1$, is

$$f(\boldsymbol{x}_2 | \boldsymbol{X}_1 = \boldsymbol{x}_1) = \frac{f(\boldsymbol{x}_1, \boldsymbol{x}_2)}{f_1(\boldsymbol{x}_1)}. \tag{2.8}$$

(It is assumed that $f_1(\boldsymbol{x}_1)$ is nonzero.) The conditional p.d.f. of $\boldsymbol{X}_1$, given $\boldsymbol{X}_2 = \boldsymbol{x}_2$, can be defined similarly.

In general, two random variables can each have the same marginal distribution, even when their joint distributions are different. For instance, the marginal p.d.f.s of the following joint p.d.f.s,

$$f(x_1, x_2) = 1, \qquad 0 < x_1, x_2 < 1, \tag{2.9}$$

and (Morgenstern, 1956)

$$f(x_1, x_2) = 1 + \alpha(2x_1 - 1)(2x_2 - 1), \qquad 0 < x_1, x_2 < 1,\ -1 \le \alpha \le 1, \tag{2.10}$$

are both uniform, although the two joint distributions are different.

Independence When the conditional p.d.f. $f(\boldsymbol{x}_2|\boldsymbol{X}_1 = \boldsymbol{x}_1)$ is the same for all values of $\boldsymbol{x}_1$, then we say that $\boldsymbol{X}_1$ and $\boldsymbol{X}_2$ are *statistically independent* of each other. In such situations, $f(\boldsymbol{x}_2|\boldsymbol{X}_1 = \boldsymbol{x}_1)$ must be $f_2(\boldsymbol{x}_2)$. Hence, the joint density must equal the product of the marginals, as stated in the following theorem.

Theorem 2.1.1 *If* $\boldsymbol{X}_1$ and $\boldsymbol{X}_2$ *are statistically independent, then*

$$f(\boldsymbol{x}) = f_1(\boldsymbol{x}_1)f_2(\boldsymbol{x}_2).$$

Note that for the p.d.f. given by (2.9), the variables x_1 and x_2 are independent, whereas the variables in (2.10) are dependent.

2.2 Population Moments

In this section, we give the population analogues of the sample moments, which were discussed in Section 1.4.

2.2.1 Expectation and Correlation

If $\boldsymbol{x}$ is a random vector with p.d.f. $f(\boldsymbol{x})$, then the *expectation* or *mean* of a scalar-valued function $g(\boldsymbol{x})$ is defined as

$$\mathsf{E}[g(\boldsymbol{x})] = \int_{-\infty}^{\infty} g(\boldsymbol{x})f(\boldsymbol{x})\,d\boldsymbol{x}. \tag{2.11}$$

We assume that all necessary integrals converge, so that the expectations are finite. Expectations have the following properties:

(1) Linearity,

$$\mathsf{E}[a_1g_1(\boldsymbol{x}) + a_2g_2(\boldsymbol{x})] = a_1\mathsf{E}[g_1(\boldsymbol{x})] + a_2\mathsf{E}[g_2(\boldsymbol{x})]. \tag{2.12}$$

(2) Partition, $\boldsymbol{x}' = (\boldsymbol{x}_1', \boldsymbol{x}_2')$. The expectation of a function of $\boldsymbol{x}_1$ may be written in terms of the marginal distribution of $\boldsymbol{x}_1$ as follows:

$$\mathsf{E}[g(\boldsymbol{x}_1)] = \int_{-\infty}^{\infty} g(\boldsymbol{x}_1)f(\boldsymbol{x})\,d\boldsymbol{x} = \int_{-\infty}^{\infty} g(\boldsymbol{x}_1)f_1(\boldsymbol{x}_1)\,d\boldsymbol{x}_1. \tag{2.13}$$

When f_1 is known, the second expression is useful for computation.

(3) If $\boldsymbol{x}_1$ and $\boldsymbol{x}_2$ are independent and $g_i(\boldsymbol{x}_i)$ is a function of $\boldsymbol{x}_i$ *only* ($i = 1, 2$), then

$$\mathsf{E}[g_1(\boldsymbol{x}_1)g_2(\boldsymbol{x}_2)] = \mathsf{E}[g_1(\boldsymbol{x}_1)]\mathsf{E}[g_2(\boldsymbol{x}_2)]. \tag{2.14}$$

More generally, the expectation of a *matrix*-valued (or vector-valued) function of $\boldsymbol{x}$, $\boldsymbol{G}(\boldsymbol{x}) = (g_{ij}(\boldsymbol{x}))$, is defined to be the matrix

$$\mathsf{E}[\boldsymbol{G}(\boldsymbol{x})] = (\mathsf{E}[g_{ij}(\boldsymbol{x})]).$$

2.2.2 Population Mean Vector and Covariance Matrix

The vector $\mathsf{E}(\boldsymbol{x}) = \boldsymbol{\mu}$ is called the *population mean vector* of $\boldsymbol{x}$. Thus,

$$\mu_i = \int_{-\infty}^{\infty} x_i f(\boldsymbol{x})\,d\boldsymbol{x}, \quad i = 1, \dots, p.$$

The population mean possesses the linearity property

$$\mathsf{E}(\boldsymbol{Ax}+\boldsymbol{b}) = \boldsymbol{A}\mathsf{E}(\boldsymbol{x})+\boldsymbol{b}, \tag{2.15}$$

where $\boldsymbol{A}(q\times p)$ and $\boldsymbol{b}(q\times 1)$ are constant.

The matrix

$$\mathsf{E}[(\boldsymbol{x}-\boldsymbol{\mu})(\boldsymbol{x}-\boldsymbol{\mu})'] = \boldsymbol{\Sigma} = \text{var}(\boldsymbol{x}) \tag{2.16}$$

is called the *covariance matrix* of $\boldsymbol{x}$ (also known as the variance–covariance, variance or dispersion matrix). For conciseness, write

$$\boldsymbol{x} \sim (\boldsymbol{\mu}, \boldsymbol{\Sigma}) \tag{2.17}$$

to describe a random vector with mean vector $\boldsymbol{\mu}$ and covariance matrix $\boldsymbol{\Sigma}$.

More generally, we can define the covariance between two vectors, $\boldsymbol{x}(p\times 1)$ and $\boldsymbol{y}(q\times 1)$, by the $(p\times q)$ matrix

$$\text{cov}(\boldsymbol{x},\boldsymbol{y}) = \mathsf{E}[(\boldsymbol{x}-\boldsymbol{\mu})(\boldsymbol{y}-\boldsymbol{\nu})'], \tag{2.18}$$

where $\boldsymbol{\mu}=\mathsf{E}(\boldsymbol{x}), \boldsymbol{\nu}=\mathsf{E}(\boldsymbol{y})$. Notice the following simple properties of covariances. Let $\text{var}(\boldsymbol{x}) = \boldsymbol{\Sigma} = (\sigma_{ij})$.

(1) $\sigma_{ij} = \text{cov}(x_i, x_j),\ i\neq j;\quad \sigma_{ii} = \text{var}(x_i) = \sigma_i^2,\quad$ say.

(2) $\boldsymbol{\Sigma} = \mathsf{E}(\boldsymbol{xx}') - \boldsymbol{\mu\mu}'$. (2.19)

(3) $\text{var}(\boldsymbol{a}'\boldsymbol{x}) = \boldsymbol{a}'\text{var}(\boldsymbol{x})\boldsymbol{a} = \sum a_i a_j \sigma_{ij}$ for all constant vectors $\boldsymbol{a}$. (2.20)

Since the left-hand side of (2.20) is always nonnegative, we get the following result:

(4) $\boldsymbol{\Sigma} \geq 0$.

(5) $\text{var}(\boldsymbol{Ax}+\boldsymbol{b}) = \boldsymbol{A}\,\text{var}(\boldsymbol{x})\boldsymbol{A}'$. (2.21)

(6) $\text{cov}(\boldsymbol{x},\boldsymbol{x}) = \text{var}(\boldsymbol{x})$. (2.22)

(7) $\text{cov}(\boldsymbol{x},\boldsymbol{y}) = \text{cov}(\boldsymbol{y},\boldsymbol{x})'$.

(8) $\text{cov}(\boldsymbol{x}_1+\boldsymbol{x}_2,\boldsymbol{y}) = \text{cov}(\boldsymbol{x}_1,\boldsymbol{y}) + \text{cov}(\boldsymbol{x}_2,\boldsymbol{y})$. (2.23)

(9) If $p=q$, $\text{var}(\boldsymbol{x}+\boldsymbol{y}) = \text{var}(\boldsymbol{x}) + \text{cov}(\boldsymbol{x},\boldsymbol{y}) + \text{cov}(\boldsymbol{y},\boldsymbol{x}) + \text{var}(\boldsymbol{y})$. (2.24)

(10) $\text{cov}(\boldsymbol{Ax},\boldsymbol{By}) = \boldsymbol{A}\text{cov}(\boldsymbol{x},\boldsymbol{y})\boldsymbol{B}'$. (2.25)

(11) If $\boldsymbol{x}$ and $\boldsymbol{y}$ are independent, then $\text{cov}(\boldsymbol{x},\boldsymbol{y}) = \boldsymbol{0}$. However the converse is *not* true. See Exercise 2.2.2. (2.26)

Example 2.2.1 Let

$$f(x_1,x_2) = \begin{cases} x_1+x_2, & 0\leq x_1,x_2\leq 1,\\ 0, & \text{otherwise}. \end{cases} \tag{2.27}$$

Then,

$$\boldsymbol{\mu} = \begin{pmatrix}\mathsf{E}(x_1)\\ \mathsf{E}(x_2)\end{pmatrix} = \begin{pmatrix}7/12\\ 7/12\end{pmatrix},\qquad \boldsymbol{\Sigma} = \begin{bmatrix}\sigma_{11} & \sigma_{12}\\ \sigma_{21} & \sigma_{22}\end{bmatrix} = \begin{bmatrix}11/144 & -1/144\\ -1/144 & 11/144\end{bmatrix}.$$ □

Correlation matrix The population correlation matrix is defined in a manner similar to its sample counterpart. Let us denote the correlation coefficient between the ith and jth variables by ρ_{ij}, so that

$$\rho_{ij} = \sigma_{ij}/(\sigma_i\sigma_j),\quad i\neq j.$$

The matrix

$$P = (\rho_{ij}) \tag{2.28}$$

with $\rho_{ii} = 1$ is called the *population correlation matrix*. Taking $\boldsymbol{\Delta} = \text{diag}(\sigma_i)$, we have

$$P = \boldsymbol{\Delta}^{-1}\boldsymbol{\Sigma}\boldsymbol{\Delta}^{-1}.$$

The matrix $P \geq 0$ because $\boldsymbol{\Sigma} \geq 0$, and $\boldsymbol{\Delta}$ is symmetric.

Generalized variance By analogy with Section 1.4.2, we may also define the population *generalized variance* and *total variation* as $|\boldsymbol{\Sigma}|$ and $\text{tr}\,\boldsymbol{\Sigma}$, respectively.

2.2.3 Mahalanobis Space

We now turn to the population analogue of the Mahalanobis distance given by (1.24). If $\boldsymbol{x}$ and $\boldsymbol{y}$ are two points in space, then the *Mahalanobis distance* between $\boldsymbol{x}$ and $\boldsymbol{y}$, with $\boldsymbol{\Sigma}$, is the square root of

$$\Delta^2_{\boldsymbol{\Sigma}}(\boldsymbol{x},\boldsymbol{y}) = (\boldsymbol{x}-\boldsymbol{y})'\boldsymbol{\Sigma}^{-1}(\boldsymbol{x}-\boldsymbol{y}). \tag{2.29}$$

(The subscript $\boldsymbol{\Sigma}$ may be omitted when there is no risk of confusion.) The matrix $\boldsymbol{\Sigma}$ is usually selected to be some convenient covariance matrix. Some examples are as follows:

(1) Let $\boldsymbol{x} \sim (\boldsymbol{\mu}_1, \boldsymbol{\Sigma})$ and $\boldsymbol{y} \sim (\boldsymbol{\mu}_2, \boldsymbol{\Sigma})$. Then, $\Delta(\boldsymbol{\mu}_1, \boldsymbol{\mu}_2)$ is a Mahalanobis distance between the parameters. It is invariant under transformations of the form

$$\boldsymbol{x} \to \boldsymbol{A}\boldsymbol{x} + \boldsymbol{b}, \qquad \boldsymbol{y} \to \boldsymbol{A}\boldsymbol{y} + \boldsymbol{b}, \qquad \boldsymbol{\Sigma} \to \boldsymbol{A}\boldsymbol{\Sigma}\boldsymbol{A}',$$

where $\boldsymbol{A}$ is a nonsingular matrix.

(2) Let $\boldsymbol{x} \sim (\boldsymbol{\mu}, \boldsymbol{\Sigma})$. The Mahalanobis distance between $\boldsymbol{x}$ and $\boldsymbol{\mu}$, $\boldsymbol{\Delta}(\boldsymbol{x}, \boldsymbol{\mu})$, is here a random variable.

(3) Let $\boldsymbol{x} \sim (\boldsymbol{\mu}_1, \boldsymbol{\Sigma}), \boldsymbol{y} \sim (\boldsymbol{\mu}_2, \boldsymbol{\Sigma})$. The Mahalanobis distance between $\boldsymbol{x}$ and $\boldsymbol{y}$ is $\Delta(\boldsymbol{x}, \boldsymbol{y})$.

2.2.4 Higher Moments

Following Section 1.8, a kth-order central moment for the variables $x_{i_1}, \ldots, x_{i_s}$ is

$$\mu^{(j_1,\ldots,j_s)}_{i_1,\ldots,i_s} = \mathsf{E}\left\{\prod_{t=1}^{s}(x_{i_t} - \mu_{i_t})^{j_t}\right\}, \tag{2.30}$$

where $j_1 + \cdots + j_s = k$, $j_t \neq 0$, $t = 1, \ldots, s$. Further, suitable population counterparts of the measures of multivariate skewness and kurtosis for independent and identically distributed random vectors $\boldsymbol{x}, \boldsymbol{y}$ with mean $\boldsymbol{\mu}$ and covariance $\boldsymbol{\Sigma}$ are, respectively,

$$\beta_{1,p} = \mathsf{E}[(\boldsymbol{x} - \boldsymbol{\mu})'\boldsymbol{\Sigma}^{-1}(\boldsymbol{y} - \boldsymbol{\mu})]^3, \tag{2.31}$$

$$\beta_{2,p} = \mathsf{E}[(\boldsymbol{x} - \boldsymbol{\mu})'\boldsymbol{\Sigma}^{-1}(\boldsymbol{x} - \boldsymbol{\mu})]^2, \tag{2.32}$$

(see Mardia (1970b)). The quantities given in (2.31) and (2.32) are known as Mardia's coefficients of multivariate skewness and kurtosis, respectively. It can be seen that these measures are invariant under linear transformations. When $p = 2$, we may simplify the notation in (2.30) and write

$$\mu^{(r,s)}_{1,2} = \mathsf{E}\{(x_1 - \mu_1)^r(x_2 - \mu_2)^s\} = \mu_{rs}, \qquad \text{say}$$

and also denote $\gamma_{rs} = \mu_{rs}/(\sigma_1^r \sigma_2^s)$ as the standardized bivariate moment of order (r, s). Mardia (1970b) shows that (see also the sample counterpart in Exercise 1.8.1)

$$\beta_{1,2} = (1-\rho^2)^{-3}[\gamma_{30}^2 + \gamma_{03}^2 + 3(1+2\rho^2)(\gamma_{12}^2 + \gamma_{21}^2) - 2\rho^3\gamma_{30}\gamma_{03} + 6\rho\{\gamma_{30}(\rho\gamma_{12} - \gamma_{21}) + \gamma_{03}(\rho\gamma_{21} - \gamma_{12}) - (2+\rho^2)\gamma_{12}\gamma_{21}\}] \tag{2.33}$$

and

$$\beta_{2,2} = (1-\rho^2)^{-2}[\gamma_{40} + \gamma_{04} + 2\gamma_{22} + 4\rho(\rho\gamma_{22} - \gamma_{13} - \gamma_{31})]. \tag{2.34}$$

In particular, if the correlation is zero, we have

$$\beta_{1,2} = \gamma_{30}^2 + 3\gamma_{21}^2 + 3\gamma_{12}^2 + \gamma_{03}^2, \tag{2.35}$$

$$\beta_{2,2} = \gamma_{40} + 2\gamma_{22} + \gamma_{04}. \tag{2.36}$$

Example 2.2.2 Consider the Morgenstern (1956) distribution on the square given by

$$f(x_1, x_2) = 1 + \alpha(2x_1 - 1)(2x_2 - 1), \qquad 0 < x_1, x_2 < 1, -1 \le \alpha \le 1,$$

which has uniform marginals on $(0, 1)$. We have

$$\mathsf{E}(x_1) = \mathsf{E}(x_2) = \frac{1}{2}, \qquad \mu_{r,s} = \mathsf{E}\left\{\left(x_1 - \frac{1}{2}\right)^r \left(x_2 - \frac{1}{2}\right)^s\right\} = \mu_r\mu_s + 4\alpha\mu_{r+1}\mu_{s+1},$$

where $\mu_{r,s} = 0$ when r or s is an odd integer, and

$$\mu_r = 2^{-r-1}\{1 + (-1)^r\}/(r+1),$$

which is the rth central moment for the uniform distribution on $(0, 1)$. Then,

$$\sigma_1^2 = \sigma_2^2 = \frac{1}{12}, \ \rho = \frac{1}{3}\alpha, \ \gamma_{12} = \gamma_{03} = 0, \ \gamma_{22} = 1, \gamma_{13} = \frac{9}{5}\rho, \ \gamma_{40} = \frac{9}{5}.$$

Consequently, using (2.33) and (2.34), we have

$$\beta_{1,2} = 0, \quad \beta_{2,2} = 4(7 - 13\rho^2)/\{5(1-\rho^2)^2\}.$$

Note that for $\rho = 0$, $\beta_{2,2} = 5.60$, and for $\rho = 1/3$, $\beta_{2,2} = 5.62$, so in both the extreme cases, the values are smaller than $\beta_{2,2} = 8$ for the normal distribution. Also note that for the uniform spherical case vs. this square case, we have from Example 3.3.5, $\beta_{2,2} = 5.33$, so the kurtosis values are lower for both of these nonnormal cases. □

2.2.5 Conditional Moments

Moments of $\boldsymbol{x}_1 \mid \boldsymbol{x}_2$ are called conditional moments. In particular, $\mathsf{E}(\boldsymbol{x}_1 \mid \boldsymbol{x}_2)$ and $\mathrm{var}(\boldsymbol{x}_1 \mid \boldsymbol{x}_2)$ are the conditional mean vector and the conditional variance–covariance matrix of $\boldsymbol{x}_1$, given $\boldsymbol{x}_2$. The *regression curve* of $\boldsymbol{x}_1$ on $\boldsymbol{x}_2$ is defined by the conditional expectation function

$$\mathsf{E}(\boldsymbol{x}_1 \mid \boldsymbol{x}_2),$$

defined on the support of $\boldsymbol{x}_2$. If this function is linear in $\boldsymbol{x}_2$, then the regression is called *linear*. The conditional variance function

$$\mathrm{var}(\boldsymbol{x}_1 \mid \boldsymbol{x}_2)$$

defines the *scedastic curve* of $\boldsymbol{x}_1$ on $\boldsymbol{x}_2$. The regression of $\boldsymbol{x}_1$ on $\boldsymbol{x}_2$ is called *homoscedastic* if $\mathrm{var}(\boldsymbol{x}_1 | \boldsymbol{x}_2)$ is a constant matrix.

Example 2.2.3 Consider the p.d.f. given by (2.27). The marginal density of x_2 is

$$f_2(x_2) = \int_0^1 (x_1 + x_2)\, dx_1 = x_2 + \frac{1}{2}, \qquad 0 < x_2 < 1.$$

Hence, the regression curve of x_1 on x_2 is

$$\mathsf{E}(x_1 \mid x_2) = \int_0^1 x_1 f(x_1 \mid x_2)\, \mathrm{d}x_1 = \int_0^1 \frac{x_1(x_1 + x_2)}{x_2 + \frac{1}{2}}\, \mathrm{d}x_1 = \frac{(3x_2 + 2)}{3(1 + 2x_2)}, \quad 0 < x_2 < 1.$$

This is a decreasing function of x_2, and the regression is not linear. Similarly,

$$\mathsf{E}(x_1^a \mid x_2) = \left(\frac{1}{a+2} + \frac{1}{a+1}x_2\right) \Big/ \left(\frac{1}{2} + x_2\right),$$

so that

$$\operatorname{var}(x_1 \mid x_2) = (1 + 6x_2 + 6x_2^2) / \{18(1 + 2x_2)^2\}, \quad 0 < x_2 < 1.$$

Hence, the regression is not homoscedastic. □

In general, if all the specified expectations exist, then

$$\mathsf{E}(\boldsymbol{x}_1) = \mathsf{E}[\mathsf{E}(\boldsymbol{x}_1 \mid \boldsymbol{x}_2)]. \tag{2.37}$$

However, note that the conditional expectations $\mathsf{E}(\boldsymbol{x}_1 \mid \boldsymbol{x}_2)$ may all be finite, even when $\mathsf{E}(\boldsymbol{x})$ is infinite (see Exercise 2.2.6).

2.3 Characteristic Functions

Let $\boldsymbol{x}$ be a random p-vector. Then, the characteristic function (c.f.) of $\boldsymbol{x}$ is defined as the function

$$\phi_{\boldsymbol{x}}(\boldsymbol{t}) = \mathsf{E}(\mathrm{e}^{i\boldsymbol{t}'\boldsymbol{x}}) = \int \mathrm{e}^{i\boldsymbol{t}'\boldsymbol{x}} f(\boldsymbol{x})\, \mathrm{d}\boldsymbol{x}, \quad \boldsymbol{t} \in \mathbb{R}^p, \tag{2.38}$$

where $i = \sqrt{-1}$. As in the univariate case, we have the following properties:

(1) The characteristic function always exists, $\phi_{\boldsymbol{x}}(\boldsymbol{0}) = 1$ and $|\phi_{\boldsymbol{x}}(\boldsymbol{t})| \le 1$.
(2) (Uniqueness theorem.) Two random vectors have the same c.f. if and only if they have the same distribution.
(3) (Inversion theorem.) If the c.f. $\phi_{\boldsymbol{x}}(t)$ is absolutely integrable, then $\boldsymbol{x}$ has a p.d.f. given by

$$f(\boldsymbol{x}) = \frac{1}{(2\pi)^p} \int_{-\infty}^{\infty} \mathrm{e}^{-i\boldsymbol{t}'\boldsymbol{x}} \phi_{\boldsymbol{x}}(\boldsymbol{t})\, \mathrm{d}\boldsymbol{t}. \tag{2.39}$$

(4) Partition $\boldsymbol{x}' = (\boldsymbol{x}_1', \boldsymbol{x}_2')$. The random vectors $\boldsymbol{x}_1$ and $\boldsymbol{x}_2$ are independent if and only if their joint c.f. factorizes into the product of their respective marginal c.f.s, that is if

$$\phi_{\boldsymbol{x}}(\boldsymbol{t}) = \phi_{\boldsymbol{x}_1}(\boldsymbol{t}_1)\phi_{\boldsymbol{x}_2}(\boldsymbol{t}_2), \tag{2.40}$$

where $\boldsymbol{t}' = (\boldsymbol{t}_1', \boldsymbol{t}_2')$.
(5)

$$\mathsf{E}(x_1^{j_1} \cdots x_p^{j_p}) = \frac{1}{i^{j_1+\cdots+j_p}} \left\{ \frac{\partial^{j_1+\cdots+j_p}}{\partial t_1^{j_1} \cdots \partial t_p^{j_p}} \phi_{\boldsymbol{x}}(\boldsymbol{t}) \right\}_{\boldsymbol{t}=\boldsymbol{0}}$$

when this moment exists. (For a proof, differentiate both sides of (2.38) and put $\boldsymbol{t} = \boldsymbol{0}$.)
(6) The c.f. of the marginal distribution of $\boldsymbol{x}_1$ is simply $\phi_{\boldsymbol{x}}(\boldsymbol{t}_1, \boldsymbol{0})$.
(7) If $\boldsymbol{x}$ and $\boldsymbol{y}$ are independent random p-vectors, then the c.f. of the *sum* $\boldsymbol{x} + \boldsymbol{y}$ is the *product* of the c.f.s of $\boldsymbol{x}$ and $\boldsymbol{y}$,

$$\phi_{\boldsymbol{x}+\boldsymbol{y}}(\boldsymbol{t}) = \phi_{\boldsymbol{x}}(\boldsymbol{t})\phi_{\boldsymbol{y}}(\boldsymbol{t}).$$

(To prove this, notice that independence implies $\mathsf{E}(\mathrm{e}^{i\boldsymbol{t}'(\boldsymbol{x}+\boldsymbol{y})}) = \mathsf{E}(\mathrm{e}^{i\boldsymbol{t}'\boldsymbol{x}})\mathsf{E}(\mathrm{e}^{i\boldsymbol{t}'\boldsymbol{y}})$.)

Example 2.3.1 Let $a(x_1, x_2)$ be a continuous positive function on an open set $S \subseteq \mathbb{R}^2$. Let

$$f(x_1, x_2) = a(x_1, x_2) q(\theta_1, \theta_2) \exp\{x_1\theta_1 + x_2\theta_2\}, \qquad \boldsymbol{x} \in S,$$

be a density defined for $(\theta_1, \theta_2) \in \{(\theta_1, \theta_2) : 1/q(\theta_1, \theta_2) < \infty\}$, where

$$1/q(\theta_1, \theta_2) = \int a(x_1, x_2) \exp\{x_1\theta_1 + x_2\theta_2\}\, \mathrm{d}x_1\, \mathrm{d}x_2$$

is a normalization constant. Since this integral converges absolutely and uniformly over compact sets of (θ_1, θ_2), q can be extended by analytic continuation to give

$$\phi_{x_1, x_2}(t_1, t_2) = \int \mathrm{e}^{it_1x_1 + it_2x_2} f(x_1, x_2)\, \mathrm{d}x_1 \mathrm{d}x_2 = q(\theta_1, \theta_2)/q(\theta_1 + it_1, \theta_2 + it_2).$$

□

We end this section with an important result.

Theorem 2.3.1 **(*Cramér-Wold*)** *The distribution of a random p-vector $\boldsymbol{x}$ is completely determined by the set of all one-dimensional distributions of linear combinations $\boldsymbol{t}'\boldsymbol{x}$ where $\boldsymbol{t} \in \mathbb{R}^p$ ranges through all fixed p-vectors.*

Proof: Let $y = \boldsymbol{t}'\boldsymbol{x}$, and let the c.f. of y be

$$\phi_y(s) = \mathsf{E}[\mathrm{e}^{isy}] = \mathsf{E}[\mathrm{e}^{ist'\boldsymbol{x}}].$$

Clearly, for $s = 1$, $\phi_y(1) = \mathsf{E}[\mathrm{e}^{i\boldsymbol{t}'\boldsymbol{x}}]$, which, regarded as a function of $\boldsymbol{t}$, is the c.f. of $\boldsymbol{x}$. □

The Cramér-Wold theorem implies that a multivariate probability distribution can be defined completely by specifying the distribution of all its linear combinations.

2.4 Transformations

Suppose that $f(\boldsymbol{x})$ is the p.d.f. of $\boldsymbol{x}$, and let $\boldsymbol{x} = \boldsymbol{u}(\boldsymbol{y})$ be a transformation from $\boldsymbol{y}$ to $\boldsymbol{x}$, which is one to one except possibly on sets of Lebesgue measure 0 in the supports of $\boldsymbol{x}$ and $\boldsymbol{y}$. Then, the p.d.f. of $\boldsymbol{y}$ is

$$f\{\boldsymbol{u}(\boldsymbol{y})\}J, \tag{2.41}$$

where J is the Jacobian of the transformation from $\boldsymbol{y}$ to $\boldsymbol{x}$. It is defined by

$$J = \text{absolute value of } |\boldsymbol{J}|, \qquad \boldsymbol{J} = \left(\frac{\partial x_i}{\partial y_j}\right), \tag{2.42}$$

and we suppose that J is never zero or infinite except possibly on a set of Lebesgue measure 0. For some problems, it is easier to compute J from

$$J^{-1} = \text{absolute value of } \left|\frac{\partial y_j}{\partial x_i}\right|, \tag{2.43}$$

using the inverse transformation $\boldsymbol{y} = \boldsymbol{u}^{-1}(\boldsymbol{x})$, and then substitute for $\boldsymbol{x}$ in terms of $\boldsymbol{y}$.

(1) *Linear transformation.* Let

$$\boldsymbol{y} = \boldsymbol{A}\boldsymbol{x} + \boldsymbol{b}, \tag{2.44}$$

where $\boldsymbol{A}$ is a nonsingular matrix. Clearly, $\boldsymbol{x} = \boldsymbol{A}^{-1}(\boldsymbol{y} - \boldsymbol{b})$. Therefore, $\partial x_i/\partial y_j = a^{ij}$, and the Jacobian of the transformation $\boldsymbol{y}$ to $\boldsymbol{x}$ is

$$\text{absolute value of } |\boldsymbol{A}|^{-1}. \tag{2.45}$$

(2) *Polar transformation.* A generalization of the two-dimensional polar transformation,

$$x_1 = r\cos\theta, \quad x_2 = r\sin\theta, \qquad r > 0, \quad 0 \le \theta < 2\pi,$$

to p dimensions is

$$\boldsymbol{x} = r\boldsymbol{u}(\boldsymbol{\theta}), \qquad \boldsymbol{\theta} = (\theta_1, \dots, \theta_{p-1})', \tag{2.46}$$

where

$$u_i(\boldsymbol{\theta}) = \cos\theta_i \prod_{j=0}^{i-1} \sin\theta_j, \qquad \sin\theta_0 = \cos\theta_p = 1,$$

and

$$0 \le \theta_j \le \pi, \quad j = 1, \dots, p-2, \qquad 0 \le \theta_{p-1} < 2\pi, \quad r > 0.$$

The Jacobian of the transformation from $(r, \boldsymbol{\theta})$ to $\boldsymbol{x}$ is

$$J = r^{p-1} \prod_{i=2}^{p-1} \sin^{p-i}\theta_{i-1}. \tag{2.47}$$

Note that the transformation is one to one except when $r = 0$ or $\theta_i = 0$ or π, for any $i = 1, \dots, p-2$.

(3) *Rosenblatt's transformation* (Rosenblatt, 1952). Suppose that $\boldsymbol{x}$ has p.d.f. $f(\boldsymbol{x})$ and denote the conditional c.d.f. of x_i, given $x_1, \dots, x_{i-1}$ by $F(x_i \mid x_1, \dots, x_{i-1})$, $i = 1, \dots, p$. The Jacobian of the transformation $\boldsymbol{x}$ to $\boldsymbol{y}$, where

$$y_i = F(x_i | x_1, \dots, x_{i-1}), \quad i = 1, \dots, p, \tag{2.48}$$

is given by $f(x_1, \dots, x_p)$. Hence, looking at the transformation $\boldsymbol{y}$ to $\boldsymbol{x}$, we see that $y_1, \dots, y_p$ are independent identically distributed uniform variables on $(0, 1)$.

Some other Jacobians useful in multivariate analysis are listed in Table 2.1. For their proof, see Deemer and Olkin (1951).

Table 2.1 Jacobians of some transformations.

Transformation *Y* **to** *X*	**Restriction**	**Jacobian (absolute value)**
$\boldsymbol{X} = \boldsymbol{Y}^{-1}$	$\boldsymbol{Y}(p \times p)$ and nonsingular (all elements random)	$\lvert\boldsymbol{Y}\rvert^{-2p}$
$\boldsymbol{X} = \boldsymbol{Y}^{-1}$	$\boldsymbol{Y}$ symmetric and nonsingular	$\lvert\boldsymbol{Y}\rvert^{-p-1}$
$\boldsymbol{X} = \boldsymbol{A}\boldsymbol{Y} + \boldsymbol{B}$	$\boldsymbol{Y}(p \times p), \boldsymbol{A}(p \times p)$ nonsingular, $\boldsymbol{B}(p \times p)$	$\lvert\boldsymbol{A}\rvert^{p}$
$\boldsymbol{X} = \boldsymbol{A}\boldsymbol{Y}\boldsymbol{B}$	$\boldsymbol{Y}(p \times q), \boldsymbol{A}(p \times p)$ and $\boldsymbol{B}(q \times q)$ nonsingular	$\lvert\boldsymbol{A}\rvert^{q}\lvert\boldsymbol{B}\rvert^{p}$
$\boldsymbol{X} = \boldsymbol{A}\boldsymbol{Y}\boldsymbol{A}'$	$\boldsymbol{Y}(p \times p)$, symmetric $\boldsymbol{A}(p \times p)$ nonsingular	$\lvert\boldsymbol{A}\rvert^{p+1}$
$\boldsymbol{X} = \boldsymbol{Y}\boldsymbol{Y}'$	$\boldsymbol{Y}$ lower triangular	$2^p \prod_{i=1}^{p} y_{ii}^{p+1-i}$

2.5 The Multivariate Normal Distribution

2.5.1 Definition

In this section, we introduce the most important multivariate probability distribution, namely the *multivariate normal* distribution. If we write the p.d.f. of $N(\mu, \sigma^2)$, the univariate normal with mean μ and variance $\sigma^2 > 0$, as

$$f(x) = \{2\pi\sigma^2\}^{-1/2} \exp\left\{-\frac{1}{2}(x-\mu)\{\sigma^2\}^{-1}(x-\mu)\right\},$$

then a plausible extension to p variates is

$$f(\boldsymbol{x}) = |2\pi\boldsymbol{\Sigma}|^{-1/2} \exp\left\{-\frac{1}{2}(\boldsymbol{x}-\boldsymbol{\mu})'\boldsymbol{\Sigma}^{-1}(\boldsymbol{x}-\boldsymbol{\mu})\right\}, \tag{2.49}$$

where $\boldsymbol{\Sigma} > 0$. (Observe that the constant can also be written as $\{(2\pi)^{p/2}|\boldsymbol{\Sigma}|^{1/2}\}^{-1}$.) Obviously, (2.49) is positive. It will be shown in Theorem 2.5.2 that the total integral is unity, but first we give a formal definition.

Definition 2.5.1 *The random vector $\boldsymbol{x}$ is said to have a p-variate normal (or p-dimensional multivariate normal) distribution with mean vector $\boldsymbol{\mu}$ and covariance matrix $\boldsymbol{\Sigma}$ if its p.d.f. is given by (2.49). We write $\boldsymbol{x} \sim N_p(\boldsymbol{\mu}, \boldsymbol{\Sigma})$.*

The quadratic form in (2.49) is equivalent to

$$\sum_{i=1}^{p}\sum_{j=1}^{p} \sigma^{ij}(x_i-\mu_i)(x_j-\mu_j), \quad \text{where} \quad \boldsymbol{\Sigma}^{-1} = (\sigma^{ij}).$$

The p.d.f. may also be written in terms of correlations rather than covariances.

Theorem 2.5.2 *Let $\boldsymbol{x}$ have the p.d.f. given by (2.49), and let*

$$\boldsymbol{y} = \boldsymbol{\Sigma}^{-1/2}(\boldsymbol{x}-\boldsymbol{\mu}), \tag{2.50}$$

where $\boldsymbol{\Sigma}^{-1/2}$ is the symmetric positive-definite square root of $\boldsymbol{\Sigma}^{-1}$. Then, $y_1, \ldots, y_p$ are independent $N(0,1)$ variables.

Proof: From (2.50), we have

$$(\boldsymbol{x}-\boldsymbol{\mu})'\boldsymbol{\Sigma}^{-1}(\boldsymbol{x}-\boldsymbol{\mu}) = \boldsymbol{y}'\boldsymbol{y}. \tag{2.51}$$

From (2.45), the Jacobian of the transformation $\boldsymbol{y}$ to $\boldsymbol{x}$ is $|\boldsymbol{\Sigma}|^{1/2}$. Hence, using (2.49), the p.d.f. of $\boldsymbol{y}$ is

$$g(\boldsymbol{y}) = \frac{1}{(2\pi)^{p/2}} \exp\{-\sum y_i^2/2\}.$$ □

Note that since $g(\boldsymbol{y})$ integrates to 1, (2.49) is a density.

Corollary 2.5.3 *If $\boldsymbol{x}$ has the p.d.f. given by (2.49), then*

$$\mathrm{E}(\boldsymbol{x}) = \boldsymbol{\mu}, \quad \mathrm{var}(\boldsymbol{x}) = \boldsymbol{\Sigma}. \tag{2.52}$$

Proof: We have

$$\mathsf{E}(\boldsymbol{y}) = \mathbf{0}, \quad \mathrm{var}(\boldsymbol{y}) = \boldsymbol{I}. \tag{2.53}$$

From (2.50),

$$\boldsymbol{x} = \boldsymbol{\Sigma}^{1/2}\boldsymbol{y} + \boldsymbol{\mu}. \tag{2.54}$$

Using Theorem 2.3, the result follows. □

For $p = 2$, it is usual to write ρ_{12} as ρ, $-1 < \rho < 1$. In this case, the p.d.f. becomes

$$f(x_1, x_2) = \frac{1}{2\pi\sigma_1\sigma_2(1-\rho^2)^{1/2}} \times \exp\left[-\frac{1}{2(1-\rho^2)}\left\{\frac{(x_1-\mu_1)^2}{\sigma_1^2} - \frac{2\rho(x_1-\mu_1)(x_2-\mu_2)}{\sigma_1\sigma_2} + \frac{(x_2-\mu_2)^2}{\sigma_2^2}\right\}\right],$$

where $-\infty < x_1, x_2 < \infty$.

2.5.2 Geometry

We now look at some of the above ideas geometrically. The multivariate normal distribution in p dimensions has constant density on ellipses or ellipsoids of the form

$$(\boldsymbol{x} - \boldsymbol{\mu})'\boldsymbol{\Sigma}^{-1}(\boldsymbol{x} - \boldsymbol{\mu}) = c^2, \tag{2.55}$$

c being a constant. These ellipsoids are called the *contours* of the distribution or the "ellipsoids of equal concentration." For $\boldsymbol{\mu} = \mathbf{0}$, these contours are centered at $\boldsymbol{x} = \mathbf{0}$, and when $\boldsymbol{\Sigma} = \boldsymbol{I}$, the contours are circles or in higher dimensions spheres or hyperspheres. Figure 2.1

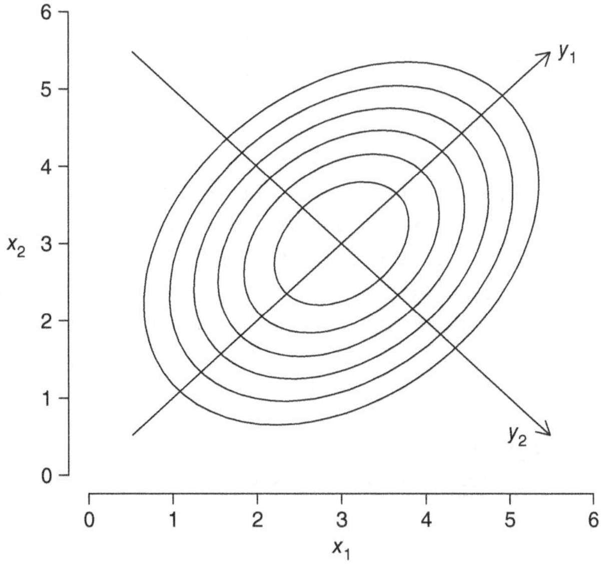

Figure 2.1 Ellipses of equal probability density for the bivariate normal distribution, showing the principal components y_1 and y_2, where $\boldsymbol{\mu}' = (3, 3)$, $\sigma_{11} = 3$, $\sigma_{12} = 1$, $\sigma_{22} = 3$.

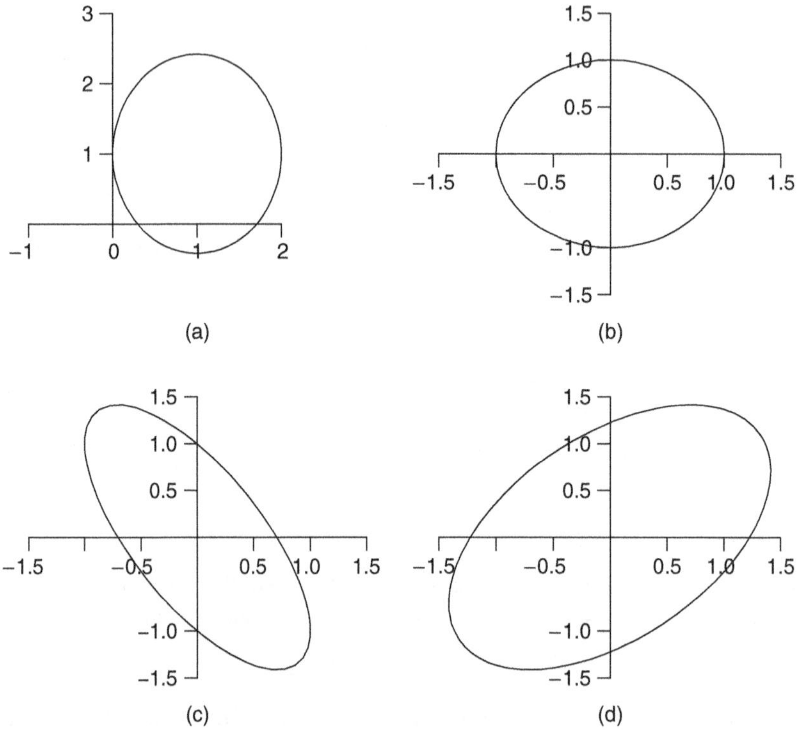

Figure 2.2 Ellipses of equal probability density for the bivariate normal distribution with $c = 1$.
(a) $\boldsymbol{\mu} = \begin{pmatrix} 1 \\ 1 \end{pmatrix}$, $\boldsymbol{\Sigma} = \begin{bmatrix} 1 & 0 \\ 0 & 2 \end{bmatrix}$, (b) $\boldsymbol{\mu} = \begin{pmatrix} 0 \\ 0 \end{pmatrix}$, $\boldsymbol{\Sigma} = \begin{bmatrix} 1 & 0 \\ 0 & 1 \end{bmatrix}$,
(c) $\boldsymbol{\mu} = \begin{pmatrix} 0 \\ 0 \end{pmatrix}$, $\boldsymbol{\Sigma} = \begin{bmatrix} 1 & -1 \\ -1 & 2 \end{bmatrix}$, (d) $\boldsymbol{\mu} = \begin{pmatrix} 0 \\ 0 \end{pmatrix}$, $\boldsymbol{\Sigma} = \begin{bmatrix} 2 & 1 \\ 1 & 2 \end{bmatrix}$.

shows a family of such contours for selected values of c for the bivariate case, and Figure 2.2 shows various types of contours for differing $\boldsymbol{\mu}$ and $\boldsymbol{\Sigma}$.

The principal component transformation facilitates interpretation of the ellipsoids of equal probability density. Using the spectral decomposition theorem (Theorem A.5), write $\boldsymbol{\Sigma} = \boldsymbol{\Gamma}\boldsymbol{\Lambda}\boldsymbol{\Gamma}'$, where $\boldsymbol{\Lambda} = \text{diag}(\lambda_1, \ldots, \lambda_p)$ is the matrix of eigenvalues of $\boldsymbol{\Sigma}$, and $\boldsymbol{\Gamma}$ is an orthogonal matrix whose columns are the corresponding eigenvectors. As in Section 1.5.3, define the *principal component transformation* by $\boldsymbol{y} = \boldsymbol{\Gamma}'(\boldsymbol{x} - \boldsymbol{\mu})$. In terms of $\boldsymbol{y}$, (2.55) becomes

$$\sum_{i=1}^{p} \frac{y_i^2}{\lambda_i} = c^2,$$

so the components of $\boldsymbol{y}$ represent axes of the ellipsoid. This property is illustrated in Figure 2.1, where y_1 and y_2 represent the major and minor semiaxes of the ellipse, respectively.

2.5.3 Properties

If $\boldsymbol{x} \sim N_p(\boldsymbol{\mu}, \boldsymbol{\Sigma})$, the following results may be derived.

Theorem 2.5.4

$$U = (\boldsymbol{x} - \boldsymbol{\mu})'\boldsymbol{\Sigma}^{-1}(\boldsymbol{x} - \boldsymbol{\mu}) \sim \chi^2_p. \tag{2.56}$$

Proof: From (2.51), the left-hand side is $\sum y_i^2$, where the y_is are independent $N(0, 1)$ by Theorem 2.5.2. Hence, the result follows. □

Using this theorem, we can calculate the probability of a point $\boldsymbol{x}$ falling within an ellipsoid (2.55), from chi-square tables, since it amounts to calculating $\mathsf{P}(U \leq c^2)$.

Theorem 2.5.5 *The c.f. of* $\boldsymbol{x}$ *is*

$$\phi_{\boldsymbol{x}}(\boldsymbol{t}) = \exp\left(i\boldsymbol{t}'\boldsymbol{\mu} - \frac{1}{2}\boldsymbol{t}'\boldsymbol{\Sigma}\boldsymbol{t}\right). \tag{2.57}$$

Proof: Using (2.54), we find that

$$\phi_{\boldsymbol{x}}(\boldsymbol{t}) = \mathsf{E}(\mathrm{e}^{i\boldsymbol{t}'\boldsymbol{x}}) = \mathrm{e}^{i\boldsymbol{t}'\boldsymbol{\mu}}\mathsf{E}(\mathrm{e}^{i\boldsymbol{u}'\boldsymbol{y}}), \tag{2.58}$$

where

$$\boldsymbol{u}' = \boldsymbol{t}'\boldsymbol{\Sigma}^{1/2}. \tag{2.59}$$

Since the y_j are independent $N(0, 1)$ from Theorem 2.5.2,

$$\mathsf{E}(\mathrm{e}^{i\boldsymbol{u}'\boldsymbol{y}}) = \prod_{j=1}^{p} \phi_{y_j}(u_j) = \prod_{j=1}^{p} \mathrm{e}^{-u_j^2/2} = \mathrm{e}^{-\boldsymbol{u}'\boldsymbol{u}/2}. \tag{2.60}$$

Substituting (2.60) and (2.59) in (2.58), we obtain the required result. □

As an example of the use of c.f.s, we prove the following result.

Theorem 2.5.6 *All nontrivial linear combinations of the elements of* $\boldsymbol{x}$ *are univariate normal.*

Proof: Let $\boldsymbol{a} \neq \boldsymbol{0}$ be a p-vector. The c.f. of $y = \boldsymbol{a}'\boldsymbol{x}$ is

$$\phi_y(t) = \phi_{\boldsymbol{x}}(t\boldsymbol{a}) = \exp\left(it\boldsymbol{a}'\boldsymbol{\mu} - \frac{1}{2}t^2\boldsymbol{a}'\boldsymbol{\Sigma}\boldsymbol{a}\right),$$

which is the c.f. of a normal random variable with mean $\boldsymbol{a}'\boldsymbol{\mu}$ and variance $\boldsymbol{a}'\sum\boldsymbol{a} > 0$. □

Theorem 2.5.7 $\beta_{1,p} = 0, \quad \beta_{2,p} = p(p+2).$

Proof: Let $V = (\boldsymbol{x} - \boldsymbol{\mu})'\boldsymbol{\Sigma}^{-1}(\boldsymbol{y} - \boldsymbol{\mu})$, where $\boldsymbol{x}$ and $\boldsymbol{y}$ are i.i.d. $N_p(\boldsymbol{\mu}, \boldsymbol{\Sigma})$. Then, from (2.31), $\beta_{1,p} = \mathsf{E}(V^3)$. However, the distribution of V is symmetric about $V = 0$, and therefore, $\mathsf{E}(V^3) = 0$. From (2.32) and (2.56),

$$\beta_{2,p} = \mathsf{E}[(\chi^2_p)^2] = p(p+2).$$ □

Theorem 2.5.8 *Let $\boldsymbol{x} \sim N_p(\boldsymbol{\mu}, \boldsymbol{\Sigma})$, be partitioned as $\boldsymbol{x} = (\boldsymbol{x}_1', \boldsymbol{x}_2')'$, and conformably partition $\boldsymbol{\mu}$ and $\boldsymbol{\Sigma}, \boldsymbol{\Sigma}^{-1}$ by*

$$\boldsymbol{\Sigma} = \begin{bmatrix} \boldsymbol{\Sigma}_{11} & \boldsymbol{\Sigma}_{12} \\ \boldsymbol{\Sigma}_{21} & \boldsymbol{\Sigma}_{22} \end{bmatrix}, \qquad \boldsymbol{\Sigma}^{-1} = \begin{bmatrix} \boldsymbol{\Sigma}^{11} & \boldsymbol{\Sigma}^{12} \\ \boldsymbol{\Sigma}^{21} & \boldsymbol{\Sigma}^{22} \end{bmatrix}.$$

Then,

$$\boldsymbol{x}_2 \mid \boldsymbol{x}_1 \sim N_{p_2}\left(\boldsymbol{\mu}_2 + \boldsymbol{\Sigma}_{21}\boldsymbol{\Sigma}_{11}^{-1}(\boldsymbol{x}_1 - \boldsymbol{\mu}_1), \boldsymbol{\Sigma}_{22.1}\right), \tag{2.61}$$

where $\boldsymbol{\Sigma}_{22.1} = \boldsymbol{\Sigma}_{22} - \boldsymbol{\Sigma}_{21}\boldsymbol{\Sigma}_{11}^{-1}\boldsymbol{\Sigma}_{12}$.

Proof: The conditional density is proportional to the joint density and can be simplified as follows:

$$\begin{aligned} f(\boldsymbol{x}_2 \mid \boldsymbol{x}_1) &\propto f(\boldsymbol{x}_2, \boldsymbol{x}_1) \qquad (2.62) \\ &\propto \exp\left\{-\frac{1}{2}(\boldsymbol{x} - \boldsymbol{\mu})'\boldsymbol{\Sigma}^{-1}(\boldsymbol{x} - \boldsymbol{\mu})\right\} \\ &\propto \exp\left\{-\frac{1}{2}\left[(\boldsymbol{x}_2 - \boldsymbol{\mu}_2)'\boldsymbol{\Sigma}^{22}(\boldsymbol{x}_2 - \boldsymbol{\mu}_2) + 2(\boldsymbol{x}_2 - \boldsymbol{\mu}_2)'\boldsymbol{\Sigma}^{21}(\boldsymbol{x}_1 - \boldsymbol{\mu}_1)\right]\right\} \\ &\propto \exp\left\{-\frac{1}{2}\left[(\boldsymbol{x}_2 - \boldsymbol{\mu}_2) + (\boldsymbol{\Sigma}^{22})^{-1}\boldsymbol{\Sigma}^{21}(\boldsymbol{x}_1 - \boldsymbol{\mu}_1)\right]' \times \right. \\ &\qquad \left. \boldsymbol{\Sigma}^{22}\left[(\boldsymbol{x}_2 - \boldsymbol{\mu}_2) + (\boldsymbol{\Sigma}^{22})^{-1}\boldsymbol{\Sigma}^{21}(\boldsymbol{x}_1 - \boldsymbol{\mu}_1)\right]\right\} \\ &\qquad \text{— completing the square} \\ &\propto \text{p.d.f. of } N_{p_2}\left(\boldsymbol{\mu}_2 - (\boldsymbol{\Sigma}^{22})^{-1}\boldsymbol{\Sigma}^{21}(\boldsymbol{x}_1 - \boldsymbol{\mu}_1), (\boldsymbol{\Sigma}^{22})^{-1}\right). \qquad (2.63) \end{aligned}$$

In the third line of (2.63), the quadratic form has been split into four terms, where the two cross-product terms are equal, and the final term $-\frac{1}{2}(\boldsymbol{x}_1 - \boldsymbol{\mu}_1)'\boldsymbol{\Sigma}^{11}(\boldsymbol{x}_1 - \boldsymbol{\mu}_1)$ has been absorbed into the proportionality constant. The fourth line involves adding and subtracting the term $\frac{1}{2}(\boldsymbol{x}_1 - \boldsymbol{\mu}_1)'\boldsymbol{\Sigma}^{12}(\boldsymbol{\Sigma}^{22})^{-1}\boldsymbol{\Sigma}^{21}(\boldsymbol{x}_1 - \boldsymbol{\mu}_1)$ within the square brackets. The minus term is used to complete the square; the plus term is absorbed into the proportionality constant. The final line can be recognized as the density of the stated normal distribution. To complete the proof, recall the identities for a partitioned covariance matrix (Section A.2.4), $(\boldsymbol{\Sigma}^{22})^{-1} = \boldsymbol{\Sigma}_{22.1}$ and $\boldsymbol{\Sigma}^{21} = -\boldsymbol{\Sigma}^{22}\boldsymbol{\Sigma}_{21}\boldsymbol{\Sigma}_{11}^{-1}$, so $(\boldsymbol{\Sigma}^{22})^{-1}\boldsymbol{\Sigma}^{21} = -\boldsymbol{\Sigma}_{21}\boldsymbol{\Sigma}_{11}^{-1}$. □

Example 2.5.1 For the bivariate normal distribution, $p = 2, \boldsymbol{x} = (x_1, x_2)', \boldsymbol{\mu} = (\mu_1, \mu_2)'$, and $\boldsymbol{\Sigma} = \begin{bmatrix} \sigma_1^2 & \rho\sigma_1\sigma_2 \\ \rho\sigma_1\sigma_2 & \sigma_2^2 \end{bmatrix}$. Then, $\boldsymbol{\Sigma}_{22.1} = \sigma_2^2(1 - \rho^2)$ and (2.61) gives

$$x_2 \mid x_1 \sim N_1\left(\mu_2 + \rho\frac{\sigma_2}{\sigma_1}(x_1 - \mu_1), \sigma_2^2(1 - \rho^2)\right). \tag{2.64}$$

Note that if $\rho > 0$ and $x_1 > \mu_1$, then the conditional mean of x_2 is higher than μ_2; conversely, if $x_1 < \mu_1$ (with ρ still positive), then the conditional mean of x_2 is lower than μ_2. Also, $\sigma_2^2(1 - \rho^2) < \sigma_2^2$; i.e., the conditional variance is less than the marginal variance. Knowing the value of x_1 reduces the variability in x_2. □

Two further applications of (2.61) are of special interest in the p-dimensional case:

(1) Suppose $\boldsymbol{x}_2 = x_i$ contains just one variable, and $\boldsymbol{x}_1 = (x_j,\ j \neq i)$ contains the remaining $p-1$ variables. Then, the conditional variance of x_i given the remaining variables is

$$\text{var}\{\boldsymbol{x}_2 \mid \boldsymbol{x}_1\} = 1/\sigma^{ii}, \tag{2.65}$$

the reciprocal of the (i, i)th element of the concentration matrix.

(2) Suppose $\boldsymbol{x}_2 = (x_i, x_j)'$ contains just two variables, and $\boldsymbol{x}_1 = (x_k,\ k \neq i, j)'$ contains the remaining $p-2$ variables. Set

$$\boldsymbol{P} = \begin{bmatrix} \sigma^{ii} & \sigma^{ij} \\ \sigma^{ji} & \sigma^{jj} \end{bmatrix}, \quad \boldsymbol{Q} = \boldsymbol{P}^{-1} = \frac{1}{\sigma^{11}\sigma^{22} - (\sigma^{12})^2} \begin{bmatrix} \sigma^{jj} & -\sigma^{ij} \\ -\sigma^{ji} & \sigma^{ii} \end{bmatrix}. \tag{2.66}$$

Then, $\boldsymbol{Q}$ represents the conditional covariance matrix of $(x_i, x_j)'$, given the remaining variables. In particular, note that the conditional correlation between x_i and x_j is 0 if and only if the (i, j)th element of the concentration matrix vanishes, $\sigma^{ij} = 0$.

The multivariate normal distribution is explored in greater detail in Chapter 4 using a density-free approach.

2.5.4 Singular Multivariate Normal Distribution

The p.d.f. of $N_p(\boldsymbol{\mu}, \boldsymbol{\Sigma})$ involves $\boldsymbol{\Sigma}^{-1}$. However, if $\text{rank}(\boldsymbol{\Sigma}) = k < p$, we can define the (singular) density of $\boldsymbol{x}$ as

$$\frac{(2\pi)^{-k/2}}{(\lambda_1 \cdots \lambda_k)^{1/2}} \exp\left\{-\frac{1}{2}(\boldsymbol{x}-\boldsymbol{\mu})'\boldsymbol{\Sigma}^{-}(\boldsymbol{x}-\boldsymbol{\mu})\right\}, \tag{2.67}$$

where

(1) $\boldsymbol{x}$ lies on the hyperplane $\boldsymbol{N}'(\boldsymbol{x}-\boldsymbol{\mu}) = \boldsymbol{0}$, where $\boldsymbol{N}$ is a $p \times (p-k)$ matrix such that

$$\boldsymbol{N}'\boldsymbol{\Sigma} = \boldsymbol{0}, \quad \boldsymbol{N}'\boldsymbol{N} = \boldsymbol{I}_{p-k} \tag{2.68}$$

(2) $\boldsymbol{\Sigma}^{-}$ is a g-inverse of $\boldsymbol{\Sigma}$ (see Section A.8), and $\lambda_1, \dots, \lambda_k$ are the nonzero eigenvalues of $\boldsymbol{\Sigma}$.

There is a close connection between the singular density (2.67) and the nonsingular multivariate normal distribution in k dimensions.

Theorem 2.5.9 *Let $\boldsymbol{y} \sim N_k(\boldsymbol{0}, \boldsymbol{\Lambda}_1)$, where $\boldsymbol{\Lambda}_1 = \text{diag}\,(\lambda_1, \dots, \lambda_k)$. Then, there exists a $(p \times k)$ column orthonormal matrix $\boldsymbol{B}$ (that is, $\boldsymbol{B}'\boldsymbol{B} = \boldsymbol{I}_k$) such that*

$$\boldsymbol{x} = \boldsymbol{B}\boldsymbol{y} + \boldsymbol{\mu} \tag{2.69}$$

has the p.d.f. (2.67).

Proof: The change of variable formula (2.41) has a generalization applicable to hypersurfaces. If $\boldsymbol{x} = \boldsymbol{\phi}(\boldsymbol{y})$, $\boldsymbol{y} \in \mathbb{R}^k$, $\boldsymbol{x} \in \mathbb{R}^p$, is a parameterization of a k-dimensional hypersurface in $\mathbb{R}^p$ $(k \leq p)$, and $g(\boldsymbol{y})$ is a p.d.f. on $\mathbb{R}^k$, then $\boldsymbol{x}$ has a p.d.f. on the hypersurface given by

$$f(\boldsymbol{x}) = g(\boldsymbol{\phi}^{-1}(\boldsymbol{x}))|\boldsymbol{D}'\boldsymbol{D}|^{-1/2}, \tag{2.70}$$

where

$$D = D(x) = \left(\frac{\partial \phi_i(y)}{\partial y_j}\right)\Bigg|_{y=\phi^{-1}(x)}$$

is a $(p \times k)$ matrix evaluated at $\boldsymbol{y} = \boldsymbol{\phi}^{-1}(\boldsymbol{x})$, and we suppose $|\boldsymbol{D}'\boldsymbol{D}|$ is never zero.

By the spectral decomposition theorem, we can write $\boldsymbol{\Sigma} = \boldsymbol{\Gamma}\boldsymbol{\Lambda}\boldsymbol{\Gamma}'$, where $\boldsymbol{\Lambda} = \text{diag}(\lambda_1, \dots, \lambda_k, 0, \dots, 0)$, and $\boldsymbol{\Gamma} = (\boldsymbol{B} : \boldsymbol{N})$ is an orthonormal matrix partitioned, so $\boldsymbol{B}$ is a $(p \times k)$ matrix. Then, $\boldsymbol{\Sigma} = \boldsymbol{B}\boldsymbol{\Lambda}_1\boldsymbol{B}'$, $\boldsymbol{B}'\boldsymbol{B} = \boldsymbol{I}_k, \boldsymbol{B}'\boldsymbol{N} = \boldsymbol{0}$, and $\boldsymbol{N}'\boldsymbol{N} = \boldsymbol{I}_{p-k}$. Notice that $\boldsymbol{N}$ can be taken to be the same as in (2.68).

The transformation $\boldsymbol{x} = \boldsymbol{B}\boldsymbol{y} + \boldsymbol{\mu}$ parameterizes the hyperplane $\boldsymbol{N}'(\boldsymbol{x} - \boldsymbol{\mu}) = \boldsymbol{0}$. The p.d.f. of $\boldsymbol{y}$ is given by

$$(2\pi)^{-k/2}(\lambda_1 \cdots \lambda_k)^{-1/2} \exp\left(-\frac{1}{2}\sum \frac{y_i^2}{\lambda_i}\right).$$

Now $\sum y_i^2/\lambda_i = \boldsymbol{y}'\boldsymbol{\Lambda}_1^{-1}\boldsymbol{y} = (\boldsymbol{x} - \boldsymbol{\mu})'\boldsymbol{B}\boldsymbol{\Lambda}_1^{-1}\boldsymbol{B}'(\boldsymbol{x} - \boldsymbol{\mu})$, and $\boldsymbol{B}\boldsymbol{\Lambda}_1^{-1}\boldsymbol{B}'$ is a g-inverse of $\boldsymbol{\Sigma}$. Also, for this transformation, $\boldsymbol{D} = \boldsymbol{B}$ and $|\boldsymbol{B}'\boldsymbol{B}|^{1/2} = |\boldsymbol{I}_k|^{1/2} = 1$. Thus, the p.d.f. of $f(\boldsymbol{x})$ takes the form given in (2.67). □

Using Theorem 2.5.9, it is easy to show that many properties for the nonsingular multivariate normal carry over to the singular case.

Corollary 2.5.10 $\text{E}(\boldsymbol{x}) = \boldsymbol{\mu}$, $\text{var}(\boldsymbol{x}) = \boldsymbol{\Sigma}$.

Corollary 2.5.11

$$\phi_x(\boldsymbol{t}) = \exp\left(i\boldsymbol{t}'\boldsymbol{\mu} - \frac{1}{2}\boldsymbol{t}'\boldsymbol{\Sigma}\boldsymbol{t}\right). \tag{2.71}$$

Corollary 2.5.12 *If $\boldsymbol{a}$ is a p-vector and $\boldsymbol{a}'\boldsymbol{\Sigma}\boldsymbol{a} > 0$, then $\boldsymbol{a}'\boldsymbol{x}$ has a univariate normal distribution.*

2.5.5 The Matrix Normal Distribution

Let $\boldsymbol{X}(n \times p)$ be a matrix whose n rows, $\boldsymbol{x}_1', \dots, \boldsymbol{x}_n'$, are independently distributed as $N_p(\boldsymbol{\mu}, \boldsymbol{\Sigma})$. Then, $\boldsymbol{X}$ has a *matrix normal distribution* and represents a random sample from $N_p(\boldsymbol{\mu}, \boldsymbol{\Sigma})$. Using (2.49), we find that the p.d.f. of $\boldsymbol{X}$ is

$$\begin{aligned} f(\boldsymbol{X}) &= |2\pi\boldsymbol{\Sigma}|^{-n/2} \exp\left\{-\frac{1}{2}\sum_{r=1}^{n}(\boldsymbol{x}_r - \boldsymbol{\mu})'\boldsymbol{\Sigma}^{-1}(\boldsymbol{x}_r - \boldsymbol{\mu})\right\} \\ &= |2\pi\boldsymbol{\Sigma}|^{-n/2} \exp\left\{-\frac{1}{2}\text{tr}\,[\boldsymbol{\Sigma}^{-1}(\boldsymbol{X} - \boldsymbol{1}\boldsymbol{\mu}')'(\boldsymbol{X} - \boldsymbol{1}\boldsymbol{\mu}')]\right\}. \end{aligned}$$

A more general formulation is to allow a $np \times np$ dimensional covariance matrix. In this case, we can simply combine the n observations into a single vector of length np, with a corresponding $np \times np$ covariance matrix, which we call $\boldsymbol{\Sigma}^*$. However, this then requires specifying and estimating an $np \times np$ matrix.

A model in between these two extremes – leading to the *matrix factored normal distribution* – is to consider a "between-observation" $(n \times n)$ covariance, say $\boldsymbol{U}$, and a "within-observation" $(p \times p)$ covariance, say $\boldsymbol{V}$, and then to take $\boldsymbol{\Sigma}^* = \boldsymbol{V} \otimes \boldsymbol{U}$, with $\otimes$

denoting the *Kronecker product* – see A.2.5. This formulation is designed for the situation in which the $\boldsymbol{x}_r'$ (i.e. the rows of $\boldsymbol{X}$) are correlated.

In this case, the probability density function of $\boldsymbol{X}$ is given by (Mardia and Goodall, 1993)

$$f(\boldsymbol{X}) = |\boldsymbol{V}|^{-n/2}|2\pi\boldsymbol{U}|^{-p/2}\exp\left\{-\frac{1}{2}\text{tr}\,[\boldsymbol{V}^{-1}(\boldsymbol{X}-\boldsymbol{M})'\boldsymbol{U}^{-1}(\boldsymbol{X}-\boldsymbol{M})]\right\}, \tag{2.72}$$

where $\boldsymbol{M}$ is an $n \times p$ matrix denoting the expected value of $\boldsymbol{X}$. Note that the overall scale of the matrices $\boldsymbol{U}$ and $\boldsymbol{V}$ is not identifiable since

$$\alpha\boldsymbol{V} \otimes \frac{1}{\alpha}\boldsymbol{U} = \boldsymbol{V} \otimes \boldsymbol{U}$$

for a scalar $\alpha > 0$, but the ambiguity can be solved by, e.g. tr $\boldsymbol{U} = n$, or simply by fixing one of the diagonal entries of $\boldsymbol{U}$ or $\boldsymbol{V}$.

This can be extended from two-way arrays (matrices) to higher orders. For example, the three-factor case (but with $\boldsymbol{M}$ only depending on two factors) is described in detail in Mardia and Goodall (1993). In this situation, it may help to use vectorization – see Exercise 2.5.3 for the simplest case.

The estimation of the parameters is discussed in Section 5.2.3.

2.6 Random Samples

In Section 2.5.5, we met the idea of a random sample from the $N_p(\boldsymbol{\mu}, \boldsymbol{\Sigma})$ distribution. We now consider a more general situation.

Suppose that $\boldsymbol{x}_1, \dots, \boldsymbol{x}_n$ is a random sample from a population with p.d.f. $f(\boldsymbol{x}; \boldsymbol{\theta})$, where $\boldsymbol{\theta}$ is a parameter vector; that is, $\boldsymbol{x}_1, \dots, \boldsymbol{x}_n$ are independently and identically distributed (i.i.d.), where the p.d.f. of $\boldsymbol{x}_r$ is $f(\boldsymbol{x}_r; \boldsymbol{\theta})$, $r = 1, \dots, n$. (The function f is the same for each value of r).

We now obtain the moments of $\overline{\boldsymbol{x}}, \boldsymbol{S}$ and the Mahalanobis distances, under the assumption of random sampling from a population with mean $\boldsymbol{\mu}$ and covariance matrix $\boldsymbol{\Sigma}$. No assumption of normality is made in this section.

Theorem 2.6.1 $\mathsf{E}(\overline{\boldsymbol{x}}) = \boldsymbol{\mu}$ *and* $\text{var}(\overline{\boldsymbol{x}}) = (1/n)\boldsymbol{\Sigma}$.

Proof: Since $\mathsf{E}(\boldsymbol{x}_r) = \boldsymbol{\mu}$, we have

$$\mathsf{E}(\overline{\boldsymbol{x}}) = \frac{1}{n}\sum \mathsf{E}(\boldsymbol{x}_r) = \boldsymbol{\mu}.$$

Because $\text{var}(\boldsymbol{x}_r) = \boldsymbol{\Sigma}$ and $\text{cov}(\boldsymbol{x}_r, \boldsymbol{x}_s) = \boldsymbol{0}$, $r \neq s$,

$$\text{var}(\overline{\boldsymbol{x}}) = \frac{1}{n^2}\left[\sum_{r=1}^{n}\text{var}(\boldsymbol{x}_r) + \sum_{r\neq s}\text{cov}(\boldsymbol{x}_r, \boldsymbol{x}_s)\right] = \frac{1}{n}\boldsymbol{\Sigma}.$$ □

Theorem 2.6.2 $\mathsf{E}(\boldsymbol{S}) = \{(n-1)/n\}\boldsymbol{\Sigma}$.

Proof: Since

$$\boldsymbol{S} = \frac{1}{n}\sum_{r=1}^{n}\boldsymbol{x}_r\boldsymbol{x}_r' - \overline{\boldsymbol{x}}\,\overline{\boldsymbol{x}}' = \frac{1}{n}\sum_{r=1}^{n}(\boldsymbol{x}_r - \boldsymbol{\mu})(\boldsymbol{x}_r - \boldsymbol{\mu})' - (\overline{\boldsymbol{x}} - \boldsymbol{\mu})(\overline{\boldsymbol{x}} - \boldsymbol{\mu})',$$

taking expectations yields

$$\mathsf{E}(\boldsymbol{S}) = \frac{1}{n}\sum_{r=1}^{n}\text{var}(\boldsymbol{x}_r) - \text{var}(\overline{\boldsymbol{x}}) = \frac{n}{n}\boldsymbol{\Sigma} - \frac{1}{n}\boldsymbol{\Sigma} = \frac{n-1}{n}\boldsymbol{\Sigma}. \qquad \square$$

Note that

$$\mathsf{E}(\boldsymbol{S}_u) = \frac{n}{n-1}\mathsf{E}(\boldsymbol{S}) = \boldsymbol{\Sigma},$$

so $\boldsymbol{S}_u = (n/(n-1))\boldsymbol{S}$ is an unbiased estimate of $\boldsymbol{\Sigma}$.

Theorem 2.6.3 *Let* $\boldsymbol{G} = (g_{rs})$, *where* $g_{rs} = (\boldsymbol{x}_r - \overline{\boldsymbol{x}})'\boldsymbol{S}^{-1}(\boldsymbol{x}_s - \overline{\boldsymbol{x}})$. *Then,*

$$\mathsf{E}(\boldsymbol{G}) = \frac{np}{n-1}\boldsymbol{H}, \tag{2.73}$$

where

$$\boldsymbol{H} = \boldsymbol{I} - \frac{1}{n}\mathbf{1}\mathbf{1}'.$$

Proof: We have the identities (see Exercise 1.6.1)

$$\sum_r g_{rs} = 0 \text{ and } \sum_r g_{rr} = np.$$

Under random sampling, the variables g_{rr} are identically distributed, as are the variables $g_{rs}(r \neq s)$. Let their expectations be a and b, respectively. From the above identities, we see that $(n-1)b + a = 0$ and $na = np$. Therefore, $a = p$ and $b = -p/(n-1)$, as stated in the theorem. □

From (1.24) note that $D_{rs}^2 = g_{rr} + g_{ss} - 2g_{rs}$. Therefore, under the assumption of random sampling,

$$\mathsf{E}(D_{rs}^2) = a + a - 2b = 2pn/(n-1), \quad r \neq s. \tag{2.74}$$

Example 2.6.1 Using the student data from Table 1.1.1, where $n = 5$ and $p = 3$, we see that, under the assumption of random sampling, $\mathsf{E}(D_{rs}^2) = 2pn/(n-1) = 7.5$. In other words, each D_{rs} should be around 2.74. Calculations show the matrix of Mahalanobis distances D_{rs} to be

$$\begin{bmatrix} 0.00 & 2.53 & 2.94 & 3.14 & 3.07 \\ & 0.00 & 0.76 & 2.79 & 2.95 \\ & & 0.00 & 2.73 & 2.50 \\ & & & 0.00 & 3.14 \\ & & & & 0.00 \end{bmatrix}.$$

It can be seen that most of the observed values of D_{rs} are indeed near 2.74, although D_{23} is substantially lower. □

2.7 Limit Theorems

Although a random sample may come from a nonnormal population, the *sample mean vector* $\overline{\boldsymbol{x}}$ will often be approximately normally distributed for large sample size n.

Theorem 2.7.1 *(Central Limit Theorem) Let $\boldsymbol{x}_1, \boldsymbol{x}_2, \ldots$ be an infinite sequence of independent identically distributed random vectors from a distribution with mean $\boldsymbol{\mu}$ and variance $\boldsymbol{\Sigma}$. Then,*

$$n^{-1/2} \sum_{r=1}^{n} (\boldsymbol{x}_r - \boldsymbol{\mu}) = n^{1/2}(\overline{\boldsymbol{x}} - \boldsymbol{\mu}) \xrightarrow{D} N_p(\boldsymbol{0}, \boldsymbol{\Sigma}) \qquad \text{as } n \to \infty,$$

where $\xrightarrow{D}$ denotes "convergence in distribution." By an abuse of notation, we also write, asymptotically,

$$\overline{\boldsymbol{x}} \sim N_p\left(\boldsymbol{\mu}, \frac{1}{n}\boldsymbol{\Sigma}\right).$$

The following theorem shows that a transformation of an asymptotically normal random vector with small variances is also asymptotically normal.

Theorem 2.7.2 *If $\boldsymbol{t}$ is asymptotically normal with mean $\boldsymbol{\mu}$ and covariance matrix $\boldsymbol{V}/n$, and if $\boldsymbol{f} = (f_1, \ldots, f_q)'$ are real-valued functions that are differentiable at $\boldsymbol{\mu}$, then $\boldsymbol{f}(\boldsymbol{t})$ is asymptotically normal with mean $\boldsymbol{f}(\boldsymbol{\mu})$ and covariance matrix $\boldsymbol{D}'\boldsymbol{V}\boldsymbol{D}/n$, where $d_{ij} = \partial f_j / \partial t_i$ evaluated at $\boldsymbol{t} = \boldsymbol{\mu}$.*

Proof: First, we need the following notation. Given a sequence of random variables $\{g_n\}$ and a sequence of positive numbers $\{b_n\}$, we say that $g_n = O_p(b_n)$ as $n \to \infty$ if

$$\lim_{n\to\infty} \sup_{m \geq n} \mathrm{P}(|b_m^{-1} g_m| > c) \to 0 \qquad \text{as } c \to \infty; \tag{2.75}$$

that is if, for all n large enough, $|b_n^{-1} g_n|$ will, with large probability, not be too big. Similarly, say that $g_n = o_p(b_n)$ if

$$\lim_{n\to\infty} \sup_{m \geq n} \mathrm{P}(|b_m^{-1} g_m| > c) = 0 \qquad \text{for all } c > 0; \tag{2.76}$$

that is, if $g_n / b_n \xrightarrow{P} 0$ as $n \to \infty$. Note that $O_p(\cdot)$ and $o_p(\cdot)$ are probabilistic versions of the $O(\cdot)$ and $o(\cdot)$ notations used for sequences of constants. Since each f_i is differentiable at $\boldsymbol{\mu}$, we can write

$$\boldsymbol{f}(\boldsymbol{t}) - \boldsymbol{f}(\boldsymbol{\mu}) = \boldsymbol{D}'(\boldsymbol{t} - \boldsymbol{\mu}) + \|\boldsymbol{t} - \boldsymbol{\mu}\| \boldsymbol{\delta}(\boldsymbol{t} - \boldsymbol{\mu}),$$

where $\|\boldsymbol{t} - \boldsymbol{\mu}\|^2 = \sum (t_i - \mu_i)^2$, and where $\boldsymbol{\delta}(\boldsymbol{t} - \boldsymbol{\mu}) \to \boldsymbol{0}$ as $\boldsymbol{t} \to \boldsymbol{\mu}$.

The assumptions about $\boldsymbol{t}$ imply that $n^{1/2}\|\boldsymbol{t} - \boldsymbol{\mu}\| = O_p(1)$ and that $\|\boldsymbol{\delta}(\boldsymbol{t} - \boldsymbol{\mu})\| = o_p(1)$ as $n \to \infty$. Thus,

$$n^{1/2}[\boldsymbol{f}(\boldsymbol{t}) - \boldsymbol{f}(\boldsymbol{\mu})] = n^{1/2}\boldsymbol{D}'(\boldsymbol{t} - \boldsymbol{\mu}) + o_p(1) \xrightarrow{D} \boldsymbol{N}_q(\boldsymbol{0}, \boldsymbol{D}'\boldsymbol{V}\boldsymbol{D}). \qquad \square$$

More general theorems of this type can be found in Rao (1973, p. 387).

Example 2.7.1 Let $\boldsymbol{x}$ be multinomially distributed with parameters n and $\boldsymbol{a}$, where $a_i > 0, i = 1, \ldots, p$, and set $\boldsymbol{z} = (1/n)\boldsymbol{x}$. Let $\boldsymbol{b} = (a_1^{1/2}, \ldots, a_p^{1/2})'$ and $\boldsymbol{w} = (z_1^{1/2}, \ldots, z_p^{1/2})'$. From Exercise 3.2.7, $\boldsymbol{x}$ can be written as a sum of n i.i.d. random vectors with mean $\boldsymbol{a}$ and variance matrix $\mathrm{diag}(\boldsymbol{a}) - \boldsymbol{a}\boldsymbol{a}'$. Thus, by the central limit theorem,

$$\boldsymbol{z} \sim N_p\left(\boldsymbol{a}, \frac{1}{n}[\mathrm{diag}(\boldsymbol{a}) - \boldsymbol{a}\boldsymbol{a}']\right),$$

asymptotically. Consider the transformation given by

$$f_i(\boldsymbol{z}) = z_i^{1/2}, \quad i = 1, \ldots, p.$$

Then,

$$\left.\frac{\partial f_i}{\partial z_j}\right|_{\boldsymbol{z}=\boldsymbol{a}} = \begin{cases} \frac{1}{2}b_i^{-1}, & i = j, \\ 0, & i \neq j, \end{cases}$$

so, asymptotically,

$$\boldsymbol{w} \sim N_p\left(\boldsymbol{b}, \frac{1}{n}\boldsymbol{\Sigma}\right),$$

where

$$\boldsymbol{\Sigma} = \frac{1}{4}\text{diag}(b_i^{-1})\left[\text{diag}(\boldsymbol{a}) - \boldsymbol{a}\boldsymbol{a}'\right]\text{diag}(b_i^{-1}) = \frac{1}{4}[\boldsymbol{I} - \boldsymbol{b}\boldsymbol{b}'].$$

Note that since $\sum b_i^2 = 1,\ \boldsymbol{\Sigma}\boldsymbol{b} = \boldsymbol{0}$, and hence, $\boldsymbol{\Sigma}$ is singular. □

Exercises and Complements

2.1.1 Consider the bivariate p.d.f.

$$f(x_1, x_2) = 2, \qquad 0 < x_1 < x_2 < 1.$$

Calculate $f_1(x_1)$ and $f_2(x_2)$ to show that x_1 and x_2 are dependent.

2.1.2 As an extension of (2.27), let

$$f(x_1, x_2) = \begin{cases} \gamma(x_1^\alpha + x_2^\beta), & 0 < x_1 < x_2 < 1, \\ 0, & \text{otherwise}, \end{cases}$$

where $\alpha > 0,\ \beta > 0$. Show that γ must equal $[1/(\alpha+1) + 1/(\beta+1)]^{-1}$ for $f(\cdot)$ to integrate to one. Calculate the corresponding c.d.f. and the probability of the following events:

(i) $0 < x_1, x_2 < \frac{1}{2}$, (ii) $0 < x_1 < \frac{1}{2} < 1$,

(iii) $\frac{1}{2} < x_1, x_2 < 1$, (iv) $0 < x_2 < \frac{1}{2} < x_1 < 1$.

2.2.1 (Mardia, 1962,1970a, p. 91) Consider the bivariate Pareto distribution defined by

$$f(x, y) = c(x + y - 1)^{-p-2}, \quad x, y > 1,$$

where $p > 0$. Show that c must equal $p(p+1)$. Calculate the joint and marginal c.d.f.s of x and y. If $p > 1$, show that x and y each have mean $p/(p-1)$, and if $p > 2$ show that the covariance matrix is given by

$$\{(p-1)^2(p-2)\}^{-1}\begin{bmatrix} p & 1 \\ 1 & p \end{bmatrix}.$$

What happens to the expectations if $0 < p \leq 1$? If $p > 2$, show that $\text{cor}(x, y) = 1/p$, and that the generalized variance and total variation are

$$(p+1)/\{(p-1)^3(p-2)^2\} \quad \text{and} \quad 2p/\{(p-1)^2(p-2)\}.$$

Calculate the regression and scedastic curves of x on y.

2.2.2 For the random variables x and y with p.d.f.

$$(2\pi^3)^{-1/2}(x^2+y^2)^{-1/2}e^{-(x^2+y^2)/2}, \qquad -\infty < x, y < \infty,$$

show that $\rho = 0$. Hence, conclude that $\rho = 0$ does not imply independence.

2.2.3 If $p \geq 2$ and $\boldsymbol{\Sigma}(p \times p)$ is the *equicorrelation* matrix, $\boldsymbol{\Sigma} = (1-\alpha)\boldsymbol{I} + \alpha\mathbf{1}\mathbf{1}'$, show that $\boldsymbol{\Sigma} \geq \mathbf{0}$ if and only if $-(p-1)^{-1} \leq \alpha \leq 1$. If $-(p-1)^{-1} < \alpha < 1$, show that

$$\boldsymbol{\Sigma}^{-1} = (1-\alpha)^{-1}[\boldsymbol{I} - \alpha\{1+(p-1)\alpha\}^{-1}\mathbf{1}\mathbf{1}'].$$

If $\boldsymbol{\mu}_1' = (\delta, \mathbf{0}')$ and $\boldsymbol{\mu}_2 = \mathbf{0}$, show that the Mahalanobis distance between them is given by

$$\Delta(\boldsymbol{\mu}_1, \boldsymbol{\mu}_2) = \delta\left[\frac{1+(p-2)\alpha}{(1-\alpha)\{1+(p-1)\alpha\}}\right]^{1/2}.$$

2.2.4 (Fréchet, 1951; Mardia, 1970c, Fréchet inequalities) Let x and y be random variables with joint c.d.f. $F(x,y)$ and marginal c.d.f.s $F_x(x)$ and $F_y(y)$. Show that

$$\max(F_x + F_y - 1, 0) \leq F \leq \min(F_x, F_y).$$

2.2.5 (Mardia, 1967b; Mardia and Thompson, 1972) Let x and y be random variables with joint c.d.f. $F(x,y)$ and marginal c.d.f.s $F_x(x)$ and $F_y(y)$. Show that the covariance between x^r and y^s is given by

$$\text{cov}(x^r, y^s) = rs\int_{-\infty}^{\infty}\int_{-\infty}^{\infty} x^{r-1}y^{s-1}\{F(x,y) - F_x(x)F_y(y)\}\,\text{d}x\text{d}y$$

for $r, s > 0$. Hence, show that

$$\text{cov}(x,y) = \int_{-\infty}^{\infty}\int_{-\infty}^{\infty}\{F(x,y) - F_x(x)F_y(y)\}\,\text{d}x\text{d}y.$$

2.2.6 (Enis, 1973) Let x and y be random variables such that the p.d.f. of y is

$$g(y) = \frac{1}{\sqrt{2}}y^{-1/2}e^{-y/2}, \quad y > 0,$$

and the conditional p.d.f. of x, given y, is

$$f(x \mid y) = (2\pi)^{-1/2}y^{1/2}e^{-yx^2/2}.$$

Show that $\mathsf{E}[x \mid y] = 0$, and hence, that $\mathsf{E}_y\{\mathsf{E}[x \mid y]\} = 0$, but that nevertheless the unconditional mean of x does not exist.

2.5.1 Consider the bivariate distribution of x and y defined as follows: let u and v be independent $N(0,1)$ random variables. Set $x = u$ if $uv > 0$, while $x = -u$ if $uv < 0$, and set $y = v$. Show that

(i) x and y are each $N(0,1)$, but their joint distribution is not bivariate normal;
(ii) x^2 and y^2 are statistically independent, but x and y are not.

2.5.2 **Sheppard's formula of orthant probability** (Kepner et al., 1989). Assume that random variables x and y are distributed as standardized bivariate normal with correlation ρ. Show that

(i) the conditional distribution of $x \mid y$ is $N(\rho y, (1-\rho^2))$.

(ii) $\mathsf{P}(\rho) = \mathsf{P}(x > 0, y > 0) = \mathsf{P}(x < 0, y < 0) = 1/4 + (2\pi)^{-1}\arcsin(\rho)$.

(iii) Show that the same result holds for the bivariate normal with zero means but with different variances.

Hint: From (i) obtain

$$\mathsf{P}(\rho) = \mathsf{P}(x < 0, y < 0) = \int_{-\infty}^{0} \phi(x)\Phi\left(\frac{-\rho x}{\sqrt{1-\rho^2}}\right)\,\mathrm{d}x,$$

where $\phi(\cdot)$ and $\Phi(\cdot)$ are the p.d.f. and c.d.f. of $N(0,1)$, respectively. Now,

$$\frac{\mathrm{d}\mathsf{P}(\rho)}{\mathrm{d}\rho} = \frac{1}{2\pi\sqrt{1-\rho^2}}\int_{-\infty}^{0} x\exp\left(-\frac{1}{2}x^2\right)\,\mathrm{d}x = \frac{1}{(2\pi)\sqrt{1-\rho^2}}.$$

Integrating out $\mathrm{d}\mathsf{P}(\rho)/\mathrm{d}\rho$, and using $\mathsf{P}(0) = 1/4$, we get (ii).

2.5.3 Show that if Equation (2.72) holds, then a vectorized $\boldsymbol{X}$, say $\boldsymbol{X}^V$ (see Definition A.6), has a multivariate normal distribution

$$\boldsymbol{X}^V \sim N_{np}(\boldsymbol{M}^V, \boldsymbol{V} \otimes \boldsymbol{U}).$$

2.6.1 In the terminology of Section 2.6, suppose $\mathsf{E}(g_{rr}^2) = c$, $\mathsf{E}(g_{rs}^2) = d, r \neq s$. Using (1.27) and $nc = \sum_{i=1}^{n}\mathsf{E}(g_{rr})^2$, show that $c = \mathsf{E}(b_{2,p})$. Further, using

$$nc + \frac{n(n-1)d}{2} = \mathsf{E}\left\{\sum_{r=1}^{n}\sum_{s=1}^{n} g_{rs}^2\right\} = n,$$

show that

$$d = \frac{2}{n-1}\{1 - \mathsf{E}(b_{2,p})\}.$$

2.6.2 (Mardia, 1964a) Let $\boldsymbol{U} = [\boldsymbol{u}_1, \ldots, \boldsymbol{u}_n]'$ be an $(n \times p)$ data matrix from a p.d.f. $f(\boldsymbol{u})$ and set

$$x_i = \min_{r=1,\ldots,n}(u_{ri}), \quad y_i = \max_{r=1,\ldots,n}(u_{ri}).$$

(i) Show that the joint p.d.f. of $(\boldsymbol{x}, \boldsymbol{y})$ is

$$(-1)^p \frac{\partial^{2p}}{\partial x_1 \cdots \partial x_p \partial y_1 \cdots \partial y_p}\left\{\int_{\boldsymbol{x}}^{\boldsymbol{y}} f(\boldsymbol{u})\,\mathrm{d}\boldsymbol{u}\right\}^n.$$

(ii) Let $R_i = y_i - x_i$ denote the *range* of the ith variable. Show that the joint distribution of $\boldsymbol{R} = (R_1, \ldots, R_p)'$ is

$$\int_{-\infty}^{\infty}\left[(-1)^p \frac{\partial^{2p}}{\partial x_1 \cdots \partial x_p \partial y_1 \cdots \partial y_p}\left\{\int_{\boldsymbol{x}}^{\boldsymbol{y}} f(\boldsymbol{u})\,\mathrm{d}\boldsymbol{u}\right\}^n\right]_{\boldsymbol{y}=\boldsymbol{x}+\boldsymbol{R}} \mathrm{d}\boldsymbol{x}.$$

2.7.1 Karl Pearson-type inequality Let $\psi(\boldsymbol{x})$ be a nonnegative function. Show that

$$\mathsf{P}(\psi(\boldsymbol{x}) < \epsilon^2) \geq 1 - \frac{\mathsf{E}[\psi(\boldsymbol{x})^s]}{\epsilon^{2s}}, \qquad s > 0.$$

In particular, for $s = 2$, we have

$$\mathsf{P}[(\boldsymbol{x} - \boldsymbol{\mu})'\boldsymbol{\Sigma}^{-1}(\boldsymbol{x} - \boldsymbol{\mu}) < \epsilon^2] \geq 1 - \beta_{2,p}/\epsilon^4,$$

where $\beta_{2,p}$ is the measure of kurtosis for $\boldsymbol{x}$.

3

Nonnormal Distributions

3.1 Introduction

In Chapter 2, we have given the properties of the multivariate normal distribution and have described its related distributions. In this chapter, we give some important nonnormal distributions that have become prominent in multivariate analysis. These include generalizations of univariate distributions. Some of these are direct extensions and others have a more distinctive multivariate character, such as spherical and elliptic distributions and examples. Also, some families of distributions are given such as the exponential family, spherical family, and elliptic distributions; some inference problems for the normal case extend easily for the elliptic distributions. We finally give a section on copulas, which have applications in a variety of different areas.

3.2 Some Multivariate Generalizations of Univariate Distributions

We give below three common techniques that are used to generalize univariate distributions to higher dimensions. However, caution must be exercised because in some cases there is more than one plausible way to carry out the generalization.

3.2.1 Direct Generalizations

Often, a property used to derive a univariate distribution has a plausible (though not necessarily unique) extension to higher dimensions.

(1) The simplest example is the multivariate normal where the squared term $(x-\mu)^2/\sigma^2$ in the exponent of the one-dimensional probability density function (p.d.f.) is generalized to the quadratic form $(\boldsymbol{x}-\boldsymbol{\mu})'\boldsymbol{\Sigma}^{-1}(\boldsymbol{x}-\boldsymbol{\mu})$.
(2) If $\boldsymbol{x} \sim \boldsymbol{N}_p(\boldsymbol{\mu}, \boldsymbol{\Sigma})$, and $\boldsymbol{u}$ is a vector with elements $u_i = \exp(x_i),\ i = 1, \dots, p$, then $\boldsymbol{u}$ is said to have a *multivariate log-normal distribution* with parameters $\boldsymbol{\mu}, \boldsymbol{\Sigma}$.
(3) Let $\boldsymbol{x} \sim \boldsymbol{N}_p(\boldsymbol{\mu}, \boldsymbol{\Sigma})$ and $y \sim \chi^2_\nu$ be independent, and let $\boldsymbol{u}$ be a vector with elements $u_i = x_i/(y/\nu)^{1/2}, i = 1, \dots, p$. Then, $\boldsymbol{u}$ has a *multivariate Student's t-distribution* with parameters $\boldsymbol{\mu}$ (location), $\boldsymbol{\Sigma}$ (scale or scatter matrix), and ν (index) (Cornish, 1954;

Multivariate Analysis, Second Edition. Kanti V. Mardia, John T. Kent, and Charles C. Taylor.

Dunnett and Sobel, 1954). The case $\nu = 1$ is termed the *multivariate Cauchy distribution*. See Exercise 3.2.5.

(4) The *Wishart distribution* defined in Chapter 4 is a matrix generalization of the χ^2 distribution. The p.d.f. is given in (4.48).

(5) The multivariate Pareto distribution (Mardia, 1962, 1964b) has p.d.f.

$$f(\boldsymbol{x}) = \prod_{i=1}^{p} \left(\frac{a+i-1}{b_i} \right) \left[1 - p + \sum_{i=1}^{p} \frac{x_i}{b_i} \right]^{-(a+p)} \quad x_i > b_i, \quad i = 1, \dots, p, \tag{3.1}$$

with parameters $b_i > 0$, $i = 1, \dots, p$, and $a > 0$. It generalizes the univariate Pareto distribution because its p.d.f. is given by a linear function of $\boldsymbol{x}$ raised to some power. See Exercise 3.2.4.

(6) The *Dirichlet distribution* is a generalization to p dimensions of the beta distribution (see Appendix B) with density

$$f(x_1, \dots, x_p) = \frac{\Gamma\left(\sum_{i=0}^{p} \alpha_i\right)}{\prod_{i=0}^{p} \Gamma(\alpha_i)} \left(1 - \sum_{i=1}^{p} x_i \right)^{\alpha_0 - 1} \prod_{i=1}^{p} x_i^{\alpha_i - 1}, \tag{3.2}$$

where $\Gamma(\cdot)$ denotes the gamma function (see B.6),

$$x_i > 0, \quad i = 1, \dots, p, \qquad \sum_{i=1}^{p} x_i < 1,$$

$$\text{and } \alpha_i > 0, \quad i = 0, 1, \dots, p, \text{ are parameters.}$$

(7) The *multinomial distribution* is a discrete generalization to p dimensions of the binomial distribution. For parameters $a_1, \dots, a_p$ ($a_i > 0$, $i = 1, \dots, p$, $\sum_{i=1}^{p} a_i = 1$) and n (a positive integer), the probabilities are given by

$$\mathsf{P}(\boldsymbol{x} = \boldsymbol{n}) = \begin{cases} \frac{n!}{n_1! \cdots n_p!} \prod_{i=1}^{p} a_i^{n_i}, & n_i \geq 0, \quad i = 1, \dots, p, \quad \sum_{i=1}^{p} n_i = n, \\ 0, & \text{otherwise.} \end{cases} \tag{3.3}$$

Note that the above equation has a vector $\boldsymbol{n}$ and a scalar n. The mean and covariance matrix of this distribution are given in Exercise 3.2.7, and its limiting behavior for large n is given in Example 2.7.1.

3.2.2 Common Components

Let $\mathcal{F}$ be some family of distributions in one dimension, and let $u_0, u_1, \dots, u_p$ denote independent members of $\mathcal{F}$. Set

$$x_i = u_i + u_0, \quad i = 1, \dots, p.$$

Then, the distribution of $\boldsymbol{x}$ is one possible generalization of $\mathcal{F}$. This approach has been used to construct a multivariate Poisson distribution (see Exercise 3.2.2) and a multivariate gamma distribution (Ramabhadran, 1951).

A similar construction was proposed by Azzalini and Dalla Valle (1996) in a generalization of the skew-normal distribution to the multivariate case. Suppose that $\boldsymbol{u} = (u_1, \dots, u_p)' \sim N(\boldsymbol{0}, \boldsymbol{\Sigma}_1)$ in which the diagonal elements of $\boldsymbol{\Sigma}_1$ are all equal to 1, and that, independently, $u_0 \sim N(0, 1)$. Also, let $\boldsymbol{\delta}$ be a vector of skew parameters with

$-1 < \delta_1, \ldots, \delta_p < 1$. Then, Azzalini and Dalla Valle (1996) show that $\boldsymbol{x} = (x_1, \ldots, x_p)'$, which is defined by

$$x_j = \delta_j |u_0| + (1 - \delta_j^2)^{1/2} u_j, \quad j = 1, \ldots, p,$$

has a multivariate skew-normal distribution with density given by

$$2f(\boldsymbol{x}; \boldsymbol{0}, \boldsymbol{\Omega})F(\boldsymbol{\alpha}'\boldsymbol{x}), \quad \boldsymbol{x} \in \mathbb{R}^p. \tag{3.4}$$

Here,

$f(\boldsymbol{x}; \boldsymbol{0}, \boldsymbol{\Omega})$ is the p.d.f. of a multivariate normal distribution with mean $\boldsymbol{0}$ and covariance $\boldsymbol{\Omega}$;
$F(\cdot)$ is the cumulative distribution function (c.d.f.) of a standard (univariate) normal distribution;

$$\begin{aligned}
\boldsymbol{\alpha}' &= \frac{\boldsymbol{\lambda}'\boldsymbol{\Sigma}_1^{-1}\boldsymbol{\Delta}^{-1}}{(1 + \boldsymbol{\lambda}'\boldsymbol{\Sigma}_1^{-1}\boldsymbol{\lambda})^{1/2}}, \\
\boldsymbol{\Delta} &= \text{diag}\,((1 - \delta_1^2)^{1/2}, \ldots, (1 - \delta_p^2)^{1/2}) \\
\boldsymbol{\lambda} &= \boldsymbol{\Delta}\boldsymbol{\delta}'; \\
\boldsymbol{\Omega} &= \boldsymbol{\Delta}(\boldsymbol{\Sigma}_1 + \boldsymbol{\lambda}\boldsymbol{\lambda}')\boldsymbol{\Delta}.
\end{aligned}$$

Azzalini and Dalla Valle (1996) show that the marginal distribution of x_j has density given by

$$2f(x_j)F\left(\frac{\delta_j x_j}{(1 - \delta_j^2)^{1/2}}\right),$$

which is known as the skew-normal distribution. Here, $f(\cdot)$ is the p.d.f. of a standard normal – given by (B.1) – and $F(\cdot)$ as above.

Further developments of multivariate skew distributions – for example, the skew multivariate t-distribution – are given in Azzalini (2005).

3.2.3 Stochastic Generalizations

Sometimes, a probabilistic argument used to construct a distribution in one dimension can be generalized to higher dimensions. The simplest example is the use of the multivariate normal distribution. As another example, consider a multivariate exponential distribution (Johnson and Kotz, 1972; Weinman, 1966). A system has p identical components with times to failure $x_1, \ldots, x_p$. Initially, each component has an independent failure rate $\lambda_0 > 0$. (If the failure rate were constant for all time, the life time of each component would have an exponential distribution,

$$f(x, \lambda_0) = \lambda_0^{-1} \exp(-x/\lambda_0), \quad x > 0.)$$

However, once a component fails, the failure rate of the remaining components changes. More specifically, conditional on exactly k components having failed by time $t > 0$, each of the remaining $p - k$ components has a failure rate $\lambda_k > 0$ (until the $(k + 1)$th component fails). It can be shown that the joint p.d.f. is given by

$$\left(\prod_{i=0}^{p-1} \lambda_i^{-1}\right) \exp\left\{-\sum_{i=0}^{p-1} (p - i)\frac{(x_{(i+1)} - x_{(i)})}{\lambda_i}\right\}, \quad x_i > 0,\ i = 1, \ldots, p, \tag{3.5}$$

where $x_{(0)} = 0$ and $x_{(1)} \leq x_{(2)} \leq \ldots \leq x_{(p)}$ are the order statistics.

A different multivariate exponential distribution has been described by Marshall and Olkin (1967), using a different underlying probability model (see Exercise 3.2.3).

3.3 Families of Distributions

3.3.1 The Exponential Family

The random vector $\boldsymbol{x}$ belongs to the *general exponential family* if its p.d.f. is of the form

$$f(\boldsymbol{x};\boldsymbol{\theta}) = \exp\left[a_0(\boldsymbol{\theta}) + b_0(\boldsymbol{x}) + \sum_{i=1}^{q} a_i(\boldsymbol{\theta})b_i(\boldsymbol{x})\right], \quad \boldsymbol{x} \in S, \tag{3.6}$$

where $\boldsymbol{\theta}' = (\theta_1, \dots, \theta_r)$ is the vector of parameters, $\exp[a_0(\boldsymbol{\theta})]$ is the normalizing constant, and S is the support. If $r = q$ and $a_i(\boldsymbol{\theta}) = \theta_i (i \neq 0)$, we say that $\boldsymbol{x}$ belongs to the *canonical exponential family*; see (3.9). The general exponential family (3.6) includes most of the important univariate distributions as special cases, for instance the normal, Poisson, negative binomial, and gamma distributions (Rao, 1973, p. 195).

Example 3.3.1 Putting

$$(b_1, \dots, b_q) = (x_1, \dots, x_p, x_1^2, \dots, x_p^2, x_1x_2, x_1x_3, \dots, x_{p-1}x_p)$$

in (3.6), it is seen that $N_p(\boldsymbol{\mu}, \boldsymbol{\Sigma})$, whose p.d.f. is given by (2.49), belongs to the general exponential family. □

Example 3.3.2 A discrete example is the *logit model* (Cox 1972) defined by

$$\log \mathsf{P}(x_1, \dots, x_p) = a_0 + \sum a_i x_i + \sum a_{ij} x_i x_j + \sum a_{ijk} x_i x_j x_k + \cdots, \tag{3.7}$$

for $x_i = \pm 1$, $i = 1, \dots, p$. If $a_{ij} = 0, a_{ijk} = 0, \dots,$ the variables are independent. The logit model plays an important role in contingency tables where the variables $z_i = (x_i + 1)/2$, taking values 0 or 1, are of interest. The interpretation of parameters in (3.7) is indicated by

$$\frac{1}{2}\log\frac{\mathsf{P}(x_1 = 1 \mid x_2, \dots, x_p)}{\mathsf{P}(x_1 = -1 \mid x_2, \dots, x_p)} = a_1 + a_{12}x_2 + \dots + a_{1p}x_p + a_{123}x_2x_3 + \cdots \tag{3.8}$$

since a_{23}, etc., do not appear in this expression. □

Properties of exponential families

(1) For the canonical exponential family,

$$f(\boldsymbol{x};\boldsymbol{\theta}) = \exp\left[a_0(\boldsymbol{\theta}) + b_0(\boldsymbol{x}) + \sum_{i=1}^{p} \theta_i b_i(\boldsymbol{x})\right], \quad \boldsymbol{x} \in S, \tag{3.9}$$

the vector $(b_1(\boldsymbol{x}), \dots, b_q(\boldsymbol{x}))$ is a sufficient and complete statistic for $\boldsymbol{\theta}$.

(2) Consider the set of all p.d.f.s $g(\boldsymbol{x})$ with support S satisfying the constraints

$$\mathsf{E}[b_i(\boldsymbol{x})] = c_i, \quad i = 1, \dots, q, \tag{3.10}$$

where the c_i are fixed. Then, the *entropy* $\mathsf{E}\{-\log g(\boldsymbol{x})\}$ is maximized by the density $f(\boldsymbol{x};\boldsymbol{\theta})$ in (3.9), provided there exists $\boldsymbol{\theta}$ for which the constraints (3.10) are satisfied. If such a $\boldsymbol{\theta}$ exists, the maximum is unique (see Kagan et al. (1973, p. 409), Mardia (1975)).

The above maximum entropy property is very powerful. For example, if we fix the expectations of $x_1, \ldots, x_p$, $x_1^2, \ldots, x_p^2$, $x_1x_2, \ldots, x_{p-1}x_p$, $\boldsymbol{x} \in \mathbb{R}^p$, then the maximum entropy distribution is the multivariate normal distribution.

Extended exponential family Let us assume that a random vector $\boldsymbol{y}$ is paired with an observable vector $\boldsymbol{x}$. Then, an extended family can be defined (Dempster, 1971, 1972) as

$$f(\boldsymbol{y}|\boldsymbol{x}, \boldsymbol{\theta}) = \exp\left(a_0 + \sum_{i=1}^{p} \sum_{j=1}^{q} \theta_{ij} x_i y_j \right), \tag{3.11}$$

where a_0 depends on the x_i and the parameters θ_{ij}. If $(\boldsymbol{x}', \boldsymbol{y}')'$ has a multivariate normal distribution, then the conditional density of $\boldsymbol{y} \mid \boldsymbol{x}$ follows a modified version of (3.11) with extra terms in the exponent, $\sum_{1 \le j \le k \le q} \phi_{jk} y_j y_k$. Obviously, the conditional distribution (3.8) for the logit model is also a particular case.

3.3.2 The Spherical Family

Any nonzero vector $\boldsymbol{x} \in \mathbb{R}^p$ can be written in polar coordinates as

$$\boldsymbol{x} = r\boldsymbol{\ell}, \quad \boldsymbol{\ell}'\boldsymbol{\ell} = 1, \tag{3.12}$$

where $r > 0$ and the unit vector

$$\boldsymbol{\ell} = \boldsymbol{u}(\boldsymbol{\theta}), \tag{3.13}$$

is expressed in terms of a vector of angles $\boldsymbol{\theta} = (\theta_1 \ldots, \theta_{p-1})'$ using (2.46). Using the Jacobian (2.47), the probability element $\mathrm{d}\boldsymbol{x}$ can be factored as

$$\mathrm{d}\boldsymbol{x} = \{r^{p-1}\,\mathrm{d}r\} \left\{ \left(\prod_{i=1}^{p-1} \sin^{p-i}\theta_{i-1} \right) \mathrm{d}\boldsymbol{\theta} \right\}. \tag{3.14}$$

Definition 3.3.1 *Let $\boldsymbol{x}$ be a random vector with p.d.f. $f(\boldsymbol{x})$. If $f(\boldsymbol{x})$ takes the form*

$$f(\boldsymbol{x}) = g(\boldsymbol{x}'\boldsymbol{x}), \tag{3.15}$$

then the distribution of $\boldsymbol{x}$ is said to belong to the spherical family *since it is spherically symmetric.*

Under spherical symmetry, the p.d.f. can be factorized as

$$f(\boldsymbol{x})\,\mathrm{d}\boldsymbol{x} = \{c_p r^{p-1} g(r^2)\,\mathrm{d}r\} \left\{ c_p^{-1} \left(\prod_{i=1}^{p-1} \sin^{p-i}\theta_{i-1} \right) \mathrm{d}\boldsymbol{\theta} \right\}, \tag{3.16}$$

where $c_p = 2\pi^{p/2}/\Gamma(p/2)$ is the surface area of the unit hypersphere in $\mathbb{R}^p$, and Γ is the gamma function given in (B.6). Thus the radial component r, with p.d.f. $c_p r^{p-1} g(r^2)\,\mathrm{d}r$, is independent of the angular component $\boldsymbol{\ell}$, which has a uniform distribution on the unit hypersphere in $\mathbb{R}^p$. It can be shown that

$$\mathsf{E}(\boldsymbol{\ell}) = \boldsymbol{0}, \quad \mathsf{E}(\boldsymbol{\ell}\boldsymbol{\ell}') = \frac{1}{p}\boldsymbol{I}_p. \tag{3.17}$$

We immediately have the following important property of spherically symmetric distributions.

Theorem 3.3.2 *The value of $r = (\boldsymbol{x}'\boldsymbol{x})^{1/2}$ is statistically independent of any scale-invariant function of $\boldsymbol{x}$.*

Note that a function $h(\boldsymbol{x})$ is scale invariant if $h(\boldsymbol{x}) = h(\alpha\boldsymbol{x})$ for all $\alpha > 0$. Setting $\alpha = 1/r$, we see that $h(\boldsymbol{x}) = h\{r\boldsymbol{u}(\boldsymbol{\theta})\} = h\{\boldsymbol{u}(\boldsymbol{\theta})\}$ depends only on $\boldsymbol{\theta}$. Thus, $h(\boldsymbol{x})$ is independent of r.

Example 3.3.3 If $\boldsymbol{x} \sim N_p(\boldsymbol{0}, \sigma^2\boldsymbol{I})$, then the p.d.f. of $\boldsymbol{x}$ is

$$(2\pi\sigma^2)^{-p/2} \exp\left(-\frac{1}{2}r^2/\sigma^2\right).$$

Hence, $N_p(\boldsymbol{0}, \sigma^2\boldsymbol{I})$ is spherically symmetric. Further, we note that $r^2/\sigma^2 = \boldsymbol{x}'\boldsymbol{x}/\sigma^2 \sim \chi_p^2$. □

Example 3.3.4 The multivariate Cauchy distribution with parameters $\boldsymbol{\theta}$ and $\boldsymbol{I}$ has p.d.f.

$$f(\boldsymbol{x}) = \pi^{-(p+1)/2}\Gamma((p+1)/2)(1 + \boldsymbol{x}'\boldsymbol{x})^{-(p+1)/2} \tag{3.18}$$

and belongs to the spherical family (see Exercise 3.2.5), where Γ is given by (B.6). □

Skewness and kurtosis for the spherical family. We know that Mardia's kurtosis is defined as

$$\beta_{2,p} = \mathsf{E}[(\boldsymbol{x} - \boldsymbol{\mu})'\boldsymbol{\Sigma}^{-1}(\boldsymbol{x} - \boldsymbol{\mu})]^2. \tag{3.19}$$

Theorem 3.3.3 *We have for the spherical family*

$$\beta_{1,p} = 0, \qquad \beta_{2,p} = p^2\frac{\mathsf{E}(r^4)}{\mathsf{E}(r^2)^2}. \tag{3.20}$$

Proof: We have

$$\boldsymbol{\mu} = \mathsf{E}(r\boldsymbol{\ell}) = \mathsf{E}(r)\mathsf{E}(\boldsymbol{\ell}) = \boldsymbol{0}. \tag{3.21}$$

$$\boldsymbol{\Sigma} = \mathsf{E}(r\boldsymbol{\ell}r\boldsymbol{\ell}') = \mathsf{E}(r^2)\mathsf{E}(\boldsymbol{\ell}\boldsymbol{\ell}') = \frac{\mathsf{E}(r^2)}{p}\boldsymbol{I}. \tag{3.22}$$

We have from (3.21) and (3.22), $\boldsymbol{\mu} = 0$, $\boldsymbol{\Sigma} = \mathsf{E}(r^2)\boldsymbol{I}/p$, and since

$$\mathsf{E}(\boldsymbol{x}'\boldsymbol{x})^4 = \mathsf{E}(r^2\boldsymbol{\ell}'\boldsymbol{\ell})^2 = \mathsf{E}(r^4),$$

the result follows. □

Example 3.3.5 Consider three cases for a distribution $\boldsymbol{x}$ in the spherical family.

a) If $\boldsymbol{x}$ is uniformly distributed on the hypersphere, then $r = 1$ is constant, so that $\mathsf{E}(r^2) = \mathsf{E}(r^4) = 1$. The formula for kurtosis simplifies to $\beta_{2,p}^{\text{sphere}} = p^2$. In fact, since for any distribution $\{\mathsf{E}(r^2)\}^2 \le \mathsf{E}(r^4)$, this is the smallest possible kurtosis.
b) If $\boldsymbol{x}$ is uniformly distributed on the unit ball $\{\boldsymbol{x} : \boldsymbol{x}'\boldsymbol{x} \le 1\}$, then the radial component has p.d.f. pr^{p-1}, $0 \le r \le 1$, with moments $\mathsf{E}(r^2) = p/(p+2)$ and $\mathsf{E}(r^4) = p/(p+4)$. The formula for kurtosis simplifies to $\beta_{2,p}^{\text{ball}} = p(p+2)^2/(p+4)$.
c) Finally, the kurtosis for the multivariate normal distribution is $\beta_{2,p}^{\text{normal}} = p(p+2)$.

These three values satisfy the inequalities

$$\beta_{2,p}^{\text{sphere}} < \beta_{2,p}^{\text{ball}} < \beta_{2,p}^{\text{normal}}.$$

Thus, the uniform distribution on the hypersphere and the uniform distribution on the ball have shorter tails than the multivariate normal distribution. □

Skewness for spherical family. Noting that $\beta_{1,p} = (p+2)^{-3}\mathsf{E}(\boldsymbol{x}'\boldsymbol{y})^3$ and recognizing that each term is a third-order moment, so equal to zero by a symmetry argument, we have

$$\beta_{1,p} = 0.$$

Theorem 3.3.2 can also be proved in a more general setting, which will be useful later.

Theorem 3.3.4 *Let $\boldsymbol{X}(n \times p)$ be a random matrix which when thought of as an np-vector $\boldsymbol{X}^v$ is spherically symmetric. If $h(\boldsymbol{X})$ is any column-scale-invariant function of $\boldsymbol{X}$, then $h(\boldsymbol{X})$ is independent of $(r_1, \ldots, r_p)$, where*

$$r_j^2 = \sum_{i=1}^{n} x_{ij}^2, \quad j = 1, \ldots, p.$$

Proof: Write $R^2 = \sum r_j^2$. Then, the density of $\boldsymbol{X}$ can be written as $g(R^2)\prod \mathrm{d}x_{ij}$. Transforming each column to polar coordinates, we get

$$\left[g(R^2)\prod_{j=1}^{p} r_j^{n-1}\,\mathrm{d}r_j\right]\left[\prod_{j=1}^{p}\left(\prod_{i=2}^{n-1}\sin^{n-i}\theta_{i-1,j}\right)\,\mathrm{d}\theta_{1j}\ldots\mathrm{d}\theta_{n-1,j}\right].$$

Thus, $(r_1, \ldots, r_p)$ is independent of $\boldsymbol{\Theta} = (\theta_{ij})$.

A function $h(\boldsymbol{X})$ is *column-scale invariant* if $h(\boldsymbol{XD}) = h(\boldsymbol{X})$ for all diagonal matrices $\boldsymbol{D} > \boldsymbol{0}$. Setting $\boldsymbol{D} = \text{diag}\,(r_1^{-1}, \ldots, r_p^{-1})$, we see that

$$\begin{aligned} h(\boldsymbol{X}) &= h(r_1\boldsymbol{u}_1(\boldsymbol{\theta}_{(1)}), \ldots, r_p\boldsymbol{u}_p(\boldsymbol{\theta}_{(p)})) \\ &= h(\boldsymbol{u}_1(\boldsymbol{\theta}_{(1)}), \ldots, \boldsymbol{u}_p(\boldsymbol{\theta}_{(p)})) \end{aligned}$$

depends only on $\boldsymbol{\Theta}$, where each $\boldsymbol{u}_j(\boldsymbol{\theta}_{(j)})$ is the function of the jth column of $\boldsymbol{\Theta}$ defined as in (2.46). Thus, $h(\boldsymbol{X})$ is statistically independent of $(r_1, \ldots, r_p)$. □

Often, results proved for the multivariate normal distribution $N_p(\boldsymbol{0}, \sigma^2\boldsymbol{I})$ are true for *all* spherically symmetric distributions. For example, if the vector $(x_1, \ldots, x_{m+n})$ is spherically symmetric, then

$$n\sum_{j=1}^{m} x_j^2 \Bigg/ \left(m\sum_{j=m+1}^{m+n} x_j^2\right) \sim F_{m,n}, \tag{3.23}$$

where $F_{m,n}$ is the F-variable; see Dempster (1969).

3.3.3 Elliptical Distributions

By noting that $\boldsymbol{y} = \boldsymbol{\Sigma}^{-1/2}(\boldsymbol{x} - \boldsymbol{\mu})$ is spherically symmetric, the above discussion can be extended to elliptically symmetric distributions of the form

$$f(\boldsymbol{x}) = c_p|\boldsymbol{\Sigma}|^{-1/2}g((\boldsymbol{x} - \boldsymbol{\mu})'\boldsymbol{\Sigma}^{-1}(\boldsymbol{x} - \boldsymbol{\mu})), \tag{3.24}$$

where c_p is defined as in Equation (3.16). This generalization is akin to that of the skew distributions (3.4). Equation (3.24) depends on two parameters, $\boldsymbol{\mu}, \boldsymbol{\Sigma}$, and on the function g, which is referred to as the *density generator*. It is obvious that, if $\boldsymbol{x}$ is such that $(\boldsymbol{x} - \boldsymbol{\mu})'\boldsymbol{\Sigma}^{-1}(\boldsymbol{x} - \boldsymbol{\mu})$ is constant, then $f(\boldsymbol{x})$ is also constant.

If $\boldsymbol{x}$ has an elliptically contoured distribution with density given by (3.24), then it is easy to show that

(1) if $\boldsymbol{A}$ is a square matrix of full rank, then $\boldsymbol{y} = \boldsymbol{c} + \boldsymbol{A}'\boldsymbol{x}$ has an elliptical distribution with parameters $\boldsymbol{c} + \boldsymbol{A}'\boldsymbol{\mu}, \boldsymbol{A}'\boldsymbol{\Sigma}\boldsymbol{A}$ and g.
(2) there exists a random p-vector $\boldsymbol{s}$, uniformly distributed on the unit sphere, and a continuous positive random variable r, independent of $\boldsymbol{s}$ with p.d.f. given by (3.19) such that

$$\boldsymbol{x} = \boldsymbol{\mu} + r\boldsymbol{L}'\boldsymbol{s}, \tag{3.25}$$

where $\boldsymbol{L}'\boldsymbol{L} = \boldsymbol{\Sigma}$.
(3) if $\boldsymbol{x}$ is partitioned into $\boldsymbol{x}' = (\boldsymbol{x}_1', \boldsymbol{x}_2')$, with $\boldsymbol{\mu}$ and $\boldsymbol{\Sigma}$ correspondingly partitioned, then
   (a) $\boldsymbol{x}_1$ has an elliptical distribution with parameters $\boldsymbol{\mu}_1, \boldsymbol{\Sigma}_{11}$ and g;
   (b) the distribution of $\boldsymbol{x}_1$ conditional on $\boldsymbol{x}_2$ has an elliptical distribution with parameters $\boldsymbol{\mu}_1 + \boldsymbol{\Sigma}_{12}\boldsymbol{\Sigma}_{22}^{-1}(\boldsymbol{x}_2 - \boldsymbol{\mu}_2)$ and $\boldsymbol{\Sigma}_{11} - \boldsymbol{\Sigma}_{12}\boldsymbol{\Sigma}_{22}^{-1}\boldsymbol{\Sigma}_{21}$ and a density generator that depends on $\boldsymbol{x}_2$ only through $(\boldsymbol{x}_2 - \boldsymbol{\mu}_2)'\boldsymbol{\Sigma}_{22}^{-1}(\boldsymbol{x}_2 - \boldsymbol{\mu}_2)$.

In the special case that $\boldsymbol{\mu} = \boldsymbol{0}$ and $\boldsymbol{\Sigma} = \boldsymbol{I}$, (3.24) becomes the spherical or isotropic distribution,

$$f(\boldsymbol{x}) = c_p g(\boldsymbol{x}'\boldsymbol{x}) = c_p g(r^2),$$

and result (2) means that $\boldsymbol{x}$ can be written as the product of two independent random variables: r, with density (3.19) and $\boldsymbol{\ell}$, uniform on the sphere, i.e. $\boldsymbol{x} = r\boldsymbol{\ell}$.

In the general case, the moments of $\boldsymbol{x}$ can be found from the characteristic function, which is of the form $\phi_{\boldsymbol{x}}(\boldsymbol{t}) = \exp(i\boldsymbol{t}'\boldsymbol{\mu})g^*(\boldsymbol{t}'\boldsymbol{\Sigma}\boldsymbol{t})$, or from the representation given by Equation (3.25). It is then clear that the kth order moment of $\boldsymbol{x}$ exists if and only if the kth order moment of r exists. For example, it can be shown that $\mathsf{E}(\boldsymbol{x}) = \boldsymbol{\mu}$, and $\mathsf{E}\{(\boldsymbol{x} - \boldsymbol{\mu})(\boldsymbol{x} - \boldsymbol{\mu})'\} = \{\mathsf{E}(r^2)/p\}\boldsymbol{\Sigma}$; see Anderson (2003, sec. 2.7.4) for more details. Further properties of elliptical distributions can be found in Kelker (1970) and Cambanis et al. (1981).

An obvious subclass of the family of elliptical distributions is the multivariate normal in which $f(\boldsymbol{x}) \propto \exp(-\boldsymbol{x}'\boldsymbol{x}/2)$; see Equation (2.49). Mardia and Holmes (1980), using maximum entropy as a criterion, proposed the elliptically symmetric density

$$f(\boldsymbol{x}) = C(\kappa)|\boldsymbol{\Sigma}|^{-1/2}\exp\left(-\frac{1}{2}\kappa\left[(\boldsymbol{x} - \boldsymbol{a})'\boldsymbol{\Sigma}^{-1}(\boldsymbol{x} - \boldsymbol{a}) - 1\right]^2\right), \tag{3.26}$$

where $\kappa > 0$ can be described as a concentration parameter, and $C(\kappa)$ is a normalizing constant. They used this distribution to investigate the hypothesis that megalithic sites were constructed in an elliptic pattern. See Exercise 3.3.3 for some properties of this density. Another subclass is given by the Kotz-type distribution with density

$$f(\boldsymbol{x}) = c_p(\boldsymbol{x}'\boldsymbol{x})^{N-1}\exp(\alpha r(\boldsymbol{x}'\boldsymbol{x})^{\beta}), \qquad \alpha, \beta > 0, \quad 2N + p > 2,$$

where

$$c_p = \frac{\beta\Gamma(p/2)}{\pi^{p/2}\Gamma((2N+p-2)/(2\beta))}\alpha^{(2N+p-2)/(2\beta)}.$$

This was proposed by Kotz (1975) with $\beta = 1$. It can be seen that, if $N = 1$, $\beta = 1$, and $\alpha = 1/2$, we obtain the multivariate normal as a special case. For further details, and applications, see Nadarajah (2003).

3.3.4 Stable Distributions

A univariate random variable x with c.d.f. $F(x)$ is called *stable* if, for all $b_1 > 0, b_2 > 0$, there exists $b > 0$ and c real such that

$$F(x/b_1) * F(x/b_2) = F((x-c)/b),$$

where $*$ denotes convolution. This concept has been generalized to higher dimensions by Lévy (1937). A *random vector* $\boldsymbol{x}$ is *stable* if every linear combination of its components is univariate stable. Call $\boldsymbol{x}$ *symmetric* about $\boldsymbol{a}$ if $\boldsymbol{x} - \boldsymbol{a}$ and $-(\boldsymbol{x} - \boldsymbol{a})$ have the same distribution. Then, a useful subclass of the symmetric stable laws is the set of random vectors whose c.f. is of the form

$$\log \phi_{\boldsymbol{x}}(\boldsymbol{t}) = i\boldsymbol{a}'\boldsymbol{t} - \frac{1}{2}\sum_{j=1}^{m}(\boldsymbol{t}'\boldsymbol{\Omega}_j\boldsymbol{t})^{\alpha/2}, \qquad (3.27)$$

where $0 < \alpha \leq 2$, $m \geq 1$, and $\boldsymbol{\Omega}_j$ is a $(p \times p)$ positive semidefinite matrix of constants for $j = 1, \ldots, m$. Equation (3.27) gives the c.f. of a nonsingular multivariate stable distribution if $\sum \boldsymbol{\Omega}_j > 0$ (Press, 1972, p. 155).

The Cauchy distribution and the multivariate normal distribution are the two most important stable distributions. For further details and applications in stable portfolio analysis, see Press (1972, Chapter 12).

3.4 Insights into Skewness and Kurtosis

Recall that the standard multivariate normal distribution in p dimensions has no skewness and a kurtosis of $p(p+2)$. In this section, we use bivariate nonnormal examples to get some insights for distributions that have skewness different from 0 and excess kurtosis different from $\beta_{2,2} - 8$. On the one hand, we describe the skewness and kurtosis for a bivariate mixture of normal distributions, using examples that have skewness with either some kurtosis or almost zero excess kurtosis. Note that some properties of the multivariate normal mixture are explored in Exercise 3.2.1. Here we use the case of a mixture of bivariate normal distributions (writing (x, y) in place of $\boldsymbol{x}$):

$$f(x,y) = \frac{\exp -y^2/2)}{2\pi}\left\{\lambda \exp\left(-\frac{(x-\mu)^2}{2}\right) + \lambda' \exp\left(-\frac{x^2}{2}\right)\right\}, 0 \leq \lambda \leq 1, \lambda' = 1 - \lambda, \qquad (3.28)$$

i.e. both components have $\boldsymbol{\Sigma} = \boldsymbol{I}_2$, and the means are $(\mu, 0)'$ and $(0, 0)'$, respectively. We note from Exercise 3.2.1 that this is an appropriate reduction of two multivariate normals with general means. It can be seen that in this case, the departure from normality is captured by the univariate variable x. In another example, we examine a bivariate t-distribution that has no skewness but high kurtosis, thus complementing the mixture examples.

The multivariate t-distribution is spherically symmetric, which is in contrast to the mixture p.d.f. Note that some properties of the multivariate t-distribution are explored in Exercises 3.2.1, 3.2.5, and 3.2.6. The bivariate ($p = 2$) standard t-distribution considered here has the p.d.f. given by (see Exercise 3.2.5)

$$g(x, y) = \frac{1}{2\pi}\left(1 + \frac{x^2 + y^2}{\nu}\right)^{-(1+\nu/2)}, \tag{3.29}$$

where we assume that $\nu > 4$.

We now work out the details of these cases. First, consider the mixture. In order to focus on the effect of the third and the fourth moments, we standardize the density (3.28), so its mean vector is zero, and the covariance matrix is the identity, using

$$x \to \frac{x - \lambda\mu}{\sqrt{1 + \lambda(1-\lambda)\mu^2}} = \frac{x - \lambda\mu}{a}, \quad \text{say.}$$

This leads to the joint density function

$$\frac{a \exp(-y^2/2)}{2\pi}\left\{\lambda \exp(-(ax - \lambda'\mu)^2/2) + \lambda' \exp(-(ax + \lambda\mu)^2/2)\right\}, \tag{3.30}$$

where $a = \sqrt{1 + \lambda\lambda'\mu^2}$ $(0 \le \lambda \le 1)$. To obtain the skewness and kurtosis of the density given by (3.30), we can use Exercise 3.2.1, with $p = 2$ and $\Delta = \mu$ to find that

$$\beta_{1,2} = \{\lambda\lambda'(\lambda - \lambda')\mu^3\}^2/(1 + \lambda\lambda'\mu^2)^3 \tag{3.31}$$

and

$$\beta_{2,2} = \{\lambda\lambda'(1 - 6\lambda\lambda')\mu^4\}/(1 + \lambda\lambda'\mu^2)^2 + 8. \tag{3.32}$$

For the t distribution, we can obtain the skewness and kurtosis for general p (see Exercise 3.2.6) as

$$\beta_{1,p} = 0, \qquad \beta_{2,p} = \frac{p(p+2)(\nu-2)}{\nu-4}, \quad \nu > 4.$$

Hence, for our case

$$\beta_{2,2} = \frac{8(\nu-2)}{\nu-4}, \quad \nu > 4.$$

We note that the mean of (3.29) is $\mathbf{0}$, and the covariance matrix is $\nu/(\nu-2)\boldsymbol{I}$, so after standardizing we have the density

$$g(x, y) = \frac{\nu}{2\pi(\nu-2)}\left(1 + \frac{x^2 + y^2}{\nu - 2}\right)^{-(1+\nu/2)} \quad (\nu > 4). \tag{3.33}$$

Example 3.4.1 We now give our specific selection for μ and λ to illustrate (a) no skewness or kurtosis (normal), (b) moderate skewness and negligible kurtosis, (c) both skewness

Table 3.1 Selected examples (shown in Figure 3.1) for a variety of $\beta_{1,2}$ and $\beta_{2,2}$ with the corresponding distributions.

Case	Distribution	$\beta_{1,2}$	$\beta_{2,2}$	Comments
a	Bivariate normal	0	8	No skewness or excess kurtosis
b	Mixture ($\mu = 3, \lambda = 0.2$)	0.462	8.087	Skewness and negligible kurtosis
c	Mixture ($\mu = 3, \lambda = 0.1$)	0.637	9.024	Skewness and small kurtosis
d	*t*-Distribution with 5 d.f.	0	24	No skewness but large kurtosis

and kurtosis, and, in contrast, (d) a multivariate t-distribution, which has no skewness but substantial kurtosis, as shown in Table 3.1, to obtain a range of examples of the coefficients of skewness and kurtosis.

In Figure 3.1, we have plotted the quantiles of the p.d.f.s given by (3.30) and (3.33), choosing the quantile levels to highlight the tails of the distributions, and so aid comparisons.

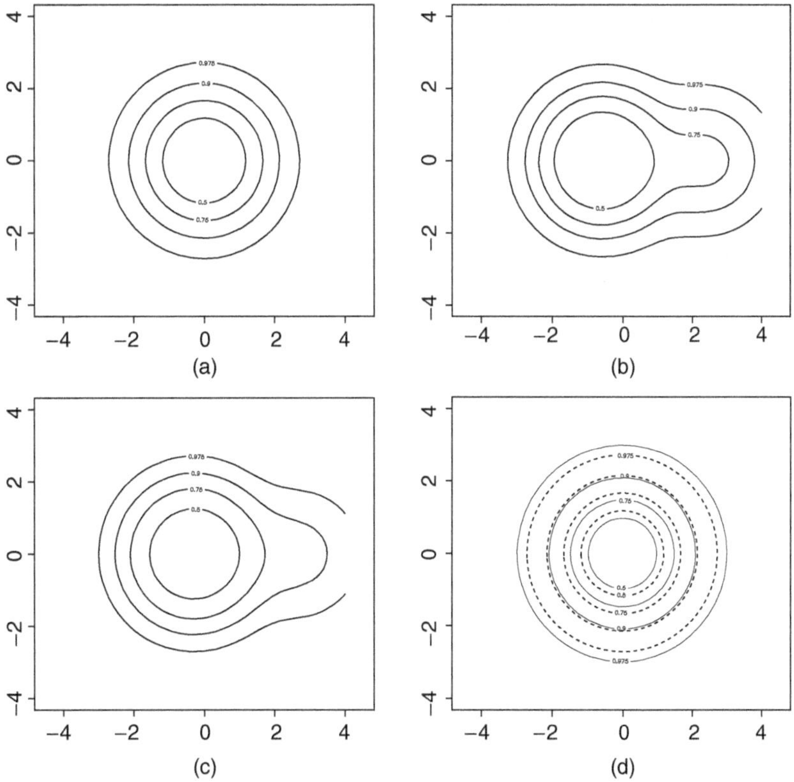

Figure 3.1 Contour plots of the quantiles for various choices of λ and μ in the probability density function given by (3.30): (a) bivariate normal; (b) mixture ($\mu = 3, \lambda = 0.2$); and (c) mixture ($\mu = 3, \lambda = 0.1$). (d) Contour plot of bivariate t-distribution with degrees of freedom $\nu = 5$ (continuous) and contours from panel (a) (dashed).

The most obvious differences are that (a) and (d) have circular contours, indicating no skewness, whereas (b) and (c) have a longer tail on the positive x-axis, indicating skewed distributions. Note that the main difference between panels (a) and (d) is in the spacing of the outer contours, whereas the spacing of the inner contours are very similar; this pattern is consistent with excess kurtosis. In comparing panels (b) and (c), we again note the differences in the spacing of the contours. □

3.5 Copulas

One way to generate a broad class of multivariate distributions is through the use of copulas. The motivation behind copulas is to separate the marginal distributions of the variables from their dependence structure. For simplicity, we restrict attention here to the continuous case – that is, multivariate distributions possessing density functions. In this section we again use capital letters to denote random variables.

Definition 3.5.1 *A copula function (or, simply, copula) is a multivariate distribution function with uniform marginal distributions. A copula function in p-dimensions is typically written as* $C(\boldsymbol{u}),\ \boldsymbol{u} = (u_1, \dots, u_p)' \in [0,1]^p$.

All continuous multivariate distributions can be written in terms of copulas using Sklar's Theorem.

Theorem 3.5.2 *(Sklar, 1959) Given a continuous random vector* $\boldsymbol{X} = (X_1, \dots, X_p)'$ *with joint c.d.f.* $F(\boldsymbol{x})$ *and marginal c.d.f.s* $F_1, \dots, F_p$*, there exists a unique copula* $C(\cdot)$ *such that for all* $\boldsymbol{x} = (x_1, \dots, x_p)'$*, we have*

$$F(\boldsymbol{x}) = C(F_1(x_1), \dots, F_p(x_p)). \tag{3.34}$$

Conversely, for given any one-dimensional c.d.f.s $F_1, \dots, F_p$ *and a copula* $C(\cdot)$*, Eq. (3.34) defines a p-variate c.d.f.* F *with marginal c.d.f.s* $F_1, \dots, F_p$.

Copulas can also be defined for discrete distributions, but the construction is less important.

For this section it is convenient to use an upper case letter X to denote a random variable and a lower case letter x to denote a possible value. Similarly, use a bold upper case letter $\boldsymbol{X}$ to denote a random vector and a lower case bold letter $\boldsymbol{x}$ to denote a possible value. The notation here conflicts with the convention in the rest of the book where bold upper case letters are reserved for matrices, and where lower case letters are used for both random vectors and their realizations.

To understand the logic behind Sklar's Theorem, recall the probability integral transformation for a one-dimensional random variable X with c.d.f. $F(x)$ and p.d.f. $f(x)$. Then, X can be transformed to a uniform random variable $U \sim \text{Unif}(0,1)$ by setting $U = F(X)$. This result is straightforward to verify since for any choice of u, $0 < u < 1$,

$$\mathsf{P}(U \le u) = \mathsf{P}(X = F^{-1}(U) \le F^{-1}(u)) = F(F^{-1}(u)) = u.$$

Once the marginal distributions of a random vector have been transformed to uniform, the dependence between the components can be represented as follows.

Let $\boldsymbol{X}$ be a p-dimensional continuous random vector with multivariate c.d.f.

$$F(\boldsymbol{x}) = \mathsf{P}(X_1 \le x_1, \ldots, X_p \le x_p),$$

and let $F_i(x_i)$, $i = 1, \ldots, p$, denote the corresponding marginal c.d.f.s. Set $U_i = F_i(X_i)$. Then, the corresponding random vector $\boldsymbol{U}$ has marginal uniform distributions with c.d.f.

$$\begin{aligned} \mathsf{P}(U_1 \le u_1, \ldots, U_p \le u_p) &= \mathsf{P}(X_1 \le F_1^{-1}(u_1), \ldots, X_p \le F_p^{-1}(u_p)) \\ &= F(F_1^{-1}(u_1), \ldots, F_p^{-1}(u_p)) \qquad (3.35) \\ &= C(\boldsymbol{u}), \text{ say,} \end{aligned}$$

where $C(\cdot)$ in the final line is the same as $C(\cdot)$ in (3.34).

The p.d.f. $c(\boldsymbol{u})$, say, of a copula $C(\boldsymbol{u})$ is given by differentiation $c(\boldsymbol{u}) = \partial^p C(\boldsymbol{u})/\partial u_1 \cdots \partial u_p$ and is straightforward to compute using the chain rule,

$$c(\boldsymbol{u}) = \frac{f(\boldsymbol{x})}{\prod_{i=1}^p f_i(x_i)}, \qquad (3.36)$$

where $x_i = F_i^{-1}(u_i)$, $i = 1, \ldots, p$, and $f_i(x_i) = \mathrm{d}F_i(x_i)/\mathrm{d}x_i = \{\mathrm{d}F_i^{-1}(u_i)/\mathrm{d}u_i\}^{-1}$ is the marginal p.d.f. of X_i.

The concept of a copula can also be defined in terms of survival functions. Let $\boldsymbol{X}$ be a continuous random vector with marginal survival functions $\overline{F}_i(x_i) = \mathsf{P}(X_i > x_i)$ and joint survival function $H(\boldsymbol{x}) = \mathsf{P}(X_i > x_i,\ i = 1, \ldots, p)$. The function $\hat{C}$ defined by $H(x) = \hat{C}(\overline{F}_1(x_1), \ldots \overline{F}_p(x_p))$ is called the *survival copula* of $\boldsymbol{X}$. In $p = 2$ dimensions, it is related to the ordinary copula by

$$\hat{C}(u_1, u_2) = u_1 + u_2 - 1 + C(1 - u_1, 1 - u_2).$$

Next, we give some examples and some properties of copulas.

3.5.1 The Gaussian Copula

An important example of the copula construction is given by the multivariate normal or Gaussian distribution, $\boldsymbol{X} \sim N_p(\boldsymbol{\mu}, \boldsymbol{\Sigma})$. Since a copula is unchanged if the component random variables undergo a monotone transformation, it suffices to limit attention to the case where $\boldsymbol{\mu} = \boldsymbol{0}$ and $\sigma_{ii} = 1$, $i = 1, \ldots, p$, so that the marginal distributions are standard normal, $X_i \sim N(0, 1)$. Let $\Phi(\cdot)$ denote the c.d.f. of $N(0, 1)$, with p.d.f. $\phi(x) = (2\pi)^{-1/2}\exp(-x^2/2)$.

The corresponding p.d.f. of $\boldsymbol{u}$ is given by

$$\begin{aligned} c(\boldsymbol{u}; \boldsymbol{\Sigma}) &= \frac{|2\pi\boldsymbol{\Sigma}|^{-1/2}\exp\left\{-\frac{1}{2}\boldsymbol{x}'\boldsymbol{\Sigma}^{-1}\boldsymbol{x}\right\}}{\prod_{i=1}^p (2\pi)^{-1/2}\exp\{-x_i^2/2\}} \\ &= |\boldsymbol{\Sigma}|^{-1/2}\exp\left\{-\frac{1}{2}\boldsymbol{x}'(\boldsymbol{\Sigma}^{-1} - \boldsymbol{I}_p)\boldsymbol{x}\right\}, \qquad (3.37) \end{aligned}$$

where $x_i = \Phi^{-1}(u_i)$, $i = 1, \ldots, p$. The most popular choice for $\boldsymbol{\Sigma}$ is the equicorrelation matrix, $\boldsymbol{\Sigma} = (1 - \rho)\boldsymbol{I} + \rho\boldsymbol{J}$, with $\boldsymbol{J} = \boldsymbol{1}\boldsymbol{1}'$ (see Section A.3.2), where the correlation parameter ρ lies in the interval $-1/(p - 1) \le \rho \le 1$. For this choice,

$$|\boldsymbol{\Sigma}| = (1 - \rho)^{p-1}(1 + (p - 1)\rho)$$

and

$$\boldsymbol{x}'(\boldsymbol{\Sigma}^{-1} - \boldsymbol{I}_p)\boldsymbol{x} = \frac{\rho\{(1 + (p-1)\rho)\boldsymbol{x}'\boldsymbol{x} - p^2\overline{x}^2\}}{(1-\rho)(1+(p-1)\rho)}, \tag{3.38}$$

where $\overline{x}$ is the mean of the vector $\boldsymbol{x}$. In particular, in the bivariate Gaussian case, $-1 < \rho < 1$, and the copula p.d.f. is given by

$$c(\boldsymbol{u};\rho) = \frac{1}{(1-\rho^2)^{1/2}} \exp\left\{-\frac{\rho^2(x_1^2 + x_2^2) - 2\rho x_1 x_2}{2(1-\rho^2)}\right\}. \tag{3.39}$$

In quantitative finance, Gaussian copulas have been applied to risk management, portfolio management and optimization, and derivatives pricing; see, for example, McNeil et al. (2015); Miller (2018). More specifically, the marginal distributions in a multivariate data set are transformed to uniform, and the Gaussian copula is then used to model the dependence structure.

However, the data sets in these applications often have the property that the underlying random variables have very skew distributions with long right-hand tails. The underlying symmetry of the Gaussian distribution means that the Gaussian copula often produces a poor model of dependence for such data. Indeed, the Gaussian copula has been described as "the formula that killed Wall Street" (Salmon, 2012). Some better copulas for financial modeling are described below.

3.5.2 The Clayton–Mardia Copula

It is shown in Exercise 3.2.4 (ii) that for the multivariate Pareto distribution given there, the multivariate survivor function, also called the upper distribution function, is given by

$$\mathsf{P}(\boldsymbol{X} > \mathbf{c}) = \left(\sum_{i=1}^{p} b_i^{-1} c_i - p + 1\right)^{-a},$$

where $x_i > b_i > 0, i = 1, \ldots, p$, and $a > 0$. Further, the marginal survival functions (just set $c_j = b_j$ for $j \neq i$) are given by

$$\mathsf{P}(X_i > c_i) = (c_i/b_i)^{-a}, \; i = 1, \ldots, p.$$

Let $b_i = \theta, \; i = 1, \ldots, p$, and $a = 1/\theta$. Then, transform $\boldsymbol{X}$ to a uniform random vector $\boldsymbol{U}$ by setting

$$U_i = (X_i/\theta)^{-1/\theta}, \quad 0 \le U_i \le 1, \; i = 1, \ldots, p, \quad \theta > 0.$$

From (3.34), the survival function of $\boldsymbol{u}$ is

$$C(\boldsymbol{u};\theta) = \left[\sum_{i=1}^{p} u_i^{-\theta} - (p-1)\right]^{-1/\theta}. \tag{3.40}$$

Equation (3.40) defines a survival copula that is known by various names including the Clayton copula, the Clayton–Mardia copula, and the Pareto copula. Joe (2014) has named it the Mardia–Takahasi–Clayton–Cook–Johnson copula, which shows its long evolution. This copula has some attractive properties, which are studied in Exercise 3.5.1. Also, it allows nonelliptical covariance patterns when the marginals are transformed to Gaussian; see, for example, Joe (2014, pp. 168–169).

Table 3.2 Some examples of Archimedean copulas.

Name	$\phi(u)$	$\psi(v)$	θ
Independence	$-\log u$	$\exp(-v)$	—
Clayton–Mardia	$\theta^{-1}(u^{-\theta}-1)$	$(1+\theta v)^{-1/\theta}$	$\theta > 0$
Frank	$-\log\left\{\dfrac{e^{-\theta u}-1}{e^{-\theta}-1}\right\}$	$-\dfrac{\log\{1+(e^{-\theta}-1)e^{-v}\}}{\theta}$	$\theta \neq 0$

3.5.3 Archimedean Copulas

Let $\phi(u)$, $0 < u < 1$, be a monotone decreasing function of u with $\phi(0+) = \infty$, $\phi(1-) = 0$. Set $\psi = \phi^{-1}$ to be the inverse function. Then, $\psi(v)$, $0 < v < \infty$, is a monotone decreasing function of v with $\psi(0+) = 1$, $\psi(\infty-) = 0$. Assume that $\psi(v)$ is a *completely monotone function*; that is, suppose that ϕ is infinitely differentiable, and that the derivatives alternate in sign

$$(-1)^k \psi^{(k)}(v) > 0, \quad 0 < v < \infty,\ k \geq 1,$$

where $\psi^{(k)}(v)$ denotes the kth derivative of $\psi(v)$.

Definition 3.5.3 *The p-variate Archimedean copula generated by $\phi(\cdot)$ is defined by*

$$C(\boldsymbol{u}) = \psi(\phi(u_1) + \phi(u_2) + \cdots + \phi(u_p)). \tag{3.41}$$

It can be shown that C defines a valid c.d.f. with p.d.f.

$$c(\boldsymbol{u}) = \psi^{(k)}(v)\prod_{i=1}^{p}\phi'(u_i), \tag{3.42}$$

where $v_i = \phi(u_i)$, $v = v_1 + \cdots + v_p$, and $\phi'(v_i)$ is the first derivative of ϕ. Table 3.2 gives various examples, including another popular copula called the Frank copula, in each case indexed by a parameter θ. The Frank copula allows positive and negative dependence.

Example 3.5.1 We make comparisons of some of these copulas using plots. According to Joe (2014, sec. 1.4), the best way to make visual comparisons is to convert to standard $N(0,1)$ margins. In Figure 3.2, we show contour plots of $c(\Phi(z_1), \Phi(z_2))\phi(z_1)\phi(z_2)$ for three copulas (Gaussian, Clayton–Mardia, and Frank), with parameters chosen so that the product-moment correlation is 0.5 after converting to standard $N(0,1)$ margins. We can see a variety of dependence behaviors and reflection asymmetries, which can be compared to the familiar ellipse of the bivariate normal distribution. □

3.5.4 Fréchet–Höffding Bounds

Every multivariate copula is bounded by the following bounds given by Fréchet (1951) and Höffding (1940):

$$W(\boldsymbol{u}) = \max\{u_1 + \cdots + u_p - (p-1), 0\} < C(\boldsymbol{u}) < \min\{u_1, \ldots, u_p\} = M(\boldsymbol{u}). \tag{3.43}$$

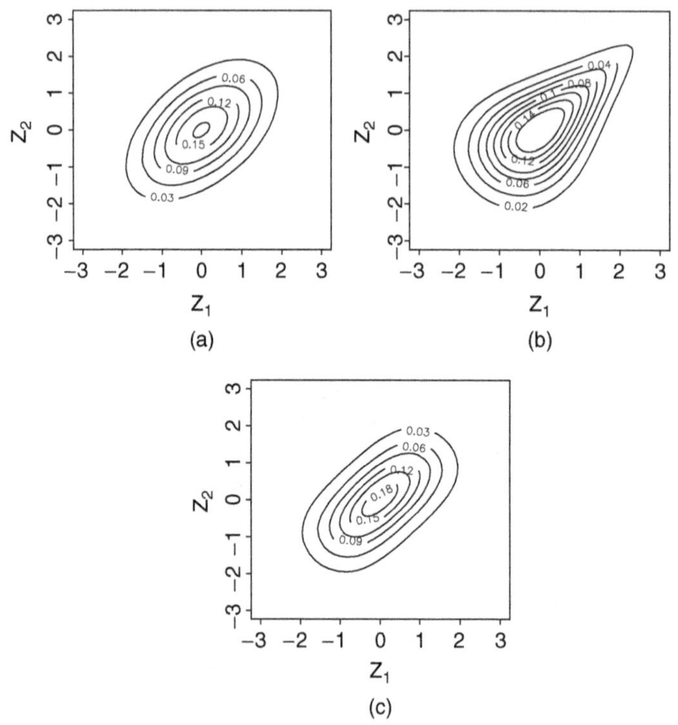

Figure 3.2 Copulas combined with $N(0,1)$ margins. (a) Gaussian ($\rho = 0.5$); (b) Clayton–Mardia ($\theta = 0.993$); (c) Frank ($\theta = 3.567$).

For the bivariate case, we can rewrite these bounds as

$$W(u_1, u_2) = \max(u_1 + u_2 - 1, 0) < C(u_1, u_2) < \min(u_1, u_2) = M(u_1, u_2) \tag{3.44}$$

(see also Exercise 2.2.4).

A one-parameter family of bivariate copulas that can achieve these bounds is given by

$$C_\theta(u_1, u_2) = \frac{1 + (\theta - 1)(u_1 + u_2) - \sqrt{\{1 + (\theta - 1)(u_1 + u_2)\}^2 - 4\theta(\theta - 1)u_1 u_2}}{2(\theta - 1)}, \tag{3.45}$$

where $0 \leq \theta < \infty$. The choice $\theta = 1$ yields the independence copula, and the limiting choices $\theta \to 0$ and $\theta \to \infty$ achieve the bounds $W(u_1, u_2)$ and $M(u_1, u_2)$, respectively. This copula, known as the Plackett copula or the Plackett–Mardia copula, was developed by Plackett (1965) and Mardia (1967b). One motivation is the fact that it is the unique copula with constant cross-ratios. That is, for all $0 < u_1, u_2 < 1$, the cross-ratios all take the same value θ,

$$\begin{aligned} \theta &= \frac{\mathrm{P}(0 < U_1 < u_1,\ 0 < U_2 < u_2)\mathrm{P}(u_1 < U_1 < 1,\ u_2 < U_2 < 1)}{\mathrm{P}(0 < U_1 < u_1,\ u_2 < U_2 < 1)\mathrm{P}(u_1 < U_1 < 1,\ 0 < U_2 < u_2)} \\ &= \frac{C(1 - u_1 - u_2 + C)}{(u_1 - C)(u_2 - C)}, \end{aligned} \tag{3.46}$$

where $C = C_\theta(u_1, u_2)$. If the unit square is divided into four rectangles at the point (u_1, u_2), then this cross-ratio is the area of the lower left rectangle times the upper right rectangle, divided by the area of the lower right rectangle times the upper left rectangle. Equation (3.46) can be rephrased as a quadratic equation for C, which has one solution defining a valid copula. Exercise 3.5.1 explores the details.

Another one-parameter family of bivariate copulas that can achieve the Fréchet–Höffding bounds was introduced by Mardia (1970c). It has parameter $-1 \le \theta \le 1$ and is given by

$$C_\theta(u_1, u_2) = \frac{\theta^2(1+\theta)}{2} M(u_1, u_2) + (1-\theta^2)u_1u_2 + \frac{\theta^2(1-\theta)}{2} W(u_1, u_2).$$

It is perhaps the simplest copula that can achieve these bounds. Also, the independence copula arises when $\theta = 0$.

Exercises and Complements

3.2.1 **Mixture of normals** (Mardia, 1974). Let $\phi(\boldsymbol{x}; \boldsymbol{\mu}, \boldsymbol{\Sigma})$ be the p.d.f. of $N_p(\boldsymbol{\mu}, \boldsymbol{\Sigma})$. Consider the mixture given by the p.d.f. $g_1(\boldsymbol{x}) = \lambda\phi(\boldsymbol{x}; \boldsymbol{\mu}_1, \boldsymbol{\Sigma}) + \lambda'\phi(\boldsymbol{x}; \boldsymbol{\mu}_2, \boldsymbol{\Sigma})$, where $0 < \lambda < 1$ and $\lambda' = 1 - \lambda$. Find a nonsingular linear transformation, $\boldsymbol{y} = \boldsymbol{A}\boldsymbol{x} + \boldsymbol{b}$, for which the p.d.f. of $\boldsymbol{y}$ is given by

$$g_2(\boldsymbol{y}) = \{\lambda\phi(y_1 - \Delta) + \lambda'\phi(y_1)\}\prod_{j=2}^{p}\phi(y_j), \quad -\infty < \boldsymbol{y} < \infty,$$

where $\phi(\cdot)$ is the p.d.f. of $N(0,1)$, and $\Delta^2 = (\boldsymbol{\mu}_1 - \boldsymbol{\mu}_2)'\boldsymbol{\Sigma}^{-1}(\boldsymbol{\mu}_1 - \boldsymbol{\mu}_2)$.
Let $y = y_1$ and $\boldsymbol{z} = (y_2, y_3, \ldots, y_p)'$. Noting that y and $\boldsymbol{z}$ are independent, y is a mixture of $N(\Delta, 1)$ and $N(0,1)$, and $\boldsymbol{z}$ is $N_{p-1}(\boldsymbol{0}, \boldsymbol{I})$ show that now

$$\boldsymbol{\mu} = \mathsf{E}(\boldsymbol{y}) = (\lambda\Delta, 0, \ldots, 0)' \quad \text{and} \quad \boldsymbol{\Sigma} = \text{cov}(\boldsymbol{y}) = \text{diag}(1 + \lambda\lambda'\Delta^2, 1, 1, \ldots, 1).$$

The multivariate measures of skewness and kurtosis given by (1.26) and (1.27) for $\boldsymbol{y}$ (and hence also for $\boldsymbol{x}$) are

$$\beta_{1,p} = \beta_{1,1}, \qquad \beta_{2,p} = \beta_{2,1} + (p-1)(p+1),$$

so that these depend only on the skewness and kurtosis for the univariate mixture y, which can be obtained using the following steps.
First, prove that the joint cumulant generating function $\log \mathsf{E}\{\exp(ty)\}$ for y is given by

$$K(t) = \frac{1}{2}t^2 + \log(\lambda' + \lambda\exp(\Delta t)),$$

and that this satisfies

$$\frac{\partial K(t)}{\partial t} = t + \Delta\lambda + \Delta\lambda\lambda'\frac{\partial K(t)}{\partial \lambda}.$$

By substituting the general form

$$K(t) = \sum_{r=0}^{\infty} \kappa_r \frac{t^r}{r!}$$

and considering the coefficient of t^r, obtain the recurrence relation given by:

$$\kappa_{r+1} = \lambda\lambda'\Delta\frac{d\kappa_r}{d\lambda}, \quad r > 1.$$

Using $\kappa_2 = \sigma^2 = 1 + \lambda\lambda'\Delta^2$, deduce that (for $\lambda > \lambda'$ without any loss of generality)

$$\kappa_3 = \lambda\lambda'(\lambda - \lambda')\Delta^3 = \sigma^3\sqrt{\beta_{1,1}}, \qquad \kappa_4 = \lambda\lambda'(1 - 6\lambda\lambda')\Delta^4 = \sigma^4\beta_{2,1} - 3.$$

Thus, conclude that

$$\beta_{1,p} = \{\lambda\lambda'(\lambda - \lambda')\Delta^3\}^2/(1 + \lambda\lambda'\Delta^2)^3$$

and

$$\beta_{2,p} = \{\lambda\lambda'(1 - 6\lambda\lambda')\Delta^4\}/(1 + \lambda\lambda'\Delta^2)^2 + p(p+2).$$

Hence, show that there is skewness unless $\lambda = 1/2$ or $\Delta = 0$, and that, if $\Delta \to \infty$, then $\beta_{1,p} = (\lambda - \lambda')^2/(\lambda\lambda')$. For $\lambda = 1/2$, $\beta_{2,p} = -2\{\Delta^4/(4 + \Delta^2)^2\} + p(p+2)$ and, for any λ as $\Delta \to \infty$, $\beta_{2,p} = 1/(\lambda\lambda') + p(p+2) - 6$. Interpret these values.

3.2.2 **A multivariate Poisson distribution** (Holgate, 1964; Krishnamoorthy, 1951). Let $u_0, u_1, \ldots, u_p$ be independent Poisson variables with parameters λ_0, $\lambda_1 - \lambda_0, \ldots, \lambda_p - \lambda_0$, respectively. Write down the joint distribution of $x_i = u_0 + u_i$, $i = 1, \ldots, p$, and show that the marginal distributions of $x_1, \ldots, x_p$ are all Poisson. For $p = 2$ and writing $(x_1, x_2) = (x, y)$, show that the (discrete) p.d.f. is given by

$$f(x,y) = \exp(-\lambda_1 - \lambda_2 + \lambda_0)\frac{a^x b^y}{x!y!}\sum_{r=0}^{s}\frac{x^{(r)}}{a^r}\cdot\frac{y(r)}{b^r}\cdot\frac{\lambda_0^{(r)}}{r!},$$

where $s = \min(x,y)$, $a = \lambda_1 - \lambda_0$, $b = \lambda_2 - \lambda_0$, $\lambda_1 > \lambda_0 > 0$, $\lambda_2 > \lambda_0 > 0$, and $x^{(r)} = x(x-1)\cdots(x-r+1)$. Furthermore,

$$\mathrm{E}(y \mid x) = b + (\lambda_0/\lambda_1)x, \quad \mathrm{var}(y \mid x) = b + \{a\lambda_0/\lambda_1^2\}x.$$

3.2.3 **A multivariate exponential distribution** (Marshall and Olkin, 1967). In a system of p components, there are $2^p - 1$ types of shocks, each of which is fatal to one of the subsets $(i_1, \ldots, i_r)$ of the p components. A component dies when one of the subsets containing it receives a shock. The different types of shocks have independent timings, and the shock that is fatal to the subset $(i_1, \ldots, i_r)$ has an exponential distribution, with parameter $\lambda_{i_1,\ldots,i_r}^{-1}$ representing the expected value of the distribution. Show that the lifetimes $x_1, \ldots, x_p$ of the components are distributed as

$$-\log \mathrm{P}(\boldsymbol{x} > \boldsymbol{a}) = \sum_{i=1}^{p}\lambda_i a_i + \sum_{i_1<i_2}\lambda_{i_1i_2}\max(a_{i_1}, a_{i_2}) + \cdots + \lambda_{1,2,\ldots,p}\max(a_1, \ldots, a_p).$$

3.2.4 **A multivariate Pareto distribution** (Mardia, 1962, 1964b). Consider the p.d.f.

$$f(\boldsymbol{x}) = \prod_{i=1}^{p}\left(\frac{a+i-1}{b_i}\right)\left(\sum_{i=1}^{p}b_i^{-1}x_i - p + 1\right)^{-(a+p)},$$

$$\text{where } x_i > b_i, \quad i = 1, \ldots, p, \qquad a_i > 0, b_i > 0.$$

(i) Show that any subset of the components of $\boldsymbol{x}$ has the same type of distribution.

(ii) Show that

$$\mathsf{P}(\boldsymbol{x} > \mathbf{c}) = \left(\sum_{i=1}^{p} b_i^{-1} c_i - p + 1 \right)^{-a}.$$

(iii) Let $\boldsymbol{x}_r$, $r = 1, \ldots, n$, be a random sample from this population. Show that $\boldsymbol{u} = \min(\boldsymbol{x}_r)$, where the minimum is taken elementwise, has the above distribution with the parameter a replaced by na.

3.2.5 **Multivariate Student's *t*-and Cauchy distribution** (Cornish, 1954; Dunnett and Sobel, 1954). A random vector has a multivariate t distribution if its p.d.f. is of the form

$$g_\nu(\boldsymbol{t}; \boldsymbol{\mu}, \boldsymbol{\Sigma}) = \frac{c_p |\boldsymbol{\Sigma}|^{-1/2}}{[1 + \nu^{-1}(\boldsymbol{t} - \boldsymbol{\mu})'\boldsymbol{\Sigma}^{-1}(\boldsymbol{t} - \boldsymbol{\mu})]^{(\nu+p)/2}},$$

where

$$c_p = \frac{\Gamma((\nu + p)/2)}{(\pi\nu)^{p/2}\Gamma(\nu/2)},$$

and ν is known as the number of degrees of freedom of the distribution.

(i) Let $x_j = y_j/(S/\sqrt{\nu})$, $j = 1, \ldots, p$, where $\boldsymbol{y} \sim N_p(\boldsymbol{0}, \boldsymbol{I})$, $S^2 \sim \chi^2_\nu$, and $\boldsymbol{y}$ and S are independent. Show that $\boldsymbol{x}$ has the p.d.f. $g_\nu(\boldsymbol{x}; \boldsymbol{0}, \boldsymbol{I})$.

(ii) For $\nu = 1$, the distribution is known as a *multivariate Cauchy distribution*. Show that its c.f. is

$$\exp\{i\boldsymbol{\mu}'\boldsymbol{t} - (\boldsymbol{t}'\boldsymbol{\Sigma}\boldsymbol{t})^{1/2}\}.$$

3.2.6 Setting $\boldsymbol{\mu} = \boldsymbol{0}$ and $\boldsymbol{\Sigma} = \boldsymbol{I}$ in the multivariate t distribution (see above), substitute $\boldsymbol{t} = r\boldsymbol{\ell}$ (cf. (3.12)) and show that

$$f(r) = \frac{2}{\nu^{p/2}} \frac{1}{\mathrm{B}(\nu/2, p/2)} \frac{r^{p-1}}{(1 + r^2/\nu)^{(\nu+p)/2}}, \qquad r > 0,$$

where $\mathrm{B}(\cdot, \cdot)$ is defined in (B.13). Hence, using the fact that B is a normalizing constant in (B.12), show that

$$\mathsf{E}(r^s) = \nu^{s/2} \frac{\mathrm{B}((\nu - s)/2, (p + s)/2)}{\mathrm{B}(\nu/2, p/2)} \qquad \text{for } s < \nu.$$

Now, substitute $s = 2$ and $s = 4$ in this expression and use (3.20) to show that (for $\nu > 4$)

$$\beta_{2,p} = \frac{p(p + 2)(\nu - 2)}{\nu - 4},$$

so that the *excess* kurtosis is $2p(p + 2)/(\nu - 4)$. Finally, show that the limiting case $\nu \to \infty$ gives a value that is consistent with that of the multivariate normal.

3.2.7 Let $\boldsymbol{e}_i$ denote the unit vector with 1 in the ith place and 0 elsewhere. Let $\boldsymbol{y}_j$, $j = 1, 2, \ldots$, be a sequence of independent identically distributed random vectors such that $\mathsf{P}(\boldsymbol{y} = \boldsymbol{e}_i) = a_i$, $i = 1, \ldots, p$, where $a_i > 0$, $\sum a_i = 1$.

(i) Show that $\mathsf{E}(\boldsymbol{y}) = \boldsymbol{a}$, $\operatorname{var}(\boldsymbol{y}) = \operatorname{diag}(\boldsymbol{a}) - \boldsymbol{a}\boldsymbol{a}'$.

(ii) Let $\boldsymbol{x} = \sum_{j=1}^{n} \boldsymbol{y}_j$. Show that $\boldsymbol{x}$ has the multinomial distribution given in (3.3).

(iii) Show that $\mathsf{E}(\boldsymbol{x}) = n\boldsymbol{a}$, $\text{var}(\boldsymbol{x}) = n[\text{diag}\,(\boldsymbol{a}) - \boldsymbol{a}\boldsymbol{a}']$.
(iv) Verify that $[\text{diag}(\boldsymbol{a}) - \boldsymbol{a}\boldsymbol{a}']\mathbf{1} = \mathbf{0}$, and hence, deduce that $\text{var}(\boldsymbol{x})$ is singular.

3.3.1 Show that the only spherical distribution for which the components of $\boldsymbol{x}$ are independent is the spherical multivariate normal distribution.

3.3.2 Consider the "dipped multivariate normal distribution," denoted by $DN_p(\alpha, \beta)$, with p.d.f. given by

$$f(\boldsymbol{x}) = c_p(\boldsymbol{x}'\boldsymbol{x} + \alpha)\exp(-\boldsymbol{x}'\boldsymbol{x}/(2\beta)),$$

where α and β are the shape and scale parameters, respectively.
(i) Show that the normalizing constant is

$$c_p = 1/\{(2\pi\beta)^{p/2}(p\beta + \alpha)\}.$$

(ii) Sketch the density when $p = \alpha = \beta = 1$.
(iii) When $p = 2$ and $\boldsymbol{x}' = (x_1, x_2)$:
(a) Show that the marginal distribution of x_1 follows $DN_1(\alpha + \beta, \beta)$.
(b) Show that the conditional distribution of $x_2 \mid x_1$ follows $DN_1(\alpha + x_1^2, \beta)$.
(c) Describe the conditions under which x_1 and x_2 are uncorrelated or independent.

This distribution can be viewed as a mixture of Kotz-type distributions. It is unusual in that both the conditional and marginal distributions belong to the same family (also true of the multivariate normal).

3.3.3 (Mardia and Holmes, 1980). For the density given by Equation (3.26), show that
(i) $\mathsf{E}(\boldsymbol{x}) = \boldsymbol{a}$.
(ii) the mode is given by the ellipsoid of $\boldsymbol{x}$ values such that $(\boldsymbol{x} - \boldsymbol{a})'\boldsymbol{\Sigma}^{-1}(\boldsymbol{x} - \boldsymbol{a}) = 1$.
(iii) $\mathsf{E}(\boldsymbol{x} - \boldsymbol{a})'(\boldsymbol{x} - \boldsymbol{a}) = \alpha\boldsymbol{\Sigma}$, where α is a function of κ.

3.4.1 Using Equations (3.31) and (3.32), find the maximum and minimum values of $\beta_{1,2}$ and $\beta_{2,2}$ in the cases that $\lambda = 1/2$ (as a function of μ) and for $\mu = 2$ (as a function of λ). Plot the corresponding contours of (3.30).

3.5.1 (Mardia, 1967b; Nelsen, 2007; Plackett, 1965). Let $C(\boldsymbol{u})$ be a bivariate copula, and suppose that for some fixed parameter $0 < \theta < \infty$, the copula satisfies the constant cross-ratio condition in (3.46) for all $0 < u_1, u_2 < 1$. Show that for each $\boldsymbol{u}$, this condition can be rewritten as a quadratic equation for $C(\boldsymbol{u})$ with roots

$$C_\theta(u_1, u_2) = \frac{1 + (\theta - 1)(u_1 + u_2) \pm \sqrt{\{1 + (\theta - 1)(u_1 + u_2)\}^2 - 4\theta(\theta - 1)u_1u_2}}{2(\theta - 1)}.$$

(i) Show that $C_\theta(u_1, 0)$ and $C_\theta(u_1, 1)$ do not satisfy the boundary conditions with the "+" sign in $\pm$. Further, check with "−" sign that $C_\theta(u_1, 0) = 0$ and $C_\theta(u_1, 1) = u_1$, and that the corresponding p.d.f. is nonnegative. Hence, deduce that a copula satisfying (3.46) exists and is unique and is given by the Plackett–Mardia copula in (3.45).

(ii) Show that $C_0(u_1, u_2) = W(u_1, u_2)$ and $C_\infty(u_1, u_2) = M(u_1, u_2)$, so that the Fréchet–Höffding bounds are achieved. Show further that the choice $\theta = 1$ yields the independence copula $C_1(u_1, u_2) = u_1 u_2$.

3.5.2 Consider a bivariate continuous random vector $(x, y)'$ with joint c.d.f. $F(x, y)$ and marginal c.d.f.s $F_x(x)$ and $F_y(y)$. Let $u = F_x(x)$, and $v = F_y(y)$, so that the random vector $(u, v)'$ has uniform marginals. Let $C(u, v)$ denote the corresponding copula function.

- The population version of Spearman's correlation between x and y, $\rho_{\text{spear}}(x, y)$, say, is defined as the population Pearson, or product moment, correlation between u and v. Show that u and v have mean $1/2$ and variance $1/12$. Hence, deduce that

$$\rho_{\text{spear}}(x, y) = \rho_{\text{spear}}(u, v) = 12 \iint (uv - 1/4)\, \mathrm{d}C(u, v).$$

Using Exercise 2.2.5, show that Spearman's correlation can be reexpressed in the form

$$\rho_{\text{spear}}(x, y) = 12 \iint (C(u, v) - uv)\, \mathrm{d}u\mathrm{d}v.$$

- Let (x_i, y_i), $i = 1, 2, 3$, be three independent bivariate random vectors from $F(x, y)$. Set $z = x_1 - x_2$ and $w = y_1 - y_3$. Show that Spearman's correlation between x and y is equal to (Kruskal, 1958)

$$\rho_{\text{spear}}(x, y) = 3[\mathsf{P}(zw > 0) - \mathsf{P}(zw < 0)]. \tag{3.47}$$

Also, show that

$$\rho_{\text{spear}}(x, y) = 3[2\, \mathsf{P}(z > 0, w > 0) + 2\, \mathsf{P}(z < 0, w < 0) - 1]. \tag{3.48}$$

(i) If $F(x, y)$ denotes the c.d.f. for the bivariate normal distribution with correlation ρ, note that (z, w) is bivariate normal with correlation $\rho/2$. Using equation (3.47) for $\rho_{\text{spear}}(x, y)$ and Sheppard's formula of Exercise 2.5.2, show that

$$\rho_{\text{spear}}(x, y) = \frac{6}{\pi} \arcsin(\rho/2).$$

(ii) If $F(x, y)$ denotes the Plackett–Mardia copula with parameter θ, show that

$$\rho_{\text{spear}}(x, y; \theta) = \frac{\theta + 1}{\theta - 1} - \frac{2\theta \log \theta}{(\theta - 1)^2}$$

(Mardia 1967b). For $\theta = 0, 1, \infty$, show that $\rho_{\text{spear}}(x, y; \theta) = -1, 0, +1$, respectively. Further, by taking the first two derivatives of $\theta^2 - 1 - 2\theta \log \theta$, show that $\rho_{\text{spear}}(x, y; \theta)$ is positive and increasing for $\theta > 1$, and then, using $\rho_{\text{spear}}(x, y; \theta) = -\rho_{\text{spear}}(x, y; 1/\theta)$, show that $\rho_{\text{spear}}(x, y; \theta)$ is negative and increasing for $\theta < 1$.

3.5.3 Let $F(x, y)$ be a bivariate c.d.f. and (x_i, y_i), $i = 1, 2$, be two independent bivariate random vectors from this distribution. Set $z = x_1 - x_2$ and $w = y_1 - y_2$. Kendall's τ is a measure of dependence defined by

$$\tau(x, y) = \mathsf{P}(zw > 0) - \mathsf{P}(zw < 0).$$

Prove that

$$\tau(x, y) = 2\ \mathsf{P}(z > 0, w > 0) + 2\ \mathsf{P}(z < 0, w < 0) - 1.$$

Show that if $C(u, v)$ is the copula corresponding to $F(x, y)$, then

$$\tau(u, v) = 4\mathsf{E}[C(u, v)] - 1. \tag{3.49}$$

(i) If $F(x, y)$ denotes the c.d.f. for the bivariate normal distribution with correlation ρ, note that (z, w) is bivariate normal with correlation ρ. Using equation (3.49) for $\tau(x, y)$ and Sheppard's formula of Exercise 2.5.2, show that

$$\tau(u, v) = (2/\pi) \arcsin(\rho).$$

(ii) If $F(x, y)$ denotes the Clayton–Mardia copula with parameter θ, show that

$$\tau(u, v) = 1/(1 + 2\theta).$$

4

Normal Distribution Theory

4.1 Introduction and Characterization

4.1.1 Introduction

There has been a tendency in multivariate analysis to assume that all random vectors come from the multivariate normal family of distributions. Among the reasons for its preponderance in the multivariate context are the following:

(1) The multivariate normal distribution is an easy generalization of its univariate counterpart, and the multivariate analysis runs almost parallel to the corresponding analysis based on univariate normality. The same cannot be said of other multivariate generalizations: different authors have given different extensions of the gamma, Poisson, and exponential distributions, and attempts to derive entirely suitable definitions have not yet proved entirely successful (see Section 3.2).
(2) The multivariate normal distribution is entirely defined by its first and second moments – a total of only $p(p+3)/2$ parameters in all. This compares with $2^p - 1$ for the multivariate binary or logit distribution (3.7). This economy of parameters simplifies the problems of estimation.
(3) In the case of normal variables, zero correlation implies independence, and pairwise independence implies total independence. Again, other distributions do not necessarily have these properties (see Exercise 2.2.2).
(4) Linear functions of a multivariate normal vector are themselves univariate normal. This opens the door to an extremely simple derivation of multivariate normal theory, as developed here. Again, other distributions may not have this property; e.g. linear functions of multivariate binary variables are not themselves binary.
(5) Even when the original data is not multivariate normal, one can often appeal to central limit theorems which prove that certain functions such as the sample mean are normal for large samples (Section 2.7).
(6) The equiprobability contours of the multivariate normal distribution are simple ellipses, which by a suitable change of coordinates can be made into circles (or, in the general case, hyperspheres). This geometric simplicity, together with the associated invariance properties, allows us to derive many crucial properties through intuitively appealing arguments.

Multivariate Analysis, Second Edition. Kanti V. Mardia, John T. Kent, and Charles C. Taylor.

In this chapter, unlike Section 2.5, we use a density-free approach and try to emphasize the interrelationships between different distributions without using their actual probability density functions (p.d.f.s).

4.1.2 A Definition by Characterization

In this chapter, we define the multivariate normal distribution with the help of the Cramér–Wold theorem (Theorem 2.3.1). This states that the multivariate distribution of any random p-vector $\boldsymbol{x}$ is completely determined by the univariate distributions of linear functions such as $\boldsymbol{a}'\boldsymbol{x}$, where $\boldsymbol{a}$ may be any nonrandom p-vector.

Definition 4.1.1 *We say that $\mathbf{x}$ has a p-variate normal distribution if and only if $\mathbf{a}'\mathbf{x}$ is univariate normal for all fixed p-vectors $\mathbf{a}$.*

To allow for the case $\boldsymbol{a} = \mathbf{0}$, we regard constants as degenerate forms of the normal distribution.

The above definition of multivariate normality has a useful geometric interpretation. If $\boldsymbol{x}$ is visualized as a random point in p-dimensional space, then linear combinations such as $\boldsymbol{a}'\boldsymbol{x}$ can be regarded as projections of $\boldsymbol{x}$ onto a one-dimensional subspace. Definition 4.1.1 therefore implies that the projection of $\boldsymbol{x}$ onto all one-dimensional subspaces has a univariate normal distribution. This geometric interpretation makes it clear that even after $\boldsymbol{x}$ is transformed by any arbitrary shift, rotation, or projection, it will still have the property of normality. In coordinate-dependent terms, this may be stated more precisely as follows. (In this theorem and others that follow, we assume that matrices and vectors such as $\boldsymbol{A}$, $\boldsymbol{b}$, and $\boldsymbol{c}$ are nonrandom unless otherwise stated.)

Theorem 4.1.2 *If $\mathbf{x}$ has a p-variate normal distribution, and if $\mathbf{y} = \mathbf{A}\mathbf{x} + \mathbf{c}$, where $\mathbf{A}$ is any $(q \times p)$ matrix and $\mathbf{c}$ is any q-vector, then $\mathbf{y}$ has a q-variate normal distribution.*

Proof: Let $\boldsymbol{b}$ be any fixed q-vector. Then, $\boldsymbol{b}'\boldsymbol{y} = \boldsymbol{a}'\boldsymbol{x} + \boldsymbol{b}'\boldsymbol{c}$, where $\boldsymbol{a} = \boldsymbol{A}'\boldsymbol{b}$. Since $\boldsymbol{x}$ is multivariate normal, $\boldsymbol{a}'\boldsymbol{x}$ is univariate normal by Definition 4.1.1. Therefore, $\boldsymbol{b}'\boldsymbol{y}$ is also univariate normal for all fixed vectors $\boldsymbol{b}$, and therefore, $\boldsymbol{y}$ is multivariate normal by virtue of Definition 4.1.1. □

Corollary 4.1.3 *Any subset of elements of a multivariate normal vector itself has a multivariate normal distribution. In particular, the individual elements each have univariate normal distributions.*

Note that the above theorem and corollary need not assume that the covariance matrix $\boldsymbol{\Sigma}$ is of full rank. Therefore, these results also apply to the singular multivariate normal distribution (Section 2.5.4). Also, the proofs do not use any intrinsic property of normality. Therefore, similar results hold in principle for any other multivariate distribution defined in a similar way. That is, if we were to say that $\boldsymbol{x}$ has a p-variate "M" distribution whenever $\boldsymbol{a}'\boldsymbol{x}$ is univariate "M" for all fixed $\boldsymbol{a}$ ("M" could be "Cauchy"), then results analogous to Theorem 4.1.2 and Corollary 4.1.3 could be derived.

However, before proceeding further, we must prove the *existence* of the multivariate normal distribution. This is done by showing that Definition 4.1.1 leads to the characteristic function (c.f.) which has already been referred to in (2.57) and (2.71).

Theorem 4.1.4 *If $\boldsymbol{x}$ is multivariate normal with mean vector $\boldsymbol{\mu}$ and covariance matrix $\boldsymbol{\Sigma}$ ($\boldsymbol{\Sigma} \geq 0$), then its c.f. is given by*

$$\phi_x(\boldsymbol{t}) = \exp\left(i\boldsymbol{t}'\boldsymbol{\mu} - \frac{1}{2}\boldsymbol{t}'\boldsymbol{\Sigma}\boldsymbol{t}\right). \tag{4.1}$$

Proof: We follow the lines of the Cramér–Wold theorem (Theorem 2.3.1) and note that if $y = \boldsymbol{t}'\boldsymbol{x}$, then y has mean $\boldsymbol{t}'\boldsymbol{\mu}$ and variance $\boldsymbol{t}'\boldsymbol{\Sigma}\boldsymbol{t}$. Since y is univariate normal, $y \sim N(\boldsymbol{t}'\boldsymbol{\mu}, \boldsymbol{t}'\boldsymbol{\Sigma}\boldsymbol{t})$. Therefore, from the c.f. of the univariate normal distribution, the c.f. of y is

$$\phi_y(s) = \mathsf{E}[\exp(isy)] = \exp\left(is\boldsymbol{t}'\boldsymbol{\mu} - \frac{1}{2}s^2\boldsymbol{t}'\boldsymbol{\Sigma}\boldsymbol{t}\right).$$

Hence, the c.f. of $\boldsymbol{x}$ must be given by

$$\phi_x(\boldsymbol{t}) = \mathsf{E}[\exp(i\boldsymbol{t}'\boldsymbol{x})] = \mathsf{E}(\exp iy) = \phi_y(1) = \exp\left(i\boldsymbol{t}'\boldsymbol{\mu} - \frac{1}{2}\boldsymbol{t}'\boldsymbol{\Sigma}\boldsymbol{t}\right).$$

From Section 2.5, we see that (4.1) is indeed the c.f. of a multivariate distribution. Hence, the multivariate normal distribution with mean $\boldsymbol{\mu}$ and covariance matrix $\boldsymbol{\Sigma}$ exists, and its c.f. has the stated form. □

As in Section 2.5, we may summarize the statement, "$\boldsymbol{x}$ is p-variate normal with mean $\boldsymbol{\mu}$ and covariance matrix $\boldsymbol{\Sigma}$," by writing $\boldsymbol{x} \sim N_p(\boldsymbol{\mu}, \boldsymbol{\Sigma})$. When the dimension is clear, we may omit the subscript p. We can obtain the p.d.f. when $\boldsymbol{\Sigma} > 0$ using the inversion formula (2.39), and it is given by (2.49).

Theorem 4.1.5

(a) *Two jointly multivariate normal vectors are independent if and only if they are uncorrelated.*
(b) *For two jointly multivariate normal vectors, pairwise independence of their components implies complete independence.*

Proof: The c.f. given in Theorem 4.1.4 factorizes as required only when the corresponding submatrix of $\boldsymbol{\Sigma}$ is zero. This happens only when the vectors are uncorrelated. □

4.2 Linear Forms

Theorem 4.1.2 proved that if $\boldsymbol{x} \sim N_p(\boldsymbol{\mu}, \boldsymbol{\Sigma})$ and $\boldsymbol{y} = \boldsymbol{A}\boldsymbol{x} + \boldsymbol{c}$, where $\boldsymbol{A}$ is any $(q \times p)$ matrix, then $\boldsymbol{y}$ has a q-variate normal distribution. Now from Section 2.2.2 we know that the moments of $\boldsymbol{y}$ are $\boldsymbol{A}\boldsymbol{\mu} + \boldsymbol{c}$ and $\boldsymbol{A}\boldsymbol{\Sigma}\boldsymbol{A}'$. Hence, we deduce immediately the following results.

Theorem 4.2.1 *If $\boldsymbol{x} \sim N_p(\boldsymbol{\mu}, \boldsymbol{\Sigma})$ and $\boldsymbol{y} = \boldsymbol{A}\boldsymbol{x} + \boldsymbol{c}$, then $\boldsymbol{y} \sim N_q(\boldsymbol{A}\boldsymbol{\mu} + \boldsymbol{c}, \boldsymbol{A}\boldsymbol{\Sigma}\boldsymbol{A}')$.*

Corollary 4.2.2 *If* $\boldsymbol{x} \sim N_p(\boldsymbol{\mu}, \boldsymbol{\Sigma})$ *and* $\boldsymbol{\Sigma} > 0$, *then* $\boldsymbol{y} = \boldsymbol{\Sigma}^{-1/2}(\boldsymbol{x} - \boldsymbol{\mu}) \sim N_p(\boldsymbol{0}, \boldsymbol{I})$ *and* $(\boldsymbol{x} - \boldsymbol{\mu})'\boldsymbol{\Sigma}^{-1}(\boldsymbol{x} - \boldsymbol{\mu}) = \boldsymbol{y}'\boldsymbol{y} = \sum_i y_i^2 \sim \chi_p^2$.

Corollary 4.2.3 *If* $\boldsymbol{x} \sim N_p(\boldsymbol{\mu}, \sigma^2\boldsymbol{I})$ *and* $\boldsymbol{G}(q \times p)$ *is any row-orthonormal matrix, i.e. satisfying* $\boldsymbol{G}\boldsymbol{G}' = \boldsymbol{I}_q$, *then* $\boldsymbol{G}\boldsymbol{x} \sim N_q(\boldsymbol{G}\boldsymbol{\mu}, \sigma^2\boldsymbol{I})$.

Corollary 4.2.4 *If* $\boldsymbol{x} \sim N_p(\boldsymbol{0}, \boldsymbol{I})$ *and* $\boldsymbol{a}$ *is any nonzero p-vector, then* $\boldsymbol{a}'\boldsymbol{x}/\sqrt{\boldsymbol{a}'\boldsymbol{a}}$ *has the standard univariate normal distribution.*

Corollary 4.2.2 shows that any normal vector can easily be converted into standard form. It also gives an important quadratic expression which has a chi-squared distribution. From Corollary 4.2.3, we note that the standard multivariate normal distribution has a certain invariance under orthogonal transformations. Note that Corollary 4.2.4 also applies if $\boldsymbol{a}$ is a random vector independent of $\boldsymbol{x}$ (see Exercise 4.2.4). A further direct result of Theorem 4.1.5 is the following.

Theorem 4.2.5 *If* $\boldsymbol{x} \sim N_p(\boldsymbol{\mu}, \boldsymbol{\Sigma})$, *then* $\boldsymbol{A}\boldsymbol{x}$ *and* $\boldsymbol{B}\boldsymbol{x}$ *are independent if and only if* $\boldsymbol{A}\boldsymbol{\Sigma}\boldsymbol{B}' = \boldsymbol{0}$.

Corollary 4.2.6 *If* $\boldsymbol{x} \sim N_p(\boldsymbol{\mu}, \sigma^2\boldsymbol{I})$ *and* $\boldsymbol{G}$ *is any row-orthonormal matrix, then* $\boldsymbol{G}\boldsymbol{x}$ *is independent of* $(\boldsymbol{I} - \boldsymbol{G}'\boldsymbol{G})\boldsymbol{x}$.

If $\boldsymbol{x}$ is partitioned into two subvectors, with r and s elements, respectively, then by noting two particular matrices that satisfy the conditions of Theorem 4.2.5, we may prove the following.

Theorem 4.2.7 *If* $\boldsymbol{x} = (\boldsymbol{x}_1', \boldsymbol{x}_2')' \sim N_p(\boldsymbol{\mu}, \boldsymbol{\Sigma})$, *then* $\boldsymbol{x}_1$ *and* $\boldsymbol{x}_{2.1} = \boldsymbol{x}_2 - \boldsymbol{\Sigma}_{21}\boldsymbol{\Sigma}_{11}^{-1}\boldsymbol{x}_1$ *have the following distributions and are statistically independent:*

$$\boldsymbol{x}_1 \sim N_r(\boldsymbol{\mu}_1, \boldsymbol{\Sigma}_{11}), \; \boldsymbol{x}_{2.1} \sim N_s(\boldsymbol{\mu}_{2.1}, \boldsymbol{\Sigma}_{22.1}),$$

where

$$\boldsymbol{\mu}_{2.1} = \boldsymbol{\mu}_2 - \boldsymbol{\Sigma}_{21}\boldsymbol{\Sigma}_{11}^{-1}\boldsymbol{\mu}_1, \; \boldsymbol{\Sigma}_{22.1} = \boldsymbol{\Sigma}_{22} - \boldsymbol{\Sigma}_{21}\boldsymbol{\Sigma}_{11}^{-1}\boldsymbol{\Sigma}_{12}. \tag{4.2}$$

Proof: We may write $\boldsymbol{x}_1 = \boldsymbol{A}\boldsymbol{x}$, where $\boldsymbol{A} = [\boldsymbol{I}, \boldsymbol{0}]$, and $\boldsymbol{x}_{2.1} = \boldsymbol{B}\boldsymbol{x}$, where $\boldsymbol{B} = [-\boldsymbol{\Sigma}_{21}\boldsymbol{\Sigma}_{11}^{-1}, \boldsymbol{I}]$. Therefore, by Theorem 4.2.1, $\boldsymbol{x}_1$ and $\boldsymbol{x}_{2.1}$ are normal. Their moments are $\boldsymbol{A}\boldsymbol{\mu}, \boldsymbol{A}\boldsymbol{\Sigma}\boldsymbol{A}', \boldsymbol{B}\boldsymbol{\mu}$, and $\boldsymbol{B}\boldsymbol{\Sigma}\boldsymbol{B}'$, which simplify to the given expressions. To prove independence, note that $\boldsymbol{A}\boldsymbol{\Sigma}\boldsymbol{B}' = 0$ and use Theorem 4.2.5. □

Similar results hold (using g-inverses) in the case of singular distributions. The above theorem can now be used to find the conditional distribution of $\boldsymbol{x}_2$ when $\boldsymbol{x}_1$ is known.

Note that the following theorem was already proved in Theorem 2.5.8 using a different approach.

Theorem 4.2.8 *Using the assumptions and notation of Theorem 4.2.7, the conditional distribution of* $\boldsymbol{x}_2$ *for a given value of* $\boldsymbol{x}_1$ *is*

$$\boldsymbol{x}_2 \mid \boldsymbol{x}_1 \sim N_s(\boldsymbol{\mu}_2 + \boldsymbol{\Sigma}_{21}\boldsymbol{\Sigma}_{11}^{-1}(\boldsymbol{x}_1 - \boldsymbol{\mu}_1), \boldsymbol{\Sigma}_{22.1}).$$

Proof: Since $\boldsymbol{x}_{2.1}$ is independent of $\boldsymbol{x}_1$, its conditional distribution for a given value of $\boldsymbol{x}_1$ is the same as its marginal distribution, which was stated in Theorem 4.2.7. Now, $\boldsymbol{x}_2$ is simply $\boldsymbol{x}_{2.1}$ plus $\boldsymbol{\Sigma}_{21}\boldsymbol{\Sigma}_{11}^{-1}\boldsymbol{x}_1$, and this term is constant when $\boldsymbol{x}_1$ is given. Therefore, the conditional distribution of $\boldsymbol{x}_2 \mid \boldsymbol{x}_1$ is normal, and its conditional mean is

$$\mathsf{E}[\boldsymbol{x}_2 \mid \boldsymbol{x}_1] = \boldsymbol{\mu}_{2.1} + \boldsymbol{\Sigma}_{21}\boldsymbol{\Sigma}_{11}^{-1}\boldsymbol{x}_1 = \boldsymbol{\mu}_2 + \boldsymbol{\Sigma}_{21}\boldsymbol{\Sigma}_{11}^{-1}(\boldsymbol{x}_1 - \boldsymbol{\mu}_1). \tag{4.3}$$

The conditional covariance matrix of $\boldsymbol{x}_2$ is the same as that of $\boldsymbol{x}_{2.1}$, namely $\boldsymbol{\Sigma}_{22.1}$. □

If the assumption of normality is dropped from Theorem 4.2.7, then $\boldsymbol{x}_1$ and $\boldsymbol{x}_{2.1}$ still have the means and covariances stated. Instead of being independent of each other, however, all that can be said in general is that $\boldsymbol{x}_1$ and $\boldsymbol{x}_{2.1}$ are uncorrelated.

Recall that Example 2.5.1 considered a case of Theorem 4.2.8 when $p = 2$.

Example 4.2.1 If $\boldsymbol{\Sigma}$ is the equicorrelation matrix $\boldsymbol{\Sigma} = (1-\rho)\boldsymbol{I} + \rho\boldsymbol{1}\boldsymbol{1}'$, then the conditional distributions take a special form. Note that $\boldsymbol{\Sigma}_{11}$ and $\boldsymbol{\Sigma}_{22}$ are equicorrelation matrices of orders $(r \times r)$ and $(s \times s)$, respectively. Also, $\boldsymbol{\Sigma}_{12} = \rho\boldsymbol{1}_r\boldsymbol{1}_s'$ and $\boldsymbol{\Sigma}_{21} = \rho\boldsymbol{1}_s\boldsymbol{1}_r'$. Furthermore, we know from (A.35) that

$$\boldsymbol{\Sigma}_{11}^{-1} = (1-\rho)^{-1}[\boldsymbol{I} - \alpha\boldsymbol{1}\boldsymbol{1}'], \ \alpha = \frac{\rho}{1+(r-1)\rho}.$$

Therefore, $\boldsymbol{\Sigma}_{21}\boldsymbol{\Sigma}_{11}^{-1} = \rho\boldsymbol{1}\boldsymbol{1}'\boldsymbol{\Sigma}_{11}^{-1} = \rho(1-\rho)^{-1}(1-\alpha r)\boldsymbol{1}\boldsymbol{1}' = \alpha\boldsymbol{1}\boldsymbol{1}'$. Substituting in (4.3), we find that the conditional mean is

$$\mathsf{E}[\boldsymbol{x}_2 \mid \boldsymbol{x}_1] = \boldsymbol{\mu}_2 + \alpha\boldsymbol{1}_r'(\boldsymbol{x}_1 - \boldsymbol{\mu}_1)\boldsymbol{1}_s,$$

and the conditional covariance matrix is

$$\boldsymbol{\Sigma}_{22.1} = (1-\rho)\boldsymbol{I} + \rho\boldsymbol{1}_s\boldsymbol{1}_s' - \rho\alpha\boldsymbol{1}_s\boldsymbol{1}_r'\boldsymbol{1}_r\boldsymbol{1}_s' = (1-\rho)\boldsymbol{I} + \rho(1-r\alpha)\boldsymbol{1}\boldsymbol{1}'.$$

Note that the conditional mean is just the original mean $\boldsymbol{\mu}_2$ with each element altered by the same amount. Moreover, this amount is proportional to $\boldsymbol{1}'(\boldsymbol{x}_1 - \boldsymbol{\mu}_1)$, the sum of the deviations of the elements of $\boldsymbol{x}_1$ from their respective means. If $r = 1$, then the conditional mean is

$$\mathsf{E}(\boldsymbol{x}_2 \mid x_1) = \boldsymbol{\mu}_2 + \rho(x_1 - \mu_1)\boldsymbol{1}.$$ □

4.3 Transformations of Normal Data Matrices

Let $\boldsymbol{x}_1, \dots, \boldsymbol{x}_n$ be a random sample from $N_p(\boldsymbol{\mu}, \boldsymbol{\Sigma})$. We call $\boldsymbol{X} = (\boldsymbol{x}_1, \dots, \boldsymbol{x}_n)'$ a data matrix from $N(\boldsymbol{\mu}, \boldsymbol{\Sigma})$, or simply a "normal data matrix." In this section, we consider linear functions such as $\boldsymbol{Y} = \boldsymbol{AXB}$, where $\boldsymbol{A}(m \times n)$ and $\boldsymbol{B}(p \times q)$ are fixed matrices of real numbers. The most important linear function is the sample mean $\overline{\boldsymbol{x}}' = n^{-1}\boldsymbol{1}'\boldsymbol{X}$, where $\boldsymbol{A} = n^{-1}\boldsymbol{1}'$ and $\boldsymbol{B} = \boldsymbol{I}_p$. The following result is immediate from Theorem 2.6.1.

Theorem 4.3.1 *If $\boldsymbol{X}(n \times p)$ is a data matrix from $N_p(\boldsymbol{\mu}, \boldsymbol{\Sigma})$, and if $n\overline{\boldsymbol{x}} = \boldsymbol{X}'\boldsymbol{1}$, then $\overline{\boldsymbol{x}}$ has the $N_p(\boldsymbol{\mu}, n^{-1}\boldsymbol{\Sigma})$ distribution.*

We may ask under what conditions $\boldsymbol{Y} = \boldsymbol{AXB}$ is itself a normal data matrix. Since

$$y_{ij} = \sum_{k,l} a_{ik} x_{kl} b_{lj},$$

clearly each element of $\boldsymbol{Y}$ is univariate normal. However, for $\boldsymbol{Y}$ to be a normal data matrix, we also require (i) that the rows of $\boldsymbol{Y}$ should be independent of each other and (ii) that each row should have the same distribution. The following theorem gives necessary and sufficient conditions on $\boldsymbol{A}$ and $\boldsymbol{B}$.

Theorem 4.3.2 *If $\boldsymbol{X}$ is a normal data matrix from $N_p(\boldsymbol{\mu}, \boldsymbol{\Sigma})$, and if $\boldsymbol{Y}(m \times q) = \boldsymbol{AXB}$, then $\boldsymbol{Y}$ is a normal data matrix if and only if*

(a) *$\boldsymbol{A1} = \alpha\boldsymbol{1}$ for some scalar α, or $\boldsymbol{B}'\boldsymbol{\mu} = \boldsymbol{0}$, and*
(b) *$\boldsymbol{AA}' = \beta\boldsymbol{I}$ for some scalar β, or $\boldsymbol{B}'\boldsymbol{\Sigma}\boldsymbol{B} = \boldsymbol{0}$.*

When both these conditions are satisfied, then $\boldsymbol{Y}$ is a normal data matrix from $N_q(\alpha\boldsymbol{B}'\boldsymbol{\mu}, \beta\boldsymbol{B}'\boldsymbol{\Sigma}\boldsymbol{B})$.

Proof: See Exercise 4.3.4. □

To understand this theorem, note that postmultiplication of $\boldsymbol{X}$ involves adding weighted *variables*, while premultiplication of $\boldsymbol{X}$ adds weighted *objects*. Since the original objects (rows of $\boldsymbol{X}$) are independent, the transformed objects (rows of $\boldsymbol{Y}$) are also independent unless the premultiplication by $\boldsymbol{A}$ has introduced some interdependence. This clearly cannot happen when $\boldsymbol{A}$ is $k\boldsymbol{I}$, since then $\alpha = k$ and $\beta = k^2$, and both conditions of the theorem are satisfied. Similarly, all permutation matrices satisfy the conditions required on $\boldsymbol{A}$.

We may also investigate the correlation structure between two linear transformations of $\boldsymbol{X}$. Conditions for independence are stated in the following theorem.

Theorem 4.3.3 *If $\boldsymbol{X}$ is a data matrix from $N(\boldsymbol{\mu}, \boldsymbol{\Sigma})$, and if $\boldsymbol{Y} = \boldsymbol{AXB}$ and $\boldsymbol{Z} = \boldsymbol{CXD}$, then the elements of $\boldsymbol{Y}$ are independent of the elements of $\boldsymbol{Z}$ if and only if either (i) $\boldsymbol{B}'\boldsymbol{\Sigma}\boldsymbol{D} = \boldsymbol{0}$ or (ii) $\boldsymbol{AC}' = \boldsymbol{0}$.*

Proof: See Exercise 4.3.5. □

This result is also valid in the situation where the rows of $\boldsymbol{X}$ do not have the same mean; see Exercise 4.4.20.

Corollary 4.3.4 *Under the conditions of the theorem, if $\boldsymbol{X} = (\boldsymbol{X}_1, \boldsymbol{X}_2)$, then $\boldsymbol{X}_1$ is independent of $\boldsymbol{X}_{2.1} = \boldsymbol{X}_2 - \boldsymbol{X}_1\boldsymbol{\Sigma}_{11}^{-1}\boldsymbol{\Sigma}_{12}$. Also, $\boldsymbol{X}_1$ is a data matrix from $N(\boldsymbol{\mu}_1, \boldsymbol{\Sigma}_{11})$, and $\boldsymbol{X}_{2.1}$ is a data matrix from $N(\boldsymbol{\mu}_{2.1}, \boldsymbol{\Sigma}_{22.1})$, where $\boldsymbol{\mu}_{2.1}$ and $\boldsymbol{\Sigma}_{22.1}$ are defined in (4.2).*

Proof: We have $\boldsymbol{X}_1 = \boldsymbol{XB}$, where $\boldsymbol{B}' = (\boldsymbol{I}, \boldsymbol{0})$, and $\boldsymbol{X}_{2.1} = \boldsymbol{XD}$, where $\boldsymbol{D}' = (-\boldsymbol{\Sigma}_{21}\boldsymbol{\Sigma}_{11}^{-1}, \boldsymbol{I})$. Since $\boldsymbol{B}'\boldsymbol{\Sigma}\boldsymbol{D} = \boldsymbol{0}$, the result follows from part (i) of the theorem. □

Corollary 4.3.5 *Under the conditions of the theorem, $\bar{\mathbf{x}} = n^{-1}\mathbf{X}'\mathbf{1}$ is independent of $\mathbf{HX}$, and therefore, $\bar{\mathbf{x}}$ is independent of $\mathbf{S} = n^{-1}\mathbf{X}'\mathbf{HX}$.*

Proof: Put $\mathbf{A} = n^{-1}\mathbf{1}'$ and $\mathbf{C} = \mathbf{H} = \mathbf{I} - n^{-1}\mathbf{11}'$ in the theorem. Since $\mathbf{AC}' = \mathbf{0}$, the result follows. □

4.4 The Wishart Distribution

4.4.1 Introduction

We now turn from linear functions to consider matrix-valued quadratic functions of the form $\mathbf{X}'\mathbf{CX}$, where $\mathbf{C}$ is a symmetric matrix. Among such functions, the most important special case is the sample covariance matrix $\mathbf{S}$ obtained by putting $\mathbf{C} = n^{-1}\mathbf{H}$, where $\mathbf{H}$ is the centering matrix (see Section A.3.3). (However, other quadratic functions can also be used, for instance, in permuting the rows of $\mathbf{X}$ or in finding within- and between-group covariance matrices in regression analysis). These quadratic forms often lead to the Wishart distribution, which constitutes a matrix generalization of the univariate chi-squared distribution and has many similar properties.

Definition 4.4.1 *If $\mathbf{M}(p \times p)$ can be written as $\mathbf{M} = \mathbf{X}'\mathbf{X}$, where $\mathbf{X}(m \times p)$ is a data matrix from $N_p(\mathbf{0}, \mathbf{\Sigma})$, then $\mathbf{M}$ is said to have a Wishart distribution with scale matrix $\mathbf{\Sigma}$ and degrees of freedom parameter m. We write $\mathbf{M} \sim \mathbf{W}_p(\mathbf{\Sigma}, m)$. When $\mathbf{\Sigma} = \mathbf{I}_p$, the distribution is said to be in standard form.*

Note when $p = 1$, the $\mathbf{W}_1(\sigma^2, m)$ distribution is given by $\mathbf{x}'\mathbf{x}$, where the elements of $\mathbf{x}(m \times 1)$ are independent and identically distributed (i.i.d.) $N_1(0, \sigma^2)$ variables; that is, the $W_1(\sigma^2, m)$ distribution is the same as the $\sigma^2\chi^2_m$ distribution.

The scale matrix $\mathbf{\Sigma}$ plays the same role in the Wishart distribution as σ^2 does in the $\sigma^2\chi^2_m$ distribution. We usually suppose $\mathbf{\Sigma} > 0$.

Note that the first moment of $\mathbf{M}$ is given by

$$\mathsf{E}[\mathbf{M}] = \sum_{r=1}^{m} \mathsf{E}[\mathbf{x}_r\mathbf{x}_r'] = m\mathbf{\Sigma}.$$

4.4.2 Properties of Wishart Matrices

Theorem 4.4.2 *If $\mathbf{M} \sim W_p(\mathbf{\Sigma}, m)$ and $\mathbf{B}$ is a $(p \times q)$ matrix, then $\mathbf{B}'\mathbf{MB} \sim W_q(\mathbf{B}'\mathbf{\Sigma B}, m)$.*

Proof: The theorem follows directly from Definition 4.4.1, since $\mathbf{B}'\mathbf{MB} = \mathbf{B}'\mathbf{X}'\mathbf{XB} = \mathbf{Y}'\mathbf{Y}$, where $\mathbf{Y} = \mathbf{XB}$, and the rows of $\mathbf{X}$ are i.i.d. $N_p(\mathbf{0}, \mathbf{\Sigma})$. From Theorem 4.3.2, the rows of $\mathbf{Y}$ are i.i.d. $N_q(\mathbf{0}, \mathbf{B}'\mathbf{\Sigma B})$. Therefore, using Definition 4.4.1 again, $\mathbf{Y}'\mathbf{Y}$ has the stated distribution. □

Corollary 4.4.3 *Diagonal submatrices of $\mathbf{M}$ themselves have a Wishart distribution.*

Corollary 4.4.4 $\boldsymbol{\Sigma}^{-1/2}\boldsymbol{M}\boldsymbol{\Sigma}^{-1/2} \sim W_p(\boldsymbol{I}, m)$.

Corollary 4.4.5 *If* $\boldsymbol{M} \sim W_p(\boldsymbol{I}, m)$, *and* $\boldsymbol{B}(p \times q)$ *satisfies* $\boldsymbol{B}'\boldsymbol{B} = \boldsymbol{I}_q$, *then* $\boldsymbol{B}'\boldsymbol{M}\boldsymbol{B} \sim W_q(\boldsymbol{I}, m)$.

The corollaries follow by inserting particular values of $\boldsymbol{B}$ in the theorem. Note from Corollary 4.4.3 that the ith diagonal element of $\boldsymbol{M}$ has a $\sigma_i^2\chi_m^2$ distribution. The following theorem generalizes and emphasizes this important relationship between the chi-squared and Wishart distributions.

Theorem 4.4.6 *If* $\boldsymbol{M} \sim W_p(\boldsymbol{\Sigma}, m)$, *and* $\boldsymbol{a}$ *is any fixed p-vector such that* $\boldsymbol{a}'\boldsymbol{\Sigma}\boldsymbol{a} \neq 0$, *then*

$$\boldsymbol{a}'\boldsymbol{M}\boldsymbol{a}/\boldsymbol{a}'\boldsymbol{\Sigma}\boldsymbol{a} \sim \chi_m^2.$$

Proof: From Theorem 4.4.2, we see that $\boldsymbol{a}'\boldsymbol{M}\boldsymbol{a} \sim W_1(\boldsymbol{a}'\boldsymbol{\Sigma}\boldsymbol{a}, m)$, which is equivalent to the stated result. □

Note that the converse of Theorem 4.4.6 is untrue – see Exercise 4.4.4.

Corollary 4.4.7 $m_{ii} \sim \sigma_i^2\chi_m^2$.

Theorem 4.4.2 shows that the class of Wishart matrices is closed under the transformation $\boldsymbol{M} \to \boldsymbol{B}'\boldsymbol{M}\boldsymbol{B}$. The Wishart family is also closed under addition.

Theorem 4.4.8 *If* $\boldsymbol{M}_1 \sim W_p(\boldsymbol{\Sigma}, m_1)$ *and* $\boldsymbol{M}_2 \sim W_p(\boldsymbol{\Sigma}, m_2)$, *and if* $\boldsymbol{M}_1$ *and* $\boldsymbol{M}_2$ *are independent, then* $\boldsymbol{M}_1 + \boldsymbol{M}_2 \sim W_p(\boldsymbol{\Sigma}, m_1 + m_2)$.

Proof: We may write $\boldsymbol{M}_i$ as $\boldsymbol{X}_i'\boldsymbol{X}_i$, where $\boldsymbol{X}_i$ has m_i independent rows taken from $N_p(\boldsymbol{0}, \boldsymbol{\Sigma})$, $i = 1, 2$. But

$$\boldsymbol{M}_1 + \boldsymbol{M}_2 = \boldsymbol{X}_1'\boldsymbol{X}_1 + \boldsymbol{X}_2'\boldsymbol{X}_2 = \boldsymbol{X}'\boldsymbol{X}.$$

Now, $\boldsymbol{X}_1$ and $\boldsymbol{X}_2$ may be chosen so as to be independent, in which case all the $(m_1 + m_2)$ rows of $\boldsymbol{X}$ are i.i.d. $N_p(\boldsymbol{0}, \boldsymbol{\Sigma})$ variables. The result then follows from Definition 4.4.1. □

So far, we have taken a normal data matrix $\boldsymbol{X}$ with zero mean and derived a Wishart distribution based on $\boldsymbol{X}'\boldsymbol{X}$. However, it is possible that other functions of $\boldsymbol{X}$ apart from $\boldsymbol{X}'\boldsymbol{X}$ also have a Wishart distribution. Clearly, any matrix containing the sums of squares and cross-products from a subset of the rows of $\boldsymbol{X}$ also has a Wishart distribution. Such a matrix equals $\boldsymbol{X}'\boldsymbol{C}\boldsymbol{X}$, where $c_{rr} = 1$ whenever the rth row of $\boldsymbol{X}$ is in the subset, and all other elements of $\boldsymbol{C}$ are zero. This matrix is symmetric and idempotent, which suggests the following theorem.

Theorem 4.4.9 ***(Cochran 1934)*** *If* $\boldsymbol{X}(n \times p)$ *is a data matrix from* $N_p(\boldsymbol{0}, \boldsymbol{\Sigma})$, *and if* $\boldsymbol{C}(n \times n)$ *is a symmetric matrix, then*

(a) $\boldsymbol{X}'\boldsymbol{C}\boldsymbol{X}$ *has the same distribution as a weighted sum of independent* $W_p(\boldsymbol{\Sigma}, 1)$ *matrices, where the weights are eigenvalues of* $\boldsymbol{C}$;

(b) $\boldsymbol{X}'\boldsymbol{C}\boldsymbol{X}$ *has a Wishart distribution if and only if* $\boldsymbol{C}$ *is idempotent, in which case* $\boldsymbol{X}'\boldsymbol{C}\boldsymbol{X} \sim W_p(\boldsymbol{\Sigma}, r)$*, where* $r = \operatorname{tr}\boldsymbol{C} = \operatorname{rank}\boldsymbol{C}$*;*

(c) If $\boldsymbol{S} = n^{-1}\boldsymbol{X}'\boldsymbol{H}\boldsymbol{X}$ *is the sample covariance matrix, then* $n\boldsymbol{S} \sim W_p(\boldsymbol{\Sigma}, n-1)$*.*

Proof: Using the spectral decomposition theorem (Theorem A.5), write

$$\boldsymbol{C} = \sum_{i=1}^{n} \lambda_i \boldsymbol{\gamma}_{(i)}\boldsymbol{\gamma}'_{(i)}, \qquad \text{where} \quad \boldsymbol{\gamma}'_{(i)}\boldsymbol{\gamma}_{(j)} = \delta_{ij}, \tag{4.4}$$

and λ_i and $\boldsymbol{\gamma}_{(i)}$ are the ith eigenvalue and eigenvector of $\boldsymbol{C}$, respectively. Using (4.4), we see that

$$\boldsymbol{X}'\boldsymbol{C}\boldsymbol{X} = \sum \lambda_i \boldsymbol{y}_i\boldsymbol{y}'_i, \qquad \text{where} \qquad \boldsymbol{y}_i = \boldsymbol{X}'\boldsymbol{\gamma}_{(i)}. \tag{4.5}$$

Writing $\boldsymbol{Y} = \boldsymbol{\Gamma}'\boldsymbol{X}$, where $\boldsymbol{\Gamma}$ is orthogonal, it is easily seen from Theorem 4.3.2 that $\boldsymbol{Y}$ is a data matrix from $N_p(\boldsymbol{0}, \boldsymbol{\Sigma})$. Therefore, $\boldsymbol{y}_1\boldsymbol{y}'_1, \dots, \boldsymbol{y}_n\boldsymbol{y}'_n$ in (4.5) are a set of independent Wishart matrices, each having rank 1. This proves part (a) of the theorem. For part (b), note that if $\boldsymbol{C}$ is idempotent and of rank r, then exactly r of the λ_i are nonzero, and each nonzero λ_i equals 1. Also, $r = \operatorname{tr}\boldsymbol{C}$. Hence, $\boldsymbol{X}'\boldsymbol{C}\boldsymbol{X} \sim W_p(\boldsymbol{\Sigma}, \boldsymbol{r})$ as required.

To prove (c), we note that $\boldsymbol{H}$ is idempotent and of rank $(n-1)$.

For the proof in (b) it can be shown that idempotence of $\boldsymbol{C}$ is in fact a *necessary* condition for $\boldsymbol{X}'\boldsymbol{C}\boldsymbol{X}$ to have a Wishart distribution; we refer the reader to Anderson (1958, p. 164). □

The above theorem is also valid when $\boldsymbol{\mu} \neq \boldsymbol{0}$ if $\boldsymbol{C}$ is a symmetric idempotent matrix whose row sums are 0; see Exercise 4.4.5. For an extension to the case where the rows of $\boldsymbol{X}$ do not have the same mean, see Exercise 4.4.20.

Using Exercise 4.4.5, and also results from Theorem 4.3.1 and Corollary 4.3.5, we emphasize the following important results concerning the sample mean and covariance matrix of a random sample $\boldsymbol{x}_1, \dots, \boldsymbol{x}_n$ from $N_p(\boldsymbol{\mu}, \boldsymbol{\Sigma})$:

$$\bar{\boldsymbol{x}} \sim N_p(\boldsymbol{\mu}, n^{-1}\boldsymbol{\Sigma}), \;\; n\boldsymbol{S} \sim W_p(\boldsymbol{\Sigma}, n-1), \;\; \bar{\boldsymbol{x}} \text{ and } \boldsymbol{S} \text{ are independent.} \tag{4.6}$$

For an alternative proof of (4.6), see Exercise 4.4.12.

We turn now to consider pairs of functions such as $\boldsymbol{X}'\boldsymbol{C}\boldsymbol{X}$ and $\boldsymbol{X}'\boldsymbol{D}\boldsymbol{X}$ and investigate the conditions under which they are independent.

Theorem 4.4.10 ***(Craig, 1943; Lancaster, 1969, p. 23).*** *If the rows of* $\boldsymbol{X}$ *are i.i.d.* $N_p(\boldsymbol{\mu}, \boldsymbol{\Sigma})$*, and if* $\boldsymbol{C}_1, \dots, \boldsymbol{C}_k$ *are symmetric matrices, then* $\boldsymbol{X}'\boldsymbol{C}_1\boldsymbol{X}, \dots, \boldsymbol{X}'\boldsymbol{C}_k\boldsymbol{X}$ *are jointly independent if* $\boldsymbol{C}_r\boldsymbol{C}_s = \boldsymbol{0}$ *for all* $r \neq s$*.*

Proof: First, consider the case $k = 2$, and let $\boldsymbol{C}_1 = \boldsymbol{C}, \boldsymbol{C}_2 = \boldsymbol{D}$. As in (4.5), we can write

$$\boldsymbol{X}'\boldsymbol{C}\boldsymbol{X} = \sum \lambda_i \boldsymbol{y}_i\boldsymbol{y}'_i, \;\; \boldsymbol{X}'\boldsymbol{D}\boldsymbol{X} = \sum \psi_j \boldsymbol{z}_j\boldsymbol{z}'_j,$$

where $\boldsymbol{y}_i = \boldsymbol{X}'\boldsymbol{\gamma}_{(i)}$ and $\boldsymbol{z}_j = \boldsymbol{X}'\boldsymbol{\delta}_{(j)}$, $\boldsymbol{\gamma}_{(i)}$, and $\boldsymbol{\delta}_{(j)}$ being eigenvectors of $\boldsymbol{C}$ and $\boldsymbol{D}$, respectively, with λ_i and ψ_j the corresponding eigenvalues. From Theorem 4.3.3, we note that $\boldsymbol{y}_i$ and $\boldsymbol{z}_j$ are independent if and only if $\boldsymbol{\gamma}'_{(i)}\boldsymbol{\delta}_{(j)} = 0$. Thus, the np-dimensional normal random vectors $(\lambda_1^{1/2}\boldsymbol{y}'_1, \dots, \lambda_n^{1/2}\boldsymbol{y}'_n)'$ and $(\psi_1^{1/2}\boldsymbol{z}'_1, \dots, \psi_n^{1/2}\boldsymbol{z}'_n)'$ will be independent if $\boldsymbol{\gamma}'_{(i)}\boldsymbol{\delta}_{(j)} = 0$ whenever $\lambda_i\psi_j$ is nonzero; that is, if

$$\lambda_i\psi_j\boldsymbol{\gamma}'_{(i)}\boldsymbol{\delta}_{(j)} = 0 \quad \text{for all } \; i, j. \tag{4.7}$$

But

$$CD = \sum \lambda_i \psi_j \gamma_{(i)} \gamma'_{(i)} \delta_{(j)} \delta'_{(j)}.$$

If $CD = 0$, then premultiplying by $\gamma'_{(u)}$ and postmultiplying by $\delta_{(v)}$ gives $\lambda_u \psi_v \gamma'_{(u)} \delta_{(v)} = 0$. This holds for all u and v, and therefore, (4.7) holds. Thus, since $X'CX$ and $X'DX$ are functions of independent normal np-vectors, they are independent.

To deal with the case $k > 2$, notice that normal np-vectors that are pairwise independent are also jointly independent. (This may be easily proved in the same way as Theorem 4.1.5(b).) Hence, the matrices $X'C_rX$, $r = 1, \ldots, k$, are jointly independent. □

The converse of the theorem also holds. For a proof, the reader is referred to Ogawa (1949). A similar theorem gives the condition for $X'CX$ to be independent of a linear function such as AXB (see Exercise 4.4.7). This theorem is also valid when the rows of X have different means; see Exercise 4.4.20.

Note that Craig's theorem does not require the C_r to be idempotent, although if they are, and if $\mu = 0$, then by Cochran's theorem, the quadratic forms to which they lead are not only independent, but also each have the Wishart distribution. An important decomposition of $X'X$ when $\mu = 0$ (and of $X'HX$ for general μ) into a sum of independent Wishart matrices is described in Exercise 4.4.6. This decomposition forms the basis for the multivariate analysis of variance (see Chapter 13).

These theorems can be easily extended to cover $(n \times n)$ matrices such as XCX' (in contrast to $(p \times p)$ matrices such as $X'CX$). This is done by noting that if the rows of X are i.i.d. $N_p(\mu, \Sigma)$, then in general, the rows of X' (that is the columns of X) are *not* i.i.d. However, when $\mu = 0$ and $\Sigma = I_p$, the rows of X' are i.i.d. $N_n(0, I)$ vectors, since in this case all the np elements of X are i.i.d. Hence, we can get the necessary extensions for the standard case (see Exercise 4.4.9) and can thereby derive the relevant properties for general Σ (Exercise 4.4.10). The special case where $n = 1$ leads to quadratic forms proper and is discussed in Exercise 4.4.11.

4.4.3 Partitioned Wishart Matrices

If $M \sim W_p(\Sigma, m)$, it is often useful to partition M into submatrices in the usual way. For instance, we may want to divide the p original variables into two subgroups, consisting of say a and b variables, respectively, where $a + b = p$. Suppose then that M_{11} is $(a \times a)$, and M_{22} is $(b \times b)$, where $a + b = p$. We have already noted (Corollary 4.4.3) that M_{11} and M_{22} have Wishart distributions, although these distributions are not in general independent. However, M_{11} *is* independent of

$$M_{22.1} = M_{22} - M_{21} M_{11}^{-1} M_{12}. \tag{4.8}$$

Here, $M_{22.1}$ is in fact just m times the sample analog of $\Sigma_{22.1}$ defined in (4.2).

Note that when p equals 2, the matrix M may be written as

$$M = m \begin{bmatrix} s_1^2 & rs_1s_2 \\ rs_1s_2 & s_2^2 \end{bmatrix}. \tag{4.9}$$

The matrix $M_{22.1}$ simplifies in this context to

$$M_{22.1} = ms_2^2(1 - r^2). \tag{4.10}$$

Various properties concerning the joint distribution of $\boldsymbol{M}_{11}, \boldsymbol{M}_{12}$, and $\boldsymbol{M}_{22.1}$ are proved in the following theorem.

Theorem 4.4.11 *Let $\boldsymbol{M} \sim W_p(\boldsymbol{\Sigma}, m), m > a$. Then,*

(a) $\boldsymbol{M}_{22.1}$ has the $W_b(\boldsymbol{\Sigma}_{22.1}, m-a)$ distribution and is independent of $(\boldsymbol{M}_{11}, \boldsymbol{M}_{12})$ and
(b) if $\boldsymbol{\Sigma}_{12} = \boldsymbol{0}$, then $\boldsymbol{M}_{22} - \boldsymbol{M}_{22.1} = \boldsymbol{M}_{21}\boldsymbol{M}_{11}^{-1}\boldsymbol{M}_{12}$ has the $W_b(\boldsymbol{\Sigma}_{22}, a)$ distribution, and $\boldsymbol{M}_{21}\boldsymbol{M}_{11}^{-1}\boldsymbol{M}_{12}, \boldsymbol{M}_{11}$, and $\boldsymbol{M}_{22.1}$ are jointly independent.

Proof: Write $\boldsymbol{M} = \boldsymbol{X}'\boldsymbol{X}$, where the rows of $\boldsymbol{X}(m \times p)$ are i.i.d. $N_p(\boldsymbol{0}, \boldsymbol{\Sigma})$ random vectors. Then, $\boldsymbol{M}_{22.1}$ may be written as

$$\boldsymbol{M}_{22.1} = \boldsymbol{X}_2'\boldsymbol{X}_2 - \boldsymbol{X}_2'\boldsymbol{X}_1\boldsymbol{M}_{11}^{-1}\boldsymbol{X}_1'\boldsymbol{X}_2 = \boldsymbol{X}_2'\boldsymbol{P}\boldsymbol{X}_2, \tag{4.11}$$

where $\boldsymbol{P}$ is the symmetric, idempotent matrix defined by $\boldsymbol{P} = \boldsymbol{I} - \boldsymbol{X}_1\boldsymbol{M}_{11}^{-1}\boldsymbol{X}_1'$. Note that $\boldsymbol{P}$ has rank $(m-a)$ and is a function of $\boldsymbol{X}_1$ alone. Since $\boldsymbol{X}_1'\boldsymbol{P} = \boldsymbol{0}$, we see that $\boldsymbol{X}_2'\boldsymbol{P}\boldsymbol{X}_2 = \boldsymbol{X}_{2.1}'\boldsymbol{P}\boldsymbol{X}_{2.1}$, where $\boldsymbol{X}_{2.1} = \boldsymbol{X}_2 - \boldsymbol{X}_1\boldsymbol{\Sigma}_{11}^{-1}\boldsymbol{\Sigma}_{12}$. By Corollary 4.3.4, $\boldsymbol{X}_{2.1} \mid \boldsymbol{X}_1$ is distributed as a data matrix from $N_p(\boldsymbol{0}, \boldsymbol{\Sigma}_{22.1})$. Therefore, using Theorem 4.4.9(b), for any given value of $\boldsymbol{X}_1$, the conditional distribution of $\boldsymbol{M}_{22.1} \mid \boldsymbol{X}_1$ is $W_b(\boldsymbol{\Sigma}_{22.1}, m-a)$. But this conditional distribution is free of $\boldsymbol{X}_1$. Therefore, it is the unconditional (marginal) distribution, and moreover, $\boldsymbol{M}_{22.1}$ is independent of $\boldsymbol{X}_1$.

Since $\boldsymbol{P}(\boldsymbol{I} - \boldsymbol{P}) = \boldsymbol{0}$, we see from Theorem 4.3.3 that, given $\boldsymbol{X}_1$, the matrices $\boldsymbol{P}\boldsymbol{X}_{2.1}$ and $(\boldsymbol{I} - \boldsymbol{P})\boldsymbol{X}_{2.1} = \boldsymbol{X}_1\boldsymbol{M}_{11}^{-1}\boldsymbol{M}_{12} - \boldsymbol{X}_1\boldsymbol{\Sigma}_{11}^{-1}\boldsymbol{\Sigma}_{12}$ are independent. Hence, given $\boldsymbol{X}_1$, the matrices $\boldsymbol{M}_{22.1} = (\boldsymbol{P}\boldsymbol{X}_{2.1})'(\boldsymbol{P}\boldsymbol{X}_{2.1})$ and $(\boldsymbol{I} - \boldsymbol{P})\boldsymbol{X}_{2.1}$ are independent. But from the above paragraph, $\boldsymbol{M}_{22.1}$ is independent of $\boldsymbol{X}_1$. Hence, $\boldsymbol{M}_{22.1}$ is independent of $(\boldsymbol{X}_1, (\boldsymbol{I} - \boldsymbol{P})\boldsymbol{X}_{2.1})$. Since $\boldsymbol{M}_{11}$ and $\boldsymbol{M}_{12}$ can be expressed in terms of $\boldsymbol{X}_1$ and $(\boldsymbol{I} - \boldsymbol{P})\boldsymbol{X}_{2.1}$, we see that $\boldsymbol{M}_{22.1}$ is independent of $(\boldsymbol{M}_{11}, \boldsymbol{M}_{12})$. Thus, the proof of part (a) of the theorem is completed.

For part (b), note that

$$\begin{aligned}\boldsymbol{M}_{22} - \boldsymbol{M}_{22.1} &= \boldsymbol{X}_2'(\boldsymbol{I} - \boldsymbol{P})\boldsymbol{X}_2 \\ &= \boldsymbol{X}_{2.1}'(\boldsymbol{I} - \boldsymbol{P})\boldsymbol{X}_{2.1} + \boldsymbol{\Sigma}_{21}\boldsymbol{\Sigma}_{11}^{-1}\boldsymbol{M}_{12} \\ &\quad + \boldsymbol{M}_{21}\boldsymbol{\Sigma}_{11}^{-1}\boldsymbol{\Sigma}_{12} - \boldsymbol{\Sigma}_{21}\boldsymbol{\Sigma}_{11}^{-1}\boldsymbol{M}_{11}\boldsymbol{\Sigma}_{11}^{-1}\boldsymbol{\Sigma}_{12}\end{aligned}$$

(see Exercise 4.3.2). Now when $\boldsymbol{\Sigma}_{12}$ is zero, the last three terms of this expression disappear, leaving just $\boldsymbol{X}_{2.1}'(\boldsymbol{I} - \boldsymbol{P})\boldsymbol{X}_{2.1}$. Because $(\boldsymbol{I} - \boldsymbol{P})$ is symmetric, idempotent, and has rank a, the distribution of $\boldsymbol{M}_{22} - \boldsymbol{M}_{22.1}$ conditional upon a given value of $\boldsymbol{X}_1$ is $W_b(\boldsymbol{\Sigma}_{22}, a)$. Moreover, Craig's theorem (Theorem 4.4.10) implies that $\boldsymbol{M}_{22.1}$ and $\boldsymbol{M}_{22} - \boldsymbol{M}_{22.1}$ are independent for any given value of $\boldsymbol{X}_1$, since $\boldsymbol{P}(\boldsymbol{I} - \boldsymbol{P}) = \boldsymbol{0}$. As the conditional distributions of $\boldsymbol{M}_{22.1}$ and $\boldsymbol{M}_{12} - \boldsymbol{M}_{22.1}$ do not involve $\boldsymbol{X}_1$, we see that $\boldsymbol{M}_{22.1}$ and $\boldsymbol{M}_{22} - \boldsymbol{M}_{22.1}$ are (unconditionally) independent, and further, $(\boldsymbol{M}_{22.1}, \boldsymbol{M}_{22} - \boldsymbol{M}_{22.1})$ is independent of $\boldsymbol{X}_1$ (and hence independent of $\boldsymbol{M}_{11}$). Therefore, in their unconditional joint distribution, the three matrices $\boldsymbol{M}_{11}, \boldsymbol{M}_{22.1}$, and $\boldsymbol{M}_{22} - \boldsymbol{M}_{22.1}$ are independent of one another. □

Recall from (A.31) that $\boldsymbol{M}^{22} = \boldsymbol{M}_{22.1}^{-1}$. Hence, parts (a) and (b) of the theorem may also be written as follows.

Corollary 4.4.12

(a) $(\boldsymbol{M}^{22})^{-1} \sim W_b((\boldsymbol{\Sigma}^{22})^{-1}, m-a)$ and is independent of $(\boldsymbol{M}_{11}, \boldsymbol{M}_{12})$.

(b) *If* $\boldsymbol{\Sigma}_{12} = \mathbf{0}$, *then* $\boldsymbol{M}_{22} - (\boldsymbol{M}^{22})^{-1} \sim W_b(\boldsymbol{\Sigma}_{22}, a)$, *and* $\boldsymbol{M}_{22} - (\boldsymbol{M}^{22})^{-1}, \boldsymbol{M}_{11}$, *and* $(\boldsymbol{M}^{22})^{-1}$ *are jointly independent.*

Hence, when $\boldsymbol{\Sigma}_{12} = \mathbf{0}$, the Wishart matrix $\boldsymbol{M}_{22}$ can be decomposed into the sum $(\boldsymbol{M}_{22} - \boldsymbol{M}_{22.1}) + (\boldsymbol{M}_{22.1})$, where the two components of the sum have independent Wishart distributions. Moreover, the degrees of freedom are additive in a similar manner.

If the population correlation coefficient is zero, then, for the bivariate case, part (b) of Theorem 4.4.11 leads to the following results with the help of (4.9) and (4.10):

$$m_{11} \sim \sigma_1^2 \chi_m^2, \tag{4.12}$$

$$m_{22.1} = m s_2^2 (1 - r^2) \sim \sigma_2^2 \chi_{m-1}^2, \tag{4.13}$$

$$m_{22} - m_{22.1} = m s_2^2 r^2 \sim \sigma_2^2 \chi_1^2. \tag{4.14}$$

Moreover, these three chi-squared variables are jointly statistically independent.

Theorem 4.4.13 *If* $\boldsymbol{M} \sim W_p(\boldsymbol{\Sigma}, m), m > p$, *then*

(a) *the ratio* $\boldsymbol{a}'\boldsymbol{\Sigma}^{-1}\boldsymbol{a}/\boldsymbol{a}'\boldsymbol{M}^{-1}\boldsymbol{a}$ *has the* χ^2_{m-p+1} *distribution for any fixed p-vector* $\boldsymbol{a}$, *and in particular,* $\sigma^{ii}/m^{ii} \sim \chi^2_{m-p+1}$ *for* $i = 1, \ldots, p$;
(b) m^{ii} *is independent of all the elements of* $\boldsymbol{M}$ *except* m_{ii}.

Proof: From Corollary 4.4.12, putting $b = 1$ and $a = p - 1$, we get

$$(m^{pp})^{-1} \sim (\sigma^{pp})^{-1} \chi^2_{m-p+1}. \tag{4.15}$$

This proves part (a) for the special case where $\boldsymbol{a} = (0, \ldots, 0, 1)'$.

For general $\boldsymbol{a}$, let $\boldsymbol{A}$ be a nonsingular matrix whose last column equals $\boldsymbol{a}$, and set $\boldsymbol{N} = \boldsymbol{A}^{-1}\boldsymbol{M}(\boldsymbol{A}^{-1})'$. Then, $\boldsymbol{N} \sim W_p(\boldsymbol{A}^{-1}\boldsymbol{\Sigma}(\boldsymbol{A}^{-1})', m)$. Since $\boldsymbol{N}^{-1} = \boldsymbol{A}'\boldsymbol{M}^{-1}\boldsymbol{A}$, we see from (4.15) that

$$\begin{aligned}(n^{pp})^{-1} &= (\boldsymbol{a}'\boldsymbol{M}^{-1}\boldsymbol{a})^{-1} \sim [(\boldsymbol{A}'\boldsymbol{\Sigma}^{-1}\boldsymbol{A})_{pp}]^{-1}\chi^2_{m-p+1} \\ &= (\boldsymbol{a}'\boldsymbol{\Sigma}^{-1}\boldsymbol{a})^{-1}\chi^2_{m-p+1}.\end{aligned}$$

For part (b), note that $(\boldsymbol{M}^{22})^{-1}$ in Theorem 4.4.11 is independent of $(\boldsymbol{M}_{11}, \boldsymbol{M}_{12})$. Hence, for this particular case, m^{pp} is independent of all the elements of $\boldsymbol{M}$ except m_{pp}. By a suitable permutation of the rows and columns, a similar result can be proved for all i. □

For a generalization of this theorem, see Exercise 4.4.19.

The following theorem describes the distribution of $|\boldsymbol{M}|$.

Theorem 4.4.14 *If* $\boldsymbol{M} \sim \boldsymbol{W}_p(\boldsymbol{\Sigma}, m)$ *and* $m \geq p$, *then* $|\boldsymbol{M}|$ *is* $|\boldsymbol{\Sigma}|$ *times p independent chi-squared random variables with degrees of freedom* $m, m-1, \ldots, m-p+1$.

Proof: We proceed by induction on p. Clearly, the theorem is true if $p = 1$. For $p > 1$, partition $\boldsymbol{M}$ with $a = p - 1$ and $b = 1$. Suppose by the induction hypothesis that $|\boldsymbol{M}_{11}|$ can be written as $|\boldsymbol{\Sigma}_{11}|$ times $p - 1$ independent chi-squared random variables with degrees of freedom $m, m-1, \ldots, m-p+2$. By Theorem 4.4.11, $\boldsymbol{M}_{11}$ is independent of the scalar

$\boldsymbol{M}_{22.1} \sim \boldsymbol{\Sigma}_{22.1}\chi^2_{m-p+1}$. Since $|\boldsymbol{M}| = |\boldsymbol{M}_{11}||\boldsymbol{M}_{22.1}|$ and $|\boldsymbol{\Sigma}| = |\boldsymbol{\Sigma}_{11}||\boldsymbol{\Sigma}_{22.1}|$ (see Eq. A.18), the theorem follows. □

Corollary 4.4.15 *If* $\boldsymbol{M} \sim W_p(\boldsymbol{\Sigma}, m), \boldsymbol{\Sigma} > 0$, *and* $m > p$, *then* $\boldsymbol{M} > 0$ *with probability 1.*

Proof: Since a chi-squared variate is strictly positive with probability 1 and $|\boldsymbol{\Sigma}| > 0$, it follows that $|\boldsymbol{M}| > 0$. Hence, since by construction $\boldsymbol{M}$ is positive semi-definite (p.s.d.), all of the eigenvalues of $\boldsymbol{M}$ are strictly positive with probability 1. □

Diaz Garcia et al. (1997) have unified all Wishart and pseudo-Wishart distributions, including central and noncentral, singular or non-singular, and applied these in shape analysis.

4.5 The Hotelling T^2 Distribution

We now turn to functions such as $\boldsymbol{d}'\boldsymbol{M}^{-1}\boldsymbol{d}$, where $\boldsymbol{d}$ is normal, $\boldsymbol{M}$ is Wishart, and $\boldsymbol{d}$ and $\boldsymbol{M}$ are independent. For instance, $\boldsymbol{d}$ may be the sample mean, and $\boldsymbol{M}$ proportional to the sample covariance matrix (see Eq. (4.6)). This important special case is examined in Corollary 4.5.3.

We now derive the general distribution of quadratic forms such as the above. This work was initiated by Hotelling (1931).

Definition 4.5.1 *If* α *can be written as* $\alpha = m\boldsymbol{d}'\boldsymbol{M}^{-1}\boldsymbol{d}$, *where* $\boldsymbol{d}$ *and* $\boldsymbol{M}$ *are independently distributed as* $N_p(\boldsymbol{0}, \boldsymbol{I})$ *and* $W_p(\boldsymbol{I}, m)$, *respectively, then we say that* α *has the Hotelling* T^2 *distribution with parameters p and m. We write* $\alpha \sim T^2(p, m)$.

Theorem 4.5.2 *If* $\boldsymbol{x}$ *and* $\boldsymbol{M}$ *are independently distributed as* $N_p(\boldsymbol{\mu}, \boldsymbol{\Sigma})$ *and* $W_p(\boldsymbol{\Sigma}, m)$, *respectively, then*

$$m(\boldsymbol{x} - \boldsymbol{\mu})'\boldsymbol{M}^{-1}(\boldsymbol{x} - \boldsymbol{\mu}) \sim T^2(p, m). \tag{4.16}$$

Proof: If $\boldsymbol{d}^* = \boldsymbol{\Sigma}^{-1/2}(\boldsymbol{x} - \boldsymbol{\mu})$ and $\boldsymbol{M}^* = \boldsymbol{\Sigma}^{-1/2}\boldsymbol{M}\boldsymbol{\Sigma}^{-1/2}$, we see that $\boldsymbol{d}^*$ and $\boldsymbol{M}^*$ satisfy the requirements of Definition 4.5.1. Therefore, $\alpha \sim T^2(p, m)$, where

$$\alpha = m\boldsymbol{d}^{*\prime}\boldsymbol{M}^{*-1}\boldsymbol{d}^* = m(\boldsymbol{x} - \boldsymbol{\mu})'\boldsymbol{M}^{-1}(\boldsymbol{x} - \boldsymbol{\mu}).$$

Hence, the theorem is proved. □

Corollary 4.5.3 *If* $\overline{\boldsymbol{x}}$ *and* $\boldsymbol{S}$ *are the mean vector and covariance matrix of a sample of size n from* $N_p(\boldsymbol{\mu}, \boldsymbol{\Sigma})$, *and* $\boldsymbol{S}_u = (n/(n-1))\boldsymbol{S}$, *then*

$$(n-1)(\overline{\boldsymbol{x}} - \boldsymbol{\mu})'\boldsymbol{S}^{-1}(\overline{\boldsymbol{x}} - \boldsymbol{\mu}) = n(\overline{\boldsymbol{x}} - \boldsymbol{\mu})'\boldsymbol{S}_u^{-1}(\overline{\boldsymbol{x}} - \boldsymbol{\mu}) \sim T^2(p, n-1). \tag{4.17}$$

Proof: Substituting $\boldsymbol{M} = n\boldsymbol{S}, m = n-1$, and $\boldsymbol{x} - \boldsymbol{\mu}$ for $n^{1/2}(\overline{\boldsymbol{x}} - \boldsymbol{\mu})$ in the theorem, the result follows immediately. □

Corollary 4.5.4 *The* T^2 *statistic is invariant under any nonsingular linear transformation* $\boldsymbol{x} \to \boldsymbol{A}\boldsymbol{x} + \boldsymbol{b}$.

Of course, the univariate t statistic also has this property of invariance mentioned above. Indeed, the square of the univariate t_m variable has the $T^2(1, m)$ distribution (see Exercise 4.5.1). In other words, the $F_{1,m}$ distribution and the $T^2(1, m)$ distribution are the same. The following theorem extends the result.

Theorem 4.5.5 *We have*

$$T^2(p, m) = \{mp/(m - p + 1)\}F_{p,m-p+1}. \tag{4.18}$$

Proof: We use the characterization $\alpha = m\mathbf{d}'\mathbf{M}^{-1}\mathbf{d}$ given in Definition 4.5.1. Write

$$\alpha = m(\mathbf{d}'\mathbf{M}^{-1}\mathbf{d}/\mathbf{d}'\mathbf{d})\mathbf{d}'\mathbf{d}.$$

Since $\mathbf{M}$ is independent of $\mathbf{d}$, we see from Theorem 4.4.13(a) that the conditional distribution of $\beta = \mathbf{d}'\mathbf{d}/\mathbf{d}'\mathbf{M}^{-1}\mathbf{d}$, given $\mathbf{d}$ is χ^2_{m-p+1}. Since this conditional distribution does not depend on $\mathbf{d}$, it is also the marginal distribution of β, and furthermore, β is independent of $\mathbf{d}$. By Theorem 2.5.4, $\mathbf{d}'\mathbf{d} \sim \chi^2_p$, so we can express α as a ratio of independent χ^2 variables:

$$\alpha = m\chi^2_p/\chi^2_{m-p+1} = \{mp/(m - p + 1)\}F_{p,m-p+1}.$$

□

Corollary 4.5.6 *If $\overline{\mathbf{x}}$ and $\mathbf{S}$ are the mean and covariance of a sample of size n from $N_p(\boldsymbol{\mu}, \boldsymbol{\Sigma})$, then*

$$\{(n - p)/p\}(\overline{\mathbf{x}} - \boldsymbol{\mu})'\mathbf{S}^{-1}(\overline{\mathbf{x}} - \boldsymbol{\mu}) \sim F_{p,n-p}. \tag{4.19}$$

Proof: The result follows immediately from (4.17). □

Corollary 4.5.7 $|\mathbf{M}|/|\mathbf{M} + \mathbf{d}\mathbf{d}'| \sim B((m - p + 1)/2, p/2)$, *where $B(\cdot, \cdot)$ is a beta distribution.*

Proof: From Eq. (A.23), the given ratio equals $m/(m + \alpha)$. Using the F distribution of α, and the univariate relationship between F and beta distributions, the result follows. (See also Exercise 4.5.2.) □

Since $\mathbf{d}'\mathbf{d}$ and β are independent chi-squared statistics, as shown in the proof of Theorem 4.5.5, their ratio is independent of their sum, which also has a chi-squared distribution (see Theorem B.4.1). The sum is

$$\mathbf{d}'\mathbf{d} + \beta = \mathbf{d}'\mathbf{d}\left(1 + \frac{1}{\mathbf{d}'\mathbf{M}^{-1}\mathbf{d}}\right).$$

Hence, we have the following:

Theorem 4.5.8 *If $\mathbf{d}$ and $\mathbf{M}$ are independently distributed as $N_p(\mathbf{0}, \mathbf{I})$ and $W_p(\mathbf{I}, m)$, respectively, then*

$$\mathbf{d}'\mathbf{d}\left(1 + \frac{1}{\mathbf{d}'\mathbf{M}^{-1}\mathbf{d}}\right) \sim \chi^2_{m+1}$$

and is distributed independently of $\mathbf{d}'\mathbf{M}^{-1}\mathbf{d}$.

Theorem 4.5.8 is one extension of the univariate result that if $d \sim N(0, 1)$ and $u^2 \sim \chi^2$, then $d^2 + u^2$ is independent of d^2/u^2. Another generalization of the same result is given in Theorem 4.5.9, which requires the following lemma.

Lemma Let $\boldsymbol{W}$ be a square symmetric $(p \times p)$ random matrix, and let $\boldsymbol{x}$ be a random p-vector. If $\boldsymbol{x}$ is independent of $(\boldsymbol{g}_1'\boldsymbol{W}\boldsymbol{g}_1, \ldots, \boldsymbol{g}_p'\boldsymbol{W}\boldsymbol{g}_p)$ for all nonrandom orthogonal matrices $\boldsymbol{G} = (\boldsymbol{g}_1, \ldots, \boldsymbol{g}_p)'$, then $\boldsymbol{x}$ is independent of $\boldsymbol{W}$.

Proof: Using the joint c.f. of $\boldsymbol{W}$ and $\boldsymbol{x}$, it is seen that if $\text{tr}\boldsymbol{A}\boldsymbol{W}$ is independent of $\boldsymbol{x}$ for every square symmetric matrix $\boldsymbol{A}$, then $\boldsymbol{W}$ is independent of $\boldsymbol{x}$. Now, $\boldsymbol{A}$ may be written in canonical form as $\sum \lambda_i \boldsymbol{g}_i \boldsymbol{g}_i'$ say, where $\boldsymbol{G} = (\boldsymbol{g}_1, \ldots, \boldsymbol{g}_p)'$ is orthogonal. Then,

$$\text{tr}\boldsymbol{A}\boldsymbol{W} = \text{tr}\left(\sum \lambda_i \boldsymbol{g}_i \boldsymbol{g}_i'\right)\boldsymbol{W} = \sum \lambda_i \boldsymbol{g}_i' \boldsymbol{W} \boldsymbol{g}_i.$$

If the conditions of the lemma are satisfied, then all of the terms in the summation, and the summation itself, are independent of $\boldsymbol{x}$. Thus, $\text{tr}\boldsymbol{A}\boldsymbol{W}$ is independent of $\boldsymbol{x}$ for all $\boldsymbol{A}$, and therefore, $\boldsymbol{W}$ is independent of $\boldsymbol{x}$. □

Theorem 4.5.9 *If* $\mathbf{d}$ *and* $\mathbf{M}$ *are independently distributed as* $N_p(\mathbf{0}, \boldsymbol{I})$ *and* $W_p(\boldsymbol{I}, m)$, *respectively, then* $\mathbf{d}'\mathbf{M}^{-1}\mathbf{d}$ *is independent of* $\mathbf{M} + \mathbf{d}\mathbf{d}'$.

Proof: Let $\boldsymbol{G} = (\boldsymbol{g}_1, \ldots, \boldsymbol{g}_p)'$ be an orthogonal matrix, and consider the quadratic forms $q_j = \boldsymbol{g}_j'(\boldsymbol{M} + \boldsymbol{d}\boldsymbol{d}')\boldsymbol{g}_j, j = 1, \ldots, p$. Write

$$\boldsymbol{M} = \boldsymbol{X}_1'\boldsymbol{X}_1 = \sum_{r=1}^{m} \boldsymbol{x}_r \boldsymbol{x}_r' \qquad \text{and} \qquad \boldsymbol{d} = \boldsymbol{x}_{m+1},$$

where $\boldsymbol{X} = (\boldsymbol{X}_1', \boldsymbol{x}_{m+1})'$ is a data matrix from $N_p(\mathbf{0}, \boldsymbol{I})$. Set $\boldsymbol{Y} = \boldsymbol{X}\boldsymbol{G}'$, so that $\boldsymbol{Y}_1 = \boldsymbol{X}_1\boldsymbol{G}'$ and $\boldsymbol{y}_{m+1} = \boldsymbol{G}\boldsymbol{x}_{m+1}$. Then,

$$q_j = \boldsymbol{g}_j'(\boldsymbol{M} + \boldsymbol{d}\boldsymbol{d}')\boldsymbol{g}_j = \sum_{r=1}^{m+1} (\boldsymbol{g}_j'\boldsymbol{x}_r)^2 = \sum_{r=1}^{m+1} y_{rj}^2.$$

Now, $\boldsymbol{Y}$ is also a data matrix from $N_p(\mathbf{0}, \boldsymbol{I})$), so, thought of as a $p(m+1)$-vector $\boldsymbol{Y}^V$, it has the $N_{p(m+1)}(\mathbf{0}, \boldsymbol{I})$ distribution and hence is spherically symmetric. Therefore, by Theorem 3.3.4, $(q_1, \ldots, q_p)$ is statistically independent of any column-scale-invariant function of $\boldsymbol{Y}$. In particular, $(q_1, \ldots, q_p)$ is independent of

$$\mathbf{d}'\mathbf{M}^{-1}\mathbf{d} = \boldsymbol{y}_{m+1}'(\boldsymbol{Y}_1'\boldsymbol{Y}_1)^{-1}\boldsymbol{y}_{m+1}.$$

Since this result holds for all orthogonal matrices $\boldsymbol{G}$, we see from the above lemma that the theorem is proved. □

4.6 Mahalanobis Distance

4.6.1 The Two-Sample Hotelling T^2 Statistic

The so-called Mahalanobis distance between two populations with means $\boldsymbol{\mu}_1$ and $\boldsymbol{\mu}_2$ and common covariance matrix $\boldsymbol{\Sigma}$ has already been defined in Section 2.2.3. It is given by Δ, where

$$\Delta^2 = (\boldsymbol{\mu}_1 - \boldsymbol{\mu}_2)'\boldsymbol{\Sigma}^{-1}(\boldsymbol{\mu}_1 - \boldsymbol{\mu}_2). \tag{4.20}$$

Of course, only rarely are the population parameters known, and it is usual for them to be estimated by the corresponding sample values. Suppose that we have two samples of size n_1 and n_2, where $n_1 + n_2 = n$. Then, the sample Mahalanobis distance, D, can be defined by

$$D^2 = (\overline{\mathbf{x}}_1 - \overline{\mathbf{x}}_2)' \mathbf{S}_u^{-1} (\overline{\mathbf{x}}_1 - \overline{\mathbf{x}}_2), \tag{4.21}$$

where $\mathbf{S}_u = (n_1\mathbf{S}_1 + n_2\mathbf{S}_2)/(n-2)$ is an unbiased estimate of $\mathbf{\Sigma}$. (The sample mean and covariance matrix for sample $i, i = 1, 2$, are denoted $\overline{\mathbf{x}}_i$ and $\mathbf{S}_i$.) The statistical distribution of D^2 under one particular set of assumptions is given by the following theorem.

Theorem 4.6.1 *If $\mathbf{X}_1$ and $\mathbf{X}_2$ are independent data matrices, and if the n_i rows of $\mathbf{X}_i$ are i.i.d. $N_p(\boldsymbol{\mu}_i, \mathbf{\Sigma}_i), i = 1, 2$, then when $\boldsymbol{\mu}_1 = \boldsymbol{\mu}_2$ and $\mathbf{\Sigma}_1 = \mathbf{\Sigma}_2$, $(n_1 n_2/n)D^2$ is a $T^2(p, n-2)$ variable.*

Proof: Since $\overline{\mathbf{x}}_i \sim N_p(\boldsymbol{\mu}_i, n_i^{-1}\mathbf{\Sigma}_i), i = 1, 2$, the general distribution of $\mathbf{d} = \overline{\mathbf{x}}_1 - \overline{\mathbf{x}}_2$ is normal with mean $\boldsymbol{\mu}_1 - \boldsymbol{\mu}_2$ and covariance matrix $n_1^{-1}\mathbf{\Sigma}_1 + n_2^{-1}\mathbf{\Sigma}_2$. When $\boldsymbol{\mu}_1 = \boldsymbol{\mu}_2$ and $\mathbf{\Sigma}_1 = \mathbf{\Sigma}_2 = \mathbf{\Sigma}$, $\mathbf{d} \sim N_p(\mathbf{0}, c\mathbf{\Sigma})$, where $c = n/n_1 n_2$.

If $\mathbf{M}_i = n_i \mathbf{S}_i$, then $\mathbf{M}_i \sim W_p(\mathbf{\Sigma}_i, n_i - 1)$. Thus, when $\mathbf{\Sigma}_1 = \mathbf{\Sigma}_2 = \mathbf{\Sigma}$,

$$\mathbf{M} = (n-2)\mathbf{S}_u = \mathbf{M}_1 + \mathbf{M}_2 \sim W_p(\mathbf{\Sigma}, n-2).$$

So, $c\mathbf{M} \sim W_p(c\mathbf{\Sigma}, n-2)$. Moreover, $\mathbf{M}$ is independent of $\mathbf{d}$ since $\overline{\mathbf{x}}_i$ is independent of $\mathbf{S}_i$ for $i = 1, 2$, and the two samples are independent of one another. Therefore,

$$(n-2)\mathbf{d}'(c\mathbf{M})^{-1}\mathbf{d} \sim T^2(p, n-2). \tag{4.22}$$

Simplifying the left-hand side gives the required result. □

The quantity

$$(n_1 n_2/n)D^2 = (n_1 n_2/n)(\overline{\mathbf{x}}_1 - \overline{\mathbf{x}}_2)' \mathbf{S}_u^{-1} (\overline{\mathbf{x}}_1 - \overline{\mathbf{x}}_2) \tag{4.23}$$

is known as *Hotelling's two-sample T^2 statistic*. Using the relationship (Theorem 4.5.5) between T^2 and F statistics, we may also deduce that, under the stated conditions,

$$\frac{n_1 n_2 (n-p-1)}{n(n-2)p} D^2 \sim F_{p, n-p-1}. \tag{4.24}$$

4.6.2 A Decomposition of Mahalanobis Distance

The Mahalanobis distance of $\boldsymbol{\mu}$ from $\mathbf{0}$ is

$$\Delta_p^2 = \boldsymbol{\mu}' \mathbf{\Sigma}^{-1} \boldsymbol{\mu}. \tag{4.25}$$

Partition $\boldsymbol{\mu}' = (\boldsymbol{\mu}_1', \boldsymbol{\mu}_2')$, where $\boldsymbol{\mu}_1$ contains the first k variables. Then, using (A.29) and (A.31) to partition $\mathbf{\Sigma}^{-1}$, we can write

$$\Delta_p^2 = \boldsymbol{\mu}_1' \mathbf{\Sigma}_{11}^{-1} \boldsymbol{\mu}_1 + \boldsymbol{\mu}_{2.1}' \mathbf{\Sigma}_{22.1}^{-1} \boldsymbol{\mu}_{2.1}, \tag{4.26}$$

where $\boldsymbol{\mu}_{2.1} = \boldsymbol{\mu}_2 - \mathbf{\Sigma}_{21}\mathbf{\Sigma}_{11}^{-1}\boldsymbol{\mu}_1$ and $\mathbf{\Sigma}_{22.1} = \mathbf{\Sigma}_{22} - \mathbf{\Sigma}_{21}\mathbf{\Sigma}_{11}^{-1}\mathbf{\Sigma}_{12}$. Note that, if $\Delta_k^2 = \boldsymbol{\mu}_1' \mathbf{\Sigma}_{11}^{-1} \boldsymbol{\mu}_1$, then Δ_k is the Mahalanobis distance based on the first k variables, so the condition $\boldsymbol{\mu}_{2.1} = \mathbf{0}$ is equivalent to $\Delta_k^2 = \Delta_p^2$.

Similarly, if $\boldsymbol{u} \sim N_p(\boldsymbol{\mu}, \boldsymbol{\Sigma})$ and $\boldsymbol{M} \sim W_p(\boldsymbol{\Sigma}, m)$, then the sample Mahalanobis distance

$$D_p^2 = m\boldsymbol{u}'\boldsymbol{M}^{-1}\boldsymbol{u} \tag{4.27}$$

can be partitioned as

$$D_p^2 = D_k^2 + m\boldsymbol{z}'\boldsymbol{M}_{22.1}^{-1}\boldsymbol{z}, \tag{4.28}$$

where $D_k^2 = m\boldsymbol{u}_1\boldsymbol{M}_{11}^{-1}\boldsymbol{u}_1$, $\boldsymbol{M}_{22.1} = \boldsymbol{M}_{22} - \boldsymbol{M}_{21}\boldsymbol{M}_{11}^{-1}\boldsymbol{M}_{12}$, and

$$\boldsymbol{z} = \boldsymbol{u}_2 - \boldsymbol{M}_{21}\boldsymbol{M}_{11}^{-1}\boldsymbol{u}_1 \tag{4.29}$$

(see Exercise 4.6.1).

The following theorem gives the distribution of $D_p^2 - D_k^2$ when $\boldsymbol{\mu}_{2.1} = \boldsymbol{0}$.

Theorem 4.6.2 *If D_p^2 and D_k^2 are as defined above and $\boldsymbol{\mu}_{2.1} = \boldsymbol{0}$, then*

$$\frac{D_p^2 - D_k^2}{m + D_k^2} \sim \frac{p-k}{m-p+1}F_{p-k,m-p+1}$$

and is independent of D_k^2.

Proof: Suppose that $\boldsymbol{M} = \boldsymbol{X}'\boldsymbol{X}$, where $\boldsymbol{X}$ is a data matrix from $N_p(\boldsymbol{0}, \boldsymbol{\Sigma})$ independent of $\boldsymbol{u}$. By Theorem 4.4.11, $\boldsymbol{M}_{22.1} \sim W_{p-k}(\boldsymbol{\Sigma}_{22.1}, m-k)$ and is independent of $(\boldsymbol{M}_{21}, \boldsymbol{M}_{11})$. By hypothesis, $\boldsymbol{u}$ is independent of $\boldsymbol{M}$. Thus, $\boldsymbol{M}_{22.1}, (\boldsymbol{M}_{21}, \boldsymbol{M}_{11})$, and $\boldsymbol{u}$ are jointly independent and hence

$$\boldsymbol{M}_{22.1} \text{is independent of } (\boldsymbol{M}_{21}, \boldsymbol{M}_{11}, \boldsymbol{u}). \tag{4.30}$$

Let us examine the distribution of $\boldsymbol{z}$ in (4.29) conditional on $\boldsymbol{u}_1$ and $\boldsymbol{X}_1$. First write

$$\boldsymbol{u}_2 = \boldsymbol{u}_{2.1} + \boldsymbol{\Sigma}_{21}\boldsymbol{\Sigma}_{11}^{-1}\boldsymbol{u}_1, \tag{4.31}$$

where $\boldsymbol{u}_{2.1} = \boldsymbol{u}_2 - \boldsymbol{\Sigma}_{21}\boldsymbol{\Sigma}_{11}^{-1}\boldsymbol{u}_1$ is normally distributed with mean $\boldsymbol{\mu}_{2.1} = \boldsymbol{0}$ and covariance matrix $\boldsymbol{\Sigma}_{22.1}$. By Theorem 4.2.7, $\boldsymbol{u}_{2.1}$ is independent of $(\boldsymbol{u}_1, \boldsymbol{X})$ and hence

$$\boldsymbol{u}_{2.1} \text{is independent of } (\boldsymbol{M}_{21}\boldsymbol{M}_{11}^{-1}\boldsymbol{u}_1, \boldsymbol{u}_1, \boldsymbol{X}_1). \tag{4.32}$$

The second term in (4.31), $\boldsymbol{\Sigma}_{21}\boldsymbol{\Sigma}_{11}^{-1}\boldsymbol{u}_1$, is a constant, given $\boldsymbol{u}_1, \boldsymbol{X}_1$.

To study $\boldsymbol{M}_{21}\boldsymbol{M}_{11}^{-1}\boldsymbol{u}_1 \mid \boldsymbol{u}_1, \boldsymbol{X}_1$, write

$$\boldsymbol{X}_2 = \boldsymbol{X}_{2.1} + \boldsymbol{X}_1\boldsymbol{\Sigma}_{11}^{-1}\boldsymbol{\Sigma}_{12},$$

where $\boldsymbol{X}_{2.1}$ is a data matrix from $N_{p-k}(\boldsymbol{0}, \boldsymbol{\Sigma}_{22.1})$ independent of $(\boldsymbol{X}_1, \boldsymbol{u}_1)$. Then,

$$\boldsymbol{M}_{21}\boldsymbol{M}_{11}^{-1}\boldsymbol{u}_1 = \boldsymbol{X}_2'\boldsymbol{X}_1\boldsymbol{M}_{11}^{-1}\boldsymbol{u}_1 = \boldsymbol{X}_{2.1}'\boldsymbol{b} + \boldsymbol{\Sigma}_{21}\boldsymbol{\Sigma}_{11}^{-1}\boldsymbol{X}_1'\boldsymbol{b},$$

where $\boldsymbol{b} = \boldsymbol{X}_1\boldsymbol{M}_{11}^{-1}\boldsymbol{u}_1$ and $\boldsymbol{\Sigma}_{21}\boldsymbol{\Sigma}_{11}^{-1}\boldsymbol{X}_1'\boldsymbol{b} = \boldsymbol{\Sigma}_{21}\boldsymbol{\Sigma}_{11}^{-1}\boldsymbol{u}_1$ are constants, given $(\boldsymbol{X}_1, \boldsymbol{u}_1)$. Given $(\boldsymbol{u}_1, \boldsymbol{X}_1)$, $\boldsymbol{X}_{2.1}'\boldsymbol{b}$ is a linear combination of the (statistically independent) rows of $\boldsymbol{X}_{2.1}$, so that $\boldsymbol{X}_{2.1}'\boldsymbol{b} \mid \boldsymbol{u}_1, \boldsymbol{X}_1$ is normally distributed with mean $\boldsymbol{0}$ and covariance matrix $(\boldsymbol{b}'\boldsymbol{b})\boldsymbol{\Sigma}_{22.1} = (\boldsymbol{u}_1'\boldsymbol{M}_{11}^{-1}\boldsymbol{u}_1)\boldsymbol{\Sigma}_{22.1}$.

From (4.32), $\boldsymbol{u}_{2.1} \mid \boldsymbol{u}_1, \boldsymbol{X}_1$ is independent of $\boldsymbol{X}_{2.1}'\boldsymbol{b}$. Thus, adding the two terms of $\boldsymbol{z}$ together, we see that

$$\boldsymbol{z} \mid \boldsymbol{u}_1, \boldsymbol{X}_1 \sim N_{p-k}(\boldsymbol{0}, (1 + \boldsymbol{u}_1'\boldsymbol{M}_{11}^{-1}\boldsymbol{u}_1)\boldsymbol{\Sigma}_{22.1}).$$

Let $\boldsymbol{y} = (1 + D_k^2/m)^{-1/2}\boldsymbol{z}$. Since $D_k^2 = m\boldsymbol{u}_1'\boldsymbol{M}_{11}^{-1}\boldsymbol{u}_1$ is a function of $\boldsymbol{u}_1$ and $\boldsymbol{X}_1$ only,

$$\boldsymbol{y} \mid \boldsymbol{u}_1, \boldsymbol{X}_1 \sim N_{p-k}(\boldsymbol{0}, \boldsymbol{\Sigma}_{22.1}).$$

As this conditional distribution does not depend on $(\boldsymbol{u}_1, \boldsymbol{X}_1)$, it is also the marginal distribution. Furthermore, $\boldsymbol{y}$ is independent of $(\boldsymbol{u}_1, \boldsymbol{X}_1)$ and hence independent of D_k^2.

Now, $\boldsymbol{y}$ and D_k^2 are functions of $(\boldsymbol{M}_{11}, \boldsymbol{M}_{12}, \boldsymbol{u})$, so, by (4.30), $\boldsymbol{M}_{22.1}$, $\boldsymbol{y}$ and D_k^2 are jointly independent. Thus, from Theorem 4.5.2,

$$\boldsymbol{y}'\boldsymbol{M}_{22.1}^{-1}\boldsymbol{y} = \frac{D_p^2 - D_k^2}{m + D_k^2} \sim (m-k)^{-1}T^2(p-k, m-k) = \frac{p-k}{m-p+1}F_{p-k,m-p+1},$$

and further, this quantity is independent of D_k^2. □

4.7 Statistics Based on the Wishart Distribution

In univariate analysis, many tests are based on statistics having independent chi-squared distributions. A particular hypothesis may imply, say, that $a \sim \sigma^2\chi_\alpha^2$ and $b \sim \sigma^2\chi_\beta^2$, where a and b are statistics based on the data. If a and b are independent, then, as is well known, a/b is α/β times an $F_{\alpha,\beta}$ variable, and $a/(a+b)$ has a beta distribution with parameters $\alpha/2$ and $\beta/2$. Neither of these functions involves the parameter σ, which in general is unknown.

In the multivariate case, many statistics are based on independent Wishart distributions. Let $\boldsymbol{A} \sim W_p(\boldsymbol{\Sigma}, m)$ be independent of $\boldsymbol{B} \sim W_p(\boldsymbol{\Sigma}, n)$, where $m \geq p$. Since $m \geq p$, $\boldsymbol{A}^{-1}$ exists, and the nonzero eigenvalues of the matrix $\boldsymbol{A}^{-1}\boldsymbol{B}$ are the quantities of interest. Note that since $\boldsymbol{A}^{-1}\boldsymbol{B}$ is similar to the p.s.d. matrix $\boldsymbol{A}^{-1/2}\boldsymbol{B}\boldsymbol{A}^{-1/2}$, all of the nonzero eigenvalues will be positive. Also, with probability 1, the number of nonzero eigenvalues equals $\min(n, p)$. Further, the scale matrix $\boldsymbol{\Sigma}$ has no effect on the distribution of these eigenvalues, so without loss of generality, we may suppose that $\boldsymbol{\Sigma} = \boldsymbol{I}$ (see Exercise 4.7.1).

For convenience, denote the joint distribution of the $\min(n, p)$ nonzero roots of $\boldsymbol{A}^{-1}\boldsymbol{B}$ by $\Psi(p, m, n)$. Then, the following theorem gives an important relationship between the Ψ distributions for different values of the parameters. See also Exercise 4.7.3.

Theorem 4.7.1 *For $m \geq p$ and $n, p \geq 1$, the $\Psi(p, m, n)$ distribution is identical to the $\Psi(n, m+n-p, p)$ distribution.*

Proof: First, note that the number of nonzero eigenvalues is the same for each distribution. Note also that $m \geq p$ implies $m + n - p \geq n$, so the latter distribution makes sense.

Suppose that $n \leq p$. The $\Psi(p, m, n)$ distribution is the joint distribution of the nonzero eigenvalues of $\boldsymbol{A}^{-1}\boldsymbol{B}$, where $\boldsymbol{A} \sim W_p(\boldsymbol{I}, m)$ independently of $\boldsymbol{B} \sim W_p(\boldsymbol{I}, n)$. Write $\boldsymbol{B} = \boldsymbol{X}\boldsymbol{X}'$, where $\boldsymbol{X}(n \times p)$ is a data matrix from $N_p(\boldsymbol{0}, \boldsymbol{I})$ independent of $\boldsymbol{A}$. Because $n \leq p$, $\boldsymbol{X}\boldsymbol{X}'$ is a nonsingular $(n \times n)$ matrix (and so $\boldsymbol{X}\boldsymbol{X}' > 0$) with probability 1. Define

$$\boldsymbol{G} = (\boldsymbol{X}\boldsymbol{X}')^{-1/2}\boldsymbol{X}. \tag{4.33}$$

Then, $\boldsymbol{G}$ is a row orthonormal matrix ($\boldsymbol{G}\boldsymbol{G}' = \boldsymbol{I}_n$) and also $\boldsymbol{X}\boldsymbol{G}'\boldsymbol{G} = \boldsymbol{X}$. Thus,

$$\boldsymbol{A}^{-1}\boldsymbol{B} = \boldsymbol{A}^{-1}\boldsymbol{X}'\boldsymbol{X} = \boldsymbol{A}^{-1}\boldsymbol{G}'\boldsymbol{G}\boldsymbol{X}'\boldsymbol{X}\boldsymbol{G}'\boldsymbol{G},$$

which has the same eigenvalues as

$$(GA^{-1}G')(GX'XG') = C^{-1}D, \quad \text{say.} \tag{4.34}$$

We now show that $C \sim W_n(I, m+n-p)$ independently of $D \sim W_n(I, p)$. The theorem will then follow.

Since G is a function of X, and A is independent of X, we find (see Exercise 4.4.19) that

$$(GA^{-1}G') \mid X = C \mid X \sim W_n(I, m+n-p).$$

Since this distribution does not depend on X, it is also the unconditional distribution, and C is independent of X. Hence, C is also independent of $D = GX'XG'$. Finally, since all np elements of X are independent $N(0, 1)$ random variables, $X'(p \times n)$ can be considered as a data matrix from $N_n(\mathbf{0}, I)$, so that

$$D = GX'XG' = XX' \sim W_n(I, p).$$

Thus, the result is proved when $n \leq p$. If $n > p$, then start with the $\Psi(n, m+n-p, p)$ distribution instead of the $\Psi(p, m, n)$ distribution in the above discussion. □

The following result describes the distribution of a generalization of the F statistic.

Theorem 4.7.2 *If $A \sim W_p(\Sigma, m)$ and $B \sim W_p(\Sigma, n)$ are independent, and if $m \geq p$ and $n \geq p$, then*

$$\phi = |A^{-1}B| = |B|/|A| \tag{4.35}$$

is proportional to the product of p independent F variables of which the ith *has degrees of freedom $(n-i+1)$ and $(m-i+1)$.*

Proof: From Theorem 4.4.14, $|A|$ and $|B|$ are each $|\Sigma|$ times the product of p independent chi-squared variables. Therefore, ϕ is the product of p ratios of independent chi-squared statistics. The ith ratio is $\chi^2_{n-i+1}/\chi^2_{m-i+1}$, i.e. $(n-i+1)/(m-i+1)$ times an $F_{n-i+1,m-i+1}$ statistic. Allowing i to vary from 1 to p, the result follows. □

The multivariate extension of the beta variable will now be defined.

Definition 4.7.3 *When $A \sim W_p(I, m)$ and $B \sim W_p(I, n)$ are independent, $m \geq p$, we say that*

$$\Lambda = |A|/|A+B| = |I + A^{-1}B|^{-1} \sim \Lambda(p, m, n) \tag{4.36}$$

has a Wilks' lambda distribution *with parameters p, m, and n.*

Unfortunately, the notation for this statistic is far from standard. The Λ family of distributions occurs frequently in the context of likelihood ratio tests. The parameter m usually represents the "error" degrees of freedom, and n the "hypothesis" degrees of freedom. Thus, $m+n$ represents the "total" degrees of freedom. Similar to the T^2 statistic, Wilks' lambda distribution is invariant under changes of the scale parameters of A and B (see Exercise 4.7.4). The distribution of Λ is given in the following theorem.

Theorem 4.7.4 *We have*

$$\Lambda(p,m,n) \sim \prod_{r=1}^{n} u_r, \tag{4.37}$$

where $u_1, \dots, u_n$ are n independent variables, and $u_r \sim B\{(m+r-p)/2, p/2\}$, $r = 1, \dots, n$.

Proof: Write $\boldsymbol{B} = \boldsymbol{X}'\boldsymbol{X}$, where the n rows of $\boldsymbol{X}$ are i.i.d. $N_p(\boldsymbol{0}, \boldsymbol{I})$ variables. Let $\boldsymbol{X}_r$ be the $(r \times p)$ matrix consisting of the first r rows of $\boldsymbol{X}$, and let

$$\boldsymbol{M}_r = \boldsymbol{A} + \boldsymbol{X}_r'\boldsymbol{X}_r, \quad r = 1, \dots, n.$$

Note that $\boldsymbol{M}_0 = \boldsymbol{A}$, $\boldsymbol{M}_n = \boldsymbol{A} + \boldsymbol{B}$, and $\boldsymbol{M}_r = \boldsymbol{M}_{r-1} + \boldsymbol{x}_r\boldsymbol{x}_r'$. Now, write

$$\Lambda(p,m,n) = \frac{|\boldsymbol{A}|}{|\boldsymbol{A}+\boldsymbol{B}|} = \frac{|\boldsymbol{M}_0|}{|\boldsymbol{M}_n|} = \frac{|\boldsymbol{M}_0||\boldsymbol{M}_1|}{|\boldsymbol{M}_1||\boldsymbol{M}_2|} \cdots \frac{|\boldsymbol{M}_{n-1}|}{|\boldsymbol{M}_n|}.$$

This product may be written as $u_1 u_2 \cdots u_n$, where

$$u_r = \frac{|\boldsymbol{M}_{r-1}|}{|\boldsymbol{M}_r|}, \qquad r = 1, \dots, n.$$

Now, $\boldsymbol{M}_r = \boldsymbol{M}_{r-1} + \boldsymbol{x}_r\boldsymbol{x}_r'$, and therefore, by Corollary 4.5.7, with $\boldsymbol{M}_{r-1}$ corresponding to $\boldsymbol{M}$ and $\boldsymbol{x}_r$ corresponding to $\boldsymbol{d}$,

$$u_r \sim B\{(m+r-p)/2, p/2\}, \quad r = 1, \dots, n.$$

It remains to be shown that the u_r are statistically independent. From Theorem 4.5.9, $\boldsymbol{M}_r$ is independent of

$$1 + \boldsymbol{x}_r'\boldsymbol{M}_{r-1}^{-1}\boldsymbol{x}_r = |\boldsymbol{M}_r|/|\boldsymbol{M}_{r-1}| = u_r^{-1}.$$

Since u_r is independent of $\boldsymbol{x}_{r+1}, \dots, \boldsymbol{x}_n$, and

$$\boldsymbol{M}_{r+s} = \boldsymbol{M}_r + \sum_{k=1}^{s} \boldsymbol{x}_{r+k}\boldsymbol{x}_{r+k}',$$

it follows that u_r is also independent of $\boldsymbol{M}_{r+1}, \boldsymbol{M}_{r+2}, \dots, \boldsymbol{M}_n$ and hence independent of $u_{r+1}, \dots, u_n$. The result follows. □

Theorem 4.7.5 *The $\Lambda(p,m,n)$ and $\Lambda(n, m+n-p, p)$ distributions are the same.*

Proof: Let $\lambda_1 \geq \cdots \geq \lambda_p$ denote the eigenvalues of $\boldsymbol{A}^{-1}\boldsymbol{B}$ in Definition 4.7.3, and let $k = \min(n,p)$ denote the number of nonzero such eigenvalues. Then, using Section A.6, we can write

$$\Lambda(p,m,n) = |\boldsymbol{I} + \boldsymbol{A}^{-1}\boldsymbol{B}|^{-1} = \prod_{i=1}^{p}(1+\lambda_i)^{-1} = \prod_{i=1}^{k}(1+\lambda_i)^{-1}. \tag{4.38}$$

Thus, Λ is a function of the nonzero eigenvalues of $\boldsymbol{A}^{-1}\boldsymbol{B}$, so the result follows from Theorem 4.7.1. □

Special cases of these results are as follows:

(a) The statistics $\Lambda(p,m,1)$ and $\Lambda(1, m+1-p, p)$ are equivalent, and each corresponds to a single $B\{(m-p+1)/2, p/2\}$ statistic.
(b) $\Lambda(p,m,2)$ and $\Lambda(2, m+2-p, p)$ are equivalent and correspond to the product of a $B\{(m-p+1)/2, p/2\}$ statistic with an independent $B\{(m-p+2)/2, p/2\}$ statistic.

From the relationship between β and F variables, functions of $\Lambda(p, m, 1)$ and $\Lambda(p, m, 2)$ statistics can also be expressed in terms of the F distribution as follows:

$$\frac{1 - \Lambda(p, m, 1)}{\Lambda(p, m, 1)} \sim \frac{p}{m - p + 1} F_{p, m-p+1}, \tag{4.39}$$

$$\frac{1 - \Lambda(1, m, n)}{\Lambda(1, m, n)} \sim \frac{n}{m} F_{n,m}, \tag{4.40}$$

$$\frac{1 - \sqrt{\Lambda(p, m, 2)}}{\sqrt{\Lambda(p, m, 2)}} \sim \frac{p}{m - p + 1} F_{2p, 2(m-p+1)}, \tag{4.41}$$

$$\frac{1 - \sqrt{\Lambda(2, m, n)}}{\sqrt{\Lambda(2, m, n)}} \sim \frac{n}{m - 1} F_{2n, 2(m-1)}. \tag{4.42}$$

Formulae (4.39) and (4.40) are easy to verify (see Exercise 4.7.5), but (4.41) and (4.42) are more complicated (see Anderson (1958, pp. 195–196)).

For other values of n and p, provided m is large, we may use Bartlett's approximation:

$$-\{m - \frac{1}{2}(p - n + 1)\} \log \Lambda(p, m, n) \sim \chi^2_{np} \tag{4.43}$$

asymptotically as $m \to \infty$. Pearson and Hartley (1972, p. 333) tabulate values of a constant factor $C(p, n, m - p + 1)$, which improves the approximation.

An approximation that uses the F distribution with noninteger degrees of freedom is discussed in Mardia and Zemroch (1978). Another approximation based on Theorem 4.7.4 is given in Exercise 4.7.6.

The Λ statistic arises naturally in likelihood ratio tests, and this property explains the limiting χ^2 distribution in (4.43). Another important statistic in hypothesis testing is the greatest root statistic defined below:

Definition 4.7.6 *Let $\boldsymbol{A} \sim W_p(\boldsymbol{I}, m)$ be independent of $\boldsymbol{B} \sim W_p(\boldsymbol{I}, n)$, where $m \geq p$. Then, the largest eigenvalue θ of $(\boldsymbol{A} + \boldsymbol{B})^{-1}\boldsymbol{B}$ is called the greatest root statistic, and its distribution is denoted $\theta(p, m, n)$.*

Note that $\theta(p, m, n)$ can also be defined as the largest root of the determinantal equation

$$|\boldsymbol{B} - \theta(\boldsymbol{A} + \boldsymbol{B})| = 0.$$

If λ is an eigenvalue of $\boldsymbol{A}^{-1}\boldsymbol{B}$, then $\lambda/(1 + \lambda)$ is an eigenvalue of $(\boldsymbol{A} + \boldsymbol{B})^{-1}\boldsymbol{B}$ (Exercise 4.7.8). Since this is a monotone function of λ, θ is given by

$$\theta = \lambda_1/(1 + \lambda_1), \tag{4.44}$$

where λ_1 denotes the largest eigenvalue of $\boldsymbol{A}^{-1}\boldsymbol{B}$. Since $\lambda_1 > 0$, we see that $0 < \theta < 1$.

Using Theorem 4.7.1 and (4.38), we easily get the following properties:

(1) $\theta(p, m, n)$ and $\theta(n, m + n - p, p)$ have the same distribution; (4.45)

(2)
$$\frac{\theta(1, m, n)}{1 - \theta(1, m, n)} = \frac{1 - \Lambda(1, m, n)}{\Lambda(1, m, n)} \sim \frac{n}{m} F_{n,m}; \tag{4.46}$$

(3)
$$\frac{\theta(p, m, 1)}{1 - \theta(p, m, 1)} = \frac{1 - \Lambda(p, m, 1)}{\Lambda(p, m, 1)} \sim \frac{p}{m - p + 1} F_{p, m-p+1}. \tag{4.47}$$

For $p \geq 2$, critical values of the θ statistic must be found from tables. Table D.7 in Appendix D gives upper percentage points for $p = 2$. Pearson and Hartley (1972, pp. 98–104, 336–350) give critical values for general values of p. Of course, the relation (4.45) can be used to extend the tables. Efficient computational expressions, based on Pfaffian representations, are derived by Gupta and Richard (1985), which has led to numerical algorithms (Butler and Paige, 2011), further developed for arbitrary p, n, m in an iterative approach (Chiani, 2016). This is now available in the R library `rootWishart`.

As above, note that $p =$ dimension, $m =$ "error" degrees of freedom, and $n =$ "hypothesis" degrees of freedom (ν_1 and ν_2, respectively, in Table D.7).

4.8 Other Distributions Related to the Multivariate Normal

Wishart Distribution The Wishart distribution $W_p(\boldsymbol{\Sigma}, m)$ has already been described. For reference, its p.d.f. (when $\boldsymbol{\Sigma} > 0$ and $m \geq p$) is given by

$$f(\boldsymbol{M}) = \frac{|\boldsymbol{M}|^{(m-p-1)/2} \exp\left(-\frac{1}{2}\mathrm{tr}\boldsymbol{\Sigma}^{-1}\boldsymbol{M}\right)}{2^{mp/2}\pi^{p(p-1)/4}|\boldsymbol{\Sigma}|^{m/2}\prod_{i=1}^{p}\Gamma\left(\frac{1}{2}(m+1-i)\right)}, \tag{4.48}$$

with respect to Lebesgue measure $\prod_{i \leq j} dm_{ij}$ in $\mathbb{R}^{p(p+1)/2}$ restricted to the set where $\boldsymbol{M} > 0$ (see Anderson (1958, p. 154)).

Inverted Wishart Distribution (Siskind, 1972) If $\boldsymbol{M} \sim W_p(\boldsymbol{\Sigma}, m)$, where $\boldsymbol{\Sigma} > 0$ and $m \geq p$, then $\boldsymbol{U} = \boldsymbol{M}^{-1}$ is said to have an inverted Wishart distribution $W_p^{-1}(\boldsymbol{\Sigma}, m)$. Using the Jacobian from Table 2.1, we see that its p.d.f. is

$$g(\boldsymbol{U}) = \frac{|\boldsymbol{U}|^{-(m+p+1)/2} \exp\left(-\frac{1}{2}\mathrm{tr}\boldsymbol{\Sigma}^{-1}\boldsymbol{U}^{-1}\right)}{2^{mp/2}\pi^{p(p-1)/4}|\boldsymbol{\Sigma}|^{m/2}\prod_{i=1}^{p}\Gamma\left(\frac{1}{2}(m+1-i)\right)}. \tag{4.49}$$

The expected value of $\boldsymbol{U}$ is given (see Exercise 4.4.13) by

$$\mathrm{E}(\boldsymbol{U}) = \boldsymbol{\Sigma}^{-1}/(m-p-1). \tag{4.50}$$

Complex Multivariate Normal Distribution (Goodman, 1963; Khatri, 1965; Wooding, 1956) Let $\boldsymbol{z} = (\boldsymbol{x}', \boldsymbol{y})' \sim N_{2p}((\boldsymbol{\mu}_1', \boldsymbol{\mu}_2')', \boldsymbol{\Sigma})$, where $\boldsymbol{\Sigma}_{11} = \boldsymbol{\Sigma}_{22}$ and $\boldsymbol{\Sigma}_{12}(p \times p)$ is a skew-symmetric matrix ($\boldsymbol{\Sigma}_{12} = -\boldsymbol{\Sigma}_{21}$). Then, the distribution of $\boldsymbol{w} = \boldsymbol{x} + i\boldsymbol{y}$ is known as complex multivariate normal.

Noncentral Wishart Distribution (James, 1964) Let $\boldsymbol{X}$ be a data matrix from $N_p(\boldsymbol{\mu}, \boldsymbol{\Sigma})$, $\boldsymbol{\mu} \neq \boldsymbol{0}$. Then, $\boldsymbol{M} = \boldsymbol{X}'\boldsymbol{X}$ has a noncentral Wishart distribution.

Matrix T Distribution (Dickey, 1967; Kshirsagar, 1960) Let $\boldsymbol{X}(n \times p)$ be a data matrix from $N_p(\boldsymbol{0}, \boldsymbol{Q})$, which is independent of $\boldsymbol{M} \sim W_n(\boldsymbol{P}, \nu)$. Then, $\boldsymbol{T} = \boldsymbol{X}'\boldsymbol{M}^{-1/2}$ has the matrix T distribution.

Matrix Beta Type I Distribution (Khatri and Pillai, 1965; Kshirsagar, 1961; Mitra, 1969) Let $\boldsymbol{M}_i \sim W_p(\boldsymbol{\Sigma}, \nu_i)$, $i = 1, 2$. Then, $(\boldsymbol{M}_1 + \boldsymbol{M}_2)^{-1/2}\boldsymbol{M}_1(\boldsymbol{M}_1 + \boldsymbol{M}_2)^{-1/2}$ has the matrix beta type I distribution.

Matrix Beta Type II Distribution (Khatri and Pillai, 1965; Kshirsagar, 1960) Let $\boldsymbol{M}_i \sim W_p(\boldsymbol{\Sigma}, \nu_i)$, $i = 1, 2$. Then, $\boldsymbol{M}_2^{-1/2}\boldsymbol{M}_1\boldsymbol{M}_2^{-1/2}$ has the matrix beta type II distribution.

Exercises and Complements

4.2.1 If the rows of $\boldsymbol{X}$ are i.i.d. $N_p(\boldsymbol{0}, \boldsymbol{I})$, then using Theorem 3.3.2 show that $\text{tr}\boldsymbol{X}'\boldsymbol{X}$ is independent of any scale-invariant function of $\boldsymbol{X}$.

4.2.2 If $\boldsymbol{x} \sim N_p(\boldsymbol{\mu}, \boldsymbol{\Sigma})$, show that $\boldsymbol{x}$ and $\boldsymbol{Gx}$ have the same distribution for all orthogonal matrices $\boldsymbol{G}$ if and only if $\boldsymbol{\mu} = \boldsymbol{0}$ and $\boldsymbol{\Sigma} = \sigma^2\boldsymbol{I}$. How does this relate to the property of spherical symmetry?

4.2.3 If $\boldsymbol{x} \sim N(\boldsymbol{0}, \sigma^2\boldsymbol{I})$, show that $\boldsymbol{Ax}$ and $(\boldsymbol{I} - \boldsymbol{A}^-\boldsymbol{A})\boldsymbol{x}$, where $\boldsymbol{A}^-$ is a generalized inverse satisfying $\boldsymbol{AA}^-\boldsymbol{A} = \boldsymbol{A}$, are independent, and each has a normal distribution.

4.2.4 (a) If $\boldsymbol{x} \sim N_p(\boldsymbol{\mu}, \boldsymbol{\Sigma})$, and $\boldsymbol{a}$ is any fixed vector, show that

$$f = \frac{\boldsymbol{a}'(\boldsymbol{x} - \boldsymbol{\mu})}{\sqrt{\boldsymbol{a}'\boldsymbol{\Sigma}\boldsymbol{a}}} \sim N(0, 1).$$

(b) If $\boldsymbol{a}$ is now a random vector independent of $\boldsymbol{x}$ for which $\mathsf{P}(\boldsymbol{a}'\boldsymbol{\Sigma}\boldsymbol{a} = 0) = 0$, show that $f \sim N(0, 1)$ and is independent of $\boldsymbol{a}$.
(c) Hence, show that if $\boldsymbol{x} \sim N_3(\boldsymbol{0}, \boldsymbol{I})$, then

$$\frac{x_1 e^{x_3} + x_2 \log|x_3|}{[e^{2x_3} + (\log|x_3|)^2]^{1/2}} \sim N(0, 1).$$

4.2.5 (a) The ordinary least-squares coefficient for a line passing through the origin is given by $b = \sum x_r y_r / \sum x_r^2 = \boldsymbol{x}'\boldsymbol{y}/\boldsymbol{x}'\boldsymbol{x}$. If $\boldsymbol{y} \sim N_n(\boldsymbol{0}, \boldsymbol{I})$, and if $\boldsymbol{x}$ is statistically independent of $\boldsymbol{y}$, then show that

$$\alpha = \boldsymbol{x}'\boldsymbol{y}/(\boldsymbol{x}'\boldsymbol{x})^{1/2} \sim N(0, 1)$$

and is independent of $\boldsymbol{x}$.
(b) Suppose that $\boldsymbol{x} \sim N_n(\boldsymbol{0}, \boldsymbol{I})$. Since $b = \alpha/(\boldsymbol{x}'\boldsymbol{x})^{1/2}$ where $\alpha \sim N(0, 1)$ and is independent of $\boldsymbol{x}$, and since $\boldsymbol{x}'\boldsymbol{x} \sim \chi_n^2$, deduce that $n^{1/2}b \sim t_n$. This is the null distribution of the ordinary least-squares coefficient for the bivariate normal model.

4.2.6 Using the covariance matrix,

$$\boldsymbol{\Sigma} = \begin{bmatrix} 1 & \rho & \rho^2 \\ & 1 & 0 \\ & & 1 \end{bmatrix},$$

show that the conditional distribution of (x_1, x_2), given x_3, has mean vector $[\mu_1 + \rho^2(x_3 - \mu_3), \mu_2]'$ and covariance matrix

$$\begin{bmatrix} 1 - \rho^4 & \rho \\ \rho & 1 \end{bmatrix}.$$

4.2.7 If $\boldsymbol{x}_1, \boldsymbol{x}_2$, and $\boldsymbol{x}_3$ are i.i.d. random variables, and if $\boldsymbol{y}_1 = \boldsymbol{x}_1 + \boldsymbol{x}_2, \boldsymbol{y}_2 = \boldsymbol{x}_2 + \boldsymbol{x}_3$, and $\boldsymbol{y}_3 = \boldsymbol{x}_1 + \boldsymbol{x}_3$, then obtain the conditional distribution of $\boldsymbol{y}_1$, given $\boldsymbol{y}_2$, and of $\boldsymbol{y}_1$, given $\boldsymbol{y}_2$ and $\boldsymbol{y}_3$.

4.3.1 If $\boldsymbol{x} \sim N_p(\boldsymbol{\mu}, \boldsymbol{\Sigma})$ and $\boldsymbol{Q\Sigma Q}'(q \times q)$ is nonsingular, then, given that $\boldsymbol{Qx} = \boldsymbol{q}$, show that the conditional distribution of $\boldsymbol{x}$ is normal with mean $\boldsymbol{\mu} + \boldsymbol{\Sigma Q}'(\boldsymbol{Q\Sigma Q}')^{-1}(\boldsymbol{q} - \boldsymbol{Q\mu})$ and (singular) covariance matrix $\boldsymbol{\Sigma} - \boldsymbol{\Sigma Q}'(\boldsymbol{Q\Sigma Q}')^{-1}\boldsymbol{Q\Sigma}$.

4.3.2 If $\boldsymbol{X}_{2.1}$ is as defined in Corollary 4.3.4, and if $\boldsymbol{Q} = \boldsymbol{X}_1(\boldsymbol{X}_1'\boldsymbol{X}_1)^{-1}\boldsymbol{X}_1'$, show that

(i) $\boldsymbol{QX}_{2.1} = \boldsymbol{QX}_2 - \boldsymbol{X}_1\boldsymbol{\Sigma}_{11}^{-1}\boldsymbol{\Sigma}_{12}$,

(ii) $(\boldsymbol{I} - \boldsymbol{Q})\boldsymbol{X}_{2.1} = (\boldsymbol{I} - \boldsymbol{Q})\boldsymbol{X}_2$.

Hence, prove that

(iii) $\boldsymbol{X}_{2.1}'(\boldsymbol{I} - \boldsymbol{Q})\boldsymbol{X}_{2.1} = \boldsymbol{X}_2'(\boldsymbol{I} - \boldsymbol{Q})\boldsymbol{X}_2$,

(iv) $\boldsymbol{X}_{2.1}'\boldsymbol{QX}_{2.1} = \boldsymbol{X}_2'\boldsymbol{QX}_2 - \boldsymbol{\Sigma}_{21}\boldsymbol{\Sigma}_{11}^{-1}\boldsymbol{M}_{12} - \boldsymbol{M}_{21}\boldsymbol{\Sigma}_{11}^{-1}\boldsymbol{\Sigma}_{12} + \boldsymbol{\Sigma}_{21}\boldsymbol{\Sigma}_{11}^{-1}\boldsymbol{M}_{11}\boldsymbol{\Sigma}_{11}^{-1}\boldsymbol{\Sigma}_{12}$, where $\boldsymbol{M}_{ij} = \boldsymbol{X}_i'\boldsymbol{X}_j$.

(see also Exercise 4.4.15.)

4.3.3 Suppose that $\boldsymbol{x} \sim (\boldsymbol{\mu}, \boldsymbol{\Sigma})$, and $\boldsymbol{a}$ is a fixed vector. If r_i is the correlation between x_i and $\boldsymbol{a}'\boldsymbol{x}$, show that $\boldsymbol{r} = (c\boldsymbol{D})^{-1/2}\boldsymbol{\Sigma a}$, where $c = \boldsymbol{a}'\boldsymbol{\Sigma a}$ and $\boldsymbol{D} = \text{Diag}(\boldsymbol{\Sigma})$. When does $\boldsymbol{r} = \boldsymbol{\Sigma a}$?

4.3.4 (Proof of Theorem 4.3.2) (a) Let $\boldsymbol{X}^V$ be the np-vector obtained by stacking the columns of $\boldsymbol{X}$ on top of one another (Section A.2.5). Then, $\boldsymbol{X}(n \times p)$ is a random data matrix from $N_p(\boldsymbol{\mu}, \boldsymbol{\Sigma})$ if and only if

$$\boldsymbol{X}^V \sim N_{np}(\boldsymbol{\mu} \otimes \boldsymbol{1}, \boldsymbol{\Sigma} \otimes \boldsymbol{I}),$$

where $\otimes$ denotes Kronecker multiplication (see Section A.2.5).

(b) Using the fact that

$$(\boldsymbol{AXB})^V = (\boldsymbol{B}' \otimes \boldsymbol{A})\boldsymbol{X}^V, \qquad (*)$$

we deduce that $(\boldsymbol{AXB})^V$ is multivariate normal with mean

$$(\boldsymbol{B}' \otimes \boldsymbol{A})(\boldsymbol{\mu} \otimes \boldsymbol{1}) = \boldsymbol{B}'\boldsymbol{\mu} \otimes \boldsymbol{A1}$$

and with covariance matrix

$$(\boldsymbol{B}' \otimes \boldsymbol{A})(\boldsymbol{\Sigma} \otimes \boldsymbol{I})(\boldsymbol{B}' \otimes \boldsymbol{A})' = \boldsymbol{B}'\boldsymbol{\Sigma B} \otimes \boldsymbol{AA}'.$$

(c) Therefore, $\boldsymbol{AXB}$ is a normal data matrix if and only if

(i) $\boldsymbol{A1} = \alpha\boldsymbol{1}$ for some scalar α or $\boldsymbol{B}'\boldsymbol{\mu} = \boldsymbol{0}$, and

(ii) $\boldsymbol{AA}' = \beta\boldsymbol{I}$ for some scalar β or $\boldsymbol{B}'\boldsymbol{\Sigma B} = \boldsymbol{0}$.

(d) If both the above conditions are satisfied then $\boldsymbol{AXB}$ is a random data matrix from $N_q(\alpha\boldsymbol{B}'\boldsymbol{\mu}, \beta\boldsymbol{B}'\boldsymbol{\Sigma B})$.

4.3.5 (Proof of Theorem 4.3.3) Suppose that $\boldsymbol{X}(n \times p)$ is a data matrix from $N_p(\boldsymbol{\mu}, \boldsymbol{\Sigma})$, and that $\boldsymbol{Y} = \boldsymbol{AXB}$ and $\boldsymbol{Z} = \boldsymbol{CXD}$. Then, using (*) from Exercise 4.3.4, show that

$$\boldsymbol{Y}^V\boldsymbol{Z}^{V\prime} = (\boldsymbol{B}' \otimes \boldsymbol{A})\boldsymbol{X}^V\boldsymbol{X}^{V\prime}(\boldsymbol{D}' \otimes \boldsymbol{C})'.$$

But using Exercise 4.3.4, we know that $\text{var}(\boldsymbol{X}^V) = \boldsymbol{\Sigma} \otimes \boldsymbol{I}$. Therefore

$$\text{cov}[\boldsymbol{Y}^V, \boldsymbol{Z}^V] = (\boldsymbol{B}' \otimes \boldsymbol{A})(\boldsymbol{\Sigma} \otimes \boldsymbol{I})(\boldsymbol{D} \otimes \boldsymbol{C}') = \boldsymbol{B}'\boldsymbol{\Sigma}\boldsymbol{D} \otimes \boldsymbol{A}\boldsymbol{C}'.$$

The elements of $\boldsymbol{Y}$ and $\boldsymbol{Z}$ are uncorrelated if and only if the above matrix is the zero matrix, i.e. if and only if either $\boldsymbol{B}'\boldsymbol{\Sigma}\boldsymbol{D} = \boldsymbol{0}$ or $\boldsymbol{A}\boldsymbol{C}' = \boldsymbol{0}$.

4.4.1 If $\boldsymbol{M} \sim W_p(\boldsymbol{\Sigma}, m)$, and $\boldsymbol{a}$ is any random p-vector that satisfies $\boldsymbol{a}'\boldsymbol{\Sigma}\boldsymbol{a} \neq 0$ with probability 1 and is independent of $\boldsymbol{M}$, then $\boldsymbol{a}'\boldsymbol{M}\boldsymbol{a}/\boldsymbol{a}'\boldsymbol{\Sigma}\boldsymbol{a}$ has the χ^2_m distribution and is independent of $\boldsymbol{a}$.

4.4.2 If $\boldsymbol{M} \sim W_p(\boldsymbol{\Sigma}, m)$, show that $\boldsymbol{b}'\boldsymbol{M}\boldsymbol{b}$ and $\boldsymbol{d}'\boldsymbol{M}\boldsymbol{d}$ are statistically independent if $\boldsymbol{b}'\boldsymbol{\Sigma}\boldsymbol{d} = 0$. (Hint: use Theorem 4.3.3.) Hence, show that m_{ii} and m_{jj} are independent if $\sigma_{ij} = 0$, and that when $\boldsymbol{\Sigma} = \boldsymbol{I}$, $\text{tr}\boldsymbol{M}$ has the χ^2_{mp} distribution. Give an alternative proof of this result that follows directly from the representation of $\boldsymbol{M}$ as $\boldsymbol{X}'\boldsymbol{X}$.

4.4.3 (Mitra, 1969) Show that the following conditions taken together are necessary (and sufficient) for $\boldsymbol{M}$ to have the $W_p(\boldsymbol{\Sigma}, m)$ distribution:
(a) $\boldsymbol{M}$ is symmetric, and if $\boldsymbol{a}'\boldsymbol{\Sigma}\boldsymbol{a} = 0$, then $\boldsymbol{a}'\boldsymbol{M}\boldsymbol{a} = 0$ with probability 1;
(b) for every $(q \times p)$ matrix $\boldsymbol{L}$ that satisfies $\boldsymbol{L}\boldsymbol{\Sigma}\boldsymbol{L}' = \boldsymbol{I}$, the diagonal elements of $\boldsymbol{L}\boldsymbol{M}\boldsymbol{L}'$ are independent χ^2_m variables.

4.4.4 (Mitra, 1969) The converse of Theorem 4.4.6 does not hold. That is, if $\boldsymbol{a}'\boldsymbol{T}\boldsymbol{a}/\boldsymbol{a}'\boldsymbol{\Sigma}\boldsymbol{a} \sim \chi^2_f$ for all $\boldsymbol{a}$, then $\boldsymbol{T}$ does not necessarily have a Wishart distribution.
(a) *Construction.* Consider $\boldsymbol{T} = \alpha\boldsymbol{M}$, where $\boldsymbol{M} \sim W_p(\boldsymbol{\Sigma}, n)$, $\alpha \sim B(f/2, (n-f)/2)$, and α and $\boldsymbol{M}$ are independent. From Theorem 4.4.6, $\boldsymbol{a}'\boldsymbol{T}\boldsymbol{a}/\boldsymbol{a}'\boldsymbol{\Sigma}\boldsymbol{a}$ is the product of independent $B(f/2, (n-f)/2)$ and χ^2_n variables. Hence, using the hint, show that $\boldsymbol{a}'\boldsymbol{T}\boldsymbol{a}/\boldsymbol{a}'\boldsymbol{\Sigma}\boldsymbol{a} \sim \chi^2_f$. Thus, $\boldsymbol{T} = \alpha\boldsymbol{M}$ satisfies the required property. (Hint: If $x_1, \ldots, x_n$ are i.i.d. $N(0, 1)$ variables, then using the fact that $\boldsymbol{x}'\boldsymbol{x}$ is independent of any scale-invariant function (see Exercise 4.2.1) note that

$$\sum_{r=1}^{f} x_r^2 \Big/ \sum_{r=1}^{n} x_r^2 \quad \text{and} \quad \sum_{r=1}^{n} x_r^2$$

are independent variables with $B(f/2, (n-f)/2)$ and χ^2_n distributions, respectively. Hence, the product of the above variables is $\sum_{r=1}^{f} x_r^2$, which has a χ^2_f distribution.)
(b) *Contradiction.* But $\boldsymbol{T}$ cannot have a Wishart distribution. For if it does, it must have f degrees of freedom. In that case, $r_{ij} = t_{ij}/(t_{ii}t_{jj})^{1/2}$ would have the distribution of a sample correlation coefficient based on a normal sample of size $(f+1)$. But r_{ij} is also $m_{ij}/(m_{ii}m_{jj})^{1/2}$, which has the distribution of a sample correlation coefficient based on a normal sample of size $(n+1)$. Hence, we have a contradiction, and $\boldsymbol{T}$ cannot have a Wishart distribution.

4.4.5 (a) If the rows of $\boldsymbol{X}$ are i.i.d. $N_p(\boldsymbol{\mu}, \boldsymbol{\Sigma})$, and $\boldsymbol{Y} = \boldsymbol{X} - \boldsymbol{1}\boldsymbol{\mu}'$, then the rows of $\boldsymbol{Y}$ are i.i.d. $N_p(\boldsymbol{0}, \boldsymbol{\Sigma})$. Now

$$\boldsymbol{X}'\boldsymbol{C}\boldsymbol{X} = \boldsymbol{Y}'\boldsymbol{C}\boldsymbol{Y} + \boldsymbol{Y}'\boldsymbol{C}\boldsymbol{1}\boldsymbol{\mu}' + \boldsymbol{\mu}\boldsymbol{1}'\boldsymbol{C}\boldsymbol{Y} + (\boldsymbol{1}'\boldsymbol{C}\boldsymbol{1})\boldsymbol{\mu}\boldsymbol{\mu}'.$$

If $\boldsymbol{C}$ is symmetric and idempotent, then $\boldsymbol{Y}'\boldsymbol{C}\boldsymbol{Y}$ has a Wishart distribution by virtue of Theorem 4.4.9, and $\boldsymbol{X}'\boldsymbol{C}\boldsymbol{X}$ is the sum of a Wishart distribution, two nonindependent normal distributions, and a constant. However, a special case arises when the rows of $\boldsymbol{C}$ sum to zero, since then $\boldsymbol{C}\boldsymbol{1} = \boldsymbol{0}$, and the final three terms above are all zero. Thus, we have the following result that generalizes Theorem 4.4.9.
(b) If the rows of $\boldsymbol{X}$ are i.i.d. $N_p(\boldsymbol{\mu}, \boldsymbol{\Sigma})$, and $\boldsymbol{C}$ is a symmetric matrix, then $\boldsymbol{X}'\boldsymbol{C}\boldsymbol{X}$ has a $W_p(\boldsymbol{\Sigma}, r)$ distribution if (and only if): (i) $\boldsymbol{C}$ is idempotent; and either (ii) $\boldsymbol{\mu} = \boldsymbol{0}$ or (iii) $\boldsymbol{C}\boldsymbol{1} = \boldsymbol{0}$. In either case, $r = \operatorname{tr} \boldsymbol{C}$.
(c) (Proof of $\boldsymbol{X}'\boldsymbol{H}\boldsymbol{X} \sim W_p(\boldsymbol{\Sigma}, n-1)$.) As a corollary to the above, we deduce that if the rows of $\boldsymbol{X}(n \times p)$ are i.i.d. $N_p(\boldsymbol{\mu}, \boldsymbol{\Sigma})$, then $\boldsymbol{X}'\boldsymbol{H}\boldsymbol{X} \sim W_p(\boldsymbol{\Sigma}, n-1)$, and $\boldsymbol{S} = n^{-1}\boldsymbol{X}'\boldsymbol{H}\boldsymbol{X} \sim W_p(n^{-1}\boldsymbol{\Sigma}, n-1)$ because the centering matrix $\boldsymbol{H}$ is idempotent of rank $(n-1)$, and all its row sums are zero.

4.4.6 (a) Let $\boldsymbol{C}_1, \dots, \boldsymbol{C}_k$ be $(n \times n)$ symmetric idempotent matrices such that $\boldsymbol{C}_1 + \cdots + \boldsymbol{C}_k = \boldsymbol{I}$, and let $\boldsymbol{X}(n \times p)$ be a data matrix from $N_p(\boldsymbol{\mu}, \boldsymbol{\Sigma})$. Show that $\boldsymbol{C}_i\boldsymbol{C}_j = \boldsymbol{0}$ for $i \neq j$. Hence, if $\boldsymbol{M}_i = \boldsymbol{X}'\boldsymbol{C}_i\boldsymbol{X}$, deduce that $\boldsymbol{X}'\boldsymbol{X} = \boldsymbol{M}_1 + \cdots + \boldsymbol{M}_k$ is a decomposition of $\boldsymbol{X}'\boldsymbol{X}$ into a sum of independent matrices.
(b) If $\boldsymbol{\mu} = \boldsymbol{0}$, show that $\boldsymbol{M}_i \sim W_p(\boldsymbol{\Sigma}, r_i)$ for $i = 1, \dots, k$, where $r_i = \operatorname{rank}(\boldsymbol{C}_i)$.
(c) For general $\boldsymbol{\mu}$, if $\boldsymbol{C}_1 = n^{-1}\boldsymbol{1}\boldsymbol{1}'$, use Exercise 4.4.5 to show that $\boldsymbol{M}_i \sim W_p(\boldsymbol{\Sigma}, r_i)$ for $i = 2, \dots, k$.

4.4.7 If the rows of $\boldsymbol{X}$ are i.i.d. $N_p(\boldsymbol{\mu}, \boldsymbol{\Sigma})$, and if $\boldsymbol{C}$ is a symmetric $(n \times n)$ matrix, then $\boldsymbol{X}'\boldsymbol{C}\boldsymbol{X}$ and $\boldsymbol{A}\boldsymbol{X}\boldsymbol{B}$ are independent if either $\boldsymbol{B}'\boldsymbol{\Sigma} = \boldsymbol{0}$ or $\boldsymbol{A}\boldsymbol{C} = \boldsymbol{0}$.
(Hint: As in (4.5), we have $\boldsymbol{X}'\boldsymbol{C}\boldsymbol{X} = \sum \lambda_i \boldsymbol{y}_i \boldsymbol{y}_i'$, where $\boldsymbol{y}_i = \boldsymbol{X}'\boldsymbol{\gamma}_{(i)}$, and $\boldsymbol{C} = \sum \lambda_i \boldsymbol{\gamma}_{(i)}\boldsymbol{\gamma}_{(i)}'$. Now, by Theorem 4.3.3, $\boldsymbol{A}\boldsymbol{X}\boldsymbol{B}$ is independent of $\lambda_i^{1/2}\boldsymbol{\gamma}_{(i)}'\boldsymbol{X}$ if and only if either $\boldsymbol{B}'\boldsymbol{\Sigma} = \boldsymbol{0}$ or $\lambda_i^{1/2}\boldsymbol{A}\boldsymbol{\gamma}_{(i)} = \boldsymbol{0}$. The second condition holds for all i if and only if $\boldsymbol{A}\boldsymbol{C} = \boldsymbol{0}$. Hence, the result follows.)

4.4.8 If the rows of $\boldsymbol{X}$ are i.i.d., and $n\boldsymbol{S} = \boldsymbol{X}'\boldsymbol{H}\boldsymbol{X}$ is statistically independent of $n\bar{\boldsymbol{x}} = \boldsymbol{X}'\boldsymbol{1}$, then the rows of $\boldsymbol{X}$ must have a multivariate normal distribution (Kagan et al., 1973). However, note that $\bar{\boldsymbol{x}}$ is not independent of $\boldsymbol{X}'\boldsymbol{X}$.

4.4.9 If the rows of $\boldsymbol{X}(n \times p)$ are i.i.d. $N_p(\boldsymbol{0}, \boldsymbol{I})$, and if $\boldsymbol{C}$ and $\boldsymbol{D}$ are symmetric $(p \times p)$ matrices, then the following results may be derived from Theorems 4.4.9 and 4.4.10:
(a) $\boldsymbol{X}\boldsymbol{C}\boldsymbol{X}' \sim W_n(\boldsymbol{I}, r)$ if and only if $\boldsymbol{C}$ is idempotent, and $r = \operatorname{tr} \boldsymbol{C}$;
(b) $\boldsymbol{X}\boldsymbol{C}\boldsymbol{X}'$ and $\boldsymbol{X}\boldsymbol{D}\boldsymbol{X}'$ are independent if and only if $\boldsymbol{C}\boldsymbol{D} = \boldsymbol{0}$;
(c) $\boldsymbol{X}\boldsymbol{C}\boldsymbol{X}'$ and $\boldsymbol{A}\boldsymbol{X}'\boldsymbol{B}$ are independent if (and only if) $\boldsymbol{A}\boldsymbol{C} = \boldsymbol{0}$ or $\boldsymbol{B} = \boldsymbol{0}$.
(Hint: the columns of $\boldsymbol{X}$ are the rows of $\boldsymbol{X}'$, and these also are i.i.d.)

4.4.10 Extend the results of Exercise 4.4.9 to the case of general $\boldsymbol{\Sigma} > 0$ and show that if the rows of $\boldsymbol{X}$ are i.i.d. $N_p(\boldsymbol{0}, \boldsymbol{\Sigma})$, then:
(a) $\boldsymbol{X}\boldsymbol{C}\boldsymbol{X}' \sim W_n(\boldsymbol{I}, r)$ if and only if $\boldsymbol{C}\boldsymbol{\Sigma}\boldsymbol{C} = \boldsymbol{C}$, in which case $r = \operatorname{tr} \boldsymbol{C}\boldsymbol{\Sigma}$;
(b) $\boldsymbol{X}\boldsymbol{C}\boldsymbol{X}'$ and $\boldsymbol{X}\boldsymbol{D}\boldsymbol{X}'$ are independent if and only if $\boldsymbol{C}\boldsymbol{\Sigma}\boldsymbol{D} = \boldsymbol{0}$;
(c) $\boldsymbol{X}\boldsymbol{C}\boldsymbol{X}'$ and $\boldsymbol{A}\boldsymbol{X}'\boldsymbol{B}$ are independent if (and only if) $\boldsymbol{A}\boldsymbol{\Sigma}\boldsymbol{C} = \boldsymbol{0}$ or $\boldsymbol{B} = \boldsymbol{0}$.
(Hint: note that the rows of $\boldsymbol{X}\boldsymbol{\Sigma}^{-1/2}$ are i.i.d. $N_p(\boldsymbol{0}, \boldsymbol{I})$ and use Exercise 4.4.9.)

4.4.11 From Exercise 4.4.9 show that if $\boldsymbol{x} \sim N_p(\boldsymbol{\mu}, \boldsymbol{\Sigma})$, then
(a) $\boldsymbol{x}'\boldsymbol{C}\boldsymbol{x} \sim \chi_r^2$ if and only if $\boldsymbol{\mu} = \boldsymbol{0}$ and $\boldsymbol{C\Sigma C} = \boldsymbol{C}$, in which case $r = \text{tr}\,\boldsymbol{C\Sigma}$ (as a special case, $(\boldsymbol{x}-\boldsymbol{\mu})'\boldsymbol{\Sigma}^{-1}(\boldsymbol{x}-\boldsymbol{\mu}) \sim \chi_p^2$);
(b) $\boldsymbol{x}'\boldsymbol{C}\boldsymbol{x}$ and $\boldsymbol{x}'\boldsymbol{D}\boldsymbol{x}$ are independent if and only if $\boldsymbol{C\Sigma D} = \boldsymbol{0}$;
(c) $\boldsymbol{x}'\boldsymbol{C}\boldsymbol{x}$ and $\boldsymbol{A}\boldsymbol{x}$ are independent if (and only if) $\boldsymbol{A\Sigma C} = \boldsymbol{0}$.

4.4.12 (Alternative proof that $\overline{\boldsymbol{x}}$ and $\boldsymbol{S}$ are independent.) Let $\boldsymbol{X}(n \times p)$ be a data matrix from $N_p(\boldsymbol{\mu}, \boldsymbol{\Sigma})$, and let $\boldsymbol{A}(n \times n)$ be an orthogonal matrix whose last row is given by $\boldsymbol{a}_n = n^{-1/2}\boldsymbol{1}$. If $\boldsymbol{Y} = \boldsymbol{AX}$, show that
(a) the rows of $\boldsymbol{Y}$ are independent;
(b) $\boldsymbol{y}_n = n^{1/2}\overline{\boldsymbol{x}} \sim N_p(n^{1/2}\boldsymbol{\mu}, \boldsymbol{\Sigma})$;
(c) $\boldsymbol{y}_r \sim N_p(\boldsymbol{0}, \boldsymbol{\Sigma})$ for $r = 1, \dots, n-1$;
(d) $n\boldsymbol{S} = \boldsymbol{X}'\boldsymbol{H}\boldsymbol{X} = \sum_{r=1}^{n-1} \boldsymbol{y}_r\boldsymbol{y}_r' \sim W_p(\boldsymbol{\Sigma}, n-1)$ independently of $\overline{\boldsymbol{x}}$.

4.4.13 (Expectation of the inverted Wishart distribution.) If $\boldsymbol{M} \sim W_p(\boldsymbol{\Sigma}, m)$, $m \geq p+2$, show that $\mathsf{E}(\boldsymbol{M}^{-1}) = \boldsymbol{\Sigma}^{-1}/(m-p-1)$.
(Hint: If $x \sim \chi_n^2$, $n \geq 3$, then use its p.d.f. to show that $\mathsf{E}(x^{-1}) = 1/(n-2)$. Also note that $\boldsymbol{a}'\boldsymbol{\Sigma}^{-1}\boldsymbol{a}/\boldsymbol{a}'\boldsymbol{M}^{-1}\boldsymbol{a} \sim \chi_{m-p+1}^2$, for all constant vectors $\boldsymbol{a}$.)

4.4.14 If $c \sim \chi_m^2$, then, using the central limit theorem (Theorem 2.7.1) and the transformation theorem (Theorem 2.7.2), it is easy to see that $\log c$ has an asymptotic $N(\log m, 2/m)$ distribution as $m \to \infty$. Deduce a corresponding result for the asymptotic distribution of $\log |\boldsymbol{M}|$, where $\boldsymbol{M} \sim W(\boldsymbol{\Sigma}, m)$.

4.4.15 (a) Let $\boldsymbol{X}$ be a data matrix from $N_p(\boldsymbol{0}, \boldsymbol{\Sigma})$. If $\boldsymbol{X}_{2.1} = \boldsymbol{X}_2 - \boldsymbol{X}_1\boldsymbol{\Sigma}_{11}^{-1}\boldsymbol{\Sigma}_{12}$, then

$$\boldsymbol{X}_{2.1}'\boldsymbol{X}_{2.1} = \boldsymbol{F} = \boldsymbol{M}_{22} + \boldsymbol{\Sigma}_{21}\boldsymbol{\Sigma}_{11}^{-1}\boldsymbol{M}_{11}\boldsymbol{\Sigma}_{11}^{-1}\boldsymbol{\Sigma}_{12} - \boldsymbol{\Sigma}_{21}\boldsymbol{\Sigma}_{11}^{-1}\boldsymbol{M}_{12} - \boldsymbol{M}_{21}\boldsymbol{\Sigma}_{11}^{-1}\boldsymbol{\Sigma}_{12},$$

and $\boldsymbol{F} \sim W(\boldsymbol{\Sigma}_{22.1}, m)$.
(Hint: use Theorem 4.4.2; see also Exercise 4.3.2.)
(b) $\boldsymbol{F}$ is independent of $\boldsymbol{X}_1$ and of $\boldsymbol{M}_{11}$.
(c) If $\hat{\boldsymbol{F}}$ is the matrix obtained from $\boldsymbol{F}$ by substituting $n^{-1}\boldsymbol{M}$ for $\boldsymbol{\Sigma}$, then $\hat{\boldsymbol{F}} = \boldsymbol{M}_{22.1}$, defined in (4.8).

4.4.16 If $\boldsymbol{A}$ and $\boldsymbol{B}$ are $(p \times p)$ symmetric idempotent matrices of rank r and s, and if $\boldsymbol{AB} = \boldsymbol{0}$, show that, for $\boldsymbol{x} \sim N_p(\boldsymbol{0}, \sigma^2\boldsymbol{I})$,

$$\frac{\boldsymbol{x}'\boldsymbol{A}\boldsymbol{x}/r}{\boldsymbol{x}'\boldsymbol{B}\boldsymbol{x}/s} \sim F_{r,s}, \qquad \frac{\boldsymbol{x}'\boldsymbol{A}\boldsymbol{x}}{\boldsymbol{x}'(\boldsymbol{A}+\boldsymbol{B})\boldsymbol{x}} \sim B(\frac{1}{2}r, \frac{1}{2}s)$$

and

$$\frac{(p-r)\boldsymbol{x}'\boldsymbol{A}\boldsymbol{x}}{r\boldsymbol{x}'(\boldsymbol{I}-\boldsymbol{A})\boldsymbol{x}} \sim F_{r,p-r}.$$

4.4.17 (a) Suppose that the elements of $\boldsymbol{x}$ are i.i.d. with mean 0, variance 1, and third and fourth moments μ_3 and μ_4. Consider the matrix $\boldsymbol{M} = \boldsymbol{x}\boldsymbol{x}'$ and show that

$$\mathsf{E}(m_{ij}) = \delta_{ij}, \qquad \text{cov}(m_{ij}, m_{kl}) = (\mu_4 - 3)\delta_{ijkl} + (\delta_{ik}\delta_{jl} + \delta_{il}\delta_{jk}),$$
$$\text{var}(m_{ij}) = (\mu_4 - 2)\delta_{ij} + 1,$$

where δ_{ij} is the Kronecker delta (equal to 1 for $i = j$, and 0 otherwise) and $\delta_{ijkl} = 1$ if and only if $i = j = k = l$, and is 0 otherwise.
(b) Suppose that the elements of $\boldsymbol{X}(n \times p)$ are i.i.d. with the moments given above. If $\boldsymbol{M} = \boldsymbol{X}'\boldsymbol{X}$, show that

$$\operatorname{cov}(m_{ij}, m_{kl}) = n[(\mu_4 - 3)\delta_{ijkl} + \delta_{ik}\delta_{jl} + \delta_{il}\delta_{jk}].$$

(c) Using the fact that $\mu_4 = 3$ for $N(0, 1)$, show that if $\boldsymbol{M} \sim W_p(\boldsymbol{\Sigma}, n)$, then

$$\operatorname{var}(m_{ij}) = n(\sigma_{ij}^2 + \sigma_{ii}\sigma_{jj}), \qquad \operatorname{cov}(m_{ij}, m_{kl}) = n(\sigma_{ik}\sigma_{jl} + \sigma_{il}\sigma_{jk}).$$

4.4.18 (Alternative proof of Corollary 4.4.15.) Let $f(\boldsymbol{x})$ be a p.d.f. on $\mathbb{R}^p$, and let $\boldsymbol{X} = (\boldsymbol{x}_1, \dots, \boldsymbol{x}_n)'$ be a random sample from $f(\boldsymbol{x})$. Show that $\operatorname{rank}(\boldsymbol{X}) = \min(n, p)$ with probability 1, and hence using (A.41), show that if $n \geq p$, then $\boldsymbol{X}'\boldsymbol{X} > 0$ with probability 1.

4.4.19 (Eaton 1972, §8.26) (a) Note the following generalization of Theorem 4.4.13(a): if $\boldsymbol{M} \sim W_p(\boldsymbol{\Sigma}, m)$, and $\boldsymbol{A}(n \times p)$ has rank k, then

$$(\boldsymbol{A}\boldsymbol{M}^{-1}\boldsymbol{A}')^{-1} \sim W_k((\boldsymbol{A}\boldsymbol{\Sigma}^{-1}\boldsymbol{A}')^{-1}, m - p + k).$$

(Hint: If $\boldsymbol{A} = [\boldsymbol{I}, \boldsymbol{0}]$, then this theorem is simply a statement that $(\boldsymbol{M}^{11})^{-1} \sim W_k((\boldsymbol{\Sigma}^{11})^{-1}, m - p + k)$, which has already been proved in Corollary 4.4.12. Now any $(k \times p)$ matrix of rank k can be written as $\boldsymbol{A} = \boldsymbol{B}[\boldsymbol{I}_k, \boldsymbol{0}]\boldsymbol{\Gamma}\boldsymbol{\Sigma}^{1/2}$, where $\boldsymbol{B}(k \times k)$ is nonsingular, and $\boldsymbol{\Gamma} = [\boldsymbol{\Gamma}_1', \boldsymbol{\Gamma}_2']'$ is orthogonal. (The first k rows, $\boldsymbol{\Gamma}_1$, of $\boldsymbol{\Gamma}$ form an orthonormal basis for the rows of $\boldsymbol{A}\boldsymbol{\Sigma}^{-1/2}$. The rows of $\boldsymbol{\Gamma}_2$ are orthogonal to the rows of $\boldsymbol{A}\boldsymbol{\Sigma}^{-1/2}$, so $\boldsymbol{A}\boldsymbol{\Sigma}^{-1/2}\boldsymbol{\Gamma}_2' = \boldsymbol{0}$.) Let $\boldsymbol{V} = \boldsymbol{\Gamma}\boldsymbol{\Sigma}^{-1/2}\boldsymbol{M}\boldsymbol{\Sigma}^{-1/2}\boldsymbol{\Gamma}'$. Clearly $\boldsymbol{V} \sim W_p(\boldsymbol{I}, m)$. Now, $\boldsymbol{A}\boldsymbol{M}^{-1}\boldsymbol{A}' = \boldsymbol{B}[\boldsymbol{I}, \boldsymbol{0}]\boldsymbol{V}^{-1}[\boldsymbol{I}, \boldsymbol{0}]'\boldsymbol{B}' = \boldsymbol{B}\boldsymbol{V}^{11}\boldsymbol{B}'$, where $\boldsymbol{V}^{11}$ is the upper left-hand submatrix of $\boldsymbol{V}^{-1}$. Now, $(\boldsymbol{A}\boldsymbol{M}^{-1}\boldsymbol{A}')^{-1} = (\boldsymbol{B}')^{-1}(\boldsymbol{V}^{11})^{-1}\boldsymbol{B}^{-1}$. But $(\boldsymbol{V}^{11})^{-1} \sim W_k(\boldsymbol{I}, m - p + k)$. Therefore, $(\boldsymbol{A}\boldsymbol{M}^{-1}\boldsymbol{A}')^{-1} \sim W_k((\boldsymbol{B}\boldsymbol{B}')^{-1}, m - p + k)$. But $\boldsymbol{B} = \boldsymbol{A}\boldsymbol{\Sigma}^{-1/2}\boldsymbol{\Gamma}'[\boldsymbol{I}, \boldsymbol{0}]' = \boldsymbol{A}\boldsymbol{\Sigma}^{-1/2}\boldsymbol{\Gamma}_1'$. Thus, since $\boldsymbol{A}\boldsymbol{\Sigma}^{-1/2}\boldsymbol{\Gamma}_2' = \boldsymbol{0}$ and $\boldsymbol{I} = \boldsymbol{\Gamma}_1'\boldsymbol{\Gamma}_1 + \boldsymbol{\Gamma}_2'\boldsymbol{\Gamma}_2$, $(\boldsymbol{B}\boldsymbol{B}')^{-1} = (\boldsymbol{A}\boldsymbol{\Sigma}^{-1}\boldsymbol{A}')^{-1}$.)
(b) Hence, deduce Theorem 4.4.13(a) as a special case.

4.4.20 Let the rows of $\boldsymbol{X}(n \times p)$ be independently distributed $\boldsymbol{x}_r \sim N_p(\boldsymbol{\mu}_r, \boldsymbol{\Sigma})$ for $r = 1, \dots, n$, where the $\boldsymbol{\mu}_r$ are not necessarily equal.
(a) Show that Theorem 4.3.3 remains valid with this assumption on $\boldsymbol{X}$.
(b) Show that Craig's theorem (Theorem 4.4.10) remains valid with this assumption on $\boldsymbol{X}$.
(c) If $\mathsf{E}(\boldsymbol{C}\boldsymbol{X}) = \boldsymbol{0}$, show that parts (a) and (b) of Cochran's theorem (Theorem 4.4.9) remain valid with this assumption on $\boldsymbol{X}$.

4.4.21 Let $\boldsymbol{x} \sim N_p(\boldsymbol{0}, \boldsymbol{\Sigma})$. Using the characteristic function of $\boldsymbol{x}$, show that

$$\mathsf{E}(x_i x_j x_k x_l) = \sigma_{ij}\sigma_{kl} + \sigma_{ik}\sigma_{jl} + \sigma_{il}\sigma_{jk},$$

for $i, j, k, l = 1, \dots, p$. Hence, prove that

$$\operatorname{cov}(\boldsymbol{x}'\boldsymbol{A}\boldsymbol{x}, \boldsymbol{x}'\boldsymbol{B}\boldsymbol{x}) = 2\operatorname{tr}(\boldsymbol{A}\boldsymbol{\Sigma}\boldsymbol{B}\boldsymbol{\Sigma}),$$

where $\boldsymbol{A}$ and $\boldsymbol{B}$ are $p \times p$ symmetric matrices. Hint: prove the result for the transformed variables $\boldsymbol{y} = \boldsymbol{\Sigma}^{-1/2}\boldsymbol{x}$.

4.5.1 (a) Examine the case $p = 1$ in Theorem 4.5.2. Putting $d \sim N_1(0, \sigma^2)$ and $ms^2 \sim \sigma^2 \chi^2_m$ to correspond to $\boldsymbol{d}$ and $\boldsymbol{M}$, respectively, show that α corresponds to $(d/s)^2$, which is the square of a t_m variable.
(b) If $\boldsymbol{d} \sim N_p(\boldsymbol{\mu}, \boldsymbol{\Sigma})$ and $\boldsymbol{M} \sim W_p(\boldsymbol{\Sigma}, m)$, then we say that $\alpha = m\boldsymbol{d}'\boldsymbol{M}^{-1}\boldsymbol{d}$ has a *noncentral* T^2 distribution. Show that

$$\alpha = \beta + 2m\boldsymbol{\mu}'\boldsymbol{M}^{-1}\boldsymbol{x} + m\boldsymbol{\mu}'\boldsymbol{M}^{-1}\boldsymbol{\mu},$$

where β has a central T^2 distribution, and $\boldsymbol{x} \sim N_p(\boldsymbol{0}, \boldsymbol{\Sigma})$.
(c) Under the conditions stated in part (b), show that α is proportional to a noncentral F statistic, where the non centrality parameter depends on $\boldsymbol{\mu}'\boldsymbol{\Sigma}^{-1}\boldsymbol{\mu}$, the Mahalanobis distance from $\boldsymbol{\mu}$ to the origin.

4.5.2 Using the assumptions of Theorem 4.5.2, show that if $\alpha = m\boldsymbol{d}'\boldsymbol{M}^{-1}\boldsymbol{d}$, then:

(a)
$$\frac{\alpha}{\alpha + m} = \frac{\boldsymbol{d}'\boldsymbol{M}^{-1}\boldsymbol{d}}{1 + \boldsymbol{d}'\boldsymbol{M}^{-1}\boldsymbol{d}} \sim B\left(\frac{1}{2}p, \frac{1}{2}(m - p + 1)\right),$$

(b)
$$\frac{m}{\alpha + m} = \frac{1}{1 + \boldsymbol{d}'\boldsymbol{M}^{-1}\boldsymbol{d}} \sim B\left(\frac{1}{2}(m - p + 1), \frac{1}{2}p\right),$$

(c)
$$\frac{|\boldsymbol{M}|}{|\boldsymbol{M} + \boldsymbol{d}\boldsymbol{d}'|} = \frac{m}{\alpha + m} \sim B\left(\frac{1}{2}(m - p + 1), \frac{1}{2}p\right);$$

(Hint: use Eq. (A.23).)

4.5.3 Show that if $m\alpha \sim T^2(p, m)$, then $(m + 1)\log(1 + \alpha)$ has an asymptotic χ^2_p distribution as $m \to \infty$.

4.6.1 Partition $\boldsymbol{\mu} = (\boldsymbol{\mu}_1', \boldsymbol{\mu}_2')'$ and suppose that $\boldsymbol{\Sigma}$ is also partitioned. Using Eqs. (A.31) and (A.29), show that

$$\boldsymbol{\mu}'\boldsymbol{\Sigma}^{-1}\boldsymbol{\mu} = \boldsymbol{\mu}_1\boldsymbol{\Sigma}_{11}^{-1}\boldsymbol{\mu}_1 + \boldsymbol{\mu}_{2.1}'\boldsymbol{\Sigma}_{22.1}^{-1}\boldsymbol{\mu}_{2.1},$$

where

$$\boldsymbol{\mu}_{2.1} = \boldsymbol{\mu}_2 - \boldsymbol{\Sigma}_{21}\boldsymbol{\Sigma}_{11}^{-1}\boldsymbol{\mu}_1 \quad \text{and} \quad \boldsymbol{\Sigma}_{22.1} = \boldsymbol{\Sigma}_{22} - \boldsymbol{\Sigma}_{21}\boldsymbol{\Sigma}_{11}^{-1}\boldsymbol{\Sigma}_{12}.$$

4.7.1 If $\boldsymbol{A} \sim W_p(\boldsymbol{\Sigma}, m)$ independently of $\boldsymbol{B} \sim W_p(\boldsymbol{\Sigma}, n)$ and $m \geq p$, show that $\boldsymbol{A}^{-1}\boldsymbol{B}$ has the same eigenvalues as $\boldsymbol{A}^{*^{-1}}\boldsymbol{B}^*$, where $\boldsymbol{A}^* = \boldsymbol{\Sigma}^{-1/2}\boldsymbol{A}\boldsymbol{\Sigma}^{-1/2} \sim W_p(\boldsymbol{I}, m)$ independently of $\boldsymbol{B}^* = \boldsymbol{\Sigma}^{-1/2}\boldsymbol{B}\boldsymbol{\Sigma}^{-1/2} \sim W_p(\boldsymbol{I}, n)$. Hence, deduce that the joint distribution of the eigenvalues of $\boldsymbol{A}^{-1}\boldsymbol{B}$ does not depend on $\boldsymbol{\Sigma}$.

4.7.2 In Theorem 4.7.1, show that $\boldsymbol{G} = (\boldsymbol{X}\boldsymbol{X}')^{-1/2}\boldsymbol{X}$ is row orthonormal. Also show that $\boldsymbol{X}\boldsymbol{G}'\boldsymbol{G} = \boldsymbol{X}$ and that $\boldsymbol{G}\boldsymbol{X}'\boldsymbol{X}\boldsymbol{G}' = \boldsymbol{X}\boldsymbol{X}'$.

4.7.3 (Alternative proof of Theorem 4.7.1.) Let $\boldsymbol{X}((m+n)\times(n+p))$, where $m \geq p$, be a data matrix from $N_{n+1}(\boldsymbol{0}, \boldsymbol{I})$ and partition

$$\boldsymbol{X}'\boldsymbol{X} = \boldsymbol{M} = \begin{bmatrix} \boldsymbol{M}_{11} & \boldsymbol{M}_{12} \\ \boldsymbol{M}_{21} & \boldsymbol{M}_{22} \end{bmatrix} \begin{matrix} n \\ p \end{matrix}.$$

Using Theorem 4.4.11, show that $(\boldsymbol{M}^{22})^{-1} = \boldsymbol{M}_{22.1} = \boldsymbol{M}_{22} - \boldsymbol{M}_{21}\boldsymbol{M}_{11}^{-1}\boldsymbol{M}_{12} \sim W_p(\boldsymbol{I}, m)$ independently of $\boldsymbol{M}_{22} - \boldsymbol{M}_{22.1} = \boldsymbol{M}_{21}\boldsymbol{M}_{11}^{-1}\boldsymbol{M}_{12} \sim W_p(\boldsymbol{I}, n)$. Thus, using Eq. (A.31), the nonzero roots $\boldsymbol{M}^{22}\boldsymbol{M}_{21}\boldsymbol{M}_{11}^{-1}\boldsymbol{M}_{12} = -\boldsymbol{M}^{21}\boldsymbol{M}_{12}$ have the $\Psi(p, m, n)$ distribution. Since $\boldsymbol{M}$ and $\boldsymbol{M}^{-1}$ are symmetric, $-\boldsymbol{M}^{21}\boldsymbol{M}_{12} = -(\boldsymbol{M}_{21}\boldsymbol{M}^{12})'$, which has the same nonzero eigenvalues as $-\boldsymbol{M}^{12}\boldsymbol{M}_{21}$ (since $\boldsymbol{AB}$ and $(\boldsymbol{AB})' = \boldsymbol{B}'\boldsymbol{A}'$ have the same nonzero eigenvalues). Interchanging the roles of 1 and 2 above, and applying Theorem 4.4.11 again, show that the roots of $-\boldsymbol{M}^{12}\boldsymbol{M}_{21}$ have the $\Psi(n, m+n-p, p)$ distribution. Thus, the $\Psi(p, m, n)$ and $\Psi(n, m+n-p, p)$ distributions are the same.

4.7.4 If $\boldsymbol{A} \sim W_p(\boldsymbol{\Sigma}, m)$ and $\boldsymbol{B} \sim W_p(\boldsymbol{\Sigma}, n)$ are independent Wishart matrices, show that $|\boldsymbol{A}|/|\boldsymbol{A}+\boldsymbol{B}|$ has the $\Lambda(p, m, n)$ distribution.

4.7.5 Verify the F distribution given in (4.39) and (4.40).

4.7.6 (a) (Bartlett, 1947) Show that $-\{t - (p+q+1)/2\} \log \Lambda(p, t-q, q)$ has approximately a χ^2_{pq} distribution for large t.
(b) (Rao 1951, 1973, p. 556) If $\Lambda \sim \Lambda(p, t-q, q)$, then an alternative approximation uses

$$R = \frac{ms - 2\lambda}{pq} \frac{1 - \Lambda^{1/s}}{\Lambda^{1/s}},$$

where $m = t - \frac{1}{2}(p+q+1)$, $\lambda = \frac{1}{4}(pq-2)$, and $s^2 = (p^2q^2 - 4)/(p^2 + q^2 - 5)$. Then, R has an asymptotic F distribution with pq and $(ms - 2\lambda)$ degrees of freedom, even though $(ms - 2\lambda)$ need not be an integer.

4.7.7 Show that if m or n is less than p, then the ϕ statistic defined in Theorem 4.7.2 cannot be used.

4.7.8 If $\boldsymbol{A} > \boldsymbol{0}$ and $\boldsymbol{B} \geq 0$ and $\boldsymbol{Bx} = \lambda \boldsymbol{Ax}$ for some $\boldsymbol{x} \neq \boldsymbol{0}$, then show that $\boldsymbol{Bx} = (\lambda/(1+\lambda))(\boldsymbol{A}+\boldsymbol{B})\boldsymbol{x}$. Hence, deduce that if λ is an eigenvalue of $\boldsymbol{A}^{-1}\boldsymbol{B}$, then $\lambda/(1+\lambda)$ is an eigenvalue of $(\boldsymbol{A}+\boldsymbol{B})^{-1}\boldsymbol{B}$.

5

Estimation

Introduction

Finding estimates of location and scatter is a fundamental task in statistics. Three general approaches are considered in this chapter, based on likelihood, robustness considerations and Bayesian ideas, respectively. Maximum likelihood estimation for the multivariate normal distribution leads to the sample mean vector and the sample covariance matrix, which are natural extensions of their one-dimensional counterparts. Robust estimators seek to limit the influence of outlying observations, though the concept of an outlier is more complicated in the multivariate setting than in the univariate setting. Bayesian inference incorporates prior distributions for the unknown parameters with corresponding estimates typically defined in terms of posterior means.

5.1 Likelihood and Sufficiency

5.1.1 The Likelihood Function

Suppose that $\boldsymbol{x}_1, \dots, \boldsymbol{x}_n$ is a random sample from a population with probability density function (p.d.f.) $f(\boldsymbol{x}; \boldsymbol{\theta})$, where $\boldsymbol{\theta}$ is a parameter vector. The likelihood function of the whole sample is

$$L(\boldsymbol{X}; \boldsymbol{\theta}) = \prod_{r=1}^{n} f(\boldsymbol{x}_r; \boldsymbol{\theta}). \tag{5.1}$$

The log-likelihood function is

$$l(\boldsymbol{X}; \boldsymbol{\theta}) = \log L(\boldsymbol{X}; \boldsymbol{\theta}) = \sum_{r=1}^{n} \log f(\boldsymbol{x}_r; \boldsymbol{\theta}). \tag{5.2}$$

Given a sample $\boldsymbol{X}$, both $L(\boldsymbol{X}; \boldsymbol{\theta})$ and $l(\boldsymbol{X}; \boldsymbol{\theta})$ are considered as functions of the parameter $\boldsymbol{\theta}$. For the special case $n = 1$, $L(\boldsymbol{x}; \boldsymbol{\theta}) = f(\boldsymbol{x}; \boldsymbol{\theta})$, and the distinction between the p.d.f and the likelihood function is to be noted: $f(\boldsymbol{x}; \boldsymbol{\theta})$ is interpreted as a p.d.f. when $\boldsymbol{\theta}$ is fixed and $\boldsymbol{x}$ is allowed to vary, and it is interpreted as the likelihood function when $\boldsymbol{x}$ is fixed and $\boldsymbol{\theta}$ is allowed to vary. As we know, the p.d.f. plays a key role in probability theory, whereas the likelihood is central to the theory of statistical inference.

Multivariate Analysis, Second Edition. Kanti V. Mardia, John T. Kent, and Charles C. Taylor.

Example 5.1.1 Suppose that $\boldsymbol{x}_1, \ldots, \boldsymbol{x}_n$ is a random sample from $N_p(\boldsymbol{\mu}, \boldsymbol{\Sigma})$. Then, Equations (5.1) and (5.2) become, respectively,

$$L(\boldsymbol{X}; \boldsymbol{\mu}, \boldsymbol{\Sigma}) = |2\pi\boldsymbol{\Sigma}|^{-n/2} \exp\left\{ -\frac{1}{2}\sum_{r=1}^{n}(\boldsymbol{x}_r - \boldsymbol{\mu})'\boldsymbol{\Sigma}^{-1}(\boldsymbol{x}_r - \boldsymbol{\mu}) \right\} \tag{5.3}$$

and

$$l(\boldsymbol{X}; \boldsymbol{\mu}, \boldsymbol{\Sigma}) = \log L(\boldsymbol{X}; \boldsymbol{\theta}) = -\frac{n}{2}\log|2\pi\boldsymbol{\Sigma}| - \frac{1}{2}\sum_{r=1}^{n}(\boldsymbol{x}_r - \boldsymbol{\mu})'\boldsymbol{\Sigma}^{-1}(\boldsymbol{x}_r - \boldsymbol{\mu}). \tag{5.4}$$

Equation (5.4) can be simplified as follows. When the identity

$$\begin{aligned}(\boldsymbol{x}_r - \boldsymbol{\mu})'\boldsymbol{\Sigma}^{-1}(\boldsymbol{x}_r - \boldsymbol{\mu}) &= (\boldsymbol{x}_r - \overline{\boldsymbol{x}})'\boldsymbol{\Sigma}^{-1}(\boldsymbol{x}_r - \overline{\boldsymbol{x}}) \\ &\quad + (\overline{\boldsymbol{x}} - \boldsymbol{\mu})'\boldsymbol{\Sigma}^{-1}(\overline{\boldsymbol{x}} - \boldsymbol{\mu}) + 2(\overline{\boldsymbol{x}} - \boldsymbol{\mu})'\boldsymbol{\Sigma}^{-1}(\boldsymbol{x}_r - \overline{\boldsymbol{x}})\end{aligned} \tag{5.5}$$

is summed over the index $r = 1, \ldots, n$, the final term on the right-hand side vanishes, yielding

$$\sum_{r=1}^{n}(\boldsymbol{x}_r - \boldsymbol{\mu})'\boldsymbol{\Sigma}^{-1}(\boldsymbol{x}_r - \boldsymbol{\mu}) = \sum_{r=1}^{n}(\boldsymbol{x}_r - \overline{\boldsymbol{x}})'\boldsymbol{\Sigma}^{-1}(\boldsymbol{x}_r - \overline{\boldsymbol{x}}) + n(\overline{\boldsymbol{x}} - \boldsymbol{\mu})'\boldsymbol{\Sigma}^{-1}(\overline{\boldsymbol{x}} - \boldsymbol{\mu}). \tag{5.6}$$

Since each term $(\boldsymbol{x}_r - \overline{\boldsymbol{x}})'\boldsymbol{\Sigma}^{-1}(\boldsymbol{x}_r - \overline{\boldsymbol{x}})$ is a scalar, it equals the trace of itself. Hence (see Section A.2.2),

$$(\boldsymbol{x}_r - \overline{\boldsymbol{x}})'\boldsymbol{\Sigma}^{-1}(\boldsymbol{x}_r - \overline{\boldsymbol{x}}) = \operatorname{tr}\boldsymbol{\Sigma}^{-1}(\boldsymbol{x}_r - \overline{\boldsymbol{x}})(\boldsymbol{x}_r - \overline{\boldsymbol{x}})'. \tag{5.7}$$

Summing (5.7) over the index r and substituting in (5.6) yields

$$\sum_{r=1}^{n}(\boldsymbol{x}_r - \boldsymbol{\mu})'\boldsymbol{\Sigma}^{-1}(\boldsymbol{x}_r - \boldsymbol{\mu}) = \operatorname{tr}\boldsymbol{\Sigma}^{-1}\left\{\sum_{r=1}^{n}(\boldsymbol{x}_r - \overline{\boldsymbol{x}})(\boldsymbol{x}_r - \overline{\boldsymbol{x}})'\right\} + n(\overline{\boldsymbol{x}} - \boldsymbol{\mu})'\boldsymbol{\Sigma}^{-1}(\overline{\boldsymbol{x}} - \boldsymbol{\mu}). \tag{5.8}$$

Writing

$$\sum_{r=1}^{n}(\boldsymbol{x}_r - \overline{\boldsymbol{x}})(\boldsymbol{x}_r - \overline{\boldsymbol{x}})' = n\boldsymbol{S},$$

and using Eq. (5.8) in Eq. (5.4) gives

$$l(\boldsymbol{X}; \boldsymbol{\mu}, \boldsymbol{\Sigma}) = -\frac{n}{2}\log|2\pi\boldsymbol{\Sigma}| - \frac{n}{2}\operatorname{tr}\boldsymbol{\Sigma}^{-1}\boldsymbol{S} - \frac{n}{2}(\overline{\boldsymbol{x}} - \boldsymbol{\mu})'\boldsymbol{\Sigma}^{-1}(\overline{\boldsymbol{x}} - \boldsymbol{\mu}). \tag{5.9}$$

For the special case when $\boldsymbol{\Sigma} = \boldsymbol{I}$ and $\boldsymbol{\mu} = \boldsymbol{0}$, Eq. (5.9) becomes

$$l(\boldsymbol{X}; \boldsymbol{\theta}) = \frac{np}{2}\log 2\pi - \frac{n}{2}\operatorname{tr}\boldsymbol{S} - \frac{n}{2}(\overline{\boldsymbol{x}} - \boldsymbol{0})'(\overline{\boldsymbol{x}} - \boldsymbol{0}). \tag{5.10}$$

□

5.1.2 Efficient Scores and Fisher's Information

The *score function* or efficient score $\boldsymbol{s} = \boldsymbol{s}(\boldsymbol{X}; \boldsymbol{\theta})$ is defined as

$$\boldsymbol{s}(\boldsymbol{X}; \boldsymbol{\theta}) = \frac{\partial}{\partial\boldsymbol{\theta}} l(\boldsymbol{X}; \boldsymbol{\theta}) = \frac{1}{L(\boldsymbol{X}; \boldsymbol{\theta})}\frac{\partial}{\partial\boldsymbol{\theta}} L(\boldsymbol{X}; \boldsymbol{\theta}). \tag{5.11}$$

Of course, $\boldsymbol{s}(\boldsymbol{X}; \boldsymbol{\theta})$ is a random vector, which we may write simply as $\boldsymbol{s}$. The covariance matrix of $\boldsymbol{s}$ is called *Fisher's information matrix*, which we denote by $\boldsymbol{F}$.

Theorem 5.1.1 *If $s(X;\theta)$ is the score of a likelihood function and t is any function of X and θ, then, under certain regularity conditions,*

$$\mathsf{E}(st') = \frac{\partial}{\partial\theta}\mathsf{E}(t') - \mathsf{E}\left(\frac{\partial t'}{\partial\theta}\right). \tag{5.12}$$

Proof: We have

$$\mathsf{E}(t') = \int t'L \, dX.$$

On differentiating both sides with respect to θ and taking the differentiation under the integral sign in the right-hand side (assuming this operation is valid), we obtain

$$\frac{\partial}{\partial\theta}\mathsf{E}(t') = \int \frac{\partial \log L}{\partial\theta} t'L \, dX + \int \frac{\partial t'}{\partial\theta} L \, dX.$$

The result follows on simplifying and rearranging this expression. □

Corollary 5.1.2 *If $s(X;\theta)$ is the score corresponding to a regular likelihood function, then* $\mathsf{E}(s) = \mathbf{0}$.

Proof: Choose t as any constant vector. Then, $\mathsf{E}(s)t' = \mathbf{0}$ for all t. Thus,

$$\mathsf{E}(s) = \mathbf{0}. \tag{5.13}$$

□

Corollary 5.1.3 *If $s(X;\theta)$ is the score corresponding to a regular likelihood function and t is any unbiased estimator of θ, then*

$$\mathsf{E}(st') = I. \tag{5.14}$$

Proof: We have $\mathsf{E}(t) = \theta$ and, since t does not involve θ, $\partial t'/\partial\theta = \mathbf{0}$. □

Corollary 5.1.4 *If $s(X;\theta)$ is the score corresponding to a regular likelihood function and t is an estimator such that $\mathsf{E}(t) = \theta + b(\theta)$, then $\mathsf{E}(st') = I + B$, where $(B)_{ij} = \partial b_j/\partial\theta_i$.*

Proof: Use $\partial/\partial\theta \, \mathsf{E}(t') = I + B$. □

From Corollary 5.1.2, we see that $F = \text{var}(s) = \mathsf{E}(ss')$. Applying Theorem 5.1.1 with $t = s$ gives

$$F = \mathsf{E}(ss') = -\mathsf{E}\left(\frac{\partial s'}{\partial\theta}\right) = -\mathsf{E}\left(\frac{\partial^2 l}{\partial\theta\partial\theta'}\right). \tag{5.15}$$

Example 5.1.2 Suppose that $x_1, \ldots, x_n$ is a random sample from $N_p(\theta, I)$. From (5.10),

$$s(X;\theta) = \frac{\partial}{\partial\theta} l(X;\theta) = n(\bar{x} - \theta), \qquad \frac{\partial s'}{\partial\theta} = -nI.$$

Hence, from (5.15), $F = nI$. Alternatively, F can be obtained as the covariance matrix of $n(\bar{x} - \theta)$. □

Example 5.1.3 For $n = 1$, the simple exponential family defined in Section 3.3.1 has the likelihood function

$$f(\boldsymbol{x}; \boldsymbol{\theta}) = \exp\left\{ a_0(\boldsymbol{\theta}) + b_0(\boldsymbol{x}) + \sum_{i=1}^{q} \theta_i b_i(\boldsymbol{x}) \right\}.$$

Hence,

$$l(\boldsymbol{x}; \boldsymbol{\theta}) = a_0(\boldsymbol{\theta}) + b_0(\boldsymbol{x}) + \sum_{i=1}^{q} \theta_i b_i(\boldsymbol{x}),$$

$$\boldsymbol{s}(\boldsymbol{x}; \boldsymbol{\theta}) = \frac{\partial a_0(\boldsymbol{\theta})}{\partial \boldsymbol{\theta}} + \boldsymbol{b},$$

where $\boldsymbol{b}' = (b_1(\boldsymbol{x}), \ldots, b_q(\boldsymbol{x}))$. Using (5.15),

$$\boldsymbol{F} = -\mathrm{E}\left(\frac{\partial \boldsymbol{s}}{\partial \boldsymbol{\theta}'}\right) = -\frac{\partial^2 a_0(\boldsymbol{\theta})}{\partial \boldsymbol{\theta} \partial \boldsymbol{\theta}'}. \tag{5.16}$$

□

5.1.3 The Cramér–Rao Lower Bound

Theorem 5.1.5 *If $\boldsymbol{t} = t(\boldsymbol{X})$ is an unbiased estimator of $\boldsymbol{\theta}$ based on a regular likelihood function, then*

$$\mathrm{var}(\boldsymbol{t}) \geq \boldsymbol{F}^{-1}, \tag{5.17}$$

where $\boldsymbol{F}$ is the Fisher information matrix.

Proof: Consider $\mathrm{cor}^2(\alpha, \gamma)$, where $\alpha = \boldsymbol{a}'\boldsymbol{t}, \gamma = \boldsymbol{c}'\boldsymbol{s}$, and $\boldsymbol{s}$ is the score function. From Equations (5.13) and (5.14), we have

$$\mathrm{cov}(\alpha, \gamma) = \boldsymbol{a}'\mathrm{cov}(\boldsymbol{t}, \boldsymbol{s})\boldsymbol{c} = \boldsymbol{a}'\boldsymbol{c}, \qquad \mathrm{var}(\gamma) = \boldsymbol{c}'\mathrm{var}(\boldsymbol{s})\boldsymbol{c} = \boldsymbol{c}'\boldsymbol{F}\boldsymbol{c}.$$

Hence,

$$\mathrm{cor}^2(\alpha, \gamma) = (\boldsymbol{a}'\boldsymbol{c})^2 / \{\boldsymbol{a}'\mathrm{var}(\boldsymbol{t})\boldsymbol{a}\boldsymbol{c}'\boldsymbol{F}\boldsymbol{c}\} \leq 1. \tag{5.18}$$

Maximizing the left-hand side of (5.18) with respect to $\boldsymbol{c}$ with the help of (A.99) gives, for all $\boldsymbol{a}$,

$$\boldsymbol{a}'\boldsymbol{F}^{-1}\boldsymbol{a}/\boldsymbol{a}'\mathrm{var}(\boldsymbol{t})\boldsymbol{a} \leq 1;$$

that is $\boldsymbol{a}'\{\mathrm{var}(\boldsymbol{t}) - \boldsymbol{F}^{-1}\}\boldsymbol{a} \geq 0$ for all $\boldsymbol{a}$, which is equivalent to (5.17). □

Note that the Cramér–Rao lower bound is attained if and only if the estimator is a linear function of the score vector.

Example 5.1.4 A genetic example presented by Fisher (1970, p. 305) has four outcomes with probabilities $(2 + \theta)/4$, $(1 - \theta)/4$, $(1 - \theta)/4$, and $\theta/4$, respectively. If in n trials, one observes x_i results for outcome i, where $\sum x_i = n$, then $\boldsymbol{x} = (x_1, x_2, x_3, x_4)'$ has a multinomial distribution, and the likelihood function is

$$L(\boldsymbol{x}; \theta) = c(2 + \theta)^{x_1}(1 - \theta)^{x_2 + x_3}\theta^{x_4}/4^n,$$

where $c = n!/(x_1!x_2!x_3!x_4!)$. Then,

$$l(\boldsymbol{x};\theta) = \log c - n\log 4 + x_1\log(2+\theta) + (x_2+x_3)\log(1-\theta) + x_4\log\theta.$$

The score function is given by

$$s(\boldsymbol{x};\theta) = \frac{\partial l}{\partial\theta} = \frac{x_1}{2+\theta} - \frac{x_2+x_3}{1-\theta} + \frac{x_4}{\theta}.$$

From Equation (5.15),

$$F = -\mathsf{E}\left(\frac{-x_1}{(2+\theta)^2} - \frac{(x_2+x_3)}{(1-\theta)^2} - \frac{x_4}{\theta^2}\right) = \frac{n(1+2\theta)}{2\theta(1-\theta)(2+\theta)}.$$

The Cramér–Rao lower bound is F^{-1}. Here, $4x_4/n$ is an unbiased estimator of θ with variance $\theta(4-\theta)/n$, which exceeds F^{-1}. Hence, $4x_4/n$ does not attain the Cramér–Rao lower bound. Since the lower bound can only be obtained by a linear function of the score, any such function (which will involve θ) cannot be a statistic. Hence, no unbiased estimator can attain the lower bound in this example. □

5.1.4 Sufficiency

Suppose that $\boldsymbol{X} = (\boldsymbol{x}_1,\ldots,\boldsymbol{x}_n)'$ is a sequence of independent identically distributed random vectors whose distribution depends on the parameter $\boldsymbol{\theta}$, and let $\boldsymbol{t}(\boldsymbol{X})$ be any statistic. The statistic $\boldsymbol{t}$ is said to be sufficient for $\boldsymbol{\theta}$ if $L(\boldsymbol{X};\boldsymbol{\theta})$ can be factorized as

$$L(\boldsymbol{X};\boldsymbol{\theta}) = g(\boldsymbol{t};\boldsymbol{\theta})h(\boldsymbol{X}), \tag{5.19}$$

where h is a nonnegative function not involving $\boldsymbol{\theta}$, and g is a function of $\boldsymbol{\theta}$ and $\boldsymbol{t}$.

Note that Equation (5.19) implies that the efficient score $\boldsymbol{s}$ depends on the data only through the sufficient statistic.

Example 5.1.5 For the multivariate normal case, we have from Equation (5.9), taking exponentials,

$$L(\boldsymbol{X};\boldsymbol{\theta},\boldsymbol{\Sigma}) = |2\pi\boldsymbol{\Sigma}|^{-n/2}\exp\left\{-\frac{n}{2}\operatorname{tr}\boldsymbol{\Sigma}^{-1}\boldsymbol{S} - \frac{n}{2}(\overline{\boldsymbol{x}}-\boldsymbol{\mu})'\boldsymbol{\Sigma}^{-1}(\overline{\boldsymbol{x}}-\boldsymbol{\mu})\right\}. \tag{5.20}$$

In Equation (5.19), taking $h(\boldsymbol{X}) = 1$, we see that $\overline{\boldsymbol{x}}$ and $\boldsymbol{S}$ are sufficient for $\boldsymbol{\mu}$ and $\boldsymbol{\Sigma}$. □

Example 5.1.6 From Example 5.1.4,

$$L(\boldsymbol{x};\theta) = c(2+\theta)^{x_1}(1-\theta)^{x_2+x_3}\theta^{x_4}/4^{x_1+x_2+x_3+x_4}.$$

Taking $g(\boldsymbol{t};\theta) = (2+\theta)^{x_1}(1-\theta)^{x_2+x+3}\theta^{x_4}$ and $\boldsymbol{t}' = (x_1, x_2+x_3, x_4)$ in (5.19), we see that $\boldsymbol{t}$ is sufficient for θ. Since $n = x_1+x_2+x_3+x_4$ is fixed, $\boldsymbol{t}^+ = (x_1,x_4)'$ is also sufficient, as x_2+x_3 can be written in terms of $\boldsymbol{t}^+$. □

Example 5.1.7 For a sample size of n from the simple exponential family as in Example 5.1.3, the likelihood can be written as

$$f(\boldsymbol{X};\boldsymbol{\theta}) = \exp\left\{na_0(\boldsymbol{\theta}) + \sum_{i=1}^{q}\theta_i\sum_{r=1}^{n}b_i(\boldsymbol{x}_r)\right\}\exp\left\{\sum_{r=1}^{n}b_0(\boldsymbol{x}_r)\right\}.$$

In Equation (5.19), take $g(\boldsymbol{t};\boldsymbol{\theta})$ as the first factor in this expression and $h(\boldsymbol{X})$ as the second. The vector $\boldsymbol{t}$, where

$$t_i = \sum_{r=1}^{n} b_i(\boldsymbol{x}_r),$$

is sufficient for $\boldsymbol{\theta}$. □

A sufficient statistic is said to minimal sufficient if it is a function of every other sufficient statistic. The Rao–Blackwell theorem states that if the minimal sufficient statistic is *complete*, then any unbiased estimator which is a function of the minimal sufficient statistic must necessarily be the unique minimum variance unbiased estimator (MVUE); that is, it will have a smaller covariance matrix than any other unbiased estimator.

Example 5.1.8 Suppose that the n rows of $\boldsymbol{X}$ are independent and identically distributed (i.i.d.) $N_p(\boldsymbol{\mu}, \boldsymbol{I})$ vectors, and we seek the best unbiased estimate of the quadratic function $\theta = \boldsymbol{\mu}'\boldsymbol{\mu} + \mathbf{1}'\boldsymbol{\mu}$. The mean $\overline{\boldsymbol{x}}$ is minimal sufficient for $\boldsymbol{\mu}$, and we consider a quadratic function of $\overline{\boldsymbol{x}}$, $t = a\overline{\boldsymbol{x}}'\overline{\boldsymbol{x}} + b\,\mathbf{1}'\overline{\boldsymbol{x}} + c$. Then,

$$\mathrm{E}(t) = a\left(\boldsymbol{\mu}'\boldsymbol{\mu} + \frac{p}{n}\right) + b\,\mathbf{1}'\boldsymbol{\mu} + c.$$

If t is to be an unbiased estimator of θ, then $a = 1, b = 1$, and $c = -ap/n$.

Since this unbiased estimator is a function of the minimal sufficient statistic $\overline{\boldsymbol{x}}$, which is complete, then, from the Rao–Blackwell theorem, t is the minimum variance unbiased estimate of θ. □

5.2 Maximum-likelihood Estimation

5.2.1 General Case

The maximum-likelihood estimate (m.l.e.) of an unknown parameter is that value of the parameter which maximizes the likelihood of the given observations. In regular cases, this maximum may be found by differentiation, and since $l(\boldsymbol{X};\boldsymbol{\theta})$ is at a maximum when $L(\boldsymbol{X};\boldsymbol{\theta})$ is at a maximum, the equation $(\partial l/\partial \boldsymbol{\theta}) = \boldsymbol{s} = \mathbf{0}$ is solved for $\boldsymbol{\theta}$. The m.l.e. $\hat{\boldsymbol{\theta}}$ of $\boldsymbol{\theta}$ is that value which provides an overall maximum. Since $\boldsymbol{s}$ is a function of a sufficient statistic, then so is the m.l.e. Also, if the density $f(\boldsymbol{x};\boldsymbol{\theta})$ satisfies certain regularity conditions, and if $\hat{\boldsymbol{\theta}}_n$ is the m.l.e. of $\boldsymbol{\theta}$ for a random sample of size n, then $\hat{\boldsymbol{\theta}}_n$ is asymptotically normally distributed with mean $\boldsymbol{\theta}$ and covariance matrix $\boldsymbol{F}_n^{-1} = (n\boldsymbol{F})^{-1}$, where $\boldsymbol{F}$ is the Fisher information matrix for a *single* observation (see, for example, Rao (1973, p. 416)). In particular, since $\mathrm{var}(\hat{\boldsymbol{\theta}}_n) \to \mathbf{0}$, $\hat{\boldsymbol{\theta}}_n$ is a *consistent* estimate of $\boldsymbol{\theta}$, i.e. plim $\hat{\boldsymbol{\theta}}_n = \boldsymbol{\theta}$, where plim denotes the limit in probability.

Example 5.2.1 The m.l.e. of θ in Example 5.1.4 is found by solving the equation $\boldsymbol{s} = \mathbf{0}$, which becomes

$$n\theta^2 + (2x_2 + 2x_3 - x_1 + x_4)\theta - 2x_4 = 0.$$

This quadratic equation gives two roots, one positive and one negative. Only the positive root is admissible. Writing $p_i = x_i/n, i = 1, 2, 3, 4$, the quadratic equation becomes

$$\theta^2 + \theta(2 - 3p_1 - p_4) - 2p_4 = 0.$$

Discarding the negative root of this quadratic equation, $\hat{\theta} = \alpha + (\alpha^2 + 2p_4)^{1/2}$, where $\alpha = (3p_1 + p_4)/2 - 1$. □

Example 5.2.2 If $(\boldsymbol{x}_1, \dots, \boldsymbol{x}_n)$ is a random sample from Weinman's p-variate exponential distribution defined in Section 3.2.3, then the log likelihood is given by

$$l = -n\sum_{j=0}^{p-1}\left(\log \lambda_j + \frac{\delta_j}{\lambda_j}\right), \tag{5.21}$$

where

$$\delta_j = \frac{p-j}{n}\sum_{r=1}^{n}\{x_{r(j+1)} - x_{r(j)}\},$$

with $x_{r(j)}$ the jth smallest element of $\boldsymbol{x}_r$ $(r = 1, \dots, n; j = 1, \dots, p-1)$, and $x_{r(0)} = 0$. Note that although the elements of $\boldsymbol{x}_r$ are not independent, the elements $\delta_1, \dots, \delta_{p-1}$ are independent. Differentiating the log-likelihood gives the elements s_j of the score $\boldsymbol{s}$ as

$$s_j = \partial l/\partial \lambda_j = -n(\lambda_j - \delta_j)/\lambda_j^2.$$

The m.l.e.s are therefore $\hat{\lambda}_j = \delta_j$. Since $\mathsf{E}(s_j) = 0$, we see that $\hat{\lambda}_j$ is an unbiased estimate of λ_j. In the special case when all the parameters are equal, say λ, then Equation (5.21) becomes

$$l = -pn\log\lambda - \frac{n}{\lambda}\sum_{j=1}^{p-1}\delta_j. \tag{5.22}$$

Differentiating with respect to λ gives

$$s = \frac{\partial l}{\partial \lambda} = -\frac{np}{\lambda} + \frac{n}{\lambda^2}\sum_{j=0}^{p-1}\delta_j.$$

Solving $s = 0$ gives

$$\hat{\lambda} = \frac{1}{p}\sum_{j=0}^{p-1}\delta_j.$$

Again, since $\mathsf{E}(s) = 0$, we see that $\hat{\lambda}$ is unbiased for λ. □

5.2.2 Multivariate Normal Case

5.2.2.1 Unconstrained Case

For the multivariate normal distribution, we have the log-likelihood function from (5.9):

$$l(\boldsymbol{X}; \boldsymbol{\mu}, \boldsymbol{\Sigma}) = -\frac{n}{2}\log|2\pi\boldsymbol{\Sigma}| - \frac{n}{2}\operatorname{tr}\boldsymbol{\Sigma}^{-1}\boldsymbol{S} - \frac{n}{2}\operatorname{tr}\boldsymbol{\Sigma}^{-1}(\overline{\boldsymbol{x}} - \boldsymbol{\mu})(\overline{\boldsymbol{x}} - \boldsymbol{\mu})'.$$

We now show that

$$\hat{\boldsymbol{\mu}} = \overline{\boldsymbol{x}}, \qquad \hat{\boldsymbol{\Sigma}} = \boldsymbol{S} \tag{5.23}$$

if $n \geq p + 1$. Note from Corollary 4.4.15 that $\boldsymbol{S} > 0$ with probability 1.

The parameters here are $\boldsymbol{\mu}$ and $\boldsymbol{\Sigma}$. Using new parameters $\boldsymbol{\mu}$ and $\boldsymbol{V}$ with $\boldsymbol{V} = \boldsymbol{\Sigma}^{-1}$, we first calculate $\partial l/\partial \boldsymbol{\mu}$ and $\partial l/\partial \boldsymbol{V}$. Equation (5.9) becomes

$$l(\boldsymbol{X}; \boldsymbol{\mu}, \boldsymbol{V}) = -\frac{np}{2}\log 2\pi + \frac{n}{2}\log|\boldsymbol{V}| - \frac{n}{2}\operatorname{tr}\boldsymbol{V}\boldsymbol{S} - \frac{n}{2}\operatorname{tr}\boldsymbol{V}(\overline{\boldsymbol{x}} - \boldsymbol{\mu})(\overline{\boldsymbol{x}} - \boldsymbol{\mu})'. \tag{5.24}$$

Using Section A.9,

$$\frac{\partial l}{\partial \boldsymbol{\mu}} = n\boldsymbol{V}(\overline{\boldsymbol{x}} - \boldsymbol{\mu}). \tag{5.25}$$

To calculate $\partial l/\partial \boldsymbol{V}$, consider separately each term on the right-hand side of (5.24). From (A.89),

$$\frac{\partial}{\partial v_{ij}} \log|\boldsymbol{V}| = \begin{cases} 2V_{ij}/|\boldsymbol{V}|, & i \neq j, \\ V_{ii}/|\boldsymbol{V}|, & i = j, \end{cases}$$

where V_{ij} is the ijth cofactor of $\boldsymbol{V}$. Since $\boldsymbol{V}$ is symmetric, the matrix with elements $V_{ij}/|\boldsymbol{V}|$ equals $\boldsymbol{V}^{-1} = \boldsymbol{\Sigma}$. Thus,

$$\frac{\partial \log|\boldsymbol{V}|}{\partial \boldsymbol{V}} = 2\boldsymbol{\Sigma} - \text{Diag}\,\boldsymbol{\Sigma}.$$

From Equation (A.90),

$$\frac{\partial \text{tr}\,\boldsymbol{VS}}{\partial \boldsymbol{V}} = 2\boldsymbol{S} - \text{Diag}\,\boldsymbol{S}$$

and

$$\frac{\partial \text{tr}\boldsymbol{V}(\overline{\boldsymbol{x}} - \boldsymbol{\mu})(\overline{\boldsymbol{x}} - \boldsymbol{\mu})'}{\partial \boldsymbol{V}} = 2(\overline{\boldsymbol{x}} - \boldsymbol{\mu})(\overline{\boldsymbol{x}} - \boldsymbol{\mu})' - \text{Diag}\,(\overline{\boldsymbol{x}} - \boldsymbol{\mu})(\overline{\boldsymbol{x}} - \boldsymbol{\mu})'.$$

Combining these equations, we see that

$$\frac{\partial l}{\partial \boldsymbol{V}} = \frac{n}{2}(2\boldsymbol{M} - \text{Diag}\,\boldsymbol{M}), \tag{5.26}$$

where $\boldsymbol{M} = \boldsymbol{\Sigma} - \boldsymbol{S} - (\overline{\boldsymbol{x}} - \boldsymbol{\mu})(\overline{\boldsymbol{x}} - \boldsymbol{\mu})'$,

To find the m.l.e.s of $\boldsymbol{\mu}$ and $\boldsymbol{\Sigma}$, we must solve

$$\frac{\partial l}{\partial \boldsymbol{\mu}} = \boldsymbol{0} \quad \text{and} \quad \frac{\partial l}{\partial \boldsymbol{V}} = \boldsymbol{0}.$$

From Equation (5.25), we see that $\boldsymbol{V}(\overline{\boldsymbol{x}} - \boldsymbol{\mu}) = \boldsymbol{0}$. Hence, the m.l.e. of $\boldsymbol{\mu}$ is $\hat{\boldsymbol{\mu}} = \overline{\boldsymbol{x}}$, and from Eq. (5.26) $\partial l/\partial \boldsymbol{V} = \boldsymbol{0}$ gives $2\boldsymbol{M} - \text{Diag}\,\boldsymbol{M} = \boldsymbol{0}$. This implies $\boldsymbol{M} = \boldsymbol{0}$, i.e. the m.l.e. of $\boldsymbol{\Sigma}$ is given by

$$\hat{\boldsymbol{\Sigma}} = \boldsymbol{S} + (\overline{\boldsymbol{x}} - \boldsymbol{\mu})(\overline{\boldsymbol{x}} - \boldsymbol{\mu})'. \tag{5.27}$$

Since $\hat{\boldsymbol{\mu}} = \overline{\boldsymbol{x}}$, Eq. (5.27) gives $\hat{\boldsymbol{\Sigma}} = \boldsymbol{S}$.

Strictly speaking, the above argument only tells us that $\overline{\boldsymbol{x}}$ and $\boldsymbol{S}$ give a stationary point of the likelihood. In order to show that $\overline{\boldsymbol{x}}$ and $\boldsymbol{S}$ give the overall maximum value, consider the following theorem.

Theorem 5.2.1 *For any fixed* $(p \times p)$ *matrix* $\boldsymbol{A} > 0$,

$$f(\boldsymbol{\Sigma}) = |\boldsymbol{\Sigma}|^{-n/2} \exp\left(-\frac{1}{2}\text{tr}\,\boldsymbol{\Sigma}^{-1}\boldsymbol{A}\right)$$

is maximized over $\boldsymbol{\Sigma} > 0$ *by* $\boldsymbol{\Sigma} = n^{-1}\boldsymbol{A}$, *and* $f(n^{-1}\boldsymbol{A}) = |n^{-1}\boldsymbol{A}|^{-n/2}e^{-np/2}$.

Proof: It is easily seen that $f(n^{-1}\boldsymbol{A})$ takes the form given above. Then, we can write

$$\log f(n^{-1}\boldsymbol{A}) - \log f(\boldsymbol{\Sigma}) = \frac{1}{2}np(a - 1 - \log g),$$

where $a = \text{tr}\,\boldsymbol{\Sigma}^{-1}\boldsymbol{A}/np$ and $g = |n^{-1}\boldsymbol{\Sigma}^{-1}\boldsymbol{A}|^{1/p}$ are the arithmetic and geometric means of the eigenvalues of $n^{-1}\boldsymbol{\Sigma}^{-1}\boldsymbol{A}$. Note that all of these eigenvalues are positive from Corollary A.9. From Exercise 5.2.3, $a - 1 - \log g \geq 0$, and hence, $f(n^{-1}\boldsymbol{A}) \geq f(\boldsymbol{\Sigma})$ for all $\boldsymbol{\Sigma} > 0$. □

If we maximize the likelihood (5.9) over $\boldsymbol{\Sigma}$ for fixed $\boldsymbol{\mu}$, we find from Theorem 5.2.1 that $\hat{\boldsymbol{\Sigma}}$ is given by (5.27). Then, since from (A.22)

$$\begin{aligned} l(\boldsymbol{X}; \boldsymbol{\mu}, \boldsymbol{S} + (\overline{\boldsymbol{x}} - \boldsymbol{\mu})(\overline{\boldsymbol{x}} - \boldsymbol{\mu})') &= -\frac{n}{2}|\boldsymbol{S} + (\overline{\boldsymbol{x}} - \boldsymbol{\mu})(\overline{\boldsymbol{x}} - \boldsymbol{\mu})'| \\ &= -\frac{n}{2}|\boldsymbol{S}|\{1 + (\overline{\boldsymbol{x}} - \boldsymbol{\mu})'\boldsymbol{S}^{-1}(\overline{\boldsymbol{x}} - \boldsymbol{\mu})\} \\ &\leq -\frac{n}{2}|\boldsymbol{S}|, \end{aligned}$$

we see that the log-likelihood is maximized by $\hat{\boldsymbol{\mu}} = \overline{\boldsymbol{x}}$. Alternatively, we could maximize over $\boldsymbol{\mu}$ first (Exercise 5.2.11). (In later examples, we in general leave it to the reader to verify that the stationary points obtained are in fact overall maxima.)

Note that the m.l.e. of $\boldsymbol{\mu}$ could have been deduced from (5.9) directly, whether $\boldsymbol{\Sigma}$ is known or not. Since $\boldsymbol{\Sigma}^{-1} > 0, -(\overline{\boldsymbol{x}} - \boldsymbol{\mu})'\boldsymbol{\Sigma}^{-1}(\overline{\boldsymbol{x}} - \boldsymbol{\mu}) \leq 0$ and is maximized when $\boldsymbol{\mu} = \overline{\boldsymbol{x}}$. Thus, $l(\boldsymbol{X}; \boldsymbol{\mu}, \boldsymbol{\Sigma})$ in (5.9) is maximized at $\boldsymbol{\mu} = \overline{\boldsymbol{x}}$ whether $\boldsymbol{\Sigma}$ is constrained or not. Consequently, $\hat{\boldsymbol{\mu}} = \overline{\boldsymbol{x}}$ when $\boldsymbol{\Sigma}$ is known. However, in cases where $\boldsymbol{\mu}$ is constrained, the m.l.e. of $\boldsymbol{\Sigma}$ will be affected, e.g. when $\boldsymbol{\mu}$ is known *a priori*, the m.l.e. of $\boldsymbol{\Sigma}$ is found by solving $\partial l/\partial \boldsymbol{V} = \boldsymbol{0}$ and is given by (5.27).

In the case where $n = 1$ and $\boldsymbol{\Sigma} = \boldsymbol{I}$ is known, Stein (1956) showed that when $p \geq 3$, the m.l.e. $\hat{\boldsymbol{\mu}} = \overline{\boldsymbol{x}} = \boldsymbol{x}$ is an *inadmissible* estimator of $\boldsymbol{\mu}$ under quadratic loss. For further discussion of this remarkable fact, see Cox and Hinkley (1974, p. 447).

5.2.2.2 Constraints on the Mean Vector $\boldsymbol{\mu}$

Consider the case where $\boldsymbol{\mu}$ is known to be proportional to a known vector, so $\boldsymbol{\mu} = k\boldsymbol{\mu}_0$. For example, the elements of $\boldsymbol{x}$ could represent a sample of repeated measurements, in which case $\boldsymbol{\mu} = k\boldsymbol{1}$. With $\boldsymbol{\Sigma}$ also known, Equation (5.9) becomes

$$l(\boldsymbol{X}; k) = -\frac{n}{2}\log|2\pi\boldsymbol{\Sigma}| - \frac{n}{2}\mathrm{tr}\,\boldsymbol{\Sigma}^{-1}\boldsymbol{S} - \frac{n}{2}(\overline{\boldsymbol{x}} - k\boldsymbol{\mu}_0)'\,\boldsymbol{\Sigma}^{-1}(\overline{\boldsymbol{x}} - k\boldsymbol{\mu}_0).$$

To find the m.l.e. of k, we solve $\partial l/\partial k = 0$. This gives

$$n\hat{\boldsymbol{\mu}}_0'\boldsymbol{\Sigma}^{-1}(\overline{\boldsymbol{x}} - k\boldsymbol{\mu}_0) = 0, \tag{5.28}$$

i.e. the m.l.e. $\hat{k}$ of k is

$$\hat{k} = \hat{\boldsymbol{\mu}}_0'\boldsymbol{\Sigma}^{-1}\overline{\boldsymbol{x}}/\boldsymbol{\mu}_0'\boldsymbol{\Sigma}^{-1}\boldsymbol{\mu}_0. \tag{5.29}$$

Then, $\hat{k}$ is unbiased with variance $(n\boldsymbol{\mu}_0'\boldsymbol{\Sigma}^{-1}\boldsymbol{\mu}_0)^{-1}$; see Exercise 5.2.1.

When $\boldsymbol{\Sigma}$ is unknown, the two equations to solve for $\hat{\boldsymbol{\Sigma}}$ and $\hat{k}$ are (5.27) and (5.29). Pre- and postmultiplying (5.27) by $\hat{\boldsymbol{\Sigma}}^{-1}$ and $\boldsymbol{S}^{-1}$, respectively, gives

$$\boldsymbol{S}^{-1} = \hat{\boldsymbol{\Sigma}}^{-1} + \hat{\boldsymbol{\Sigma}}^{-1}(\overline{\boldsymbol{x}} - \boldsymbol{\mu})(\overline{\boldsymbol{x}} - \boldsymbol{\mu})'\boldsymbol{S}^{-1}. \tag{5.30}$$

Premultiplying (5.30) by $\boldsymbol{\mu}_0'$ and using (5.28) gives

$$\boldsymbol{\mu}_0'\boldsymbol{S}^{-1} = \boldsymbol{\mu}_0'\hat{\boldsymbol{\Sigma}}^{-1}.$$

Thus, from (5.29),

$$\hat{k} = \boldsymbol{\mu}_0'\boldsymbol{S}^{-1}\overline{\boldsymbol{x}}/\boldsymbol{\mu}_0'\boldsymbol{S}^{-1}\boldsymbol{\mu}_0. \tag{5.31}$$

A further type of constraint is $\boldsymbol{R\mu} = \boldsymbol{r}$, where $\boldsymbol{R}$ and $\boldsymbol{r}$ are prespecified. Maximizing the log-likelihood subject to this constraint may be achieved by augmenting the log-likelihood with a Lagrangian expression; thus, we maximize

$$l^{+} = l - n\boldsymbol{\lambda}'(\boldsymbol{R\mu} - \boldsymbol{r}),$$

where $\boldsymbol{\lambda}$ is a vector of Lagrangian multipliers, and l is as given in (5.9). With $\boldsymbol{\Sigma}$ assumed known, to find the m.l.e. of $\boldsymbol{\mu}$, we are required to find a $\boldsymbol{\lambda}$ for which the solution to

$$\frac{\partial l^{+}}{\partial \boldsymbol{\mu}} = \boldsymbol{0} \tag{5.32}$$

satisfies the constraint $\boldsymbol{R\mu} = \boldsymbol{r}$. From (5.25),

$$\frac{\partial l^{+}}{\partial \boldsymbol{\mu}} = n\boldsymbol{\Sigma}^{-1}(\overline{\boldsymbol{x}} - \boldsymbol{\mu}) - n\boldsymbol{R}'\boldsymbol{\lambda}.$$

Thus,

$$\overline{\boldsymbol{x}} - \boldsymbol{\mu} = \boldsymbol{\Sigma R}'\boldsymbol{\lambda}.$$

Premultiplying by $\boldsymbol{R}$ gives $\boldsymbol{R}\overline{\boldsymbol{x}} - \boldsymbol{r} = (\boldsymbol{R\Sigma R}')\boldsymbol{\lambda}$ if the constraint is to be satisfied. Thus, we take $\boldsymbol{\lambda} = (\boldsymbol{R\Sigma R})^{-1}(\boldsymbol{R}\overline{\boldsymbol{x}} - \boldsymbol{r})$, so

$$\hat{\boldsymbol{\mu}} = \overline{\boldsymbol{x}} - \boldsymbol{\Sigma R}'\boldsymbol{\lambda} = \overline{\boldsymbol{x}} - \boldsymbol{\Sigma R}'(\boldsymbol{R\Sigma R}')^{-1}(\boldsymbol{R}\overline{\boldsymbol{x}} - \boldsymbol{r}). \tag{5.33}$$

When $\boldsymbol{\Sigma}$ is unknown, the m.l.e. of $\boldsymbol{\mu}$ becomes

$$\hat{\boldsymbol{\mu}} = \overline{\boldsymbol{x}} - \boldsymbol{SR}'(\boldsymbol{RSR}')^{-1}(\boldsymbol{R}\overline{\boldsymbol{x}} - \boldsymbol{r}). \tag{5.34}$$

See Exercise 5.2.8.

5.2.2.3 Constraints on $\boldsymbol{\Sigma}$

First, we consider the case where $\boldsymbol{\Sigma} = k\boldsymbol{\Sigma}_0$, where $\boldsymbol{\Sigma}_0$ is known. From (5.9),

$$2n^{-1}l(\boldsymbol{X}; \boldsymbol{\mu}, k) = -p\log k - \log|2\pi\boldsymbol{\Sigma}_0| - k^{-1}\alpha, \tag{5.35}$$

where $\alpha = \operatorname{tr}\boldsymbol{\Sigma}_0^{-1}\boldsymbol{S} + (\overline{\boldsymbol{x}} - \boldsymbol{\mu})'\boldsymbol{\Sigma}_0^{-1}(\overline{\boldsymbol{x}} - \boldsymbol{\mu})$ is independent of k. If $\boldsymbol{\mu}$ is known, then to obtain the m.l.e. of k, we solve $\partial l/\partial k = 0$. This gives

$$-p/k + \alpha/k^2 = 0.$$

Thus, the m.l.e. of k is $\hat{k} = \alpha/p$. If $\boldsymbol{\mu}$ is unconstrained, then we solve $\partial l/\partial \boldsymbol{\mu} = \boldsymbol{0}$ and $\partial l/\partial k = 0$. These give $k = \alpha/p$ and $\hat{\boldsymbol{\mu}} = \overline{\boldsymbol{x}}$. Together, these give

$$\hat{k} = \operatorname{tr}\boldsymbol{\Sigma}_0^{-1}\boldsymbol{S}/p. \tag{5.36}$$

The constraint $\boldsymbol{\Sigma}_{12} = \boldsymbol{0}$, where $\boldsymbol{\Sigma}_{12}$ is an off-diagonal submatrix of $\boldsymbol{\Sigma}$, implies that the two corresponding groups of variables are independent. The m.l.e. of $\boldsymbol{\Sigma}$ can be found by considering each subgroup separately and is

$$\hat{\boldsymbol{\Sigma}} = \begin{bmatrix} \boldsymbol{S}_{11} & \boldsymbol{0} \\ \boldsymbol{0} & \boldsymbol{S}_{22} \end{bmatrix}.$$

Other constraints on $\boldsymbol{\Sigma}$, in which some entries of $\boldsymbol{\Sigma}^{-1}$ are constrained to be zero or constrained to be equal, are considered in Section 8.3.

Example 5.2.3 For 47 female cats, the body weight (kg) and heart weight (g) were recorded; see Fisher (1947) and Exercise 1.4.3. The sample mean vector and covariance matrix are

$$\bar{\boldsymbol{x}} = (2.36, 9.20)', \qquad \boldsymbol{S} = \begin{bmatrix} 0.0735 & 0.1937 \\ 0.1937 & 1.8040 \end{bmatrix}.$$

Thus, $\bar{\boldsymbol{x}}$ and $\boldsymbol{S}$ are the unconstrained m.l.e.s for $\boldsymbol{\mu}$ and $\boldsymbol{\Sigma}$. However, if from other information, we know that $\boldsymbol{\mu} = (2.5, 10.0)'$, then $\hat{\boldsymbol{\Sigma}}$ is given by (5.27),

$$\begin{aligned} \bar{\boldsymbol{x}} - \boldsymbol{\mu} &= (-0.14, -0.80)', \\ \hat{\boldsymbol{\Sigma}} = \boldsymbol{S} + (\bar{\boldsymbol{x}} - \boldsymbol{\mu})(\bar{\boldsymbol{x}} - \boldsymbol{\mu})' &= \begin{bmatrix} 0.0931 & 0.3057 \\ 0.3057 & 2.4440 \end{bmatrix}. \end{aligned}$$

If instead we assume $\boldsymbol{\mu} = k(2.5, 10.0)'$, then from (5.31),

$$\hat{k} = (2.5, 10.0)\boldsymbol{S}^{-1}(2.36, 9.20)'/(2.5, 10.0)\boldsymbol{S}^{-1}(2.5, 10.0)'.$$

This gives $\hat{k} = 0.937$, so the m.l.e. of $\boldsymbol{\mu}$ is $(2.34, 9.37)'$, and the m.l.e. for $\boldsymbol{\Sigma}$ is now $\hat{\boldsymbol{\Sigma}} = \boldsymbol{S} + (0.02, -0.17)'(0.02, -0.17)$, which is quite close to $\boldsymbol{S}$.

If the covariance was known to be proportional to

$$\boldsymbol{\Sigma}_0 = \begin{bmatrix} 0.1 & 0.2 \\ 0.2 & 2.0 \end{bmatrix}$$

with $\boldsymbol{\mu}$ unconstrained, then from (5.36), $\hat{k} = \operatorname{tr} \boldsymbol{\Sigma}_0^{-1}\boldsymbol{S}/p = 0.781$. Hence,

$$\hat{\boldsymbol{\Sigma}} = \begin{bmatrix} 0.078 & 0.156 \\ 0.156 & 1.562 \end{bmatrix}.$$

□

5.2.2.4 Samples with Linked Parameters

We turn now to consider situations with several normal samples, where we know *a priori* that some relationship exists between their parameters. For instance, we may have independent data matrices $\boldsymbol{X}_1, \dots, \boldsymbol{X}_k$, where the rows of $\boldsymbol{X}_i(n_i \times p)$ are i.i.d. $N_p(\boldsymbol{\mu}_i, \boldsymbol{\Sigma}_i)$, $i = 1, \dots, k$. The most common constraints are

(a) $\boldsymbol{\Sigma}_1 = \cdots = \boldsymbol{\Sigma}_k$ or
(b) $\boldsymbol{\Sigma}_1 = \cdots = \boldsymbol{\Sigma}_k$ and $\boldsymbol{\mu}_1 = \cdots = \boldsymbol{\mu}_k$.

Of course, if (b) holds, we can treat all the data matrices as constituting one sample from a single population.

To calculate the m.l.e.s if (a) holds, note that, from (5.9), the log-likelihood function is given by

$$l = -\frac{1}{2}\sum_i \left\{ n_i \log |2\pi\boldsymbol{\Sigma}| + n_i \operatorname{tr} \boldsymbol{\Sigma}^{-1}(\boldsymbol{S}_i + \boldsymbol{d}_i\boldsymbol{d}_i') \right\}, \tag{5.37}$$

where $\boldsymbol{S}_i$ is the covariance matrix of the ith sample, $i = 1, \dots, k$, and $\boldsymbol{d}_i = \bar{\boldsymbol{x}}_i - \boldsymbol{\mu}_i$. Since there is no restriction on the population means, the m.l.e. of $\boldsymbol{\mu}_i$ is $\bar{\boldsymbol{x}}_i$, and setting $n = \sum n_i$, (5.37) becomes

$$l = -\frac{n}{2} \log |2\pi\boldsymbol{\Sigma}| - \frac{1}{2} \operatorname{tr} \boldsymbol{\Sigma}^{-1}\boldsymbol{W}, \qquad \text{where} \quad \boldsymbol{W} = \sum_i n_i \boldsymbol{S}_i. \tag{5.38}$$

Differentiating (5.38) with respect to $\boldsymbol{\Sigma}$ and equating to zero gives $\boldsymbol{\Sigma} = n^{-1}\boldsymbol{W}$. Therefore, $\hat{\boldsymbol{\Sigma}} = n^{-1}\boldsymbol{W}$ is the m.l.e. of $\boldsymbol{\Sigma}$ under the conditions stated.

These results will be used in the derivation of Wilks' Λ test (Section 6.3.3), and are also relevant to the Hotelling T^2 distribution (Section 4.5).

5.2.3 Matrix Normal Distribution

Given N $(n \times p)$ matrices $\boldsymbol{X}_1, \dots, \boldsymbol{X}_N$, each with a matrix normal distribution (see Equation (2.72)), the log-likelihood function $l(\boldsymbol{X}_1, \dots, \boldsymbol{X}_N; \boldsymbol{M}, \boldsymbol{U}, \boldsymbol{V})$ is given by

$$-\frac{Nn}{2}\log|2\pi\boldsymbol{V}| - \frac{Np}{2}\log|2\pi\boldsymbol{U}| - \frac{1}{2}\sum_{i=1}^{N}\operatorname{tr}\left[\boldsymbol{V}^{-1}(\boldsymbol{X}_i - \boldsymbol{M})'\boldsymbol{U}^{-1}(\boldsymbol{X}_i - \boldsymbol{M})\right],$$

where $\boldsymbol{M}$ is an $n \times p$ matrix. Using Section A.9, it can be seen that the likelihood is maximized when

$$\hat{\boldsymbol{M}} = \frac{1}{N}\sum_{i=1}^{N}\boldsymbol{X}_i. \tag{5.39}$$

Also, using Theorem 5.2.1, it is easy to see that (Mardia and Goodall, 1993) with fixed $\boldsymbol{A} = \sum_{i=1}^{N}(\boldsymbol{X}_i - \boldsymbol{M})'\boldsymbol{U}^{-1}(\boldsymbol{X}_i - \boldsymbol{M})$, the likelihood is maximized when

$$\boldsymbol{V} = \frac{1}{Nn}\sum_{i=1}^{N}(\boldsymbol{X}_i - \boldsymbol{M})'\boldsymbol{U}^{-1}(\boldsymbol{X}_i - \boldsymbol{M}). \tag{5.40}$$

Similarly, beginning with (A.5) and using Theorem 5.2.1 with fixed $\boldsymbol{A} = \sum_{i=1}^{N}(\boldsymbol{X}_i - \boldsymbol{M})\boldsymbol{V}^{-1}(\boldsymbol{X}_i - \boldsymbol{M})'$, the log-likelihood is maximized when

$$\boldsymbol{U} = \frac{1}{Np}\sum_{i=1}^{N}(\boldsymbol{X}_i - \boldsymbol{M})\boldsymbol{V}^{-1}(\boldsymbol{X}_i - \boldsymbol{M})'. \tag{5.41}$$

Equations (5.40) and (5.41) can be solved in an iterative procedure, after setting $\boldsymbol{M} = \hat{\boldsymbol{M}}$. It has been shown that the m.l.e. exists (with probability 1) if $N > \max\{p/n, n/p\} + 1$, but that the iterative solution for $\hat{\boldsymbol{\Sigma}} = \hat{\boldsymbol{V}} \otimes \hat{\boldsymbol{U}}$ depends on the starting values and so is not unique (Dutilleul, 1999; Roś et al., 2016).

This iterative algorithm has been extended to the three-factor case (but with $\boldsymbol{M}$ only depending on two factors) in Mardia and Goodall (1993). In this case, each covariance matrix can be determined conditional on the other two being known, and the algorithm is again iteratively applied.

5.3 Robust Estimation of Location and Dispersion for Multivariate Distributions

In classical statistics, it is assumed that the data form an i.i.d. sample from an $N_p(\boldsymbol{\mu}, \boldsymbol{\Sigma})$ distribution. However, in many practical examples, a data set may include some outliers not coming from this distribution. That is, a data set may consist mainly of "good" data from a multivariate normal distribution, but that there may also be some contamination by "bad" data, and the bad data may be outliers. The theory of robust statistics has been developed to deal with this situation. Here, we consider estimators of location and dispersion that are resistant in some aspects to the presence of outliers. In a data set of size n, suppose that h values are "good." To develop the theory, suppose that h is known. Typically, $h > n/2$, so that more than half the data are good; in addition Rousseeuw & van Zomeren (1990) recommend $h \geq 5p$.

5.3.1 M-Estimates of Location

An estimator based on each variable separately is the M-estimator (maximum-likelihood-type estimator). For a univariate sample $x_1 \ldots, x_n$, it is defined implicitly by

$$\sum_{r=1}^{n} \psi(x_r - T_n) = 0,$$

where ψ is a given function. If the sample comes from a symmetric distribution F, and if one takes $\psi(x) = \max[-k, \min(x, k)]$, where $F(-k) = \alpha$, then this estimator has the same asymptotic behavior as the α-trimmed mean (Huber, 1972). (Note that if $\psi(x) = x$, then $T_n = \bar{x}$). A similar strategy for a multivariate version is to find a location $\boldsymbol{T}$ to minimize

$$\sum_{r=1}^{n} \psi(||\boldsymbol{x}_r - \boldsymbol{T}||),$$

but such estimators are not affine equivariant (see Section 1.8). Maronna (1976) proposed affine equivariant M-estimates of both location and scatter, but they have poor properties and have been shown not to work very well in many situations (Wisnowski et al., 2002).

5.3.2 Minimum Covariance Determinant

The minimum covariance determinant (MCD) method (Rousseeuw, 1984) looks for the h observations whose sample covariance matrix has the lowest possible determinant. The MCD estimate of location is then the simple average of these h observations, and the MCD estimate of scatter is a multiple of their covariance matrix. These estimates are affine equivariant, and this makes the analysis independent of the units of measurement as well as translations or rotations of the data. An efficient algorithm to compute the MCD is described in Rousseeuw and van Driessen (1999).

One way to characterize the robustness of an estimator is the breakdown point, which can be roughly described as the smallest proportion of observations that can make the estimator arbitrarily badly behaved. A more formal definition (Donoho and Huber, 1983) of the breakdown value $\varepsilon^*(\boldsymbol{T}, \boldsymbol{X})$ of an estimator of location $\boldsymbol{T}$ for the data set $\boldsymbol{X}$ is as follows. Consider all possible contaminated data sets $\tilde{\boldsymbol{X}}$ obtained by replacing any m of the original observations by arbitrary points. Then, the breakdown value of the location estimator $\boldsymbol{T}$ is the smallest fraction m/n of outliers that can make the estimate arbitrarily large, i.e.

$$\varepsilon^*(\boldsymbol{T}, \boldsymbol{X}) = \min_{m} \left\{ \frac{m}{n} \sup_{\tilde{X}} ||\boldsymbol{T}(\tilde{\boldsymbol{X}}) - T(\boldsymbol{X})|| = \infty \right\}.$$

Similarly, the breakdown value of a covariance estimator $\hat{\boldsymbol{\Sigma}}$ is defined as the smallest fraction of outliers that can make the largest eigenvalue tend to infinity, or the smallest eigenvalue tend to zero. For the MCD estimates of location and scatter, the breakdown values are both $\varepsilon^* \approx (n - h)/n$, with a highest possible breakdown value when $h = (n + p + 1)/2$ (Lopuhaä and Rousseeuw, 1991).

A limitation of the method is that the number of observations selected h should be larger than the number of variables p, so high-dimensional data sets will need to be reduced first by variable selection or principal components (see Chapter 9).

5.3.3 Multivariate Trimming

One possible way to estimate the location of a multivariate sample is to treat each variable separately. For example, one can use the α-trimmed mean for each variable (i.e. the mean after omitting a proportion α of the smallest and a proportion α of the largest observations on each variable — note that this convention for α differs from the univariate case). Coordinate-wise trimming results in an uncertain overall fraction of data which is discarded, and it is not affine invariant. A better version for multivariate data is to use the intersection of all closed halfspaces that contain at least $100(1 - \alpha)\%$ of the sample. The resulting α-trimmed region is a convex set, the centroid of which could be used as a location estimator (Donoho and Huber, 1983).

Tukey has proposed a multivariate analogue to trimming called "peeling" (Barnett, 1976). This consists of deleting extreme points of the convex hull of the sample and either repeating this a fixed number of times or until a fixed percentage of the points has been removed. A closely related method, based on projection depth, was proposed by Zuo (2006).

Example 5.3.1 We illustrate the above techniques on the iris data of Table 1.3 for *Iris versicolor* with x_1 = petal length and x_2 = sepal length. Figure 5.1 shows the set of convex hulls for this data. The mean vector after "peeling once", i.e. excluding the extreme points (nine in all) of the convex hull, is $\bar{\boldsymbol{x}}_p^* = (4.26, 5.91)'$. This is close to the sample mean vector $\bar{\boldsymbol{x}} = (4.26, 5.94)'$. The excluded points are listed in Table 5.1.

The coordinate-wise α-trimmed values for $\alpha = 0.04$ are also shown in Table 5.1, i.e. the two largest and two smallest extreme values are trimmed for each variable. The trimmed vector is $\bar{\boldsymbol{x}}_{(\alpha)}^* = (4.27, 5.93)'$. This is also very close to the untrimmed mean $\bar{\boldsymbol{x}}$. Hence on these data the values of the three estimates are very similar. However, the properties of the estimators are vastly different, especially on samples which are contaminated in some way. □

Table 5.1 Extreme values to be excluded from robust estimates in the iris data of Table 1.3.

Values excluded after peeling once (x_1, x_2)	α-Trimming, $\alpha = 0.04$
(4.7, 7.0), (4.9, 6.9),	
(3.3, 4.9), (4.4, 6.7)	x_1 : 3.3, 5.0, 5.1, 3.0
(5.0, 6.7), (3.5, 5.7),	x_2 : 7.0, 6.9, 4.9, 5.0
(5.1, 6.0), (4.5, 5.4),	
(3.0, 5.1)	

5.3.4 Stahel–Donoho Estimator

The first high-breakdown location and scatter estimator was proposed by Stahel (1981) and Donoho (1982). These are essentially a weighted mean, with weights based on a measure of *outlyingness*, given by

$$u_r = \sup_{||\boldsymbol{v}||=1} \frac{|\boldsymbol{x}_r'\boldsymbol{v} - m(\boldsymbol{v})|}{\text{med}_k |\boldsymbol{x}_k'\boldsymbol{v} - m(\boldsymbol{v})|}, \qquad r = 1, \dots, n, \tag{5.42}$$

where $m(\boldsymbol{v}) = \text{med}_j(\boldsymbol{x}_j'\boldsymbol{v})$ is the median of projections of all data points $\boldsymbol{x}_j$, $j = 1, \dots, n$, in the direction of the vector $\boldsymbol{v}$. The location and scatter are then estimated using weights that are a strictly positive and decreasing function of $w(u) \geq 0$.

The Stahel–Donoho estimator (SDE) is related to projection pursuit (Section 9.9) in that there is a search over all possible projections $\boldsymbol{v}$ with u_r as a projection index. The estimator has good robustness properties – these properties are essentially determined by the denominator in Equation (5.42) – but it has relatively high computational cost.

5.3.5 Minimum Volume Estimator

Another affine equivariant estimator, also introduced by Rousseeuw (1984), is the minimum volume estimator (MVE). The objective of MVE is to find the minimal volume ellipsoid covering at least a fraction h/n of the data points, where h/n can be taken as low as $1/2 + 1/n$. The location estimator is then given as the center of the ellipsoid, and the corresponding covariance estimator is defined as the covariance matrix of the ellipsoid multiplied by a suitable correction factor, which makes the estimate consistent for the multivariate normal distribution.

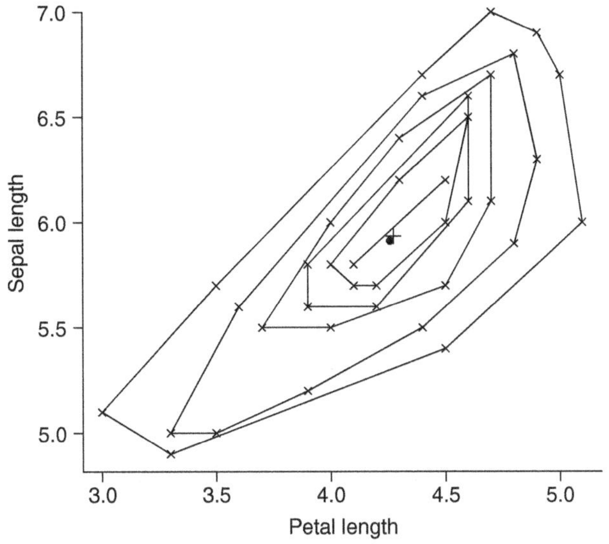

Figure 5.1 Convex hull for the iris data. (*I. versicolor* variety) with • = mean after "peeling" once and + = α-trimmed mean $\alpha = 0.04$.

According to Hubert et al. (2008), MVE has efficiency zero due to its low rate of convergence ($n^{-2/3}$), and the algorithms suffer from inefficiency and high computational complexity, which make them hard to use for large data sets.

Example 5.3.2 We use simulated data to illustrate the performance of MCD, MVE, and SDE compared with the classical maximum-likelihood estimator (5.23). We initially simulate 1500 observations from a standard multivariate normal, with four independent components. These data are then corrupted in one of two ways to create two contaminated data sets. "Outlying data" are 500 simulated observations generated from a bivariate normal with means $\mu_1 = -\mu_2 = 10$, variance $\sigma_1 = \sigma_2 = 1$, and correlation $\rho = 0.9$. In the first contamination, the first two variables of a random sample of size 500 of the initial 1500 observations are substituted with the outlying data ("Random contamination"). In the second contamination, we substitute the 500 observations with the largest $\boldsymbol{x}'\boldsymbol{x}$ with the outlying data ("Contaminated edge").

The results are shown in Table 5.2. It can be seen that the location and scatter estimates of the m.l.e. are greatly affected by the outliers. The other three (robust) estimates are quite similar for both data sets. The estimate of the mean is good for both data sets, but the three robust estimates of the (co)variances are somewhat better for the second contamination. □

Table 5.2 Estimates of location and scatter for each of four methods: maximum-likelihood estimate (MLE); minimum volume estimator (MVE); minimum covariance determinant (MCD); and Stahel–Donoho estimator (SDE).

	Random contamination					Contaminated edge				
	$\bar{\boldsymbol{x}}$	$\hat{\boldsymbol{\Sigma}}$				$\bar{\boldsymbol{x}}$	$\hat{\boldsymbol{\Sigma}}$			
	3.34	23.44	−21.94	0.02	0.02	3.34	23.21	−22.05	0.01	−0.07
M	−3.34	−21.94	23.02	−0.11	0.02	−3.32	−22.05	22.93	0.05	−0.01
L	−0.01	0.02	−0.11	1.01	0.04	−0.01	0.01	0.05	1.01	0.04
E	0.00	0.02	0.02	0.04	1.02	0.00	−0.07	−0.01	0.04	1.02
	−0.01	1.67	0.07	0.00	−0.01	−0.01	1.03	0.02	0.00	−0.03
M	−0.01	0.07	1.54	−0.14	−0.01	0.02	0.02	0.94	−0.02	0.00
V	−0.01	0.00	−0.14	1.71	0.14	−0.01	0.00	−0.02	0.96	−0.03
E	0.00	−0.01	−0.01	0.14	1.68	0.00	−0.03	0.00	−0.03	1.03
	−0.01	1.67	0.06	0.04	−0.01	−0.01	1.03	0.02	0.00	−0.03
M	−0.01	0.06	1.54	−0.13	−0.01	0.02	0.02	0.94	−0.02	0.00
C	−0.02	0.04	−0.13	1.71	0.12	−0.01	0.00	−0.02	0.96	−0.03
D	0.00	−0.01	−0.01	0.12	1.68	0.00	−0.03	0.00	−0.03	1.03
	0.00	1.68	−0.04	0.07	0.02	−0.01	1.02	0.02	0.00	−0.03
S	−0.03	−0.04	1.58	−0.12	0.01	0.02	0.02	0.93	−0.02	0.00
D	0.01	0.07	−0.12	1.51	0.06	−0.01	0.00	−0.02	0.95	−0.03
E	0.02	0.02	0.01	0.06	1.61	0.00	−0.03	0.00	−0.03	1.01

See text for models used to create the two data sets.

5.3.6 Tyler's Estimate of Scatter

The preceding suggestions for robustness assume that a data set consists of mainly "good" data from a multivariate normal distribution, but that there may be some contamination by "bad" data, and the bad data may be outliers. This section takes a slightly different perspective. It is assumed that the data $\boldsymbol{x}_r$, $r = 1, \ldots, n$, are i.i.d. from an elliptical distribution centered at the origin (Section 3.3.3), but that the distribution may be long-tailed. The objective is to estimate the scale matrix $\boldsymbol{\Sigma}$. In this setting, the scale matrix is only determined up to a scalar constant, so that $\boldsymbol{\Sigma}$ and $c\boldsymbol{\Sigma}$, $c > 0$, are equivalent.

Let $\boldsymbol{x}$ be a random vector from this distribution, and set $\boldsymbol{y} = \boldsymbol{\Sigma}^{-1/2}\boldsymbol{x}$. Then, $\boldsymbol{y}$ has an isotropic distribution, so that $\boldsymbol{y}/||\boldsymbol{y}||$ has a uniform distribution on the unit sphere in $\mathbb{R}^p$. Hence,

$$\mathsf{E}\{\boldsymbol{y}\boldsymbol{y}'/||\boldsymbol{y}||^2\} = \mathsf{E}\{\boldsymbol{\Sigma}^{-1/2}\boldsymbol{x}\boldsymbol{x}'\boldsymbol{\Sigma}^{-1/2}/\boldsymbol{x}'\boldsymbol{\Sigma}^{-1}\boldsymbol{x}\} = \frac{1}{p}\boldsymbol{I}_p,$$

or equivalently,

$$\mathsf{E}\{\boldsymbol{x}\boldsymbol{x}'/\boldsymbol{x}'\boldsymbol{\Sigma}^{-1}\boldsymbol{x}\} = \frac{1}{p}\boldsymbol{\Sigma}.$$

The sample version of this equation is given by

$$\sum \boldsymbol{x}_r\boldsymbol{x}_r'/\boldsymbol{x}_r'\boldsymbol{\Sigma}^{-1}\boldsymbol{x}_r = \frac{n}{p}\boldsymbol{\Sigma}. \tag{5.43}$$

Under mild conditions on the data (e.g. $n \geq p$ and no linear dependencies in the data), it turns out that (5.43) has a unique solution $\hat{\boldsymbol{\Sigma}}$, say, defined up to a scalar multiple (Tyler, 1987; Kent and Tyler, 1988).

5.4 Bayesian Inference

Until now the vector $\boldsymbol{\theta}$ has been regarded as fixed but unknown. However, a Bayesian approach regards $\boldsymbol{\theta}$ as a random variable whose distribution reflects prior information or subjective beliefs in what the value of $\boldsymbol{\theta}$ is likely to be. A fairly complete exposition of the multivariate aspects of Bayesian inference appears in chapter 8 of Box and (Tiao 1973).

Suppose that before observing the vector $\boldsymbol{x}$, our beliefs about the value of $\boldsymbol{\theta}$ are represented by a prior distribution $\pi(\boldsymbol{\theta})$. For instance, if θ is a scalar which is believed to lie between 1 and 5 with probability 95%, then we may take the prior distribution to be normal with mean 3 and variance 1, since about 95% of this distribution falls between 1 and 5. This would lead to

$$\pi(\theta) = \frac{1}{\sqrt{2\pi}} \exp\left\{-\frac{(\theta-3)^2}{2}\right\}.$$

Alternatively, if our beliefs were not symmetric about 3, some other prior distribution might be more suitable. For instance, if θ is nonnegative, then a gamma distribution could be used.

The prior density, given by $\pi(\boldsymbol{\theta})$, and the likelihood of the observed data $f(\boldsymbol{x}; \boldsymbol{\theta})$ together define the so-called posterior distribution, which is the conditional distribution of $\boldsymbol{\theta}$, given $\boldsymbol{x}$. By Bayes' theorem, the posterior density is given by

$$\pi(\boldsymbol{\theta} \mid \boldsymbol{x}) = \{\pi(\boldsymbol{\theta})f(\boldsymbol{x}; \boldsymbol{\theta})\} \bigg/ \int \pi(\boldsymbol{\theta})f(\boldsymbol{x}; \boldsymbol{\theta})\, \mathrm{d}\boldsymbol{\theta} \propto \pi(\boldsymbol{\theta})f(\boldsymbol{x}; \boldsymbol{\theta}) . \tag{5.44}$$

All Bayesian inference is based on the posterior probability function. In particular, the Bayesian estimate of $\boldsymbol{\theta}$ is given by the mean of the posterior density $\pi(\boldsymbol{\theta} \mid \boldsymbol{x})$.

Sometimes, it is convenient to use an improper prior density ($\int \pi(\theta)\, d\theta = \infty$). Such priors are allowable as long as $\int \pi(\theta) f(\boldsymbol{x}; \theta)\, d\theta < \infty$, so that the posterior is defined.

Example 5.4.1 Suppose that $\boldsymbol{X} = (\boldsymbol{x}_1, \dots, \boldsymbol{x}_n)'$ is a random sample from $N_p(\boldsymbol{\mu}, \boldsymbol{\Sigma})$, and that there is no prior knowledge concerning $\boldsymbol{\mu}$ and $\boldsymbol{\Sigma}$. Then, a natural choice of prior distribution for $\theta = (\boldsymbol{\mu}, \boldsymbol{\Sigma})$ is the noninformative (or vague) prior based on Jeffreys' principle of invariance (Jeffreys 1961, p. 179). This principle assumes that the location parameter $\boldsymbol{\mu}$ and scale parameter $\boldsymbol{\Sigma}$ are independent, so $\pi(\boldsymbol{\mu}, \boldsymbol{\Sigma}) = \pi(\boldsymbol{\mu})\pi(\boldsymbol{\Sigma})$, and second that

$$\pi(\boldsymbol{\mu}) \propto |\boldsymbol{F}(\boldsymbol{\mu})|^{1/2}, \qquad \pi(\boldsymbol{\Sigma}) \propto |\boldsymbol{F}(\boldsymbol{\Sigma})|^{1/2},$$

where $\boldsymbol{F}(\boldsymbol{\mu})$ and $\boldsymbol{F}(\boldsymbol{\Sigma})$ are the information matrices for $\boldsymbol{\mu}$ and $\boldsymbol{\Sigma}$. Hence, using Exercise 5.1.12, we see that

$$\pi(\boldsymbol{\mu}, \boldsymbol{\Sigma}) \propto |\boldsymbol{\Sigma}|^{-(p+1)/2}. \tag{5.45}$$

The marginal posterior of $\boldsymbol{\mu}$ can now be obtained by substituting for $\pi(\boldsymbol{\mu}, \boldsymbol{\Sigma})$ in (5.44) and integrating $\pi(\boldsymbol{\mu}, \boldsymbol{\Sigma} \mid \boldsymbol{X})$ with respect to $\boldsymbol{\Sigma}$ to give

$$\pi(\boldsymbol{\mu} \mid \boldsymbol{X}) \propto \int |\boldsymbol{\Sigma}|^{-(n+p+1)/2} \exp\left(-\frac{1}{2} \text{tr } \boldsymbol{\Sigma}^{-1}\boldsymbol{V}\right) d\boldsymbol{\Sigma}, \tag{5.46}$$

where $\boldsymbol{V} = n\boldsymbol{S} + n(\overline{\boldsymbol{x}} - \boldsymbol{\mu})(\overline{\boldsymbol{x}} - \boldsymbol{\mu})'$. Although the prior density is improper, with an infinite integral, the posterior density defines a proper distribution.

The function to be integrated here is similar to the p.d.f. of the inverted Wishart distribution given in (4.49). Using the normalizing constant of that expression, it is clear from (5.46) that

$$\pi(\boldsymbol{\mu} \mid \boldsymbol{X}) \propto |\boldsymbol{V}|^{-n/2}.$$

But

$$|\boldsymbol{V}| = n^p |\boldsymbol{S}| \left\{1 + (\overline{\boldsymbol{x}} - \boldsymbol{\mu})'\boldsymbol{S}^{-1}(\overline{\boldsymbol{x}} - \boldsymbol{\mu})\right\}.$$

Therefore,

$$\pi(\boldsymbol{\mu} \mid \boldsymbol{X}) \propto \left\{1 + (\overline{\boldsymbol{x}} - \boldsymbol{\mu})'\boldsymbol{S}^{-1}(\overline{\boldsymbol{x}} - \boldsymbol{\mu})\right\}^{-n/2}.$$

Thus, the posterior distribution of $\boldsymbol{\mu}$ is a multivariate t distribution with $(n - p)$ degrees of freedom and parameters $\overline{\boldsymbol{x}}$ and $\boldsymbol{S}$ (see Exercise 3.2.5). The mean of this posterior distribution is the sample mean $\overline{\boldsymbol{x}}$, so in this case the Bayes' estimate of $\boldsymbol{\mu}$ is the same as the m.l.e.

The posterior distribution of $\boldsymbol{\Sigma}$ is obtained by integrating $\pi(\boldsymbol{\mu}, \boldsymbol{\Sigma} \mid \boldsymbol{X})$ with respect to $\boldsymbol{\mu}$. This leads to an inverted Wishart posterior distribution, with a mean of $n\boldsymbol{S}/(n - p - 2)$ (see Exercises 4.4.14 and 5.4.1). Note that in this case the Bayes' estimate differs from the m.l.e. □

Example 5.4.2 When one wishes to incorporate *informative* prior information into the analysis, then a useful prior distribution is the "natural conjugate prior" (see, for example, Press (1972), p. 76). For example, in the case of a random sample from the $N_p(\boldsymbol{\mu}, \boldsymbol{\Sigma})$ distribution, the density of $\overline{\boldsymbol{x}}$ and $\boldsymbol{S}$ from (4.6) and (4.48) is

$$\begin{aligned} f(\overline{\boldsymbol{x}}, \boldsymbol{S}; \boldsymbol{\mu}, \boldsymbol{\Sigma}) \propto &\left\{|\boldsymbol{\Sigma}|^{-1/2} \exp\left[-\frac{1}{2} n(\overline{\boldsymbol{x}} - \boldsymbol{\mu})'\boldsymbol{\Sigma}^{-1}(\overline{\boldsymbol{x}} - \boldsymbol{\mu})\right]\right\} \\ &\times \left\{|\boldsymbol{\Sigma}|^{-(n-1)/2} \exp\left[-\frac{1}{2} n \text{ tr}(\boldsymbol{\Sigma}^{-1}\boldsymbol{S})\right]\right\}. \end{aligned} \tag{5.47}$$

Then, the conjugate prior for $\boldsymbol{\mu}$ and $\boldsymbol{\Sigma}$ is obtained by thinking of (5.47) as a density in $\boldsymbol{\mu}$ and $\boldsymbol{\Sigma}$ after interpreting the quantities $\boldsymbol{x}$, $n\boldsymbol{S}$, and $n-1$ as parameters of the prior distribution $\boldsymbol{\phi}$, $\boldsymbol{G} > 0$, and $m > 2p-1$ and adjusting the proportionality constant, so that the density integrates to 1. Thus,

$$\pi(\boldsymbol{\mu}, \boldsymbol{\Sigma}) \propto \left\{ |\boldsymbol{\Sigma}|^{-m/2} \exp\left(-\frac{1}{2}\text{tr}\, \boldsymbol{\Sigma}^{-1}\boldsymbol{G}\right)\right\} \left\{ |\boldsymbol{\Sigma}|^{-1/2} \exp\left[-\frac{1}{2}(\boldsymbol{\mu}-\boldsymbol{\phi})'\boldsymbol{\Sigma}^{-1}(\boldsymbol{\mu}-\boldsymbol{\phi})\right]\right\}. \quad (5.48)$$

The Bayes' estimates of $\boldsymbol{\mu}$ and $\boldsymbol{\Sigma}$ are, respectively (see Exercise 5.4.2),

$$(\boldsymbol{\phi} + n\overline{\boldsymbol{x}})/(1+n), \qquad \left\{ n\boldsymbol{S} + \boldsymbol{G} + \frac{n}{1+n}(\overline{\boldsymbol{x}}-\boldsymbol{\phi})(\overline{\boldsymbol{x}}-\boldsymbol{\phi})' \right\} /(n+m-2p-2) \quad (5.49)$$

Neither of these Bayes' estimates is the same as the corresponding m.l.e., although they can each be expressed as a weighted average of the prior mean and the m.l.e. □

The Bayesian theory in this chapter has focused on prior distributions related to the multivariate normal distribution, for which posterior distributions can be computed in closed form. More modern approaches to Bayesian statistics relax this restriction by using simulation-based inference methods such as Markov chain Monte Carlo (MCMC) when closed-form expressions are not available. Further, this chapter has not emphasized the philosophy behind Bayesian inference. From the Bayesian perspective, parameters are viewed as random in the same way as random vectors and often reflect personal beliefs. Further, the use of Bayesian methods allows statisticians to address more complicated problems than would otherwise be possible. See, for example, O'Hagan and Forster (2004) and Green et al. (2003).

Exercises and Complements

5.1.1 Let $\boldsymbol{x}$ come from the general exponential family with p.d.f

$$f(\boldsymbol{x}; \boldsymbol{\theta}) = \exp\left\{ a_0(\boldsymbol{\theta}) + b_0(\boldsymbol{x}) + \sum_{i=1}^{q} a_i(\boldsymbol{\theta})b_i(\boldsymbol{x}) \right\}.$$

(a) Show that the score function is

$$\boldsymbol{s} = \frac{\partial a_0}{\partial \boldsymbol{\theta}} + \sum_{i=1}^{q} \frac{\partial a_i}{\partial \boldsymbol{\theta}} b_i(\boldsymbol{x}) = \frac{\partial a_0}{\partial \boldsymbol{\theta}} + \boldsymbol{A}\boldsymbol{b},$$

where $\boldsymbol{A}$ is the matrix whose ith column is $\partial a_i/\partial \boldsymbol{\theta}$, and $\boldsymbol{b}'$ is the vector $(b_1(\boldsymbol{x}), \ldots, b_q(\boldsymbol{x}))$.

(b) Note that $\partial a_0/\partial \boldsymbol{\theta}$ and $\boldsymbol{A}$ are both free of $\boldsymbol{x}$. Therefore, since the score has mean zero, prove that

$$\boldsymbol{A}\text{E}[\boldsymbol{b}] = -\frac{\partial a_0}{\partial \boldsymbol{\theta}}.$$

If $\boldsymbol{A}$ is a nonsingular matrix, then

$$\text{E}[\boldsymbol{b}] = -\boldsymbol{A}^{-1}\frac{\partial a_0}{\partial \boldsymbol{\theta}}.$$

(c) Comparing (a) and (b), prove that the value of $\boldsymbol{\theta}$ for which $\boldsymbol{s}$ is zero (i.e. the m.l.e) is also the value that equates $\boldsymbol{b}$ to its expected value.

(d) Show that the information matrix is given by

$$\boldsymbol{F} = -\mathsf{E}\left(\frac{\partial \boldsymbol{s}'}{\partial \boldsymbol{\theta}}\right) = -\frac{\partial^2 a_0}{\partial\boldsymbol{\theta}\partial\boldsymbol{\theta}'} - \sum_{i=1}^{q} \frac{\partial^2 a_i}{\partial\boldsymbol{\theta}\partial\boldsymbol{\theta}'}\mathsf{E}[b_i(\boldsymbol{x})].$$

5.1.2 (a) If $\boldsymbol{x}$ has the multinomial distribution, so that $\text{var}(\boldsymbol{x}) = \text{diag}\,(\boldsymbol{a}) - \boldsymbol{a}\boldsymbol{a}'$, as shown in Section 3.2.1, then show that

$$\text{var}(\boldsymbol{\beta}'\boldsymbol{x}) = \boldsymbol{\beta}'\text{var}(\boldsymbol{x})\boldsymbol{\beta} = \sum \beta_i^2 a_i - \left(\sum \beta_i a_i\right)^2.$$

(b) Note that the score s defined in Example 5.1.4 may be written as $s = \boldsymbol{\beta}'\boldsymbol{x}$, where $a_1 = (2+\theta)/4$, $a_2 = a_3 = (1-\theta)/4$, and $a_4 = \theta/4$ and $\beta_1 = 1/(4a_1)$, $\beta_2 = \beta_3 = -1/(4a_2)$, and $\beta_4 = 1/(4a_4)$. Hence, deduce that $\text{var}(s) = \frac{1}{16}\sum a_i^{-1}$.

5.1.3 Show from (5.2) that the score of a set of independent random variables equals the sum of their individual scores.

5.1.4 If $l(\boldsymbol{X};\boldsymbol{\theta})$ is defined in (5.10) with $\boldsymbol{t} = \overline{\boldsymbol{x}}$, verify Corollary 5.1.3.

5.1.5 If $\boldsymbol{x}_1, \ldots, \boldsymbol{x}_n$ are independent variables whose distributions depend on the parameter $\boldsymbol{\theta}$, and if $\boldsymbol{F}_r = \mathsf{E}(\boldsymbol{s}_r\boldsymbol{s}_r')$ is the Fisher information matrix corresponding to $\boldsymbol{x}_r$, then show that $\mathsf{E}(\boldsymbol{s}_r\boldsymbol{s}_j') = \boldsymbol{0}$ for $r \neq j$, and hence that $\boldsymbol{F} = \boldsymbol{F}_1 + \cdots + \boldsymbol{F}_n$ is the Fisher information matrix corresponding to the matrix $\boldsymbol{X} = (\boldsymbol{x}_1, \ldots, \boldsymbol{x}_n)'$.

5.1.6 Obtain the Fisher information corresponding to the genetic experiment of Example 5.1.4 by squaring the score and evaluating $\mathsf{E}(s^2)$. Check that the result is equal to that obtained in Example 5.1.4.

5.1.7 Show that if $\mathsf{E}(\boldsymbol{t}) = \boldsymbol{\theta} + \boldsymbol{b}(\boldsymbol{\theta})$, where $\boldsymbol{t}$ does not depend on $\boldsymbol{\theta}$, then equation (5.18) becomes

$$\text{cor}^2(\alpha,\gamma) = \{\mathbf{c}'(\boldsymbol{I}+\boldsymbol{B})\boldsymbol{a}\}^2/(\boldsymbol{a}'\boldsymbol{V}\boldsymbol{a}\;\mathbf{c}'\boldsymbol{F}\mathbf{c}), \qquad \boldsymbol{B} = (\partial b_j(\boldsymbol{\theta})/\partial\theta_i),$$

so

$$\max_{\boldsymbol{c}} \text{cor}^2(\alpha,\gamma) = \boldsymbol{a}'(\boldsymbol{I}+\boldsymbol{B})'\boldsymbol{F}^{-1}(\boldsymbol{I}+\boldsymbol{B})\boldsymbol{a}/\boldsymbol{a}'\boldsymbol{V}\boldsymbol{a}.$$

Hence, show that the Cramér–Rao lower bound can be generalized.

5.1.8 If $\boldsymbol{A} \sim W_p(\boldsymbol{\Sigma}, m)$, show that its likelihood is given by

$$L(\boldsymbol{A};\boldsymbol{\Sigma}) = c|\boldsymbol{\Sigma}|^{-m/2}|\boldsymbol{A}|^{(m-p-1)/2}\exp\left(-\frac{1}{2}\text{tr}\,\boldsymbol{\Sigma}^{-1}\boldsymbol{A}\right),$$

where c includes all the terms that depend only on m and p. Taking $\boldsymbol{V} = \boldsymbol{\Sigma}^{-1}$ as the parameter as in Section 5.2.2.1, the log-likelihood is therefore

$$l(\boldsymbol{A};\boldsymbol{V}) = \log c + \frac{1}{2}m\log|\boldsymbol{V}| + \frac{1}{2}(m-p-1)\log|\boldsymbol{A}| - \frac{1}{2}\text{tr}\,\boldsymbol{V}\boldsymbol{A}.$$

Writing $\boldsymbol{K} = \frac{1}{2}m\boldsymbol{\Sigma} - \frac{1}{2}\boldsymbol{A}$, use the results of Section 5.2.2.1 to show that the score is the matrix

$$\boldsymbol{U} = \frac{\partial l}{\partial \boldsymbol{V}} = 2\boldsymbol{K} - \text{Diag}\,\boldsymbol{K}.$$

Note that, since $\mathsf{E}(\boldsymbol{A}) = m\boldsymbol{\Sigma}$, it is clear that $\boldsymbol{U}$ has mean zero.

5.1.9 If $\boldsymbol{x} \sim N_p(\mu\boldsymbol{a}, \sigma^2\boldsymbol{I})$, where $\boldsymbol{a}$ is known, then show that $\boldsymbol{x}$ enters the likelihood function only through the bivariate vector $\boldsymbol{t} = (\boldsymbol{a}'\boldsymbol{x}, \boldsymbol{x}'\boldsymbol{x})$. Hence, show that $\boldsymbol{t}$ is sufficient for (μ, σ^2), and that when the elements of $\boldsymbol{x}$ are i.i.d. $N(\mu, \sigma^2)$, then $(\boldsymbol{x}'\mathbf{1}, \boldsymbol{x}'\boldsymbol{x})$ is sufficient for (μ, σ).

5.1.10 (Generalization of Example 5.1.4 and Exercise 5.1.2). If $\boldsymbol{x} = (x_1, \dots, x_p)'$ has a multinomial distribution with parameters $a_1(\boldsymbol{\theta}), \dots, a_p(\boldsymbol{\theta})$, and $N = \sum x_i$,

(a) show that the score function is

$$\boldsymbol{s} = \sum \frac{x_i}{a_i}\frac{\partial a_i}{\partial \boldsymbol{\theta}};$$

(b) show that the condition $\mathsf{E}[\boldsymbol{s}] = \mathbf{0}$ is equivalent to the condition $\sum(\partial a_i/\partial \boldsymbol{\theta}) = \mathbf{0}$, which is implied by the condition $\sum a_i = 1$;

(c) show that

$$\frac{\partial \boldsymbol{s}'}{\partial \boldsymbol{\theta}} = \sum \frac{x_i}{a_i^2}\left(\frac{\partial^2 a_i}{\partial \boldsymbol{\theta}\partial \boldsymbol{\theta}'} - \frac{\partial a_i \partial a_i}{\partial \boldsymbol{\theta}\partial \boldsymbol{\theta}'}\right),$$

and that the information matrix is

$$\boldsymbol{F} = -\mathsf{E}\left(\frac{\partial \boldsymbol{s}'}{\partial \boldsymbol{\theta}}\right) = N\sum\left(\frac{1}{a_i}\frac{\partial a_i}{\partial \boldsymbol{\theta}}\frac{\partial a_i}{\partial \boldsymbol{\theta}'} - \frac{\partial^2 a_i}{\partial \boldsymbol{\theta}\partial \boldsymbol{\theta}'}\right).$$

Confirm that, when $\boldsymbol{\theta}$ is a scalar, this simplifies to

$$f = N\sum\left(\frac{1}{a_i}\left(\frac{\partial a_i}{\partial \theta}\right)^2 - \frac{\partial^2 a_i}{\partial \theta^2}\right).$$

5.1.11 Suppose that $\boldsymbol{s}(\boldsymbol{x}; \boldsymbol{\theta})$ and $\boldsymbol{F}(\boldsymbol{\theta})$ are the score and vector information matrix with respect to a particular parameter vector $\boldsymbol{\theta}$. Let $\boldsymbol{\phi} = \boldsymbol{\phi}(\boldsymbol{\theta})$, where $\boldsymbol{\phi}(\cdot)$ is a one-to-one differentiable transformation. Then, show that the new score vector and information matrix are, respectively,

$$\boldsymbol{s}(\boldsymbol{x}; \boldsymbol{\phi}) = \frac{\partial \boldsymbol{\theta}'}{\partial \boldsymbol{\phi}}\boldsymbol{s}(\boldsymbol{x}; \boldsymbol{\theta}), \qquad \boldsymbol{F}(\boldsymbol{\phi}) = \frac{\partial \boldsymbol{\theta}'}{\partial \boldsymbol{\phi}}\boldsymbol{F}(\boldsymbol{\theta})\frac{\partial \boldsymbol{\theta}}{\partial \boldsymbol{\phi}'}.$$

5.1.12 Let $\boldsymbol{x}_1, \dots, \boldsymbol{x}_n$ be a sample from $N_p(\boldsymbol{\mu}, \boldsymbol{\Sigma})$ with $\boldsymbol{\mu}$ known and $p = 2$. For simplicity, take $\boldsymbol{V} = \boldsymbol{\Sigma}^{-1}$ as the parameter matrix. Put $\boldsymbol{v}' = (v_{11}, v_{22}, v_{12})$. Then, using (5.24), show that

$$\boldsymbol{F}(\boldsymbol{V}) = -\mathsf{E}\left[\frac{\partial^2 l}{\partial \boldsymbol{v}\partial \boldsymbol{v}'}\right] = \frac{n}{2(v_{11}v_{22} - v_{12}^2)^2}\begin{bmatrix} v_{22}^2 & v_{12}^2 & -2v_{22}v_{12} \\ v_{12}^2 & v_{11}^2 & -2v_{11}v_{12} \\ -2v_{22}v_{12} & -2v_{11}v_{12} & 2(v_{11}v_{22} + v_{12}^2) \end{bmatrix}.$$

Hence, show that

$$|\boldsymbol{F}(\boldsymbol{V})| = \frac{1}{4}n^3|\boldsymbol{V}|^{-3} = \frac{1}{4}n^3|\boldsymbol{\Sigma}|^3.$$

This result can be generalized to the general case to give

$$|\boldsymbol{F}(\boldsymbol{V})| \propto |\boldsymbol{V}|^{-(p+1)} \propto |\boldsymbol{\Sigma}|^{p+1}$$

(see, for example, Box and Tiao (1973)). Hence, using Exercise 5.1.11 and the Jacobian from Table 2.1, show that

$$|\boldsymbol{F}(\boldsymbol{\Sigma})| \propto |\boldsymbol{\Sigma}|^{-(p+1)}.$$

5.1.13 Show that the information matrix derived in Exercise 5.1.12 for $p = 2$ can also be written as

$$F(\boldsymbol{V}) = \frac{n}{2}\begin{bmatrix} \sigma_{11}^2 & \sigma_{12}^2 & 2\sigma_{11}\sigma_{12} \\ \sigma_{12}^2 & \sigma_{22}^2 & 2\sigma_{22}\sigma_{12} \\ 2\sigma_{11}\sigma_{12} & 2\sigma_{22}\sigma_{12} & 2(\sigma_{22}\sigma_{11} + \sigma_{12}^2) \end{bmatrix}.$$

5.2.1 If $\boldsymbol{\mu} = k\boldsymbol{\mu}_0$, and $\hat{k}$ is the m.l.e. of k given by (5.29), show that $\mathsf{E}(\hat{k}) = k$ and $\text{var}(\hat{k}) = (n\boldsymbol{\mu}_0'\boldsymbol{\Sigma}^{-1}\boldsymbol{\mu}_0)^{-1}$, where n is the sample size.

5.2.2 Since $\overline{\boldsymbol{x}}$ is independent of $\boldsymbol{S}$ and has expectation $k\boldsymbol{\mu}_0$, show that the estimator $\hat{k}$ given in (5.31) is unbiased, and that its variance is

$$n^{-1}\mathsf{E}\{\boldsymbol{\mu}_0'\boldsymbol{S}^{-1}\boldsymbol{\Sigma}\boldsymbol{S}^{-1}\boldsymbol{\mu}_0/(\boldsymbol{\mu}_0'\boldsymbol{S}^{-1}\boldsymbol{\mu}_0)^2\}.$$

5.2.3 If $x > 0$, then show that

$$f(x) = x - 1 - \log x$$

has a minimum of 0 when $x = 1$. (The first derivative of x is positive when $x > 1$, negative when $x < 1$, and zero when $x = 1$.) Therefore, show that

$$x \geq \log x + 1.$$

If a and g are the arithmetic and geometric means of a set of positive numbers, then establish that $a \geq 1 + \log g$, and that equality holds only when each positive number equals unity.

5.2.4 Using the data in Example 5.2.3, show that if $\boldsymbol{\mu}'$ was known to be proportional to $(2, 8)$, then the constant of proportionality would be $k = 1.17$.

5.2.5 If $\boldsymbol{x} \sim N_p(\boldsymbol{\mu}, \sigma^2\boldsymbol{I})$, where $\boldsymbol{\mu}$ is known to lie on the unit sphere (i.e. $\boldsymbol{\mu}'\boldsymbol{\mu} = 1$), show that the m.l.e. of $\boldsymbol{\mu}$ is $\boldsymbol{x}/(\boldsymbol{x}'\boldsymbol{x})^{1/2}$.

5.2.6 If $\boldsymbol{R}\boldsymbol{\mu} = \boldsymbol{r}$, and $\boldsymbol{\mu}$ is given by (5.33), show that $\mathsf{E}(\hat{\boldsymbol{\mu}}) = \boldsymbol{\mu}$ and $\text{var}(\hat{\boldsymbol{\mu}}) = n^{-1}(\boldsymbol{\Sigma} - \boldsymbol{\Sigma}\boldsymbol{R}'(\boldsymbol{R}\boldsymbol{\Sigma}\boldsymbol{R}')^{-1}\boldsymbol{R}\boldsymbol{\Sigma})$. Note that $\text{var}(\hat{\boldsymbol{\mu}})$ is less than $\text{var}(\overline{\boldsymbol{x}})$ by a nonnegative definite matrix.

5.2.7 If $\boldsymbol{\Sigma} = k\boldsymbol{\Sigma}_0$, and $\boldsymbol{\mu}$ is unconstrained, so that $\hat{k}$ is given by $p^{-1}\text{tr } \boldsymbol{\Sigma}_0^{-1}\boldsymbol{S}$, as in (5.36), show that $\mathsf{E}[\hat{k}] = (n-1)k/n$.

5.2.8 (Proof of (5.34)) If $\boldsymbol{R\mu} = \boldsymbol{r}$, and $\boldsymbol{\Sigma}$ is unknown, then the m.l.e. of $\boldsymbol{\mu}$ is obtained by solving simultaneously

$$\frac{\partial l^+}{\partial \boldsymbol{\mu}} = \mathbf{0} \qquad \text{and} \qquad \boldsymbol{R\mu} = \boldsymbol{r}.$$

Note that $\hat{\boldsymbol{\Sigma}} = \boldsymbol{S} + \boldsymbol{dd}'$, where $\boldsymbol{d} = \hat{\boldsymbol{\Sigma}}\boldsymbol{R}'(\boldsymbol{R}\hat{\boldsymbol{\Sigma}}\boldsymbol{R}')^{-1}\boldsymbol{a}$ and $\boldsymbol{a} = \boldsymbol{R}\overline{\boldsymbol{x}} - \boldsymbol{r}$. Also, note that $\boldsymbol{Rd} = \boldsymbol{a}$, and therefore, $\hat{\boldsymbol{\Sigma}}\boldsymbol{R}' = \boldsymbol{SR}' + \boldsymbol{da}'$ and $\boldsymbol{R}\hat{\boldsymbol{\Sigma}}\boldsymbol{R}' = \boldsymbol{RSR}' + \boldsymbol{aa}'$. Thus,

$$\boldsymbol{d} = (\boldsymbol{SR}' + \boldsymbol{da}')(\boldsymbol{aa}' + \boldsymbol{RSR}')^{-1}\boldsymbol{a} = \boldsymbol{SR}'(\boldsymbol{RSR}')^{-1}\boldsymbol{a},$$

where we have used (A.29) with $\boldsymbol{A} = \boldsymbol{RSR}'$, $\boldsymbol{B} = \boldsymbol{D}' = \boldsymbol{a}$, and $\boldsymbol{C} = 1$ for the second part, plus some simplification. But $\boldsymbol{d} = \overline{\boldsymbol{x}} - \hat{\boldsymbol{\mu}}$. Therefore, $\hat{\boldsymbol{\mu}} = \overline{\boldsymbol{x}} - \boldsymbol{SR}'(\boldsymbol{RSR}')^{-1}(\boldsymbol{R}\overline{\boldsymbol{x}} - \boldsymbol{r})$. This is the same as the expression obtained in (5.33) when $\boldsymbol{\Sigma}$ is known, with the substitution of $\boldsymbol{S}$ for $\boldsymbol{\Sigma}$.

5.2.9 Use (5.33) to show that when $\boldsymbol{\mu}$ is known *a priori* to be of the form $(\boldsymbol{\mu}_1', \mathbf{0}')'$, then the m.l.e. of $\boldsymbol{\mu}_1$ is $\overline{\boldsymbol{x}}_1 = \boldsymbol{\Sigma}_{12}\boldsymbol{\Sigma}_{22}^{-1}\overline{\boldsymbol{x}}_2$.

5.2.10 (Mardia, 1962, 1964b) (a) Show that the log-likelihood of a random sample $(x_1, y_1)', \ldots, (x_n, y_n)'$ from the bivariate Pareto distribution of Exercise 2.2.1 satisfies

$$n^{-1}l = \log p(p+1) - (p+2)\overline{\alpha},$$

where

$$n\overline{\alpha} = \sum_{r=1}^{n} \log(x_r + y_r - 1).$$

(b) Show that the maximum-likelihood estimator of p is given by

$$\hat{p} = \frac{1}{\overline{\alpha}} - \frac{1}{2} + \left(\frac{1}{\overline{\alpha}^2} + \frac{1}{4}\right)^{1/2}.$$

(c) By evaluating $\partial^2 l/\partial p^2$, show that the Fisher information matrix is $n(2p^2 + 2p + 1)/\{p(p+1)\}^2$, and hence, show that the asymptotic variance of $\hat{p}$ is $n^{-1}\{p(p+1)\}^2/(2p^2 + 2p + 1)$.

5.2.11 Show that for any value of $\boldsymbol{\Sigma} > 0$, the multivariate normal log-likelihood function (5.9) is maximized when $\boldsymbol{\mu} = \overline{\boldsymbol{x}}$. Then, using Theorem 5.2.1, show that if $\boldsymbol{\mu} = \overline{\boldsymbol{x}}$, then (5.9) is maximized when $\boldsymbol{\Sigma} = \boldsymbol{S}$.

5.2.12 If $\boldsymbol{X}$ is a data matrix from $N_p(\boldsymbol{\mu}, \boldsymbol{\Sigma})$, where $\boldsymbol{\Sigma} = \sigma^2[(1-\rho)\boldsymbol{I} + \rho\mathbf{11}']$ is proportional to the equicorrelation matrix, show that the m.l.e.s are given by

$$\hat{\boldsymbol{\mu}} = \overline{\boldsymbol{x}}, \qquad \hat{\sigma}^2 = p^{-1}\operatorname{tr}\boldsymbol{S}, \qquad \hat{\rho} = \frac{2}{p(p-1)\hat{\sigma}^2}\sum_{i<j} s_{ij}.$$

(Hint: using (A.35), write $\boldsymbol{\Sigma}^{-1} = a\boldsymbol{I} + b\mathbf{11}'$ and differentiate the log-likelihood with respect to a and b.)

5.2.13 Show that the maximum of the likelihood function $l(\boldsymbol{X}_1, \ldots, \boldsymbol{X}_n; \boldsymbol{M}, \boldsymbol{U}, \boldsymbol{V})$ is satisfied by the conditions given in Equations (5.39)–(5.41).

5.4.1 (Press, 1972, p. 167) Show that if $\boldsymbol{x}_1, \ldots, \boldsymbol{x}_n$ are i.i.d. $N_p(\boldsymbol{\mu}, \boldsymbol{\Sigma})$, then $\overline{\boldsymbol{x}}$ and $\boldsymbol{M} = n\boldsymbol{S}$ have the joint density

$$f(\overline{\boldsymbol{x}}, \boldsymbol{S}; \boldsymbol{\mu}, \boldsymbol{\Sigma}) \propto \frac{|\boldsymbol{S}|^{(n-p-2)/2}}{|\boldsymbol{\Sigma}|^{n/2}} \exp\left\{-\frac{1}{2} n \operatorname{tr} \boldsymbol{\Sigma}^{-1} \left[\boldsymbol{S} + (\overline{\boldsymbol{x}} - \boldsymbol{\mu})(\overline{\boldsymbol{x}} - \boldsymbol{\mu})'\right]\right\}.$$

If the vague prior $\pi(\boldsymbol{\mu}, \boldsymbol{\Sigma}) \propto |\boldsymbol{\Sigma}|^{-(p+1)/2}$ is taken, show that the posterior density can be written as

$$\pi(\boldsymbol{\mu}, \boldsymbol{\Sigma} \mid \boldsymbol{X}) \propto |\boldsymbol{\Sigma}|^{-(\nu+p+1)/2} \exp\left\{-\frac{1}{2} \operatorname{tr} \boldsymbol{V}^{-1}\boldsymbol{\Sigma}^{-1}\right\},$$

where $\nu = n$ and $\boldsymbol{V} = [\boldsymbol{S} + (\overline{\boldsymbol{x}} - \boldsymbol{\mu})(\overline{\boldsymbol{x}} - \boldsymbol{\mu})']^{-1} n^{-1}$. By integrating $\pi(\boldsymbol{\mu}, \boldsymbol{\Sigma} \mid \boldsymbol{X})$ with respect to $\boldsymbol{\Sigma}$ and using the normalizing constant of the inverse Wishart distribution, show that the posterior density of $\boldsymbol{\mu}$ is

$$\pi(\boldsymbol{\mu} \mid \boldsymbol{X}) \propto \{1 + (\overline{\boldsymbol{x}} - \boldsymbol{\mu})'\boldsymbol{S}^{-1}(\overline{\boldsymbol{x}} - \boldsymbol{\mu})\}^{-n/2},$$

i.e. $\boldsymbol{\mu}$ has a multivariate t distribution with parameters $\overline{\boldsymbol{x}}$, $\boldsymbol{S}$, and $\nu = n - p$. By integrating $\pi(\boldsymbol{\mu}, \boldsymbol{\Sigma} \mid \boldsymbol{X})$ with respect to $\boldsymbol{\mu}$, show that the posterior density of $\boldsymbol{\Sigma}$ is inverted Wishart, $W_p^{-1}(n\boldsymbol{S}, n - 1)$ (see (4.49)). Hence, show that the posterior mean of $\boldsymbol{\mu}$ is $\overline{\boldsymbol{x}}$, and if $n > p - 2$, show that the posterior mean of $\boldsymbol{\Sigma}$ is $n\boldsymbol{S}/(n - p - 2)$ (see Exercise 4.4.13).

5.4.2 (Press, 1972, p. 168) (a) If the informative prior from (5.48) for $\boldsymbol{\mu}$ and $\boldsymbol{\Sigma}$ is taken, then show that the posterior density is

$$\pi(\boldsymbol{\mu}, \boldsymbol{\Sigma} \mid \boldsymbol{X}) \propto |\boldsymbol{\Sigma}|^{-(n+m+1)/2} \exp\{-\frac{1}{2}Q\},$$

where

$$Q = \operatorname{tr} \boldsymbol{\Sigma}^{-1}[n\boldsymbol{S} + \boldsymbol{G} + (\boldsymbol{\mu} - \boldsymbol{\phi})(\boldsymbol{\mu} - \boldsymbol{\phi})' + n(\boldsymbol{\mu} - \overline{\boldsymbol{x}})(\boldsymbol{\mu} - \overline{\boldsymbol{x}})'].$$

(b) Show that the posterior density of $\boldsymbol{\mu}$ is multivariate t with parameters

$$\boldsymbol{a} = \frac{\boldsymbol{\phi} + n\overline{\boldsymbol{x}}}{1 + n}, \qquad \boldsymbol{C} = \frac{n\boldsymbol{S} + \boldsymbol{G}}{1 + n} + \frac{n}{(1 + n)^2}(\boldsymbol{\phi} - \overline{\boldsymbol{x}})(\boldsymbol{\phi} - \overline{\boldsymbol{x}})', \qquad \nu = n + m - p.$$

(c) Show that the posterior density of $\boldsymbol{\Sigma}$ is inverted Wishart, $W_p^{-1}((n + 1)\boldsymbol{C}, n + m - p - 1)$ (see Equation (4.49)).
(d) Hence, show that the posterior mean of $\boldsymbol{\mu}$ is $\boldsymbol{a}$, and if $n + m - 2p - 2 > 0$, show that the posterior mean of $\boldsymbol{\Sigma}$ is $(1 + n)\boldsymbol{C}/(n + m + 2p - 2)$. (see Exercise 4.4.13).

6

Hypothesis Testing

6.1 Introduction

The fundamental problems of multivariate hypothesis testing may be attributed to two sources: the sheer number of different hypotheses that exist and the difficulty in choosing between various plausible test statistics. In this chapter, we present two general approaches based on the likelihood ratio test (LRT) and union intersection test (UIT), respectively. On some occasions, the LRT and UIT both lead to the same test statistics, but on other occasions, they lead to different statistics.

The sheer number of possible hypotheses is well illustrated by the p-dimensional normal distribution. This has $p(p+3)/2$ parameters, and therefore, quite apart from other hypotheses, there are $2^{p(p+3)/2}$ hypotheses that only specify values for a subset of these parameters. This function of p is large even for p as small as 3. Other hypotheses could specify the values of ratios between parameters or other functions, and further hypotheses could test linear or nonlinear restrictions. Of course, in practice, the number of hypotheses of interest is much smaller than this.

The second source of fundamental problems mentioned in the opening paragraph concerns the difficulty of choosing between certain plausible test statistics, and the key question of whether to use a sequence of univariate tests or whether some new multivariate test would be better. These problems are apparent even in the bivariate case, as is shown in the following example. (See also Exercise 6.3.6.)

Example 6.1.1 To illustrate the difference between multivariate hypothesis testing and the corresponding univariate alternatives, consider only the variables x_1 and x_3, where

$$x_1 = \text{head length of first son} \qquad \text{and} \qquad x_3 = \text{head length of second son},$$

from Fret's head data given in Table 6.1. For the complete data of Table 6.1, $n = 25$,

$$\bar{\boldsymbol{x}} = (185.72, 151.12, 183.84, 149.24)'$$

and

$$\boldsymbol{S} = \begin{bmatrix} 91.482 & 50.754 & 66.875 & 44.267 \\ & 52.186 & 49.259 & 33.651 \\ & & 96.774 & 54.278 \\ & & & 43.222 \end{bmatrix}.$$

Multivariate Analysis, Second Edition. Kanti V. Mardia, John T. Kent, and Charles C. Taylor.

Table 6.1 The measurements on the first and second adult sons in a sample of 25 families.

First son		Second son	
Head length	**Head breadth**	**Head length**	**Head breadth**
191	155	179	145
195	149	201	152
181	148	185	149
183	153	188	149
176	144	171	142
208	157	192	152
189	150	190	149
197	159	189	152
188	152	197	159
192	150	187	151
179	158	186	148
183	147	174	147
174	150	185	152
190	159	195	157
188	151	187	158
163	137	161	130
195	155	183	158
186	153	173	148
181	145	182	146
175	140	165	137
192	154	185	152
174	143	178	147
176	139	176	143
197	167	200	158
190	163	187	150

Data used is adapted from a small subset (Table VI) taken from Frets (1921).

We assume initially that x_1 and x_3 are independent, and that each is normally distributed with a variance of 100. (The assumption of independence is woefully unrealistic and is advanced here purely for pedagogic reasons.) Suppose that we wish to test the univariate hypotheses (simultaneously) that both means are 182, i.e.

$$H_1 : x_1 \sim N(182, 100), \qquad H_2 : x_3 \sim N(182, 100).$$

They may be tested using the following z statistics:

$$z_1 = \frac{\overline{x}_1 - 182}{10/\sqrt{25}} = 1.86, \qquad z_2 = \frac{\overline{x}_3 - 182}{10/\sqrt{25}} = 0.92.$$

Since $|z_i| < 1.96$ for $i = 1, 2$, neither of the univariate hypotheses H_1 nor H_2 would be rejected at the 5% level of significance.

Now, let us consider the bivariate hypothesis:

$$H_3 : x_1 \sim N(182, 100) \quad \text{and} \quad x_3 \sim N(182, 100).$$

This hypothesis is true if and only if both H_1 and H_2 are true. There are various ways of testing H_3.

One way would be to accept H_3 if and only if the respective z tests led to acceptance of both H_1 and H_2. Another approach would be to note that if both H_1 and H_2 are true, then

$$z_3 = \frac{1}{\sqrt{2}}(z_1 + z_2) \sim N(0, 1).$$

An acceptance region based on z_3 would lie between the two lines $z_1 + z_2 = \pm c$. Note that the observed value of z_3 equals 1.966, which is just significant at the 5% level, even though neither z_1 nor z_3 were significant. Hence, a multivariate hypothesis may be rejected even when each of it univariate components is accepted.

A third approach is based on the observation that if H_1 and H_2 are true, then

$$z_1^2 + z_2^2 \sim \chi_2^2.$$

This would lead to a circular acceptance region. In the above example, $z_1^2 + z_2^2 = 4.306$, which is beyond the 95th percentile of the χ_2^2 distribution.

A final possibility, to be considered in detail below, would be to use information on the correlation between z_1 and z_2 to derive a test statistic based on a quadratic form in z_1 and z_2. This leads to an elliptical acceptance region. Thus, depending on which test statistic is chosen, our acceptance region can be rectangular, linear, circular, or elliptical. These different acceptance regions can easily lead to conflicting results. □

We concentrate on developing general strategies and particular test statistics that can be used in many frequently occurring situations. Section 6.2 gives two systematic strategies, based on the LRT and UIT, respectively. The tests are introduced briefly in Section 6.2, with several examples of each. We find that for some hypotheses, the LRT and UIT lead to identical test statistics, while for other hypotheses, they lead to different statistics. Sections 6.3 and 6.5 discuss more detailed applications of the LRT and UIT. Section 6.4 considers the construction of simultaneous confidence intervals (SCIs), while Section 6.6 makes some points of comparison between different types of hypothesis testing. Throughout the chapter, we use the tables in Appendix D and also those in Pearson and Hartley (1972, pp. 98–116, 333–358). Sections 6.7 and 6.8 consider nonnormal situations and nonparametric tests.

6.2 The Techniques Introduced

This section introduces the LRT and UIT and gives a few applications.

6.2.1 The Likelihood Ratio Test (LRT)

We assume that the reader is already familiar with the likelihood ratio (LR) procedure from his or her knowledge of univariate statistics. The general strategy of the LRT is to maximize the likelihood under the null hypothesis H_0 and also to maximize the likelihood under the alternative hypothesis H_1. These main results are given in the following definitions and theorem.

Definition 6.2.1 *If the distribution of the random sample $\boldsymbol{X} = (\boldsymbol{x}_1, \dots, \boldsymbol{x}_n)'$ depends on a parameter vector $\boldsymbol{\theta}$, and if $H_0 : \boldsymbol{\theta} \in \Omega_0$ and $H_1 : \boldsymbol{\theta} \in \Omega_1$ are any two hypotheses, then the LR statistic for testing H_0 against H_1 is defined as*

$$\lambda(\boldsymbol{x}) = L_0^*/L_1^*, \tag{6.1}$$

where L_i^* is the largest value that the likelihood function takes in the region $\Omega_i, i = 0, 1$.

Equivalently, we may use the statistic

$$-2\log\lambda = 2(l_1^* - l_0^*), \tag{6.2}$$

where $l_1^* = \log L_1^*$ and $l_0^* = \log L_0^*$.

In the case of simple hypotheses, where Ω_0 and Ω_1 each contain only a single point, the optimal properties of the LR statistic are proved in the well-known Neyman–Pearson lemma. For LR properties when H_0 and H_1 are composite hypotheses, see Kendall and Stuart (1973, p. 195). In general, one tends to favor H_1 when the LR statistic is low, and H_0 when it is high. A test procedure based on the LR statistic may be defined as follows.

Definition 6.2.2 *The LRT of size α for testing H_0 against H_1 has as its rejection region*

$$R = \{\boldsymbol{x} \mid \lambda(\boldsymbol{x}) < c\},$$

where c is determined, so that

$$\sup_{\boldsymbol{\theta}\in\Omega_0} \mathrm{P}_{\boldsymbol{\theta}}(\boldsymbol{x} \in R) = \alpha.$$

For the hypotheses we are interested in, the distribution of λ does not, in fact, depend on the particular value of $\boldsymbol{\theta} \in \Omega_0$, so the above supremum is unnecessary. The LRT has the following very important asymptotic property.

Theorem 6.2.3 *In the notation of Definition 6.2.1, if Ω_1 is a region in $\mathbb{R}^q$, and if Ω_0 is an r-dimensional subregion of Ω_1, then under suitable regularity conditions, for each $\boldsymbol{\theta} \in \Omega_0$, $-2\log\lambda$ has an asymptotic χ^2_{q-r} distribution as $n \to \infty$.*

Proof: See, for example, Silvey (1970, p. 113). □

To illustrate the LRT, we examine three hypotheses assuming that $\boldsymbol{X} = (\boldsymbol{x}_1, \dots, \boldsymbol{x}_n)'$ is a random sample from $N_p(\boldsymbol{\mu}, \boldsymbol{\Sigma})$.

6.2.1.1 The Hypothesis $H_0 : \boldsymbol{\mu} = \boldsymbol{\mu}_0$, $\boldsymbol{\Sigma}$ Known

When $\boldsymbol{\Sigma}$ is known, then H_0 is a simple hypothesis, and from (5.9), the maximized log likelihood under H_0 is

$$l_0^* = l(\boldsymbol{\mu}_0, \boldsymbol{\Sigma}) = -\frac{1}{2}\log|2\pi\boldsymbol{\Sigma}| - \frac{1}{2}n\text{tr}\,\boldsymbol{\Sigma}^{-1}\boldsymbol{S} - \frac{1}{2}n(\overline{\boldsymbol{x}} - \boldsymbol{\mu}_0)'\boldsymbol{\Sigma}^{-1}(\overline{\boldsymbol{x}} - \boldsymbol{\mu}_0).$$

Since H_1 places no constraints on $\boldsymbol{\mu}$, the maximum likelihood estimate (m.l.e.) of $\boldsymbol{\mu}$ is $\overline{\boldsymbol{x}}$, and

$$l_1^* = l(\overline{\boldsymbol{x}}, \boldsymbol{\Sigma}) = -\frac{1}{2}n\log|2\pi\boldsymbol{\Sigma}| - \frac{1}{2}n\text{tr}\,\boldsymbol{\Sigma}^{-1}\boldsymbol{S}.$$

Therefore, using (6.2), we get

$$-2\log\lambda = 2(l_1^* - l_0^*) = n(\overline{\boldsymbol{x}} - \boldsymbol{\mu}_0)'\boldsymbol{\Sigma}^{-1}(\overline{\boldsymbol{x}} - \boldsymbol{\mu}_0). \tag{6.3}$$

Now from Theorem 2.5.4, we know that this function has an exact χ_p^2 distribution under H_0. Hence, we have a statistic whose distribution is known. It can therefore be used to test the null hypothesis. (Note that in this case the asymptotic distribution given by Theorem 2.5.2 is also the small sample distribution.)

Example 6.2.1 Consider the first and third variables of Fret's head measurement data from Table 6.1 and the hypothesis considered in Example 6.1.1, namely that $(x_1, x_3)' \sim N(\boldsymbol{\mu}_0, \boldsymbol{\Sigma})$, where

$$\boldsymbol{\mu}_0 = \begin{bmatrix}182\\182\end{bmatrix} \quad \text{and} \quad \boldsymbol{\Sigma} = \begin{bmatrix}100 & 0\\0 & 100\end{bmatrix}.$$

Using (6.3), we deduce that

$$-2\log\lambda = 25(3.72, 1.84)\begin{bmatrix}0.01 & 0\\0 & 0.01\end{bmatrix}\begin{pmatrix}3.72\\1.84\end{pmatrix} = 4.31.$$

Since this is below 5.99, the 95th percentile of χ_2^2, we accept the hypothesis that $\boldsymbol{\mu} = \boldsymbol{\mu}_0$.

In this case, we can find a confidence region for μ_1 and μ_3 using Theorem 2.5.4. A 95% confidence region for the means of x_1 and x_3 can be given using the inequality

$$25(185.72 - \mu_1, 183.84 - \mu_3)\ \text{diag}\ (0.01, 0.01)\begin{pmatrix}185.72 - \mu_1\\183.84 - \mu_3\end{pmatrix} < 5.99,$$

i.e.

$$(185.72 - \mu_1)^2 + (183.84 - \mu_3)^2 < 23.96.$$

Because $\boldsymbol{\Sigma}$ is assumed to be proportional to $\boldsymbol{I}$, this gives a confidence region that is circular in the parameter space.

A more useful way to express this confidence region in terms of simultaneous confidence intervals is given in Exercise 6.4.2. See also Section 6.4. □

6.2.1.2 The Hypothesis $H_0 : \boldsymbol{\mu} = \boldsymbol{\mu}_0$, $\boldsymbol{\Sigma}$ Unknown (Hotelling's One-Sample T^2 Test)

In this case, $\boldsymbol{\Sigma}$ must be estimated under H_0 and also under H_1. Therefore, both hypotheses are composite. Using results from Section 5.2.2.1, we know that the m.l.e.s are as follows:

$$\text{under } H_0,\quad \hat{\boldsymbol{\mu}} = \boldsymbol{\mu}_0 \quad \text{and} \quad \hat{\boldsymbol{\Sigma}} = \boldsymbol{S} + \boldsymbol{dd}', \quad \text{where} \quad \boldsymbol{d} = \overline{\boldsymbol{x}} - \boldsymbol{\mu}_0,$$

whereas

$$\text{under } H_1, \quad \hat{\boldsymbol{\mu}} = \overline{\boldsymbol{x}} \quad \text{and} \quad \hat{\boldsymbol{\Sigma}} = \boldsymbol{S}.$$

Now from (5.9), we deduce that

$$l_0^* = l(\boldsymbol{\mu}_0, \boldsymbol{S} + \boldsymbol{dd}') = -\frac{1}{2}n\{p \log 2\pi + \log |\boldsymbol{S}| + \log(1 + \boldsymbol{d}'\boldsymbol{S}^{-1}\boldsymbol{d}) + p\},$$

whereas $l_1^* = l(\overline{\boldsymbol{x}}, \boldsymbol{S})$ is obtained by putting $\boldsymbol{d} = \boldsymbol{0}$ in this expression. Thus,

$$-2 \log \lambda = 2(l_1^* - l_0^*) = n \log(1 + \boldsymbol{d}'\boldsymbol{S}^{-1}\boldsymbol{d}). \tag{6.4}$$

This statistic depends on $(n-1)\boldsymbol{d}'\boldsymbol{S}^{-1}\boldsymbol{d}$, which is known from Corollary 4.5.3 to be a $T^2(p, n-1)$ statistic, often known as the Hotelling one-sample T^2 statistic. Further, $\{(n-p)/p\}\boldsymbol{d}'\boldsymbol{S}^{-1}\boldsymbol{d}$ has an $F_{p,n-p}$ distribution.

Example 6.2.2 Consider again the first and third variables of Frets' head data, and let us test the hypothesis $H_0 : (x_1, x_3)' \sim N_2(\boldsymbol{\mu}_0, \boldsymbol{\Sigma})$, where $\boldsymbol{\mu}_0 = (182, 182)'$, and $\boldsymbol{\Sigma}$ is assumed unknown. Using the numbers given in Example 6.1.1, we test

$$\frac{n-p}{p}\boldsymbol{d}'\boldsymbol{S}^{-1}\boldsymbol{d} = \frac{23}{2}(3.72, 1.84)\begin{bmatrix} 91.482 & 66.875 \\ 66.875 & 96.774 \end{bmatrix}^{-1}\begin{pmatrix} 3.72 \\ 1.84 \end{pmatrix} = 1.93$$

against the $F_{2,23}$ distribution. Since $F_{2,23}(0.95) = 3.44$, we accept the null hypothesis. □

6.2.1.3 The Hypothesis $H_0 : \boldsymbol{\Sigma} = \boldsymbol{\Sigma}_0$, $\boldsymbol{\mu}$ Unknown

The m.l.e.s for this case under H_0 are $\hat{\boldsymbol{\mu}} = \overline{\boldsymbol{x}}$ and $\boldsymbol{\Sigma} = \boldsymbol{\Sigma}_0$, and under H_1, $\hat{\boldsymbol{\mu}} = \overline{\boldsymbol{x}}$ and $\hat{\boldsymbol{\Sigma}} = \boldsymbol{S}$. Therefore,

$$l_0^* = l(\overline{\boldsymbol{x}}, \boldsymbol{\Sigma}_0) = -\frac{1}{2}n \log |2\pi\boldsymbol{\Sigma}_0| - \frac{1}{2}\text{tr } \boldsymbol{\Sigma}_0^{-1}\boldsymbol{S}$$

and

$$l_1^* = l(\overline{\boldsymbol{x}}, \boldsymbol{S}) = -\frac{1}{2}n \log |2\pi\boldsymbol{S}| - \frac{1}{2}np.$$

Therefore,

$$-2 \log \lambda = 2(l_1^* - l_0^*) = n\text{tr } \boldsymbol{\Sigma}_0^{-1}\boldsymbol{S} - n \log |\boldsymbol{\Sigma}_0^{-1}\boldsymbol{S}| - np. \tag{6.5}$$

Note that this statistic is a function of the eigenvalues of $\boldsymbol{\Sigma}_0^{-1}\boldsymbol{S}$ and also that, as $\boldsymbol{S}$ approaches $\boldsymbol{\Sigma}_0$, $-2 \log \lambda$ approaches zero. In fact, if we write a and g for the arithmetic and geometric mean, respectively, of the eigenvalues of $\boldsymbol{\Sigma}_0^{-1}\boldsymbol{S}$, so that $\text{tr } \boldsymbol{\Sigma}_0^{-1}\boldsymbol{S} = pa$ and $|\boldsymbol{\Sigma}_0^{-1}\boldsymbol{S}| = g^p$, then (6.5) becomes

$$-2 \log \lambda = np(a - \log g - 1). \tag{6.6}$$

One problem with the statistic given by (6.5) is that its distribution is far from simple. Anderson (1958, p. 265) finds a formula for the moments of λ and also derives the characteristic function of $-2 \log \lambda$, under both H_0 and H_1. Korin (1968) has expressed (6.6) as an infinite sum of chi-squared variables for certain small values of n and p (Pearson and Hartley, 1972, pp. 111, 358). However, these results are not easy to use. The general result cited in Theorem 6.2.3 indicates, however, that $-2 \log \lambda$ given by (6.6) has an asymptotic χ_m^2 distribution under H_0, where m equals the number of independent parameters in $\boldsymbol{\Sigma}$, i.e. $m = p(p+1)/2$.

Example 6.2.3 Using the first and third variables of Frets' data from Table 6.1, we may test the hypothesis $\boldsymbol{\Sigma} = \text{diag}\,(100, 100)$, which was assumed in Example 6.1.1. Here,

$$\boldsymbol{\Sigma}_0^{-1}\boldsymbol{S} = \begin{bmatrix} 0.01 & 0 \\ 0 & 0.01 \end{bmatrix} \begin{bmatrix} 91.482 & 66.875 \\ 66.875 & 96.774 \end{bmatrix} = \begin{bmatrix} 0.9148 & 0.6688 \\ 0.6688 & 0.9677 \end{bmatrix}.$$

The eigenvalues of this matrix are given by $\lambda_1 = 1.611$ and $\lambda_2 = 0.272$. Therefore, $a = 0.9413$ and $g = 0.6619$. Hence, from (6.6), we find that

$$-2\log\lambda = 17.70.$$

This must be compared asymptotically with a χ_3^2 distribution, and we see quite clearly that the hypothesis is rejected; that is, our original assumption regarding the covariance matrix was false. This might perhaps have been expected from a cursory look at the sample covariance matrix, which shows the presence of a strong correlation.

Using the same data, and having rejected the hypothesis that $\boldsymbol{\Sigma} = \text{diag}\,(100, 100)$, we might now examine the hypothesis that

$$\boldsymbol{\Sigma} = \boldsymbol{\Sigma}_0 = \begin{bmatrix} 100 & 50 \\ 50 & 100 \end{bmatrix}.$$

Certainly judging from the sample covariance matrix $\boldsymbol{S}$, this hypothesis seems distinctly more plausible. It is found that

$$\boldsymbol{\Sigma}_0^{-1}\boldsymbol{S} = \begin{bmatrix} 0.7739 & 0.2465 \\ 0.2818 & 0.8445 \end{bmatrix}.$$

The arithmetic and geometric means of the eigenvalues are $a = 0.8092$ and $g = 0.7643$. Note that these are far closer together than the values obtained previously, reflecting the fact that $\boldsymbol{\Sigma}^{-1}\boldsymbol{S}$ is closer to the form $k\boldsymbol{I}$. Inserting the values of a and g in (6.6), we find that $-2\log\lambda = 3.9021$, which is well below the 95th percentile of the χ_3^2 distribution. Hence, we deduce that this revised value of $\boldsymbol{\Sigma}$ is quite plausible under the given assumptions. □

6.2.2 The Union Intersection Test (UIT)

Consider a random vector $\boldsymbol{x}$, which has a $N_p(\boldsymbol{\mu}, \boldsymbol{I})$ distribution and a nonrandom p-vector $\boldsymbol{a}$. Then, if $y_{\boldsymbol{a}} = \boldsymbol{a}'\boldsymbol{x}$, we know that $y_{\boldsymbol{a}} \sim N_1(\boldsymbol{a}'\boldsymbol{\mu}, \boldsymbol{a}'\boldsymbol{a})$. Moreover, this is true for all p-vectors $\boldsymbol{a}$.

Now suppose that we wish to test the hypothesis $H_0 : \boldsymbol{\mu} = \boldsymbol{0}$. Then, under H_0, we know that $y_{\boldsymbol{a}} \sim N(0, \boldsymbol{a}'\boldsymbol{a})$, a hypothesis which we may call $H_{0\boldsymbol{a}}$. Moreover, $H_{0\boldsymbol{a}}$ is true for all p-vectors $\boldsymbol{a}$. In other words, the multivariate hypothesis H_0 can be written as the intersection of the set of univariate hypotheses $H_{0\boldsymbol{a}}$; that is

$$H_0 = \cap H_{0\boldsymbol{a}}. \tag{6.7}$$

The intersection sign is used here because *all* the $H_{0\boldsymbol{a}}$ must be true in order for H_0 to be true. We call $H_{0\boldsymbol{a}}$ a *component* of H_0.

Now let us consider how we would test the univariate hypothesis $H_{0\boldsymbol{a}}$. One obvious way is to use $z_{\boldsymbol{a}} = y_{\boldsymbol{a}}/\sqrt{\boldsymbol{a}'\boldsymbol{a}}$, which has the standard normal distribution. A rejection region for $H_{0\boldsymbol{a}}$ based on $z_{\boldsymbol{a}}$ would be of the form

$$R_{\boldsymbol{a}} = \{z_{\boldsymbol{a}} : |z_{\boldsymbol{a}}| > c\},$$

where c is some arbitrary critical value, say 1.96. This rejection region for $H_{0\boldsymbol{a}}$ could also be written as

$$R_{\boldsymbol{a}} = \{z_{\boldsymbol{a}} : z_{\boldsymbol{a}}^2 > c^2\}. \tag{6.8}$$

Hence, we have a rejection region for each of the univariate hypotheses, which together would imply H_0 in (6.7).

We turn now to consider a sensible rejection region for the composite hypothesis H_0. Since H_0 is true, it seems sensible to accept H_0 if and only if *every* component hypothesis $H_{0\boldsymbol{a}}$ is accepted; that is, we reject H_0 if *any* component hypothesis is rejected. This leads to a rejection region for H_0 that is the union of the rejection regions for the component hypotheses; that is

$$R = \cup R_{\boldsymbol{a}}. \tag{6.9}$$

The *union* of the rejection regions given by (6.9) and the *intersection* of component hypotheses formulated in (6.7) provide the basis of the union intersection strategy, which was due initially to Roy (1957).

Definition 6.2.4 *A UIT for the hypothesis H_0 is a test whose rejection region R can be written as in (6.9), where $R_{\boldsymbol{a}}$ is the rejection region corresponding to a component hypothesis $H_{0\boldsymbol{a}}$, and where H_0 can be written as in (6.7).*

Applied to the above example, the union intersection (UI) strategy based on (6.8) leads to a rejection of H_0 if and only if $z_{\boldsymbol{a}}^2$ exceeds c^2; that is, H_0 is *accepted* if and only if $z_{\boldsymbol{a}}^2 \leq c^2$ for *all* $z_{\boldsymbol{a}}$. This is the same as saying that H_0 is accepted if and only if

$$\max_{\boldsymbol{a}} z_{\boldsymbol{a}}^2 \leq c^2.$$

In general, the UIT often leads to the maximization or minimization of some composite test statistic such as $z_{\boldsymbol{a}}^2$. In this example,

$$z_{\boldsymbol{a}}^2 = y_{\boldsymbol{a}}^2/\boldsymbol{a}'\boldsymbol{a} = \boldsymbol{a}'\boldsymbol{x}\boldsymbol{x}'\boldsymbol{a}/\boldsymbol{a}'\boldsymbol{a}.$$

This is maximized when $\boldsymbol{a} = \boldsymbol{x}$, so that max $z_{\boldsymbol{a}}^2 = \boldsymbol{x}'\boldsymbol{x}$. Hence, in this example, the UIT statistic would be $\boldsymbol{x}'\boldsymbol{x}$, whose distribution under H_0 is known to be χ_p^2.

The method of constructing UITs has important practical consequences. If the null hypothesis is rejected, then one can ask which of the component rejection regions $R_{\boldsymbol{a}}$ is responsible, thus getting a clearer idea about the nature of the deviation from the null hypothesis. In the above example, if one rejects H_0, then one can ask which linear combinations $\boldsymbol{a}'\boldsymbol{x}$ were responsible. In particular, one can look at the variables $x_i = \boldsymbol{e}_i'\boldsymbol{x}$ on their own. For example, it might be the case that some of the variables x_i lie in $R_{\boldsymbol{e}_i}$, whereas the others are all acceptable. See Section 6.4 for further details.

Unfortunately, the LRT does not have this property. If one rejects the null hypothesis using an LRT, then one cannot ask for more details about the reasons for rejection.

We now take the hypotheses used to illustrate the LRT procedure in Sections 6.2.1.1–6.2.1.3 and study them using UITs. As in those examples, we take $\boldsymbol{X}$ to be a matrix whose n rows are independent and identically distributed (i.i.d.) $N_p(\boldsymbol{\mu}, \boldsymbol{\Sigma})$ vectors. Note that this implies that $\boldsymbol{y} = \boldsymbol{X}\boldsymbol{a}$ is a vector whose n elements are i.i.d. $N(\boldsymbol{a}'\boldsymbol{\mu}, \boldsymbol{a}'\boldsymbol{\Sigma}\boldsymbol{a})$ variables.

6.2.2.1 The Hypothesis $H_0 : \mu = \mu_0$, Σ Known (Union Intersection Approach)

Under H_0, the elements of $\boldsymbol{y}$ are i.i.d. $N(\mu_y, \sigma_y^2)$, where $\boldsymbol{y} = \boldsymbol{Xa}$, $\mu_y = \boldsymbol{a}'\boldsymbol{\mu}_0$ and $\sigma_y^2 = \boldsymbol{a}'\boldsymbol{\Sigma a}$. An obvious test statistic based on $\bar{y} = \boldsymbol{a}'\bar{\boldsymbol{x}}$ is

$$z_{\boldsymbol{a}}^2 = n\boldsymbol{a}'(\bar{\boldsymbol{x}} - \boldsymbol{\mu}_0)(\bar{\boldsymbol{x}} - \boldsymbol{\mu}_0)'\boldsymbol{a}/\boldsymbol{a}'\boldsymbol{\Sigma a}.$$

Using (6.9), we wish to reject H_0 for large values of

$$\max_{\boldsymbol{a}} z_{\boldsymbol{a}}^2 = n(\bar{\boldsymbol{x}} - \boldsymbol{\mu}_0)'\boldsymbol{\Sigma}^{-1}(\bar{\boldsymbol{x}} - \boldsymbol{\mu}_0)$$

(see Section A.9). This chi-squared statistic for the UIT has already been derived as the LR statistic in Section 6.2.1.1.

Thus, for this hypothesis, the UIT and LRT both lead to the same test statistic. However, this property is not true in general, as we see in Section 6.2.2.3. Note that the critical values for the UIT statistic should be calculated on the basis of the distribution of the UI statistic, in this case a χ_p^2 distribution, and not from the value 1.96 or $(1.96)^2$, which relates to one of the component hypotheses of H_0 rather than H_0 itself.

6.2.2.2 The Hypothesis $H_0 : \mu = \mu_0$, Σ Unknown (Hotelling One-Sample T^2 Test, Union Intersection Approach)

Once more, we note that under H_0, the elements of $\boldsymbol{y}$ are i.i.d. $N(\mu_y, \sigma_y^2)$. However, in this example, σ_y^2 is unknown and therefore must be estimated. An obvious estimator is

$$s_y^2 = \frac{1}{n}\sum (y_r - \bar{y})^2 = \frac{1}{n}\sum (\boldsymbol{x}_r'\boldsymbol{a} - \bar{\boldsymbol{x}}'\boldsymbol{a})^2 = \boldsymbol{a}'\boldsymbol{Sa}. \tag{6.10}$$

This leads to the test statistic for $H_{0\boldsymbol{a}}$, namely

$$t_{\boldsymbol{a}} = \frac{\bar{y} - \mu_y}{\sqrt{s_y^2/(n-1)}}.$$

We note that

$$t_{\boldsymbol{a}}^2 = (n-1)\frac{\boldsymbol{a}'(\bar{\boldsymbol{x}} - \boldsymbol{\mu}_0)(\bar{\boldsymbol{x}} - \boldsymbol{\mu}_0)'\boldsymbol{a}}{\boldsymbol{a}'\boldsymbol{Sa}},$$

once more a ratio of quadratic forms. This time the UIT statistic is

$$\max_{\boldsymbol{a}} t_{\boldsymbol{a}}^2 = (n-1)(\bar{\boldsymbol{x}} - \boldsymbol{\mu}_0)'\boldsymbol{S}^{-1}(\bar{\boldsymbol{x}} - \boldsymbol{\mu}_0). \tag{6.11}$$

Note that (6.11) is Hotelling's one-sample T^2 statistic, which has already been discussed in Section 6.2.1.2. Hence, once more, the UIT statistic and LRT procedures have led to the same test statistic.

6.2.2.3 The Hypothesis $H_0 : \Sigma = \Sigma_0$, μ Unknown (Union Intersection Test)

Once more the elements of $\boldsymbol{y}$ are i.i.d. $N(\mu_y, \sigma_y^2)$. However, this time we wish to examine σ_y^2 and see whether it satisfies

$$\sigma_{0y}^2 = \boldsymbol{a}'\boldsymbol{\Sigma}_0\boldsymbol{a}.$$

An obvious estimator of σ_y^2 is s_y^2 defined in (6.10). This leads to the test statistic

$$U_{\boldsymbol{a}} = ns_y^2/\sigma_{0y}^2 = n\boldsymbol{a}'\boldsymbol{Sa}/\boldsymbol{a}'\boldsymbol{\Sigma}_0\boldsymbol{a},$$

and we reject H_{0a} if $U_a \leq c_{1a}$ or $U_a \geq c_{2a}$, where c_{1a} and c_{2a} are chosen to make the size of the test equal to α. Since

$$\max_{a} \frac{n\boldsymbol{a}'\boldsymbol{S}\boldsymbol{a}}{\boldsymbol{a}'\boldsymbol{\Sigma}_0\boldsymbol{a}} = n\lambda_1(\boldsymbol{\Sigma}_0^{-1}\boldsymbol{S}), \qquad \min_{a} \frac{n\boldsymbol{a}'\boldsymbol{S}\boldsymbol{a}}{\boldsymbol{a}'\boldsymbol{\Sigma}_0\boldsymbol{a}} = n\lambda_p(\boldsymbol{\Sigma}_0^{-1}\boldsymbol{S}),$$

where λ_i denotes the ith largest eigenvalue, we see that the critical region for the UIT takes the form

$$\lambda_p(\boldsymbol{\Sigma}_0^{-1}\boldsymbol{S}) < c_1 \qquad \text{or} \qquad \lambda_1(\boldsymbol{\Sigma}_0^{-1}\boldsymbol{S}) > c_2,$$

where c_1 and c_2 are chosen to make the size of the test equal to α. However, the joint distribution of the roots of $\boldsymbol{\Sigma}^{-1}\boldsymbol{S}$ is quite complicated, and its critical values have not yet been tabulated. Note that this UIT is *not* the same as the corresponding LRT which was obtained in (6.5), although both statistics depend only on the eigenvalues of $\boldsymbol{\Sigma}_0^{-1}\boldsymbol{S}$.

6.3 The Techniques Further Illustrated

In this section, we take various further hypotheses and derive the corresponding LRTs and UITs.

6.3.1 One-sample hypotheses on μ

We have already seen that the hypothesis $\boldsymbol{\mu} = \boldsymbol{\mu}_0$ leads to the same test statistic using both the LRT and UIT principle, whether or not $\boldsymbol{\Sigma}$ is known. The following general result allows us to deduce the LRT for a wide variety of hypotheses on $\boldsymbol{\mu}$. As before, we assume that $\boldsymbol{x}_1 \ldots, \boldsymbol{x}_n$ is a random sample from $N_p(\boldsymbol{\mu}, \boldsymbol{\Sigma})$.

Theorem 6.3.1 *If H_0 and H_1 are hypotheses that lead to m.l.e.s $\hat{\boldsymbol{\mu}}$ and $\overline{\boldsymbol{x}}$, respectively, and if there are no constraints on $\boldsymbol{\Sigma}$, then the m.l.e.s of $\boldsymbol{\Sigma}$ are $\boldsymbol{S} + \boldsymbol{dd}'$ and $\boldsymbol{S}$, respectively, where $\boldsymbol{d} = \overline{\boldsymbol{x}} - \hat{\boldsymbol{\mu}}$. The LRT for testing H_0 against H_1 is given by*

$$-2\log\lambda = n\boldsymbol{d}'\boldsymbol{\Sigma}^{-1}\boldsymbol{d} \qquad \text{if } \boldsymbol{\Sigma} \text{ is known} \tag{6.12}$$

and

$$-2\log\lambda = n\log(1 + \boldsymbol{d}'\boldsymbol{S}^{-1}\boldsymbol{d}) \qquad \text{if } \boldsymbol{\Sigma} \text{ is unknown.} \tag{6.13}$$

Proof: The proof is identical to that followed in deriving (6.3) and (6.4). □

Unfortunately, no result of the corresponding generality exists concerning the UIT. Therefore, UITs will have to be derived separately for each of the following examples.

6.3.1.1 The Hypothesis $H_0 : \boldsymbol{R\mu} = \boldsymbol{r}$, $\boldsymbol{\Sigma}$ Known (Hypothesis of Linear Constraints)

LRT. Here, $\boldsymbol{R}$ and $\boldsymbol{r}$ are prespecified. The m.l.e. of $\boldsymbol{\mu}$ under $H_0 : \boldsymbol{R\mu} = \boldsymbol{r}$ is given by (5.33). The corresponding LRT is given by (6.12), where

$$\boldsymbol{d} = \overline{\boldsymbol{x}} - \boldsymbol{\mu} = \boldsymbol{\Sigma R}'(\boldsymbol{R\Sigma R}')^{-1}(\boldsymbol{R}\overline{\boldsymbol{x}} - \boldsymbol{r}).$$

The LRT is therefore given by

$$-2\log\lambda = n(\boldsymbol{R}\overline{\boldsymbol{x}} - \boldsymbol{r})'(\boldsymbol{R}\boldsymbol{\Sigma}\boldsymbol{R}')^{-1}(\boldsymbol{R}\overline{\boldsymbol{x}} - \boldsymbol{r}). \tag{6.14}$$

Under H_0, the rows of $\boldsymbol{XR}'$ are i.i.d. $N_q(\boldsymbol{r}, \boldsymbol{R}\boldsymbol{\Sigma}\boldsymbol{R}')$ random vectors, where $q < p$ is the number of elements in $\boldsymbol{r}$. Therefore, by Theorem 2.5.4, Eq. (6.14) has an exact χ_q^2 distribution under H_0.

An important special case occurs when $\boldsymbol{\mu}$ is partitioned, $\boldsymbol{\mu} = (\boldsymbol{\mu}_1', \boldsymbol{\mu}_2')'$, and we wish to test whether $\boldsymbol{\mu}_1 = \boldsymbol{0}$. This hypothesis can be expressed in the form $\boldsymbol{R\mu} = \boldsymbol{r}$ on setting $\boldsymbol{R} = (\boldsymbol{I}, \boldsymbol{0})$ and $\boldsymbol{r} = \boldsymbol{0}$.

In an obvious notation, Eq. (6.14) becomes

$$-2\log\lambda = n\overline{\boldsymbol{x}}_1'\boldsymbol{\Sigma}_{11}^{-1}\overline{\boldsymbol{x}}_1.$$

Another special case occurs when we wish to test the hypothesis

$$H_0 : \boldsymbol{\mu} = k\boldsymbol{\mu}_0 \quad \text{for some } k, \tag{6.15}$$

where $\boldsymbol{\mu}_0$ is a given vector. This hypothesis may be expressed in the form $\boldsymbol{R\mu} = \boldsymbol{0}$, if we take $\boldsymbol{R}$ to be a $((p-1) \times p)$ matrix of rank $p - 1$, whose rows are all orthogonal to $\boldsymbol{\mu}_0$. Under H_0, the m.l.e. $\hat{k}$ of k is given by (5.29), and $\hat{k}$ may be used to express the LR statistic without explicit use of the matrix $\boldsymbol{R}$. See Exercise 6.3.3.

An alternative method of deriving the LRT in (6.14) can be given using the methods of Section 6.2.1.1 to test the hypothesis $\boldsymbol{R\mu} = \boldsymbol{r}$ for the transformed data matrix $\boldsymbol{Y} = \boldsymbol{XR}'$. This approach also leads to (6.14).

UIT. A sensible UIT for this hypothesis can be obtained by applying the methods of Section 6.2.2.1 to the transformed variables $\boldsymbol{XR}'$. Since the methods of Sections 6.2.1.1 and 6.2.2.1 lead to the same test, the UIT is the same as the LRT for this hypothesis.

6.3.1.2 The Hypothesis $H_0 : \boldsymbol{R\mu} = \boldsymbol{r}$, $\boldsymbol{\Sigma}$ Unknown (Hypothesis of Linear Constraints)

LRT. Using (5.34) and (6.13), we know that

$$-2\log\lambda = n\log(1 + \boldsymbol{d}'\boldsymbol{S}^{-1}\boldsymbol{d}),$$

where

$$\boldsymbol{d} = \boldsymbol{SR}'(\boldsymbol{RSR}')^{-1}(\boldsymbol{R}\overline{\boldsymbol{x}} - \boldsymbol{r}).$$

Note that this test is based on the statistic

$$(n-1)\boldsymbol{d}'\boldsymbol{S}^{-1}\boldsymbol{d} = (n-1)(\boldsymbol{R}\overline{\boldsymbol{x}} - \boldsymbol{r})'(\boldsymbol{RSR}')^{-1}(\boldsymbol{R}\overline{\boldsymbol{x}} - \boldsymbol{r}),$$

which has a $T^2(q, n-1)$ distribution under H_0 since $\boldsymbol{R}\overline{\boldsymbol{x}} \sim N_q(\boldsymbol{r}, n^{-1}\boldsymbol{R\Sigma R}')$ independently of $n\boldsymbol{RSR}' \sim W_q(\boldsymbol{R\Sigma R}', n-1)$. For the hypothesis that $\boldsymbol{\mu}_1 = \boldsymbol{0}$, this T^2 statistic becomes

$$(n-1)\overline{\boldsymbol{x}}_1'\boldsymbol{S}_{11}^{-1}\overline{\boldsymbol{x}}_1.$$

Example 6.3.1 We may use the cork tree data from Table 1.4 to examine the hypothesis H_0 that the depth of bark deposit on $n = 28$ trees is the same in all directions. Let $\boldsymbol{x}$ denote the four directions (N, E, S, W)$'$. One way to represent the null hypothesis is $\boldsymbol{R\mu} = \boldsymbol{0}$, where

$$\boldsymbol{R} = \begin{bmatrix} 1 & -1 & 1 & -1 \\ 1 & 0 & -1 & 0 \\ 0 & 1 & 0 & -1 \end{bmatrix}.$$

Here, $q = 3$. To test H_0, we use the statistic

$$(n-1)\overline{\mathbf{y}}'\mathbf{S}_y^{-1}\overline{\mathbf{y}} = 20.74 \sim T^2(3, 27),$$

where $\overline{\mathbf{y}}$ and $\mathbf{S}_y$ are the sample mean and covariance matrix for the transformed variables $\mathbf{Y} = \mathbf{XR}'$ and were calculated in Example 1.5.1. The corresponding F statistic is given by

$$\frac{n-q}{q}\overline{\mathbf{y}}'\mathbf{S}_y^{-1}\overline{\mathbf{y}} = 6.40 \sim F_{3,25}.$$

Since $F_{3,25}(0.99) = 4.675$, we conclude that H_0 must be rejected.

Note also that the null hypothesis for this example may also be represented in the form $H_0 : \boldsymbol{\mu} = k\mathbf{1}$, as in (6.15). □

Another commonly-used constrained linear hypothesis leads to the paired Hotelling T^2 test (see Exercise 6.3.7), which can be adapted to test for bilateral symmetry (see Exercise 6.3.8).

UIT. As in the previous section, the UIT gives the same test here as the LRT.

6.3.2 One-sample hypotheses on $\boldsymbol{\Sigma}$

Theorem 6.3.2 *Let $\mathbf{x}_1, \ldots, \mathbf{x}_n$ be a random sample from $N_p(\boldsymbol{\mu}, \boldsymbol{\Sigma})$. If H_0 and H_1 are hypotheses that lead to $\hat{\boldsymbol{\Sigma}}$ and $\mathbf{S}$ as the m.l.e.s for $\boldsymbol{\Sigma}$, and if $\overline{\mathbf{x}}$ is the m.l.e. of $\boldsymbol{\mu}$ under both hypotheses, then the LRT for testing H_0 against H_1 is given by*

$$-2\log\lambda = np(a - \log g - 1), \tag{6.16}$$

where a and g are the arithmetic and geometric means of the eigenvalues of $\hat{\boldsymbol{\Sigma}}^{-1}\mathbf{S}$.

Proof: The proof is identical to that followed in deriving (6.5) and (6.6). □

A similar result holds if $\boldsymbol{\mu}$ is known. Unfortunately, however, no corresponding results are known for the UIT.

6.3.2.1 The Hypothesis $H_0 : \boldsymbol{\Sigma} = k\boldsymbol{\Sigma}_0$, $\boldsymbol{\mu}$ Unknown

LRT. The m.l.e. $\hat{k}$ was found in (5.36). The LRT is therefore given by (6.16), where a and g relate to the eigenvalues of $\boldsymbol{\Sigma}_0^{-1}\mathbf{S}/\hat{k}$. But in this particular case, Eq. (6.14) conveniently simplifies. We may write $a = a_0/\hat{k}$ and $g = g_0/\hat{k}$, where a_0 and g_0 are the arithmetic and geometric means of the eigenvalues of $\boldsymbol{\Sigma}_0^{-1}\mathbf{S}$. But from (5.36), $\hat{k} = p^{-1}\mathrm{tr}\,\boldsymbol{\Sigma}_0^{-1}\mathbf{S} = a_0$. Therefore, in fact, $a = 1$ and $g = g_0/a_0$. Hence, from (6.16), we get

$$-2\log\lambda = np\log(a_0/g_0). \tag{6.17}$$

In other words, the LR criterion for this hypothesis depends simply on the ratio between the arithmetic and geometric means of the eigenvalues of $\boldsymbol{\Sigma}_0^{-1}\mathbf{S}$. This result is intuitively appealing, since, as $\mathbf{S}$ tends to $k\boldsymbol{\Sigma}_0$, the values of a_0 and g_0 both tend to k, and the expression given by (6.17) tends to zero, just as one would expect when H_0 is satisfied. From Theorem 6.2.3, we know that (6.17) has an asymptotic χ^2 distribution with $p(p+1)/2 - 1 = (p-1)(p+2)/2$ degrees of freedom. Korin (1968) has used the technique of Box (1949) to express $-2\log\lambda$ as a sum of chi-squared variates and thereby found χ^2 and F approximations.

UIT. There seems to be no straightforward UIT for this hypothesis, but see Olkin and Tomsky (1975).

Example 6.3.2 In the so-called test for sphericity, we wish to test whether $\boldsymbol{\Sigma} = k\boldsymbol{I}$. The LRT is given by (6.17), where a_0 and g_0 now relate to the eigenvalues of $\boldsymbol{S}$ in the following way:

$$a_0 = \frac{1}{p}\sum s_{ii}, \qquad g_0^p = |\boldsymbol{S}|.$$

For the first and third variables of Frets' data, which has already been used in Example 6.1.1, $n = 25, p = 2, a_0 = 94.13, |\boldsymbol{S}| = 4380.78$, and $g_0 = 66.19$. Therefore, from (6.17), $-2\log\lambda = 17.6$. This has an asymptotic χ^2 distribution with $(p-1)(p+2)/2 = 2$ degrees of freedom. Since the upper 5% critical value is exceeded, we conclude that $\boldsymbol{\Sigma}$ does not have the stated form. □

6.3.2.2 The Hypothesis $H_0 : \boldsymbol{\Sigma}_{12} = \boldsymbol{0}$, $\boldsymbol{\mu}$ Unknown

LRT. Partition the variables into two sets with dimensions p_1 and p_2, respectively, $p_1 + p_2 = p$. If $\boldsymbol{\Sigma}_{12} = \boldsymbol{0}$, the likelihood splits into two factors. Each factor can be maximized separately over $\boldsymbol{\Sigma}_{11}, \boldsymbol{\mu}_1$ and $\boldsymbol{\Sigma}_{22}, \boldsymbol{\mu}_2$, respectively, giving $\hat{\boldsymbol{\mu}} = (\hat{\boldsymbol{\mu}}_1', \hat{\boldsymbol{\mu}}_2')' = \overline{\boldsymbol{x}}$ and

$$\hat{\boldsymbol{\Sigma}} = \begin{bmatrix} \boldsymbol{S}_{11} & \boldsymbol{0} \\ \boldsymbol{0} & \boldsymbol{S}_{22} \end{bmatrix}.$$

Since the m.l.e. of $\boldsymbol{\mu}$ under H_0 and H_1 is $\overline{\boldsymbol{x}}$, we know from (6.16) that the LRT for this hypothesis depends on the eigenvalues of

$$\hat{\boldsymbol{\Sigma}}\boldsymbol{S} = \begin{bmatrix} \boldsymbol{S}_{11} & \boldsymbol{0} \\ \boldsymbol{0} & \boldsymbol{S}_{22} \end{bmatrix}^{-1} \begin{bmatrix} \boldsymbol{S}_{11} & \boldsymbol{S}_{12} \\ \boldsymbol{S}_{21} & \boldsymbol{S}_{22} \end{bmatrix} = \begin{bmatrix} \boldsymbol{I} & \boldsymbol{S}_{11}^{-1}\boldsymbol{S}_{12} \\ \boldsymbol{S}_{22}^{-1}\boldsymbol{S}_{21} & \boldsymbol{I} \end{bmatrix}.$$

Clearly, $\operatorname{tr} \hat{\boldsymbol{\Sigma}}^{-1}\boldsymbol{S} = p$, and therefore, the arithmetic mean of the eigenvalues, a, is 1. Also, using (A.18),

$$g^p = |\hat{\boldsymbol{\Sigma}}^{-1}\boldsymbol{S}| = |\boldsymbol{S}|/(|\boldsymbol{S}_{11}||\boldsymbol{S}_{22}|) = |\boldsymbol{S}_{22} - \boldsymbol{S}_{21}\boldsymbol{S}_{11}^{-1}\boldsymbol{S}_{12}|/|\boldsymbol{S}_{22}|.$$

Hence,

$$\begin{aligned} -2\log\lambda &= -n\log(|\boldsymbol{S}_{22} - \boldsymbol{S}_{21}\boldsymbol{S}_{11}^{-1}\boldsymbol{S}_{12}|/|\boldsymbol{S}_{22}|) \\ &= -n\log|\boldsymbol{I} - \boldsymbol{S}_{22}^{-1}\boldsymbol{S}_{21}\boldsymbol{S}_{11}^{-1}\boldsymbol{S}_{12}| \\ &= -n\log\prod_{i=1}^{k}(1-\lambda_i), \end{aligned} \tag{6.18}$$

where the λ_i are the nonzero eigenvalues of $\boldsymbol{S}_{22}^{-1}\boldsymbol{S}_{21}\boldsymbol{S}_{11}^{-1}\boldsymbol{S}_{12}$, and $k = \min(p_1, p_2)$.

This result is intuitively appealing since if $\boldsymbol{S}_{12}$ is close to $\boldsymbol{0}$, as it should be if H_0 is true, then $-2\log\lambda$ also takes a small value. Note that $-2\log\lambda$ can also be written in terms of the correlation matrix as $-n\log|\boldsymbol{I} - \boldsymbol{R}_{22}^{-1}\boldsymbol{R}_{21}\boldsymbol{R}_{11}^{-1}\boldsymbol{R}_{12}|$.

To find the distribution of the LR statistic, write $\boldsymbol{M}_{11} = n\boldsymbol{S}_{11}, \boldsymbol{M}_{22} = n\boldsymbol{S}_{22}$, and $\boldsymbol{M}_{22.1} = n(\boldsymbol{S}_{22} - \boldsymbol{S}_{21}\boldsymbol{S}_{11}^{-1}\boldsymbol{S}_{12})$. Then, from Theorem 4.4.11(b), we see that under H_0, $\boldsymbol{M}_{22.1}$ and $\boldsymbol{M}_{22} - \boldsymbol{M}_{22.1}$ are independently distributed Wishart matrices, $\boldsymbol{M}_{22.1} \sim W_{p_2}(\boldsymbol{\Sigma}_{22}, n-1-p_1)$ and $\boldsymbol{M}_{22} - \boldsymbol{M}_{22.1} \sim W_{p_2}(\boldsymbol{\Sigma}_{22}, p_1)$. Hence, provided $n - 1 \geq p_1 + p_2$,

$$\lambda^{2/n} = |\boldsymbol{M}_{22.1}|/|\boldsymbol{M}_{22.1} + (\boldsymbol{M}_{22} - \boldsymbol{M}_{22.1})| \sim \Lambda(p_2, n-1-p_1, p_1). \tag{6.19}$$

The null hypothesis is rejected for small $\lambda^{2/n}$, and the test can be conducted using Wilks' Λ. For the cases $p_1 = 1, 2$ (or $p_2 = 1, 2$), the exact distribution of Λ from (4.39) to (4.42) can be used. For general values of p_1, p_2, we may use Bartlett's approximation (4.43):

$$-\left(n - \frac{1}{2}(p_1 + p_2 + 3)\right) \log |\boldsymbol{I} - \boldsymbol{S}_{22}^{-1}\boldsymbol{S}_{21}\boldsymbol{S}_{11}^{-1}\boldsymbol{S}_{12}| \sim \chi^2_{p_1 p_2}, \tag{6.20}$$

asymptotically, for large n.

As we will see in Chapter 11, the LRT can be used in canonical correlation analysis. The exact distribution of (6.18) was investigated by Hotelling (1936), Girshick (1939), and Anderson (1958, p. 237). Narain (1950) has shown that the LRT has the desirable property of being unbiased.

When one set of variables has just a single member, so that $p_1 = 1$ and $p_2 = p - 1$, these formulae simplify: in that case (6.18) is just $-n \log |\boldsymbol{I} - \boldsymbol{R}_{22}^{-1}\boldsymbol{\alpha}\boldsymbol{\alpha}'|$, where $\boldsymbol{\alpha} = \boldsymbol{R}_{21}$ is a vector. Now, using (A.20), this equals

$$\begin{aligned} -n \log(1 - \boldsymbol{\alpha}'\boldsymbol{R}_{22}^{-1}\boldsymbol{\alpha}) &= -n \log(1 - \boldsymbol{R}_{12}\boldsymbol{R}_{22}^{-1}\boldsymbol{R}_{21}) \\ &= -n \log(1 - R^2), \end{aligned}$$

where R is the multiple correlation coefficient between the first variable and the others (see Section 7.5).

UIT. If $\boldsymbol{x}_1$ and $\boldsymbol{x}_2$ are the two subvectors of $\boldsymbol{x}$, then independence between $\boldsymbol{x}_1$ and $\boldsymbol{x}_2$ implies that the scalars $\boldsymbol{a}'\boldsymbol{x}_1$ and $\boldsymbol{b}'\boldsymbol{x}_2$ are also independent whatever the vectors $\boldsymbol{a}$ and $\boldsymbol{b}$. Now, an obvious way of testing whether or not the two scalars are independent is to look at their sample correlation coefficient, r. Clearly,

$$r^2 = \frac{[\text{cov}(\boldsymbol{a}'\boldsymbol{x}_1, \boldsymbol{b}'\boldsymbol{x}_2)]^2}{\text{var}(\boldsymbol{a}'\boldsymbol{x}_1)\text{var}(\boldsymbol{b}'\boldsymbol{x}_2)} = \frac{(\boldsymbol{a}'\boldsymbol{S}_{12}\boldsymbol{b})^2}{\boldsymbol{a}'\boldsymbol{S}_{11}\boldsymbol{a}\boldsymbol{b}'\boldsymbol{S}_{22}\boldsymbol{b}}.$$

The UIT based on this decomposition uses the statistic $\max_{\boldsymbol{a},\boldsymbol{b}} r^2(\boldsymbol{a}, \boldsymbol{b})$, which equals the largest eigenvalue of $\boldsymbol{S}_{11}^{-1}\boldsymbol{S}_{12}\boldsymbol{S}_{22}^{-1}\boldsymbol{S}_{21}$, also the largest eigenvalue of $\boldsymbol{S}_{22}^{-1}\boldsymbol{S}_{21}\boldsymbol{S}_{11}^{-1}\boldsymbol{S}_{12}$. (This result is proved in Theorem 11.2.1.)

Since $\boldsymbol{S}_{22}^{-1}\boldsymbol{S}_{21}\boldsymbol{S}_{11}^{-1}\boldsymbol{S}_{12}$ can be written as $[\boldsymbol{M}_{22.1} + (\boldsymbol{M}_{22} - \boldsymbol{M}_{22.1})]^{-1}(\boldsymbol{M}_{22} - \boldsymbol{M}_{22.1})$, where $\boldsymbol{M}_{22} = n\boldsymbol{S}_{22}$ and $\boldsymbol{M}_{22} - \boldsymbol{M}_{22.1} = n\boldsymbol{S}_{21}\boldsymbol{S}_{11}^{-1}\boldsymbol{S}_{12}$ have the above mentioned independent Wishart distributions, we see that the largest eigenvalue has the greatest root distribution, $\theta(p_2, n - 1 - p_1, p_1)$ described in Section 4.7.

Writing the nonzero eigenvalues of $\boldsymbol{S}_{22}^{-1}\boldsymbol{S}_{21}\boldsymbol{S}_{11}^{-1}\boldsymbol{S}_{12}$ as $\lambda_1, \lambda_2, \ldots, \lambda_k$, note that the UIT is based on λ_1, while the LRT derived in (6.18) is based on $\prod(1 - \lambda_i)$.

Example 6.3.3 The test statistics derived above may be used to examine whether there is a correlation between head length and breadth measurements of first sons and those of second sons. The matrices $\boldsymbol{S}_{11}, \boldsymbol{S}_{22}$, and $\boldsymbol{S}_{12}$ are the relevant submatrices of the matrix $\boldsymbol{S}$ given in Example 6.1.1, and $n = 25, p_1 = 2, p_2 = 2$. The LRT and UIT both require evaluation of the eigenvalues of

$$\boldsymbol{S}_{22}^{-1}\boldsymbol{S}_{21}\boldsymbol{S}_{11}^{-1}\boldsymbol{S}_{12} = \begin{bmatrix} 0.3014 & 0.2006 \\ 0.4766 & 0.3232 \end{bmatrix}.$$

Here, $\lambda_1 = 0.6218$ and $\lambda_2 = 0.0029$.

For the LRT (6.19), we require

$$\lambda^{2/n} = (1 - \lambda_1)(1 - \lambda_2) = 0.377.$$

Using the asymptotic distribution (6.20), $-(25 - \frac{7}{2})\log(0.377) = 21.0$ is to be tested against the 5% critical value of a $\chi^2_{p_1 p_2} = \chi^2_4$ distribution. The observed value of the statistic is clearly significant, and hence, the null hypothesis is rejected. To use an exact distribution, note that $\lambda^{2/n} \sim \Lambda(p_2, n-1-p_1, p_1) = \Lambda(2, 22, 2)$, and from (4.42),

$$21(1 - \sqrt{\Lambda})/2\sqrt{\Lambda} = 6.60 \sim F_{4,42}.$$

This value is significant at the 5% level, and hence, we still reject H_0.

For the UIT, we require the largest eigenvalue of $\boldsymbol{S}_{22}^{-1}\boldsymbol{S}_{21}\boldsymbol{S}_{11}^{-1}\boldsymbol{S}_{12}$, which is $\lambda_1 = 0.6218$. From Table D.7 in Appendix D with $\nu_1 = n - p_2 - 1 = 22$ and $\nu_2 = p_2 = 2$, and $\alpha = 0.05$, the critical value is about 0.33. Hence, the null hypothesis is rejected for this test also. □

6.3.2.3 The Hypothesis $H_0 : \boldsymbol{P} = \boldsymbol{I}$, $\boldsymbol{\mu}$ Unknown

The hypothesis of the last section may be generalized to consider the hypothesis that all the variables are uncorrelated with one another; that is, $\boldsymbol{P} = \boldsymbol{I}$ or, equivalently, $\boldsymbol{\Sigma}$ is diagonal. Under H_0, the mean and variance of each variable are estimated separately, so that $\hat{\boldsymbol{\mu}} = \overline{\boldsymbol{x}}$ and $\hat{\boldsymbol{\Sigma}} = \text{diag}\,(s_{11}, \ldots, s_{pp})$.

Hence, using (6.16), it is easily seen that the LRT is given in terms of the correlation matrix $\boldsymbol{R}$ by the statistic

$$-2\log\lambda = -n\log|\boldsymbol{R}|,$$

which, by Theorem 6.2.3, has an asymptotic χ^2 distribution under H_0 with $p(p+1)/2 - p = p(p-1)/2$ degrees of freedom. Box (1949) showed that the χ^2 approximation is improved if n is replaced by

$$n' = n - 1 - \frac{1}{6}(2p + 5) = n - \frac{1}{6}(2p + 11).$$

Thus, we use the test statistic

$$-n'\log|\boldsymbol{R}| \sim \chi^2_{p(p-1)/2} \tag{6.21}$$

asymptotically.

6.3.3 Multisample Hypotheses

We now consider the situation of k independent normal samples, whose likelihood was given by (5.37). We use some results from Section 5.2.2.4 in deriving LRTs.

6.3.3.1 The Hypothesis $H_b : \boldsymbol{\mu}_1 = \cdots = \boldsymbol{\mu}_k$ Given That $\boldsymbol{\Sigma}_1 = \cdots = \boldsymbol{\Sigma}_k$ (One-Way Multivariate Analysis of Variance)

LRT (Wilks' Λ test). The LRT of H_b is easily derived from the result of Section 5.2.2.4. The m.l.e.s under H_b are $\overline{\boldsymbol{x}}$ and $\boldsymbol{S}$, since the observations can be viewed under H_b as constituting a single random sample. The m.l.e. of $\boldsymbol{\mu}_i$ under the alternative hypothesis is $\overline{\boldsymbol{x}}_i$, the ith sample mean, and the m.l.e. of the common variance matrix is $n^{-1}\boldsymbol{W}$, where $\boldsymbol{W} = \sum n_i \boldsymbol{S}_i$ is the "within-groups" sum of squares and products (SSP) matrix, and $n = \sum n_i$. Using (5.37), the LRT is given by

$$\lambda_b = \{|\boldsymbol{W}|/|n\boldsymbol{S}|\}^{n/2} = |\boldsymbol{T}^{-1}\boldsymbol{W}|^{n/2}. \tag{6.22}$$

Here, $\boldsymbol{T} = n\boldsymbol{S}$ is the "total" SSP matrix, derived by regarding all the data matrices as if they constituted a single sample. In contrast, the matrix $\boldsymbol{W}$ is the "within-groups" SSP matrix, and

$$\boldsymbol{B} = \boldsymbol{T} - \boldsymbol{W} = \sum n_i(\overline{\boldsymbol{x}}_i - \overline{\boldsymbol{x}})(\overline{\boldsymbol{x}}_i - \overline{\boldsymbol{x}})' \tag{6.23}$$

may be regarded as the "between-groups" SSP matrix. Hence, from (6.22),

$$\lambda_b^{2/n} = |\boldsymbol{W}|/|\boldsymbol{B} + \boldsymbol{W}| = |\boldsymbol{I} + \boldsymbol{W}^{-1}\boldsymbol{B}|^{-1}. \tag{6.24}$$

The matrix $\boldsymbol{W}^{-1}\boldsymbol{B}$ is an obvious generalization of the univariate variance ratio. It will tend to zero if H_0 is true.

We now find the distribution of (6.24). Write the k samples as a single data matrix,

$$\boldsymbol{X}(n \times p) = \begin{bmatrix} \boldsymbol{X}_1 \\ \vdots \\ \boldsymbol{X}_k \end{bmatrix},$$

where $\boldsymbol{X}_i(n_i \times p)$ represents the observations from the ith sample, $i = 1, \ldots, k$. Let $\boldsymbol{1}_i$ denote the n-vector with 1 in the places corresponding to the ith sample and 0 elsewhere, and set $\boldsymbol{I}_i = \text{diag}\,(\boldsymbol{1}_i)$. Then, $\boldsymbol{I} = \sum \boldsymbol{I}_i$ and $\boldsymbol{1} = \sum \boldsymbol{1}_i$. Let $\boldsymbol{H}_i = \boldsymbol{I}_i - n_i^{-1}\boldsymbol{1}_i\boldsymbol{1}_i'$ represent the centering matrix for the ith sample, so that $n_i\boldsymbol{S}_i = \boldsymbol{X}'\boldsymbol{H}_i\boldsymbol{X}$, and set

$$\boldsymbol{C}_1 = \sum \boldsymbol{H}_i, \qquad \boldsymbol{C}_2 = \sum n_i^{-1}\boldsymbol{1}_i\boldsymbol{1}_i' - n^{-1}\boldsymbol{1}\boldsymbol{1}'.$$

It is easily verified that $\boldsymbol{W} = \boldsymbol{X}'\boldsymbol{C}_1\boldsymbol{X}$ and $\boldsymbol{B} = \boldsymbol{X}'\boldsymbol{C}_2\boldsymbol{X}$. Further, $\boldsymbol{C}_1$ and $\boldsymbol{C}_2$ are idempotent matrices of ranks $n - k$ and $k - 1$, respectively, and $\boldsymbol{C}_1\boldsymbol{C}_2 = \boldsymbol{0}$.

Now, under H_0, $\boldsymbol{X}$ is a data matrix from $N_p(\boldsymbol{\mu}, \boldsymbol{\Sigma})$. Thus, by Cochran's theorem (Theorem 4.4.9) and Craig's theorem (Theorem 4.4.10),

$$\begin{aligned} \boldsymbol{W} &= \boldsymbol{X}'\boldsymbol{C}_1\boldsymbol{X} \sim W_p(\boldsymbol{\Sigma}, n - k), \\ \boldsymbol{B} &= \boldsymbol{X}'\boldsymbol{C}_2\boldsymbol{X} \sim W_p(\boldsymbol{\Sigma}, k - 1), \end{aligned}$$

and, furthermore, $\boldsymbol{W}$ and $\boldsymbol{B}$ are independent. Therefore, provided $n \geq p + k$,

$$|\boldsymbol{I} + \boldsymbol{W}^{-1}\boldsymbol{B}|^{-1} \sim \Lambda(p, n - k, k - 1)$$

under H_0, where Wilks' Λ statistic is described in Section 4.7.

UIT. The univariate analog of H_0 would be tested using the analysis of variance statistics

$$\sum n_i(\overline{x}_i - \overline{x})^2 \Big/ \sum n_i s_i^2,$$

where $\overline{x}$ is the overall mean. The corresponding formula for a linear combination $\boldsymbol{Xa}$ is $\sum n_i(\boldsymbol{a}'(\overline{\boldsymbol{x}}_i - \overline{\boldsymbol{x}}))^2 / \sum n_i\boldsymbol{a}'\boldsymbol{S}_i\boldsymbol{a}$. This expression's maximum value is the largest eigenvalue of

$$\left(\sum n_i\boldsymbol{S}_i\right)^{-1} \sum n_i(\overline{\boldsymbol{x}}_i - \overline{\boldsymbol{x}})(\overline{\boldsymbol{x}}_i - \overline{\boldsymbol{x}})' = \boldsymbol{W}^{-1}\boldsymbol{B},$$

where $\boldsymbol{W}$ and $\boldsymbol{B}$ were defined above. Thus, the UI statistic for this hypothesis is the greatest root of $\boldsymbol{W}^{-1}\boldsymbol{B}$, which is not the same as the LR statistic. However, note from (6.24) that the LRT is in fact based on $\prod (1 + \lambda_i)^{-1}$, where $\lambda_1, \ldots, \lambda_p$ are the eigenvalues of $\boldsymbol{W}^{-1}\boldsymbol{B}$. Hence again, we see the familiar pattern that the LRT and UIT are both functions of the eigenvalues of the same matrix, but they lead to different test statistics.

Two-sample Hotelling T^2 test ($k = 2$). When $k = 2$, the LRT can be simplified in terms of the two-sample Hotelling T^2 statistic given in Section 4.6. In this case,

$$\boldsymbol{B} = n_1(\overline{\boldsymbol{x}}_1 - \overline{\boldsymbol{x}})(\overline{\boldsymbol{x}}_1 - \overline{\boldsymbol{x}})' + n_2(\overline{\boldsymbol{x}}_2 - \overline{\boldsymbol{x}})(\overline{\boldsymbol{x}}_2 - \overline{\boldsymbol{x}})'.$$

However, $\overline{\boldsymbol{x}}_1 - \overline{\boldsymbol{x}} = (n_2/n)\boldsymbol{d}$, where $\boldsymbol{d} = \overline{\boldsymbol{x}}_1 - \overline{\boldsymbol{x}}_2$. Also, $\overline{\boldsymbol{x}}_2 - \overline{\boldsymbol{x}} = -n_1\boldsymbol{d}/n$. Therefore,

$$\boldsymbol{B} = (n_1 n_2/n)\boldsymbol{d}\boldsymbol{d}'$$

and

$$|\boldsymbol{I} + \boldsymbol{W}^{-1}\boldsymbol{B}| = |\boldsymbol{I} + (n_1 n_2/n)\boldsymbol{W}^{-1}\boldsymbol{d}\boldsymbol{d}'| = 1 + (n_1 n_2/n)\boldsymbol{d}'\boldsymbol{W}^{-1}\boldsymbol{d}.$$

The second term is, of course, proportional to the Hotelling two-sample T^2 statistic, and we reject H_0 for large values of this statistic.

Furthermore, $\boldsymbol{W}^{-1}\boldsymbol{B}$ is of rank 1, and its largest eigenvalue is

$$(n_1 n_2/n)\boldsymbol{d}'\boldsymbol{W}^{-1}\boldsymbol{d} \tag{6.25}$$

(Corollary A.6.1). Therefore, the UIT is also given in terms of the two-sample Hotelling T^2 statistic. Hence, although in general the LRT and UIT are different for this hypothesis, they are the same in the two-sample case. For numerical examples of these tests, see Examples 13.3.1 and 13.6.1.

6.3.3.2 The Hypothesis $H_a : \Sigma_1 = \cdots = \Sigma_k$ (Test of Homogeneity of Covariances)

LRT (Box's M test). The m.l.e. of $\boldsymbol{\mu}_i$ is $\overline{\boldsymbol{x}}_i$ under both H_a and the alternative. The m.l.e. of $\boldsymbol{\Sigma}_i$ is $\boldsymbol{S} = n^{-1}\boldsymbol{W}$ under H_a, and $\boldsymbol{S}_i$ under the alternative. Therefore,

$$-2\log\lambda_a = n\log|\boldsymbol{S}| - \sum n_i \log|\boldsymbol{S}_i| = \sum n_i \log|\boldsymbol{S}_i^{-1}\boldsymbol{S}|. \tag{6.26}$$

Using Theorem 6.2.3, this has an asymptotic chi-squared distribution with $p(p+1)(k-1)/2$ degrees of freedom. It may be argued that if n_i is small, then (6.26) gives too much weight to the contribution of $\boldsymbol{S}$. This consideration led Box (1949) to propose the test statistic M in place of that given by (6.26). Box's M is given by

$$M = \gamma \sum (n_i - 1)\log|\boldsymbol{S}_{ui}^{-1}\boldsymbol{S}_u|, \tag{6.27}$$

where

$$\gamma = 1 - \frac{2p^2 + 3p - 1}{6(p+1)(k-1)}\left(\sum \frac{1}{n_i - 1} - \frac{1}{n-k}\right),$$

and $\boldsymbol{S}_u$ and $\boldsymbol{S}_{ui}$ are the unbiased estimators:

$$\boldsymbol{S}_u = \frac{n}{n-k}\boldsymbol{S}, \qquad \boldsymbol{S}_{ui} = \frac{n_i}{n_i - 1}\boldsymbol{S}_i.$$

Box's M also has an asymptotic $\chi^2_{p(p+1)(k-1)/2}$ distribution. Box's approximation seems to be good if each n_i exceeds 20 and if k and p do not exceed 5. Box also gave an asymptotic F distribution (see Pearson and Hartley (1972, p. 108)). For $p = 1$, (6.26) reduces to the test statistic for Bartlett's test of homogeneity, namely

$$n\log s^2 - \sum n_i \log s_i^2,$$

where s_i^2 is the variance of sample i, and s^2 is the pooled estimate of the variance.

No simple UIT seems to be available for this hypothesis.

Example 6.3.4 Jolicoeur and Mosimann (1960) measured the shell length, width, and height of 48 painted turtles, 24 male and 24 female. The resulting mean vectors and covariance matrices for males and females, respectively, were as follows:

$$\overline{\boldsymbol{x}}_1 = \begin{pmatrix} 113.38 \\ 88.29 \\ 40.71 \end{pmatrix}, \qquad \boldsymbol{S}_1 = \begin{bmatrix} 132.98 & 75.85 & 35.82 \\ & 47.96 & 10.75 \\ & & 10.79 \end{bmatrix},$$

$$\overline{\boldsymbol{x}}_2 = \begin{pmatrix} 136.00 \\ 102.58 \\ 51.96 \end{pmatrix}, \qquad \boldsymbol{S}_2 = \begin{bmatrix} 432.58 & 259.88 & 161.67 \\ & 164.58 & 98.98 \\ & & 63.87 \end{bmatrix}.$$

Here, $n_1 = n_2 = 24$, $n = 48$. The overall covariance matrix $\boldsymbol{S} = (n_1\boldsymbol{S}_1 + n_2\boldsymbol{S}_2)/n$ is

$$\begin{bmatrix} 282.78 & 167.86 & 98.74 \\ & 106.27 & 59.87 \\ & & 37.33 \end{bmatrix}.$$

We have $|\boldsymbol{S}_1| = 699.77$, $|\boldsymbol{S}_2| = 11{,}138.39$, and $|\boldsymbol{S}| = 4898.73$, while the corresponding determinants of the unbiased estimators are

$$|\boldsymbol{S}_{u1}| = 795.07, \qquad |\boldsymbol{S}_{u2}| = 12{,}655.30, \qquad |\boldsymbol{S}_u| = 5565.88.$$

Further, $k = 2, p = 3$, and therefore, $\gamma = 0.9293$. Hence, from (6.27), $M = 24.04$, which is to be tested against the 5% critical value 12.592 provided by the χ_6^2 distribution. Since this is highly significant, we conclude that the male and female population covariance matrices are not equal. □

6.3.3.3 The Hypothesis $H_c : \mu_1 = \cdots = \mu_k$ and $\Sigma_1 = \cdots = \Sigma_k$ (Test of Complete Homogeneity)

LRT. Let us first discuss the relationship between this hypothesis, H_c, and the hypotheses H_a and H_b that were considered in the last two subsections. Note that H_b is a "conditional" hypothesis; that is, if L_a^*, L_b^*, and L_c^* are the maximized likelihoods under the three hypotheses, and if L^* is the unconstrained maximum likelihood, then

$$\lambda_a = L_a^*/L^*, \qquad \lambda_b = L_b^*/L_a^*, \qquad \lambda_c = L_c^*/L^*.$$

Therefore $\lambda_c = \lambda_a\lambda_b$, and this observation enables us to obtain λ_c directly from (6.22) and (6.26). We must have

$$-2\log\lambda_c = -2\log\lambda_a - 2\log\lambda_b = n\log\left|\frac{1}{n}\boldsymbol{W}\right| - \sum n_i \log|\boldsymbol{S}_i|.$$

This statistic has an asymptotic chi-squared distribution with $p(k-1)(p+3)/2$ degrees of freedom.

No simple UIT seems to be available for the hypothesis.

6.4 Simultaneous Confidence Intervals

We now show through examples how the UI principle leads to the construction of the so-called SCIs for parameters.

6.4.1 The one-sample Hotelling T^2 case

From (6.11), we have

$$t_{\boldsymbol{a}}^2 \le T^2 \qquad \text{for all } p\text{-vectors } \boldsymbol{a}, \tag{6.28}$$

where

$$t_{\boldsymbol{a}} = \boldsymbol{a}'(\overline{\boldsymbol{x}} - \boldsymbol{\mu})/[\boldsymbol{a}'\boldsymbol{S}\boldsymbol{a}/(n-1)]^{1/2}, \tag{6.29}$$

and T^2 is the one-sample Hotelling T^2 statistic. Since, from Theorem 4.5.5, T^2 is proportional to an F distribution, the upper 100α percentage point of T^2 is

$$\mathsf{P}(T^2 \ge T_\alpha^2) = \alpha, \tag{6.30}$$

where

$$T_\alpha^2 = \{(n-1)p/(n-p)\}F_{p,n-p}(1-\alpha), \tag{6.31}$$

and $F_{p,n-p}(1-\alpha)$ is the upper 100α percentage point of $F_{p,n-p}$. Now, if $T^2 \le T_\alpha^2$, then (6.28) implies

$$t_{\boldsymbol{a}}^2 \le T_\alpha^2 \qquad \text{for all } p\text{-vectors } \boldsymbol{a}. \tag{6.32}$$

On substituting (6.29) and (6.31) in (6.32), and inverting the inequality, we can rewrite (6.30) as

$$\mathsf{P}\{\boldsymbol{a}'\boldsymbol{\mu} \in (\boldsymbol{a}'\overline{\boldsymbol{x}} - b, \boldsymbol{a}'\overline{\boldsymbol{x}} + b) \text{ for all } \boldsymbol{a}\} = 1 - \alpha,$$

where

$$b = [\{\boldsymbol{a}'\boldsymbol{S}\boldsymbol{a}p/(n-p)\}F_{p,n-p}(1-\alpha)]^{1/2}.$$

Thus,

$$(\boldsymbol{a}'\overline{\boldsymbol{x}} - b, \boldsymbol{a}'\overline{\boldsymbol{x}} + b) \tag{6.33}$$

is the $100(1-\alpha)\%$ confidence interval for $\boldsymbol{a}'\boldsymbol{\mu}$. Note that *there is a probability of* $(1-\alpha)$ *that all confidence intervals for* $\boldsymbol{a}'\boldsymbol{\mu}$ *obtained by varying* $\boldsymbol{a}$ *will be true.* Hence, they are called the simultaneous confidence intervals (SCIs) for $\boldsymbol{a}'\boldsymbol{\mu}$.

(a) It can be seen that, for fixed $\boldsymbol{a}$, we could obtain $100(1-\alpha)\%$ confidence intervals for $\boldsymbol{a}'\boldsymbol{\mu}$ from $t_{\boldsymbol{a}}$ given by (6.29) since it is distributed as Student's t with $(n-1)$ d.f.; namely, in (6.33), take $b = [\{\boldsymbol{a}'\boldsymbol{S}\boldsymbol{a}/(n-1)\}F_{1,n-1}(1-\alpha)]^{1/2}$ since $F_{1,n-1} = (t_{n-1})^2$. (Of course, the width b will change if $\boldsymbol{a}$ is varied.)

(b) The SCIs may be used to study more specific hypotheses concerning $\boldsymbol{\mu}$. For example, the hypothesis $H_0 : \boldsymbol{\mu} = \boldsymbol{0}$ is accepted at the $100\alpha\%$ level of significance if and only if every $100(1-\alpha)\%$ simultaneous confidence interval contains the zero value. *Thus, if* H_0 *is rejected, there must be at least one vector* $\boldsymbol{a}$ *for which the corresponding confidence interval does not contain zero.* Hence, this method allows us to examine which particular linear combinations lead H_0 to be rejected.

Example 6.4.1 In Example 6.3.1, the differences in cork depth in the four directions were summarized by the transformed variables $\boldsymbol{Y} = \boldsymbol{X}\boldsymbol{R}'$, and $\boldsymbol{Y}$ was assumed to be a data matrix from $N_3(\boldsymbol{R}\boldsymbol{\mu}, \boldsymbol{R}\boldsymbol{\Sigma}\boldsymbol{R}')$, where $\boldsymbol{R}\boldsymbol{\mu} = \boldsymbol{\nu}$, say. Hotelling's T^2 test for the cork data led to the rejection of $H_0 : \boldsymbol{\nu} = \boldsymbol{0}$. We now examine which mean might have led to the rejection of H_0. If we put $\boldsymbol{a} = (1,0,0)'$, then $\boldsymbol{a}'\boldsymbol{S}\boldsymbol{a} = s_{11}$. Using Example 1.5.1,

$$n = 28, \qquad p = 3, \qquad \overline{y}_1 = 8.8571, \qquad F_{3,25}(0.99) = 4.675, \qquad s_{11} = 124.122,$$

so (6.33) gives the 99% confidence interval for ν_1 as

$$0.51 < \nu_1 < 17.20.$$

Similarly, using the values $\boldsymbol{a} = (0, 1, 0)'$ and $\boldsymbol{a} = (0, 0, 1)'$, respectively, gives the intervals

$$-5.01 < \nu_2 < 6.72 \quad \text{and} \quad -6.47 < \nu_3 < 8.47.$$

Hence, only the interval for ν_1 does not contain zero, and thus, it is this particular mean that led to the rejection of the null hypothesis. □

In general, when the hypothesis of zero mean is rejected, there must exist at least one $\boldsymbol{a}$ for which the associated confidence interval excludes zero. However, it is not necessary that $\boldsymbol{a}$ be one of the vectors $\boldsymbol{e}_i = (0, \dots, 0, 1, 0, \dots, 0)'$ as in the above example.

6.4.2 The two-sample Hotelling T^2 case

As in Section 6.4.1, we can construct an SCI for $\boldsymbol{a}'\boldsymbol{\delta} = \boldsymbol{a}'(\boldsymbol{\mu}_1 - \boldsymbol{\mu}_2)$ using the two sample means $\overline{\boldsymbol{x}}_1$ and $\overline{\boldsymbol{x}}_2$ based on sample sizes n_1 and n_2. A UIT for $\boldsymbol{\delta} = \boldsymbol{0}$ was developed in (6.25). Using the same argument as in Section 6.4.1, it is found that the $100(1-\alpha)\%$ SCIs for $\boldsymbol{a}'\boldsymbol{\delta}$ are given by

$$\boldsymbol{a}'(\overline{\boldsymbol{x}}_1 - \overline{\boldsymbol{x}}_2) \pm \{\boldsymbol{a}'(n_1\boldsymbol{S}_1 + n_2\boldsymbol{S}_2)\boldsymbol{a}p(n_1+n_2)F_{p,m}(1-\alpha)/n_1n_2m\}^{1/2},$$

where $m = n_1 + n_2 - p - 1$.

6.4.3 Other Examples

The UI approach can be used to give SCIs in more complicated situations. For example, SCIs can be evaluated for the one-way multivariate analysis of variance of Section 6.3.3.1 to examine which group differences on which variables are important in the rejection of the null hypothesis. These SCIs are discussed in Section 13.6.

More generally, SCIs can be calculated for the parameters of a multivariate regression. See Section 7.3.3.

6.5 The Behrens–Fisher Problem

When it is not assumed that $\boldsymbol{\Sigma}_1 = \boldsymbol{\Sigma}_2$, the problem of testing equality of means is a multivariate extension of the Behrens–Fisher problem. An excellent discussion of the univariate case appears in Kendall and Stuart (1973, pp. 146–161). The multivariate problem is as follows. Suppose that $\boldsymbol{X}_1$ and $\boldsymbol{X}_2$ are independent random data matrices, where $\boldsymbol{X}_i(n_i \times p)$ is drawn from $N_p(\boldsymbol{\mu}_i, \boldsymbol{\Sigma}_i)$, $i = 1, 2$; how can we test the hypothesis that $\boldsymbol{\mu}_1 = \boldsymbol{\mu}_2$?

A variety of test statistics and approximations to their null distributions have been proposed in the literature, and some detailed comparisons have been given by Christensen and Rencher (1997) and Gamage et al. (2004). Here, we limit attention to the more recent proposal of Krishnamoorthy and Yu (2004), which is straightforward to construct and which has good overall performance.

Replace Hotelling's two-sample statistic

$$T^2 = \frac{n_1 n_2}{n_1 + n_2} \boldsymbol{d}' \boldsymbol{S}_u^{-1} \boldsymbol{d},$$

where

$$\boldsymbol{d} = \overline{\boldsymbol{x}}_1 - \overline{\boldsymbol{x}}_2, \quad \boldsymbol{S}_u = \{(n_1 - 1)\boldsymbol{S}_1 + (n_2 - 1)\boldsymbol{S}_2\}/(n_1 + n_2 - 2)$$

by the modified statistic

$$T^{*2} = \boldsymbol{d}' \tilde{\boldsymbol{S}}^{-1} \boldsymbol{d}, \tag{6.34}$$

where

$$\tilde{\boldsymbol{S}} = \tilde{\boldsymbol{S}}_1 + \tilde{\boldsymbol{S}}_2, \qquad \tilde{\boldsymbol{S}}_i = \boldsymbol{S}_i / n_i, \ i = 1, 2.$$

The tilde notation is used to indicate that $\tilde{\boldsymbol{S}}_1$, $\tilde{\boldsymbol{S}}_2$, and $\tilde{\boldsymbol{S}}$ are estimates of the variance matrices of $\overline{\boldsymbol{x}}_1$, $\overline{\boldsymbol{x}}_2$, and $\boldsymbol{d}$, respectively.

The null distribution of T^{*2} is approximated by

$$T^{*2} \sim T^2(p, \nu^*),$$

where

$$\nu^* = \frac{p + p^2}{Q(\tilde{\boldsymbol{S}}^{-1}\tilde{\boldsymbol{S}}_1)/(n_1 - 1) + Q(\tilde{\boldsymbol{S}}^{-1}\tilde{\boldsymbol{S}}_2)/(n_2 - 1)}.$$

Here, $Q(\boldsymbol{A})$ denotes the following quadratic function of the elements of a $p \times p$ matrix $\boldsymbol{A}$:

$$Q(\boldsymbol{A}) = \text{tr}\,(\boldsymbol{A}^2) + \{\text{tr}\,\boldsymbol{A}\}^2. \tag{6.35}$$

In $p = 1$ dimension, this approach simplifies to Welch's solution to the Behrens–Fisher problem. Krishnamoorthy and Yu (2004) have carried out an extensive simulations and found that the size of this test is very close to the nominal level, provided $n_1 \geq 5p$ and $n_2 \geq 5p$.

6.6 Multivariate Hypothesis Testing: Some General Points

Having discussed the LRT and UIT, what conclusions can one draw? They are different statistical procedures, and in general, they may lead to different test statistics.

An important property of LRT is the general asymptotic chi-squared result stated in Theorem 6.2.3. No such result exists for UITs. However, the importance of this difference should not be overemphasized, since no corresponding small-sample result is known. In addition, the LRT depends on the assumption of a specific distributional form. This is not crucial to the UI procedure, which merely requires a "sensible" way of testing each of the component hypotheses. For example, the t statistic may be sensible for many nonnormal distributions.

A more important advantage of the UI approach is that, if H_0 is rejected, it is a simple matter to calculate which of the component hypotheses led to the rejection of H_0. This can suggest a way of amending the hypothesis in order to accord more with the data. For instance, if the T^2 statistic devised under both the LR and UI approaches were "significant," then the implication using the LR approach would be simply to "reject H_0". However, the UI approach would indicate more than this. As well as knowing that H_0 should be rejected, one could enquire which specific linear combinations $\boldsymbol{a}'\boldsymbol{x}$ led to its rejection. We saw this property illustrated in the discussion of simultaneous confidence intervals in Section 6.4.

Another nice property of the LRT is that it sometimes leads to convenient factorizations. For example, in Theorem 4.7.4 it was shown that Wilks' Λ can be written as the product of independent beta variates.

From extensive studies it has become clear that the UI criterion is insensitive to alternative hypotheses involving more than one nonzero eigenvalue. On the other hand, Wilks' criterion provides good protection over a wide range of alternatives (Schatzoff, 1966; Pillai and Jayachandran, 1967). However, Wilks' criterion in itself does not provide simultaneous confidence intervals, a feature that might in practice outweigh its theoretical power properties.

Gabriel (1970) has examined theoretically the relationship between LRT and UIT. Other criteria have also been proposed, including the following functions of the eigenvalues:

$$\text{Pillai (1955)} \qquad \sum \lambda_i;$$

$$\text{Giri (1968)} \qquad \sum \frac{\lambda_i}{1+\lambda_i};$$

$$\text{Kiefer and Schwartz (1965)} \qquad \prod \left(\frac{1+\lambda_i}{\lambda_i}\right).$$

Mardia (1971) has investigated the effect of nonnormality upon Pillai's criterion. He found that it is fairly robust when used for testing hypotheses on means but not when used for hypotheses on covariance matrices. These properties are reminiscent of similar results from robustness studies on univariate statistics.

Of course, there are further criteria that can be used for assessing multivariate tests – the criterion of maximal invariance is one (Cox and Hinkley, 1974, pp. 157–170).

6.7 Nonnormal Data

We have so far assumed that the parent population is multivariate normal. For nonnormal populations when the form of the density is known, we can proceed to obtain the LRT for a given testing problem. The asymptotic distribution of the LR criterion is known through Wilks' theorem (see, for example, Exercise 6.7.1). However, the exact distribution is not known in general.

Since multivariate normal theory is so developed, it is worthwhile to enquire which tests of multivariate normal theory can safely be applied to nonnormal data. Initially, of course, a test of multivariate normality should be applied. We now describe a few such tests. For a review of various tests of multivariate normality, see Mardia (1978).

In Theorem 2.5.7, it was shown that, for the multivariate normal distribution, the multivariate measures of skewness and kurtosis have values $\beta_{1,p} = 0$ and $\beta_{2,p} = p(p+2)$. Mardia (1970b) has shown that the sample counterparts of the measures given in Section 1.8, $b_{1,p}$ and $b_{2,p}$, have the following asymptotic distributions as $n \to \infty$ when the underlying population is multivariate normal:

$$U = \frac{1}{6} n b_{1,p} \sim \chi^2_f \qquad \text{where} \qquad f = \frac{1}{6} p(p+1)(p+2)$$

and

$$V = \{b_{2,p} - p(p+2)\}/\{8p(p+2)/n\}^{1/2} \sim N(0,1). \tag{6.36}$$

These statistics can be used to test the null hypothesis of multivariate normality. One rejects the null hypothesis for large U and for large values of $|V|$. Critical values of these statistics for small samples are given in Mardia (1974).

If the observations are uniformly distributed on a hypersphere , then $b_{1,p}$ will be nearly zero as we know that $\beta_{1,p} = 0$ (see Section 3.3.2) due to spherical symmetry. If, however, the observations deviate from spherical symmetry, then $b_{1,p}$ will be large; this could occur when there is an abnormal clustering . In such cases, the value of $b_{2,p}$ will also be large. Thus, both measures are needed to obtain a full view of any departure from normality. In addition, a composite measure of nonnormality T is defined below by (6.37). More details can be found in Mardia (1980, 1985) and Mardia and Kent (1991). In addition, Davis (1980, 1982) has shown analytically that the first-order effects of moderate nonnormality on the Type I error of the standard manova tests (Wilks' LR criterion and Roy's largest root test, respectively) are mainly due to Mardia's multivariate skewness and kurtosis.

Mardia's tests of multivariate normality have turned out to be very popular in practice. In particular, Cain et al. (2017) have done a systematic study of such tests. They examined several (254) published data sets in Psychological Science and the American Education Research Journal and found that 64% of studies rejected multivariate normality, commenting

> *"There are also several measures of multivariate skewness and kurtosis, though Mardia's measures Mardia (*1970b*) are by far the most common.... These are currently available in* stata, *or as add-on macros* `multnorm` *in* sas *or* `mardia` *in* spss.... *To facilitate future report of skewness and kurtosis, we provide a tutorial on how to compute univariate and multivariate skewness and kurtosis by* sas, spss, *R and a newly developed web application.*

We note that the R libraries `mnt` and `MVN` (Ebner and Henze, 2020; Korkmaz et al., 2014) have a wide choice of tests and diagnostics for multivariate data.

Example 6.7.1 Consider the cork data of Table 1.4, where $n = 28$ and $p = 4$. From (1.26) and (1.27), it is found that

$$b_{1,4} = 4.476 \qquad \text{and} \qquad b_{2,4} = 22.957.$$

Then, $nb_{1,4}/6 = 20.9$ is clearly not significant when compared to the upper 5% critical value of the χ^2_{20} distribution. From (6.36), we find that the observed value of the standard normal variate obtained from the standardized value of $b_{2,p}$ is −0.40, which again is not significant. □

Another test, which is derived from Rao's score statistic, is given by Mardia (1987). This test for multivariate normality is based on a test statistic T, which can be expressed in terms of multivariate skewness and kurtosis (see Section 1.8) as

$$T = nb_{1,p}/6 + n\{b^*_{2,p} - 6b_{2,p} + 3p(p+2)\}/24, \tag{6.37}$$

where $b_{1,p}$ and $b_{2,p}$ are given by (1.26) and (1.27), and

$$b^*_{2,p} = \frac{1}{n^2} \sum_{r,s=1}^{n} g_{rs}^4,$$

with $g_{rs} = (\boldsymbol{x}_r - \overline{\boldsymbol{x}})'\boldsymbol{S}^{-1}(\boldsymbol{x}_s - \overline{\boldsymbol{x}})$. We will call this statistic T "Mardia's omnibus skewness and kurtosis measure". A population version of this omnibus measure is given by

$$\tau = \beta_{1,p}/12 + \{\beta^*_{2,p} - 6\beta_{2,p} + 3p(p+2)\}/48, \tag{6.38}$$

where $\beta_{1,p}$ and $\beta_{2,p}$ are given by (2.31) and (2.32), respectively, and

$$\beta^*_{2,p} = \mathsf{E}\{(\boldsymbol{x} - \boldsymbol{\mu})'\boldsymbol{\Sigma}^{-1}(\boldsymbol{y} - \boldsymbol{\mu})\}^4$$

where $\boldsymbol{x}, \boldsymbol{y}$ are i.i.d. with mean vector $\boldsymbol{\mu}$ and covariance matrix $\boldsymbol{\Sigma}$ (Mardia, 1987). If the data follow a multivariate normal distribution, Mardia and Kent (1991) show that $T \sim \chi^2_{p(3)+p(4)}$, where

$$p(3) = p + p(p-1) + p(p-1)(p-2)/6$$

and

$$p(4) = p + 3p(p-1)/2 + p(p-1)(p-2)/2 + p(p-1)(p-2)(p-3)/24.$$

Example 6.7.2 Now, we illustrate all the main tests for normality using the iris data set. First, we test using the statistics U and V for $b_{1,4}$ and $b_{2,4}$ (given $p = 4$). From Table 1.6, we obtain the values as given in Table 6.2.

It can be seen that for the individual species, skewness and kurtosis are not significant at 5%, which is consistent with normality. For the pooled data, there is evidence of high skewness (see Table 6.2, footnote a), but kurtosis is still not significant.

Now, we calculate the statistics T. In the case that we pool all the data, we have $b^*_{2,4} = 78.289$, so it is found that $T = 116.497$, which we compare with χ^2_{55}, so there is clear evidence against normality. This is as expected, given that the data contains a mixture of iris species. If the test is repeated for each species separately, then the values of T are 66.07, 51.05, and 64.61 for *I. setosa*, *I. versicolor*, and *I. virginica*, respectively – which are all consistent with the hypothesis of multivariate normality. Finally, repeating the procedure for pairs of species, we obtain values of T given by 71.89, 94.44, and 56.80 for the cases when each of *I. setosa*, *I. versicolor*, and *I. virginica*, respectively, are *omitted*. Comparing these to a χ^2 distribution with 55 degrees of freedom shows that only for dataset obtained from the union of the pair of species (*I. setosa* and *I. virginica*) do we reject the hypothesis of multivariate normality. This approach can be viewed (Mardia and Kent, 1991) as a way to determine the separation of clusters – see Chapter 14. The iris data is also used in Chapters 12 and 13. □

Table 6.2 Values of the statistics U and $|V|$ for each of the iris species and the pooled data.

	setosa	versicolor	virginica	pooled
Skewness (U)	25.67	25.18	26.27	67.43[a)]
Kurtosis ($\|V\|$)	1.30	0.6	0.2	0.2

a) Significant at 1%.

The main advantage of these tests is that if the hypotheses of $\beta_{1,p} = 0$ and $\beta_{2,p} = p(p+2)$ are accepted, then we can use the normal theory tests on mean vectors and covariance matrices because Mardia (1970b, 1971, 1974, 1975) has shown the following:

(a) The size of Hotelling's T^2 test is overall robust to nonnormality, although both Hotelling's one-sample T^2 test and the two-sample test for $n_1 \neq n_2$ are more sensitive to $\beta_{1,p}$ than to $\beta_{2,p}$. If $n_1 = n_2$, the two-sample test is hardly influenced by $\beta_{1,p}$ or $\beta_{2,p}$.
(b) Tests on covariances are sensitive, and sensitivity is measured by $\beta_{2,p}$.

Thus, broadly speaking, in the presence of nonnormality, the normal theory tests on means are influenced by $\beta_{1,p}$, whereas tests on covariances are influenced by $\beta_{2,p}$.

If the hypothesis of multivariate normality is rejected, one can sometimes take remedial steps to transform the data to normality, e.g. using the log transformation. Another possibility is to use a nonparametric test. See, e.g. Oja and Randles (2004) and Oja (2010) where multivariate sign tests and rank tests are developed.

Here is an example of a multivariate sign test. Suppose that the data $\boldsymbol{x}_r$, $r = 1, \dots, n$, are i.i.d. from a p-variate elliptical distribution with location parameter $\boldsymbol{\mu}$ and scale matrix $\boldsymbol{\Sigma}$. It is desired to test $H_0 : \boldsymbol{\mu} = \boldsymbol{0}$. Let $\hat{\boldsymbol{\Sigma}}$ denote Tyler's estimate of $\boldsymbol{\Sigma}$ from Eq. (5.43) and transform the data by setting

$$\boldsymbol{y}_r = \hat{\boldsymbol{\Sigma}}^{-1/2}\boldsymbol{x}_r, \quad \boldsymbol{z}_r = \boldsymbol{y}_r/\|\boldsymbol{y}_r\|.$$

Here, $\boldsymbol{z}_r$ is called the *spatial sign* of $\boldsymbol{y}_r$ and can be viewed as the multivariate analog of the sign of a real number. Up to the sampling variability in $\hat{\boldsymbol{\Sigma}}$, the $\boldsymbol{y}_r$ follow an isotropic distribution in $\mathbb{R}^p$, and the $\boldsymbol{z}_r$ are uniformly distributed on the unit sphere. A suitable test statistic is

$$Q^2 = np\bar{\boldsymbol{z}}'\bar{\boldsymbol{z}} = np\|\bar{\boldsymbol{z}}\|^2,$$

which is asymptotically distributed as χ^2_p under H_0 (Randles, 2000).

Another example is given in the following section using a circular version of a rank test.

6.8 Mardia's Nonparametric Test for the Bivariate Two-sample Problem

In $p = 2$ dimensions suppose that $\boldsymbol{x}_{1i}, i = 1, \dots, n_1$, and $\boldsymbol{x}_{2j}, j = 1, \dots, n_2$, are independent random samples from populations with cumulative distribution functions (c.d.f.s) $F(\boldsymbol{x})$ and $G(\boldsymbol{x})$, respectively. Suppose that $\bar{\boldsymbol{x}}$ is the mean vector of the combined sample. Let the angles made with the x_1-axis in the positive direction by the vectors $(\boldsymbol{x}_{1i} - \bar{\boldsymbol{x}}), i = 1, \dots, n_1$, and $(\boldsymbol{x}_{2j} - \bar{\boldsymbol{x}}), j = 1, \dots, n_2$, be ranked in a single sequence. Let $(r_1, \dots, r_{n_1})$ and $(r'_1, \dots, r'_{n_2})$ be the ranks of the angles in the first and second samples, respectively. Let $N = n_1 + n_2$. We replace the angles of the first and second samples by

$$\boldsymbol{y}_{1i} = \{\cos(2\pi r_i/N), \sin(2\pi r_i/N)\}', \quad i = 1, \dots, n_1$$

and

$$\boldsymbol{y}_{2j} = \{\cos(2\pi r'_j/N), \sin(2\pi r'_j/N)\}', \quad j = 1, \dots, n_2.$$

This means that the angles are replaced by the "uniform scores" on a circle. For these observations, it can be shown that

$$\bar{\mathbf{y}} = \mathbf{0}, \qquad \mathbf{S}_y = \frac{1}{2}\mathbf{I}. \tag{6.39}$$

Let T^2 be Hotelling's two-sample T^2 statistic for the $\mathbf{y}$ given by (4.23), and define

$$U = \frac{N-1}{N-2}T^2 = \frac{n_1(N-1)}{n_2}(\bar{\mathbf{y}}_1 - \bar{\mathbf{y}})'\mathbf{S}_y^{-1}(\bar{\mathbf{y}}_1 - \bar{\mathbf{y}}). \tag{6.40}$$

(The second equality follows because $\bar{\mathbf{y}}_1 - \bar{\mathbf{y}}_2 = \bar{\mathbf{y}}_1 - \bar{\mathbf{y}} + \bar{\mathbf{y}} - \bar{\mathbf{y}}_2 = (N/n_2)(\bar{\mathbf{y}}_1 - \bar{\mathbf{y}})$.) Consider the hypothesis $H_0 : F(\mathbf{x}) = G(\mathbf{x})$ and $H_1 : F(\mathbf{x}) = G(\mathbf{x} + \boldsymbol{\delta})$, $\boldsymbol{\delta} \neq \mathbf{0}$. Since we expect $\bar{\mathbf{y}}_1 - \bar{\mathbf{y}}$ to be shifted in the direction $\boldsymbol{\delta}$ when the alternative hypothesis is true, U is a sensible statistic for this test. On substituting (6.39) in (6.40), we get

$$U = \frac{2(N-1)}{n_1 n_2}\left\{\left(\sum_{i=1}^{n_1}\cos\frac{2\pi r_i}{N}\right)^2 + \left(\sum_{i=1}^{n_1}\sin\frac{2\pi r_i}{N}\right)^2\right\}. \tag{6.41}$$

We reject H_0 for large values of U.

As $n_1, n_2 \to \infty$, $n_1/n_2 \to \beta$, and $0 < \beta < 1$, it is found under H_0 that, asymptotically,

$$U \sim \chi_2^2.$$

Example 6.8.1 The distance between the shoulders of the larger left valve (x_1) and the length of the specimens (x_2) of *Bairda oklahomensis* from two geological levels are given in Table 6.3. Here, $n_1 = 8$, $n_2 = 11$, and $N = 19$. It is found that the ranks r_i for level 1 are

4, 13, 14, 15, 16, 17, 18, 19,

Table 6.3 Data for the geological problem (in micrometers).

Level 1		Level 2	
x_1	x_2	x_1	x_2
631	1167	682	1257
606	1222	631	1227
682	1278	631	1237
480	1045	707	1368
606	1151	631	1227
556	1172	682	1262
429	970	707	1313
454	1166	656	1283
		682	1298
		656	1283
		672	1278

Source: Data from Shaver (1960).

and

$$\sum \cos \frac{2\pi r_i}{N} = 3.2886, \qquad \sum \sin \frac{2\pi r_i}{N} = -3.6884.$$

Consequently, U given by (6.40) is 9.99, which is significant since the 5% critical value of χ_2^2 is 5.99. □

We note the following points. Similar to T^2, U is invariant under the nonsingular transformations of the variables $\boldsymbol{x}_{1i}$ and $\boldsymbol{x}_{2i}$. The Pitman's asymptotic efficiency of this test relative to Hotelling's T^2 test for the normal model is $\pi/4$. The U test is strictly nonparametric in the sense that the null distribution does not depend on the underlying distribution of the variables. This test was proposed by Mardia (1967a). For critical values and extensions, see Mardia (1967a, 1968, 1969, 1972), Mardia and Jupp (2000) and Mardia and Spurr (1977).

Another nonparametric test can be constructed by extending the Mann–Whitney test as follows. First, combine the two groups and rank the observations for each variable separately. Then, use T^2 for the resulting ranks. This test is *not* invariant under all linear transformations, but it is invariant under monotone transformations of each variable separately. For further nonparametric tests, see Puri and Sen (1971).

Exercises and Complements

6.2.1 Derive a LRT for the hypothesis $\boldsymbol{\mu}'\boldsymbol{\mu} = 1$, based on one observation $\boldsymbol{x}$ from $N_p(\boldsymbol{\mu}, \sigma^2\boldsymbol{I})$, where σ^2 is known.

6.2.2 Show that $\boldsymbol{\Sigma}_0^{-1}\boldsymbol{S}, a, g$, and $-2\log\lambda$ take the values indicated in Example 6.2.3.

6.2.3 If $\boldsymbol{M}$ is a $(p \times p)$ matrix, distributed as $W_p(\boldsymbol{\Sigma}, n)$, where

$$\boldsymbol{\Sigma} = \begin{bmatrix} \boldsymbol{\Sigma}_{11} & \boldsymbol{0} \\ \boldsymbol{0} & \boldsymbol{\Sigma}_{22} \end{bmatrix} \begin{matrix} p_1 \\ p_2 \end{matrix},$$

and $\boldsymbol{M}$ is partitioned conformably, show that $\boldsymbol{F} = \boldsymbol{M}_{21}\boldsymbol{M}_{11}^{-1}\boldsymbol{M}_{12}$ and $\boldsymbol{G} = \boldsymbol{M}_{22} - \boldsymbol{M}_{21}\boldsymbol{M}_{11}^{-1}\boldsymbol{M}_{12}$ are distributed independently as $W_{p_2}(\boldsymbol{\Sigma}_{22}, p_1)$ and $W_{p_2}(\boldsymbol{\Sigma}_{22}, n - p_1)$.

Using the fact that $\boldsymbol{a}'\boldsymbol{F}\boldsymbol{a}$ and $\boldsymbol{a}'\boldsymbol{G}\boldsymbol{a}$ are independent χ^2 variables for all $\boldsymbol{a}$, derive a two-sided UIT, based on the F-ratio, of the hypothesis that $\boldsymbol{\Sigma}$ is in fact of the form given above.

6.2.4 (a) Show that if Example 6.2.1 is amended, so that

$$\boldsymbol{\Sigma} = \begin{bmatrix} 100 & 50 \\ 50 & 100 \end{bmatrix},$$

then $-2\log\lambda = 3.46$, and H_0 is still not rejected.

(b) Show that with $\boldsymbol{\Sigma}$ given above, the 95% confidence region for μ_1 and μ_3 is elliptical and takes the form

$$(185.72 - \mu_1)^2 + (183.84 - \mu_3)^2 - (185.72 - \mu_1)(183.84 - \mu_3) \leq 17.97.$$

6.2.5 Many of the tests in this chapter have depended on the statistic $\mathbf{d}'\mathbf{S}^{-1}\mathbf{d}$, where $\mathbf{S}$ is a $(p \times p)$ matrix. Show that in certain special cases this statistic simplifies as follows:

(a) If $\mathbf{S}$ is diagonal, then $\mathbf{d}'\mathbf{S}^{-1}\mathbf{d} = t_1^2 + \cdots + t_p^2$, where $t_i = d_i/\sqrt{s_{ii}}$.

(b) When $p = 2$,

$$\mathbf{d}'\mathbf{S}^{-1}\mathbf{d} = \frac{t_1^2 + t_2^2 - 2t_1t_2r_{12}}{1 - r_{12}^2}.$$

(c) When $p = 3$, $\mathbf{d}'\mathbf{S}^{-1}\mathbf{d}$ is given by

$$\frac{t_1^2(1 - r_{23}^2) + t_2^2(1 - r_{13}^2) + t_3^2(1 - r_{12}^2) - 2t_1t_2u_{12.3} - 2t_1t_3u_{13.2} - 2t_2t_3u_{23.1}}{1 + 2r_{12}r_{13}r_{23} - r_{23}^2 - r_{13}^2 - r_{12}^2}$$

where $u_{12.3} = r_{12} - r_{13}r_{23}$, etc.

6.3.1 (Rao, 1948, p. 63) (a) Consider the cork-tree data from Example 6.3.1 and the contrasts

$$z_1 = (\mathrm{N} + \mathrm{S}) - (\mathrm{E} + \mathrm{W}), \qquad z_2 = \mathrm{S} - \mathrm{W}, \qquad z_3 = \mathrm{N} - \mathrm{S}.$$

Show that $z_1 = y_1$, $z_3 = y_2$, and $z_2 = (y_1 - y_2 + y_3)/2$, where y_1, y_2, and y_3 were defined in Example 6.3.1.

(b) Show that the sample mean and covariance matrix for $\mathbf{z} = (z_1, z_2, z_3)'$ are

$$\bar{\mathbf{z}} = \begin{pmatrix} 8.86 \\ 4.50 \\ 0.86 \end{pmatrix} \quad \text{and} \quad \begin{bmatrix} 124.12 & 59.22 & -20.27 \\ & 54.89 & -27.29 \\ & & 61.26 \end{bmatrix}.$$

(c) Hence, show that the value of T^2 for testing the hypothesis $\mathsf{E}(\mathbf{z}) = \mathbf{0}$ is $T^2 = 20.736$, and the value of F is 6.40. Show that this is significant at the 1% level. (Note that the value of T^2 obtained in this exercise is the same as that obtained in Example 6.3.1, thus illustrating that T^2 is invariant under linear transformations.)

6.3.2 (Seal, 1964, p. 106) (a) Measurements of cranial length (x_1) and cranial breadth (x_2) on a sample of 35 mature female frogs led to the following statistics:

$$\bar{\mathbf{x}} = \begin{pmatrix} 22.860 \\ 24.397 \end{pmatrix}, \qquad \mathbf{S}_1 = \begin{bmatrix} 17.178 & 19.710 \\ & 23.710 \end{bmatrix}.$$

Test the hypothesis that $\mu_1 = \mu_2$.

(b) Similar measurements on 14 male frogs led to the statistics:

$$\bar{\mathbf{x}} = \begin{pmatrix} 21.821 \\ 22.843 \end{pmatrix}, \qquad \mathbf{S}_2 = \begin{bmatrix} 17.159 & 17.731 \\ & 19.273 \end{bmatrix}.$$

Again, test the hypothesis that $\mu_1 = \mu_2$.

(c) Show that the pooled covariance matrix obtained from the above data is

$$\mathbf{S} = \frac{n_1\mathbf{S}_1 + n_2\mathbf{S}_2}{n_1 + n_2} = \begin{bmatrix} 17.173 & 19.145 \\ & 22.442 \end{bmatrix}.$$

Use this to test the hypothesis that the covariance matrices of male and female populations are the same.

(d) Test the hypothesis that $\boldsymbol{\mu}_1 = \boldsymbol{\mu}_2$, stating your assumptions carefully.

6.3.3 (a) Consider the hypothesis $H_0 : \boldsymbol{\mu} = k\boldsymbol{\mu}_0$, where $\boldsymbol{\Sigma}$ is known. From (5.29), the m.l.e. of k under H_0 is given by $\hat{k} = \boldsymbol{\mu}_0'\boldsymbol{\Sigma}^{-1}\overline{\mathbf{x}}/\boldsymbol{\mu}_0\boldsymbol{\Sigma}^{-1}\boldsymbol{\mu}_0$. Using (6.12), show that the LRT is given by

$$-2\log\lambda = n\overline{\mathbf{x}}'\boldsymbol{\Sigma}^{-1}\{\boldsymbol{\Sigma} - (\boldsymbol{\mu}_0'\boldsymbol{\Sigma}^{-1}\boldsymbol{\mu}_0)^{-1}\boldsymbol{\mu}_0\boldsymbol{\mu}_0'\}\boldsymbol{\Sigma}^{-1}\overline{\mathbf{x}}.$$

Using Section 6.3.1.1, deduce that $-2\log\lambda \sim \chi^2_{p-1}$ exactly.

(b) Consider now the hypothesis $H_0 : \boldsymbol{\mu} = k\boldsymbol{\mu}_0$, where $\boldsymbol{\Sigma}$ is *unknown*. From (5.31), the m.l.e. of k under H_0 is given by $\hat{k} = \boldsymbol{\mu}_0'\boldsymbol{S}^{-1}\overline{\mathbf{x}}/\boldsymbol{\mu}_0'\boldsymbol{S}^{-1}\boldsymbol{\mu}_0$. Using (6.13), show that

$$-2\log\lambda = n\log(1 + \mathbf{d}'\boldsymbol{S}^{-1}\mathbf{d}),$$

where

$$\mathbf{d}'\boldsymbol{S}^{-1}\mathbf{d} = \overline{\mathbf{x}}\boldsymbol{S}^{-1}\{\boldsymbol{S} - (\boldsymbol{\mu}_0'\boldsymbol{S}^{-1}\boldsymbol{\mu}_0)^{-1}\boldsymbol{\mu}_0\boldsymbol{\mu}_0'\}\boldsymbol{S}^{-1}\overline{\mathbf{x}}.$$

Using Section 6.3.1.2, show that $(n-1)\mathbf{d}'\boldsymbol{S}^{-1}\mathbf{d} \sim T^2(p-1, n-1)$.

6.3.4 If λ_a, λ_b, and λ_c are as defined in Section 6.3.3.3, and if m_a, m_b, and m_c are the respective asymptotic degrees of freedom, show that $m_c = m_a + m_b$. Explain.

6.3.5 Let $\boldsymbol{X}$ be a data matrix from $N_p(\boldsymbol{\mu}, \boldsymbol{\Sigma})$.

(a) Show that the LRT for the hypothesis that x_1 is uncorrelated with $x_2, \ldots, x_p$ is given by

$$\lambda = (1 - R^2)^{n/2},$$

where $R^2 = \boldsymbol{R}_{12}\boldsymbol{R}_{22}^{-1}\boldsymbol{R}_{21}$ is the squared multiple correlation coefficient between x_1 and $x_2, \ldots, x_p$.

(b) Show that the LRT for the hypothesis that $\boldsymbol{\Sigma} = \sigma^2\{(1-\rho)\boldsymbol{I} + \rho\mathbf{1}\mathbf{1}'\}$ is given by

$$\lambda = \{|\boldsymbol{S}|/[v^p(1-r)^{p-1}(1+(p-1)r)]\}^{n/2},$$

where v is the average of the diagonal elements of $\boldsymbol{S}$, and vr is the average of the off-diagonal elements. (See Exercise 5.2.12 and also Wilk et al. (1962, p. 477).)

6.3.6 (a) Rao (1966, p. 91) gives an anthropometric example in which each of two variables taken separately gives a "significant" difference between two communities, whereas the T^2 or F test utilizing both variables together does not indicate a significant difference.

		Mean length (in mm) of	
Community	**Sample size**	**Femur**	**Humerus**
1	27	460.4	335.1
2	20	444.3	323.1
t statistics for each character		2.302	2.214

The pooled unbiased estimate of the common covariance matrix can be constructed from the information in Rao (1966) as

$$\boldsymbol{S}_u = \begin{bmatrix} 561.8 & 380.7 \\ & 343.2 \end{bmatrix}.$$

Show that each of the individual t statistics is significant at the 5% level (using 45 degrees of freedom). However, if one tests whether the population means of femur and humerus are the same in both communities, show that the resulting F statistic takes the value 2.685, which is not significant at the 5% level (using 2 and 44 degrees of freedom).

(b) Rao shows that the power of the T^2 test depends on p, the number of variables, and Δ_p, the Mahalanobis distance in p dimensions between the two populations. He gives charts that illustrate the power functions for values of p between 2 and 9. It can be seen from these charts that, for a given sample size, the power can decrease when the number of variables increases from p to $(p+q)$, unless the increase in squared Mahalanobis distance $\Delta^2_{p+q} - \Delta^2_p$ is of a certain order of magnitude. In the example given in part (a), the D^2 for femur alone was $D_1^2 = 0.4614$, while that for femur and humerus together was 0.4777. Such a small increase led to a loss of power with samples of the given size. Comment on this behavior.

6.3.7 (Paired Hotelling T^2 Test). Let $\boldsymbol{x}_r$, $r = 1, \ldots, n$, be a random sample from $N_{2p}(\boldsymbol{\mu}, \boldsymbol{\Sigma})$. Partition $\boldsymbol{x}_r$ and the parameters as

$$\boldsymbol{x}_r = \begin{pmatrix} \boldsymbol{x}_{Ar} \\ \boldsymbol{x}_{Br} \end{pmatrix}, \quad \boldsymbol{\mu} = \begin{pmatrix} \boldsymbol{\mu}_A \\ \boldsymbol{\mu}_B \end{pmatrix}, \quad \boldsymbol{\Sigma} = \begin{bmatrix} \boldsymbol{\Sigma}_{AA} & \boldsymbol{\Sigma}_{AB} \\ \boldsymbol{\Sigma}_{BA} & \boldsymbol{\Sigma}_{BB} \end{bmatrix}, \tag{6.42}$$

where $\boldsymbol{x}_{Ar}$ and $\boldsymbol{x}_{Br}$ are vectors of the same length p. The purpose of this exercise is to construct a test of the hypothesis $H_0 : \boldsymbol{\mu}_A = \boldsymbol{\mu}_B$ using the differences

$$\boldsymbol{d}_r = \boldsymbol{x}_{Ar} - \boldsymbol{x}_{Br}, \quad r = 1, \ldots, n.$$

(a) Show that the differences $\boldsymbol{d}_r$ have population mean $\boldsymbol{\mu}_A - \boldsymbol{\mu}_B$ and population covariance matrix $\boldsymbol{\Sigma}_{AA} + \boldsymbol{\Sigma}_{BB} - \boldsymbol{\Sigma}_{AB} - \boldsymbol{\Sigma}_{BA}$.

(b) Let $\bar{\boldsymbol{d}}$ and $\boldsymbol{S}_{d,u}$ denote the sample mean and unbiased sample covariance matrix of $\boldsymbol{d}_r$, $r = 1, \ldots, n$. Show that a test of $H_0 : \boldsymbol{\mu}_A = \boldsymbol{\mu}_B$ can be carried out using the one-sample Hotelling T^2 statistic

$$T^2 = n\bar{\boldsymbol{d}}' \boldsymbol{S}_{d,u}^{-1} \bar{\boldsymbol{d}},$$

which is distributed as $T^2_{p,n-1}$ under H_0.

(c) Using Section 6.3.1.2, show that H_0 can also formulated as a a constrained linear hypothesis

$$H_0 : \boldsymbol{C\mu} = 0$$

where $\boldsymbol{C} = (\boldsymbol{I}_p, -\boldsymbol{I}_p)$ is a $p \times 2p$ matrix.

6.3.8 (Bilateral Symmetry, Mardia et al. (2000)). The Hotelling T^2 test in Exercise 6.3.7 can be adapted to test for bilateral symmetry. If $\boldsymbol{v} = (v_1, v_2)'$ is a two-dimensional vector representing a point in the plane, and $\boldsymbol{v}^R = (-v_1, v_2)'$ denotes its reflection about the vertical axis, then the points $\boldsymbol{v}$ and $\boldsymbol{v}^R$ are said to display bilateral symmetry about the vertical axis.

Let $\boldsymbol{x}_r$, $r = 1, \dots, n$, be a random sample from $N_{2p}(\boldsymbol{\mu}, \boldsymbol{\Sigma})$, partitioned as in (6.42). Suppose that $p = 2$ and it is desired to test $H_0 : \boldsymbol{\mu}_A = \boldsymbol{\mu}_B^R$. That is, if $\boldsymbol{\mu}_A = (a, b)'$, where a and b are unknown, we wish to test if $\mu_B = (-a, b)'$.

(a) Define modified differences

$$\boldsymbol{z}_r = \boldsymbol{x}_{Ar} - \boldsymbol{x}_{Br}^R = \begin{pmatrix} x_{Ar1} + x_{Br1} \\ x_{Ar2} - x_{Br2} \end{pmatrix}, \quad r = 1, \dots, n.$$

Show that each $\boldsymbol{z}_r$ has population mean $\boldsymbol{0}$ under H_0 and adapt the argument in Exercise 6.3.7 to construct a Hotelling T^2 statistic to test for bilateral symmetry. For example, $\boldsymbol{\mu}_A$ and $\boldsymbol{\mu}_B$ might represent the coordinates of the centers of the left and right eyes, respectively, on a human face, and the data might represent multiple measurements. The aim of the test might be to assess any asymmetry in the face, though in practice more than one pair of landmarks would typically be used.

(b) Suppose the blocks of the partitioned matrix $\boldsymbol{\Sigma}$ take the form

$$\boldsymbol{\Sigma}_{AA} = \sigma^2 \begin{bmatrix} 1 & \rho_A \\ \rho_A & 1 \end{bmatrix}, \quad \boldsymbol{\Sigma}_{BB} = \sigma^2 \begin{bmatrix} 1 & \rho_B \\ \rho_B & 1 \end{bmatrix}$$

and $\boldsymbol{\Sigma}_{AB} = \boldsymbol{\Sigma}_{BA} = 0$. Show that the population covariance matrix of each $\boldsymbol{z}_r$ takes the form

$$\text{var}(\boldsymbol{z}_r) = \sigma^2 \begin{bmatrix} 2 & \rho_A - \rho_B \\ \rho_A - \rho_B & 2 \end{bmatrix}$$

with determinant

$$|\text{var}(\boldsymbol{z}_r)| = \sigma^4\{4 - (\rho_A - \rho_B)^2\}.$$

Confirm that this determinant is positive if $\sigma^2 > 0, -1 < \rho_A, \rho_B < 1$. Show that $|\text{var}(\boldsymbol{z}_r)| \leq 4\sigma^4$ and discuss the limiting case to explain why $|\text{var}(\boldsymbol{z}_r)| = 0$ when $\rho_A \rho_B = -1$.

6.4.1 Following Section 6.4.1, show that the SCIs for $\boldsymbol{a}'(\boldsymbol{\mu}_1 - \boldsymbol{\mu}_2)$ in the two-sample case are given by (6.33).

6.4.2 (a) In the one-sample case when $\boldsymbol{\Sigma}$ is known, use the approach of Section 6.4.1 to show that simultaneous $100(1-\alpha)\%$ confidence intervals for $\boldsymbol{a}'\boldsymbol{\mu}$ are given by

$$\boldsymbol{a}'\boldsymbol{\mu} \in \boldsymbol{a}'\overline{\boldsymbol{x}} \pm \left\{ \frac{1}{n} (\boldsymbol{a}'\boldsymbol{\Sigma}\boldsymbol{a}) \chi_p^2(1-\alpha) \right\}^{1/2} \quad \text{for all } \boldsymbol{a}.$$

(b) Show that $\boldsymbol{a}'\boldsymbol{\mu}$ lies in the above confidence interval for all $\boldsymbol{a}$ if and only if $\boldsymbol{\mu}$ lies in the 100(1 − α)% confidence ellipsoid of Example 6.2.1,

$$n(\boldsymbol{\mu} - \overline{\mathbf{x}})'\boldsymbol{\Sigma}^{-1}(\boldsymbol{\mu} - \overline{\mathbf{x}}) < \chi_p^2(1 - \alpha).$$

(Hint: see Corollary A.9.9).

(c) Construct a corresponding confidence ellipsoid for the one-sample case when $\boldsymbol{\Sigma}$ is unknown.

6.7.1 Consider a random sample of size n from Weinman's p-variate exponential distribution (Section 3.2.3) and consider the null hypothesis $H_0 : \lambda_0 = \lambda_1 = \cdots = \lambda_{p-1}$. The maximum log likelihood may be found from Example 5.2.2. Show that

$$-2\log\lambda = 2np\log(\tilde{a}/\tilde{g}),$$

where $\tilde{a}$ and $\tilde{g}$ are the arithmetic and geometric means, and give the asymptotic distribution of $-2\log\lambda$.

6.7.2 For the cork data of Table 1.4, show that the measures of skewness and kurtosis of Section 1.8, taking one variable at a time, are

	N	E	S	W
$b_{1,1}$	0.6542	0.6925	0.5704	0.1951
$b_{2,1}$	2.6465	2.7320	2.8070	2.2200

whereas taking two variables at a time, they are

	NE	NS	NW	ES	EW	SW
$b_{1,2}$	1.214	1.019	1.138	2.192	0.965	0.935
$b_{2,2}$	7.512	6.686	7.186	9.359	6.896	7.835

What can be concluded regarding normality of variables in the lower dimensions? Can the large values of $b_{1,2}$ and $b_{2,2}$ for ES be partially ascribed to the larger values of $b_{1,1}$ and $b_{2,1}$ for E and S, respectively?

6.7.3 For the cork data (Table 1.4), let $y_1 = \text{N} + \text{S} - \text{E} - \text{W}$, $y_2 = \text{N} - \text{S}$, and $y_3 = \text{E} - \text{W}$. Show that

$$b_{1,3} = 1.777, \qquad b_{2,3} = 13.558.$$

Test the hypothesis of normality.

6.8.1 Show that

$$\sum_{i=1}^{N} \cos \frac{2\pi i}{N} = \sum_{i=1}^{N} \sin \frac{2\pi i}{N} = 0$$

and

$$\sum_{i=1}^{N} \cos^2 \frac{2\pi i}{N} = \sum_{i=1}^{N} \sin^2 \frac{2\pi i}{N} = \frac{N}{2}, \qquad \sum_{i=1}^{N} \cos \frac{2\pi i}{N} \sin \frac{2\pi i}{N} = 0.$$

Hence, prove (6.39). Verify the second equality in (6.40) and prove (6.41).

6.8.2 Apply Hotelling's T^2 test to the data in Table 6.3 and compare the conclusions with those in Example 6.8.1.

7

Multivariate Regression Analysis

7.1 Introduction

Consider the model defined by

$$\boldsymbol{Y} = \boldsymbol{X}\boldsymbol{B} + \boldsymbol{U}, \tag{7.1}$$

where $\boldsymbol{Y}(n \times p)$ is an observed matrix of p response variables on each of n individuals, $\boldsymbol{X}(n \times q)$ is a known matrix, $\boldsymbol{B}(q \times p)$ is a matrix of unknown regression parameters, and $\boldsymbol{U}$ is a matrix of unobserved random disturbances whose rows for given $\boldsymbol{X}$ are uncorrelated, each with mean $\boldsymbol{0}$ and common covariance $\boldsymbol{\Sigma}$. In this chapter, we usually have $n \geq p + q$. The high-dimensional case (in which p may be larger than n) is considered in Chapter 16.

When $\boldsymbol{X}$ represents a matrix of q "independent" variables observed on each of the n individuals, Eq. (7.1) is called the *multivariate regression model*. (If $\boldsymbol{X}$ is a random matrix, then the distribution of $\boldsymbol{U}$ is assumed to be uncorrelated with $\boldsymbol{X}$.) When $\boldsymbol{X}$ is a design matrix (usually 0s and 1s), Eq. (7.1) is called the *general linear model*. Since the mathematics is the same in both cases, we will not emphasize the difference, and we usually suppose that the first column of $\boldsymbol{X}$ equals $\boldsymbol{1}$ (namely $\boldsymbol{X} = (\boldsymbol{1}, \boldsymbol{X}_1)$) to allow for an overall mean effect. For simplicity, we treat $\boldsymbol{X}$ as a fixed matrix throughout the chapter. If $\boldsymbol{X}$ is random, then all the likelihoods and expectations are to be interpreted as "conditional on $\boldsymbol{X}$."

The columns of $\boldsymbol{Y}$ represent "dependent" variables, which are to be explained in terms of the "independent" variables given by the columns of $\boldsymbol{X}$. Note that

$$\mathsf{E}(y_{rj}) = \boldsymbol{x}_r' \boldsymbol{\beta}_{(j)},$$

so the expected value of y_{rj} depends on the rth row of $\boldsymbol{X}$ and the j th column of the matrix of regression coefficients. (The case $p = 1$, where there is only one dependent variable, is the familiar *multiple regression model* which we write as $\boldsymbol{y} = \boldsymbol{X}\boldsymbol{\beta} + \boldsymbol{u}$.)

In most applications, we suppose that $\boldsymbol{U}$ is normally distributed, so that

$$\boldsymbol{U} \text{ is a data matrix from } N_p(\boldsymbol{0}, \boldsymbol{\Sigma}), \tag{7.2}$$

where $\boldsymbol{U}$ is independent of $\boldsymbol{X}$. Under the assumption of normal errors, the log likelihood for the data $\boldsymbol{Y}$ in terms of the parameters $\boldsymbol{B}$ and $\boldsymbol{\Sigma}$ is given from (5.4) by

$$l(\boldsymbol{B}, \boldsymbol{\Sigma}) = -\frac{1}{2} n \log |2\pi\boldsymbol{\Sigma}| - \frac{1}{2} \operatorname{tr} (\boldsymbol{Y} - \boldsymbol{X}\boldsymbol{B})\boldsymbol{\Sigma}^{-1}(\boldsymbol{Y} - \boldsymbol{X}\boldsymbol{B})'. \tag{7.3}$$

Multivariate Analysis, Second Edition. Kanti V. Mardia, John T. Kent, and Charles C. Taylor.

7.2 Maximum-likelihood Estimation

7.2.1 Maximum-likelihood estimators for B and Σ

We now give the maximum-likelihood estimates (m.l.e.s) for $\boldsymbol{B}$ and $\boldsymbol{\Sigma}$ in the multivariate regression when $n \geq p + q$. In order for the estimates of $\boldsymbol{B}$ to be unique, we suppose that $\boldsymbol{X}$ has full rank q, so that the inverse $(\boldsymbol{X}'\boldsymbol{X})^{-1}$ exists. The situation where $\text{rank}(\boldsymbol{X}) < q$ is discussed in Section 7.4.

Let

$$\boldsymbol{P} = \boldsymbol{I} - \boldsymbol{X}(\boldsymbol{X}'\boldsymbol{X})^{-1}\boldsymbol{X}'. \tag{7.4}$$

Then, $\boldsymbol{P}(n \times n)$ is a symmetric idempotent matrix of rank $n - q$ that projects onto the subspace of $\mathbb{R}^n$ orthogonal to the columns of $\boldsymbol{X}$ (see Exercise 7.2.1). In particular, note that $\boldsymbol{PX} = \boldsymbol{0}$.

Theorem 7.2.1 *For the log-likelihood function (7.3), the m.l.e.s of $\boldsymbol{B}$ and $\boldsymbol{\Sigma}$ are*

$$\hat{\boldsymbol{B}} = (\boldsymbol{X}'\boldsymbol{X})^{-1}\boldsymbol{X}'\boldsymbol{Y} \tag{7.5}$$

and

$$\hat{\boldsymbol{\Sigma}} = n^{-1}\boldsymbol{Y}'\boldsymbol{PY}. \tag{7.6}$$

Proof: Let

$$\hat{\boldsymbol{Y}} = \boldsymbol{X}\hat{\boldsymbol{B}} = \boldsymbol{X}(\boldsymbol{X}'\boldsymbol{X})^{-1}\boldsymbol{X}'\boldsymbol{Y}$$

and

$$\hat{\boldsymbol{U}} = \boldsymbol{Y} - \boldsymbol{X}\hat{\boldsymbol{B}} = \boldsymbol{PY}$$

denote the "fitted" value of $\boldsymbol{Y}$ and the "fitted" error matrix, respectively. Then, $\boldsymbol{Y} - \boldsymbol{XB} = \hat{\boldsymbol{U}} + \boldsymbol{X}\hat{\boldsymbol{B}} - \boldsymbol{XB}$, so that the second term in the right-hand side of (7.3) can be written as

$$-\frac{1}{2}\text{tr}\,\boldsymbol{\Sigma}^{-1}(\boldsymbol{Y} - \boldsymbol{XB})'(\boldsymbol{Y} - \boldsymbol{XB}) = -\frac{1}{2}\text{tr}\,\boldsymbol{\Sigma}^{-1}[\hat{\boldsymbol{U}}'\hat{\boldsymbol{U}} + (\hat{\boldsymbol{B}} - \boldsymbol{B})'\boldsymbol{X}'\boldsymbol{X}(\hat{\boldsymbol{B}} - \boldsymbol{B})].$$

Substituting in (7.3), we get

$$l(\boldsymbol{B}, \boldsymbol{\Sigma}) = -\frac{1}{2}n\log|2\pi\boldsymbol{\Sigma}| - \frac{1}{2}n\text{tr}\,\boldsymbol{\Sigma}^{-1}\hat{\boldsymbol{\Sigma}} - \frac{1}{2}\text{tr}\,\boldsymbol{\Sigma}^{-1}(\hat{\boldsymbol{B}} - \boldsymbol{B})'\boldsymbol{X}'\boldsymbol{X}(\hat{\boldsymbol{B}} - \boldsymbol{B}). \tag{7.7}$$

Only the final term here involves $\boldsymbol{B}$, and this is maximized (at the value zero) when $\boldsymbol{B} = \hat{\boldsymbol{B}}$. Then, the "reduced" log-likelihood function is given by

$$l(\hat{\boldsymbol{B}}, \boldsymbol{\Sigma}) = -\frac{1}{2}np\log 2\pi - \frac{1}{2}n(\log|\boldsymbol{\Sigma}| + \text{tr}\,\boldsymbol{\Sigma}^{-1}\hat{\boldsymbol{\Sigma}}),$$

which by Theorem 5.2.1 is maximized when $\boldsymbol{\Sigma} = \hat{\boldsymbol{\Sigma}}$.

The maximum value of the likelihood is given from (7.7) by

$$l(\hat{\boldsymbol{B}}, \hat{\boldsymbol{\Sigma}}) = -\frac{1}{2}n\log|2\pi\hat{\boldsymbol{\Sigma}}| - \frac{1}{2}np. \tag{7.8}$$

□

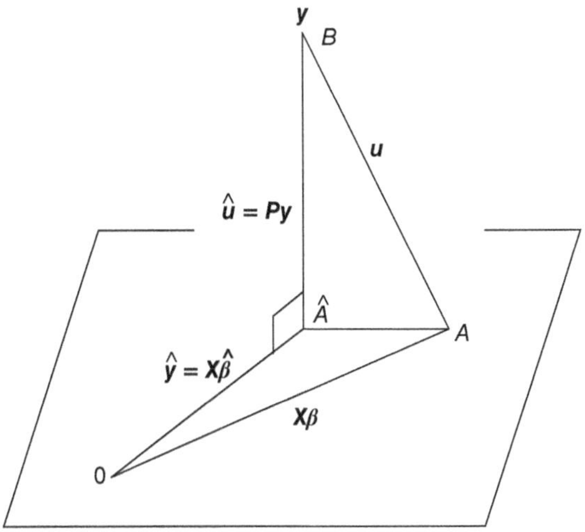

Figure 7.1 Geometry of multiple regression model, $\boldsymbol{y} = \boldsymbol{X\beta} + \boldsymbol{u}$.

The statistics $\hat{\boldsymbol{B}}$ and $\hat{\boldsymbol{\Sigma}}$ are sufficient for $\boldsymbol{B}$ and $\boldsymbol{\Sigma}$ (Exercise 7.2.2). For an efficient method of calculating $\hat{\boldsymbol{B}}$ and $\hat{\boldsymbol{\Sigma}}$, see Anderson (1958, pp. 184–187).

Note that

$$\hat{\boldsymbol{U}}'\hat{\boldsymbol{U}} = \boldsymbol{Y}'\boldsymbol{Y} - \hat{\boldsymbol{Y}}'\hat{\boldsymbol{Y}} = \boldsymbol{Y}'\boldsymbol{PY}. \tag{7.9}$$

In other words, the residual sum of squares and products (SSP) matrix $\hat{\boldsymbol{U}}'\hat{\boldsymbol{U}}$ equals the total SSP matrix $\boldsymbol{Y}'\boldsymbol{Y}$ minus the fitted SSP matrix $\hat{\boldsymbol{Y}}'\hat{\boldsymbol{Y}}$. Geometrically, each column of $\boldsymbol{Y}$ can be split into two orthogonal parts. First, $\hat{\boldsymbol{y}}_{(i)}$ is the projection of $\boldsymbol{y}_{(i)}$ onto the space spanned by the columns of $\boldsymbol{X}$. Second, $\hat{\boldsymbol{u}}_{(i)} = \boldsymbol{Py}_{(i)}$ is the projection of $\boldsymbol{y}_{(i)}$ onto the space orthogonal to the columns of $\boldsymbol{X}$. This property is illustrated for $p = 1$ in Figure 7.1. The true value of $\boldsymbol{X\beta} = \mathsf{E}(\boldsymbol{y})$ is denoted by the point A. The observed value of $\boldsymbol{y}$ (the point B) is projected onto the plane spanned by the columns of $\boldsymbol{X}$ to give the point $\hat{A}$ as an estimate of A.

If we let $\boldsymbol{M}_{11} = \boldsymbol{X}'\boldsymbol{X}$, $\boldsymbol{M}_{22} = \boldsymbol{Y}'\boldsymbol{Y}$, and $\boldsymbol{M}_{12} = \boldsymbol{X}'\boldsymbol{Y}$, then we can write

$$n\hat{\boldsymbol{\Sigma}} = \hat{\boldsymbol{U}}'\hat{\boldsymbol{U}} = \boldsymbol{M}_{22} - \boldsymbol{M}_{21}\boldsymbol{M}_{11}^{-1}\boldsymbol{M}_{12}.$$

Thus, the residual SSP matrix can be viewed as a residual Wishart matrix $\boldsymbol{M}_{22.1}$ conditioned on $\boldsymbol{X}$, as in (4.8).

7.2.2 The distribution of $\hat{B}$ and $\hat{\Sigma}$

We now consider the joint distribution of the m.l.e.s derived in Theorem 7.2.1 under the multivariate linear model with normal errors.

Theorem 7.2.2 *Under the model defined by (7.1) and (7.2)*

(a) $\hat{\boldsymbol{B}}$ *is unbiased for* $\boldsymbol{B}$,
(b) $\mathsf{E}(\hat{\boldsymbol{U}}) = \boldsymbol{0}$,
(c) $\hat{\boldsymbol{B}}$ *and* $\hat{\boldsymbol{U}}$ *are multivariate normal,*
(d) $\hat{\boldsymbol{B}}$ *is statistically independent of* $\hat{\boldsymbol{U}}$, *and hence of* $\hat{\boldsymbol{\Sigma}}$.

Proof: (a) Inserting $\boldsymbol{Y} = \boldsymbol{XB} + \boldsymbol{U}$ in the formula for $\hat{\boldsymbol{B}}$ defined in Theorem 7.2.1,

$$\hat{\boldsymbol{B}} = (\boldsymbol{X}'\boldsymbol{X})^{-1}\boldsymbol{X}'(\boldsymbol{XB} + \boldsymbol{U}) = \boldsymbol{B} + (\boldsymbol{X}'\boldsymbol{X})^{-1}\boldsymbol{X}'\boldsymbol{U}. \tag{7.10}$$

Therefore, $\mathsf{E}(\hat{\boldsymbol{B}}) = \boldsymbol{B}$, since $\mathsf{E}(\boldsymbol{U}) = \boldsymbol{0}$. Hence, (a) is proved.

(b) We have $\hat{\boldsymbol{U}} = \boldsymbol{Y} - \boldsymbol{X}\hat{\boldsymbol{B}}$. Therefore,

$$\mathsf{E}(\hat{\boldsymbol{U}}) = \mathsf{E}(\boldsymbol{Y}) - \boldsymbol{X}\mathsf{E}(\hat{\boldsymbol{B}}) = \boldsymbol{XB} - \boldsymbol{XB} = \boldsymbol{0}.$$

(c) We have

$$\hat{\boldsymbol{U}} = \boldsymbol{PY} = \boldsymbol{PU}, \tag{7.11}$$

where $\boldsymbol{P}$ was defined in (7.4). Also, $\hat{\boldsymbol{B}} = \boldsymbol{B} + (\boldsymbol{X}'\boldsymbol{X})^{-1}\boldsymbol{X}'\boldsymbol{U}$ from (7.10). Hence, both $\hat{\boldsymbol{U}}$ and $\hat{\boldsymbol{B}}$ are linear functions of $\boldsymbol{U}$. Since $\boldsymbol{U}$ has a multivariate normal distribution, the normality of $\hat{\boldsymbol{U}}$ and $\hat{\boldsymbol{B}}$ follows immediately.

(d) This follows by applying Theorem 4.3.3 to the linear functions defined in part (c) above. □

Of particular importance in the above theorem is the fact that the $\hat{\boldsymbol{B}}$ and $\hat{\boldsymbol{\Sigma}}$ are independent for normal data. More precise results concerning the distribution of $\hat{\boldsymbol{B}}$ and $\hat{\boldsymbol{\Sigma}}$ are given in the following theorem.

Theorem 7.2.3 *(a) The covariance between $\hat{\beta}_{ij}$ and $\hat{\beta}_{kl}$ is $\sigma_{jl}g_{ik}$, where $\boldsymbol{G} = (\boldsymbol{X}'\boldsymbol{X})^{-1}$.*
(b) $n\hat{\boldsymbol{\Sigma}} \sim W_p(\boldsymbol{\Sigma}, n-q)$.

Proof: (a) From (7.10), we see that $\hat{\beta}_{ij} - \beta_{ij} = \boldsymbol{a}_i'\boldsymbol{u}_{(j)}$, where $\boldsymbol{A} = (\boldsymbol{X}'\boldsymbol{X})^{-1}\boldsymbol{X}'$. Therefore,

$$\text{cov}(\hat{\beta}_{ij}, \hat{\beta}_{kl}) = \boldsymbol{a}_i'\mathsf{E}[\boldsymbol{u}_{(j)}\boldsymbol{u}_{(l)}']\boldsymbol{a}_k = \boldsymbol{a}_i'[\sigma_{jl}\boldsymbol{I}]\boldsymbol{a}_k = \sigma_{jl}(\boldsymbol{AA}')_{ik} = \sigma_{jl}g_{ik},$$

since $\boldsymbol{AA}' = \boldsymbol{G}$. Therefore, part (a) is proved.

(b) From (7.11), $n\hat{\boldsymbol{\Sigma}} = \hat{\boldsymbol{U}}'\hat{\boldsymbol{U}} = \boldsymbol{U}'\boldsymbol{PU}$, where $\boldsymbol{P}$ is an idempotent matrix of rank $(n-q)$. Since each row of $\boldsymbol{U}$ has mean $\boldsymbol{0}$, the result follows from Theorem 4.4.9. □

Note that when $\boldsymbol{X}'\boldsymbol{X} = \boldsymbol{I}$, or nearly so, the elements of a column of $\hat{\boldsymbol{\beta}}_{(i)}$ will tend to have a low correlation with one another; this property is considered desirable.

7.3 The General Linear Hypothesis

We wish to consider the hypotheses of the form $\boldsymbol{C}_1\boldsymbol{BM}_1 = \boldsymbol{D}$, where $\boldsymbol{C}_1(g \times q)$, $\boldsymbol{M}_1(p \times r_0)$, and $\boldsymbol{D}(g \times r_0)$ are given matrices, and $\boldsymbol{C}_1$ and $\boldsymbol{M}_1$ have ranks g and r_0, respectively. In many cases, $\boldsymbol{D} = \boldsymbol{0}$ and $\boldsymbol{M}_1 = \boldsymbol{I}$, so the hypothesis takes the form $\boldsymbol{C}_1\boldsymbol{B} = \boldsymbol{0}$. Note that the rows of $\boldsymbol{C}_1$ make assertions about the effect on the regression from linear combinations of the *independent* variables, whereas the columns of $\boldsymbol{M}_1$ focus attention on particular linear combinations of the *dependent* variables.

We follow the two-pronged approach begun in Chapter 6; that is, we consider the hypothesis from the point of view of the likelihood ratio test (LRT) and the union intersection test (UIT).

7.3.1 The Likelihood Ratio Test (LRT)

Consider first the hypothesis $\boldsymbol{C}_1\boldsymbol{B} = \boldsymbol{D}$ (so, $\boldsymbol{M}_1 = \boldsymbol{I}$). It is convenient to define some additional matrices before constructing the LRT. Let $\boldsymbol{C}_2((q-g)\times q)$ be a matrix such that $\boldsymbol{C}' = (\boldsymbol{C}_1', \boldsymbol{C}_2')'$ is a nonsingular matrix of order $q \times q$, and let $\boldsymbol{B}_0(q\times p)$ be any matrix satisfying $\boldsymbol{C}_1\boldsymbol{B}_0 = \boldsymbol{D}$.

Then, the model (7.1) can be written as

$$\boldsymbol{Y}_+ = \boldsymbol{Z}\boldsymbol{\Delta} + \boldsymbol{U}, \tag{7.12}$$

where $\boldsymbol{Y}_+ = \boldsymbol{Y} - \boldsymbol{X}\boldsymbol{B}_0, \boldsymbol{Z} = \boldsymbol{X}\boldsymbol{C}^{-1}$, and $\boldsymbol{\Delta} = (\boldsymbol{\Delta}_1', \boldsymbol{\Delta}_2')' = \boldsymbol{C}(\boldsymbol{B}-\boldsymbol{B}_0)$. The hypothesis $\boldsymbol{C}_1\boldsymbol{B} = \boldsymbol{D}$ becomes $\boldsymbol{\Delta}_1 = \boldsymbol{0}$. Partition $\boldsymbol{C}^{-1} = (\boldsymbol{C}^{(1)}, \boldsymbol{C}^{(2)})$, and let

$$\boldsymbol{P}_1 = \boldsymbol{I} - \boldsymbol{X}\boldsymbol{C}^{(2)}(\boldsymbol{C}^{(2)'}\boldsymbol{X}'\boldsymbol{X}\boldsymbol{C}^{(2)})^{-1}\boldsymbol{C}^{(2)'}\boldsymbol{X}'$$

represent the projection onto the subspace orthogonal to the columns of $\boldsymbol{X}\boldsymbol{C}^{(2)}$.

From (7.8), we see that the maximized likelihoods under the null and alternative hypotheses are given by

$$|2\pi n^{-1}\boldsymbol{Y}_+'\boldsymbol{P}_1\boldsymbol{Y}_+|^{-n/2}\mathrm{e}^{-np/2} \qquad \text{and} \qquad |2\pi n^{-1}\boldsymbol{Y}_+'\boldsymbol{P}\boldsymbol{Y}_+|^{-n/2}\mathrm{e}^{-np/2},$$

respectively. Thus, the likelihood ratio (LR) statistic is given by

$$\lambda^{2/n} = |\boldsymbol{Y}_+'\boldsymbol{P}\boldsymbol{Y}_+|/|\boldsymbol{Y}_+'\boldsymbol{P}_1\boldsymbol{Y}_+|.$$

Note that $\boldsymbol{P}$ is a projection onto a subspace of $\boldsymbol{P}_1$ (a vector orthogonal to all the columns of $\boldsymbol{X}$ is orthogonal to the columns of $\boldsymbol{X}\boldsymbol{C}^{(2)}$), so that $\boldsymbol{P}_2 = \boldsymbol{P}_1 - \boldsymbol{P}$ is also a projection. It can be shown that

$$\boldsymbol{P}_2 = \boldsymbol{X}(\boldsymbol{X}'\boldsymbol{X})^{-1}\boldsymbol{C}_1'[\boldsymbol{C}_1(\boldsymbol{X}'\boldsymbol{X})^{-1}\boldsymbol{C}_1']^{-1}\boldsymbol{C}_1(\boldsymbol{X}'\boldsymbol{X})^{-1}\boldsymbol{X}' \tag{7.13}$$

(see Exercise 7.3.1). Also note that $\boldsymbol{P}\boldsymbol{Y} = \boldsymbol{P}\boldsymbol{Y}_+$ because $\boldsymbol{P}\boldsymbol{X} = \boldsymbol{0}$. This observation is useful in proving the following result.

Theorem 7.3.1 *The LRT of the hypothesis $\boldsymbol{C}_1\boldsymbol{B} = \boldsymbol{D}$ for the model (7.12) has test statistic*

$$\lambda^{2/n} = |\boldsymbol{Y}'\boldsymbol{P}\boldsymbol{Y}|/|\boldsymbol{Y}'\boldsymbol{P}\boldsymbol{Y} + \boldsymbol{Y}_+'\boldsymbol{P}_2\boldsymbol{Y}_+|,$$

which has a $\Lambda(p, n-q, g)$ distribution under the null hypothesis.

Proof: The formula for $\lambda^{2/n}$ was derived above. For its distribution, note from (7.11) that $\boldsymbol{Y}'\boldsymbol{P}\boldsymbol{Y} = \boldsymbol{U}'\boldsymbol{P}\boldsymbol{U}$ under the null and alternative hypotheses. Also, under the null hypothesis, $\boldsymbol{Y}_+'\boldsymbol{P}_2\boldsymbol{Y}_+ = \boldsymbol{U}'\boldsymbol{P}_2\boldsymbol{U}$ (Exercise 7.3.2). Since $\boldsymbol{P}$ and $\boldsymbol{P}_2$ are independent projections ($\boldsymbol{P}\boldsymbol{P}_2 = \boldsymbol{0}$) of ranks $n-q$ and g, respectively, it follows from Theorems 4.4.8 and 4.4.9 that $\boldsymbol{Y}'\boldsymbol{P}\boldsymbol{Y}$ and $\boldsymbol{Y}_+'\boldsymbol{P}_2\boldsymbol{Y}_+$ have independent $W_p(\boldsymbol{\Sigma}, n-q)$ and $W_p(\boldsymbol{\Sigma}, g)$ distributions. The distribution of $\lambda^{2/n}$ follows. □

Now, let us consider the hypothesis $\boldsymbol{C}_1\boldsymbol{B}\boldsymbol{M}_1 = \boldsymbol{D}$ for the model (7.1). From Exercise 7.3.3, we see that it is equivalent to study the hypothesis $\boldsymbol{C}_1\boldsymbol{B}\boldsymbol{M}_1 = \boldsymbol{D}$ for the transformed model

$$\boldsymbol{Y}\boldsymbol{M}_1 = \boldsymbol{X}\boldsymbol{B}\boldsymbol{M}_1 + \boldsymbol{U}\boldsymbol{M}_1, \tag{7.14}$$

where $\boldsymbol{U}\boldsymbol{M}_1$ is now a data matrix from $N_p(\boldsymbol{0}, \boldsymbol{M}_1'\boldsymbol{\Sigma}\boldsymbol{M}_1)$, and $\boldsymbol{B}\boldsymbol{M}_1$ contains qr_0 parameters.

Define matrices

$$\begin{aligned} \boldsymbol{H} &= \boldsymbol{M}_1'\boldsymbol{Y}_+'\boldsymbol{P}_2\boldsymbol{Y}_+\boldsymbol{M}_1 \\ &= \boldsymbol{M}_1'\boldsymbol{Y}_+'\boldsymbol{X}(\boldsymbol{X}'\boldsymbol{X})^{-1}\boldsymbol{C}_1'[\boldsymbol{C}_1(\boldsymbol{X}'\boldsymbol{X})^{-1}\boldsymbol{C}_1']^{-1}\boldsymbol{C}_1(\boldsymbol{X}'\boldsymbol{X})^{-1}\boldsymbol{X}'\boldsymbol{Y}_+\boldsymbol{M}_1 \end{aligned} \tag{7.15}$$

and

$$\begin{aligned} \boldsymbol{E} &= \boldsymbol{M}_1'\boldsymbol{Y}'\boldsymbol{P}\boldsymbol{Y}\boldsymbol{M}_1 \\ &= \boldsymbol{M}_1'\boldsymbol{Y}'[\boldsymbol{I} - \boldsymbol{X}(\boldsymbol{X}'\boldsymbol{X})^{-1}\boldsymbol{X}']\boldsymbol{Y}\boldsymbol{M}_1 \end{aligned} \tag{7.16}$$

to be the SSP matrix due to the regression and the residual SSP matrix, respectively. Then, the LRT is given by the following theorem.

Theorem 7.3.2 *The LRT of the hypothesis $\boldsymbol{C}_1\boldsymbol{B}\boldsymbol{M}_1 = \boldsymbol{D}$ for the model (7.1) is given by the test statistic*

$$\lambda^{2/n} = |\boldsymbol{E}|/|\boldsymbol{H} + \boldsymbol{E}| \sim \Lambda(r_0, n - q, g). \tag{7.17}$$

Proof: Apply Theorem 7.3.1 to the transformed model (7.14). □

Note that the LR statistic can be written as

$$|\boldsymbol{E}|/|\boldsymbol{H} + \boldsymbol{E}| = \prod_{i=1}^{r_0} (1 + \lambda_i)^{-1}, \tag{7.18}$$

where $\lambda_1, \ldots, \lambda_{r_0}$ are the eigenvalues of $\boldsymbol{H}\boldsymbol{E}^{-1}$.

When $p = 1$, the matrices in (7.17) are scalars, and it is more usual to use the equivalent F statistic

$$\boldsymbol{H}/\boldsymbol{E} \sim [g/(n - q)]F_{g,n-q}. \tag{7.19}$$

The null hypothesis is rejected for large values of this statistic.

7.3.2 The Union Intersection Test (UIT)

The multivariate hypothesis $\boldsymbol{C}_1\boldsymbol{B}\boldsymbol{M}_1 = \boldsymbol{D}$ is true if and only if $\boldsymbol{b}'\boldsymbol{C}_1\boldsymbol{B}\boldsymbol{M}_1\boldsymbol{a} = \boldsymbol{b}'\boldsymbol{D}\boldsymbol{a}$ for all $\boldsymbol{a}$ and $\boldsymbol{b}$. Replacing $\boldsymbol{C}_1$ and $\boldsymbol{M}_1$ by $\boldsymbol{b}'\boldsymbol{C}_1$ and $\boldsymbol{M}_1\boldsymbol{a}$ in (7.15) and (7.16), we see from (7.19) that each of these univariate hypotheses is tested by the statistic

$$\frac{\{\boldsymbol{b}'\boldsymbol{C}_1(\boldsymbol{X}'\boldsymbol{X})^{-1}\boldsymbol{X}'\boldsymbol{Y}_+\boldsymbol{M}_1\boldsymbol{a}\}^2}{\{\boldsymbol{b}'\boldsymbol{C}_1(\boldsymbol{X}'\boldsymbol{X})^{-1}\boldsymbol{C}_1'\boldsymbol{b}\}\{\boldsymbol{a}'\boldsymbol{M}_1'\boldsymbol{Y}'\boldsymbol{P}\boldsymbol{Y}\boldsymbol{M}_1\boldsymbol{a}\}}, \tag{7.20}$$

which has a $(n - q)^{-1}F_{1,n-q}$ distribution under the null hypothesis for *fixed* $\boldsymbol{a}$ and $\boldsymbol{b}$. Maximizing (7.20) over $\boldsymbol{b}$ gives

$$\boldsymbol{a}'\boldsymbol{H}\boldsymbol{a}/\boldsymbol{a}'\boldsymbol{E}\boldsymbol{a}, \tag{7.21}$$

which has a $[g/(n - q)]F_{g,n-q}$ distribution for *fixed* $\boldsymbol{a}$ (because $\boldsymbol{H}$ and $\boldsymbol{E}$ have independent Wishart distributions). Finally, maximizing (7.21) over $\boldsymbol{a}$ shows the union intersection (UI) statistic to be λ_1, the largest eigenvalue of $\boldsymbol{H}\boldsymbol{E}^{-1}$. Let $\theta = \lambda_1/(1 + \lambda_1)$ denote the greatest root of $\boldsymbol{H}(\boldsymbol{H} + \boldsymbol{E})^{-1}$.

The distribution of $\theta = \theta(r_0, n - q, g)$ is described in Chapter 4. The null hypothesis is rejected for large values of θ.

If $\text{rank}(\boldsymbol{M}_1) = r_0 = 1$, then maximizing over $\boldsymbol{a}$ is unnecessary, and (7.21) becomes simply the ratio of scalars (7.19). Thus, the UIT is equivalent to the LRT in this case.

If $\text{rank}(\boldsymbol{C}_1) = g = 1$, then maximizing over $\boldsymbol{b}$ is unnecessary. Since $\boldsymbol{H}$ and $\boldsymbol{HE}^{-1}$ have rank 1 in this case, we see from (7.18) that the UIT and LRT are equivalent here also. The only nonzero eigenvalue of $\boldsymbol{HE}^{-1}$ equals its trace, so from (7.15) and (7.16),

$$\lambda_1 = \text{tr}\,\boldsymbol{HE}^{-1} = \{\boldsymbol{C}_1(\boldsymbol{X}'\boldsymbol{X})^{-1}\boldsymbol{C}_1'\}^{-1}\boldsymbol{C}_1(\boldsymbol{X}'\boldsymbol{X})^{-1}\boldsymbol{X}'\boldsymbol{Y}_+\boldsymbol{M}_1\boldsymbol{E}^{-1}\boldsymbol{M}_1'\boldsymbol{Y}_+'\boldsymbol{X}(\boldsymbol{X}'\boldsymbol{X})^{-1}\boldsymbol{C}_1'.$$

Let $\boldsymbol{d} = \boldsymbol{X}(\boldsymbol{X}'\boldsymbol{X})^{-1}\boldsymbol{C}_1'$. Under the null hypothesis, $\boldsymbol{M}_1'\boldsymbol{Y}_+'\boldsymbol{d} \sim N_p(\boldsymbol{0}, (\boldsymbol{d}'\boldsymbol{d})\boldsymbol{M}_1'\boldsymbol{\Sigma}\boldsymbol{M}_1)$ (Exercise 7.3.4), and under both the null and alternative hypotheses, $\boldsymbol{E} \sim W_{r_0}(\boldsymbol{M}_1'\boldsymbol{\Sigma}\boldsymbol{M}_1, n-q)$ independently. Thus, under the null hypothesis, λ_1 has the

$$(n-q)^{-1}T^2_{r_0,n-q} = [r_0/(n-q-r_0+1)]F_{r_0,n-q-r_0+1}$$

distribution.

7.3.3 Simultaneous Confidence Intervals

Suppose that $\boldsymbol{B}$ is the true value of the regression parameter matrix, and let $\boldsymbol{Y}_+ = \boldsymbol{Y} - \boldsymbol{XB}$. Then, the UI principle states that the probability that the ratio in (7.20) will be less than or equal to $\theta_\alpha/(1-\theta_\alpha)$ simultaneously for all $\boldsymbol{a}$ and $\boldsymbol{b}$ for which the denominator does not vanish equals $1-\alpha$. Here, θ_α denotes the upper α critical value of the $\theta(r_0, n-q, g)$ distribution. Rearranging gives the simultaneous confidence intervals (SCIs):

$$\text{P}\left(\boldsymbol{b}'\boldsymbol{C}_1\boldsymbol{BM}_1\boldsymbol{a} \in \boldsymbol{b}'\boldsymbol{C}_1(\boldsymbol{X}'\boldsymbol{X})^{-1}\boldsymbol{X}'\boldsymbol{YM}_1\boldsymbol{a} \pm \left\{\frac{\theta_\alpha}{1-\theta_\alpha}(\boldsymbol{a}'\boldsymbol{Ea})[\boldsymbol{b}'\boldsymbol{C}_1(\boldsymbol{X}'\boldsymbol{X})^{-1}\boldsymbol{C}_1'\boldsymbol{b}]\right\}^{1/2}, \text{ for all } \boldsymbol{a}, \boldsymbol{b}\right) = 1-\alpha. \tag{7.22}$$

Note that $\boldsymbol{b}'\boldsymbol{C}_1(\boldsymbol{X}'\boldsymbol{X})^{-1}\boldsymbol{X}'\boldsymbol{YM}_1\boldsymbol{a} = \boldsymbol{b}'\boldsymbol{C}_1\hat{\boldsymbol{B}}\boldsymbol{M}_1\boldsymbol{a}$ is an unbiased estimator of $\boldsymbol{b}'\boldsymbol{C}_1\boldsymbol{BM}_1\boldsymbol{a}$.

In some applications, $\boldsymbol{a}$ and/or $\boldsymbol{b}$ are given *a priori*, and the confidence limits in (7.22) can be narrowed. If $\boldsymbol{a}$ is fixed, then SCIs for $\boldsymbol{b}$ can be obtained by replacing $\theta_\alpha/(1-\theta_\alpha)$ by $[(g/(n-q)]F_{g,n-q}(1-\alpha)$. If $\boldsymbol{b}$ is fixed, then SCIs for $\boldsymbol{a}$ can be found by replacing $\theta_\alpha/(1-\theta_\alpha)$ by

$$(n-q)^{-1}T_{r_0,n-q;\alpha} = [r_0/(n-q-r_0+1)]F_{r_0,n-q-r_0+1}(1-\alpha).$$

Finally, if both $\boldsymbol{a}$ and $\boldsymbol{b}$ are known, a single confidence interval is obtained by replacing $\theta_\alpha/(1-\theta_\alpha)$ by $(n-q)^{-1}F_{1,n-q}(1-\alpha)$.

7.4 Design Matrices of Degenerate Rank

In the design of experiments, it is often convenient for symmetry reasons to consider design matrices $\boldsymbol{X}$ which are not of full rank. Unfortunately, the parameter matrix in (7.1) is not well defined in this case. The simplest way to resolve this indeterminacy is to restrict the regression to a subset of the columns of $\boldsymbol{X}$.

Suppose that $\boldsymbol{X}(n \times q)$ has rank $k < q$. Rearrange the columns of $\boldsymbol{X}$ and partition $\boldsymbol{X} = (\boldsymbol{X}_1, \boldsymbol{X}_2)$, so that $\boldsymbol{X}_1(n \times k)$ has full rank. Then, $\boldsymbol{X}_2$ can be written in terms of $\boldsymbol{X}_1$ as

$X_2 = X_1A$ for some matrix $A(k \times (q-k))$. Partition $B' = (B_1', B_2')$. If the regression model (7.1) is valid, then it can be reformulated in terms of X_1 alone as

$$Y = X_1B_1^* + U, \tag{7.23}$$

where $B_1^* = B_1 + AB_2$ is uniquely determined.

Consider a hypothesis $C_1BM_1 = D$, and partition $C_1 = (C_{11}, C_{12})$. The hypothesis matrix C_1 is called *testable* if, as a function of B,

$$XB = 0 \quad \text{implies} \quad C_1B = 0. \tag{7.24}$$

Then, although B is indeterminate in the model (7.1), $C_1B = C_{11}B_1^*$ is uniquely determined. Also, note that the condition (7.24) does not involve M_1. With a similar motivation, a linear combination $d'B$ is called *estimable* if $XB = 0$ implies $d'B = 0$.

An alternative way to describe testability was given by Roy (1957). He showed that (7.24) holds if and only if

$$C_{12} = C_{11}(X_1'X_1)^{-1}X_1'X_2. \tag{7.25}$$

(See Exercise 7.4.1.) This criterion is convenient in practice.

If $\hat{B}_1^*$ is the m.l.e. of B_1^* in the model (7.23), and if C_1 is testable, then the (unique) m.l.e. of C_1B is given by

$$C_{11}\hat{B}_1^* = C_{11}(X_1'X_1)^{-1}X_1'Y.$$

Thus, for a design matrix of nonfull rank, the estimation of estimable linear combinations and the testing of testable hypotheses are most conveniently carried out using the full model (7.23) instead of (7.1).

Example 7.4.1 (One-way multivariate analysis of variance). In Section 6.3.3.1, we considered k multivariate normal samples, $y_1, \dots, y_{n_1}, y_{n_1+1}, \dots, y_{n_1+n_2}, y_{n_1+n_2+1}, \dots, y_n$, where the ith sample consists of n_i observations from $N_p(\mu + \tau_i, \Sigma)$, $i = 1, \dots, k$, and $n = n_1 + n_2 + \cdots + n_k$. We can write this model in the form (7.1) using the design and parameter matrices

$$X = \begin{bmatrix} 1 & 0 & \cdots & 0 & 1 \\ \vdots & \vdots & & \vdots & \vdots \\ 1 & 0 & \cdots & 0 & 1 \\ 0 & 1 & \cdots & 0 & 1 \\ \vdots & \vdots & & \vdots & \vdots \\ 0 & 1 & \cdots & 0 & 1 \\ \vdots & \vdots & & \vdots & \vdots \\ 0 & 0 & \cdots & 1 & 1 \\ \vdots & \vdots & & \vdots & \vdots \\ 0 & 0 & \cdots & 1 & 1 \end{bmatrix} \quad \text{and} \quad B' = [\tau_1 \quad \cdots \quad \tau_k \quad \mu].$$

The matrix $X(n \times (k+1))$ has rank k. The indeterminacy of this parameterization is usually resolved by requiring $\sum_1^k \tau_i = 0$. However, in this section, it is more convenient to drop μ from the regression; that is, partition X so that X_1 contains the first k columns, and use the model (7.23). Since

$$(X_1'X_1)^{-1} = \text{diag}(n_i^{-1}),$$

we see that the m.l.e.s of $\tau_1^*, \ldots, \tau_k^*$ in the model (7.23) are given by

$$(\hat{\tau}_1^*, \ldots, \hat{\tau}_k^*)' = (X_1'X_1)^{-1}X_1'Y = (\bar{y}_1, \ldots, \bar{y}_k)'.$$

One way to express the hypothesis that $\tau_1 = \cdots = \tau_k$ for the model (7.1) is to use the $((k-1) \times (k+1))$ matrix:

$$C_1 = \begin{bmatrix} 1 & 0 & 0 & \cdots & 0 & -1 & 0 \\ 0 & 1 & 0 & \cdots & 0 & -1 & 0 \\ \vdots & \vdots & \vdots & & \vdots & \vdots & \vdots \\ 0 & 0 & 0 & \cdots & 1 & -1 & 0 \end{bmatrix}.$$

Partition $C_1 = (C_{11}, C_{12})$, so that C_{11} contains the first k columns. It is easily checked that the hypothesis $C_1B = 0$ is testable (and is equivalent to the hypothesis $\tau_1^* = \cdots = \tau_k^*$) for the model (7.23). The matrices H and E defined by (7.15) and (7.16), replaced by X_1 and C_{11}, can be shown to equal B and W, respectively, the "between-groups" and "within-groups" SSP matrices given in Section 6.3.3.1 (see Exercise 7.4.2). Thus, as expected, the LRT and UIT take the same forms as given there.

As b varies through all $(k-1)$-vectors, $C_{11}'b = c$ varies through all k-vectors, satisfying $\sum_1^k c_i = 0$. Such vectors are called *contrasts*, and $\sum c_i\tau_i = \sum c_i\tau_i^*$ for all μ. Thus, from (7.22), SCIs for linear combinations a of contrasts c between group means are given by

$$c'\tau a \in c'\bar{Y}a \pm \left\{ \frac{\theta_\alpha}{1-\theta_\alpha} a'Wa \sum_{i=1}^{k} \frac{c_i^2}{n_i} \right\}^{1/2},$$

where we have written $\tau' = (\tau_1, \ldots, \tau_k)$ and $\bar{Y}' = (\bar{y}_1, \ldots, \bar{y}_k)$, and θ_α is the upper α critical value of $\theta(p, n-k, k-1)$. If a and/or c are known *a priori*, the adjustments described at the end of Section 7.3.3 are appropriate. □

7.5 Multiple Correlation

7.5.1 The Effect of the Mean

In the model (7.1), it is sometimes convenient to separate the effect of the mean from the other "independent" variables. Rewrite the regression model as

$$Y = 1\mu' + XB + U, \tag{7.26}$$

where $(1, X)$ now represents an $(n \times (1+q))$ matrix of full rank. Without loss of generality, we suppose that the columns of X are each centered to have mean 0. (If $x_{(i)}$ is replaced by $x_{(i)} - \bar{x}_i 1$, then (7.26) remains valid if μ is replaced by $\mu + \sum \bar{x}_i\beta_{i\cdot}$.) Then,

$$\left[\begin{pmatrix} 1' \\ X' \end{pmatrix} (1, X)\right]^{-1} = \begin{bmatrix} n^{-1} & 0 \\ 0 & (X'X)^{-1} \end{bmatrix}$$

because $X'1 = 0$. Thus, by Theorem 7.2.3(a), $\hat{\mu} = \bar{y}$ is independent of $\hat{B} = (X'X)^{-1}X'Y = (X'X)^{-1}X'(Y - 1\bar{y}')$. If the mean of Y is estimated by $\bar{y}$, then (7.26) can be written as

$$Y - 1\bar{y}' = XB + W, \tag{7.27}$$

where $\boldsymbol{W} = \boldsymbol{U} - \mathbf{1}\bar{\boldsymbol{u}}'$ denotes the centered error matrix which, thought of as a column vector $\boldsymbol{W}^V$, has distribution $N_{np}(\mathbf{0}, \boldsymbol{\Sigma} \otimes \boldsymbol{H})$. Here, $\boldsymbol{H} = \boldsymbol{I} - n^{-1}\mathbf{1}\mathbf{1}'$ (see Section 2.5.5 and Exercise 7.5.1).

7.5.2 Multiple Correlation Coefficient

Suppose that $p = 1$, so that (7.26) becomes

$$\boldsymbol{y} - \bar{y}\mathbf{1} = \boldsymbol{X}\boldsymbol{\beta} + \boldsymbol{v},$$

where $\boldsymbol{v} \sim N_n(\mathbf{0}, \sigma^2\boldsymbol{H})$, and again, we suppose that the columns of $\boldsymbol{X}$ are each centered to have zero mean. Define

$$\boldsymbol{S} = \begin{bmatrix} s_{11} & \boldsymbol{S}_{12} \\ \boldsymbol{S}_{21} & \boldsymbol{S}_{22} \end{bmatrix} = n^{-1} \begin{bmatrix} \boldsymbol{y}' - \bar{y}\mathbf{1}' \\ \boldsymbol{X}' \end{bmatrix} (\boldsymbol{y} - \bar{y}\mathbf{1}, \boldsymbol{X}) \tag{7.28}$$

to be the sample covariance matrix for $\boldsymbol{y}$ and the columns of $\boldsymbol{X}$. Let $\boldsymbol{R}$ denote the corresponding correlation matrix. Since

$$\hat{\boldsymbol{\beta}} = (\boldsymbol{X}'\boldsymbol{X})^{-1}\boldsymbol{X}'\boldsymbol{y} = (\boldsymbol{X}'\boldsymbol{X})^{-1}\boldsymbol{X}'(\boldsymbol{y} - \bar{y}\mathbf{1}) = \boldsymbol{S}_{22}^{-1}\boldsymbol{S}_{21},$$

we see that the (sample) correlation between $\boldsymbol{y}$ and $\boldsymbol{X}\hat{\boldsymbol{\beta}}$ is given by

$$\begin{aligned} R_{y \cdot X} = \operatorname{cor}(\boldsymbol{y}, \boldsymbol{X}\hat{\boldsymbol{\beta}}) &= \boldsymbol{S}_{12}\boldsymbol{S}_{22}^{-1}\boldsymbol{S}_{21} / \{s_{11}\boldsymbol{S}_{12}\boldsymbol{S}_{22}^{-1}\boldsymbol{S}_{21}\}^{1/2} \\ &= \{\boldsymbol{S}_{12}\boldsymbol{S}_{22}^{-1}\boldsymbol{S}_{21}/s_{11}\}^{1/2}. \end{aligned}$$

Then, $R_{y \cdot X}$ is called the *multiple correlation coefficient of* $\boldsymbol{y}$ with $\boldsymbol{X}$. It can also be defined in terms of the correlations as

$$R^2_{y \cdot X} = \boldsymbol{R}_{12}\boldsymbol{R}_{22}^{-1}\boldsymbol{R}_{21}. \tag{7.29}$$

An important property of the multiple correlation is given by the following theorem.

Theorem 7.5.1 *The largest sample correlation between* $\boldsymbol{y}$ *and a linear combination of the columns of* $\boldsymbol{X}$ *is given by* $R_{y \cdot X}$ *and is attained by the linear combination* $\boldsymbol{X}\hat{\boldsymbol{\beta}}$, *where* $\hat{\boldsymbol{\beta}} = (\boldsymbol{X}'\boldsymbol{X})^{-1}\boldsymbol{X}'\boldsymbol{y}$ *is the regression coefficient of* $\boldsymbol{y}$ *on* $\boldsymbol{X}$.

Proof: Let $\boldsymbol{X}\boldsymbol{\beta}$ denote a linear combination of the columns of $\boldsymbol{X}$. Then,

$$\operatorname{cor}^2(\boldsymbol{y}, \boldsymbol{X}\boldsymbol{\beta}) = (\boldsymbol{\beta}'\boldsymbol{S}_{21})^2 / (s_{11}\boldsymbol{\beta}'\boldsymbol{S}_{22}\boldsymbol{\beta}).$$

This function is a ratio of quadratic forms in $\boldsymbol{\beta}$, and by Corollary A.4, its maximum is given by $\boldsymbol{S}_{12}\boldsymbol{S}_{22}^{-1}\boldsymbol{S}_{21}/s_{11}$ and is attained when $\hat{\boldsymbol{\beta}} = \boldsymbol{S}_{22}^{-1}\boldsymbol{S}_{21} = (\boldsymbol{X}'\boldsymbol{X})^{-1}\boldsymbol{X}'\boldsymbol{y}$. □

Another common notation, which is used for a single matrix $\boldsymbol{X}(n \times q)$ to label the multiple correlation of $\boldsymbol{x}_{(1)}$ with $\boldsymbol{x}_{(2)}, \ldots, \boldsymbol{x}_{(q)}$, is $R_{1 \cdot 2 \ldots q}$. It is necessary to exercise some caution when interpreting the multiple correlation, as shown by the following exercise.

Example 7.5.1 Consider the case when

$$r_{13} = 0, \qquad r_{12} = \cos\theta, \qquad r_{23} = \sin\theta.$$

These correlations will arise when $\boldsymbol{x}_{(1)}$, $\boldsymbol{x}_{(2)}$, and $\boldsymbol{x}_{(3)}$ lie in the same plane in $\mathbb{R}^n$, and the angle between $\boldsymbol{x}_{(1)}$ and $\boldsymbol{x}_{(2)}$ is θ. If θ is small, then r_{13} and r_{23} are both small. However, $R_{3\cdot12} = 1$. This example illustrates the seemingly paradoxical result that, for small θ, $\boldsymbol{x}_{(3)}$ has a low correlation with both $\boldsymbol{x}_{(1)}$ and $\boldsymbol{x}_{(2)}$, but the multiple correlation coefficient $R_{3\cdot12}$ is at its maximum. □

The multiple correlation coefficient can be used to partition the total sum of squares as in (7.9). If $\hat{\boldsymbol{y}} = \boldsymbol{X}\hat{\boldsymbol{\beta}}$ denotes the fitted value of $\boldsymbol{y}$, and $\hat{\boldsymbol{u}} = \boldsymbol{y} - \bar{y}\boldsymbol{1} - \boldsymbol{X}\hat{\boldsymbol{\beta}}$ denotes the fitted residual, then the total sum of squares (about the mean) can be expressed as

$$\boldsymbol{y}'\boldsymbol{y} - n\bar{y}^2 = \hat{\boldsymbol{y}}'\hat{\boldsymbol{y}} + \hat{\boldsymbol{u}}'\hat{\boldsymbol{u}},$$

which can also be written as

$$ns_y^2 = ns_y^2 R_{y\cdot\boldsymbol{X}}^2 + ns_y^2(1 - R_{y\cdot\boldsymbol{X}}). \tag{7.30}$$

Thus, the squared multiple correlation coefficient represents the proportion of the total sum of squares explained by the regression on $\boldsymbol{X}$.

7.5.3 Partial Correlation Coefficient

It is sometimes of interest to study the correlation between two variables after allowing for the effects of other variables. Consider the multivariate regression (7.27) with $p = 2$ and center the columns of $\boldsymbol{Y}$, so that $\bar{y}_1 = \bar{y}_2 = 0$. We can write each column of $\boldsymbol{Y}$ as

$$\boldsymbol{y}_{(i)} = \hat{\boldsymbol{y}}_i + \hat{\boldsymbol{u}}_{(i)}, \quad i = 1,2,$$

where $\hat{\boldsymbol{y}}_i = \boldsymbol{X}(\boldsymbol{X}'\boldsymbol{X})^{-1}\boldsymbol{X}'\boldsymbol{y}_{(i)}$ is the "fitted value" of $\boldsymbol{y}_{(i)}$. Since $\hat{\boldsymbol{u}}_{(1)}$ and $\hat{\boldsymbol{u}}_{(2)}$ are both residuals after fitting $\boldsymbol{X}$, a correlation between $\boldsymbol{y}_{(1)}$ and $\boldsymbol{y}_{(2)}$ after "eliminating" the effect of $\boldsymbol{X}$ can be defined by

$$r_{12\cdot\boldsymbol{X}} = \hat{\boldsymbol{u}}_{(1)}'\hat{\boldsymbol{u}}_{(2)}/\{||\hat{\boldsymbol{u}}_{(1)}||\ ||\hat{\boldsymbol{u}}_{(2)}||\}.$$

The coefficient $r_{12\cdot\boldsymbol{X}}$ is called the *(sample) partial correlation coefficient between* $\boldsymbol{y}_{(1)}$ *and* $\boldsymbol{y}_{(2)}$, *given* $\boldsymbol{X}$.

To calculate the partial correlation coefficient, partition the covariance matrix of $\boldsymbol{Y}$ and $\boldsymbol{X}$ as

$$\boldsymbol{S} = \begin{bmatrix} s_{11} & s_{12} & \boldsymbol{s}_1' \\ s_{21} & s_{22} & \boldsymbol{s}_2' \\ \boldsymbol{s}_1 & \boldsymbol{s}_2 & \boldsymbol{S}_{22} \end{bmatrix} = n^{-1}\begin{bmatrix} \boldsymbol{y}_{(1)}' \\ \boldsymbol{y}_{(2)}' \\ \boldsymbol{X} \end{bmatrix}[\boldsymbol{y}_{(1)} \quad \boldsymbol{y}_{(2)} \quad \boldsymbol{X}].$$

If we set

$$s_{ij}^* = s_{ij} - \boldsymbol{s}_i'\boldsymbol{S}_{22}^{-1}\boldsymbol{s}_j, \quad i,j = 1,2,$$

then

$$r_{12\cdot\boldsymbol{X}} = s_{12}^*/\{s_{11}^* s_{22}^*\}^{1/2}. \tag{7.31}$$

(See Exercise 7.5.5.)

If we partition the correlation matrix $\boldsymbol{R}$ similarly, and set $r_{ij}^* = r_{ij} - \boldsymbol{r}_i'\boldsymbol{R}_{22}^{-1}\boldsymbol{r}_j$, $i,j = 1,2$, then $r_{12\cdot\boldsymbol{X}}$ can also be written as

$$r_{12\cdot\boldsymbol{X}} = r_{12}^*/\{r_{11}^* r_{22}^*\}^{1/2}. \tag{7.32}$$

Another notation, used for a single matrix $\boldsymbol{X}(n \times q)$ to describe the partial correlation between $\boldsymbol{x}_{(1)}$ and $\boldsymbol{x}_{(2)}$, given $\boldsymbol{x}_{(3)}, \dots, \boldsymbol{x}_{(q)}$, is $r_{12\cdot3\dots q}$.

In general, if $\boldsymbol{X} = (\boldsymbol{X}_1, \boldsymbol{X}_2)$ is a partitioned data matrix with sample covariance matrix $\boldsymbol{S}$, we can define the "partial" or "residual" covariance matrix of $\boldsymbol{X}_1$ after allowing for the effect of $\boldsymbol{X}_2$ by

$$\boldsymbol{S}_{11\cdot2} = \boldsymbol{S}_{11} - \boldsymbol{S}_{12}\boldsymbol{S}_{22}^{-1}\boldsymbol{S}_{21}. \tag{7.33}$$

Note that $\boldsymbol{S}_{11\cdot2}$ can also be interpreted as the sample conditional covariance matrix of $\boldsymbol{X}_1$, given $\boldsymbol{X}_2$. (A population version of (7.33) for the multivariate normal distribution is given in (4.2).)

Example 7.5.2 Observations on the intelligence (x_1), weight (x_2), and age (x_3) of school children produced the following correlation matrix:

$$\boldsymbol{R} = \begin{bmatrix} 1.0000 & 0.6162 & 0.8267 \\ & 1.0000 & 0.7321 \\ & & 1.0000 \end{bmatrix}.$$

This in fact suggests a high dependence (0.6162) between weight and intelligence. However, calculating the partial correlation coefficient $r_{12\cdot3} = 0.0286$ shows that the correlation between weight and intelligence is very much lower, in fact almost zero, after the effect of age has been taken into account.

The multiple correlation coefficient between intelligence and the other two variables is evaluated from (7.29) as

$$R_{1\cdot23} = 0.8269.$$

However, the simple correlation r_{13} is 0.8267, so weight obviously plays little further part in explaining intelligence. □

7.5.4 Measures of Correlation Between Vectors

In the univariate model defined by (7.26), that is when $p = 1$, we have seen that the squared multiple correlation coefficient R^2 represents the "proportion of variance explained" by the model, and from (7.30),

$$1 - R^2 = \hat{\boldsymbol{u}}'\hat{\boldsymbol{u}}/\boldsymbol{y}'\boldsymbol{y}.$$

(We assume $\boldsymbol{y}$ is centered, so that $\bar{y} = 0$.)

It seems reasonable to seek a similar measure for the multivariate correlation between matrices $\boldsymbol{X}$ and $\boldsymbol{Y}$ in the model $\boldsymbol{Y} = \boldsymbol{XB} + \boldsymbol{U}$.

Let us start by writing $\boldsymbol{D} = (\boldsymbol{Y}'\boldsymbol{Y})^{-1}\hat{\boldsymbol{U}}'\hat{\boldsymbol{U}}$. (Again, we assume that $\boldsymbol{Y}$ is centered, so that the columns of $\boldsymbol{Y}$ have zero mean.) The matrix $\boldsymbol{D}$ is a simple generalization of $1 - R^2$ in the univariate case. Note that $\hat{\boldsymbol{U}}'\hat{\boldsymbol{U}} = \boldsymbol{Y}'\boldsymbol{PY}$ ($\boldsymbol{P}$ is defined in Eq. (7.4)) ranges between zero, when all the variation in $\boldsymbol{Y}$ is "explained" by the model, and $\boldsymbol{Y}'\boldsymbol{Y}$ at the other extreme, when no part of the variation in $\boldsymbol{Y}$ is explained. Therefore, $\boldsymbol{I}-\boldsymbol{D}$ varies between the identity matrix and the zero matrix (see Exercise 7.5.7). Any sensible measure of multivariate correlation should range between one and zero at these extremes, and this property is satisfied by at least two oft-used coefficients, the trace correlation r_T and the determinant correlation r_D, defined as follows (Hooper 1959, pp. 249–250):

$$r_T^2 = p^{-1}\text{tr}\,(\boldsymbol{I}-\boldsymbol{D}), \qquad r_D^2 = |\boldsymbol{I}-\boldsymbol{D}|. \tag{7.34}$$

For their population counterparts, see Exercise 7.5.8. Hotelling (1936) suggested the "vector alienation coefficient" (r_A), which in our notation is $|\boldsymbol{D}|$, and ranges from zero when $\hat{\boldsymbol{U}} = \boldsymbol{0}$ to one when $\hat{\boldsymbol{U}} = \boldsymbol{Y}$.

If $d_1, \ldots, d_p$ are the eigenvalues of $\boldsymbol{I}-\boldsymbol{D}$, then

$$r_T^2 = p^{-1}\sum d_i, \qquad r_D^2 = \prod d_i, \qquad r_A = \prod (1-d_i).$$

From this formulation, it is clear that r_D or r_A is zero if *just one* d_i is zero or one, respectively, but r_T is zero only if *all* the d_i are zero. Note that these measures of vector correlations are invariant under commutation, namely they would remain the same if we were to define $\boldsymbol{D}$ as equal to $(\hat{\boldsymbol{U}}'\hat{\boldsymbol{U}})(\boldsymbol{Y}'\boldsymbol{Y})^{-1}$.

7.6 Least-squares Estimation

7.6.1 Ordinary Least-squares (OLS) Estimation

Consider the multiple regression model

$$\boldsymbol{y} = \boldsymbol{X}\boldsymbol{\beta} + \boldsymbol{u}, \tag{7.35}$$

where $\boldsymbol{u}(n \times 1)$ is the vector of disturbances. Suppose that the matrix $\boldsymbol{X}(n \times q)$ is a known matrix of rank q, and let us relax the assumptions about the disturbances, assuming merely that

$$\mathsf{E}(\boldsymbol{u}) = \boldsymbol{0}, \qquad \text{var}(\boldsymbol{u}) = \boldsymbol{\Omega}.$$

When $\boldsymbol{u}$ is not normally distributed, then the estimation of $\boldsymbol{\beta}$ can be approached by the method of least squares.

Definition 7.6.1 *The* ordinary least-squares (OLS) estimator *of $\boldsymbol{\beta}$ is given by*

$$\hat{\boldsymbol{\beta}} = (\boldsymbol{X}'\boldsymbol{X})^{-1}\boldsymbol{X}'\boldsymbol{y}. \tag{7.36}$$

Differentiation shows that $\hat{\boldsymbol{\beta}}$ minimizes the residual sum of squares

$$(\boldsymbol{y}-\boldsymbol{X}\boldsymbol{\beta})'(\boldsymbol{y}-\boldsymbol{X}\boldsymbol{\beta}),$$

hence the terminology. Note that $\hat{\boldsymbol{\beta}}$ is exactly the same as the m.l.e. of $\boldsymbol{\beta}$ given in Theorem 7.2.1 (with $p = 1$) under the assumption of normality and the assumption $\boldsymbol{\Omega} = \sigma^2\boldsymbol{I}_n$.

Note that

$$\mathsf{E}(\hat{\boldsymbol{\beta}}) = (\boldsymbol{X}'\boldsymbol{X})^{-1}\boldsymbol{X}'(\boldsymbol{X}\boldsymbol{\beta} + \mathsf{E}\boldsymbol{u}) = \boldsymbol{\beta},$$

So that $\hat{\boldsymbol{\beta}}$ is an *unbiased* estimate of $\boldsymbol{\beta}$. Also,

$$\text{var}(\hat{\boldsymbol{\beta}}) = (\boldsymbol{X}'\boldsymbol{X})^{-1}(\boldsymbol{X}'\boldsymbol{\Omega}\boldsymbol{X})(\boldsymbol{X}'\boldsymbol{X})^{-1}. \tag{7.37}$$

In particular, if $\boldsymbol{\Omega} = \sigma^2\boldsymbol{I}_n$, this formula simplifies to

$$\text{var}(\hat{\boldsymbol{\beta}}) = \sigma^2(\boldsymbol{X}'\boldsymbol{X})^{-1}, \tag{7.38}$$

and in this case, $\hat{\boldsymbol{\beta}}$ has the following optimal property.

Theorem 7.6.2 (Gauss–Markov). *Consider the multiple regression model (7.35) and suppose that the disturbance terms are uncorrelated with one another,* $\text{var}(\boldsymbol{u}) = \sigma^2\boldsymbol{I}_n$. *Then, the OLS estimator (7.36) has a covariance matrix which is smaller than that of any other linear unbiased estimator.*

In other words, the OLS estimator is the *best* linear unbiased estimator (BLUE). The proof of this theorem is outlined in Exercise 7.6.2.

7.6.2 Generalized Least Squares

When $\boldsymbol{u}$ does not have a covariance matrix equal to $\sigma^2\boldsymbol{I}$, then, in general, the OLS estimator is *not* BLUE. However, a straightforward transformation adjusts the covariance matrix, so that the OLS can be used.

Let

$$\mathsf{E}(\boldsymbol{u}) = \boldsymbol{0}, \quad \text{var}(\boldsymbol{u}) = \boldsymbol{\Omega}.$$

Suppose that $\boldsymbol{\Omega}$ is known, and consider the transformed model

$$\boldsymbol{z} = \boldsymbol{\Omega}^{-1/2}\boldsymbol{X}\boldsymbol{\beta} + \boldsymbol{v},$$

where

$$\boldsymbol{z} = \boldsymbol{\Omega}^{-1/2}\boldsymbol{y}, \qquad \boldsymbol{v} = \boldsymbol{\Omega}^{-1/2}\boldsymbol{u}.$$

Since $\text{var}(\boldsymbol{v}) = \boldsymbol{I}$, this transformed model satisfies the assumptions of the Gauss–Markov theorem (Theorem 7.6.2). Thus, the best linear unbiased estimate of $\boldsymbol{\beta}$ is given by

$$\tilde{\boldsymbol{\beta}} = (\boldsymbol{X}'\boldsymbol{\Omega}\boldsymbol{X})^{-1}\boldsymbol{X}'\boldsymbol{\Omega}^{-1/2}\boldsymbol{z} = (\boldsymbol{X}'\boldsymbol{\Omega}^{-1}\boldsymbol{X})^{-1}\boldsymbol{X}'\boldsymbol{\Omega}^{-1}\boldsymbol{y}, \tag{7.39}$$

and the covariance matrix of $\tilde{\boldsymbol{\beta}}$ is

$$\text{var}(\tilde{\boldsymbol{\beta}}) = (\boldsymbol{X}'\boldsymbol{\Omega}^{-1}\boldsymbol{X})^{-1}.$$

The estimator defined by (7.39) is called the *generalized least squares* (GLS) estimator and is an obvious generalization of the OLS estimator. The fact that the GLS estimator is the BLUE can also be proved directly (Exercise 7.6.3).

Note that (7.39) is not an estimator if $\boldsymbol{\Omega}$ is not known. However, in some applications, although $\boldsymbol{\Omega}$ is unknown, there exists a consistent estimator $\hat{\boldsymbol{\Omega}}$ of $\boldsymbol{\Omega}$. Then, $\hat{\boldsymbol{\Omega}}$ may be used in place of $\boldsymbol{\Omega}$ in (7.39).

7.6.3 Application to Multivariate Regression

Multivariate regression is an example where OLS and GLS lead to the same estimators. Write the multivariate regression model (7.35) in vector form,

$$\boldsymbol{Y}^V = \boldsymbol{X}^* \boldsymbol{B}^V + \boldsymbol{U}^V, \tag{7.40}$$

where $\boldsymbol{Y}^V = (\boldsymbol{y}'_{(1)}, \ldots, \boldsymbol{y}'_{(p)})'$ is obtained by stacking the columns of $\boldsymbol{Y}$ on top of one another, and $\boldsymbol{X}^* = \boldsymbol{I}_p \otimes \boldsymbol{X}$. The disturbance $\boldsymbol{U}^V$ is assumed to have mean $\boldsymbol{0}$ and covariance matrix

$$\text{var}(\boldsymbol{U}^V) = \boldsymbol{\Omega} = \boldsymbol{\Sigma} \otimes \boldsymbol{I}_n.$$

Then, the GLS estimator of $\boldsymbol{B}^V$ is given by

$$\begin{aligned} \hat{\boldsymbol{B}}^V &= (\boldsymbol{X}^{*\prime}\boldsymbol{\Omega}^{-1}\boldsymbol{X}^*)^{-1}\boldsymbol{X}^{*\prime}\boldsymbol{\Omega}^{-1}\boldsymbol{Y}^V \\ &= (\boldsymbol{\Sigma} \otimes \boldsymbol{X}'\boldsymbol{X})^{-1}(\boldsymbol{\Sigma}^{-1} \otimes \boldsymbol{X}')\boldsymbol{Y}^V \\ &= [\boldsymbol{I} \otimes (\boldsymbol{X}'\boldsymbol{X})^{-1}\boldsymbol{X}']\boldsymbol{Y}^V. \end{aligned} \tag{7.41}$$

Note that (7.40) does not depend on $\boldsymbol{\Sigma}$, and (7.41) defines an estimator whether or not $\boldsymbol{\Sigma}$ is known. In particular, the OLS estimator is obtained when $\boldsymbol{\Sigma} = \boldsymbol{I}$, so the OLS and GLS estimators are the same in this case. Further, (7.41) also coincides with the m.l.e. of $\boldsymbol{B}$ obtained in Theorem 7.2.1 under the assumption of normality.

7.6.4 Asymptotic Consistency of Least-squares Estimators

Let us examine the asymptotic behavior of the GLS estimator as the sample size n tends to ∞. Let $\boldsymbol{x}_1, \boldsymbol{x}_2, \ldots$ be a sequence of (fixed) "independent" variables, and let $\boldsymbol{u}_1, \boldsymbol{u}_2, \ldots$ be a sequence of random disturbances. Writing $\boldsymbol{X} = \boldsymbol{X}_n = (\boldsymbol{x}'_1, \ldots, \boldsymbol{x}'_n)'$, $\boldsymbol{\Omega} = \boldsymbol{\Omega}_n$, and $\boldsymbol{\beta} = \boldsymbol{\beta}_n$, to emphasize the dependence on n, suppose that

$$\lim_{n\to\infty} (\boldsymbol{X}'_n \boldsymbol{\Omega}^{-1} \boldsymbol{X}_n)^{-1} = \boldsymbol{0}. \tag{7.42}$$

Then, $\mathsf{E}(\tilde{\boldsymbol{\beta}}_n) = \boldsymbol{\beta}$ for all n, and $\text{var}(\tilde{\boldsymbol{\beta}}_n) = (\boldsymbol{X}'_n\boldsymbol{\Omega}_n^{-1}\boldsymbol{X}_n)^{-1} \to \boldsymbol{0}$ as $n \to \infty$. Hence, $\tilde{\boldsymbol{\beta}}$ is *consistent*.

If the "independent" variables $\boldsymbol{x}_r$, $r = 1, \ldots, n$, are *random* rather than fixed, but still uncorrelated with random disturbances, then the results of Sections 7.6.1 and 7.6.2 remain valid, provided all expectations and variances are interpreted as "conditional on $\boldsymbol{X}$". In particular, suppose that

$$\mathsf{E}(\boldsymbol{u} \mid \boldsymbol{X}) = \boldsymbol{0} \qquad \text{and} \qquad \text{var}(\boldsymbol{u} \mid \boldsymbol{X}) = \boldsymbol{\Omega} \tag{7.43}$$

do not depend on $\boldsymbol{X}$.

If we replace condition (7.42) by

$$\text{plim}\ (\boldsymbol{X}'_n\boldsymbol{\Omega}_n^{-1}\boldsymbol{X}_n)^{-1} = \boldsymbol{0}, \tag{7.44}$$

then consistency holds in this situation also. (Here, plim denotes the limit in probability; that is, if $\boldsymbol{A}_n$ is a sequence of random matrices and $\boldsymbol{A}_0$ is a constant matrix, then $\text{plim}\ \boldsymbol{A}_n = \boldsymbol{A}_0$ or $\boldsymbol{A}_n \xrightarrow{p} \boldsymbol{A}_0$ if and only if

$$\mathsf{P}(||\boldsymbol{A}_n - \boldsymbol{A}_0|| > \epsilon) \to 0 \quad \text{as} \quad n \to \infty,$$

for each $\epsilon > 0$, where $||\boldsymbol{A}|| = \sum\sum a_{ij}^2$.) The proof is outlined in Exercise 7.6.5.

If

$$\text{plim}\; n^{-1}\mathbf{X}_n'\boldsymbol{\Omega}_n^{-1}\mathbf{X}_n = \boldsymbol{\Psi}, \quad \text{or} \quad \lim n^{-1}\mathbf{X}_n'\boldsymbol{\Omega}_n^{-1}\mathbf{X}_n = \boldsymbol{\Psi} \tag{7.45}$$

where $\boldsymbol{\Psi}(q \times q)$ is nonsingular, then $(\mathbf{X}_n'\boldsymbol{\Omega}_n^{-1}\mathbf{X}_n)^{-1}$ tends to zero at rate n^{-1}, so that (7.42) or (7.44) holds. In this situation, with some further regularity assumptions on $\boldsymbol{X}_n$ and/or $\boldsymbol{u}_n$, it can be shown that in fact $\tilde{\boldsymbol{\beta}}_n$ is asymptotically *normally* distributed, with mean $\boldsymbol{\beta}$ and covariance matrix $n^{-1}\boldsymbol{\Psi}^{-1}$. Note that (7.45) will hold, for example, if $\boldsymbol{\Omega}_n = \sigma^2\boldsymbol{I}_n$ and $\boldsymbol{X}_n$ is a random sample from some distribution with mean $\boldsymbol{\mu}$ and covariance matrix $\boldsymbol{\Sigma}$, letting $\boldsymbol{\Psi} = \boldsymbol{\Sigma} + \boldsymbol{\mu}\boldsymbol{\mu}'$.

7.7 Discarding of Variables

7.7.1 Dependence Analysis

Consider the multiple regression

$$\boldsymbol{y} = \boldsymbol{X}\boldsymbol{\beta} + \mu\mathbf{1} + \boldsymbol{u}$$

and suppose that some of the columns of $\boldsymbol{X}(n \times q)$ are "nearly" collinear. Since $\text{rank}(\boldsymbol{X})$ equals the number of linearly independent columns of $\boldsymbol{X}$ and also the number of nonzero eigenvalues of $\boldsymbol{X}'\boldsymbol{X}$, we see that in this situation some of the eigenvalues of $\boldsymbol{X}'\boldsymbol{X}$ will be "nearly" zero. Hence, some of the eigenvalues of $(\boldsymbol{X}'\boldsymbol{X})^{-1}$ will be very large. The variance of the OLS estimator $\hat{\boldsymbol{\beta}}$ equals $\sigma^2(\boldsymbol{X}'\boldsymbol{X})^{-1}$, and so, at least some of these regression estimates will have large variances.

Clearly, because of the high collinearity, some of the independent variables are contributing little to the regression. Therefore, it is of interest to ask how well the regression can be explained using a smaller number of independent variables. There are two important reasons for discarding variables here; namely

(1) to increase the precision of the regression estimates for the retained variables and
(2) to reduce the number of measurements needed on similar data in the future.

Suppose that it has been decided to retain k variables $(k < q)$. Then, a natural choice of retained variables is the subset $\boldsymbol{x}_{i_1}, \ldots, \boldsymbol{x}_{i_k}$, which maximizes the squared multiple correlation coefficient,

$$R_{y \cdot i_1 \ldots i_k}. \tag{7.46}$$

This choice will explain more of the variation in $\boldsymbol{y}$ than any other set of k variables.

Computationally, the search for the best subset of k variables is not simple, but efficient algorithms, which work well for data sets that are not too large, are described in Miller (2002).

The choice of k, the number of retained variables, is somewhat arbitrary. One rule of thumb is to retain enough variables so that the squared multiple correlation with $\boldsymbol{y}$ using k variables is at least 90% or 95% of the squared multiple correlation using all q variables (but see Thompson (1978), for some test criteria).

Further, although there is usually just one set of k variables maximizing (7.46), there are often other k-sets that are "nearly optimal," so that for practical purposes the choice of variables is not uniquely determined.

An alternative method of dealing with multicollinearity in a multiple regression can be developed using principal components (see Sections 9.8 and 16.3).

Example 7.7.1 In a study of $n = 180$ wooden pitprops cut from the Corsican pine tree, Jeffers (1967) studied the dependence of maximum compressive strength (y) on 13 other variables ($\boldsymbol{X}$) measured on each prop. The strength of pitprops is of interest because it is used to support the roofs of mine tunnels.

The physical variables measure on each prop were

x_1 = the top diameter of the prop (inches);

x_2 = the length of the prop (inches);

x_3 = the moisture content of the prop, expressed as a percentage of the dry weight;

x_4 = the specific gravity of the timber at the time of the test;

x_5 = the oven-dry specific gravity of the timber;

x_6 = the number of annual rings at the top of the prop;

x_7 = the number of annual rings at the base of the prop;

x_8 = the maximum bow (inches);

x_9 = the distance of the point of maximum bow from the top of the prop (inches);

x_{10} = the number of knot whorls;

x_{11} = the length of clear prop from the top of the prop (inches);

x_{12} = the average number of knots per whorl;

x_{13} = the average diameter of the knots (inches).

The columns of $\boldsymbol{X}$ have each been standardized to have mean 0 and variance 1, so that $\boldsymbol{X}'\boldsymbol{X}$, given in Table 7.1, represents the correlation matrix for the independent variables. The correlation between each of the independent variables and $\boldsymbol{y}$ is given in Table 7.2. The eigenvalues of $\boldsymbol{X}'\boldsymbol{X}$ are 4.22, 2.38, 1.88, 1.11, 0.91, 0.82, 0.58, 0.44, 0.35, 0.19, 0.05, 0.04, and 0.04, and clearly, the smallest ones are nearly zero.

For each value of k, the subset of k variables that maximizes (7.46) is given in Table 7.3. Note that if all the variables are retained, then 73.1% of the variance of $\boldsymbol{y}$ can be explained by the regression. If only $k = 6$ variables are retained, then we can still explain 71.6% of the variance of $\boldsymbol{y}$, which is $(0.716/0.731) \times 100\%$ or over 95% of the "explainable" variance of $\boldsymbol{y}$.

Note that for this data, the optimal $(k + 1)$-set of variables equals the optimal k-set of variables plus one new variable, for $k = 1, \ldots, 12$. Unfortunately, this property does not hold in general. Further discussion of this data appears in Example 9.8.1. □

For further discussion of methods for discarding variables and related tests, see Thompson (1978).

7.7.2 Interdependence Analysis

Consider now a situation where q explanatory variables $\boldsymbol{X}$ are considered on their own, rather than as independent variables in a regression. When the columns of $\boldsymbol{X}$ are nearly

Table 7.1 Correlation matrix for the physical properties of props.

x_1												
0.954	x_2											
0.364	0.297	x_3										
0.342	0.284	0.882	x_4									
−0.129	−0.118	−0.148	0.22	x_5								
0.313	0.291	0.153	0.381	0.364	x_6							
0.496	0.503	−0.029	0.174	0.296	0.813	x_7						
0.424	0.419	−0.054	−0.059	0.004	0.090	0.372	x_8					
0.592	0.648	0.125	0.137	−0.039	0.211	0.465	0.482	x_9				
0.545	0.569	−0.081	−0.014	0.037	0.274	0.679	0.557	0.526	x_{10}			
0.084	0.076	0.162	0.097	−0.091	−0.036	−0.113	0.061	0.085	−0.319	x_{11}		
−0.019	−0.036	0.220	0.169	−0.145	0.024	−0.232	−0.357	−0.127	−0.368	0.029	x_{12}	
0.134	0.144	0.126	0.015	−0.208	−0.329	−0.424	−0.202	−0.076	−0.291	0.007	0.184	x_{13}

Table 7.2 Correlation between the independent variables and $\boldsymbol{y}$ for pitprop data.

Variable	Correlation
1	−0.419
2	−0.338
3	−0.728
4	−0.543
5	0.247
6	0.117
7	0.110
8	−0.253
9	−0.235
10	−0.101
11	−0.055
12	−0.117
13	−0.153

collinear, it is desirable to discard some of the variables in order to reduce the number of measurements needed to describe the data effectively. In this context, we need a measure of how well a retained set of k variables, $\boldsymbol{x}_{i_1}, \ldots, \boldsymbol{x}_{i_k}$, explains the whole data set $\boldsymbol{X}$.

A measure of how well one rejected variable $\boldsymbol{x}_{j_1}$ is explained by the retained variables $\boldsymbol{x}_{i_1}, \ldots, \boldsymbol{x}_{i_k}$ is given by the squared multiple correlation coefficient $R^2_{j_1 \cdot i_1 \ldots i_k}$. Thus, an overall

Table 7.3 Variables selected in multiple regression for pitprop data.

k	Variables selected	R^2
1	3	0.530
2	3, 8	0.616
3	3, 6, 8	0.684
4	3, 6, 8, 11	0.695
5	1, 3, 6, 8, 11	0.706
6	1, 2, 3, 6, 8, 11	0.716
7	1, 2, 3, 4, 6, 8, 11	0.721
8	1, 2, 3, 4, 6, 8, 11, 12	0.725
9	1, 2, 3, 4, 5, 6, 8, 11, 12	0.727
10	1, 2, 3, 4, 5, 6, 8, 9, 11, 12	0.729
11	1, 2, 3, 4, 5, 6, 7, 8, 9, 11, 12	0.729
12	1, 2, 3, 4, 5, 6, 7, 8, 9, 10, 11, 12	0.731
13	1, 2, 3, 4, 5, 6, 7, 8, 9, 10, 11, 12, 13	0.731

measure of the ability of the retained variables to explain the data is obtained by looking at the worst possible case; namely

$$\min_{j_l} R^2_{j_l \cdot i_1 \dots i_k}, \tag{7.47}$$

where j_l runs through all the rejected variables. Then, the best choice for the retained variables is obtained by maximizing (7.47) over all k-sets of variables.

Computationally, this problem is similar to the problem of Section 7.7.1. See Beale et al. (1967) and Beale (1970).

As in Section 7.7.1, the choice of k is somewhat arbitrary. A possible rule of thumb is to retain enough variables so that the minimum R^2 with any rejected variable is at least 0.50, say. Further, although the best choice of k variables is usually well defined, there are often several "nearly optimal" choices of k variables making the selection nonunique for practical purposes.

Another method of rejecting variables in interdependence analysis based on principal components is discussed in Section 9.7.

Example 7.7.2 Let us carry out an interdependence analysis for the 13 explanatory variables of Jeffers' pitprop data of Example 7.7.1. For each value of $k = 1, \dots, 13$, the optimal k-set of variables is given in Table 7.4, together with the minimum value of the squared multiple correlation of any of the rejected variables with the retained variables. For example, eight variables, namely the variables 3, 5, 7, 8, 9, 11, 12, and 13, are needed to explain at least 50% of the variation in each of the rejected variables.

Note from Table 7.4 that the best k-set of variables is *not* in general a subset of the best $(k + 1)$-set.

Further discussion of this data is given in Example 9.7.1. □

Table 7.4 Variables selected in interdependence analysis for pitprop data.

k	Variables selected	min R^2 with any rejected variable
1	9	0.002
2	4, 10	0.050
3	1, 5, 10	0.191
4	1, 5, 8, 10	0.237
5	1, 4, 7, 11, 12	0.277
6	3, 4, 10, 11, 12, 13	0.378
7	2, 3, 4, 8, 11, 12, 13	0.441
8	3, 5, 7, 8, 9, 11, 12, 13	0.658
9	2, 4, 5, 7, 8, 9, 11, 12, 13	0.751
10	1, 3, 5, 6, 8, 9, 10, 11, 12, 13	0.917
11	2, 3, 5, 6, 7, 8, 9, 10, 11, 12, 13	0.919
12	1, 3, 4, 5, 6, 7, 8, 9, 10, 11, 12, 13	0.927
13	1, 2, 3, 4, 5, 6, 7, 8, 9, 10, 11 12, 13	—

Exercises and Complements

7.1.1 (Box and Tiao 1973, p. 439). For the multivariate regression model (7.1) with normal errors, show that a Bayesian analysis with a noninformative prior leads to a posterior probability density function (p.d.f.) for $\boldsymbol{B}$, which is proportional to

$$|n\hat{\boldsymbol{\Sigma}} + (\hat{\boldsymbol{B}} - \boldsymbol{B})'\boldsymbol{X}'\boldsymbol{X}(\hat{\boldsymbol{B}} - \boldsymbol{B})|^{-n/2},$$

where $\hat{\boldsymbol{\Sigma}}$ and $\hat{\boldsymbol{B}}$ are defined in (7.5) and (7.6). Hence, the posterior distribution of $\boldsymbol{B}$ is a matrix T distribution.

7.2.1 Show that the $(n \times n)$ matrix $\boldsymbol{P}$ defined in (7.4) is symmetric and idempotent. Show that $\boldsymbol{Pw} = \boldsymbol{w}$ if and only if $\boldsymbol{w}$ is orthogonal to all columns of $\boldsymbol{X}$, and that $\boldsymbol{Pw} = \boldsymbol{0}$ if and only if $\boldsymbol{w}$ is a linear combination of the columns of $\boldsymbol{X}$. Hence, $\boldsymbol{P}$ is a projection onto the subspace of $\mathbb{R}^n$ orthogonal to the columns of $\boldsymbol{X}$.

7.2.2 (a) From (7.7), show that $l(\boldsymbol{B}, \boldsymbol{\Sigma})$ depends on $\boldsymbol{Y}$ only through $\hat{\boldsymbol{B}}$ and $\hat{\boldsymbol{\Sigma}}$, and hence, show that $(\hat{\boldsymbol{B}}, \hat{\boldsymbol{\Sigma}})$ is sufficient for $(\boldsymbol{B}, \boldsymbol{\Sigma})$.
(b) Is $\hat{\boldsymbol{B}}$ sufficient for $\boldsymbol{B}$ when $\boldsymbol{\Sigma}$ is known?
(c) Is $\hat{\boldsymbol{\Sigma}}$ sufficient for $\boldsymbol{\Sigma}$ when $\boldsymbol{B}$ is known?

7.2.3 Show the following from Theorem 7.2.3(a):
(a) The correlation between $\hat{\beta}_{ij}$ and $\hat{\beta}_{kl}$ is $\rho_{jl}g_{ik}/(g_{ii}g_{kk})^{1/2}$, where $\rho_{jl} = \sigma_{jl}/(\sigma_{jj}\sigma_{ll})^{1/2}$.
(b) The covariance between two rows of $\hat{\boldsymbol{B}}$ is $\text{cov}(\hat{\boldsymbol{\beta}}_i, \hat{\boldsymbol{\beta}}_k) = g_{ik}\boldsymbol{\Sigma}$.
(c) The covariance between two columns of $\hat{\boldsymbol{\beta}}$ is $\text{cov}(\hat{\boldsymbol{\beta}}_{(j)}, \hat{\boldsymbol{\beta}}_{(l)}) = \sigma_{jl}\boldsymbol{G}$.

7.2.4 Show from (7.10) that

$$\hat{B} - B = AU,$$

where $A = (X'X)^{-1}X'$ and $AA' = (X'X)^{-1} = G$. Hence, following the approach adopted in Exercise 4.3.4, show that $\hat{B} - B$, thought of as a column vector, has the distribution

$$(\hat{B} - B)^V \sim N(0, \Sigma \otimes G),$$

thus confirming Theorem 7.2.3.

7.3.1 Verify formula (7.13) for the projection P_2. Hint: let $Z = (Z_1, Z_2) = XC^{-1} = (XC^{(1)}, XC^{(2)})$, and let $A = Z'Z$. Note that $CC^{-1} = I$ implies $C_1C^{(1)} = I$ and $C_1C^{(2)} = 0$. Show that $P_1 - P = ZA^{-1}Z' - Z_2A_{22}^{-1}Z_2'$, and that $P_2 = ZA^{-1}[I, 0]'(A^{11})^{-1}[I, 0]A^{-1}Z'$. Hence, using (A.31), show that $P_1 - P$ and P_2 can both be written in the form

$$[Z_1, Z_2]\begin{bmatrix} A^{11} & A^{12} \\ A^{21} & A_{21}(A^{11})^{-1}A_{12} \end{bmatrix}\begin{bmatrix} Z_1' \\ Z_2' \end{bmatrix}.$$

7.3.2 In Theorem 7.3.1, verify that $Y_+'P_2Y_+ = U'P_2U$.

7.3.3 (a) Partition $X = (X_1, X_2)$, $Y = (Y_1, Y_2)$, and

$$B = \begin{bmatrix} B_{11} & B_{12} \\ B_{21} & B_{22} \end{bmatrix}, \qquad \Sigma = \begin{bmatrix} \Sigma_{11} & \Sigma_{12} \\ \Sigma_{21} & \Sigma_{22} \end{bmatrix},$$

and consider the hypothesis $H_0 : B_{11} = 0$. Let $L_1(Y; B, \Sigma)$ denote the likelihood of the whole data set Y under the model (7.1), and let $L_2(Y_1; B_{11}, B_{21}, \Sigma_{11})$ denote the likelihood of the data Y_1 under the model (7.14). Show that

$$\frac{\begin{array}{c}\max L_1 \text{ over } (\Sigma, B) \\ \text{such that } B_{11} = 0\end{array}}{\max L_1 \text{ over } (\Sigma, B)} = \frac{\begin{array}{c}\max L_2 \text{ over } (\Sigma_{11}, B_{11}, B_{21}) \\ \text{such that } B_{11} = 0\end{array}}{\max L_2 \text{ over } (\Sigma_{11}, B_{11}, B_{21})},$$

and hence, the LRT for H_0 is the same whether one uses the model (7.1) or (7.14). (Hint: Split the maximization of L_1 into two parts by factorizing the joint density of (Y_1, Y_2) as the product of the marginal density of Y_1 (which depends on B_{11}, B_{21}, and Σ_{11} alone) times the conditional density of $Y_2 \mid Y_1$ which when maximized over $B_{12}, B_{22}, \Sigma_{12}$, and Σ_{22}, takes the same value, whatever the value of B_{11}, B_{21}, and Σ_{11}.)

(b) Carry out a similar analysis for the hypothesis $C_1BM_1 = D$.

7.3.4 If $Z(n \times p)$ is a data matrix from $N_p(0, \Sigma)$, and a is a fixed n-vector, show that $Z'a \sim N_p(0, (a'a)\Sigma)$.

7.4.1 If a hypothesis matrix C_1 is testable in the sense of (7.24), show that the parameter matrix C_1B is uniquely determined in the model (7.1) because $\mathsf{E}Y = XB$. Show that C_1 is testable if and only if Roy's criterion (7.25) holds.

7.4.2 In Example 7.4.1, show that $\boldsymbol{H} = \boldsymbol{Y}'\boldsymbol{P}_2\boldsymbol{Y}$ and $\boldsymbol{E} = \boldsymbol{Y}'\boldsymbol{P}\boldsymbol{Y}$ are the "between-groups" and "within-groups" SSP matrices, respectively, which are defined in Section 6.3.3.1. (Hint: Set $N_0 = 0$ and $N_i = n_1 + \cdots + n_i$ for $i = 1, \ldots, k$. Consider the following orthogonal basis in $\mathbb{R}^n$. Let $\boldsymbol{e} = \mathbf{1}$. Let $\boldsymbol{f}_{ij}$ have $+1/(j-1)$ in the places $N_{i-1}+1, \ldots, N_{i-1}+j-1$, -1 in the $(N_{i-1}+j)$th place, and 0 elsewhere for $j = 2, \ldots, n_i; i = 1, \ldots, k$. Let $\mathbf{g}_i$ have $1/N_{i-1}$ in the places $1, \ldots, N_{i-1}$ and $-1/n_i$ in the places $N_{i-1}+1, \ldots, N_i$, with 0 elsewhere, for $i = 2, \ldots, k$. Then, verify that $\boldsymbol{P}_2\boldsymbol{e} = \boldsymbol{P}_2\boldsymbol{f}_{ij} = \mathbf{0}$ and $\boldsymbol{P}_2\,\mathbf{g}_i = \mathbf{g}_i$, so that

$$\boldsymbol{P}_2 = \sum_{i=2}^{k} (\mathbf{g}_i'\mathbf{g}_i)^{-1}\mathbf{g}_i\mathbf{g}_i'.$$

Similarly, verify that $\boldsymbol{P}\boldsymbol{e} = \boldsymbol{P}\mathbf{g}_i = \mathbf{0}$ and $\boldsymbol{P}\boldsymbol{f}_{ij} = \boldsymbol{f}_{ij}$, so that $\boldsymbol{P} = \boldsymbol{I} - n^{-1}\mathbf{1}\mathbf{1}' - \boldsymbol{P}_2$. Thus, verify that $\boldsymbol{H}$ and $\boldsymbol{E}$ are the "between-groups" and "within-groups" SSP matrices from Section 6.3.3.1.)

7.4.3 Give the form for the SCIs in Example 7.4.1 when the contrast $\mathbf{c}$ and/or the linear combination of variables $\boldsymbol{a}$ is given *a priori*.

7.5.1 Let $\boldsymbol{U}$ be a data matrix from $N_p(\boldsymbol{\mu}, \boldsymbol{\Sigma})$, and set $\boldsymbol{W} = \boldsymbol{H}\boldsymbol{U} = \boldsymbol{U} - \mathbf{1}\bar{\boldsymbol{u}}'$. Show that $\mathsf{E}(\boldsymbol{w}_r) = \mathbf{0}$ for $r = 1, \ldots, n$, and that

$$\operatorname{cov}(\boldsymbol{w}_r, \boldsymbol{w}_s) = \mathsf{E}(\boldsymbol{w}_r\boldsymbol{w}_s') = \begin{cases} (1-n^{-1})\boldsymbol{\Sigma}, & r = s, \\ -n^{-1}\boldsymbol{\Sigma}, & r \neq s. \end{cases}$$

Hence, writing $\boldsymbol{W}$ as a vector (see Section A.2.5), deduce $\boldsymbol{W}^V \sim N_{np}(\mathbf{0}, \boldsymbol{\Sigma} \otimes \boldsymbol{H})$.

7.5.2 (Population multiple correlation coefficient). If $\boldsymbol{x}$ is a random p-vector with covariance matrix $\boldsymbol{\Sigma}$ partitioned as (7.28), show that the largest correlation between x_1 and a linear combination of $x_2, \ldots, x_p$ equals $\sigma_{11}^{-1}\boldsymbol{\Sigma}_{12}\boldsymbol{\Sigma}_{22}^{-1}\boldsymbol{\Sigma}_{21}$.

7.5.3 (a) If $\boldsymbol{R}$ is a (2×2) correlation matrix, show that $R_{1\cdot 2} = |r_{12}|$.
(b) If $\boldsymbol{R}$ is a (3×3) correlation matrix, show that

$$R_{1\cdot 23}^2 = (r_{12}^2 + r_{13}^2 - 2r_{12}r_{13}r_{23})/(1 - r_{23}^2).$$

7.5.4 Let $\boldsymbol{R}$ be the correlation matrix of $\boldsymbol{X}(n \times p)$. Show that
(a) $0 \leq R_{1\cdot 2\ldots p} \leq 1$;
(b) $R_{1\cdot 2\ldots p} = 0$ if and only if $\boldsymbol{x}_{(1)}$ is uncorrelated with each of $\boldsymbol{x}_{(2)}, \ldots, \boldsymbol{x}_{(p)}$;
(c) $R_{1\cdot 2\ldots p} = 1$ if and only if $\boldsymbol{x}_{(1)}$ is a linear combination of $\boldsymbol{x}_{(2)}, \ldots, \boldsymbol{x}_{(p)}$.

7.5.5 Verify formula (7.31) for the partial correlation coefficient and show that it can be expressed in the form (7.32).

7.5.6 Let $\boldsymbol{R}$ be a (3×3) correlation matrix. Show that
(a) $r_{12\cdot 3} = (r_{12} - r_{13}r_{23})/\{(1 - r_{13}^2)(1 - r_{23}^2)\}^{1/2}$;
(b) $r_{12\cdot 3}$ may be nonzero when $r_{12} = 0$ and vice versa;

(c) $r_{12\cdot3}$ and r_{12} have different signs when $r_{13}r_{23} > r_{12} > 0$ or $r_{13}r_{23} < r_{12} < 0$.

7.5.7 (a) Show that the eigenvalues $d_1, \ldots, d_p$ of $\boldsymbol{I} - \boldsymbol{D}$ in (7.34) all lie between 0 and 1.
(b) If a and g are arithmetic and geometric means of $d_1, \ldots, d_p$, show that

$$r_T^2 = a, \qquad r_D^2 = g^p.$$

(c) If a_1 and g_1 are arithmetic and geometric means of $1 - d_1, \ldots, 1 - d_p$, show that

$$a_1 = 1 - a \quad \text{and} \quad r_A = g_1^p.$$

(d) Hence, show that $r_D^2 \leq r_T^{2p}$ and $r_A \leq (1 - r_T^2)^p$.

7.5.8 Let $\boldsymbol{x} = (\boldsymbol{x}_1, \boldsymbol{x}_2)$ be a normally distributed random $(p + q)$-vector with known mean $\boldsymbol{\mu}$ and covariance matrix $\boldsymbol{\Sigma}$ with $\boldsymbol{\Sigma}$ conformably partitioned as $\boldsymbol{x}$. Let $\boldsymbol{\Sigma}_{22.1} = \boldsymbol{\Sigma}_{22} - \boldsymbol{\Sigma}_{21}\boldsymbol{\Sigma}_{11}^{-1}\boldsymbol{\Sigma}_{12}$ denote the residual covariance matrix after regressing $\boldsymbol{x}_2$ on $\boldsymbol{x}_1$ and set

$$\boldsymbol{\Sigma}_{22}^{-1}\boldsymbol{\Sigma}_{22\cdot1} = \boldsymbol{I} - \boldsymbol{\Sigma}_{22}^{-1}\boldsymbol{\Sigma}_{21}\boldsymbol{\Sigma}_{11}^{-1}\boldsymbol{\Sigma}_{12} = \boldsymbol{\Delta}, \quad \text{say.}$$

Show that the corresponding population values of ρ_T, the trace correlation, and ρ_D, the determinant correlation, are given by

$$\rho_T^2 = \frac{1}{q}\operatorname{tr}(\boldsymbol{I} - \boldsymbol{\Delta}) = \frac{1}{q}\operatorname{tr}\boldsymbol{\Sigma}_{22}^{-1}\boldsymbol{\Sigma}_{21}\boldsymbol{\Sigma}_{11}^{-1}\boldsymbol{\Sigma}_{12}$$

and

$$\rho_D^2 = |\boldsymbol{I} - \boldsymbol{\Delta}| = |\boldsymbol{\Sigma}_{21}\boldsymbol{\Sigma}_{11}^{-1}\boldsymbol{\Sigma}_{12}|/|\boldsymbol{\Sigma}_{22}|.$$

7.6.1 If (7.35) holds, show that $\mathrm{E}(\boldsymbol{y}) = \boldsymbol{X}\boldsymbol{\beta}$. Show that the residual sum of squares $(\boldsymbol{y} - \boldsymbol{X}\boldsymbol{\beta})'(\boldsymbol{y} - \boldsymbol{X}\boldsymbol{\beta})$ is minimized when $\boldsymbol{\beta} = \hat{\boldsymbol{\beta}}$, given by (7.36).

7.6.2 Consider the regression model (7.35) for $\boldsymbol{y}$ and let $\hat{\boldsymbol{y}}$ denote the OLS estimator in (7.36). If $\boldsymbol{t} = \boldsymbol{B}\boldsymbol{y}$ is any linear estimator of $\boldsymbol{\beta}$, show that
(a) $\boldsymbol{t}$ is unbiased if $\boldsymbol{B}\boldsymbol{X} = \boldsymbol{I}$, and hence, confirm that the OLS estimator is unbiased;
(b) the covariance of $\boldsymbol{t}$ is $\boldsymbol{B}\boldsymbol{\Omega}\boldsymbol{B}'$, and hence, confirm (7.37) and (7.38);
(c) $\boldsymbol{t} = \hat{\boldsymbol{\beta}} + \boldsymbol{C}\boldsymbol{y}$, where $\boldsymbol{C} = \boldsymbol{B} - (\boldsymbol{X}'\boldsymbol{X})^{-1}\boldsymbol{X}'$, and hence, show that $\boldsymbol{t}$ is unbiased if and only if $\boldsymbol{C}\boldsymbol{X} = \boldsymbol{0}$.
If now $\boldsymbol{t}$ is an unbiased estimator and $\boldsymbol{\Omega} = \sigma^2\boldsymbol{I}_n$, show that the covariance matrix of $\boldsymbol{t}$ is

$$\mathrm{E}[(\boldsymbol{t} - \boldsymbol{\beta})(\boldsymbol{t} - \boldsymbol{\beta})'] = \mathrm{E}[(\boldsymbol{t} - \hat{\boldsymbol{\beta}})(t - \hat{\boldsymbol{\beta}})'] + \mathrm{E}[(\hat{\boldsymbol{\beta}} - \boldsymbol{\beta})(\hat{\boldsymbol{\beta}} - \boldsymbol{\beta})'],$$

and hence, that $\hat{\boldsymbol{\beta}}$ has smaller covariance matrix than any other linear unbiased estimator. (This proves the Gauss–Markov result stated as Theorem 7.6.2.)

7.6.3 If $\boldsymbol{t} = \boldsymbol{B}\boldsymbol{y}$ is a linear unbiased estimator of $\boldsymbol{\beta}$ where $\boldsymbol{y}$ satisfies (7.35) and $\tilde{\boldsymbol{\beta}}$ is given by (7.39), show that $\boldsymbol{t} = \tilde{\boldsymbol{\beta}} + \tilde{\boldsymbol{C}}\boldsymbol{y}$, where

$$\tilde{\boldsymbol{C}} = \boldsymbol{B} - (\boldsymbol{X}'\boldsymbol{\Omega}^{-1}\boldsymbol{X})^{-1}\boldsymbol{X}'\boldsymbol{\Omega}^{-1}.$$

Hence, show that $\boldsymbol{t}$ is unbiased if and only if $\tilde{\boldsymbol{C}}\boldsymbol{X} = \boldsymbol{0}$, and that, if $\boldsymbol{t}$ is unbiased, then its covariance matrix is not less than that of $\tilde{\boldsymbol{\beta}}$.

7.6.4 If $\boldsymbol{y} \sim N_p(\boldsymbol{X\beta}, \boldsymbol{\Omega})$ with known covariance matrix $\boldsymbol{\Omega}$, then show that the GLS estimator is the maximum-likelihood estimator of $\boldsymbol{\beta}$, and that it is normal with mean $\boldsymbol{\beta}$ and variance $(\boldsymbol{X}'\boldsymbol{\Omega}^{-1}\boldsymbol{X})^{-1}$.

7.6.5 To prove the consistency of the GLS estimator $\tilde{\boldsymbol{\beta}}$ under assumptions (7.43) and (7.44), it is necessary and sufficient to prove that plim $\tilde{\beta}_{n,i} = \beta_i$ for each component, $i = 1, \dots, q$.

(a) Show that

$$\mathsf{E}(\tilde{\boldsymbol{\beta}} \mid \boldsymbol{X}_n) = \boldsymbol{\beta}, \qquad \mathrm{var}(\tilde{\boldsymbol{\beta}} \mid \boldsymbol{X}_n) = (\boldsymbol{X}_n'\boldsymbol{\Omega}_n^{-1}\boldsymbol{X}_n)^{-1}.$$

(b) Let σ_n^{ii} denote the (i, i)th element of $(\boldsymbol{X}_n'\boldsymbol{\Omega}_n^{-1}\boldsymbol{X}_n)^{-1}$, which is a function of the random matrix $\boldsymbol{X}_n$. Then, by (7.44), plim $\sigma_n^{ii} = 0$. Using Chebyshev's inequality, show that

$$\mathsf{P}(|\tilde{\beta}_{n,i} - \beta_i| > \epsilon \mid \boldsymbol{X}_n) < \sigma_n^{ii}/\epsilon^2.$$

(c) Show that for all $\epsilon > 0, \delta > 0$

$$\mathsf{P}(|\tilde{\beta}_{n,i} - \beta_i| > \epsilon) \leq \mathsf{P}(|\tilde{\beta}_{n,i} - \beta_i| > \epsilon \mid \sigma_n^{ii} < \delta) + \mathsf{P}(\sigma_n^{ii} \geq \delta).$$

Using (b) and (7.44), deduce that for any fixed $\epsilon > 0$, the right-hand side of the above equation can be made arbitrarily small by choosing δ small enough and restricting n to be sufficiently large. Hence, $\mathsf{P}(|\tilde{\beta}_{n,i} - \beta_i| > \epsilon) \to 0$ as $n \to \infty$, so that plim $\tilde{\beta}_{n,i} = \beta_i$.

8

Graphical Models

8.1 Introduction

A *graphical model* is a partial representation of a multivariate distribution; it describes the *conditional dependence* structure of the variables. Such representations are usually much more concise than a description of the full joint probability distribution. Their flexibility for learning and inference and their ease of interpretation have led to widespread usage for both discrete and continuous multivariate data.

Let $\boldsymbol{x}$ be a p-dimensional random vector. A graphical model to describe the distribution of $\boldsymbol{x}$ contains two components: a set of p *vertices* (or nodes) to represent the variables and a set of *edges* between some of the pairs of vertices to describe the relationship between the variables. In particular, the absence of an edge between two vertices i and j indicates the conditional independence of x_i and x_j, given the remaining variables, written as

$$x_i \perp\!\!\!\perp x_j \mid (x_k, k \neq i, j). \tag{8.1}$$

Thus, the graphical model partially specifies the joint distribution of $\boldsymbol{x}$.

In the case of a multivariate normal distribution, $\boldsymbol{x} \sim N_p(\boldsymbol{\mu}, \boldsymbol{\Sigma})$, conditional independence is straightforward to specify in terms of the *concentration matrix* or inverse covariance matrix $\boldsymbol{K} = \boldsymbol{\Sigma}^{-1}$. Variables x_i and x_j are conditionally independent, given the remaining variables, if $k_{ij} = \sigma^{ij} = 0$. Further details are given in Section 8.3.

Example 8.1.1 Consider the open/closed-book data given in Table 1.2. Recall that there are $p = 5$ variables: Mechanics (C), Vectors (C), Algebra (O), Analysis (O), and Statistics (O), where O indicates open book, and C indicates closed book. Figure 8.1 shows a graph reminiscent of a butterfly (Whittaker, 1990), and it can be used to define a Gaussian graphical model. The variables are represented by the five vertices, and the presence of an edge joining two vertices represents conditional dependence between the two variables, given the remaining variables (similarly, the absence of an edge represents conditional independence). For example, Vectors and Statistics are conditionally independent, given the remaining variables. It also follows for this model that Vectors and Statistics are conditionally independent, given Algebra (that is, knowing Algebra renders Vectors irrelevant for predicting Statistics). Further details are given in Example 8.3.1. □

Multivariate Analysis, Second Edition. Kanti V. Mardia, John T. Kent, and Charles C. Taylor.

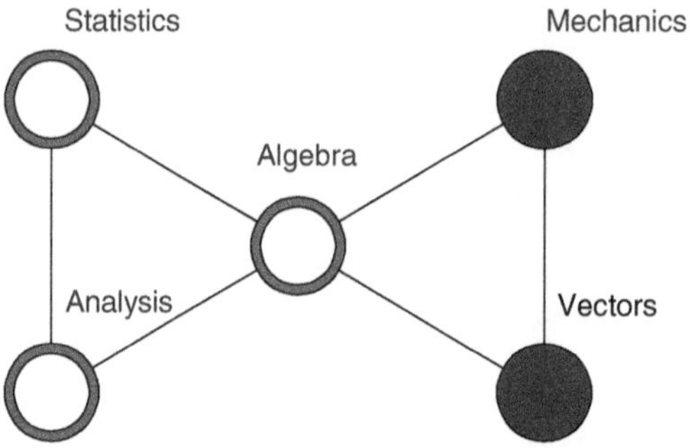

Figure 8.1 Graphical model for open/closed-book data given in Table 1.2. The open-book exams are indicated by open circles, and the closed-book exams are indicated by filled circles.

Section 8.2 sets out the basic definitions and concepts for a graphical model. Gaussian graphical models can be used to model continuous data; these are studied in Section 8.2, including problems of estimation and model selection. Log-linear graphical models can be used to model discrete categorical data; these are discussed in Section 8.4.

Throughout the chapter, we assume that all the probability densities or probability mass functions $f(\boldsymbol{x})$ are defined on a product domain $\mathcal{X} = \mathcal{X}_1 \times \cdots \times \mathcal{X}_p$, and that they satisfy the *positivity condition*,

$$f(\boldsymbol{x}) > 0 \text{ for all } \boldsymbol{x} \in \mathcal{X}. \tag{8.2}$$

This condition is very mild and is satisfied by most models of practical interest, including Gaussian and log-linear graphical models.

Most of this chapter is restricted to the case where the edges are *undirected*. However, a wider class of graphical models can be obtained by allowing the edges to have directions, both unidirectional and bidirectional. A brief discussion is given in Section 8.5.

Entire books are devoted to the topic of graphical models, so the exposition in this chapter is inevitably condensed. For an accessible introduction, see Whittaker (1990). A more theoretical treatment is given by Lauritzen (1996), and more applied texts include Højsgaard et al. (2012), which is designed specifically for programming with R.

8.2 Graphs and Conditional Independence

A *graph* $\mathcal{G}$ is defined by a set of vertices V and a set of edges E, written as $\mathcal{G} = (V, E)$. Single vertices are indicated by lower case subscripts, e.g. i and a. Multiple vertices are indicated by upper case subscripts, e.g. A, where $A = \{a, b\}$. This notation also includes the case where A is a singleton, e.g. $A = \{a\}$. Since conditional independence properties depend on the underlying graph rather than the specific distribution, we sometimes write $i \perp\!\!\!\perp j \mid (k, k \neq i, j)$ as a shorthand for $x_i \perp\!\!\!\perp x_j \mid (x_k, k \neq i, j)$.

The graph can be represented in various ways. For example, if there are vertices labeled $1, \ldots, p$, then a $p \times p$ symmetric indicator matrix – known as an *adjacency matrix* – can be used to define which vertices are connected by an edge. Alternatively, the set of edges can be specified as a list, for each vertex, of the adjacent vertices.

Example 8.2.1 To illustrate some different representations of a graph, consider a graph with vertices $V = \{a, b, \dots, e\}$ and edges $E = \{\{a, b\}, \{b, c\}, \{b, d\}, \{c, d\}\}$. The adjacency list can be represented in several ways. One way is a table,

a	b		
b	a	c	d
c	b	d	
d	b	c	
e			

This list can also be represented as an adjacency matrix or a plot:

	a	b	c	d	e
a	0	1	0	0	0
b	1	0	1	1	0
c	0	1	0	1	0
d	0	1	1	0	0
e	0	0	0	0	0

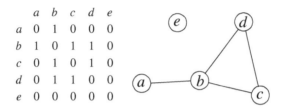

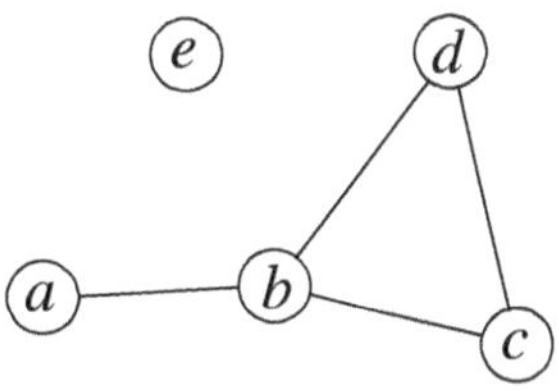

In this example, if the vertices represent variables, then variable x_e is independent of all other variables, but there are dependencies among the others. □

Since any pair of vertices may (or may not) be joined, and since there are

$$\binom{p}{2} = \frac{p(p-1)}{2}$$

pairs of vertices, there are $2^{p(p-1)/2}$ possible graphs with p vertices. For example, if there are 20 variables, then there are about 1.5×10^{57} possible graphs. Choosing between these possible graphs is one of the challenges in selecting a suitable graphical model for a set of data.

Denoting by $f_1, f_2, \dots$ some generic probability densities, and given disjoint sets of indices A, B, C in $\{1, \dots, p\}$, say that the sets of variables x_A and x_B are *conditionally independent*, given the set of variables x_C, if

$$f(x_A, x_B \mid x_C) = f_1(x_A \mid x_C) f_2(x_B \mid x_C) \qquad \text{for some } f_1, f_2 \tag{8.3}$$

or

$$f(x_A, x_B, x_C) = g(x_A, x_C) h(x_B, x_C) \qquad \text{for some functions } g, h, \tag{8.4}$$

and, when this condition holds, we write $A \perp\!\!\!\perp B \mid C$. These conditions and definitions also apply when A, B, and C are individual variables, a, b, c, say, and this interchangeable notation will apply throughout this chapter. Conditional independence models can be represented by an undirected graph, but undirected graphs can also be used for interaction models (see, for example, Section 8.4). As an example, the following statements will hold for the graph shown in Example 8.2.1:

$$a \perp\!\!\!\perp \{c, d, e\} \mid b, \qquad b \perp\!\!\!\perp e, \qquad \{c, d\} \perp\!\!\!\perp \{a, e\} \mid b, \qquad e \perp\!\!\!\perp \{a, b, c, d\}.$$

Given a graph $\mathcal{G}$, a graphical model is defined to have the conditional independence properties in (8.1), and from this starting point, further conditional independence properties can be derived.

Given sets of vertices A, B, C, and D, the following properties are straightforward to verify:

$$\text{If } A \perp\!\!\!\perp B \mid C, \text{ then } B \perp\!\!\!\perp A \mid C. \tag{8.5}$$

$$\text{If } A \perp\!\!\!\perp B \mid C, \text{ and } D \text{ is a function of } B, \text{ then } A \perp\!\!\!\perp D \mid C. \tag{8.6}$$

$$\text{If } A \perp\!\!\!\perp B \mid C, \text{ and } D \text{ is a function of } B, \text{ then } A \perp\!\!\!\perp B \mid \{C, D\}. \tag{8.7}$$

$$\text{If } A \perp\!\!\!\perp B \mid C, \text{ and } A \perp\!\!\!\perp D \mid \{B, C\}, \text{ then } A \perp\!\!\!\perp \{B, D\} \mid C. \tag{8.8}$$

If A denotes a set of vertices in a graph, then the *neighbors* of A are defined to be those vertices that are not in A but are adjacent to at least one vertex in A. The neighbors of A are also the *boundary* of A, which we denote as bd(A).

Another useful concept is that of a *separator*, which depends on the concept of a path. A sequence of vertices $u_0, u_1, \dots, u_k$ ($k > 0$) is called a *path* if there is an edge connecting each pair u_i and u_{i+1} for $i = 0, \dots, k-1$. Given three disjoint sets of vertices A, B, and C, C is said to *separate* A from B if for every $a \in A,\ b \in B$, every path connecting a and b must pass through some $c \in C$. Note that, in some of what follows, it is important to remember that a separator set may be the empty set (see Example 8.2.1).

Starting with a graph (V, E), Speed and Kiiveri (1986) prove that the following three conditions (known as *Markov properties*) are equivalent:

1. for every vertex $v \in V$, $x_v \perp\!\!\!\perp x_B \mid x_{\text{bd}(v)}$, where $B = V \backslash \{\{v\} \cup \text{bd}(v)\}$ (local Markov property). Here, the standard set notation $A \backslash B$ has been used for the relative complement of B in A, i.e. the set of all elements that are members of A but not B;
2. for every A, B, and C, with C separating A and B, $x_A \perp\!\!\!\perp x_B \mid x_C$ (global Markov property);
3. if vertices i and j are not adjacent, then $x_i \perp\!\!\!\perp x_j \mid x_{V\backslash\{i,j\}}$ (pairwise Markov property).

Further, Geiger and Pearl (1993) have formally proved that for every graph there exists a probability model over the vertices (satisfying the positivity condition (8.2)), such that for every choice of disjoint sets A, B, and C,

$$C \text{ separates } A \text{ from B in the graph} \Leftrightarrow x_A \perp\!\!\!\perp x_B \mid x_C.$$

For the rest of the chapter, we assume that given a graph (V, E), any probability model satisfies the Markov conditions.

A subset $A \subseteq V$ is said to be *complete* if all vertex pairs in A are *connected*. A *clique* is a maximal complete subset (i.e. not a subset of a larger complete subset). Note that some authors, e.g. Kent and Mardia (2022), use the term clique as a synonym for a complete subset.

Example 8.2.2 Using the graph of Example 8.2.1, the subsets $\{a, b\}$, $\{b, c\}$, $\{b, d\}$, $\{c, d\}$, and $\{b, c, d\}$ are complete, but $\{a, b, c, d\}$ is not complete since a and d are not connected. The cliques are the subsets $\{a, b\}$, $\{b, c, d\}$, and $\{e\}$. The boundaries of these cliques are $\{c, d\}$, $\{a\}$, and $\emptyset$, respectively; the boundary of $\{b, c\}$ is the set of neighbors $\{a, d\}$, and the boundary of $\{a\}$ is $\{b\}$. □

If the vertices V can be partitioned into disjoint sets A, B, and C such that $A \perp\!\!\!\perp B \mid C$ with C complete, then the partition (A, B, C) is said to form a *decomposition* of the graph. The *components* of this decomposition are $A \cup C$ and $B \cup C$. A graph is said to be *prime* if no decomposition exists in which $A \neq \emptyset$ and $B \neq \emptyset$. An important result (Diestel, 1987) is that any graph can be recursively decomposed into its maximal prime components.

A graph is (fully) *decomposable* if its prime components are cliques. Alternatively, we can say that a graph is fully decomposable if it is complete or admits a proper decomposition into decomposable components (note that this is a recursive definition). Thus, for a fully decomposable graph, the maximal prime components are cliques (Lauritzen, 1996).

A *chord* of a set C is defined to be an edge that is not in the edge set of C but whose endpoints lie in the vertex set of C (West, 2001, p. 225). A *circuit* or cycle is defined to be a path along the edges of a graph that begins and ends at the same vertex. A graph is said to be a *chordal graph* if it possesses no chordless circuits. These definitions are illustrated in the following example, in which it can be seen that a graph can be prime but not chordal.

Example 8.2.3 We consider two graphs with eight vertices. The first example, shown in Figure 8.2, can be decomposed into its maximal prime components. But it is not fully decomposable, since the subgraph with vertices $\{a, c, f, g\}$ is not a chordal graph because the chords $\{a, f\}$ and $\{c, g\}$ are not in the edge set of the subgraph. The second example is shown in Figure 8.3. Note that the circuit formed by the vertices $\{a, e, f, h\}$ has a chord $\{a, h\}$, so the graph does not have any chordless circuits. It can be decomposed into six cliques, as shown in Figure 8.3. □

Equation (8.4) gives an example of a factorization of the distribution f. More generally, we can say that the distribution f *factorizes* with respect to the graph $\mathcal{G}$ if

$$f(\boldsymbol{x}) = \prod_{A \in \mathcal{A}} \psi_A(\boldsymbol{x}_A),$$

where $\mathcal{A}$ denotes the collection of complete subsets of $\mathcal{G}$, and $\psi_A(\boldsymbol{x}_A)$ denotes a function that depends only on $\boldsymbol{x}_A$. Recall that the complete subsets of a graph are sets in which all elements are pairwise neighbors.

Under the positivity condition (8.2), it can be shown that this factorization property is equivalent to the global Markov property (and hence also to the local and pairwise

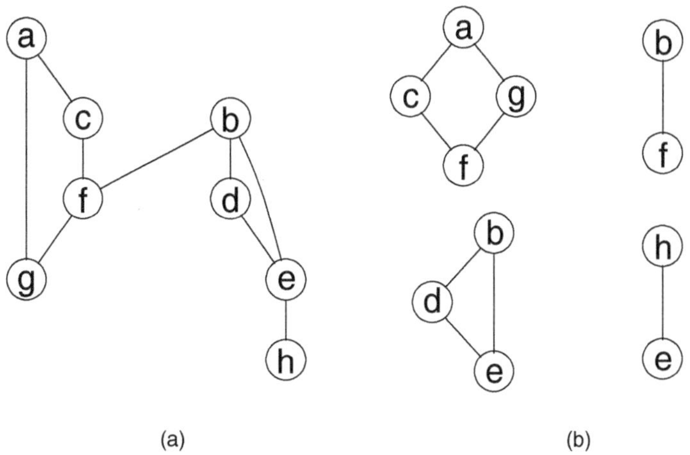

Figure 8.2 (a) Example of a graph with eight vertices; (b) decomposition into four subgraphs, the first of which is not a clique.

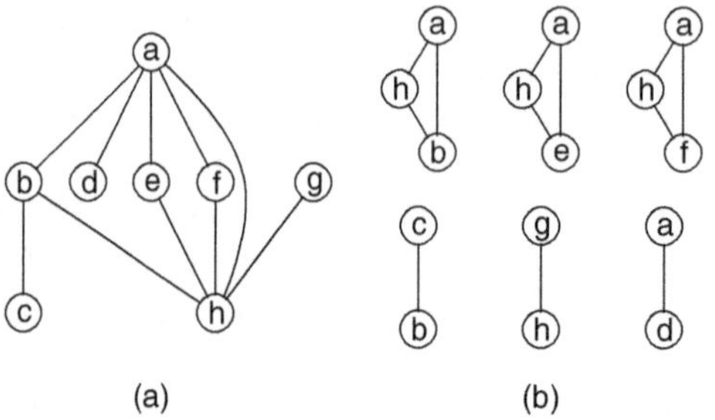

Figure 8.3 (a) Example of a graph with eight vertices; (b) decomposition into six subgraphs, which are all cliques.

Markov properties). This result has become known as the Hammersley–Clifford Theorem (Clifford and Hammersley, 1971); see also Besag (1974) and Kent and Mardia (2022).

We note that for any undirected graph $\mathcal{G}$, the following are equivalent (Green and Thomas, 2013; Lauritzen, 1996, p. 18):

- $\mathcal{G}$ is chordal
- $\mathcal{G}$ is decomposable
- all maximal prime subgraphs of $\mathcal{G}$ are cliques
- the cliques of $\mathcal{G}$ can be ordered using the running intersection property (see Exercise 8.2.5).

Decomposability is an important property because it allows for a clique-separator factorization of the joint distribution; in the following section, Eq. (8.10) gives an example of this factorization for the normal distribution, but these factorizations hold more generally. Such decompositions make it easier to represent prior knowledge about a plausible model. They also give computational advantages in model fitting (see Section 8.3.1), simplify prediction, and make it easier to check the fit of the model (Green and Thomas, 2013).

8.3 Gaussian Graphical Models

Graphical models are particularly simple to describe for the multivariate Gaussian distribution because the conditional independence between two variables, given the rest, can be expressed in terms of a zero entry in the inverse covariance matrix. The following theorem gives the key result.

Theorem 8.3.1 *Let* $\boldsymbol{x} \sim N_p(\boldsymbol{\mu}, \boldsymbol{\Sigma})$ *follow a multivariate normal distribution, and let* $\boldsymbol{K} = \boldsymbol{\Sigma}^{-1}$ *denote the concentration matrix. Consider two variables* x_i *and* x_j, $i \neq j$, $1 \leq i,j \leq p$. *Then,* x_i *and* x_j *are conditionally independent of one another, given the remaining variables, if and only if* $k_{ij} = 0$.

Proof: Partition the variables into two sets, $A = \{i,j\}$ and $B = I \backslash \{i,j\}$, where $I = \{1, \dots, p\}$ indexes all the variables. From Theorem 4.2.7, the conditional distribution of $\boldsymbol{x}_A$ given $\boldsymbol{x}_B$ is

$$N_2(\boldsymbol{\mu}_A + \boldsymbol{\Sigma}_{AB}\boldsymbol{\Sigma}_{BB}^{-1}\boldsymbol{\mu}_B, \boldsymbol{\Sigma}_{AA.B}),$$

where

$$\mathbf{\Sigma}_{AA.B} = \mathbf{\Sigma}_{AA} - \mathbf{\Sigma}_{AB}\mathbf{\Sigma}_{BB}^{-1}\mathbf{\Sigma}_{BA}.$$

From Eq. (A.31), the conditional variance matrix can be expressed in terms of the concentration matrix,

$$\mathbf{\Sigma}_{AA.B} = \boldsymbol{K}_{AA}^{-1} = \begin{bmatrix} k_{ii} & k_{ij} \\ k_{ij} & k_{jj} \end{bmatrix}^{-1} = \frac{1}{k_{ii}k_{jj} - k_{ij}^2} \begin{bmatrix} k_{jj} & -k_{ij} \\ -k_{ij} & k_{ii} \end{bmatrix}.$$

Hence, the conditional correlation between x_i and x_j, given the remaining variables, can be expressed in terms of the elements of $\boldsymbol{Q} = \mathbf{\Sigma}_{AA.B}$ as

$$\rho_{ij.\mathcal{I}\setminus\{i,j\}} = \frac{q_{ij}}{\sqrt{q_{ii}q_{jj}}} = -\frac{k_{ij}}{\sqrt{k_{ii}k_{jj}}}.$$

This conditional correlation is also known as the partial correlation between x_i and x_j, given the remaining variables (Section 7.5.3). Clearly, $\rho_{ij.\mathcal{I}\setminus\{i,j\}} = 0$ if and only if $k_{ij} = 0$. □

For later use, let $\mathbf{\Phi} = (\phi_{ij})$ denote the *standardized concentration matrix* with entries

$$\phi_{ij} = \frac{k_{ij}}{\sqrt{k_{ii}k_{jj}}}, \tag{8.9}$$

so the diagonal entries are $\phi_{ii} = 1$, and the off-diagonal entries are $\phi_{ij} = -\rho_{ij.\mathcal{I}\setminus\{i,j\}}$.

Thus, the following statements about the distribution of $\boldsymbol{x}$ are equivalent:

- the variables x_i and x_j are conditionally independent, given $\boldsymbol{x}_{\mathcal{I}\setminus\{i,j\}}$.
- the (i,j)th (population) partial correlation vanishes, $\rho_{ij.\mathcal{I}\setminus\{i,j\}} = 0$.
- the (i,j)th element of the concentration matrix vanishes, $k_{ij} = \sigma^{ij} = 0$.

8.3.1 Estimation

Let $\boldsymbol{X}$ denote an $n \times p$ data matrix from the $N_p(\boldsymbol{\mu}, \mathbf{\Sigma})$ distribution, with sample covariance matrix $\boldsymbol{S}$. Let $\boldsymbol{K}^{(\text{sample})} = \boldsymbol{S}^{-1}$ denote the sample concentration matrix, and let $\mathbf{\Phi}^{(\text{sample})}$ denote the sample version of the standardized concentration matrix in (8.9). It is of interest to identify a suitable graphical model for the data. Further, once a suitable graphical model has been identified, it is of interest to estimate the corresponding $\mathbf{\Sigma}$. Use the notation $\hat{\mathbf{\Sigma}}$ and $\hat{\boldsymbol{K}} = \hat{\mathbf{\Sigma}}^{-1}$ to denote an estimate of the covariance matrix and its inverse under a graphical model. Estimation, especially maximum-likelihood estimation, is covered in this section, and model selection is discussed in Section 8.3.2.

Thus, suppose that a suitable Gaussian graphical model has been identified, and it is desired to estimate $\mathbf{\Sigma}$. In certain cases (chordal graphs), it is possible to write the maximum-likelihood estimator of $\mathbf{\Sigma}$ in closed form. In other cases, it is necessary to use an iterative procedure.

Consider first a decomposable (i.e. chordal) graph. Given a decomposition (A, B, S), the joint density factorizes in the form $\psi_{A\cup S}(\boldsymbol{x}_{A\cup S})\psi_{B\cup S}(\boldsymbol{x}_{B\cup S})$, and such factorizations can be used repeatedly. In the case of the multivariate normal density, the final result is the factorization

$$f(\boldsymbol{x}; \boldsymbol{\mu}, \mathbf{\Sigma}) = \frac{\prod_{j=1}^{Q} f_j(\boldsymbol{x}_{C_j}; \boldsymbol{\mu}_{C_j}, \mathbf{\Sigma}_{C_j})}{\prod_{k=1}^{R} f_k(\boldsymbol{x}_{S_k}; \boldsymbol{\mu}_{S_k}, \mathbf{\Sigma}_{S_k})^{\nu(S_k)}}, \tag{8.10}$$

where $C_1, \ldots, C_Q$ are the cliques of the graph, $S_1, \ldots, S_R$ are the separators, with $\nu(S_j)$ the number of times that S_j appears between the neighboring cliques, and $f_j, j = 1, \ldots, Q$ are functions of only the variables associated with the respective clique. The notation $\boldsymbol{x}_{C_j}$ is used to denote the components of $\boldsymbol{x} = (x_1, \ldots, x_p)'$ that belong to the clique C_j, and similarly for $\boldsymbol{x}_{S_j}$. Note that a separator may be the empty set.

Now suppose that we have a data matrix $\boldsymbol{X}$ with observations $\boldsymbol{x}_r$, $r = 1, \ldots, n$. Let $\boldsymbol{x}_r^{(C_j)}$ be a vector of length $|C_j|$, which consists of components of $\boldsymbol{x}_r$ corresponding to the ordered elements of C_j, and similarly, define $\boldsymbol{x}_r^{(S_j)}$. Let

$$\boldsymbol{M}^{(j)} = \sum_{r=1}^{n} \boldsymbol{x}_r^{(C_j)} \boldsymbol{x}_r^{(C_j)'} \quad j = 1, \ldots, Q \quad \text{and} \quad \boldsymbol{N}^{(k)} = \sum_{r=1}^{n} \boldsymbol{x}_r^{(S_k)} \boldsymbol{x}_r^{(S_k)'} \quad k = 1, \ldots, R$$

be the corresponding $(|C_j| \times |C_j|)$ and $(|S_k| \times |S_k|)$ sums of squares and products (SSP) matrices – see (1.10) – for the clique and separator sets, respectively.

In this case, the maximum-likelihood estimate (m.l.e.) of $\boldsymbol{K}$ under the graphical model is given by

$$\hat{\boldsymbol{K}} = n \left\{ \sum_{j=1}^{Q} \left[\left(\boldsymbol{M}^{(j)}\right)^{-1} \right]^{[p]} - \sum_{k=1}^{R} \nu(S_k) \left[\left(\boldsymbol{N}^{(k)}\right)^{-1} \right]^{[p]} \right\}, \tag{8.11}$$

with determinant

$$|\hat{\boldsymbol{K}}| = n^p \frac{\prod_{k=1}^{R} |\boldsymbol{N}^{(k)}|^{\nu(S_k)}}{\prod_{j=1}^{Q} |\boldsymbol{M}^{(j)}|}, \tag{8.12}$$

where the square bracket notation $\left[\boldsymbol{A}^{(j)}\right]^{[p]}$ is used for a $p \times p$ "padded" matrix which has elements $\boldsymbol{A}^{(j)}$ in the rows and columns corresponding to elements of C_j, and zero otherwise, with a similar definition for the separator sets. See Lauritzen (1996, sec. 5.3) for more details.

Example 8.3.1 Recall the "butterfly" graph proposed in Example 8.1.1 to model the open/closed-book data. Here, we look at an empirical justification for this choice of graph. The sample covariance matrix and the sample version of the standardized concentration matrix in (8.9) are given by

$$\boldsymbol{S} = \begin{bmatrix} 302.3 & 125.8 & 100.4 & 105.1 & 116.1 \\ & 170.9 & 84.2 & 93.6 & 97.9 \\ & & 111.6 & 110.8 & 120.5 \\ & & & 217.9 & 153.8 \\ & & & & 294.4 \end{bmatrix} \tag{8.13}$$

and

$$\boldsymbol{\Phi}^{(\text{sample})} = \begin{bmatrix} 1.00 & -0.33 & -0.23 & 0.00 & -0.02 \\ & 1.00 & -0.28 & -0.08 & -0.02 \\ & & 1.00 & -0.43 & -0.36 \\ & & & 1.00 & -0.25 \\ & & & & 1.00 \end{bmatrix}, \tag{8.14}$$

respectively. It can be seen that some of the entries of $\boldsymbol{\Phi}^{(\text{sample})}$ are nearly zero (namely, the matrix entries $(1,4), (1,5), (2,4)$, and $(2,5)$, which form a 2×2 block above the diagonal in $\boldsymbol{\Phi}^{(\text{sample})}$); so, we infer that the butterfly graph in Figure 8.1 defines a reasonable model. Replacing the variable names (Mechanics, Vectors, Algebra, Analysis, and Statistics) by numeric labels (1, 2, 3, 4, and 5), respectively, it can be seen that the cliques are $C_1 = \{1,2,3\}$ and $C_2 = \{3,4,5\}$, with the separator $S_1 = \{3\}$. The separator appears once, so $\nu(S_1) = 1$. Then,

$$\boldsymbol{M}^{(1)} = 88\begin{bmatrix} 302.3 & 125.8 & 100.4 \\ & 170.9 & 84.2 \\ & & 111.6 \end{bmatrix}, \quad \boldsymbol{M}^{(2)} = 88\begin{bmatrix} 111.6 & 110.8 & 120.5 \\ & 217.9 & 153.8 \\ & & 294.4 \end{bmatrix},$$

and $\boldsymbol{N}^{(1)}$ is a 1×1 matrix with value 9821.1. Using (8.11), the m.l.e. under the graphical model is

$$\begin{aligned} \hat{\boldsymbol{K}}^c \times 10^4 &= 88\left\{\begin{bmatrix} 0.6 & -0.3 & -0.3 & 0 & 0 \\ & 1.2 & -0.6 & 0 & 0 \\ & & 1.8 & 0 & 0 \\ & & & 0 & 0 \\ & & & & 0 \end{bmatrix}\right. \\ &\quad \left. + \begin{bmatrix} 0 & 0 & 0 & 0 & 0 \\ & 0 & 0 & 0 & 0 \\ & & 2.5 & -0.9 & -0.6 \\ & & & 1.1 & -0.2 \\ & & & & 0.7 \end{bmatrix} - \begin{bmatrix} 0 & 0 & 0 & 0 & 0 \\ & 0 & 0 & 0 & 0 \\ & & 1.0 & 0 & 0 \\ & & & 0 & 0 \\ & & & & 0 \end{bmatrix}\right\} \\ &= \begin{bmatrix} 53.0 & -24.7 & -29.1 & 0 & 0 \\ & 104.6 & -56.7 & 0 & 0 \\ & & 288.2 & -76.4 & -49.9 \\ & & & 99.3 & -20.6 \\ & & & & 65.1 \end{bmatrix}. \end{aligned}$$

The determinant, which can be directly computed as

$$|\hat{\boldsymbol{K}}| = 2.67 \times 10^{-11},$$

can also be obtained using Eq. (8.12),

$$|\hat{\boldsymbol{K}}| = n^p \frac{\prod_{k=1}^{R} |\boldsymbol{N}^{(k)}|^{\nu(S_k)}}{\prod_{j=1}^{Q} |\boldsymbol{M}^{(j)}|} = n^p \frac{|\boldsymbol{N}^{(1)}|}{|\boldsymbol{M}^{(1)}| \cdot |\boldsymbol{M}^{(2)}|} = 88^5 \frac{9821.1}{1.54 \times 10^{12} \cdot 1.54 \times 10^{12}}.$$

Standardizing $\hat{\boldsymbol{K}}$ gives

$$\hat{\boldsymbol{\Phi}} = \begin{bmatrix} 1.00 & -0.33 & -0.24 & 0.00 & 0.00 \\ & 1.00 & -0.33 & 0.00 & 0.00 \\ & & 1.00 & -0.45 & -0.36 \\ & & & 1.00 & -0.26 \\ & & & & 1.00 \end{bmatrix}.$$

Note the exact zeroes as required by the graphical model. (Due to the labeling of the vertices, these zeroes lie in a 2×2 block, but this need not be the case.) Inverting $\hat{\boldsymbol{K}}$ gives the (constrained) m.l.e. of $\hat{\boldsymbol{\Sigma}}$,

$$\hat{\boldsymbol{\Sigma}} = \begin{bmatrix} 302.3 & 125.8 & 100.4 & 99.7 & 108.4 \\ & 170.9 & 84.2 & 83.6 & 90.9 \\ & & 111.6 & 110.8 & 120.5 \\ & & & 217.9 & 153.8 \\ & & & & 294.4 \end{bmatrix}.$$

Both $\hat{\boldsymbol{\Sigma}}$ and $\hat{\boldsymbol{\Phi}}$ are close to the empirical matrices $\boldsymbol{S}$ and $\boldsymbol{\Phi}^{(\text{sample})}$ shown at the beginning of this example. □

If the graph is not decomposable, then in general, the m.l.e. $\hat{\boldsymbol{K}}$ is not available in closed form. Instead, an iterative algorithm must be used.

The iterative method usually used is iterative proportional scaling (IPS), which was first introduced into the statistics literature by Deming and Stephan (1940) for contingency tables; see, e.g. Haberman (1974) for further discussion. Two main versions of the algorithm are given by Whittaker (1990, pp. 182–185), who proposes updates to the constrained m.l.e. of $\boldsymbol{\Sigma}$, and Lauritzen (1996), who proposes an update to the estimate of $\boldsymbol{K}$. An algorithmic comparison between these approaches, as well as a proposed improvement in efficiency, is given by Xu et al. (2011).

The IPS procedure successively updates the covariance matrix for each of the cliques as follows (Speed and Kiiveri, 1986). For an undirected Gaussian graphical model, the joint distribution can be factorized as

$$f(\boldsymbol{x}) = \prod_{j=1}^{Q} g_j(x_{C_j}),$$

where $C_1, \ldots, C_Q$ are the cliques of the graph, and $g_j, j = 1, \ldots, Q$, are functions of only the variables associated with the respective clique. To avoid a notation of nested subscripts, consider one of the cliques and call this C, with all the remaining variables denoted by D. Now reorder and partition the sample covariance matrix (1.7) $\boldsymbol{S}$ as

$$\boldsymbol{S} = \begin{bmatrix} \boldsymbol{S}_{CC} & \boldsymbol{S}_{CD} \\ \boldsymbol{S}_{DC} & \boldsymbol{S}_{DD}, \end{bmatrix}$$

with the corresponding reordering and partitioning of the current estimate of $\boldsymbol{\Sigma}$. The estimate of $\boldsymbol{\Sigma}$ is updated to

$$\begin{bmatrix} \boldsymbol{S}_{CC} & \boldsymbol{S}_{CC}\boldsymbol{\Sigma}_{CC}^{-1}\boldsymbol{\Sigma}_{CD} \\ \boldsymbol{\Sigma}_{DC}\boldsymbol{\Sigma}_{CC}^{-1}\boldsymbol{S}_{CC} & \boldsymbol{\Sigma}_{DD} - \boldsymbol{\Sigma}_{DC}\boldsymbol{\Sigma}_{CC}^{-1}\boldsymbol{\Sigma}_{CD} + \boldsymbol{\Sigma}_{DC}\boldsymbol{\Sigma}_{CC}^{-1}\boldsymbol{S}_{CC}\boldsymbol{\Sigma}_{CC}^{-1}\boldsymbol{\Sigma}_{CD} \end{bmatrix}.$$

Starting from an initial value $\hat{\boldsymbol{\Sigma}} = \boldsymbol{I}$, the process cycles through all cliques (in any order). Then, the process is repeated until convergence to some desired tolerance.

To gain insight into the way that the IPS procedure works, consider the multivariate normal probability density function (p.d.f.) for a single choice of clique C and a single observation $\boldsymbol{x}$. The p.d.f. can be written as

$$f(\boldsymbol{x}; \boldsymbol{\Sigma}) = f(\boldsymbol{x}_C; \boldsymbol{\Sigma}_{CC}) f(\boldsymbol{x}_D | \boldsymbol{x}_C; \boldsymbol{B}, \boldsymbol{\Sigma}_{DD.C}),$$

where for notational convenience it is assumed that $\mathsf{E}(\boldsymbol{x}) = \boldsymbol{0}$. The conditional density has been parameterized by $\boldsymbol{B} = \boldsymbol{\Sigma}_{CC}^{-1}\boldsymbol{\Sigma}_{CD}$, the regression coefficient matrix of $\boldsymbol{x}_D$ on $\boldsymbol{x}_C$, and $\boldsymbol{\Sigma}_{DD.C}$, the residual variance matrix. Since C is a clique, $\boldsymbol{\Sigma}_{CC}$ is unrestricted in the first factor. Hence, holding $\boldsymbol{B}$ and $\boldsymbol{\Sigma}_{DD.C}$ fixed and replacing $\boldsymbol{\Sigma}_{CC}$ by $\boldsymbol{S}_{CC}$ increases the likelihood (for all observations) for the first factor and keeps the likelihood constant for the second factor. This gives the update as given above.

Example 8.3.2 We consider the turtle data of Example 6.3.4 with the sex of the turtle denoting a binary variable. Given the presence of a binary variable, it is clear that a Gaussian model is not really appropriate here, but its use will allow us to illustrate an IPS algorithm on a small data set. The covariance matrix, $\boldsymbol{S}$, and the standardized inverse correlation matrix $\Phi^{(\text{sample})}$ are given by:

$$\begin{bmatrix} 0.25 & -5.66 & -3.57 & -2.81 \\ & 410.76 & 248.70 & 162.38 \\ & & 157.33 & 100.06 \\ & & & 68.97 \end{bmatrix} \text{ and } \begin{bmatrix} 1.00 & -0.33 & -0.08 & 0.63 \\ & 1.00 & -0.64 & -0.52 \\ & & 1.00 & -0.29 \\ & & & 1.00 \end{bmatrix},$$

with sex, length, width, and height as the four variables (in order). We consider fitting the model described by Figure 8.4, which has two cliques: $C_1 = \{\text{sex, length, height}\}$, and $C_2 = \{\text{length, width, height}\}$.

This graph is decomposable, and so, there is a closed-form solution for $\hat{\boldsymbol{\Sigma}}$. However, for illustrative purposes, we demonstrate the use of the IPS algorithm. We start the algorithm using the identity matrix $\hat{\boldsymbol{\Sigma}} = \boldsymbol{I}$. Setting $C = C_2$ for the first iteration gives the updated matrix:

$$\hat{\boldsymbol{\Sigma}} = \begin{bmatrix} 1 & 0 & 0 & 0 \\ & 410.76 & 248.70 & 162.38 \\ & & 157.33 & 100.06 \\ & & & 68.97 \end{bmatrix}.$$

Setting $C = C_1$ for the second iteration gives the updated matrix:

$$\hat{\boldsymbol{\Sigma}} = \begin{bmatrix} 0.25 & -5.66 & -3.64 & -2.81 \\ & 410.76 & 248.70 & 162.38 \\ & & 157.33 & 100.06 \\ & & & 68.97 \end{bmatrix}. \tag{8.15}$$

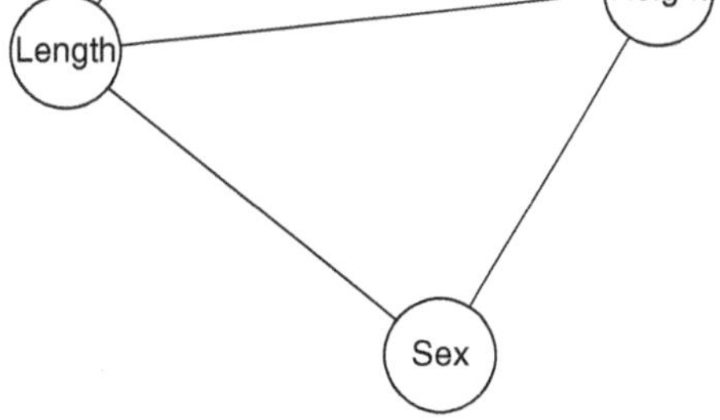

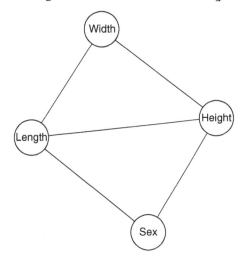

Figure 8.4 Proposed graphical model for turtle data.

The algorithm then repeats the updating steps. For this data set, the algorithm converges after only a few iterations to a final estimate $\hat{\boldsymbol{\Sigma}}$ that is very close to (8.15).

The final estimate $\hat{\boldsymbol{\Sigma}}$ differs from $\boldsymbol{S}$ only in entry (1, 3), and only the edge (1, 3) is missing in the graph of Figure 8.4. This behavior is consistent with a rule proposed by Dempster (1972). Dempster's rule states that $\hat{\boldsymbol{\Sigma}}$ should be chosen to be the positive definite symmetric matrix such that $\boldsymbol{S}$ and $\hat{\boldsymbol{\Sigma}}$ are identical for all entries, except those positions (i, j) [here (1,3)] for which $\hat{\boldsymbol{\Sigma}}^{-1}$ is identically 0.

Next, consider a likelihood ratio test of this graphical model against the unconstrained model for $\boldsymbol{\Sigma}$ (with edges between all pairs of vertices). Given the numerical similarity between $\hat{\boldsymbol{\Sigma}}$ under the graphical model and the unconstrained estimate $\boldsymbol{S}$, it seems likely that this model will not be rejected in a likelihood ratio test. Indeed, the log-likelihood,

$$-\frac{1}{2}\sum_{r=1}^{n}(\boldsymbol{x}_r - \overline{\boldsymbol{x}})'\boldsymbol{\Sigma}^{-1}(\boldsymbol{x}_r - \overline{\boldsymbol{x}}) - \frac{n}{2}\log(|2\pi\boldsymbol{\Sigma}|),$$

of the full model is $l_1^* = -443.1$ and of the graphical model is $l_0^* = -443.2$. The log-likelihood ratio test statistic is $-2\log\lambda = 0.3$, which is far below the 95th percentile of χ_1^2. □

Højsgaard and Lauritzen (2008) have investigated graphical Gaussian models with edge and vertex "symmetries"; that is, some of the diagonal elements of $\hat{\boldsymbol{K}}^c$ are constrained to equal one another, and some of the off-diagonal elements of $\hat{\boldsymbol{K}}^c$ are constrained to equal one another. Algorithms for the maximum-likelihood estimation are available in the R library `gRc` (Højsgaard and Lauritzen, 2007). However, if the number of variables is large, the computation of the maximum-likelihood estimate of $\boldsymbol{K}$ can be time consuming. More recently, Forbes and Lauritzen (2015) have developed a closed-form estimator based on score matching estimation. In score matching estimation, the term *score* refers to the derivatives of the p.d.f. of a continuous random vector $\boldsymbol{x}$; in contrast, in maximum-likelihood estimation (Section 5.1.2), the term *score* has an entirely different meaning, referring to the derivatives of the log likelihood with respect to the parameter vector $\boldsymbol{\theta}$.

8.3.2 Model Selection

When the number of variables p is small, then it is possible to consider all $2^{p(p-1)/2}$ models. Likelihood ratio tests can be used as a form of exploratory data analysis. Alternatively, Akaike's information criterion (AIC) (Akaike, 1974) can be used to choose a model. The AIC is defined by

$$\text{AIC} = -2\log(\text{maximized likelihood}) + 2 \times (\text{number of parameters fitted}),$$

and the model with the smallest AIC is the selected model.

For larger values of p, stepwise procedures may be adopted. For example, a backward procedure would start from the full model, and consider the $p(p-1)/2$ submodels in which one k_{ij} is zero. If the best (with minimum AIC) of these is less than the AIC of the full model, then this k_{ij} is set to zero. Then, the process restarts by considering the $p(p-1)/2 - 1$ submodels of the current model. Termination of this stepwise procedure occurs when the best submodel has an AIC which is not less than the current value.

Example 8.3.3 We return to the open/closed-book data. The sample covariance matrix $\boldsymbol{S}$ is given in Example 8.3.1. As before, replace the variable names (Mechanics, Vectors,

Algebra, and Analysis and Statistics) by numeric labels (1, 2, 3, 4, and 5). The full model has log likelihood equal to −1695.062. The difference in AIC corresponding to a deletion of each edge is given by the `stepwise` command in the R package `gRim` (Højsgaard, 2013):

Edge	Edge	AIC
Vectors–Mechanics	(1, 2)	8.100
Algebra–Mechanics	(1, 3)	2.800
Analysis–Mechanics	(1, 4)	−2.000
Statistics–Mechanics	(1, 5)	−1.947
Algebra–Vectors	(2, 3)	5.229
Analysis–Vectors	(2, 4)	−1.462
Statistics–Vectors	(2, 5)	−1.964
Analysis–Algebra	(3, 4)	16.164
Statistics–Algebra	(3, 5)	9.985
Statistics–Analysis	(4, 5)	3.812

and so the best choice (the smallest of the values above) is to delete edge (1, 4). At the next four steps, the edge with the largest (negative) AIC difference is removed, and the AIC differences corresponding to each edge being considered are:

	Step 2	Step 3	Step 4	Step 5
Vectors–Mechanics	8.157	8.200	8.251	8.251
Algebra–Mechanics	3.902	4.015	7.327	7.327
Analysis–Mechanics				
Statistics–Mechanics	−1.945	−1.903		
Algebra–Vectors	5.547	7.521	7.436	16.965
Analysis–Vectors	−1.404	−1.289	−1.239	
Statistics–Vectors	−1.963			
Analysis–Algebra	17.265	17.811	17.713	25.431
Statistics–Algebra	10.146	12.010	14.825	14.825
Statistics–Analysis	3.814	3.928	3.979	3.979

which means that the edges (2, 5), (1, 5), and (2, 4) are sequentially removed. At the last step, all AIC differences are positive, and so, the process terminates. The final graph is the same as the butterfly graph in Figure 8.1. Thus, the stepwise procedure here has led to the same model that was identified empirically in Example 8.3.1. □

The stepwise procedure may also be carried out in the forward direction. In addition, stepwise procedures may be restricted to just decomposable graphical models (Højsgaard et al., 2012) or to just models with graphical symmetries (Højsgaard and Lauritzen, 2008).

8.4 Log-linear Graphical Models

The Gaussian graphical models in Section 8.3 are designed for continuous data, and they allow at most pairwise interaction between variables. Log-linear graphical models (or, equivalently, graphical log-linear models) are principally designed for categorical variables, and in this case higher order interaction terms are possible.

8.4.1 Notation

Log-linear graphical models can be used for multivariate discrete data (Section 3.2). In this section, we suppose that variable x_v has m_v possible levels ($v = 1, \dots, p$). (Recall that bold capital letters are reserved for matrices, so the distinction between a random variable and its realized value must be interpreted from the context.) Associated with each combination of levels, there is a cell, denoted

$$\boldsymbol{i} = (i_1, i_2, \dots, i_p), \tag{8.16}$$

in which i_v is a level of the vth variable, and so $1 \le i_v \le m_v$. Let $\mathcal{I}$ denote the set of all cells, and let $n(\boldsymbol{i})$ and $p(\boldsymbol{i}) = \mathsf{P}(\boldsymbol{x} = \boldsymbol{i})$, respectively, denote the count (frequency) of cell $\boldsymbol{i}$ and the probability that a randomly chosen observation falls in cell $\boldsymbol{i}$, for $\boldsymbol{i} \in \mathcal{I}$. Write $\boldsymbol{n} = (n(\boldsymbol{i}))$ and $\boldsymbol{p} = (p(\boldsymbol{i}))$ for the vectors of cell counts and cell probabilities.

If the observations are independent and identically distributed (i.i.d.), and the total number of observations, n, say, is fixed, then $\boldsymbol{n}$ follows a multinomial distribution; hence, the probability of the observed contingency table is

$$\frac{n!}{\prod_{\boldsymbol{i} \in \mathcal{I}} n(\boldsymbol{i})!} \prod_{\boldsymbol{i} \in \mathcal{I}} p(\boldsymbol{i})^{n(\boldsymbol{i})}. \tag{8.17}$$

The log-likelihood

$$l(\boldsymbol{p} \mid \boldsymbol{n}) = \sum_{\boldsymbol{i} \in \mathcal{I}} n(\boldsymbol{i}) \log p(\boldsymbol{i}) + \text{constant} \tag{8.18}$$

is of interest for inference about $\boldsymbol{p}$. The multinomial coefficient, which does not depend on $\boldsymbol{p}$, has been absorbed into the "constant." For simplicity, below, we limit attention to data for which all the observed cell counts are positive $n(\boldsymbol{i}) > 0$ for all $\boldsymbol{i}$. Log-linear models can accommodate zero counts but at the cost of extra technicalities.

In most situations we can gain insight into the data by investigating simpler models, and these can give a more parsimonious way to understand the data. These simpler models are usually obtained with reference to *marginal* sums of counts.

Suppose that the set S is a subset of the p variables, i.e. $S \subset \{1, 2, \dots, p\}$. For any $\boldsymbol{i}$, let $\boldsymbol{i}_S$ be a vector containing the elements of $\boldsymbol{i}$ that correspond to the positions in S. Then, we can obtain the marginal cell counts $n^S(\boldsymbol{i}_S)$ by

$$n^S(\boldsymbol{i}_S) = \sum_{\{\boldsymbol{j} \in \mathcal{I} \mid \boldsymbol{j}_S = \boldsymbol{i}_S\}} n(\boldsymbol{j}) \qquad \text{for} \quad \boldsymbol{i}_S \in \mathcal{I}^S = \times_{v \in S} \mathcal{I}_v, \tag{8.19}$$

where $\mathcal{I}_v$, of size m_v, is the set of possible levels for the vth variable, $v = 1, \dots, p$. Similarly, define the marginal probabilities $p^S(\boldsymbol{i}_S)$.

Example 8.4.1 We use data from an investigation into satisfaction with housing conditions (Madsen, 1976) to illustrate the above notation. Each of 1681 Copenhagen residents is cross-classified according to their

- housing type (Tower blocks, Apartments, Atrium houses, Terraced houses)
- feeling of influence on apartment management (Low, Medium, High)
- degree of contact with neighbors (Low, High)
- satisfaction with housing conditions (Low, Medium, High)

Such data can be represented in a "flat table" in which a row is used to describe the combination of the four variables, together with the frequency denoting the number of times this combination was observed. Coding each of the above variables with integers, variable x_v has 4, 3, 2, and 3 levels, respectively, for $v = 1, 2, 3$, and 4. The first few rows of data are:

Housing	Influence	Contact	Satisfaction	Frequency
Tower	Low	Low	Low	21
Tower	Low	Low	Medium	21
Tower	Low	Low	High	28
Tower	Low	High	Low	14
Tower	Low	High	Medium	19
Tower	Low	High	High	37
⋮	⋮	⋮	⋮	⋮

and so (for example), $n((1, 1, 2, 3)) = 37$, and if $S = \{1, 2, 4\}$, then $(1, 1, 2, 3)_{\{1,2,4\}} = (1, 1, 3)$. Similarly, we have (for example)

$$n^{\{1,2,3\}}(1, 1, 1) = n((1, 1, 1, 1)) + n((1, 1, 1, 2)) + n((1, 1, 1, 3)) \qquad = 21 + 21 + 28$$
$$n^{\{1,2,3\}}(1, 1, 2) = n((1, 1, 2, 1)) + n((1, 1, 2, 2)) + n((1, 1, 2, 3)) \qquad = 14 + 19 + 37$$

□

8.4.2 Log-linear Models

We briefly review the fitting of standard log-linear models to count data before describing the graphical model versions. For simplicity, we assume that n is fixed, and the observed counts follow a multinomial distribution in (8.17). However, the theory carries over with little change to the case where n is random and the counts follow independent Poisson distributions.

A multinomial contingency table with total count n summarizes the values of n i.i.d. observations with probabilities $\mathsf{P}(\boldsymbol{x} = \boldsymbol{i}) = p(\boldsymbol{i})$. Under a log-linear model, the log probabilities are assumed to be linear combinations of unknown parameters. Some common models of increasing complexity are

$$\log p(\boldsymbol{i}) = \beta_0, \tag{8.20}$$

$$\log p(\boldsymbol{i}) = \beta_0 + \sum_{v=1}^{p} \beta_{i_v}^{\{v\}}, \tag{8.21}$$

$$\log p(\boldsymbol{i}) = \beta_0 + \sum_{v=1}^{p} \beta_{i_v}^{\{v\}} + \sum_{u \neq v} \beta_{i_u i_v}^{\{u,v\}}, \tag{8.22}$$

$$\log p(\boldsymbol{i}) = \beta_0 + \sum_{v=1}^{p} \beta_{i_v}^{\{v\}} + \sum_{u \neq v} \beta_{i_u i_v}^{\{u,v\}} + \sum_{u \neq v \neq w} \beta_{i_u i_v i_w}^{\{u,v,w\}}. \tag{8.23}$$

The superscript on each β is a subset of variables describing an *effect* in the model. The subscript specifies the levels of those variables. Model (8.21) is an *independence* model; the marginal counts for each of the variables are independent. It can also be called a *main effects* model. Model (8.22) is called a *pairwise interaction* or *two-way interaction* model because the highest order interactions involve two variables. Model (8.23) is called a *three-way interaction* model since the highest order interactions involve three variables. It is straightforward to write down models with higher order interactions. For all the models, the intercept term is not a free parameter but is determined from the other parameters by the constraint that the probabilities should add to 1.

As presented here, the models contain too many free parameters to be identifiable (the aliasing problem). The simplest way to resolve the problem is to set a parameter to 0 whenever at least one of the subscripts is equal to 1, the first level of a variable, e.g. set $\beta_1^{\{u\}} = 0, \beta_{2,1}^{\{u,v\}} = 0$, for all variables u, v, etc. The model with all interactions up to p-way interactions is called the *saturated model*. In this case, the number of parameters equals the number of cells.

It is also important to consider models with a limited number of effects present. For example, a model might include first-order interactions for only some of the pairs of variables $\{u, v\}$. To facilitate interpretability, we restrict attention to *hierarchical models*. For such models, if an effect is present in the model for a certain set of variables, then all simpler effects involving subsets of these variables must also be present. For example, if a model includes the three-way interaction for three variables $\{u, v, w\}$, then the model must also include the pairwise interactions for $\{u, v\}$, $\{u, w\}$, and $\{v, w\}$ and the main effects for $\{u\}$, $\{v\}$, and $\{w\}$. The most concise way to specify a hierarchical log-linear model is through a set of *generators*, that is a list of the highest order effects in a model that, together with the hierarchical property, determine all the effects in the model.

Next, consider maximum-likelihood estimation for a given hierarchical model. The log likelihood is given in (8.18). If an effect corresponding to a subset of variables S is present in the model, then it can be shown that the m.l.e. satisfies

$$\hat{p}^S(\boldsymbol{i}_S) = n^S(\boldsymbol{i}_S)/n.$$

That is, for the S-margins, the fitted probabilities match the observed proportions. For some models, the m.l.e. can be computed in closed form. For example, for the independence model, each fitted joint probability is a product of fitted marginal probabilities

$$\hat{p}(\boldsymbol{i}) = \prod_{v=1}^{p} \hat{p}_v^{\{v\}}(i_v), \quad \hat{p}_v^{\{v\}}(i_v) = \hat{n}_v^{\{v\}}(i_v)/n,$$

and for the saturated model,

$$\hat{p}(\boldsymbol{i}) = n(\boldsymbol{i})/n, \ \boldsymbol{i} \in \mathcal{I}.$$

However, in general, an iterative algorithm is needed. In such cases, maximization is usually carried out by the IPS algorithm (Deming and Stephan, 1940). This algorithm was also used in Section 8.3.1.

Next, consider the comparison between models. The goodness of fit of a model is measured by the *deviance*. For a given model, the deviance is defined to be twice the difference between the log likelihoods for the given model and the saturated model,

$$D = 2\sum_{\boldsymbol{i}\in\mathcal{I}} n(\boldsymbol{i}) \log\left(\frac{n(\boldsymbol{i})}{\hat{n}(\boldsymbol{i})}\right).$$

In particular, the saturated model has deviance zero. A comparison of two (nested) models is given by the difference in deviance

$$D_0 - D_1 = 2\sum_{i \in \mathcal{I}} n(\boldsymbol{i}) \log \left(\frac{\hat{n}^{(1)}(\boldsymbol{i})}{\hat{n}^{(0)}(\boldsymbol{i})} \right),$$

and this can be compared to a χ^2 distribution with degrees of freedom equal to the difference in the number of parameters between the models under the assumption (null hypothesis) that the smaller model is correct.

Example 8.4.2 Using the data set introduced in Example 8.4.1, we begin with a saturated model, which has a total of $4 \times 3 \times 2 \times 3 = 72$ levels. There is one constraint, so this model has 71 degrees of freedom. Omitting the four-way interaction term (Housing.Influence.Contact.Satisfaction) which has $(4-1) \times (3-1) \times (2-1) \times (3-1) = 12$ parameters (allowing for constraints) gives a difference in deviance of 5.944, so a formal test would accept the simpler model. There are four three-way interaction terms. Models that omit one of these terms in turn have degrees of freedom and deviances given by

Interaction term	DF	Deviance
Housing.Influence.Contact	6	10.01
Influence.Contact.Satisfaction	4	6.83
Housing.Influence.Satisfaction	12	27.00
Housing.Contact.Satisfaction	6	15.23

and using sequential χ^2 tests in a stepwise (greedy) fashion, we end up with a model with only one three-way interaction: Housing.Influence.Satisfaction and all six two-way interactions (three of which are implied by the hierarchical constraint). Thus, the generators of the model are: {Housing, Influence, Satisfaction}, {Contact, Housing}, {Contact, Influence}, and {Contact, Satistfaction}. □

8.4.3 Log-linear Models with a Graphical Interpretation

The class of graphical log-linear models is a subclass of the standard hierarchical log-linear models of Section 8.4.2. Let $\mathcal{G}$ be a graph on the vertices $\{1, \ldots, p\}$. If the effects specified by the cliques of $\mathcal{G}$ are used to generate a hierarchical log-linear model, then the log-linear model is called a *graphical log-linear model with respect to* $\mathcal{G}$. In particular, if $S \subset \{1, \ldots, p\}$ is a clique containing k vertices, then the corresponding log-linear model includes the k-way interaction corresponding to the vertices in S, together with all the effects corresponding to subsets of S.

Not all log-linear models can be given a graphical interpretation. As a simple example, consider a log-linear model on $p = 3$ variables generated by the three pairwise effects $\{1, 2\}$, $\{1, 3\}$, and $\{2, 3\}$. If the model were a graphical model, then the corresponding graph would need to include these edges. The only such graph with three vertices is shown in Figure 8.5. But since all three vertices are connected to one another, the full set of vertices $\{1, 2, 3\}$ is

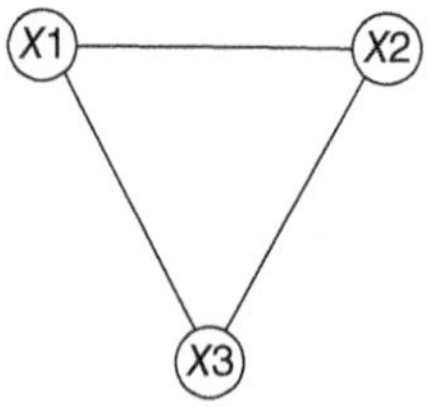

Figure 8.5 This graph shows a graphical interaction model since the three two-way interactions will imply the higher order interaction. (Christensen, 2006)/Springer Nature.

a clique; hence, the corresponding log-linear model for this graph is the saturated model generated by the three-way interaction. In other words, the log-linear model generated by the three pairwise effects cannot be viewed as a graphical log-linear model.

However, graphical log-linear models are still very useful. The aims of a log-linear graphical model are very similar to those of the standard log-linear model applied to count data: to obtain fitted values that are similar to the observed values using the variables in a parsimonious way (and so fewer estimated parameters), leading to an analysis of the estimated probabilities and an understanding of the way that the variables interact. In addition, limiting attention to graphical log-linear models (or to the even more restricted class to graphical models for which the graph is decomposable) can make it easier to interpret the conditional independence structure.

Example 8.4.3 Returning to the housing data of Example 8.4.2, starting from a saturated graphical model, we use AIC to search backward for the best model, restricting attention to decomposable graphical models. The final model, shown in Figure 8.6, has two cliques: Housing.Contact.Satisfaction and Influence.Contact.Satisfaction. It can be seen that the variables Housing (housing type) and Influence (feeling of influence on

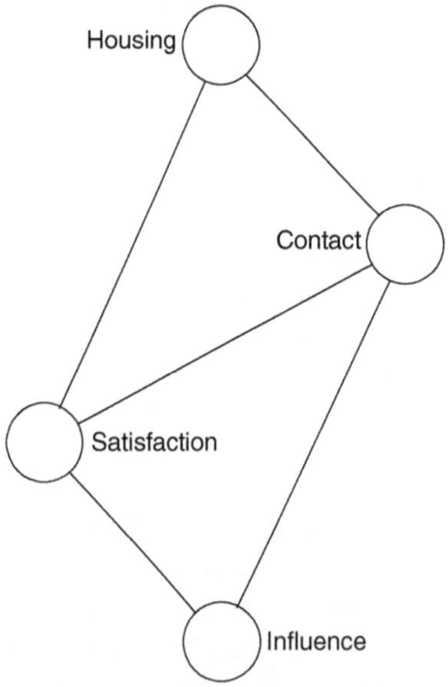

Figure 8.6 Graphical model for the Copenhagen housing data.

apartment management) are independent, given the variables Satisfaction and Contact. The AIC for this model is 456.7, which is larger than the AIC (451.1) for the model selected in Example 8.4.2, although this latter model is not a proper graphical model. The respective deviances are 43.8 for the graphical model and 22.1 for the log-linear model with 36 and 28 degrees of freedom, respectively. The graphical model is easier to visualize than the log-linear model of Example 8.4.2 and possibly simpler to interpret with only two generators. □

We conclude this section with an analysis of a larger data set on vehicle accidents.

Example 8.4.4 Road safety data is available from the UK Department of Transport (data.gov.uk/dataset/road-accidents-safety-data) and includes information about casualties, accidents, and vehicles, with each accident recorded by date, together with associated road, weather and lighting condition. The data only covers personal injury accidents on public roads that are reported to the police, and subsequently recorded, using the STATS19 accident reporting form. Data is available from 2005, but we have selected only accidents from the month of April 2013 and considered the following variables:

- Accident severity (Fatal, Serious, Slight)
- Number of vehicles (recoded as 1 or more than 1)
- Number of casualties (recoded as 1 or more than 1)
- Road classification (Motorway, A(M), A, B, C, Unclassified)
- Road type (Roundabout, One-way street, Dual carriageway, Single carriageway, Slip road, Unknown)
- Light conditions (recoded as daylight or not)
- Weather conditions (recoded as fine or not)
- Road surface conditions (recoded as dry or not)

There were $n = 9948$ observations in this month, cross-classified into $3 \times 2 \times 2 \times 6 \times 6 \times 2 \times 2 \times 2 = 3456$ cells, with the most frequent occurring 1092 times (Accident Severity = Slight, Number of vehicles > 1, Number of casualties = 1, "A" Road, Unknown road type, daylight, fine weather, and dry road conditions). A large (82%) proportion – 2839 – of the cells are empty.

Starting from a saturated model, we use a stepwise procedure using the AIC criterion to select models that are decomposable. (Recall that such representations are useful since the likelihood can be represented as ratios of products of terms indexed by cliques and separators.) The final graphical model is shown in Figure 8.7, with annotations to highlight the five cliques:

1. Accident severity, Number of vehicles, Number of casualties
2. Number of vehicles, Road classification, Road type
3. Number of vehicles, Weather conditions, Road surface conditions
4. Number of vehicles, Light conditions, Road surface conditions
5. Number of vehicles, Number of casualties, Road classification, Light conditions

Informally, we can view each clique as capturing the possible dependencies between the included variables. Note that the variable "Number of vehicles" appears in all five cliques and is a separator between cliques 1 and 3 and between cliques 3 and 5. In the first clique it

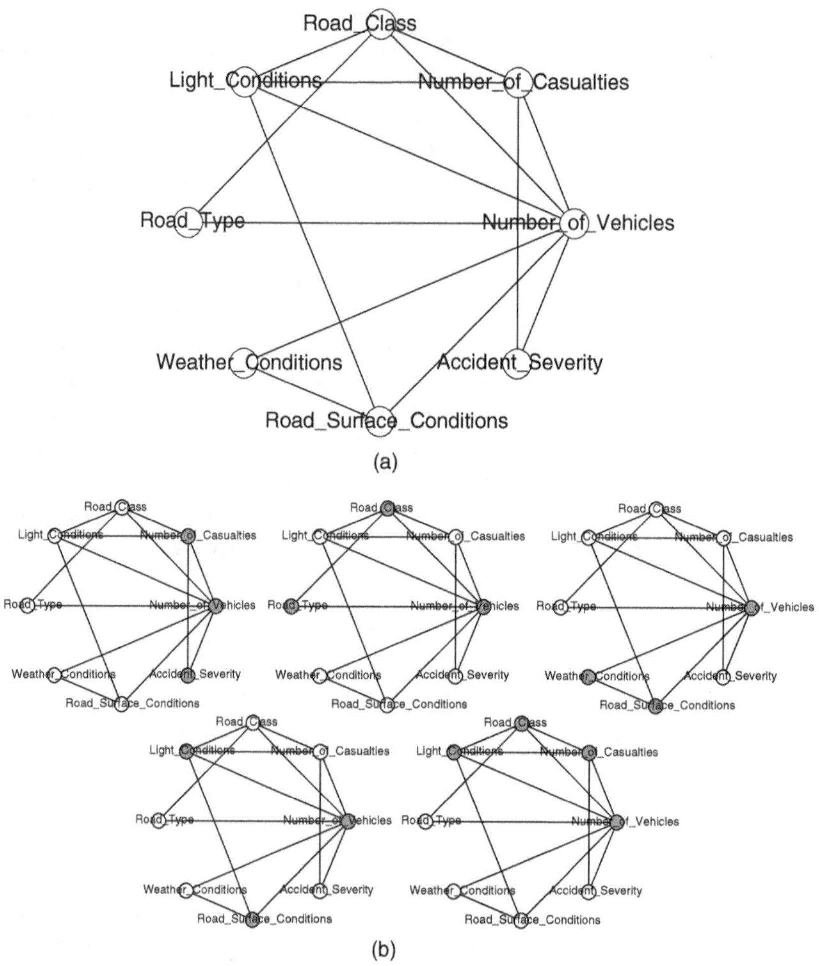

Figure 8.7 (a) Decomposable graphical model for the vehicle accident data. (b) The five cliques (colored in gray).

seems natural that the severity of the accident should be related to both the number of vehicles and the number of casualties. That Road type and Road classification are connected is to be expected since (for example) Motorways are generally dual carriageway, and slip roads are likely to occur with Motorways. So, there is an interaction between road classification and road type. □

8.5 Directed and Mixed Graphs

The focus of this chapter so far has been on *undirected graphs*. In this final section, we briefly mention other types of graphs that can be used to address questions of *causality*, when the behavior of some variables is thought to have an effect on the behavior of others.

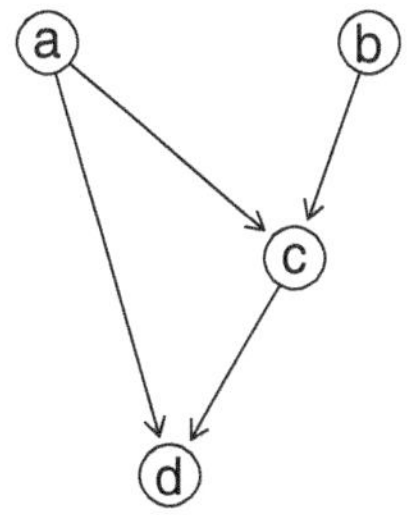

Figure 8.8 Example of directed acyclic graph (DAG).

So, we now broaden the concept of a graph to a pair $\mathcal{G} = (V, E)$, where V is as before, but now there is a wider set of choices for the edges in E. Each edge can be undirected (as before), unidirected (pointing in either direction), or *bidirected* (pointing in both directions).

A *directed (oriented) graph* is one in which all the edges are unidirected. A *mixed graph* is one in which there are at least two types of edge.

A sequence of vertices $u_0, u_1, \dots, u_k$ $(k > 0)$ is called a *directed path* if there is a directed edge connecting each pair u_i and u_{i+1} for $i = 0, \dots, k-1$, i.e. $u_i \to u_{i+1}$ for each i, where the notation $u \to v$ means there is a directed edge from u to v. Note that, if $u_0 = u_k$, then this path is a cycle. A *directed acyclic graph* (often abbreviated to DAG or ADG) is a directed graph with no cycles, i.e. no paths with arrows pointing in the same direction, which begin and end at the same vertex.

A *chain graph* is a generalization of an undirected graph and a DAG; it is a mixed graph with no bidirected edges and no semidirected cycles (a cycle with at least one directed edge).

A directed graphical model is also referred to as a *belief network* or Bayesian network; see Nielsen and Jensen (2009). Figure 8.8 shows an example of a DAG. From the graph in the figure, we say that the vertex a is a *parent* (pa) of c and d, and d is a *child* (ch) of a and c. In general, for vertices u and v, this can be formally defined by

$$\mathrm{pa}(v) = \{w \in V \backslash v \mid w \to v\}$$

and

$$\mathrm{ch}(v) = \{w \in V \backslash v \mid v \to w\}.$$

When the vertices represent variables of $\boldsymbol{x}$, then the joint probability density of a DAG can be factorized as

$$p(\boldsymbol{x}) = \prod_{v \in V} p(x_v \mid \boldsymbol{x}_{\mathrm{pa}(v)}), \tag{8.24}$$

where $p(x_v \mid \boldsymbol{x}_{\mathrm{pa}(v)})$ is the conditional probability density of x_v, given its parents.

The design of directed graphs is often based on possible *causal* relationships. However, it can also be informed by temporal information, for example the events "rain tomorrow" and "rain the next day." Given such usage, it is clear that a directed graph does not necessarily *imply* causality. The following simple example illustrates the need for external information in constructing a Bayesian network from the joint distribution.

Example 8.5.1 Suppose that x_A and x_B are binary variables that can be described by events A and B with $\mathsf{P}(A) = p_A$, $\mathsf{P}(B) = p_B$, and $\mathsf{P}(A \cap B) = p_C$. A standard Venn diagram is shown

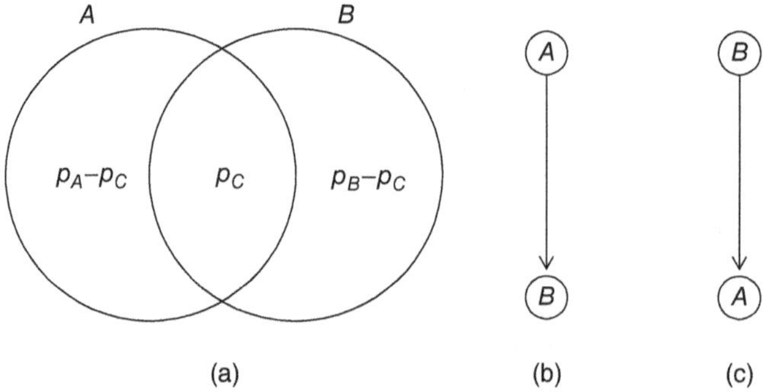

Figure 8.9 Venn diagram depicting two events A and B (a) with $P(A \cap B) = p_C$ and other probabilities also shown. (b, c) Two possible (equivalent) Bayesian network representations, either of which could be used to encode the probability distribution.

in Figure 8.9, and the symmetry gives no indication of any causality. Nevertheless, we can compute $\mathrm{P}(B \mid A) = p_C/p_A$ and $\mathrm{P}(B \mid A^c) = (p_B - p_C)/(1 - p_A)$ and depict the Bayesian network of $A \rightarrow B$ or alternatively we can compute $\mathrm{P}(A \mid B) = p_C/p_B$ and $\mathrm{P}(A \mid B^c) = (p_A - p_C)/(1 - p_B)$ and depict this with its Bayesian network. The two networks differ only in the way they encode the joint probability distribution. Thus, from a directed or mixed graph, we can deduce properties of the joint probability distribution but in general not the other way round.

Now, if A is the event "had previous contact with person who has tested positive for COVID-19" and B is the event "now have COVID-19," then a natural choice would be the graphical model in panel (b) of Figure 8.9. However, given only a set of probabilities, there would be no way to identify this representation over that of the network in panel (c). □

Two DAGs are said to be *equivalent* if the set of distributions that can be represented with one of the graphs is the same as the set of distributions that can be represented with the other. The above example – in which the two Bayesian networks belong to an *equivalence class* – illustrates one of the challenges of *learning* the structure of a directed graph from data. Other difficulties include the size of the search space (the number of possible graphs) as well as determining whether numerical correlations are sufficiently large to warrant an edge in the graph. There are various approaches to obtain the *equivalence class* of a graph – this is the best that can be hoped for, given the identifiability issue illustrated above. These approaches include constraint-based structure learning, score-based structure learning, and Bayesian model averaging methods. See Koller and Friedman (2009) for a thorough treatment of DAGs.

Exercises and Complements

8.2.1 Prove that the following identities are equivalent:
(a) $\mathrm{P}(A, B \mid C) = \mathrm{P}(A \mid C)\mathrm{P}(B \mid C)$. (b) $\mathrm{P}(A \mid B, C) = \mathrm{P}(A \mid C)$.

8.2.2 Prove the properties listed in Eqs. (8.5)–(8.8).

8.2.3 Show that

$$a \perp\!\!\!\perp b, c \mid d \Rightarrow a \perp\!\!\!\perp c \mid d.$$

8.2.4 Given six vertices, labeled a–f with (a) $c \perp\!\!\!\perp f \mid b$; (b) $\{a, e\} \perp\!\!\!\perp b \mid c$; and (c) $c \perp\!\!\!\perp d \mid \{a, e\}$, draw the (undirected) graph and list the separating sets.

8.2.5 Show that an undirected graph is decomposable if and only if there exists an ordering $C_1, \ldots, C_Q$ of its cliques such that for all $j = 1, \ldots, Q$,

$$S_j = C_j \cap (\cup_{i<j} C_i)$$

is a subset of a clique C_i with $i < j$, and S_j is complete. This is known as the *running intersection property*. (Hint: use proof by induction.)

8.3.1 Suppose that $\boldsymbol{x}' = (x_1, x_2, x_3, x_4)$ with $0 < x_i < 1, i = 1, \ldots, 4$, has p.d.f. proportional to $\exp(x_2 + x_2x_3 + x_1x_2x_4)$. Draw the variables in an undirected graph to illustrate the conditional independence relationships.

8.3.2 If $\boldsymbol{x}$ is normally distributed with a graphical structure as shown in Figure 8.3, describe the structure of $\boldsymbol{\Sigma}^{-1}$.

8.3.3 Suppose that $\boldsymbol{x} = (x_1, x_2, x_3)' \sim N_3(\boldsymbol{0}, \boldsymbol{\Sigma})$, with

$$\boldsymbol{\Sigma} = \begin{bmatrix} 1 & \sqrt{\delta} & 1/2 \\ & 2\delta & \sqrt{\delta} \\ & & 1 \end{bmatrix}.$$

Show that $x_1 \perp\!\!\!\perp x_3 \mid x_2$. Sketch the undirected graphical model and investigate what happens to the global Markov property as $\delta \to 0$.

8.3.4 Let $\boldsymbol{x} \sim N_3(\boldsymbol{0}, \boldsymbol{\Sigma})$ and $\boldsymbol{K} = \boldsymbol{\Sigma}^{-1}$. Writing k_{ij} for the (i, j)th entry of $\boldsymbol{K}$, use basic conditional probability to show that the conditional density of $\boldsymbol{x}_A \mid \boldsymbol{x}_B$ (with $\boldsymbol{x}'_A = (x_1, x_2)$ and $\boldsymbol{x}_B = x_3$) satisfies

$$\log f(x_1, x_2 \mid x_3) = \text{constant} - \frac{1}{2}\left\{ k_{11}x_1^2 + k_{22}x_2^2 + k_{33}x_3^2 \right.$$
$$\left. + 2(k_{12}x_1x_2 + k_{13}x_1x_3 + k_{23}x_2x_3) - \frac{x_3^2}{\sigma_{33}} \right\},$$

and relate this expression to Theorem 8.3.1.

8.4.1 Show that the maximum-likelihood estimates of $\boldsymbol{p}$ for the model associated with Eq. (8.18) are given by $\hat{p}(\boldsymbol{i}) = n(\boldsymbol{i})/n, \quad \boldsymbol{i} \in \mathcal{I}$.

8.4.2 Given a three-way contingency table with categorical variables x_1, x_2 and x_3 list all nine hierarchical log-linear models that may be considered. For each case, write down any conditional independence implied and state – with reasons – which are graphical models.

8.5.1 Prove Eq. (8.24) by induction.

8.5.2 For the graph in Figure 8.8, obtain the joint probability of $\boldsymbol{x}' = (x_a, x_b, x_c, x_d)$ as a joint factorization of the conditional probabilities, i.e. using (8.24).
Suppose that each variable is binary (encoded as 0/1) in which case the full joint probability distribution of $\boldsymbol{x}$ is defined by $2^4 = 16$ probabilities. From a computational perspective, compare the number of operations required to obtain the distribution of $\mathsf{P}(x_a \mid x_c = 0)$ using (i) the full joint probability distribution, with (ii) the factorization you have obtained in the first part of this question.

8.5.3 Consider the following DAG:

$$x_1 \longrightarrow x_2 \longrightarrow x_3,$$

with the following distributional information:

$$x_1 \sim N(0,1), \quad x_2 \mid x_1 \sim N(x_1, 1), \quad x_3 \mid x_2 \sim N(x_2, 1).$$

Obtain the distribution of $\boldsymbol{x} = (x_1, x_2, x_3)'$ and plot the conditional independence (undirected) graph. (Hint: consider Equation (8.3)).

9

Principal Component Analysis

9.1 Introduction

Chapter 1 has already introduced the open/closed-book data, which involves the scores of 88 students on five examinations. This data is presented in Table 1.2 as a matrix having 88 rows and 5 columns. One question that can be asked concerning these data is how the results on the five different examinations should be combined to produce an overall score. Various answers are possible. One obvious answer would be to use the overall mean, that is the linear combination $(x_1 + x_2 + x_3 + x_4 + x_5)/5$, or, equivalently, $\boldsymbol{l}'\boldsymbol{x}$, where $\boldsymbol{l}$ is the vector of weights $\boldsymbol{l}/5 = (1/5, 1/5, 1/5, 1/5, 1/5)'$. We call a linear combination $\boldsymbol{l}'\boldsymbol{x}$ a *standardized linear combination* (SLC) if $\sum l_i^2 = 1$. However, can one do better than an arithmetic mean? That is one of the questions that principal component analysis (PCA) seeks to answer. This technique was developed after Hotelling (1933) after its origin by Pearson (1901).

As a first objective, PCA seeks the SLC of the original variables which has *maximal* variance. In the examination situation, this might seem sensible – a large variance "separates out" the candidates, thereby easing consideration of differences between them. The students can then be ranked with respect to this SLC. Similar considerations apply in other situations, such as constructing an index of the cost of living.

More generally, PCA looks for a *few* linear combinations that can be used to summarize the data, losing in the process as little information as possible. This attempt to reduce the dimensionality can be described as "parsimonious summarization" of the data. For example, we might ask whether the important features of the open/closed-book data can be summarized by, say, two linear combinations. If this were possible, we would be reducing the dimensionality from $p = 5$ to $p = 2$.

9.2 Definition and Properties of Principal Components

9.2.1 Population Principal Components

Section 1.5.3 introduces the so-called principal component transformation in the context of sample data. This was an orthogonal transformation that transforms any set of variables into a set of new variables that are uncorrelated with each other. Here, we give its population counterpart.

Multivariate Analysis, Second Edition. Kanti V. Mardia, John T. Kent, and Charles C. Taylor.

Definition 9.2.1 *If $\boldsymbol{x}$ is a random vector with mean $\boldsymbol{\mu}$ and covariance matrix $\boldsymbol{\Sigma}$, then the* principal component transformation *is the transformation*

$$\boldsymbol{x} \to \boldsymbol{y} = \boldsymbol{\Gamma}'(\boldsymbol{x} - \boldsymbol{\mu}), \tag{9.1}$$

where $\boldsymbol{\Gamma}$ is orthogonal, $\boldsymbol{\Gamma}'\boldsymbol{\Sigma}\boldsymbol{\Gamma} = \boldsymbol{\Lambda}$ is diagonal, and $\lambda_1 \geq \lambda_2 \geq \cdots \geq \lambda_p \geq 0$. The strict positivity of the eigenvalues λ_i is guaranteed if $\boldsymbol{\Sigma}$ is positive definite. This representation of $\boldsymbol{\Sigma}$ follows from the spectral decomposition theorem (Theorem A.5). The ith principal component *of $\boldsymbol{x}$ may be defined as the* ith *element of vector $\boldsymbol{y}$, namely as*

$$y_i = \boldsymbol{\gamma}'_{(i)}(\boldsymbol{x} - \boldsymbol{\mu}). \tag{9.2}$$

Here, $\boldsymbol{\gamma}_{(i)}$ is the ith *column of $\boldsymbol{\Gamma}$ and may be called the* ith *vector of* principal component loadings. *The function y_p may be called the last principal component of $\boldsymbol{x}$.*

Figure A.2 gives a pictorial representation in two dimensions of the principal components for $\boldsymbol{\mu} = \boldsymbol{0}$.

Example 9.2.1 Suppose that x_1 and x_2 are standardized to have mean 0, variance 1, and have correlation ρ. The eigenvalues of this covariance matrix are $1 \pm \rho$. If ρ is positive, then the first eigenvalue is $\lambda_1 = 1 + \rho$, and the first eigenvector is $(1, 1)'$. Hence, the first principal component of $\boldsymbol{x}$ is

$$y_1 = 2^{-1/2}(x_1 + x_2),$$

which is proportional to the mean of the elements of $\boldsymbol{x}$. Note that y_1 has variance $(1 + \rho)$, which equals the first eigenvalue. The second principal component corresponding to the second eigenvalue $1 - \rho$ is

$$y_2 = 2^{-1/2}(x_1 - x_2),$$

which is proportional to the difference between the elements of $\boldsymbol{x}$ and has variance $(1 - \rho)$, which equals the second eigenvalue. If ρ is negative, then the order of the principal components is reversed. For a p-variate extension of this example, see Example 9.2.2. □

In the above example, y_1 and y_2 are uncorrelated. This is a special case of a more general result, which is proved in part (c) of the following theorem.

Theorem 9.2.2 *If $\boldsymbol{x} \sim (\boldsymbol{\mu}, \boldsymbol{\Sigma})$, and $\boldsymbol{y}$ is as defined in (9.1), then*

(a) $\mathrm{E}(y_i) = 0$;
(b) $\mathrm{var}(y_i) = \lambda_i$;
(c) $\mathrm{cov}(y_i, y_j) = 0, i \neq j$;
(d) $\mathrm{var}(y_1) \geq \mathrm{var}(y_2) \geq \cdots \geq \mathrm{var}(y_p) \geq 0$;
(e) $\sum_{i=1}^{p} \mathrm{var}(y_i) = \mathrm{tr}\, \boldsymbol{\Sigma}$;
(f) $\prod_{i=1}^{p} \mathrm{var}(y_i) = |\boldsymbol{\Sigma}|$.

Proof: Parts (a)–(d) follow from Definition 9.2.1 and the properties of the expectation operator. Part (e) follows from part (b) and the fact that $\mathrm{tr}\, \boldsymbol{\Sigma}$ is the sum of the eigenvalues. Part (f) follows from part (b) and the fact that $|\boldsymbol{\Sigma}|$ is the product of the eigenvalues. □

It is instructive to confirm the results of the above theorem with respect to Example 9.2.1. Note particularly that $\text{var}(y_1) + \text{var}(y_2) = 2$, and that $\text{var}(y_1) \times \text{var}(y_2) = (1 - \rho^2)$, thus confirming parts (e) and (f) of the theorem. Section (d) of Theorem 9.2.2 states that y_1 has a larger variance than any of the other principal components. (Here and elsewhere, we use "larger" somewhat loosely to mean "not smaller.") However, we now prove a stronger result.

Theorem 9.2.3 *No SLC of $\boldsymbol{x}$ has a variance larger than λ_1, the variance of the first principal component.*

Proof: Let the SLC be $\boldsymbol{a}'\boldsymbol{x}$, where $\boldsymbol{a}'\boldsymbol{a} = 1$. We may write

$$\boldsymbol{a} = c_1\boldsymbol{\gamma}_{(1)} + \cdots + c_p\boldsymbol{\gamma}_{(p)}, \tag{9.3}$$

where $\boldsymbol{\gamma}_{(1)}, \ldots, \boldsymbol{\gamma}_{(p)}$ are the eigenvectors of $\boldsymbol{\Sigma}$. (Any vector can be written in this form since the eigenvectors constitute a basis for $\mathbb{R}^p$.) Now, if $\alpha = \boldsymbol{a}'\boldsymbol{x}$, then

$$\text{var}(\alpha) = \boldsymbol{a}'\boldsymbol{\Sigma}\boldsymbol{a} = \boldsymbol{a}'\left(\sum_{i=1}^{p} \lambda_i \boldsymbol{\gamma}_{(i)}\boldsymbol{\gamma}'_{(i)}\right)\boldsymbol{a}, \tag{9.4}$$

using the spectral decomposition of $\boldsymbol{\Sigma}$ (Theorem A.5). Now, $\boldsymbol{\gamma}'_{(i)}\boldsymbol{\gamma}_{(j)} = \delta_{ij}$, the Kronecker delta. Inserting (9.3) into (9.4), we find that

$$\text{var}(\alpha) = \sum \lambda_i c_i^2. \tag{9.5}$$

However, we know that

$$\sum c_i^2 = 1, \tag{9.6}$$

since $\boldsymbol{a}$ is given by (9.3) and satisfies $\boldsymbol{a}'\boldsymbol{a} = 1$. Therefore, since λ_1 is the largest of the eigenvalues, the maximum of (9.5) subject to (9.6) is λ_1. This maximum is obtained when $c_1 = 1$ and all the other c_i are zero. Therefore, $\text{var}(\alpha)$ is maximized when $\boldsymbol{a} = \boldsymbol{\gamma}_{(1)}$. □

For a multivariate normal random vector, the above theorem may be interpreted geometrically by saying that the first principal component represents the major axis of the ellipsoids of the probability density function (p.d.f.) (see Figure 2.1).

A similar argument shows that the last principal component of $\boldsymbol{x}$ has a variance which is *smaller* than that of any other SLC. The intermediate components have a maximal variance property given by the following theorem.

Theorem 9.2.4 *If $\alpha = \boldsymbol{a}'\boldsymbol{x}$ is an SLC of $\boldsymbol{x}$ which is uncorrelated with the first k principal components of $\boldsymbol{x}$, then the variance of α is maximized when α is the $(k + 1)$th principal component of $\boldsymbol{x}$.*

Proof: We may write $\boldsymbol{a}$ in the form (9.3). Since α is uncorrelated with $\boldsymbol{\gamma}'_{(i)}\boldsymbol{x}$ for $i = 1, \ldots, k$, we know that $\boldsymbol{a}'\boldsymbol{\gamma}_{(i)} = 0$, and therefore, that $c_i = 0$ for $i = 1, \ldots, k$. Therefore, using the same argument as in (9.5) and (9.6), the result follows. □

Example 9.2.2 We now extend the bivariate example of Example 9.2.1. Suppose that $\boldsymbol{\Sigma}$ is the equicorrelation matrix, viz. $\boldsymbol{\Sigma} = (1 - \rho)\boldsymbol{I} + \rho\boldsymbol{1}\boldsymbol{1}'$. If $\rho > 0$, then $\lambda_1 = 1 + (p - 1)\rho$ and

$\lambda_2 = \lambda_3 = \cdots = \lambda_p = 1 - \rho$. The eigenvector corresponding to λ_1 is $\mathbf{1}$. Hence, if $\rho > 0$, the first principal component is proportional to $p^{-1}\mathbf{1}'\boldsymbol{x}$, the average of the $\boldsymbol{x}$. This may be taken as a measure of the overall "size," while "shape" may be defined in terms of the other eigenvectors which are all orthogonal to $\mathbf{1}$ (i.e. vectors containing some negative signs). (For a further discussion of size and shape, see Section 9.6.) □

It is clear that $\boldsymbol{a}'\boldsymbol{x}$ always has the same variance as $-\boldsymbol{a}'\boldsymbol{x}$, and therefore that in terms of variance, we are uninterested in the difference between the two vectors $\boldsymbol{a}$ and $-\boldsymbol{a}$. As mentioned before, we consider only SLCs, so that $\sum a_i^2 = 1$. However, some authors scale $\boldsymbol{a}$ differently, e.g. in such a way that $\sum a_i^2$ is equal to the corresponding eigenvalue.

9.2.2 Sample Principal Components

We now turn to consider the sample-based counterpart of the previous section, thus building on Section 1.5.3. Let $\boldsymbol{X} = (\boldsymbol{x}_1, \dots, \boldsymbol{x}_n)'$ be a sample data matrix, and $\boldsymbol{a}$ a standardized vector. Then, $\boldsymbol{Xa}$ gives n observations on a new variable defined as a weighted sum of the columns of $\boldsymbol{X}$. The sample variance of this new variable is $\boldsymbol{a}'\boldsymbol{Sa}$, where $\boldsymbol{S}$ is the sample covariance matrix of $\boldsymbol{X}$. We may then ask which SLC has the largest variance. Not surprisingly, the answer is that the SLC with largest variance is the first principal component defined by direct analogy with (9.2) as

$$\boldsymbol{y}_{(1)} = (\boldsymbol{X} - \mathbf{1}\overline{\boldsymbol{x}}')\boldsymbol{g}_{(1)}.$$

Here, $\boldsymbol{g}_{(1)}$ is the standardized eigenvector corresponding to the largest eigenvalue of $\boldsymbol{S}$ (i.e. $\boldsymbol{S} = \boldsymbol{GLG}'$). This result may be proved in the same way as Theorem 9.2.3. Similarly, the ith sample principal component is defined as $\boldsymbol{y}_{(i)} = (\boldsymbol{X} - \mathbf{1}\overline{\boldsymbol{x}}')\boldsymbol{g}_{(i)}$, and these components satisfy the straightforward sample extensions of Theorems 9.2.2–9.2.4 (see Exercise 9.2.8).

Putting the principal components together, we get

$$\boldsymbol{Y} = (\boldsymbol{X} - \mathbf{1}\overline{\boldsymbol{x}}')\boldsymbol{G}. \tag{9.7}$$

Also, $\boldsymbol{X} - \mathbf{1}\overline{\boldsymbol{x}}' = \boldsymbol{YG}'$ since $\boldsymbol{G}$ is orthogonal; that is, $\boldsymbol{G}$ has transformed one $(n \times p)$ matrix $(\boldsymbol{X} - \mathbf{1}\overline{\boldsymbol{x}}')$ into another one of the same order $(\boldsymbol{Y})$. The covariance matrix of $\boldsymbol{Y}$ is given by

$$\begin{aligned}\boldsymbol{S}_Y = n^{-1}\boldsymbol{Y}'\boldsymbol{HY} &= n^{-1}\boldsymbol{G}'(\boldsymbol{X} - \mathbf{1}\overline{\boldsymbol{x}}')'\boldsymbol{H}(\boldsymbol{X} - \mathbf{1}\overline{\boldsymbol{x}}')\boldsymbol{G}\\ &= n^{-1}\boldsymbol{G}'\boldsymbol{X}'\boldsymbol{HXG} = \boldsymbol{G}'\boldsymbol{SG} = \boldsymbol{L},\end{aligned}$$

where

$$\boldsymbol{H} = \boldsymbol{I} - (n^{-1}\mathbf{1}\mathbf{1}');$$

that is, the columns of $\boldsymbol{Y}$ are uncorrelated, and the variance of $\boldsymbol{y}_{(j)}$ is l_j.

The rth element of $\boldsymbol{y}_{(i)}$, y_{ri}, represents the *score* of the ith principal component on the rth individual. Similarly, the vector $\boldsymbol{g}_{(i)}$ is called the *loading* of the ith principal component. In terms of individuals, we can write the principal component transformation as

$$y_{ri} = \boldsymbol{g}'_{(i)}(\boldsymbol{x}_r - \overline{\boldsymbol{x}}).$$

Often, we omit the subscript r and write $y_i = \boldsymbol{g}'_{(i)}(\boldsymbol{x} - \overline{\boldsymbol{x}})$ (as in Eq. (9.2)) to emphasize the transformation rather than its effect on any particular individual.

Example 9.2.3 Consider the open/closed-book examination data from Table 1.2. It has covariance matrix

$$\boldsymbol{S} = \begin{bmatrix} 302.3 & 125.8 & 100.4 & 105.1 & 116.1 \\ & 170.9 & 84.2 & 83.6 & 97.9 \\ & & 111.6 & 110.8 & 120.5 \\ & & & 217.9 & 153.8 \\ & & & & 294.4 \end{bmatrix}$$

and mean $\overline{\boldsymbol{x}}' = (39.0, 50.6, 46.7, 42.3)$. A spectral decomposition of $\boldsymbol{S}$ yields principal components

$$\begin{aligned} y_1 &= 0.51x_1 + 0.37x_2 + 0.35x_3 + 0.45x_4 + 0.53x_5 - 99.5, \\ y_2 &= 0.75x_1 + 0.21x_2 - 0.08x_3 - 0.30x_4 - 0.55x_5 + 1.4, \\ y_3 &= -0.30x_1 + 0.42x_2 + 0.15x_3 + 0.60x_4 - 0.60x_5 - 19.2, \\ y_4 &= 0.30x_1 - 0.78x_2 - 0.00x_3 + 0.52x_4 - 0.18x_5 + 11.5, \\ y_5 &= 0.08x_1 + 0.19x_2 - 0.92x_3 + 0.29x_4 + 0.15x_5 + 14.4, \end{aligned}$$

with variances 679.2, 199.8, 102.6, 83.7, and 31.8, respectively. Note that the first principal component gives similar positive weight (loadings) to all the variables and thus represents an "average" grade. On the other hand, the second component represents a contrast between the open- and closed-book examinations, with the first and last examinations weighted most heavily. Of course, the difference in performance due to the two types of examination is also confounded with the differences between the individual subjects. □

9.2.3 Further Properties of Principal Components

The most important properties of principal components have already been given in Theorems 9.2.2–9.2.4, along with the corresponding statements for sample principal components. We now turn to examine several other useful properties, each of which is briefly summarized as a heading at the beginning of the relevant section.

(a) *The sum of the first k eigenvalues divided by the sum of all the eigenvalues, $(\lambda_1 + \cdots + \lambda_k)/(\lambda_1 + \cdots + \lambda_p)$, represents the "proportion of total variation" explained by the first k principal components.*

The "total variation" in this statement is tr $\boldsymbol{\Sigma}$, and the rationale for this statement is clear from part (e) of Theorem 9.2.2. A justification for the use of tr $\boldsymbol{\Sigma}$ as a measure of multivariate scatter is given in Section 1.5.3, and the proportion of total variation defined here gives a quantitative measure of the amount of information retained in the reduction from p to k dimensions. For instance, we may say in Example 9.2.1 that the first component "explains" the proportion $(1+\rho)/2$ of the total variation, and in Example 9.2.2, the proportion explained is $\{1 + (p-1)\rho\}/p$.

In Example 9.2.3, the first principal component explains $679.2/1097.1 = 61.9\%$ of the total variation, and the first two principal components explain 80.1%. (Compare this with the fact that the largest of the original variables comprised only $302.3/1097.1 = 27.6\%$ of the total variation, and the largest two only 54.4%.)

(b) *The principal components of a random vector are not scale invariant.*

One disadvantage of PCA is that it is *not* scale invariant. For example, given three variables, say weight in pounds, height in feet, and age in years, we may seek a principal component expressed, say, in ounces, inches, and decades. Two procedures seem feasible:

(i) multiply the data by 16, 12, and 1/10, respectively, and then carry out a PCA;
(ii) carry out a PCA and then multiply the elements of the relevant components by 16, 12, and 1/10.

Unfortunately, procedures (i) and (ii) do *not* generally lead to the same result. This may be seen theoretically for $p = 2$ by considering

$$\boldsymbol{\Sigma} = \begin{bmatrix} \sigma_1^2 & \rho\sigma_1\sigma_2 \\ \rho\sigma_1\sigma_2 & \sigma_2^2 \end{bmatrix},$$

where $\rho > 0$. The larger eigenvalue is $\lambda = (\sigma_1^2 + \sigma_2^2)/2 + \Delta/2$, where $\Delta = \{(\sigma_1^2 - \sigma_2^2)^2 + 4\sigma_1^2\sigma_2^2\rho^2\}^{1/2}$, with eigenvector proportional to

$$(a_1, a_2) + (\sigma_1^2 - \sigma_2^2 + \Delta, 2\rho\sigma_1\sigma_2) \tag{9.8}$$

(see Exercise 9.2.1). When $\sigma_1/\sigma_2 = 1$, the ratio a_2/a_1 given by (9.8) is unity. If $\sigma_1 = \sigma_2$ and the first variable is multiplied by a factor k, then for scale invariance, we would like the new ratio a_2/a_1 to be k. However, changing σ_1 to $k\sigma_1$ in (9.8) shows that this is not the case.

Alternatively, the lack of scale invariance can be examined empirically, as shown in the following example (see also Exercise 9.2.7.)

Example 9.2.4 From the correlation matrix of the open/closed-book data, the first principal component (after setting the sample mean to be $\mathbf{0}$) is found to be

$$0.40u_1 + 0.43u_2 + 0.50u_3 + 0.46u_4 + 0.44u_5.$$

Here, $u_1, \ldots, u_5$ are the standardized variables, so that $u_i = x_i/s_i$. Reexpressing the above components in terms of the original variables and standardizing, we get

$$y_1 = 0.023x_1 + 0.033x_2 + 0.047x_3 + 0.031x_4 + 0.025x_5$$

or

$$y_1^* = 13.56y_1 = 0.31x_1 + 0.45x_2 + 0.64x_3 + 0.42x_4 + 0.34x_5.$$

The first component of the original variables obtained from the covariance matrix is given in Example 9.2.3 and has different coefficients. (The eigenvalues are also different – they account for 61.9, 18.2, 9.3, 7.6, and 2.9% of the variance when the covariance matrix is used, and for 63.6, 14.8, 8.9, 7.8, and 4.9% when the correlation matrix is used.) □

Algebraically, the lack of scale invariance can be explained as follows. Let $\boldsymbol{S}$ be the sample covariance matrix. Then, if the ith variable is divided by d_i, the covariance matrix of the new variables is $\boldsymbol{DSD}$, where $\boldsymbol{D} = \text{diag}(d_i^{-1})$. However, if $\boldsymbol{x}$ is an eigenvector of $\boldsymbol{S}$, then $\boldsymbol{D}^{-1}\boldsymbol{x}$ is *not* an eigenvector of $\boldsymbol{DSD}$. In other words, the eigenvectors are not scale invariant.

The lack of scale invariance illustrated above implies a certain sensitivity to the way scales are chosen. Two ways out of this dilemma are possible. First, one may seek the so-called natural units, by ensuring that all variables measured are of the same type (for instance, all heights and all weights). Alternatively, one can standardize all variables so that they have unit variance and find the principal components of the correlation matrix rather than the covariance matrix. The second option is the one most commonly employed, although this does complicate the question of hypothesis testing – see Section 9.4.4.

(c) *If the covariance matrix of* $\mathbf{x}$ *has rank* $r < p$, *then the total variation of* $\mathbf{x}$ *can be entirely explained by the first r principal components.*

This follows from the fact that if $\boldsymbol{\Sigma}$ has rank r, then the last $(p - r)$ eigenvalues of $\boldsymbol{\Sigma}$ are identically zero. Hence, the result follows from (a).

(d) *The vector subspace spanned by the first k principal components* $(1 \le k < p)$ *has smaller mean square deviation from the population (or sample) variables than any other k-dimensional subspace.*

If $\mathbf{x} \sim (\mathbf{0}, \boldsymbol{\Sigma})$, and a subspace $H \subset \mathbb{R}^p$ is spanned by orthonormal vectors $\mathbf{h}_{(1)}, \ldots, \mathbf{h}_{(k)}$, then by projecting $\mathbf{x}$ onto this subspace we see that the squared distance d^2 from $\mathbf{x}$ to H has expectation

$$\mathsf{E}(d^2) = \mathsf{E}(\mathbf{x}'\mathbf{x}) - \sum_{j=1}^{k} \mathsf{E}(\mathbf{h}'_{(j)}\mathbf{x}).$$

Let $\mathbf{f}_{(j)} = \boldsymbol{\Gamma}'\mathbf{h}_{(j)}, j = 1, \ldots, k$. We have

$$\begin{aligned} \mathsf{E}(d^2) &= \operatorname{tr} \boldsymbol{\Sigma} - \sum_{j=1}^{k} \mathsf{E}(\mathbf{f}'_{(j)}\mathbf{y}) \\ &= \operatorname{tr} \boldsymbol{\Sigma} - \sum_{i=1}^{p} \sum_{j=1}^{k} f_{ij}^2 \lambda_i, \end{aligned} \tag{9.9}$$

since the $\mathbf{f}_{(j)}$ are also orthonormal. Exercise 9.2.10 then shows that (9.9) is minimized when $f_{ij} = 0, i = k + 1, \ldots, p$, for each $j = 1, \ldots, k$, that is when the $\mathbf{f}_{(j)}$ span the subspace of the first k principal components. Note that the case $k = 1$ is just a reformulation of Theorem 9.2.3.

(e) *As a special case of (d) for* $k = p - 1$, *the plane perpendicular to the last principal component has a smaller mean square deviation from the population (or sample) variables than any other plane.*

(Of course, the plane perpendicular to the last principal component equals the subspace spanned by the first $(p - 1)$ principal components.)

Consider the multiple regression $x_1 = \boldsymbol{\beta}'\mathbf{x}_2 + u$ of x_1 on $(x_2, \ldots, x_p)' = \mathbf{x}_2$, say, where u is uncorrelated with $\mathbf{x}_2$. The linear relationship $x_1 = \boldsymbol{\beta}'\mathbf{x}_2$ defines a plane in $\mathbb{R}^p$, and the parameter $\boldsymbol{\beta}$ may be found by minimizing $\mathsf{E}(d^2)$, where $d = |x_1 - \boldsymbol{\beta}'\mathbf{x}_2|$ is the distance between the point $\mathbf{x}$ and the plane, with d measured *in the direction of the "dependent variable."* On the other hand, the plane perpendicular to the last principal component can be found by minimizing $\mathsf{E}(d^2)$, where d is measured perpendicular to the plane. Thus, PCA represents "orthogonal" regression in contrast to the "simple" regression of Chapter 7.

As an example where this point of view is helpful, consider the following bivariate regression-like problem, where both variables are subject to errors.

Example 9.2.5 Suppose that an observed vector $\mathbf{x}'_r = (x_{r1}, x_{r2}), r = 1, \ldots, n$, is modeled by

$$x_{r1} = \xi_r + \varepsilon_{r1}, \qquad x_{r2} = \alpha + \beta\xi_r + \varepsilon_{r2}, \tag{9.10}$$

where ξ_r is the (unknown) "independent" variable, $\varepsilon_{r1} \sim N(0, \tau_1^2), \varepsilon_{r2} \sim N(0, \tau_2^2)$, and the εs are independent of one another and of the ξs. The problem is to find the linear relationship (9.10) between x_1 and x_2, that is to estimate α and β. However, because x_1 is subject to error, a simple regression of x_2 on x_1 is *not* appropriate.

There are two ways to approach this problem: First, the ξs can be viewed as n unknown additional parameters to be estimated, in which case (9.10) is called a *functional relationship*. Alternatively, the ξs can be regarded as a sample from $N(\mu, \sigma^2)$, where μ and σ^2 are two additional parameters to be estimated, in which case (9.10) is called a *structural relationship* (Kendall and Stuart 1967, chap. 29).

If the ratio between the error variances $\lambda = \tau_2/\tau_1$ is known, then the parameters of both the functional and structural relationships can be estimated using maximum likelihood. In both cases it is appropriate to use the line through the sample mean $(\overline{x}_1, \overline{x}_2)$ whose slope is given by the first principal component of the covariance matrix of $(\lambda^{1/2}x_1, x_2)$. (With this scaling, the error variance is spherically symmetric.) Explicit formulae for $\hat{\alpha}$ and $\hat{\beta}$ are given in Exercise 9.2.11. □

Remarks (1) Unfortunately, if the error variances are completely unknown, it is *not possible* to estimate the parameters. In particular, maximum-likelihood estimation breaks down in both cases. In the functional relationship case, the likelihood becomes infinite for a wide choice of parameter values, whereas in the structural relationship case, there is an inherent indeterminacy in the parameters; that is, different parameter values can produce the same distribution for the observations.

Thus, in this situation, more information is needed before estimation can be carried out. For example, if it is possible to make replicated observations, then information can be obtained about the error variances.

Alternatively, it may be possible to make observations on new variables which also depend linearly on the unknown ξ.

(2) More generally, the linear relationship in (9.10) can be extended to any number of dimensions $p > 2$ to express variables $x_1, \ldots, x_p$ each as a linear function of an unknown ξ plus an error term, where all the error terms are independent of one another. If the ratios between the error variances are known, then the above comments on the use of the first principal component remain valid, but again, with unknown error variances, the estimation of the parameters of the functional relationship breaks down here also. (However, the structural relationship with unknown error variances can be considered as a special case of factor analysis, and estimation of the parameters is now possible. See Chapter 10, especially Exercise 10.2.7.)

9.2.4 Correlation Structure

We now examine the correlations between $\boldsymbol{x}$ and its vector of principal components, $\boldsymbol{y}$, defined in Definition 9.2.1. For simplicity, we assume that $\boldsymbol{x}$ (and therefore $\boldsymbol{y}$) has mean zero. The covariance between $\boldsymbol{x}$ and $\boldsymbol{y}$ is then

$$\mathsf{E}(\boldsymbol{x}\boldsymbol{y}') = \mathsf{E}(\boldsymbol{x}\boldsymbol{x}'\boldsymbol{\Gamma}) = \boldsymbol{\Sigma\Gamma} = \boldsymbol{\Gamma\Lambda\Gamma}'\boldsymbol{\Gamma} = \boldsymbol{\Gamma\Lambda}.$$

Therefore, the covariance between x_i and y_j is $\gamma_{ij}\lambda_j$. Now, x_i and y_j have variances σ_{ii} and λ_j, respectively, so if their correlation is ρ_{ij}, then

$$\rho_{ij} = \gamma_{ij}\lambda_j/(\sigma_{ii}\lambda_j)^{1/2} = \gamma_{ij}(\lambda_j/\sigma_{ii})^{1/2}. \quad (9.11)$$

When $\boldsymbol{\Sigma}$ is a correlation matrix, $\sigma_{ii} = 1$, so $\rho_{ij} = \gamma_{ij}\sqrt{\lambda_j}$.

We may say that the proportion of variation of x_i "explained" by y_j is ρ_{ij}^2. Then, since the elements of $\mathbf{y}$ are uncorrelated, any set G of components explains a proportion

$$\rho_{iG}^2 = \sum_{j\in G} \rho_{ij}^2 = \frac{1}{\sigma_{ii}} \sum_{j\in G} \lambda_j \gamma_{ij}^2 \tag{9.12}$$

of the variation in x_i. The denominator of this expression represents the variation explained in x_i that is to be explained, and the numerator gives the amount explained by the set G. When G includes all the components, the numerator is the (i, i)th element of $\boldsymbol{\Gamma\Lambda\Gamma}'$, which is of course just σ_{ii}, so that the ratio in (9.12) is 1.

Note that the part of the *total variation* accounted for by the components in G can be expressed as the sum over all p variables of the proportion of *variation in each variable* explained by the components in G, where each variable is weighted by its variance; that is,

$$\sum_{j\in G} \lambda_j = \sum_{i=1}^{p} \sigma_{ii} \rho_{iG}^2. \tag{9.13}$$

See Exercise 9.2.12.

9.2.5 The Effect of Ignoring Some Components

In the open/closed-book data of Example 9.2.3, the final two principal components explain barely 10% of the total variance. It therefore seems fair to ask what if anything would be lost by ignoring these components completely. This question can be broken down into several parts, by analogy with those considered in Section 9.2.4.

First, we might ask what are the values of the correlation coefficients, which by analogy with (9.11) are given by

$$r_{ij}^2 = l_j g_{ij}^2 / s_{ii}. \tag{9.14}$$

Calculations for the open/closed-book data lead to the values of r_{ij} and r_{ij}^2 shown in Table 9.1. (Note that the rows of the second table sum to 1 except for rounding error.) In other words, the first component for this data explains $0.758^2 = 57.4\%$ of x_1, 53.9% of x_2, 72.7% of x_3, 63.4% if x_4, and 66.0% of x_5. The first two components taken together explain 94.5% of x_1, etc. Note also that while the last component explains only a small proportion of the variables 1, 2, 4, and 5, it accounts for $0.493^2 = 24.3\%$ of the variation in variable 3. Hence, "throwing away" the last component is in fact rejecting much more information about variable 3 than it is about the others. Similarly, "throwing away" the last two components rejects most about variables 2 and 3 and least about variable 5.

Another example using these correlations is given in Exercise 9.2.5.

A common practical way of looking at the contributions of various principal components (due to Cattell (1966)) is to look at a "scree graph" such as Figure 9.1 (that is, one plots λ_j versus j). Such a diagram can often indicate clearly where "large" eigenvalues cease and "small" eigenvalues begin. This is often referred to as an "elbow" point, though its determination is somewhat subjective.

Various other "rules of thumb" for excluding principal components exist, including the following:

(a) include just enough components to explain say 90% of the total variation;
(b) (Kaiser, 1958) exclude those principal components whose eigenvalues are less than the average, i.e. less than 1 if a correlation matrix has been used.

Table 9.1 Values of r_{ij}^2, defined by (9.14), – and their square root – for the open/closed-book data.

	r_{ij}	Component j 1	2	3	4	5
	1	0.758	0.609	−0.175	0.156	0.026
	2	0.734	0.224	0.322	0.548	0.081
Variable i	3	0.853	−0.102	0.139	−0.003	−0.493
	4	0.796	−0.288	0.409	0.321	0.109
	5	0.812	−0.451	−0.354	−0.094	0.050

	r_{ij}^2	Component j 1	2	3	4	5
	1	0.574	0.371	0.030	0.024	0.001
	2	0.539	0.050	0.104	0.300	0.007
Variable i	3	0.727	0.010	0.019	0.000	0.243
	4	0.634	0.083	0.168	0.103	0.012
	5	0.660	0.204	0.125	0.009	0.002

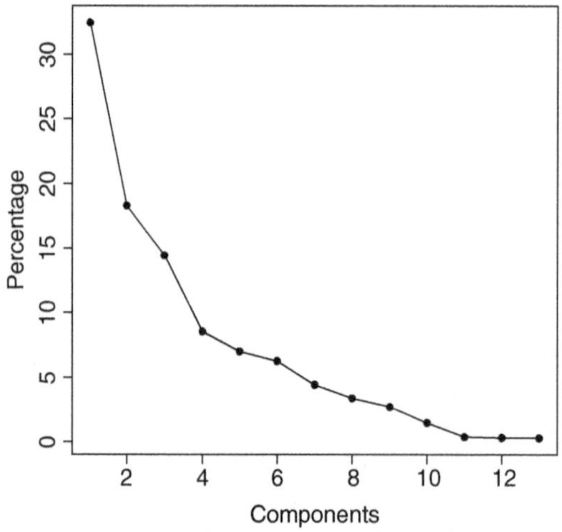

Figure 9.1 Scree graph for the Corsican pitprop data of Table 9.2.

It has been suggested that when $p \leq 20$, then Kaiser's criterion (b) tends to include too few components. Similarly, Cattell's scree test is said to include too many components. Thus, a compromise is often used in practice.

The isotropy test (Section 9.4.3) can also give some indication of the number of components to include.

Table 9.2 Eigenvectors and eigenvectors for the pitprop data.

	Variable												
Eigenvector	$g_{(1)}$	$g_{(2)}$	$g_{(3)}$	$g_{(4)}$	$g_{(5)}$	$g_{(6)}$	$g_{(7)}$	$g_{(8)}$	$g_{(9)}$	$g_{(10)}$	$g_{(11)}$	$g_{(12)}$	$g_{(13)}$
x_1	−0.40	0.22	−0.21	−0.09	−0.08	0.12	−0.11	0.14	0.33	−0.31	−0.00	0.39	−0.57
x_2	−0.41	0.19	−0.24	−0.10	−0.11	0.16	−0.08	0.02	0.32	−0.27	−0.05	−0.41	0.58
x_3	−0.12	0.54	0.14	0.08	0.35	−0.28	−0.02	0.00	−0.08	0.06	0.12	0.53	0.41
x_4	−0.17	0.46	0.35	0.05	0.36	−0.05	0.08	−0.02	−0.01	0.10	−0.02	−0.49	−0.38
x_5	−0.06	−0.17	0.48	0.05	0.18	0.63	0.42	−0.01	0.28	-0.00	0.01	0.20	0.12
x_6	−0.28	−0.01	0.48	−0.06	−0.32	0.05	−0.30	0.15	−0.41	−0.10	−0.54	0.08	0.06
x_7	−0.40	−0.19	0.25	−0.07	−0.22	0.00	−0.23	0.01	−0.13	0.19	0.76	−0.04	0.00
x_8	−0.29	−0.19	−0.24	0.29	0.19	−0.06	0.40	0.64	−0.35	−0.08	0.03	−0.05	0.02
x_9	−0.36	0.02	−0.21	0.10	−0.10	0.03	0.40	−0.70	−0.38	−0.06	−0.05	0.05	−0.06
x_{10}	−0.38	−0.25	−0.12	−0.21	0.16	−0.17	0.00	−0.01	0.27	0.71	−0.32	0.06	0.00
x_{11}	0.01	0.21	−0.07	0.80	−0.34	0.18	−0.14	0.01	0.15	0.34	−0.05	0.00	−0.01
x_{12}	0.12	0.34	0.09	−0.30	−0.60	−0.17	0.54	0.21	0.08	0.19	0.05	0.00	0.00
x_{13}	0.11	0.31	−0.33	−0.30	0.08	0.63	−0.16	0.11	−0.38	0.33	0.04	0.01	−0.01
Eigenvalue λ_i	4.22	2.38	1.88	1.11	0.91	0.82	0.58	0.44	0.35	0.19	0.05	0.04	0.04
% Eigenvalues	32.5	18.3	14.4	8.5	7.0	6.3	4.4	3.4	2.7	1.5	0.4	0.3	0.3
Cumulative %	32.5	50.7	65.2	73.7	80.7	87.0	91.4	94.8	97.5	99.0	99.4	99.7	100.0

Example 9.2.6 Consider the pine pitprop data (Jeffers, 1967) of Example 7.7.1. A PCA of the correlation matrix yields the eigenvalues and eigenvectors given in Table 9.2 together with the associated percentages and cumulative percentages of the variance explained. A look at Cattell's scree graph (Figure 9.1) suggests that three or six components should be retained, whereas Kaiser's criterion suggests the number 4. Note that seven components are needed to explain 90% of the variance. Jeffers decided to retain the first six components.

In this example, the first six principal components turn out to have natural physical interpretations. For each component j, consider those variables that have relatively high positive or negative weighting (say the variables i for which $|g_{ij}| \geq 0.70 \max_k g_{kj}$) as constituting an index of the combined action, or contrast, of the original variables. (However, it must be warned that this approach does not always lead to meaningful interpretations.) The first component weights highly variables 1, 2, 6, 7, 8, 9, and 10 and represents the overall size of the prop. The second component, weighting highly 3 and 4, measures the degree of seasoning. The third component, weighting 5 and 6, is a measure of the rate of growth of the timber. The fourth, fifth, and sixth components are most easily described as a measure of variable 11, a measure of variable 12, and an average of variables 5 and 13. □

9.2.6 Graphical Representation of Principal Components

The reduction in dimensions afforded by principal components analysis can be used graphically. Thus, if the first two principal components explain "most" of the variance,

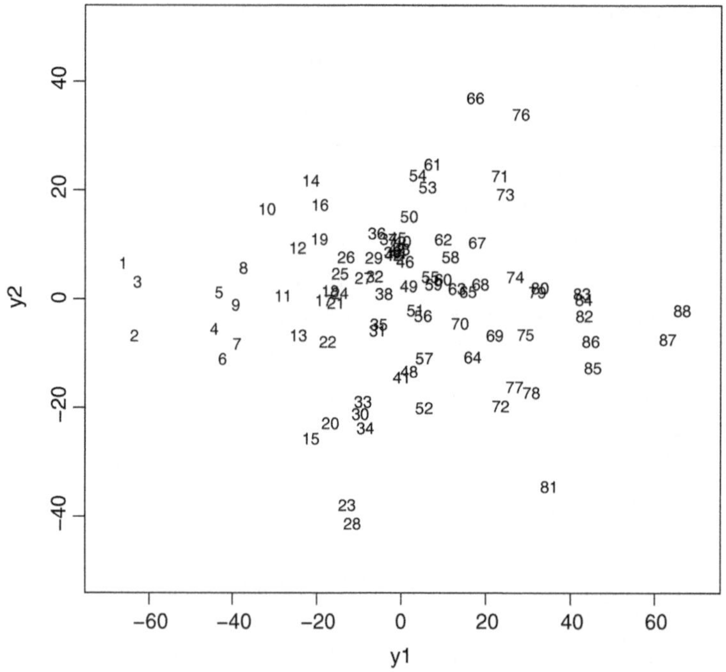

Figure 9.2 The 88 individuals of the open/closed-book data plotted on the first two principal components.

then a scatter plot showing the distribution of the objects on these two dimensions will often give a fair indication of the overall distribution of the data. Such plots can augment the graphical methods of Section 1.7 on the original variables.

Figure 9.2 shows the 88 individuals of the open/closed-book data plotted on the first two principal components of the covariance matrix (Example 9.2.3). Note that the variance along the y_1-axis is greater than the variance along the y_2-axis.

As well as showing the distribution of *objects* on the first two principal components, the *variables* can also be represented on a similar diagram, using as coordinates the correlations derived in Section 9.2.5.

Example 9.2.7 The values of r_{ij} for the open/closed-book data were calculated in Section 9.2.5. Hence, variables 1–5 can be plotted on the first two principal axes, given in Figure 9.3. Note that all of the plotted points lie inside the unit circle, and variables 1 and 5, which lie fairly close to the unit circle, are explained better by the first two principal components than variables 2, 3, and 4. □

9.2.7 Biplots

The Section 9.2.6 gave a simple graphical representation of the first two principal components. However, it is possible to convey more information in a *biplot* (Gabriel, 1971), which adds information about the variables. The motivation is as follows.

If the sample data matrix $\boldsymbol{X}$ is of rank r, then it can be written as

$$\boldsymbol{X} = \boldsymbol{G}\boldsymbol{H}', \tag{9.15}$$

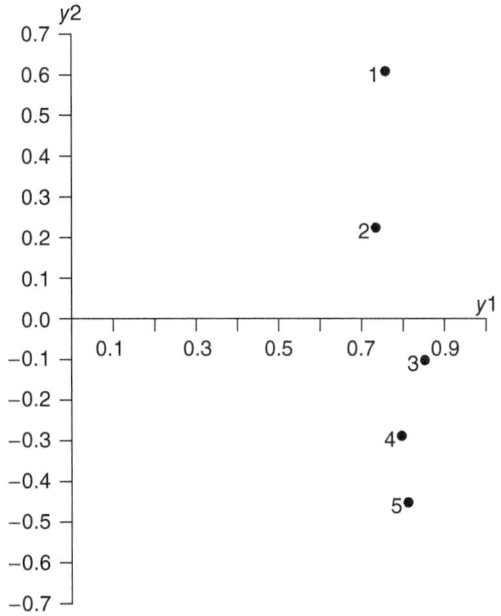

Figure 9.3 Correlations between the variables and principal components for the open/closed-book data.

where $\boldsymbol{G}$ is $n \times r$ and $\boldsymbol{H}$ is $p \times r$, both necessarily of rank r. For example, the r columns of $\boldsymbol{G}$ can be obtained as an orthonormal basis of the column space of $\boldsymbol{X}$, and then, $\boldsymbol{H} = \boldsymbol{X}'\boldsymbol{G}$, but the choice of $\boldsymbol{H}$ and $\boldsymbol{G}$ is not unique.

Equation (9.15) can also be written as

$$x_{ij} = \boldsymbol{g}_i'\boldsymbol{h}_j,$$

and so, we have represented the data by $n + p$ vectors in r-dimensional space. The vectors $\boldsymbol{g}_1, \ldots, \boldsymbol{g}_n$ may be considered as "row effects," and the $\boldsymbol{h}_j$ may be considered as "column effects." The biplot is simply a graphical display that superimposes a plot of $\boldsymbol{G}$ and $\boldsymbol{H}$ on the same graph, when the rank $r = 2$. When the rank $r > 2$, then an approximate biplot can be drawn, based on the first two principal components.

The singular value decomposition (Theorem A.8) for a matrix $\boldsymbol{X}$ of rank r is given by

$$\boldsymbol{X} = \boldsymbol{ULV}'.$$

This gives an approximation, corresponding to the two largest eigenvalues, as

$$\hat{\boldsymbol{X}} = \boldsymbol{U}_2\boldsymbol{L}_2\boldsymbol{V}_2', \tag{9.16}$$

where the subscript 2 is used to denote that the first two columns of the corresponding matrix are selected. Now, $\hat{\boldsymbol{X}}$ is of rank 2 and can be factorized as $\hat{\boldsymbol{X}} = \hat{\boldsymbol{G}}\hat{\boldsymbol{H}}'$, where

$$\begin{aligned} \hat{\boldsymbol{G}} &= \boldsymbol{U}_2\boldsymbol{L}_2^c, \\ \hat{\boldsymbol{H}}' &= \boldsymbol{L}_2^{1-c}\boldsymbol{V}_2', \end{aligned} \tag{9.17}$$

where $0 \leq c \leq 1$ is a power to be chosen. So, an approximate biplot of the data matrix $\boldsymbol{X}$ is obtained by superimposing plots of $\hat{\boldsymbol{G}}$ (which depicts the n observations) and $\hat{\boldsymbol{H}}$

(which depicts the p variables). The choice of c in (9.17) determines which distances in the plots should be preserved. That is, choosing $c = 1$ means that the distances between observations are closely approximated in the plot; choosing $c = 0$ means that the distances between the columns are nearly preserved, and choosing $c = 1/2$ gives a trade-off. Note that the principal components correspond to $c = 1$. Clearly, the effectiveness of the biplot is very dependent on how good is the approximation given by (9.16), as the following examples illustrate for $c = 0$ and $c = 1$.

Example 9.2.8 To illustrate the biplot with $c = 0$, consider the open/closed-book data. Up to the scaling of the axes, the resulting plot is almost a superposition of Figures 9.2 and 9.3. In Figure 9.2, the observations had an elliptical pattern with the major axis in the horizontal direction. In Figure 9.4, the data are isotropic. The arrows (for which the scale is shown on the top and right edges) show the plot of the variables $\hat{\boldsymbol{H}}$, which has the same overall shape (though inverted) as Figure 9.3. In this plot, the lengths of the $\boldsymbol{h}$ vectors are indicative of the relative standard deviations of the variables, and the angles between them indicate pairwise correlations. □

Example 9.2.9 Using the simulated data from the model given by Gabriel (1971), we illustrate the biplot for $c = 1$. The model uses $Z_{ij} \sim N(0, 1)$, $i = 1, \ldots, 30$, $j = 1, \ldots, 4$, with rows of $\boldsymbol{X}$ given by

$$\boldsymbol{x}_i = \begin{cases} (Z_{i1} + 10, Z_{i1} + Z_{i2}, -Z_{i1} - Z_{i2} - Z_{i3}, Z_{i1} + Z_{i2} + Z_{i3} + Z_{i4}), & i = 1, \ldots, 10, \\ (Z_{i1}, Z_{i1} + Z_{i2}, -Z_{i1} - Z_{i2} - Z_{i3}, Z_{i1} + Z_{i2} + Z_{i3} + Z_{i4}), & i = 11, \ldots, 20, \\ (Z_{i1} + 10, Z_{i1} + Z_{i2}, -Z_{i1} - Z_{i2} - Z_{i3}, Z_{i1} + Z_{i2} + Z_{i3} + Z_{i4} + 10), & i = 21, \ldots, 30. \end{cases}$$

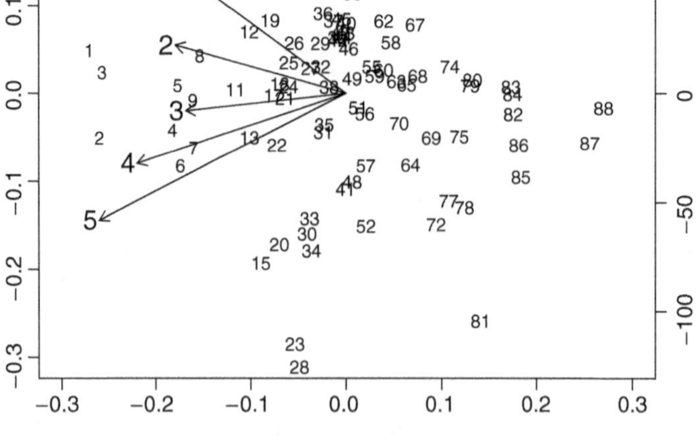

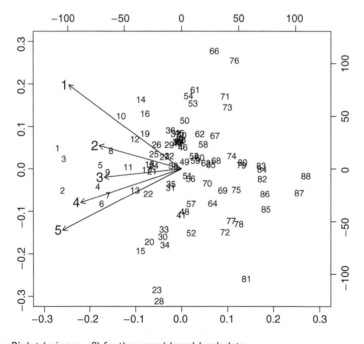

Figure 9.4 Biplot (using $c = 0$) for the open/closed-book data.

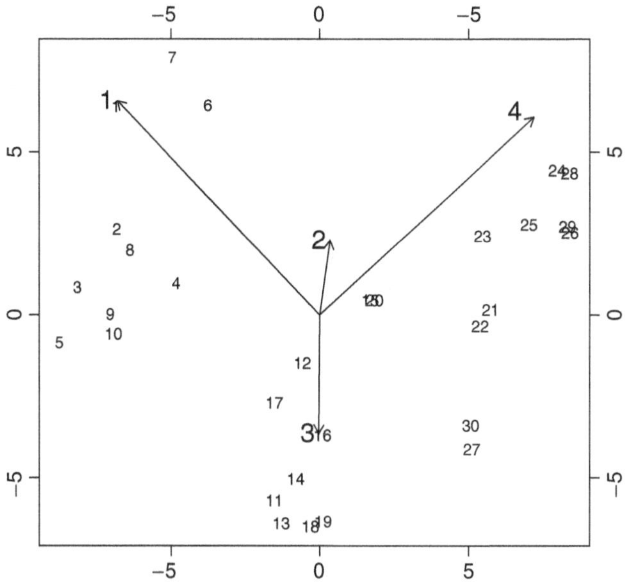

Figure 9.5 Biplot (using $c = 1$) using simulated data of Example 9.2.9.

For our simulated set of data, the first two principal components explain 96.15% of the variation in the data, and the first two principal components are:

$$y_1 = -0.80x_1 - 0.07x_2 + 0.05x_3 + 0.59x_4 + 0.59,$$
$$y_2 = -0.57x_1 - 0.15x_2 + 0.13x_3 - 0.80x_4 - 5.00.$$

In the biplot (Figure 9.5), we can clearly see the three clusters of observations corresponding to the three groups in the model. The first cluster (observations 1–10) is closely aligned with the first variable, and the third cluster (observations 21–30) with the fourth variable. The average distance between the observations within each cluster is very similar (since we used $c = 1$), and the standard deviation of variables x_1 and x_4 is much larger than those of variables x_2 and x_3. From the angles between the $\boldsymbol{h}$ vectors, we can conclude that x_2 is positively correlated with x_1 and x_4 and negatively correlated with x_3. □

9.3 Sampling Properties of Principal Components

9.3.1 Maximum-likelihood Estimation for Normal Data

The small-sample distribution of eigenvalues and eigenvectors of a covariance matrix $\boldsymbol{S}$ is extremely complicated even when all parent correlations vanish. One reason is that the eigenvalues are *nonrational* functions of the elements of $\boldsymbol{S}$. However, some large sample results are known, and many useful properties of the sample principal components for normal data stem from the following maximum-likelihood results.

Theorem 9.3.1 *For normal data when the eigenvalues of* $\boldsymbol{\Sigma}$ *are distinct, the sample principal components and eigenvalues are the maximum-likelihood estimators of the corresponding population parameters.*

Proof: The principal components and eigenvalues of $\mathbf{\Sigma}$ are related to $\mathbf{\Sigma}$ by means of a one-to-one function, except when $\mathbf{\Sigma}$ has coincident eigenvalues. Hence, the theorem follows from the invariance of maximum-likelihood estimators under one-to-one transformations. □

When the eigenvalues of $\mathbf{\Sigma}$ are not distinct, the above theorem does not hold. (This can easily be confirmed for the case $\mathbf{\Sigma} = \sigma^2\mathbf{I}$.) In such cases, there is a certain arbitrariness in defining the eigenvectors of $\mathbf{\Sigma}$. Even if this is overcome, the eigenvalues of $\mathbf{S}$ are in general distinct, so the eigenvalues and eigenvectors of $\mathbf{S}$ would not be the same function of $\mathbf{S}$ as the eigenvalues and eigenvectors of $\mathbf{\Sigma}$ are of $\mathbf{\Sigma}$. Hence, when the eigenvectors of $\mathbf{\Sigma}$ are not distinct, the sample principal components are *not* maximum-likelihood estimators of their population counterparts. However, in such cases, the following modified result may be used.

Theorem 9.3.2 *For normal data, when $k > 1$ eigenvalues of $\mathbf{\Sigma}$ are* not *distinct, but take the common value $\bar{\lambda}$, then*

(a) *the maximum-likelihood estimate (m.l.e.) of $\bar{\lambda}$ is $\bar{l}$, the arithmetic mean of the corresponding sample eigenvalues, and*
(b) *the sample eigenvectors corresponding to the repeated eigenvalue are m.l.e.s, although they are not unique m.l.e.s.*

Proof: See Anderson (1963) and Kshirsagar (1972, p. 439). □

Although we will not prove the above theorem here, we may note that (a) appears sensible in the light of Theorem 9.3.1, since if λ_i is a distinct eigenvalue, then the m.l.e. of λ_i is l_i, and since if the last k eigenvalues are equal to $\bar{\lambda}$, then

$$\operatorname{tr} \mathbf{\Sigma} = \lambda_1 + \cdots + \lambda_{p-k} + k\bar{\lambda}$$

and

$$\operatorname{tr} \mathbf{S} = l_1 + \cdots + l_{p-k} + k\bar{l}.$$

Part (b) of the theorem appears sensible in the light of Theorem 9.3.1.

Example 9.3.1 Let $\mathbf{\Sigma} = (1-\rho)\mathbf{I} + \rho\mathbf{1}\mathbf{1}'$, $\rho > 0$. We know from Example 9.2.2 that the largest eigenvalue of $\mathbf{\Sigma}$ is $\lambda_1 = 1 + (p-1)\rho$, and that the other $(p-1)$ eigenvalues all take the value $\bar{\lambda} = (1-\rho)$.

Suppose that we assume nothing about $\mathbf{\Sigma}$ except that the smallest $p-1$ eigenvalues are equal. Then, the m.l.e. of λ_1 is l_1, the first sample eigenvalue of $\mathbf{S}$, and the m.l.e. of $\bar{\lambda}$ is

$$\bar{l} = (\operatorname{tr} \mathbf{S} - l_1)/(p-1).$$

Note that the corresponding estimates of ρ, namely

$$(l_1 - 1)/(p-1) \quad \text{and} \quad 1 - \bar{l} \tag{9.18}$$

are not equal (see Exercise 9.3.3). □

9.3.2 Asymptotic Distributions for Normal Data

For large samples, the following asymptotic result, based on the central limit theorem, provides useful distributions for the eigenvalues and eigenvectors of the sample covariance matrix.

Theorem 9.3.3 *(Anderson, 1963) Let $\boldsymbol{\Sigma}$ be a positive-definite matrix with distinct eigenvalues. Let $\boldsymbol{M} \sim W_p(\boldsymbol{\Sigma}, m)$, and set $\boldsymbol{U} = m^{-1}\boldsymbol{M}$. Consider spectral decompositions $\boldsymbol{\Sigma} = \boldsymbol{\Gamma}\boldsymbol{\Lambda}\boldsymbol{\Gamma}'$ and $\boldsymbol{U} = \boldsymbol{G}\boldsymbol{L}\boldsymbol{G}'$, and let λ and $\boldsymbol{l}$ be the vectors of diagonal elements in $\boldsymbol{\Lambda}$ and $\boldsymbol{L}$. Then, the following distributions hold as $m \to \infty$:*

(a) $\boldsymbol{l} \sim N_p(\lambda, 2\boldsymbol{\Lambda}^2/m)$; that is the eigenvalues of $\boldsymbol{U}$ are asymptotically normal, unbiased, and independent, with l_i having variance $2\lambda_i^2/m$.

(b) $\boldsymbol{g}_{(i)} \sim N_p(\boldsymbol{\gamma}_i, \boldsymbol{V}_i/m)$, where

$$\boldsymbol{V}_i = \lambda_i \sum_{j \neq i} \frac{\lambda_j}{(\lambda_j - \lambda_i)^2} \boldsymbol{\gamma}_{(j)}\boldsymbol{\gamma}'_{(j)};$$

that is, the eigenvectors of $\boldsymbol{U}$ are asymptotically normal and unbiased and have the stated asymptotic covariance matrix $\boldsymbol{V}_i/m$.

(c) The covariance between the r*th element of $\boldsymbol{g}_{(i)}$ and the* s*th element of $\boldsymbol{g}_{(j)}$ is $-\lambda_i\lambda_j\gamma_{rj}\gamma_{si}/\{m(\lambda_i - \lambda_j)^2\}$.*

(d) The elements of $\boldsymbol{l}$ are asymptotically independent of the elements of $\boldsymbol{G}$.

Proof: We only give the proof of part (a) here, and we will use the $O_p(\cdot)$ and $o_p(\cdot)$ notation given in (2.75) and (2.76).

If $\boldsymbol{M} \sim W_p(\boldsymbol{\Sigma}, m)$, then $\boldsymbol{M}^* = \boldsymbol{\Gamma}'\boldsymbol{M}\boldsymbol{\Gamma} \sim W_p(\boldsymbol{\Lambda}, m)$; that is, $\boldsymbol{M}^*$ has a Wishart distribution with diagonal scale matrix. Set $\boldsymbol{U}^* = m^{-1}\boldsymbol{M}^*$. Then, using Exercise 4.4.17, it is easily seen that $\mathsf{E}(u^*_{ii}) = \lambda_i$, $\text{var}(u^*_{ii}) = 2m^{-1}\lambda_i^2$, $\mathsf{E}(u^*_{ij}) = 0$, $\text{var}(u^*_{ij}) = m^{-1}\lambda_i\lambda_j$, and $\text{cov}(u^*_{ii}, u^*_{jj}) = 0$ for $i \neq j$. Since $\boldsymbol{M}^*$ can be represented as the sum of m independent matrices from $W_p(\boldsymbol{\Lambda}, 1)$, it follows from the central limit theorem that if we write $\boldsymbol{u}^* = (u^*_{11}, \dots, u^*_{pp})'$ and $\lambda = (\lambda_1, \dots, \lambda_p)'$, then

$$m^{1/2}(\boldsymbol{u}^* - \lambda) \xrightarrow{D} N_p(\boldsymbol{0}, 2\boldsymbol{\Lambda}^2) \tag{9.19}$$

and

$$m^{1/2}u^*_{ij} \xrightarrow{D} N_1(0, \lambda_i\lambda_j), \quad i \neq j.$$

Hence,

$$u^*_{ii} = \lambda_i + O_p(m^{-1/2}), \qquad u^*_{ij} = O_p(m^{-1/2}), \quad i \neq j. \tag{9.20}$$

Thus, the roots, $l_1 > \cdots > l_p$, of the polynomial $|\boldsymbol{U}^* - l\boldsymbol{I}|$ converge in probability to the roots, $\lambda_1 > \cdots > \lambda_p$, of the polynomial $|\boldsymbol{\Lambda} - l\boldsymbol{I}|$; that is

$$u^*_{ii} - l_i \xrightarrow{P} 0, \qquad i = 1, \dots, p.$$

To get more precise bounds on $u^*_{ii} - l_i$, use (9.20) in the standard asymptotic expansion for determinants (A.8) to write

$$|\boldsymbol{U}^* - v\boldsymbol{I}| = \prod_{j=1}^{p}(u^*_{jj} - v) + O_p(m^{-1}), \tag{9.21}$$

where v is any random variable bounded above in probability $v = O_p(1)$. In particular, because $l_i \leq \text{tr}\, \boldsymbol{U}^* = \text{tr}\, \boldsymbol{U}$ for all $i = 1, \dots, p$, it is clear from (9.20) that $l_i = O_p(1)$. Since $l_1, \dots, l_p$ are the eigenvalues of $\boldsymbol{U}^*$, setting $v = l_i$ in (9.21) yields

$$\prod_{j=1}^{p}(u_{jj}^* - l_i) = O_p(m^{-1}), \qquad i = 1, \dots, p.$$

Now, the population eigenvalues λ_i are distinct, so for $i \neq j$, $u_{jj}^* - l_i$ is bounded in probability away from 0. Hence, the nearness of the above product to 0 is due to only one factor, $u_{ii} - l_i$; that is

$$u_{ii}^* = l_i + O_p(m^{-1}), \qquad i = 1, \dots, p.$$

Inserting this result in (9.19) gives

$$m^{1/2}(\boldsymbol{l} - \boldsymbol{\lambda}) = m^{1/2}(\boldsymbol{u}^* - \boldsymbol{\lambda}) + O_p(m^{-1/2}) \xrightarrow{D} N_p(\boldsymbol{0}, 2\boldsymbol{\Lambda}^2),$$

because the second term, $O_p(m^{-1/2})$, converges in probability to $\boldsymbol{0}$. Hence, part (a) is proved. □

For a single sample of size n from a multivariate normal distribution, the unbiased covariance matrix $\boldsymbol{S}_u = (n-1)^{-1}n\boldsymbol{S}$ satisfies the assumptions on $\boldsymbol{U}$ in the above theorem with $m = n - 1$ degrees of freedom. Thus, the eigenvalues $l_1 > \cdots > l_p$ of $\boldsymbol{S}_u$ with eigenvectors $\boldsymbol{g}_{(1)}, \dots, \boldsymbol{g}_{(p)}$ have the asymptotic distributions stated in the theorem. Of course, asymptotically, it does not matter whether we work with $\boldsymbol{S}$ or $\boldsymbol{S}_u$.

Example 9.3.2 Let $l_1, \dots, l_p$ denote the eigenvalues of $\boldsymbol{S}_u$ for a sample of size n. From part (a) of the theorem, we deduce that, asymptotically, $l_i \sim N(\lambda_i, 2\lambda_i^2/(n-1))$. Therefore, using Theorem 2.7.2, asymptotically,

$$\log l_i \sim N(\log \lambda_i, 2/(n-1)). \tag{9.22}$$

This leads to the asymptotic confidence limits

$$\log \lambda_i = \log l_i \pm z\sqrt{2/(n-1)}, \tag{9.23}$$

where z is the relevant critical value from the standard normal distribution.

For instance, consider the open/closed-book data from Example 9.2.3, where $n = 88$. Putting $z = 1.96$ gives $(l_i e^{-0.297}, l_i e^{0.297})$ as an asymptotic 95% confidence interval for λ_i. Use of the five sample eigenvalues of $\boldsymbol{S}_u$, 687.0, 202.1, 103.7, 84.6, and 32.2, leads to the following intervals:

$$(510, 925), \quad (150, 272), \quad (77, 140), \quad (63, 114), \quad \text{and} \quad (24, 43).$$

The sampling variability of an estimator can also be approximated using the bootstrap. The bootstrap is a powerful methodology for statistical inference which can be applied in the absence of a formal statistical model (Efron and Tibshirani, 1994). It can often be more accurate than the asymptotic distribution when the sample size n is not sufficiently large for the central limit theorem to be valid.

The bootstrap procedure works as follows. Simulate a bootstrap data matrix $\boldsymbol{X}^*$, say, of size $n \times p$ by sampling with replacement from the rows of the original $n \times p$ data matrix $\boldsymbol{X}$, and compute the estimator for $\boldsymbol{X}^*$. Repeat this process a large number of times (m^*, say).

Then, the sampling variability of the estimator is approximated by the bootstrap distribution, i.e. the empirical distribution of the estimator over the m^* bootstrap replications.

As an illustration of bootstrap methods, Diaconis and Efron (1983) used the open–closed-book data to investigate the sampling variability of the eigenvalues and eigenvectors of the sample covariance matrix. Since Theorem 9.3.3 gives the asymptotic normal distribution for these parameters, it is possible to see how closely the two approaches agree with each other. Here, we set $m^* = 100\,000$.

The standard errors using the two methods are given in Table 9.3 for the log eigenvalues, where the constant standard error $\sqrt{2/87} = 0.152$ comes from (9.23). A similar summary is given in Table 9.4 for the elements of the first two eigenvectors. It can be seen that both methods are broadly similar. The standard errors are always within 20% of each other and often within 10%.

An additional advantage of the bootstrap is that the shape of the bootstrap distribution can also be compared to the asymptotic normal distribution. Histograms are given in Figure 9.6 of the bootstrap distributions of the first two eigenvalues. Each panel also includes a normal density determined by the asymptotic theory, i.e. centered at the sample eigenvalue with standard deviation $\sqrt{2/(n-1)}$. For the first eigenvalue, the bootstrap histogram seems compatible with the asymptotic normal distribution, with a very slight (but statistically significant) negative skewness (−0.23). For the second eigenvalue, the bootstrap histogram is close to the asymptotic normal distribution, but some differences are visible. The bootstrap distribution is a bit skewed to the left, and its peak is slightly lower. For the higher order eigenvalues, the differences from the asymptotic normal distribution tend to be much more pronounced. □

Table 9.3 Open–closed-book data in Example 9.3.2: standard errors for the log eigenvalues, computed using the asymptotic formula (9.23) and using the bootstrap.

Method	PC1	PC2	PC3	PC4	PC5
Asymptotic	0.152	0.152	0.152	0.152	0.152
Bootstrap	0.162	0.172	0.132	0.136	0.138

Table 9.4 Open–closed-book data in Example 9.3.2: standard errors for the loading vectors of the first two principal components; Asymp1 and Asymp2 come from the asymptotic formula in Theorem 9.3.3; Boot1 and Boot2 come from the bootstrap.

Exam	Asymp1	Boot1	Asymp2	Boot2
Mechanics	0.065	0.056	0.072	0.086
Vectors	0.043	0.042	0.119	0.126
Algebra	0.024	0.029	0.060	0.065
Analysis	0.045	0.039	0.120	0.130
Statistics	0.055	0.046	0.107	0.129

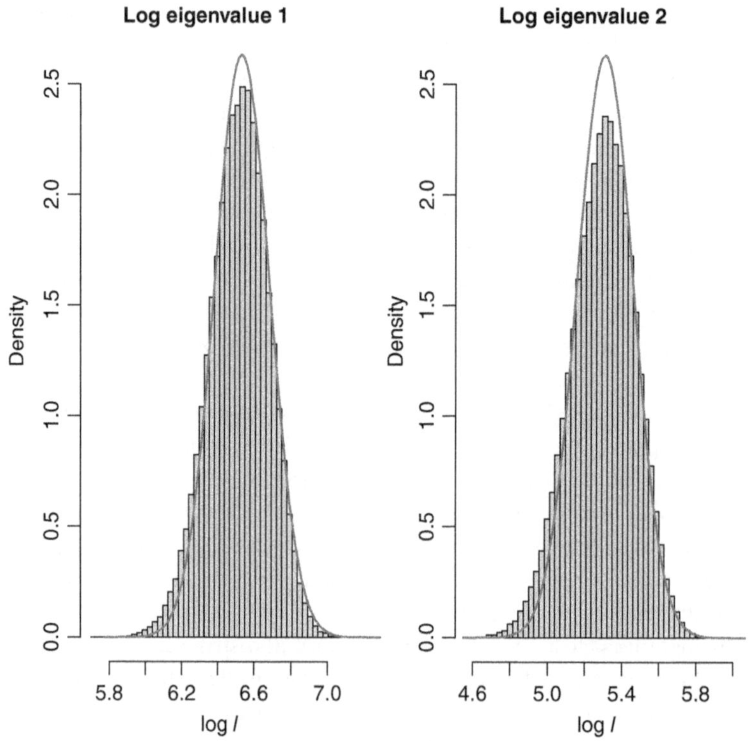

Figure 9.6 Sampling distribution of the first two log eigenvalues for the open–closed-book data in Example 9.3.2. The solid line gives the fitted asymptotic normal distribution from (9.22). The histogram gives the bootstrap distribution.

Example 9.3.3 Theorem 9.3.3 can also be used for testing hypotheses on the eigenvectors of $\mathbf{\Sigma}$. Consider, for example, the null hypothesis that the ith eigenvector $\boldsymbol{\gamma}_{(i)}$ takes a given value $\boldsymbol{r}$, where $\boldsymbol{r}'\boldsymbol{r} = 1$, and we suppose that λ_i has multiplicity 1. If the null hypothesis $\boldsymbol{\gamma}_{(i)} = \boldsymbol{r}$ is true, then, by part (b) of Theorem 9.3.3, asymptotically,

$$(n-1)^{1/2}(\mathbf{g}_{(i)} - \boldsymbol{r}) \sim N(\mathbf{0}, \boldsymbol{V}_i).$$

Note that $\boldsymbol{V}_i$ has rank $(p-1)$; its eigenvectors are $\boldsymbol{\gamma}_{(i)}$ with eigenvalue 0 and $\boldsymbol{\gamma}_{(j)}$ with eigenvalue $(\lambda_i\lambda_j^{-1} + \lambda_i^{-1}\lambda_j - 2)^{-1}$ for $j \neq i$. Thus (see Section A.8), a g-inverse $\boldsymbol{V}_i^-$ is defined by the same eigenvectors and by eigenvalues zero and $\lambda_i\lambda_j^{-1} + \lambda_i^{-1}\lambda_j - 2$; that is,

$$\boldsymbol{V}_i^- = \sum_{j\neq i}(\lambda_i\lambda_j^{-1} + \lambda_i^{-1}\lambda_j - 2)\boldsymbol{\gamma}_{(j)}\boldsymbol{\gamma}'_{(j)} = \lambda_i\mathbf{\Sigma}^{-1} + \lambda_i^{-1}\mathbf{\Sigma} - 2\boldsymbol{I}.$$

Hence, by a modified version of Theorem 2.5.4 on Mahalanobis distances (see specifically Exercise 4.4.11), asymptotically,

$$(n-1)(\mathbf{g}_{(i)} - \boldsymbol{r})'(\lambda_i\mathbf{\Sigma}^{-1} + \lambda_i^{-1}\mathbf{\Sigma} - 2\boldsymbol{I})(\mathbf{g}_{(i)} - \boldsymbol{r}) \sim \chi^2_{p-1}. \tag{9.24}$$

Let $\boldsymbol{W}_i^- = l_i\boldsymbol{S}^{-1} + l_i^{-1}\boldsymbol{S} - 2\boldsymbol{I}$ denote the sample version of $\boldsymbol{V}_i^-$. Since $(n-1)^{1/2}(\mathbf{g}_{(i)} - \boldsymbol{r}) = O_p(1)$ and $\boldsymbol{V}_i^- = \boldsymbol{W}_i^- + o_p(1)$, the limiting distribution in (9.24) remains the same if we replace $\boldsymbol{V}_i^-$ by $\boldsymbol{W}_i^-$. As $\boldsymbol{g}_{(i)}$ is an eigenvector of $\boldsymbol{W}_i^-$ with eigenvalue 0, (9.24) implies, asymptotically,

$$(n-1)(l_i\boldsymbol{r}'\boldsymbol{S}^{-1}\boldsymbol{r} + l_i^{-1}\boldsymbol{r}'\boldsymbol{S}\boldsymbol{r} - 2) \sim \chi^2_{p-1}. \tag{9.25}$$

□

9.4 Testing Hypotheses About Principal Components

9.4.1 Introduction

It is often useful to have a procedure for deciding whether the first k principal components include all the important variations in $\boldsymbol{x}$. Clearly, one would be happy to ignore the remaining $(p-k)$ components if their corresponding population eigenvalues were all zero. However, this happens only if $\boldsymbol{\Sigma}$ has rank k, in which case $\boldsymbol{S}$ must also have rank k. Hence, this situation is only trivially encountered in practice.

A second possible alternative would be to test the hypothesis that the proportion of the variance explained by the last $(p-k)$-dimensional space spanned by the last $(p-k)$ components is less than a certain value. This possibility is examined in Section 9.4.2.

A more convenient hypothesis is the question whether the last $(p-k)$ eigenvalues are equal. This would imply that the variation is equal in all directions of the $(p-k)$-dimensional space spanned by the last $(p-k)$ eigenvectors. This is the situation of *isotropic variation* and would imply that if one component is discarded, then the last k components should also be discarded. The hypothesis of isotropic variation is examined in Section 9.4.3.

In all the following subsections, we assume that we are given a random sample of normal data of size n.

9.4.2 The Hypothesis that $(\lambda_1 + \cdots + \lambda_k)/(\lambda_1 + \cdots + \lambda_p) = \psi$

Let $l_1, \ldots, l_p$ be the eigenvalues of $\boldsymbol{S}_u$, and denote the sample counterpart of ψ by

$$\hat{\psi} = (l_1 + \cdots + l_k)/(l_1 + \cdots + l_p).$$

We seek the asymptotic distribution of $\hat{\psi}$. From Theorem 9.3.3, we know that the elements of $\boldsymbol{l}$ are asymptotically normal. Using Theorem 2.7.2 with $\boldsymbol{t} = \boldsymbol{l}$, $\boldsymbol{\mu} = \lambda$, $f(\boldsymbol{t}) = (t_1 + \cdots + t_k)/(t_1 + \cdots + t_p)$, and n replaced by $n-1$, we see that $\boldsymbol{V} = 2\boldsymbol{\Lambda}^2$, and

$$d_i = \frac{\partial \hat{\psi}}{\partial l_i} = \begin{cases} (1-\psi)/\mathrm{tr}\,\boldsymbol{\Sigma} & \text{for} \quad i = 1, \ldots, k, \\ -\psi/\mathrm{tr}\,\boldsymbol{\Sigma} & \text{for} \quad i = k+1, \ldots, p. \end{cases}$$

Therefore, $\hat{\psi}$ is asymptotically normal with mean $f(\lambda) = \psi$, and with variance

$$\begin{aligned} \tau^2 &= \frac{2}{(n-1)(\mathrm{tr}\,\boldsymbol{\Sigma})^2}\{(1-\psi)^2(\lambda_1^2 + \cdots + \lambda_k^2) + \psi^2(\lambda_{k+1}^2 + \cdots + \lambda_p^2)\} \\ &= \frac{2\mathrm{tr}\,\boldsymbol{\Sigma}^2}{(n-1)(\mathrm{tr}\,\boldsymbol{\Sigma})^2}(\psi^2 - 2\alpha\psi + \alpha), \end{aligned} \tag{9.26}$$

where $\alpha = (\lambda_1^2 + \cdots + \lambda_k^2)/(\lambda_1^2 + \cdots + \lambda_p^2)$. Note that α may be interpreted as the proportion of variance explained by the first k principal components of a variable whose covariance matrix is $\boldsymbol{\Sigma}^2$. Results similar to the above have also been obtained by Sugiyama and Tong (1976) and Kshirsagar (1972, p. 454). Note that since we are only interested in ratios of eigenvalues, it makes no difference whether we work with the eigenvalues of $\boldsymbol{S}$ or of $\boldsymbol{S}_u$.

Example 9.4.1 Equation (9.26) may be used to derive approximate confidence intervals for ψ. This may be illustrated using the open/closed-book data. Let $k = 1$, so that we seek a confidence interval for the proportion of variance explained by the first principal component. Here, $n = 88$, and we have

$$\mathrm{tr}\boldsymbol{S}_u = l_1 + \cdots + l_p = 1109.6, \qquad \mathrm{tr}\ \boldsymbol{S}_u^2 = l_1^2 + \cdots + l_p^2 = 5.3176 \times 10^5,$$

$$\hat{\psi} = 687.0/\mathrm{tr}\ \boldsymbol{S}_u = 0.619.$$

If we estimate α and τ^2 by $\hat{\alpha}$ and $\hat{\tau}^2$, using the sample eigenvalues $l_1, \ldots, l_p$, then we find

$$\hat{\alpha} = 0.888 \qquad \hat{\tau}^2 = 0.148/87 = 0.0017.$$

Thus, the standard error of $\hat{\psi}$ is $0.0017^{1/2} = 0.041$, and an approximate 95% confidence interval is given by

$$0.619 \pm 1.96 \times 0.041 = (0.538, 0.700).$$

In other words, although the point estimate seemed to indicate that the first principal component explains 61% of the variation, the 95% confidence interval suggests that the true value for the population lies between 54% and 70%. □

9.4.3 The Hypothesis that $(p - k)$ Eigenvalues of $\boldsymbol{\Sigma}$ are Equal

It should be noted that any nonzero eigenvalue, however small, is "significantly" different from zero, because if $\lambda_p = 0$, then the scatter of points lies in a lower dimensional hyperplane, and so, $l_p = 0$ with probability 1. Hence, a test of the hypothesis $\lambda_p = 0$ has no meaning. However, a test of whether $\lambda_p = \lambda_{p-1}$ is meaningful. This would be testing the hypothesis that the scatter is partially isotropic in a two-dimensional subspace.

We consider the more general hypothesis that $\lambda_p = \lambda_{p-1} = \cdots = \lambda_{k+1}$. The isotropy test is useful in determining the number of principal components to use to describe the data. Suppose that we have decided to include at least k components, and we wish to decide whether or not to include any more. The acceptance of the null hypothesis implies that if we are to include more than k principal components, we should include all p of them because each of the remaining components contains the same amount of information.

Often, one conducts a sequence of isotropy tests, starting with $k = 0$ and increasing k until the null hypothesis is accepted.

The likelihood ratio for this hypothesis can be found in the same manner as that used for the one-sample hypotheses on $\boldsymbol{\Sigma}$, which is discussed in Chapter 6. As Theorem 6.3.2 indicated, the likelihood ratio test (LRT) is given by

$$-2\log\lambda = np(a - 1 - \log g),$$

where a and g are the arithmetic and geometric means of the eigenvalues of $\tilde{\boldsymbol{\Sigma}}^{-1}\boldsymbol{S}$, where $\tilde{\boldsymbol{\Sigma}}$ is the m.l.e. of $\boldsymbol{\Sigma}$ under the null hypothesis. Now, in our case, by virtue of Theorem 9.3.2, $\tilde{\boldsymbol{\Sigma}}$ and $\boldsymbol{S}$ have the same eigenvectors, and if the eigenvalues of $\boldsymbol{S}$ are $(l_1, \ldots, l_p)$, then the eigenvalues of $\tilde{\boldsymbol{\Sigma}}$ are $(l_1, \ldots, l_k, a_0, \ldots, a_0)$, where $a_0 = (l_{k+1} + \cdots + l_p)/(p-k)$ denotes the arithmetic mean of the sample estimates of the repeated eigenvalue. (In our problem, we assume that the repeated eigenvalues come last, although there is no need for this to be so.) It follows that the eigenvalues of $\tilde{\boldsymbol{\Sigma}}^{-1}\boldsymbol{S}$ are $(1, \ldots, 1, l_{k+1}/a_0, \ldots, l_p/a_0)$. Therefore, $a = 1$ and $g = (g_0/a_0)^{(p-k)/p}$, where $g_0 = (l_{k+1} \times \cdots \times l_p)^{1/(p-k)}$ is the geometric mean of the sample estimates of the repeated eigenvalue. Inserting these values in the above equation we deduce that

$$-2\log\lambda = n(p-k)\log(a_0/g_0). \tag{9.27}$$

Now, the number of degrees of freedom in the asymptotic distribution of $-2\log\lambda$ might appear to be $(p-k-1)$, the difference in the number of eigenvalues. However, this is not so because H_0 also affects the number of orthogonality conditions that $\boldsymbol{\Sigma}$ must satisfy. In fact (see Exercise 9.4.1), the relevant number of degrees of freedom is $(p-k+2)(p-k-1)/2$. Also, a better chi-squared approximation is obtained if n in (9.27) is replaced by $n' = n - (2p+11)/6$ to give Bartlett's approximation,

$$\left(n - \frac{2p+11}{6}\right)(p-k)\log\left(\frac{a_0}{g_0}\right) \sim \chi^2_{(p-k+2)(p-k-1)/2}, \tag{9.28}$$

asymptotically. Note that the ratio a_0/g_0 is the same whether we work with the eigenvalues of $\boldsymbol{S}$ or of $\boldsymbol{S}_u$.

Example 9.4.2 The three-dimensional data from Jolicoeur and Mosimann (1960) relating to $n = 24$ male turtles given in Example 6.3.4 may be used to test the hypothesis that $\lambda_2 = \lambda_3$. So, we have $p = 3$ and $k = 1$. Using $\boldsymbol{S}_1$ given in Example 6.3.4, it is found that $l_1 = 187.14$, $l_2 = 3.54$, and $l_3 = 1.06$. The values of a_0 and g_0 in (9.27) are

$$a_0 = \frac{1}{2}(3.54 + 1.06) = 2.30, \qquad g_0 = (3.54 \times 1.05)^{1/2} = 1.93.$$

Also, $n' = 24 - 17/6 = 21.17$. Thus, the test statistic is given by

$$-2\log\lambda = (21.17)(2)\log(2.30/1.93) = 7.28,$$

which under H_0 has a chi-squared distribution with $(p-k+2) \times (p-k-1)/2 = 2$ degrees of freedom. The observed value lies between the 95th percentile (= 5.99) and the 99th percentile (= 9.21). □

9.4.4 Hypotheses Concerning Correlation Matrices

When the variables have been standardized, the sample correlation matrix $\boldsymbol{R}$ may be regarded as an estimate of the population correlation matrix $\boldsymbol{P}$. The hypothesis that all eigenvalues of $\boldsymbol{P}$ are equal is equivalent to the hypothesis that $\boldsymbol{P} = \boldsymbol{I}$. In (6.21), we described a test of this hypothesis due to Box (1949). This uses the statistic $-n'\log|\boldsymbol{R}|$, where $n' = n - (2p+11)/6$. This statistic has an asymptotic chi-squared distribution, with $p(p-1)/2$ degrees of freedom.

Considerable difficulties arise if a test is required of the hypothesis that the $(p-k)$ smallest eigenvalues of $\boldsymbol{P}$ are equal, where $0 < k < p-1$. Bartlett (1951) suggested a test statistic similar to (9.27), namely

$$(n-1)(p-k)\log(a_0/g_0), \tag{9.29}$$

where a_0 and g_0 are the arithmetic and geometric means of the $(p - k)$ smallest eigenvalues of $\boldsymbol{R}$. (See also Section 6.3.2.1.) Unfortunately, this statistics does not have an asymptotic chi-squared distribution. Nevertheless, if the first k components account for a fairly high proportion of the variance, or if we are prepared to accept a conservative test, we may treat the above expression as a chi-squared statistic with $(p - k + 2)(p - k - 1)/2$ degrees of freedom (Anderson, 1963; Dagnelie, 1975, p. 181).

Example 9.4.3 The correlation matrix of the open/closed-book data has eigenvalues 3.18, 0.74, 0.44, 0.39, and 0.25. Use of (9.29) leads to the following test statistics:

$$\begin{aligned} &\text{for the hypothesis } \lambda_2 = \lambda_3 = \lambda_4 = \lambda_5, && \chi_9^2 = 25.90, \\ &\text{for the hypothesis } \lambda_3 = \lambda_4 = \lambda_5, && \chi_5^2 = 7.63. \end{aligned}$$

The former is significant at the 1% level, but the latter is not even significant at the 5% level. Thus, we accept the hypothesis that $\lambda_3 = \lambda_4 = \lambda_5$. □

9.5 Correspondence Analysis

9.5.1 Contingency Tables

Correspondence analysis can be used to interpret contingency tables. The objective is to investigate relationships between rows and columns of the table by looking for departures from independence. The methods are similar to those used in PCA. In particular, the singular value decomposition, which underlies principal coordinate analysis, also plays a key role here. For further information on correspondence analysis, see, e.g. Hill (1974) and Greenacre (1984, 2010, 2016).

The starting point is an $I \times J$ matrix $\boldsymbol{N} = (n_{ij})$ of counts. For simplicity, assume that all the counts are positive, $n_{ij} > 0$. Let $n_{i\cdot} = \sum_j n_{ij}$, $n_{\cdot j} = \sum_i n_{ij}$, and $n_{\cdot\cdot} = \sum_{i,j} n_{ij}$ denote the row sums, the column sums, and the total count, respectively, and write $n_{\cdot\cdot} = n$ for simplicity. It is also useful to scale the counts to get a matrix of proportions,

$$\boldsymbol{P} = \boldsymbol{N}/n,$$

so that $\sum_{ij} p_{ij} = 1$. (In this section, $\boldsymbol{P}$ is a matrix of proportions, whereas it usually refers to a population correlation matrix.) Denote the row and column sums of $\boldsymbol{P}$ by $\boldsymbol{r}$ and $\boldsymbol{c}$, respectively, where

$$r_i = p_{i\cdot} = n_{i\cdot}/n, \quad c_j = p_{\cdot j} = n_{\cdot j}/n, \tag{9.30}$$

with $\sum_i r_i = 1$ and $\sum_j c_j = 1$. It will be convenient to define vectors and matrices such as

$$\sqrt{\boldsymbol{r}} = (r_1^{1/2}, \dots, r_I^{1/2})', \quad \boldsymbol{D_r} = \mathrm{diag}(r_1, \dots, r_I), \quad \boldsymbol{D_r^{1/2}} = \mathrm{diag}(r_1^{1/2}, \dots, r_I^{1/2}),$$

and correspondingly for $\boldsymbol{D_c}$ and $\boldsymbol{D_c^{1/2}}$.

Other matrices of interest related to $\boldsymbol{P}$ are the *matrix of row profiles*

$$\boldsymbol{P}^{\mathrm{row}} = \boldsymbol{D_r^{-1}P}, \quad p_{ij}^{\mathrm{row}} = p_{ij}/r_i,$$

the *matrix of column profiles*

$$\boldsymbol{P}^{\mathrm{col}} = \boldsymbol{PD_c^{-1}}, \quad p_{ij}^{\mathrm{col}} = p_{ij}/c_j,$$

the *matrix of standardized proportions*

$$\boldsymbol{P}^{\text{std}} = \boldsymbol{D}_{\boldsymbol{r}}^{-1/2}\boldsymbol{P}\boldsymbol{D}_{\boldsymbol{c}}^{-1/2}, \quad p_{ij}^{\text{std}} = p_{ij}/\sqrt{r_i c_j}, \tag{9.31}$$

and the *matrix of ratios of proportions*

$$\boldsymbol{P}^{\text{rat}} = \boldsymbol{D}_{\boldsymbol{r}}^{-1}\boldsymbol{P}\boldsymbol{D}_{\boldsymbol{c}}^{-1}, \quad p_{ij}^{\text{rat}} = p_{ij}/(r_i c_j).$$

Each row of $\boldsymbol{P}^{\text{row}}$ sums to 1; similarly, each column of $\boldsymbol{P}^{\text{col}}$ sums to 1. As discussed below, the matrix $\boldsymbol{P}^{\text{std}}$ is used to construct the χ^2 test statistic of independence between the rows and columns, and the elements of $\boldsymbol{P}^{\text{rat}}$, minus 1, represent departures from the independence model.

A common model for a contingency table is to suppose that the counts come from independent Poisson distributions,

$$n_{ij} \sim P(\lambda_{ij}),$$

with the λ_{ij} representing intensities for each cell. Another way to generate this model is to start with a Poisson distribution for the total count, $n_{..} \sim P(\lambda_{..})$, and to suppose that the row and column labels of each of these $n_{..}$ observations come from a multinomial distribution with probabilities $\rho_{ij} = \lambda_{ij}/\lambda_{..}$. Recall (see Section B.6) that for a Poisson random variable, the mean and variance are given by the Poisson parameter,

$$\mathsf{E}\{n_{ij}\} = \lambda_{ij}, \quad \text{var}\{n_{ij}\} = \lambda_{ij}.$$

The simplest model for the multinomial probabilities is the *independence model between the row and column labels*, i.e.

$$\rho_{ij} = \alpha_i \beta_j,$$

say, where $\sum_i \alpha_i = 1$ and $\sum_j \beta_j = 1$. In practice, these parameters are estimated by the row and column proportions, $r_i = n_{i.}/n_{..}, c_j = n_{.j}/n_{..}$.

Residuals representing the departure from the independence model in each cell can be defined by

$$a_{ij} = (p_{ij} - r_i c_j)/(r_i c_j)^{1/2} = p_{ij}^{\text{std}} - (r_i c_j)^{1/2},$$

each with mean 0 and variance $1/n$. Let $\boldsymbol{A} = (a_{ij})$ denote the matrix containing these residuals. In matrix notation, $\boldsymbol{A}$ can be defined as

$$\boldsymbol{A} = \boldsymbol{D}_{\boldsymbol{r}}^{-1/2}\{\boldsymbol{P} - \boldsymbol{r}\boldsymbol{c}'\}\boldsymbol{D}_{\boldsymbol{c}}^{-1/2} = \boldsymbol{P}^{\text{std}} - \sqrt{\boldsymbol{r}}(\sqrt{\boldsymbol{c}})'.$$

An overall measure for the departure from independence can be obtained by the usual chi-squared test statistic,

$$\chi^2 = n\sum_{ij} a_{ij}^2 = n \text{ tr } \{\boldsymbol{D}_{\boldsymbol{r}}^{-1}(\boldsymbol{P} - \boldsymbol{r}\boldsymbol{c}')\boldsymbol{D}_{\boldsymbol{c}}^{-1}(\boldsymbol{P} - \boldsymbol{r}\boldsymbol{c}')'\} = n \text{ tr } (\boldsymbol{A}\boldsymbol{A}'), \tag{9.32}$$

which has an approximate $\chi^2_{(I-1)(J-1)}$ distribution under independence. The χ^2 statistic can also be written as weighted sum of squared distances between the row profiles and $\mathbf{c}$,

$$\chi^2 = \sum r_i d_{\text{row}}^2(\boldsymbol{p}_i^{\text{row}}, \mathbf{c}), \tag{9.33}$$

where

$$d_{\text{row}}^2(\boldsymbol{p}_i^{\text{row}}, \mathbf{c}) = (\boldsymbol{p}_i^{\text{row}} - \mathbf{c})'\boldsymbol{D}_{\boldsymbol{c}}^{-1}(\boldsymbol{p}_i^{\text{row}} - \mathbf{c})$$

is the squared *Pearson distance* between the row profiles $\boldsymbol{p}_i^{\text{row}}$ and **c**, with respect to the matrix $\boldsymbol{D}_c$. A similar Pearson distance d^2_{col} can be defined between the column profiles. See Exercise 9.5.1.

When the independence model does not hold, it is useful to describe departures from the independence model using the ratios $p_{ij}/(r_i c_j)$. If $p_{ij}/(r_i c_j) > 1$, then row i and column j are said to be positively associated. Similarly, if $p_{ij}/(r_i c_j) < 1$, then row i and column j are said to be negatively associated.

Correspondence analysis is based on the singular value decomposition of $\boldsymbol{A}$. Let $q \leq \min(I, J) - 1$ denote the rank of $\boldsymbol{A}$ and write

$$\boldsymbol{A} = \boldsymbol{U}\boldsymbol{L}\boldsymbol{V}' \tag{9.34}$$

in terms of column orthonormal matrices $\boldsymbol{U}(I \times q)$ and $\boldsymbol{V}(J \times q)$ and a diagonal matrix $\boldsymbol{L} = \text{diag}(\boldsymbol{l})$, where $l_1 \geq \cdots \geq l_q$ are the nonzero singular values. It is also useful to rescale $\boldsymbol{U}$ and $\boldsymbol{V}$ to the space of ratios by defining

$$\boldsymbol{F} = \boldsymbol{D}_r^{-1/2}\boldsymbol{U}, \quad \boldsymbol{G} = \boldsymbol{D}_c^{-1/2}\boldsymbol{V}. \tag{9.35}$$

The matrices $\boldsymbol{U}, \boldsymbol{V}, \boldsymbol{F}, \boldsymbol{G}$ have the orthogonality properties

$$\boldsymbol{U}'\boldsymbol{U} = \boldsymbol{I}_q, \quad \boldsymbol{F}'\boldsymbol{D}_r\boldsymbol{F} = \boldsymbol{I}_q,$$
$$\boldsymbol{V}'\boldsymbol{V} = \boldsymbol{I}_q, \quad \boldsymbol{G}'\boldsymbol{D}_c\boldsymbol{G} = \boldsymbol{I}_q.$$

At least one singular value of $\boldsymbol{A}$ must vanish, with the corresponding singular vectors

$$\boldsymbol{u} = \sqrt{\boldsymbol{r}}, \quad \boldsymbol{v} = \sqrt{\boldsymbol{c}},$$

since

$$\boldsymbol{A}\sqrt{\boldsymbol{c}} = \boldsymbol{D}_r^{-1/2}(\boldsymbol{P} - \boldsymbol{r}\boldsymbol{c}')\boldsymbol{D}_c^{-1/2}\sqrt{\boldsymbol{c}} = \boldsymbol{D}_r^{-1/2}(\boldsymbol{P} - \boldsymbol{r}\boldsymbol{c}')\boldsymbol{1}_J = \boldsymbol{D}_r^{-1/2}(\boldsymbol{r} - \boldsymbol{r}) = \boldsymbol{0}_I,$$

and, similarly, $\boldsymbol{A}'\sqrt{\boldsymbol{r}} = \boldsymbol{0}_J$. This component corresponds to the independence model, which has been subtracted from $\boldsymbol{P}^{\text{std}}$ to define $\boldsymbol{A}$. Further, $\sqrt{\boldsymbol{r}}$ and $\sqrt{\boldsymbol{c}}$ are orthogonal to the columns of $\boldsymbol{U}$ and $\boldsymbol{V}$, respectively,

$$\boldsymbol{U}'\sqrt{\boldsymbol{r}} = \boldsymbol{0}_q, \quad \boldsymbol{V}'\sqrt{\boldsymbol{c}} = \boldsymbol{0}_q.$$

In summary,

$$\boldsymbol{P}^{\text{std}} - \sqrt{\boldsymbol{r}}\sqrt{\boldsymbol{c}}' = \boldsymbol{U}\boldsymbol{L}\boldsymbol{V}'.$$

Pre- and postmultiplying by $\boldsymbol{D}_r^{-1/2}$ and $\boldsymbol{D}_c^{-1/2}$, respectively, yield the related decomposition for the centered ratio matrix:

$$\boldsymbol{P}^{\text{rat}} - \boldsymbol{1}_I\boldsymbol{1}_J' = \boldsymbol{F}\boldsymbol{L}\boldsymbol{G}'.$$

See Exercises 9.5.2 and 9.5.3. The results here also imply that

$$(\boldsymbol{P}^{\text{col}})'\boldsymbol{P}^{\text{row}}\boldsymbol{1}_J = \boldsymbol{1}_J, \quad (\boldsymbol{P}^{\text{col}})'\boldsymbol{P}^{\text{row}}\boldsymbol{g}_{(h)} = l_h^2\boldsymbol{g}_{(h)}, \quad h = 1, \ldots, q,$$

so that $\boldsymbol{1}_J$ and the columns of $\boldsymbol{G}$ are eigenvectors of $(\boldsymbol{P}^{\text{col}})'\boldsymbol{P}^{\text{row}} = \boldsymbol{D}_c^{-1}\boldsymbol{P}'\boldsymbol{D}_r^{-1}\boldsymbol{P}$. Similarly, $\boldsymbol{1}_I$ and $\boldsymbol{f}_{(h)}$, $h = 1, \ldots, q$ are eigenvectors of $(\boldsymbol{P}^{\text{row}})'\boldsymbol{P}^{\text{col}}$; see Exercise 9.5.4.

Next, note the identity

$$\begin{aligned}(\boldsymbol{P}^{\text{row}} - \boldsymbol{1}_I\boldsymbol{c}')\boldsymbol{D}_c^{-1}(\boldsymbol{P}^{\text{row}} - \boldsymbol{1}_I\boldsymbol{c}') &= \boldsymbol{D}_r^{-1/2}\boldsymbol{A}\boldsymbol{A}'\boldsymbol{D}_r^{-1/2} \\ &= \boldsymbol{D}_r^{-1/2}\boldsymbol{U}\boldsymbol{L}\boldsymbol{V}'\boldsymbol{V}\boldsymbol{L}\boldsymbol{U}'\boldsymbol{D}_r^{-1/2} \\ &= \boldsymbol{F}\boldsymbol{L}^2\boldsymbol{F}' = \boldsymbol{M}, \quad \text{say.}\end{aligned} \tag{9.36}$$

Then, $m_{ii} + m_{jj} - 2m_{ij}$ represents both the squared Pearson distance between rows i and j of $\boldsymbol{P}^{\text{row}}$ and the squared Euclidean distance between rows i and j of $\boldsymbol{FL}$ (Exercise 9.5.5).

Just as in PCA, we can try to approximate $\boldsymbol{P}^{\text{rat}} - \boldsymbol{1}_I\boldsymbol{1}_J'$ using a low rank approximation, often with $q_1 = 1$ or 2 terms. The nonzero singular values l_h satisfy the bounds

$$0 < l_h < 1, \quad h = 1, \ldots, q$$

(Exercise 9.5.6). The total *inertia* of the contingency table is defined by

$$\mathcal{I} = \sum l_i^2 = \sum a_{ij}^2 = \text{tr}\,(\boldsymbol{D}_r^{-1}\boldsymbol{P}\boldsymbol{D}_c^{-1}\boldsymbol{P}').$$

The proportion of total inertia explained by the first q_1 components is

$$(l_1^2 + \cdots + l_{q_1}^2)/\mathcal{I}.$$

Consider the problem of visualizing the differences between the rows (and similarly the differences between the columns). The rows of $\boldsymbol{F}$ are called *standard coordinates* of the rows, and the rows of $\boldsymbol{FL}$ are called the *principal coordinates*. As noted in (9.36), the Euclidean distances between the rows of $\boldsymbol{FL}$ are the same as the Pearson distances between the row profiles.

Coordinates for the rows and columns can be combined together in a *biplot* as in Section 9.2.7. The two most important biplots are called the *row-* and *column-principal* biplots. The choice between them depends on the motivation. The row-principal biplot is used to visualize the differences between the rows. We plot the principal coordinates $\boldsymbol{FL}$ for the rows and the standard coordinates $\boldsymbol{G}$ for the columns. The inner product between the ith row of $\boldsymbol{FL}$ and the jth row of $\boldsymbol{G}$ gives the i,jth element of the centered ratio matrix $\boldsymbol{P}^{\text{rat}} - \boldsymbol{1}_I\boldsymbol{1}_J' = \boldsymbol{FLG}'$. A positive value (when the two vectors point in the same direction) indicates positive dependence, and a negative value (when the two vectors point in opposite directions) indicates negative dependence. A similar column-principal biplot involving the rows of $\boldsymbol{GL}$ and $\boldsymbol{F}$ can be constructed to visualize the differences between the columns. The inner products between rows and columns have the same values whether a row- or a column-principal biplot is used. In practice, only the dominant $q_1 = 2$ components are plotted in each case.

The biplots in the previous paragraph are examples of *asymmetric biplots*. Another choice commonly used is the *symmetric* biplot based on $\boldsymbol{FL}$ and $\boldsymbol{GL}$, but the inner product interpretation no longer holds.

Example 9.5.1 The data in Table 9.5 were obtained from the U.S. General Social Survey (Smith et al., 2018) in 2016. The rows indicate the marital status, and the columns the religion in which the person was raised. We first carry out an initial analysis to assess whether the variables "religion" and "marital status" are independent. The chi-squared test statistic is 35.1, which when compared to the critical value $\chi^2_{12}(0.99) = 26.217$ leads to a rejection of independence.

Next, we look at various representations of the data set to explore how it differs from the independence model. The row and column sums of $\boldsymbol{P}$ are given by $\boldsymbol{r} = (0.42, 0.09, 0.21, 0.28)'$ and $\boldsymbol{c} = (0.53, 0.32, 0.02, 0.04, 0.09)'$. The row profiles and column profiles are given by

$$\boldsymbol{P}^{\text{row}} = \begin{bmatrix} 0.52 & 0.33 & 0.02 & 0.04 & 0.08 \\ 0.66 & 0.27 & 0.01 & 0.02 & 0.04 \\ 0.53 & 0.33 & 0.03 & 0.02 & 0.09 \\ 0.50 & 0.33 & 0.02 & 0.05 & 0.11 \end{bmatrix}, \qquad \boldsymbol{P}^{\text{col}} = \begin{bmatrix} 0.42 & 0.43 & 0.42 & 0.48 & 0.39 \\ 0.11 & 0.07 & 0.05 & 0.05 & 0.04 \\ 0.21 & 0.21 & 0.29 & 0.13 & 0.21 \\ 0.26 & 0.28 & 0.24 & 0.34 & 0.36 \end{bmatrix}.$$

Table 9.5 Cross-tabulation of frequencies taken from the General Social Survey (GSS) relating marital status and religion.

	Religion in which raised					
	1: Protestant	**2: Catholic**	**3: Jewish**	**4: Other**	**5: None**	**Total**
A: Married	632	401	23	53	97	1206
B: Widowed	165	67	3	6	9	250
C: Divorced/separated	315	195	16	14	53	593
D: Never married	399	260	13	37	88	797
Total	1511	923	55	110	247	2846

Source: Adapted from Smith et al. (2018).

Note that the rows of $\boldsymbol{P}^{\text{row}}$ and the columns of $\boldsymbol{P}^{\text{col}}$ sum to 1. The centered ratios $p_{ij}/(r_i c_j) - 1$ (after multiplication by 10 and rounding to the nearest integer to make the differences from the independence model stand out more prominently) are given in the following table.

	1	2	3	4	5
A	0	0	0	1	−1
B	2	−2	−4	−4	−6
C	0	0	4	−4	0
D	−1	0	−2	2	3

These values can be used to identify the cells with the largest absolute values for the centered ratios. For example, the cell value B5 = −6 indicates that there are fewer widowed/no religion cases than expected, whereas C3 = 4 indicates that there are more divorced Jews than expected.

To investigate the data more visually, we turn to correspondence analysis. The singular value decomposition of $\boldsymbol{A}$ has $q = 3$ nonzero singular values with $\boldsymbol{L} = \text{diag}(0.096, 0.048, 0.029)$. The matrices $\boldsymbol{F}$ and $\boldsymbol{G}$ are given by

$$\boldsymbol{F} = \begin{bmatrix} -0.04 & 0.46 & -1.07 \\ 2.89 & 0.80 & 1.17 \\ 0.16 & -1.94 & 0.09 \\ -0.96 & 0.49 & 1.19 \end{bmatrix}, \quad \boldsymbol{G} = \begin{bmatrix} 0.81 & 0.13 & 0.41 \\ -0.48 & -0.25 & -0.93 \\ -0.43 & -4.36 & -2.66 \\ -1.73 & 3.84 & -1.42 \\ -2.30 & -0.62 & 2.19 \end{bmatrix}.$$

The proportion of total inertia explained by the first two components is $(l_1^2 + l_2^2)/(l_1^2 + l_2^2 + l_3^2) = 93\%$, so two-dimensional plots will capture most of the total inertia. The row- and the column-principal biplots are shown in Figure 9.7. The scaling of the row variables is indicated at the left and bottom in each plot. The scaling of the column variables is indicated at the right and top in each plot. The scalings have been chosen so that the horizontal coordinates of the row variables are the same in both plots; similarly for the horizontal coordinates of the column variables. The spacing of the vertical coordinate of the row variables gets larger (by the ratio $l_1/l_2 = .096/.048 = 1.98$) from Figure 9.7(a) to Figure 9.7(b). The spacing of the vertical coordinate of the column variables gets smaller by the ratio l_2/l_1.

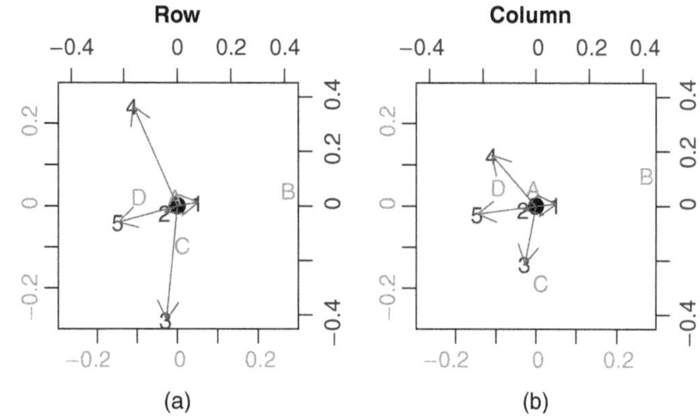

Figure 9.7 The row-principal biplot and the column-principal biplot for the marital-religious data of Example 9.5.1. The row variables are labeled A–D. The column variables are labeled 1–5 and also indicated by arrows.

The origin is at the center in each panel, both for the row variables and the column variables.

From Figure 9.7a, we see that the row variables D, A, and B lie roughly on the horizontal axis, with D being on the opposite side of the origin from A. That is, married is between never married and widowed in terms of religion. Variable C (divorced/separated) lies roughly along the vertical axis.

From Figure 9.7b, we see that the column variables 5, 2, and 1 lie roughly on the horizontal axis, with 5 being on the opposite side of the origin from 1. That is, Catholics lie between no religion and Protestants in terms of their marital status. Variables 3 and 4 (Jewish and other) lie roughly in opposite directions along the vertical axis.

The biplot inner products in either Figure 9.7a or b can be used to describe the relationships between the row and column variables. For example, C3 (divorced and Jewish) has a positive inner product, whereas B5 (widowed and no religion) has a negative inner product. These inner products approximate the values of the centered ratios described above. At the same time, caution is needed to avoid overinterpreting the data. Some of the cell counts are very low, indicating that for these cells there will be substantial sampling fluctuations. □

9.5.2 Gradient Analysis

The terms *seriation*, *ordination*, and *gradient analysis* all refer to the imposition of a one-dimensional ordering on multivariate data. For multivariate data, one way to create such an ordering is by taking the first principal component. Similarly, for contingency tables, the first principal coordinates can be used to give compatible orderings for the row and column variables. For more details beyond the methods described here, see, e.g. ter Braak and Prentice (1988).

Here is an argument, using an ecological setting, to motivate the correspondence analysis solution. Suppose that certain species of flora prefer certain types of habitat, and that their presence in a particular location can be taken as an indicator of the local conditions. Thus, one species of grass might prefer wet conditions, while another might prefer dry conditions. Other species may be indifferent. The classical approach to gradient analysis involves giving each species a "wet-preference score," according to its known preferences. Thus, a wet-loving grass may score 5, and a dry-loving grass −5, with a fickle or ambivalent grass perhaps receiving a score of 0. The conditions in a given location may now be estimated by averaging the wet-preference scores of the species that are found there.

To formalize this approach, let $\boldsymbol{N}$ be an $I \times J$ one-zero matrix representing the occurrences of I species in J locations; that is, $n_{ij} = 1$ if species i occurs in location j, and $n_{ij} = 0$ otherwise. As before, set $\boldsymbol{P} = \boldsymbol{N}/n_{\cdot\cdot}$. If f_i is the wet-preference score allocated to the ith species, then the average wet-preference score of the species found in location j is

$$g_j \propto \sum_i p_{ij} f_i / p_{\cdot j}.$$

This is the estimate of wetness in location j produced by the classical method of gradient analysis.

One drawback of the above method is that the f_i may be highly subjective. However, they themselves could be estimated by adopting the same procedure in reverse. If g_j denotes the physical conditions in location j, then f_i could be estimated as the average score of the locations in which the ith species is found; that is

$$f_i \propto \sum_j p_{ij} g_j / p_{i\cdot}.$$

Using correspondence analysis effectively takes both the above relationships simultaneously and uses them to deduce scoring vectors $\boldsymbol{f}$ and $\mathbf{g}$ that satisfy both the above equations. The vectors $\boldsymbol{f}$ and $\mathbf{g}$ are generated internally by the data, rather than being externally given.

Using the notations $\boldsymbol{P}^{\text{row}}$ and $\boldsymbol{P}^{\text{col}}$ defined earlier, the above two equations can be written in matrix form as

$$\mathbf{g} \propto (\boldsymbol{P}^{\text{col}})'\boldsymbol{f} \qquad \text{and} \qquad \boldsymbol{f} \propto \boldsymbol{P}^{\text{row}}\mathbf{g}.$$

If both these equations hold simultaneously, then

$$\boldsymbol{f} \propto \boldsymbol{P}^{\text{row}}(\boldsymbol{P}^{\text{col}})'\boldsymbol{f} \tag{9.37}$$

and

$$\mathbf{g} \propto (\boldsymbol{P}^{\text{col}})'\boldsymbol{P}^{\text{row}}\mathbf{g}. \tag{9.38}$$

Therefore, $\boldsymbol{f}$ is an eigenvector of $\boldsymbol{P}^{\text{row}}(\boldsymbol{P}^{\text{col}})'$, and $\mathbf{g}$ is an eigenvector of $(\boldsymbol{P}^{\text{col}})'\boldsymbol{P}^{\text{row}}$.

Ideally, one would like the coefficients of proportionality in each of the above equations to be unity, so that $\boldsymbol{f}$ and $\mathbf{g}$ can be interpreted as arithmetic means. This would lead one to choose $\boldsymbol{f}$ and $\mathbf{g}$ as the eigenvectors corresponding to an eigenvalue of unity. Unfortunately, although unity is indeed an eigenvalue, the corresponding eigenvectors are each $\mathbf{1}$. Hence, this requirement would involve all scores being made identical, which is not a very sensible requirement.

Therefore, we are led to consider solutions of (9.37) and (9.38) which involve an eigenvalue $\lambda < 1$. The correspondence analysis solution to this problem is to choose the largest eigenvalue less than 1, so that the largest proportion of the total inertia is explained by the seriation (see Section 15.5). Hence, the solutions for $\boldsymbol{f}$ and $\mathbf{g}$ are given by first columns of $\boldsymbol{F}$ and $\boldsymbol{G}$ in (9.35), respectively; see Section 9.5.1 for the eigenvector justification. Note that for the purpose of determining a one-dimensional ordering, $\boldsymbol{f}$ is determined only up to a proportionality constant; similarly for $\mathbf{g}$.

Example 9.5.2 In an archaeological study, there are six graves containing five types of pottery between them. It is desired to order the graves and the pottery chronologically. Let $n_{ij} = 1$ if the ith grave contains the jth pottery, and 0 otherwise. Suppose that

$$\boldsymbol{N} = \begin{array}{c} \\ A \\ B \\ C \\ D \\ E \\ F \end{array} \begin{array}{c} \begin{array}{ccccc} 1 & 2 & 3 & 4 & 5 \end{array} \\ \begin{bmatrix} 0 & 0 & 1 & 1 & 0 \\ 1 & 1 & 0 & 0 & 1 \\ 0 & 1 & 1 & 1 & 1 \\ 0 & 0 & 1 & 1 & 0 \\ 1 & 0 & 0 & 0 & 1 \\ 1 & 0 & 1 & 1 & 1 \end{bmatrix} \end{array}.$$

The largest eigenvalue of $\boldsymbol{P}^{\text{row}}(\boldsymbol{P}^{\text{col}})'$ less than 1 is 0.54, giving the corresponding eigenvectors $\boldsymbol{f} = (1.44, -1.26, 0.18, 1.44, -1.39, 0.02)'$ and $\mathbf{g} = (-1.20, -0.74, 1.05, 1.05, -0.84)'$, thus leading to chronological orders $((A, D), C, F, B, E)$ and $((3, 4), 2, 5, 1)$. Reordering the rows and columns of $\boldsymbol{N}$, we get

$$\boldsymbol{N} = \begin{array}{c} \\ A \\ D \\ C \\ F \\ B \\ E \end{array} \begin{array}{c} \begin{array}{ccccc} 3 & 4 & 2 & 5 & 1 \end{array} \\ \begin{bmatrix} 1 & 1 & 0 & 0 & 0 \\ 1 & 1 & 0 & 0 & 0 \\ 1 & 1 & 1 & 1 & 0 \\ 1 & 1 & 0 & 1 & 1 \\ 0 & 0 & 1 & 1 & 1 \\ 0 & 0 & 0 & 1 & 1 \end{bmatrix} \end{array}.$$

Columns "3" and "4" are indistinguishable, as are rows A and D. Also, the direction of time cannot be determined by this technique. Notice that the 1s are clustered about the line between the upper left-hand corner of the matrix and the lower right-hand corner, agreeing with one's intuition that later graves are associated with more recent types of pottery. Thus, the correspondence analysis seems reasonable. □

9.6 Allometry – Measurement of Size and Shape

One problem of great concern to botanists and zoologists is the question of how bodily measurements are best summarized to give overall indications of size and shape. This is the problem of "allometry," which is described in considerable detail by Hopkins (1966), Sprent (1969, 1972, p. 30), and Mosimann (1970). We will follow the approach of Rao (1971).

Define a *size factor* to be a linear function $s = \boldsymbol{a}'\boldsymbol{x}$ such that an increase in s results "on average" in an increase in each element of $\boldsymbol{x}$. Similarly, a *shape factor* is defined as a linear function $h = \boldsymbol{b}'\boldsymbol{x}$ for which an increase in h results "on average" in an increase in some of the elements of $\boldsymbol{x}$ and in a decrease in others.

The phrase "on average" is to be interpreted in the sense of a regression of $\boldsymbol{x}$ on s (and h). The covariance between s and x_i is $\mathsf{E}(\boldsymbol{a}'\boldsymbol{x}x_i) = \boldsymbol{a}'\boldsymbol{\sigma}_i$, where $\boldsymbol{\sigma}_i$ is the ith column of the covariance matrix $\boldsymbol{\Sigma} = \text{var}(\boldsymbol{x})$. Therefore, the regression coefficient of x_i on s is

$$\frac{\text{cov}(x_i, s)}{\text{var}(s)} = \frac{\boldsymbol{a}'\boldsymbol{\sigma}_i}{\boldsymbol{a}'\boldsymbol{\Sigma}\boldsymbol{a}}.$$

If ξ is a preassigned vector of positive elements that specify the ratios of increases desired in the individual measurements, then to determine $\boldsymbol{a}$ such that $\boldsymbol{a}'\boldsymbol{x}$ represents *size*, we must solve the equation

$$\boldsymbol{\Sigma a} \propto \xi.$$

Then, ξ is called a *size vector*. Note the difference between the size *vector* ξ and the size *factor* $\boldsymbol{a}'\boldsymbol{x}$. In practice, we often use standardized vectors $\boldsymbol{x}$, so that $\boldsymbol{\Sigma}$ is replaced by the sample correlation matrix $\boldsymbol{R}$. In these circumstances, it is plausible to take ξ to be a vector of ones, so that $\boldsymbol{a} \propto \boldsymbol{R}^{-1}\mathbf{1}$.

Using similar arguments, Rao suggests that a shape factor should be $h = \boldsymbol{b}'\boldsymbol{x}$, where the elements of $\boldsymbol{b}$ are given by $\boldsymbol{R}^{-1}\boldsymbol{\eta}$, where $\boldsymbol{\eta}$ is a preassigned vector of plus and minus ones, with possibly some zeros. Of course, it is possible to consider several different shape factors. Note that the interpretation of a shape factor is given by looking at the components of the shape *vector* $\boldsymbol{\eta}$, not the coefficients $\boldsymbol{b}$ of the shape *factor* $\boldsymbol{b}'\boldsymbol{x}$. If the variables are uncorrelated so that $\boldsymbol{R} = \boldsymbol{I}$, then the two vectors coincide.

Since it is only the relative magnitudes of the elements of ξ and $\boldsymbol{\eta}$ which are of interest, both s and h may be rescaled without affecting their interpretive value. If $\boldsymbol{a}$ and $\boldsymbol{b}$ are scaled so that s and h have unit variance (and if each column of the data $\boldsymbol{X}$ is scaled to have mean 0 and unit variance), then the correlation between s and h is given by

$$\rho = \frac{1}{n}\sum_{i=1}^{n} \xi'\boldsymbol{R}^{-1}\boldsymbol{x}_i\boldsymbol{x}_i'\boldsymbol{R}^{-1}\boldsymbol{\eta} = \xi'\boldsymbol{R}^{-1}\boldsymbol{\eta}.$$

A shape factor corrected for size can be obtained by considering

$$h' = h - \rho s.$$

The values of s and h' are uncorrelated and can be plotted in a scatterplot.

An alternative approach to allometry based on principal components has been suggested by Jolicoeur and Mosimann (1960). If all the coefficients of the first principal component are positive, then it can be used to measure size. In this case, the other principal components will have positive and negative coefficients, since they are orthogonal to the first principal component. Therefore, second and subsequent principal components may be taken as measures of shape. (Note that if all the elements of the covariance matrix are positive, then the Perron–Frobenius Theorem ensures that all the coefficients of the first principal component are positive – see Exercise 9.6.1.)

One advantage of principal components is that the principal component loadings have a dual interpretation in Rao's framework. First, if $\boldsymbol{g}_{(1)}$ is the first eigenvector of $\boldsymbol{R}$, that is $\boldsymbol{g}_{(1)}$ is the loading vector of the first principal component, then $\xi = \boldsymbol{g}_{(1)}$ can play the role of a size *vector*. Moreover, since $\boldsymbol{a} = \boldsymbol{R}^{-1}\boldsymbol{g}_{(1)} \propto \boldsymbol{g}_{(1)}$, $\boldsymbol{g}_{(1)}$ can also be interpreted as the coefficient vector of a size *factor* $s = \boldsymbol{g}_{(1)}'\boldsymbol{x}$. The other eigenvectors can be used to measure shape, and the sizes of the eigenvalues of $\boldsymbol{R}$ give some indication of the relative magnitude of the various size and shape variations within the data.

Unfortunately, the interpretation of principal components as size and shape vectors is not always straightforward. One's notion of size and shape is usually given in terms of simply defined ξ and $\boldsymbol{\eta}$ (usually vectors of $\pm$1s and 0s), but the principal component loadings can behave too arbitrarily to represent size and shape in a consistent way (Rao, 1964).

Example 9.6.1 In the analysis of 54 apple trees, Pearce (1965) considers the following three variables: total length of lateral shoots, circumference of the trunk, and height. Although the variables can all be measured in the same units, it is sensible in this case to use standardized variables because the variances are clearly of different orders of magnitude.

The correlation matrix is

$$\begin{bmatrix} 1 & 0.5792 & 0.2414 \\ & 1 & 0.5816 \\ & & 1 \end{bmatrix},$$

giving the principal components

$$\begin{aligned} y_1 &= 0.534x_1 + 0.654x_2 + 0.536x_3, \\ y_2 &= -0.709x_1 + 0.001x_2 + 0.705x_3, \\ y_3 &= -0.460x_1 + 0.757x_2 - 0.464x_3, \end{aligned}$$

with the corresponding variances 1.95, 0.76, and 0.29 (65%, 25%, and 10% of the total, respectively). Note that the first principal component is effectively the sum of the three variables, the second principal component represents a contrast between x_1 and x_3, and the third is a contrast between x_2 on the one hand and x_1 and x_3 on the other.

If we use a size vector $\xi = (1, 1, 1)'$, then we get a size factor

$$s = 0.739x_1 + 0.142x_2 + 0.739x_3,$$

which is similar to the first principal component y_1, although the second coefficient is perhaps low. Similarly, a shape vector $\eta_1 = (1, 0, -1)'$ contrasting x_1 with x_3 leads to a shape factor

$$h_1 = 1.32x_1 + 0.007x_2 - 1.32x_3,$$

which is similar to the second principal component. However, if we use a shape vector $\eta_2 = (1, -1, 0)'$ to study the contrast between x_1 and x_2, we obtain a shape factor

$$h_2 = 2.56x_1 - 3.21x_2 + 1.25x_3,$$

which bears no resemblance to any of the principal components. □

Example 9.6.2 Various body circumference measurements and an estimate of body fat determined by underwater weighing were obtained for 252 men (Penrose et al., 1985). In this example, we use a subset of the data with the following circumference measurements (all measured in cm) of:

(1)	Neck	(2)	Chest
(3)	Abdomen	(4)	Hip
(5)	Thigh	(6)	Knee
(7)	Ankle	(8)	Biceps (extended)
(9)	Forearm	(10)	Wrist

Although the measurements are in the same units, we again use the correlation matrix to standardize the variances. The correlation matrix is

$$\begin{bmatrix} 1.00 & 0.79 & 0.76 & 0.75 & 0.71 & 0.68 & 0.48 & 0.73 & 0.62 & 0.74 \\ & 1.00 & 0.92 & 0.83 & 0.73 & 0.72 & 0.48 & 0.73 & 0.58 & 0.66 \\ & & 1.00 & 0.87 & 0.77 & 0.74 & 0.45 & 0.68 & 0.50 & 0.63 \\ & & & 1.00 & 0.89 & 0.82 & 0.56 & 0.74 & 0.55 & 0.64 \\ & & & & 1.00 & 0.80 & 0.54 & 0.77 & 0.57 & 0.57 \\ & & & & & 1.00 & 0.61 & 0.68 & 0.56 & 0.67 \\ & & & & & & 1.00 & 0.48 & 0.42 & 0.57 \\ & & & & & & & 1.00 & 0.68 & 0.63 \\ & & & & & & & & 1.00 & 0.59 \\ & & & & & & & & & 1.00 \end{bmatrix},$$

giving eigenvalues 7.05, 0.72, 0.66, 0.48, 0.30, 0.28, 0.20, 0.17, 0.08, and 0.06. The first three eigenvectors,

$$\begin{aligned} y_1 &= 0.33x_1 + 0.34x_2 + 0.33x_3 + 0.35x_4 + 0.33x_5 \\ &\quad + 0.33x_6 + 0.25x_7 + 0.32x_8 + 0.27x_9 + 0.30x_{10}, \\ y_2 &= -0.06x_1 - 0.29x_2 - 0.39x_3 - 0.21x_4 - 0.15x_5 \\ &\quad + 0.07x_6 + 0.70x_7 - 0.01x_8 + 0.28x_9 + 0.33x_{10}, \\ y_3 &= 0.23x_1 + 0.00x_2 - 0.14x_3 - 0.23x_4 - 0.18x_5 \\ &\quad - 0.25x_6 - 0.48x_7 + 0.27x_8 + 0.67x_9 + 0.16x_{10}, \end{aligned}$$

account for 84.3% of the variability. The first eigenvector is effectively the average of the 10 variables, since all the coefficients are very similar. The second and third eigenvectors may be taken as measures of shape. The first shape variable is a contrast with approximate weights $(0, -1, -1, -1, -1, 0, 2, 0, 1, 1)$, while the second shape variable has approximate weights $(1, 0, 0, -1, -1, -1, -1, 1, 2, 0)$. In this example, the size factor corresponding to the vector $\xi = (1, 1, \ldots, 1)'$ is

$$\begin{aligned} s = {} & 0.03x_1 - 0.17x_2 + 0.65x_3 - 0.39x_4 + 0.24x_5 - 0.01x_6 \\ & +0.53x_7 + 0.09x_8 + 0.44x_9 + 0.18x_{10}, \end{aligned}$$

which does not seem to correspond to any of the main principal components. □

9.7 Discarding of Variables

In Section 7.7, we gave a method of discarding redundant variable based on regression analysis. We now give a method based on PCA.

Suppose that a PCA is performed on the correlation matrix of all the p variables. Initially, let us assume that k variables are to be retained, where k is known, and $(p - k)$ variables are to be discarded. Consider the eigenvector corresponding to the smallest eigenvalue and reject the variable with the largest coefficient (absolute value). Then, the next smallest eigenvalue is considered. This process continues until the $(p - k)$ smallest eigenvalues have been considered.

In general, we discard the variable

(a) which has the largest coefficient (absolute value) in the component and
(b) which has not previously been discarded.

This principle is consistent with the notion that we regard a component with small eigenvalue as of less importance and, consequently, the variable that dominates it should be of less importance if redundant. However, the choice of k can be made more realistic as follows. Let λ_0 be a threshold value so that eigenvalues $\lambda \leq \lambda_0$ can be regarded as contributing too little to the data. The method in essence is due to Beale et al. (1967). Jolliffe (1972) recommends $\lambda_0 = 0.70$ from various numerical examples.

Example 9.7.1 Consider Jeffers (1967)' pine pitprop data of Example 7.7.1. From the principal component analysis on the correlation matrix (see Example 9.2.6), it can be seen that the first six components account for 87% of the variability. The other seven eigenvalues are below the threshold value of $\lambda_0 = 0.70$, so seven variables are to be rejected. The variable with the largest coefficient of $\mathbf{g}_{(13)}$, the eigenvector corresponding to the smallest eigenvalue, is the second, and so x_2 is rejected. Similarly, the variables with largest coefficients in $\mathbf{g}_{(12)}, \ldots, \mathbf{g}_{(7)}$ are $x_4, x_7, x_{10}, x_6, x_9, x_{12}$, respectively, and as these are all distinct, these are successively rejected.

An alternative method of discarding variables based on the multiple correlation coefficient is given in Example 7.7.2, and the two methods give different sets of retained variables. For example, to retain six variables, the above principal component method chooses variables 1, 3, 5, 8, 11, 13, whereas from Table 7.4, the multiple correlation method chooses variables 3, 4, 10, 11, 12, 13. Variables 3, 11, and 13 are retained by both methods. □

Remarks (1) Instead of using λ_0, k may be set equal to the number of components needed to account for more than some proportion, α_0, of the total variation. Jolliffe (1972) found in his examples that λ_0 appeared to be better than α_0 as a criterion for deciding how many variables to reject. He finds $\alpha_0 = 0.80$ as the best value if it is to be used. Further, using either criterion, at least four variables should always be retained.

(2) A variant of this method of discarding variables is given by the following iterative procedure. At each stage of the procedure reject the variable associated with the smallest eigenvalue and reperform the PCA on the remaining variables. Carry on until all the eigenvalues are high.

(3) The relation between the "smallness" of $\sum l_i x_i$ and the rejection of one particular variable is rather open to criticism. However, present studies lead us to the belief that various methods of rejecting variables in order to explain the variation within $\boldsymbol{X}$ produce in practice almost the same results (Jolliffe, 1972, 1973). Of course, the method selected to deal with one's data depends on the purpose of the study, as the following section makes clear.

9.8 Principal Component Regression

In the multiple regression model of Chapter 7 it was noted that if the independent variables are highly dependent, then the estimates of the regression coefficients will be very imprecise. In this situation it is advantageous to disregard some of the variables in order to increase the stability of the regression coefficients.

An alternative way to reduce dimensionality is to use principal components (Massy, 1965). However, the choice of components in the regression context is somewhat different from the choice in Section 9.2.5. In that section the principal components with the *largest variances* are used in order to explain as much of the total variation of $\mathbf{X}$ as possible. On the other hand, in the context of multiple regression, it is sensible to take those components having the *largest correlations* with the dependent variable because the purpose in a regression is to explain the dependent variable. In a *multivariate* regression, the correlations with each of the dependent variables must be examined. Fortunately, there is often a tendency in data for the components with the largest variances to best explain the dependent variables.

If the principal components have a natural intuitive meaning, it is perhaps best to leave the regression equation expressed in terms of the components. In other cases, it is more convenient to transform back to the original variables.

More specifically, if n observations are made on a dependent variable and on p independent variables ($\mathbf{y}(n \times 1)$ and $\mathbf{X}(n \times p)$), and if the observations are centered so that $\overline{y} = \overline{x}_i = 0, i = 1, \ldots, p$, then the regression equation is

$$\mathbf{y} = \mathbf{X}\boldsymbol{\beta} + \boldsymbol{\epsilon}, \qquad \text{where} \qquad \boldsymbol{\epsilon} \sim N_n(\mathbf{0}, \sigma^2\mathbf{H}) \qquad \text{and} \qquad \mathbf{H} = \mathbf{I} - n^{-1}\mathbf{1}\mathbf{1}'. \tag{9.39}$$

If $\mathbf{W} = \mathbf{X}\mathbf{G}$ denotes the principal component transformation, then the regression equation may be written as

$$\mathbf{y} = \mathbf{W}\boldsymbol{\alpha} + \boldsymbol{\epsilon}, \tag{9.40}$$

where $\boldsymbol{\alpha} = \mathbf{G}'\boldsymbol{\beta}$. Since the columns of $\mathbf{W}$ are orthogonal, the least squares estimators $\hat{\alpha}_i$ are unaltered if some of the columns of $\mathbf{W}$ are deleted from the regression. Further, their distribution is also unaltered. The full vector of least squares estimators is given by

$$\hat{\boldsymbol{\alpha}} = (\mathbf{W}'\mathbf{W})^{-1}\mathbf{W}'\mathbf{y} \qquad \text{and} \qquad \hat{\boldsymbol{\epsilon}} = \mathbf{y} - \mathbf{W}\hat{\boldsymbol{\alpha}}$$

or

$$\hat{\alpha}_i = n^{-1}l_i^{-1}\mathbf{w}'_{(i)}\mathbf{y}, \qquad i = 1, \ldots, p, \tag{9.41}$$

where l_i is the ith eigenvalue of the covariance matrix, $n^{-1}\mathbf{X}'\mathbf{X}$. Then, $\hat{\alpha}_i$ has expectation α_i and variance $\sigma^2/(nl_i)$. The covariance between $\mathbf{w}_i$ and $\mathbf{y}$,

$$\text{cov}(\mathbf{w}_i, \mathbf{y}) = n^{-1}\mathbf{w}'_{(i)}\mathbf{y} = n^{-1}\mathbf{g}'_{(i)}\mathbf{X}'\mathbf{y}, \tag{9.42}$$

can be used to test whether or not the contribution of $\mathbf{w}_i$ to the regression is significant. Under the null hypothesis, $\alpha_i = 0$,

$$\frac{\hat{\alpha}_i(nl_i)^{1/2}}{(\hat{\boldsymbol{\epsilon}}'\hat{\boldsymbol{\epsilon}}/(n-p-1))^{1/2}} = \frac{\mathbf{w}'_{(i)}\mathbf{y}}{(l_i n\hat{\boldsymbol{\epsilon}}'\hat{\boldsymbol{\epsilon}}/(n-p-1))^{1/2}} \sim t_{n-p-1}, \tag{9.43}$$

regardless of the true values of the other α_j.

Because the components are orthogonal to one another and orthogonal to the residual vector, we can write

$$\begin{aligned} \mathbf{y}'\mathbf{y} &= \sum_{i=1}^{p} \hat{\alpha}_i^2 \mathbf{w}'_{(i)}\mathbf{w}_{(i)} + \hat{\boldsymbol{\epsilon}}'\hat{\boldsymbol{\epsilon}} \\ &= \sum_{i=1}^{p} nl_i\hat{\alpha}_i^2 + \hat{\boldsymbol{\epsilon}}'\hat{\boldsymbol{\epsilon}}. \end{aligned} \tag{9.44}$$

Thus, $nl_i\hat{\alpha}_i^2/\mathbf{y}'\mathbf{y} = \text{cor}^2(\mathbf{y}, \mathbf{w}_{(i)})$ represents the proportion of variance of $\mathbf{y}$ accounted for by component i.

Selecting those components for which the statistic (9.43) is significant is a straightforward way to choose which components to retain. Procedures that are more theoretically justifiable have also been developed (Hill et al., 1977).

Example 9.8.1 Consider the Jeffers (1967) pitprop data of Examples 7.7.1, 7.7.2, and 9.2.6. The proportions of the variance of $\mathbf{y}$ accounted for by each of the components and by the residual vector (times 100%) are given by 7.56, 33.27, 3.05, 0.26, 13.70, 6.13, 3.24, 1.28, 0.03, 0.00, 0.07, 4.56, 0.00, and 26.84, and $\hat{\boldsymbol{\alpha}}$ is given by

$$\begin{aligned}\hat{\boldsymbol{\alpha}} = {} & (0.134^{**}, -0.374^{**}, 0.128^{**}, -0.048, -0.388^{**}, 0.274^{**}, -0.237^{**},\\ & -0.170^{**}, 0.031, 0.000, -0.115, -1.054^{**}, 0.002).\end{aligned}$$

Using the distributional result (9.43), we can check from Appendix D, Table D.2, whether each $\hat{\alpha}_i$ is significantly different from zero, and $*$ $(**)$ indicates significance at the 5% (1%) level.

By a procedure similar to Example 9.2.6, Jeffers decided to retain six components in his analysis. Of these components, 1, 2, 3, 5, and 6 had significant regression coefficients, so Jeffers carried out his regression on these five components.

However, from the above discussion, a more valid analysis would be obtained by including components 7, 8, and 12 as well. Letting $J = \{1, 2, 3, 5, 6, 7, 8, 12\}$ denote the set of retained components, the restricted least squares estimator $\hat{\boldsymbol{\alpha}}$ for $\boldsymbol{\alpha}$ is given by $\tilde{\alpha}_j = \hat{\alpha}_j$ for $j \in J$ and $\tilde{\alpha}_j = 0$ for $j \notin J$. Then, $\tilde{\boldsymbol{\beta}} = \mathbf{G}\tilde{\boldsymbol{\alpha}}$. If $\alpha_j = 0$ for $j \notin J$, then the variance of each $\tilde{\beta}_i$ is given by

$$\text{var}(\tilde{\beta}_i) = \sum_{j \in J} g_{ij}^2 \sigma^2 / nl_j.$$

Notice that $l_{12} = 0.04$ is very small compared to the other retained eigenvalues. Thus, for several variables (namely 1, 2, 3, 4, and 5), $\text{var}(\tilde{\beta}_i)$ is drastically increased by including component 12. (However, these high variances are partly balanced by the resulting large correlations between these $\tilde{\beta}_i$.) Hence, for these data, the regression is more easily interpreted in terms of the principal components than in terms of the original variables.

The problem of redundant information in a multiple regression can also be approached by discarding variables, and Example 7.7.1 describes this technique applied to Jeffers' data. Note that the regression based on the eight most significant principal components explains 72.8% of the variance of y, which is slightly better than the 72.5% obtained from Table 7.3, using the optimal choice for eight retained variables. However, for fewer than eight retained variables or components, the discarding variables approach explains a higher percentage of y than the principal components approach.

In general, it is perhaps preferable to use the discarding variables approach because it is conceptually simpler. □

Another technique for dealing with redundant information in a regression analysis is the method of ridge regression (Hoerl and Kennard, 1970). See Exercises 9.8.1 and 11.2.14 and Section 16.2.

9.9 Projection Pursuit and Independent Component Analysis

9.9.1 Projection Pursuit

While PCA finds SLCs that have maximal variance and are uncorrelated with one another, *projection pursuit* techniques (Friedman and Tukey, 1974) look for "interesting" projections of the data by selecting linear combinations to maximize some *projection index*. Let the projection index be determined by some objective function Q. Then, a projection of the data matrix $\boldsymbol{X}$ given by $\boldsymbol{Xa}$ has objective function $Q(\boldsymbol{Xa})$. If Q is the variance, i.e.

$$Q(\boldsymbol{Xa}) = \|(\boldsymbol{X} - \mathbf{1}\overline{\mathbf{x}}^T)\boldsymbol{a}\|^2,$$

then we know that the maximum value of Q, subject to $\sum a_i^2 = 1$, is given by the standardized eigenvector corresponding to the largest eigenvalue of $\boldsymbol{S}$. So, we can see that PCA is a special case of projection pursuit.

However, a projection of the data is generally deemed interesting when it is non-Gaussian, which makes the above example for Q rarely appropriate. A natural starting point is to assume that Q satisfies the requirements

- Q is affine invariant, and
- if X and Y are independent random variables with finite variance, then Q satisfies $Q(X + Y) \leq \max(Q(X), Q(Y))$.

An example given by Huber (1985) is to take Q as a function of standardized absolute cumulants

$$Q(X) = |c_m(X)|/c_2(X)^{m/2}, \qquad m > 2,$$

so, with $m = 3$, the index is based on absolute skewness, and with $m = 4$, the index is based on kurtosis. Jones and Sibson (1987) considered a combination of measures based on skewness and kurtosis, leading to

$$Q(X) = \frac{1}{12}\left\{\frac{c_3(X)^2}{c_2(X)^3} + \frac{c_4(X)^2}{4c_2(X)^4}\right\}$$

as well as a bivariate moment index – for projection onto a plane – given by

$$\{\kappa_{30}^2 + 3\kappa_{21}^2 + 3\kappa_{12}^2 + \kappa_{03}^2 + (\kappa_{40}^2 + 4\kappa_{31}^2 + 6\kappa_{2}^2 + 4\kappa_{13}^2 + \kappa_{04}^2)/4\}/12,$$

where κ_{rs} is the standardized bivariate cumulant of order (r, s). This index is the same as Mardia's omnibus skewness and kurtosis measure τ (Mardia, 1987) for $p = 2$ given in (6.38).

Example 9.9.1 We illustrate projection pursuit by considering two of the variables (sepal width and petal length) from the iris data set, ignoring the species labels. We first center each variable and scale, so that the standard deviation is 1. Then, we consider three options for Q: minimum kurtosis, maximum kurtosis, and maximum variance (the first principal component). In each case, we seek a vector $\boldsymbol{a}$ such that $Q(\boldsymbol{Xa})$ is maximized for the particular choice of Q. We can visualize the transformed data using a kernel density estimate (see Exercise 12.6.2). The results are shown in Figure 9.8 where we show the centered data with the vectors $\boldsymbol{a}$ and the density estimate of $\boldsymbol{Xa}$. The maximum kurtosis is 3.57, and the minimum is 1.60. Given that the kurtosis of the standard normal is 3, we might expect the latter projection to be more interesting. Both of these can be compared to the projection given by the first principal component – also shown in Figure 9.8, in which we have used the `ica` library (Helwig, 2018) in R. □

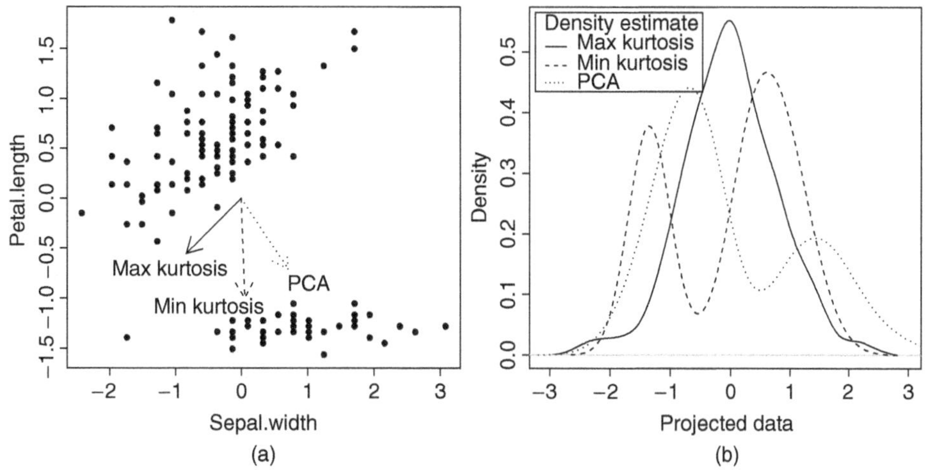

Figure 9.8 (a) Scatterplot of two of the iris data variables, with normalized direction vectors showing the projections for each choice of projection index (maximum kurtosis, minimum kurtosis, and PCA); (b) kernel density estimates of the projected data for each choice of projection index Q.

An obvious drawback of the projection pursuit is that it is computationally demanding as many choices of Q cannot be maximized analytically. Moreover, measures that are based on sample moments can find projections that are greatly influenced by outliers, though robust estimators (see Section 5.3) can be used in this case.

A modification of projection pursuit, called invariant coordinate selection (ICS), was developed by Tyler et al. (2009). It is based on two measures of scatter with different levels of robustness and can be used to explore a wide range of patterns in multivariate data. Some of the strengths and weaknesses when fourth order moments are used for clustering are illustrated in Alashwali and Kent (2016).

9.9.2 Independent Component Analysis

Independent component analysis (ICA) (Hyvärinen et al., 2001) also uses projections that attempt to maximize some measure of non-Gaussianity. However, the goal is now somewhat different. An easy way to describe the motivating application, known as the "cocktail party problem," is to consider two conversations taking place on one side of a room and two microphones placed on the other side of the room. Each microphone gives a time series recording of the sound, and the two recordings are different linear combinations of the two conversations. The goal – sometimes known as *blind source separation* – is to use these two observed times series to reconstruct both conversations. In this example, our usual notation gives $p = 2$ (microphones), and n is the length of the time series.

We can write this mixing model, in more general notation, as

$$\boldsymbol{x} = \boldsymbol{\mu} + \boldsymbol{A}\boldsymbol{s}, \tag{9.45}$$

where $\boldsymbol{x}$, $\boldsymbol{\mu}$ and $\boldsymbol{s}$ are p-dimensional vectors, and where we have now dropped the dependence on time. Here, $\boldsymbol{A}$ is an unknown $p \times p$ matrix, with columns (called the ICA basis vectors) given by $\boldsymbol{a}_{(j)}$, $j = 1, \ldots, p$, and $\boldsymbol{x}$ is observed. Also, $\boldsymbol{s}$ is to be found. Without further assumptions there would be many solutions to this problem since all three of

$\boldsymbol{A}$, $\boldsymbol{\mu}$ and $\boldsymbol{s}$ are unknown. However, the problem can be approached by assuming that the random variables $s_i, i = 1, \dots, p$, are independent if at most one of these is normally distributed. Then – subject to the ambiguities described below – $\boldsymbol{A}$ is identifiable.

Note that the order of components cannot be determined (so in the above application, we cannot determine which conversation is which). This is obvious, since if $\boldsymbol{P}$ is a permutation matrix, we can easily replace $\boldsymbol{A}$ by $\boldsymbol{AP}^{-1}$ and $\boldsymbol{s}$ by $\boldsymbol{Ps}$, so that we still have $\boldsymbol{x} = \boldsymbol{AP}^{-1}\boldsymbol{Ps}$. Secondly, we note the arbitrary scaling which can be applied to $\boldsymbol{s}$, with an inverse scaling applied to $\boldsymbol{A}$.

To reduce nonidentifiability, it is useful to do some preprocessing of $\boldsymbol{x}$. Specifically, the data are mean centered and "whitened," so that the transformed data has zero mean, unit variance, and uncorrelated components. One way to achieve this is to use an eigenvalue decomposition of the sample covariance matrix $\boldsymbol{S}$, similar to the description in Section 9.2.2.

Given n observations of $\boldsymbol{x}$, we can form an $n \times p$ data matrix $\boldsymbol{X}$. Now, define $\boldsymbol{L}$ and an orthogonal matrix $\boldsymbol{G}$ such that $\boldsymbol{S} = \boldsymbol{GLG}'$, with $\boldsymbol{L} = \text{diag}(l_1, l_2, \dots, l_p)$ the diagonal matrix of the eigenvalues of $\boldsymbol{S}$. Then, the whitened transform of $\boldsymbol{X}$ can be obtained using

$$\boldsymbol{Y} = (\boldsymbol{X} - \boldsymbol{1}\bar{\boldsymbol{x}}')\boldsymbol{M}, \tag{9.46}$$

where $\boldsymbol{M} = \boldsymbol{S}^{-1/2} = \boldsymbol{GL}^{-1/2}\boldsymbol{G}' = \boldsymbol{M}'$. Combining Eqs. (9.45) and (9.46), we have a new relationship for the transformed data:

$$\boldsymbol{y} = \boldsymbol{MAs}.$$

So, given observations $\boldsymbol{x}_1, \dots, \boldsymbol{x}_n$, we first obtain $\boldsymbol{y}_1, \dots, \boldsymbol{y}_n$, and then, we now need to find $\boldsymbol{s}_1, \dots, \boldsymbol{s}_n$, scaled so that $\mathsf{E}(\boldsymbol{ss}') = \boldsymbol{I}$, and the matrix, say $\boldsymbol{B} = \boldsymbol{MA}$, such that $\boldsymbol{y} = \boldsymbol{Bs}$. The problem is now simplified because $\boldsymbol{B}$ is orthogonal since, by construction, we have

$$\boldsymbol{I} = \mathsf{E}(\boldsymbol{yy}') = \boldsymbol{B}\,\mathsf{E}(\boldsymbol{ss}')\boldsymbol{B}' = \boldsymbol{BB}'.$$

Since $\boldsymbol{B}$ is orthogonal, we have $\boldsymbol{B}^{-1} = \boldsymbol{B}'$, so we can write $\boldsymbol{s} = \boldsymbol{B}'\boldsymbol{y}$. The columns of $\boldsymbol{B}$, say $\boldsymbol{b}_{(1)}, \dots, \boldsymbol{b}_{(p)}$, are called the transformed ICA basis vectors, with the ith component of $\boldsymbol{s}$ given by $s_i = \boldsymbol{b}'_{(i)}\boldsymbol{y}, i = 1, \dots, p$.

The goal of projection pursuit was to look for interesting projections, whereas the goal here is to use a model to obtain independent components. However, in both cases, nonnormality is the key concept. In ICA, this is motivated as follows. Suppose that the components are to be obtained in a sequential manner (it is also possible to consider parallel methods), and so, we wish to find a vector, say $\boldsymbol{w}$, so that the first component is given by the random variable $u = \boldsymbol{w}'\boldsymbol{y}$. Then, if we let $\boldsymbol{z} = \boldsymbol{B}'\boldsymbol{w}$, we can also express u in the form $u = \boldsymbol{w}'\boldsymbol{y} = \boldsymbol{w}'\boldsymbol{Bs} = \boldsymbol{z}'\boldsymbol{s}$, and we see that u is also a linear combination of the unknown $\boldsymbol{s}$. Now, by the central limit theorem, a linear combination of independent and identically distributed (i.i.d.) random variables is closer (using a measure similar to Q in Section 9.9.1) to a normal distribution than just one of the random variables, with the *least* normality obtained when only one of the elements of $\boldsymbol{z}$ is nonzero. This means that our goal can be achieved by finding the first column of $\boldsymbol{B}$ which maximizes the non-Gaussianity of linear combinations of $\boldsymbol{y}$.

In practice, the use of ICA requires two ingredients: a choice of *measure* (contrast function) of nonnormality and an algorithm to *optimize* this contrast function. Choices of contrast function include kurtosis, negentropy, and mutual information. A simple approximation to the negentropy in terms of a nonlinear function G can be defined by the contrast function

$$Q(X) = [\mathsf{E}\{G(X)\} - \mathsf{E}\{G(X_0)\}]^2,$$

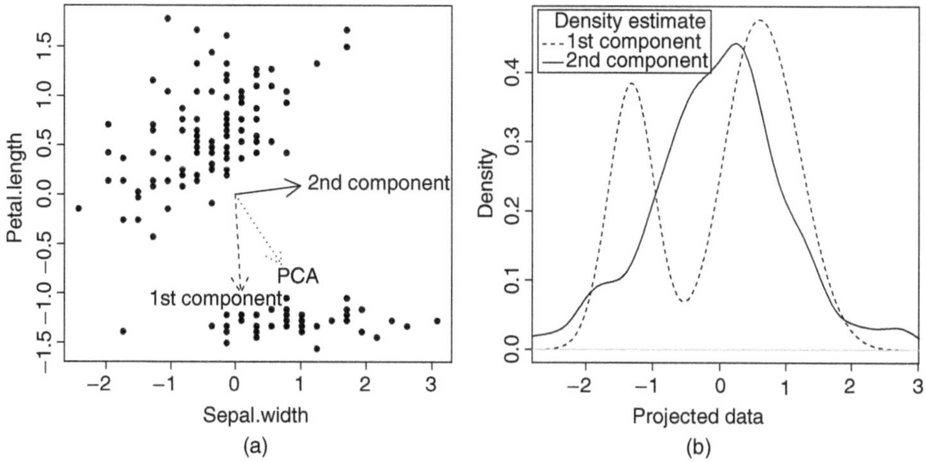

Figure 9.9 (a) Scatterplot of two of the iris data variables, with direction vectors showing the projections for PCA (to help comparison with Figure 9.8) and the two components found by FastICA; (b) kernel density estimates for each of the two components.

where X follows a distribution with mean 0 and variance 1, and X_0 follows an $N(0, 1)$ distribution. Popular choices for G include $G_1(u) = (1/a_1) \log\cosh(a_1\, u)$, where $a_1 > 0$ is a tuning parameter, and $G_2(u) = -\exp(-\tfrac{1}{2}u^2)$. Using these contrast functions, an efficient algorithm can be developed called FastICA, which can be used to obtain the components sequentially, by decorrelating the data at each stage; see Hyvärinen (1999) and Hyvärinen and Oja (2000) for more details.

Example 9.9.2 We use the same data as in Example 9.9.1 in order to look at the small differences between ICA and projection pursuit. In this example, we use the function icafast which is included in the R library `ica` (Helwig, 2018). Here, both of the components are obtained in parallel and the contrast function used is the log of hyperbolic cosine to approximate the negentropy. It can be seen that the first component is nearly identical to the projection that minimized the kurtosis, and the second component (which has kurtosis equal to 3.53) is somewhat different to the direction that maximizes the kurtosis. Similar types of plot are given in Figure 9.9, in which we have again shown the first principal component direction to help with the comparison. □

9.10 PCA in High Dimensions

In Section 9.3.2 it has been assumed that p is fixed as $n \to \infty$. The asymptotic results have a very different character if the dimension p is allowed to get large. Following Jung and Marron (2009), it is convenient to highlight three asymptotic settings:

(a) Low dimension, high sample size (LDHSS) p fixed, $n \to \infty$. This is the classic setting in multivariate analysis for asymptotic theory.
(b) High dimension, high sample size (HDHSS) $n, p \to \infty$, $n/p \to \delta$, $\delta \geq 1$. Thus, the number of observations and the sample size increase at the same rate.

(c) High dimension, low sample size (HDLSS) $p \to \infty$, n fixed. This setting is particularly suited to functional data, where an "observation" is a curve observed at a large number p of time points, but only a limited number of curves are available.

The essential problem for PCA in high dimensions is that the estimated eigenvalues and eigenvectors in principal components get noisier as p increases. In this section, we illustrate some of the surprising results that can arise. Let $\boldsymbol{X}$ be an $n \times p$ data matrix from an $N_p(\boldsymbol{\mu}, \boldsymbol{\Sigma})$ distribution, where $\boldsymbol{\Sigma}$ has eigenvalues $\lambda_1 \geq \cdots \geq \lambda_p > 0$, and let $\boldsymbol{S}$ denote the sample covariance matrix with eigenvalues $l_1 > \cdots > l_p$.

To begin with, consider the HDHSS setting with $\boldsymbol{\Sigma} = \boldsymbol{I}$, so that $p = n \to \infty$. In this case, $\boldsymbol{\Sigma}$ has just one eigenvalue, $\lambda = 1$, repeated p times. However, the sample eigenvalues of $\boldsymbol{S}$ are asymptotically spread out more widely. In particular, their empirical distribution function tends almost surely to a limiting cumulative distribution function $G(t)$, say. When $\delta = 1$, $G(\cdot)$ has p.d.f.

$$g(t) = \frac{1}{2\pi}\sqrt{(4-t)/t}, \quad 0 < t < 4,$$

and there is a similar form for $\delta > 1$. This celebrated result, due to Marchenko and Pastur (1967), can be viewed as a functional version of the law of large numbers. It is particularly surprising that the asymptotic support is the *finite* interval $(0, 4)$. In particular, the largest sample eigenvalue satisfies $l_1 \to 4$, so that l_1 is wildly inconsistent for the population eigenvalue $\lambda = 1$.

A more refined result gives the asymptotic distribution of l_1. Johnstone (2001), building on Tracy and Widom (1996), showed that when $\delta = 1$, $n^{1/3}(l_1 - 4)$ tends to a nondegenerate limiting distribution which can be described explicitly. This result is analogous to the central limit theorem, but note the smaller power $n^{1/3}$ in comparison to the power $n^{1/2}$ appearing in the central limit theorem. There is a similar result for $\delta > 1$.

The choice $\boldsymbol{\Sigma} = \boldsymbol{I}$ is essentially a "null" model with no structure in the covariance matrix. A more useful model in many applications is the *spiked covariance model* introduced by Johnstone (2001). Under the spiked covariance model, the first k (k fixed) eigenvalues are substantially greater than the remaining eigenvalues, which are assumed to be equal to one another. This model for fixed dimension p was considered earlier in the isotropy test of Section 9.4.2.

A simple version of the spiked model with $k = 1$ is given by

$$\lambda_1 > \lambda_2 = \cdots = \lambda_p = \lambda_0, \quad \text{say},$$

where $\lambda_1/\lambda_0 = cp^{\alpha}$ as $p \to \infty$ for some $c > 0$ and $\alpha > 0$. The parameter α measures the degree of spikiness. For larger values of α, the model is more strongly spiked, so that the "signal" stands out more easily from the "noise" as the dimension increases. Here, the signal is given by the first eigenvector $\boldsymbol{\gamma}_{(1)}$ of $\boldsymbol{\Sigma}$ and is estimated by the first sample eigenvector $\hat{\boldsymbol{\gamma}}_{(1)}$ of $\boldsymbol{S}$. Note that $(\hat{\boldsymbol{\gamma}}'\boldsymbol{\gamma})^2 = 1$ if the two vectors are equal to each other (up to sign), and $(\hat{\boldsymbol{\gamma}}'_{(1)}\boldsymbol{\gamma}_{(1)})^2 = 0$ if they are orthogonal to each other. Jung and Marron (2009) show the following result in the HDLSS setting, so that n is fixed as p increases. The results for the strongly spiked case ($\alpha > 1$) are strikingly different from the mildly spiked case ($0 < \alpha < 1$):

(a) If $\alpha > 1$, then $(\hat{\boldsymbol{\gamma}}'_{(1)}\boldsymbol{\gamma}_{(1)})^2 \to 1$ almost surely, so that $\hat{\boldsymbol{\gamma}}_{(1)}$ is consistent for $\boldsymbol{\gamma}_{(1)}$.
(b) If $0 < \alpha < 1$, then $(\hat{\boldsymbol{\gamma}}'_{(1)}\boldsymbol{\gamma}_{(1)})^2 \to 0$ almost surely, so that $\hat{\boldsymbol{\gamma}}_{(1)}$ is inconsistent for $\boldsymbol{\gamma}_{(1)}$. Indeed, in the limit $\hat{\boldsymbol{\gamma}}_{(1)}$ is as far away as possible from $\boldsymbol{\gamma}_{(1)}$.

Jung and Marron (2009) also give various extensions as the assumptions are relaxed.

The take-home message is that caution is needed in high dimensions when using asymptotic theory in PCA. Results for low-dimensional p do not generally carry over to high dimensions. However, all is not lost. As Jung and Marron (2009) show, if the signal stands out strongly enough from the noise, it is possible to estimate it consistently, even in high dimensions. See also Chapter 16 for some powerful and successful methods of regression analysis in high dimensions using regularization.

Exercises and Complements

9.2.1 For

$$\boldsymbol{\Sigma} = \begin{bmatrix} \sigma_1^2 & \rho\sigma_1\sigma_2 \\ \rho\sigma_1\sigma_2 & \sigma_2^2 \end{bmatrix}, \qquad \rho \neq 0,$$

show that the eigenvalues are

$$\frac{1}{2}(\sigma_1^2 + \sigma_2^2) \pm \frac{1}{2}\Delta, \qquad \text{where} \qquad \Delta = \{(\sigma_1^2 - \sigma_2^2)^2 + 4\sigma_1^2\sigma_2^2\rho^2\}^{1/2},$$

with eigenvectors proportional to $(2\rho\sigma_1\sigma_2, \sigma_2^2 - \sigma_1^2 + \Delta)$ and $(\sigma_2^2 - \sigma_1^2 + \Delta, -2\rho\sigma_1\sigma_2)$. Compare your results with those of Example 9.2.1 in the case that $\sigma_1 = \sigma_2 = 1$.

9.2.2 Find the eigenvalues and eigenvectors of

$$\boldsymbol{\Sigma} = \begin{bmatrix} \beta^2 + \delta & \beta & \beta \\ \beta & 1 + \delta & 1 \\ \beta & 1 & 1 + \delta \end{bmatrix}.$$

(Hint: $\boldsymbol{\Sigma} - \delta\boldsymbol{I} = (\beta, 1, 1)'(\beta, 1, 1)$ is of rank 1 with nonzero eigenvalue $\beta^2 + 2$ and eigenvector $(\beta, 1, 1)'$. Therefore, $\boldsymbol{\Sigma}$ has eigenvalues $\beta^2 + \delta + 2$, δ, δ, and the same eigenvectors as $\boldsymbol{\Sigma} - \delta\boldsymbol{I}$.)

9.2.3 Suppose that $\boldsymbol{x}' = (x_1, x_2)$ has a bivariate multinomial distribution with $n = 1$, so that $x_1 = 1$ with probability p, and $x_1 = 0$ with probability $q = 1 - p$, and x_2 always satisfies $x_2 = 1 - x_1$.

(a) Show that the covariance matrix of $\boldsymbol{x}$ is

$$pq \begin{bmatrix} 1 & -1 \\ -1 & 1 \end{bmatrix}.$$

(b) Show that the eigenvalues of this matrix are $2pq$ and zero, and that the eigenvalues are $(1, -1)'$ and $(1, 1)'$, respectively.

9.2.4 (Dagnelie, 1975, p. 57) Berce and Wilbaux (1935) collected measurements on five meteorological variables over an 11-year period. Hence, $n = 11$ and $p = 5$. The variables were

x_1 = rainfall in November and December (in millimeters)

x_2 = average July temperature (in degrees Celsius)

x_3 = rainfall in July (in millimeters)

x_4 = radiation in July (in milliliters of alcohol)

x_5 = average harvest yield (in quintals per hectare)

The raw data was as follows:

Year	x_1	x_2	x_3	x_4	x_5
1920–1921	87.9	19.6	1.0	1661	28.37
1921–1922	89.9	15.2	90.1	968	23.77
1922–1923	153.0	19.7	56.6	1353	26.04
1923–1924	132.1	17.0	91.0	1293	25.74
1924–1925	88.8	18.3	93.7	1153	26.68
1925–1926	220.9	17.8	106.9	1286	24.29
1926–1927	117.7	17.8	65.5	1104	28.00
1927–1928	109.0	18.3	41.8	1574	28.37
1928–1929	156.1	17.8	57.4	1222	24.96
1929–1930	181.5	16.8	140.6	902	21.66
1930–1931	181.4	17.0	74.3	1150	24.37

Show that these data lead to the mean vector, covariance matrix, and correlation matrix given below:

$$\overline{\boldsymbol{x}}' = (138.0, 17.75, 74.4, 1242, 25.66),$$

$$\boldsymbol{R}/\boldsymbol{S} = \begin{bmatrix} 1794 & -4.473 & 726.9 & -2218 & -52.01 \\ -0.087 & 1.488 & -26.62 & 197.5 & 1.577 \\ 0.490 & -0.624 & 1224 & -6203 & -56.44 \\ -0.239 & 0.738 & -0.808 & 48104 & 328.9 \\ -0.607 & 0.639 & -0.798 & 0.742 & 4.087 \end{bmatrix}.$$

(This matrix contains correlations below the main diagonal, variances on the main diagonal, and covariances above the main diagonal.)

9.2.5 (Dagnelie, 1975, p. 176) Using the correlation matrix for the meteorological variables $x_1, \ldots, x_4$ in Exercise 9.2.4, verify that the principal components $y_1, \ldots, y_4$ explain 65%, 24%, 7%, and 4%, respectively, of the total variance, and that the eigenvectors (×1000) are, respectively,

$$(291, -506, 577, -571), \qquad (871, 425, 136, 205),$$

$$(-332, 742, 418, -404), \qquad (-214, -111, 688, 685).$$

Following Section 9.2.5, show that the correlations between x_i and y_j are given by the elements of the following matrix (÷1000):

$$\begin{bmatrix} 468 & 862 & -177 & -81 \\ -815 & 420 & 397 & -42 \\ 930 & 135 & 223 & 260 \\ -919 & 202 & -216 & 259 \end{bmatrix}.$$

Hence, calculate the proportion of variance explained by y_3 and y_4 for each of x_1, x_2, x_3, and x_4.

9.2.6 (Dagnelie, 1975, p. 178) Using the covariance matrix for the meteorological variables $x_1, \ldots, x_4$ used in Exercise 9.2.5, show that the eigenvalues are 49023, 1817, 283, and 0.6, and that the eigenvectors ($\times$1000) are, respectively,

$$(-49, 4, -129, 990), \qquad (954, 3, 288, 84),$$

$$(-296, -8, 949, 109), \qquad (-5, 1000, 8, -3).$$

Interpret the principal components and compare them with those based on the correlation matrix found above. Which interpretation is more meaningful?

9.2.7 Exercise 9.2.5 gives the first principal component in terms of the standardized variables $u_i = x_i/s_i$. Express this linear combination in terms of the original variables x_i and standardize the coefficients, so their squares sum to 1. Compare this new linear combination with the first principal component using the covariance matrix (Exercise 9.2.6) to illustrate that principal component analysis is not scale invariant.

9.2.8 If $\boldsymbol{X}$ is a sample data matrix, show that no SLC of the observed variables has a variance which is larger than that of the first principal component. Also, prove results corresponding to the other parts of Theorems 9.2.2–9.2.4.

9.2.9 (a) Extending Example 9.2.3, show that the first four students in Table 1.2 have the following scores on the first two principal components:

	Principal component	
Student	**1**	**2**
1	66.4	6.5
2	63.7	−6.8
3	63.0	3.1
4	44.6	−5.6

(b) Show that the score on the first principal component is zero if and only if

$$0.51x_1 + 0.27x_2 + 0.35x_3 + 0.45x_4 + 0.53x_5 - 99.7 = 0,$$

and the score on the second principal component is zero if and only if

$$0.75x_1 + 0.21x_2 - 0.08x_3 - 0.30x_4 - 0.55x_5 + 1.5 = 0.$$

9.2.10 Suppose that $\boldsymbol{F}$ is a $(p \times k)$ matrix $(k \leq p)$ with orthonormal columns, and let $\lambda_1 \geq \lambda_2 \geq \cdots \geq \lambda_p$. Set

$$h_i = \sum_{j=1}^{k} f_{ij}^2, \qquad i = 1, \ldots, p,$$

and set

$$\phi(\boldsymbol{h}) = \sum_{i=1}^{p} h_i \lambda_i = \sum_{i=1}^{p} \sum_{j=1}^{k} f_{ij}^2 \lambda_i.$$

Show that

$$0 \leq h_1 \leq 1 \qquad \text{and} \qquad \sum_{i=1}^{p} h_i = k.$$

Hence, deduce that $\phi(\boldsymbol{h})$ is maximized when $h_1 = \cdots = h_k = 1$ and $h_{k+1} = \cdots = h_p = 0$, that is when the columns of $\boldsymbol{F}$ span the subspace of the first k rows of $\boldsymbol{F}$.

9.2.11 In Example 9.2.5, the regression with both variables subject to error, let $\boldsymbol{S}$ denote the (2×2) covariance matrix of the data $\boldsymbol{X}$. Using Exercise 9.2.1, show that the first principal component of the covariance matrix of $(\lambda x_1, x_2)$ is given in terms of (x_1, x_2) by the line with slope

$$\hat{\beta} = [s_{22} - \lambda s_{11} + \{(s_{22} - \lambda s_{11})^2 + 4\lambda s_{12}^2\}^{1/2}]/2s_{12}.$$

Show that

$$|s_{12}|/s_{11} \leq |\hat{\beta}| \leq s_{22}/|s_{12}|;$$

that is, the slope lies between the slopes of the regression lines of x_2 on x_1 and x_1 on x_2. Also show that the line goes through the sample mean $(\overline{x}_1, \overline{x}_2)$ if we take $\hat{\alpha} = \overline{x}_2 - \hat{\beta}\overline{x}_1$.

9.2.12 Verify formula (9.13) to show that the correlations between all the variables and a group G of components can be combined to give that part of the total variation accounted for by the components in G.

9.3.1 (Girshick, 1939) The p.d.f. of the eigenvalues of a $W_p(\boldsymbol{I}, m)$ distribution is

$$f(l_1, \ldots, l_p) = c \prod_{i=1}^{p} l_i^{(m-p-1)/2} \prod_{i<j} (l_i - l_j) \exp\left(-\frac{1}{2} \sum_{i=1}^{p} l_i\right).$$

Show that this may also be written as

$$f(l_1, \ldots, l_p) = c g^{p(m-p-1)/2} \exp\left(-\frac{1}{2} a p\right) \prod_{i<j} (l_i - l_j),$$

where a and g are the arithmetic and geometric means of the eigenvalues. Show that the eigenvectors of the Wishart matrix have the Haar invariant distribution on the space of orthogonal matrices.

9.3.2 Show that in the situation of Example 9.3.2, alternative confidence intervals could in principle be obtained from the fact that, asymptotically,

$$\left(\frac{1}{2}(n-1)\right)^{1/2}(l_i - \lambda_i)/\lambda_i \sim N(0,1).$$

Show that this leads to confidence intervals which can include negative values of λ_i, and hence, that the variance-stabilizing logarithmic transform used in Example 9.3.2 is preferable.

9.3.3 If $\hat{\rho}_1$ and $\hat{\rho}_2$ are the two estimates defined in (9.18), show that

$$\hat{\rho}_2 = \hat{\rho}_1 + \{(p - \text{tr}\, \boldsymbol{S})/(p-1)\}.$$

Hence, the estimates are equal if and only if tr $\boldsymbol{S} = p$.

Compare these estimators with the m.l.e. of ρ given in Exercise 5.2.12, where $\boldsymbol{\Sigma}$ is assumed to be of the form $\boldsymbol{\Sigma} = \sigma^2[(1-\rho)\boldsymbol{I} + \rho\boldsymbol{11}']$.

9.4.1 Show that $-2\log\lambda$ given by (9.27) has an asymptotic chi-squared distribution with $(p-k+2)(p-k-1)/2$ degrees of freedom. (Hint: Under H_0, $\boldsymbol{\Sigma}$ is determined by k distinct eigenvalues and one common eigenvalue and the k eigenvectors, each with p components, that correspond to the distinct eigenvalues. This gives $k+1+kp$ parameters. However, the eigenvectors are constrained by $k(k+1)/2$ orthogonality conditions, leaving $k+1+kp-k(k+1)/2$ free parameters. Under H_1, the number of free parameters is obtained by putting $k = p-1$, viz. $p + (p-1)p - (p-1)p/2 = p(p+1)/2$. By subtraction and using Theorem 6.2.3, we find that the number of degrees of freedom is $(p-k+2)(p-k-1)/2$.)

9.4.2 (Dagnelie, 1975, p. 183) An (86×4) data matrix led to the covariance matrix

$$\boldsymbol{S} = \begin{bmatrix} 0.029004 & -0.008545 & 0.001143 & -0.006594 \\ & 0.003318 & 0.000533 & 0.003248 \\ & & 0.004898 & 0.005231 \\ & & & 0.008463 \end{bmatrix}.$$

Here, the variables relate to the number of trees, height, surface area, and volume of 86 parcels of land. The eigenvalues of $\boldsymbol{S}$ are

$$l_1 = 0.033687, \quad l_2 = 0.011163, \quad l_3 = 0.000592, \quad \text{and} \quad l_4 = 0.000241.$$

Calculate the "percentage of variance explained" by the various principal components and show that

$$\text{tr}\,\boldsymbol{S} = 0.045683 \quad \text{and} \quad |\boldsymbol{S}| = 0.53641 \times 10^{-10}.$$

Show that if the hypothesis is $\lambda_1 = \lambda_2 = \lambda_3 = \lambda_4$, then using (9.28) gives as a test statistic

$$(86 - \frac{19}{6})\log 317.2 = 477,$$

which is highly significant against χ_9^2.

Show similarly that the hypothesis $\lambda_2 = \lambda_3 = \lambda_4$ gives a χ_5^2 statistic of 306, and the hypothesis $\lambda_3 = \lambda_4$ gives a χ_2^2 statistic of 16.2. Since all these are highly significant, we conclude that the eigenvalues of $\boldsymbol{\Sigma}$ are all distinct.

9.4.3 (Kendall, 1975, p. 79). Let D_1 be the determinant of $\boldsymbol{S}$, so that $D_1 = l_1 l_2 \cdots l_p$. Let D_2 be the determinant of the m.l.e. of $\boldsymbol{\Sigma}$ that would have arisen had the last $(p-k)$ eigenvalues of $\boldsymbol{\Sigma}$ been assumed equal. Show that

$$D_2 = l_1 l_2 \cdots l_k \left(\frac{l_{k+1} + \cdots + l_p}{p-k} \right)^{p-k}.$$

Hence, show that

$$D_2/D_1 = (a_0/g_0)^{p-k},$$

where a_0 and g_0 are the arithmetic and geometric means of the last $(p-k)$ eigenvalues of $\boldsymbol{S}$. Hence, show that $n \log(D_2/D_1) = -2 \log \lambda$, where $-2 \log \lambda$ is given by (9.27).

9.4.4 Consider the special case of (9.26) where $k = 1$. Show that in this case $\alpha = \psi^2 \mathrm{tr}^2\, \boldsymbol{\Sigma}/\mathrm{tr}\, \boldsymbol{\Sigma}^2$, so that

$$\tau^2 = \frac{2\psi^2}{n} \left(\frac{\mathrm{tr}\, \boldsymbol{\Sigma}^2}{\mathrm{tr}^2\, \boldsymbol{\Sigma}} - 2\psi + 1 \right).$$

Confirm that this formula leads to the same numerical results as obtained in Example 9.4.1.

9.5.1 The usual chi-squared test statistic compares observed counts with expected counts which are computed assuming a null hypothesis of independence. In the notation of this chapter, the observed counts are $n_{ij} = np_{ij}$, and the expected values given by $np_{i\cdot}p_{\cdot j}$, so the test statistic is equal to

$$\chi^2 = n \sum_{i=1}^{I} \sum_{j=1}^{J} \frac{(p_{ij} - p_{i\cdot}p_{\cdot j})^2}{p_{i\cdot}p_{\cdot j}}.$$

Show that this statistic can be rewritten in the forms given by Eqs. (9.32) and (9.33).

9.5.2 Starting from the singular value decomposition $\boldsymbol{A} = \boldsymbol{U}\boldsymbol{L}\boldsymbol{V}'$ in (9.34), show that the matrix $\boldsymbol{P}^{\mathrm{std}}$ satisfies

$$\boldsymbol{P}^{\mathrm{std}}\sqrt{\boldsymbol{c}} = \sqrt{\boldsymbol{r}}, \quad (\boldsymbol{P}^{\mathrm{std}})'\sqrt{\boldsymbol{r}} = \sqrt{\boldsymbol{c}}$$

and

$$\boldsymbol{P}^{\mathrm{std}}\boldsymbol{v}_{(h)} = l_h \boldsymbol{u}_{(h)}, \quad (\boldsymbol{P}^{\mathrm{std}})'\boldsymbol{u}_{(h)} = l_h \boldsymbol{v}_{(h)}, \quad h = 1, \dots q,$$

so that $\boldsymbol{P}^{\mathrm{std}}$ has singular values 1 and l_h, $h = 1, \dots, q$.

9.5.3 The matrices $\boldsymbol{F}$ and $\boldsymbol{G}$ are defined in (9.35). Using Exercise 9.5.2, show that

$$\boldsymbol{P}^{\mathrm{row}}\mathbf{1}_J = \mathbf{1}_I, \quad (\boldsymbol{P}^{\mathrm{col}})'\mathbf{1}_I = \mathbf{1}_J$$

and

$$\boldsymbol{P}^{\mathrm{row}}\boldsymbol{g}_{(h)} = l_h \boldsymbol{f}_{(h)}, \quad (\boldsymbol{P}^{\mathrm{col}})'\boldsymbol{f}_{(h)} = l_h \boldsymbol{g}_{(h)}, \quad h = 1, \dots q.$$

9.5.4 Using Exercise 9.5.3, show that

$$(\boldsymbol{P}^{\text{col}})'\boldsymbol{P}^{\text{row}}\boldsymbol{1}_J = \boldsymbol{1}_J, \quad \boldsymbol{P}^{\text{row}}(\boldsymbol{P}^{\text{col}})'\boldsymbol{1}_I = \boldsymbol{1}_I$$

and

$$(\boldsymbol{P}^{\text{col}})'\boldsymbol{P}^{\text{row}}\boldsymbol{g}_{(h)} = l_h^2\boldsymbol{g}_{(h)}, \quad \boldsymbol{P}^{\text{row}}(\boldsymbol{P}^{\text{col}})'\boldsymbol{f}_{(h)} = l_h^2\boldsymbol{f}_{(h)}, \quad h = 1, \dots q,$$

so that $(\boldsymbol{P}^{\text{col}})'\boldsymbol{P}^{\text{row}}$ and $\boldsymbol{P}^{\text{row}}(\boldsymbol{P}^{\text{col}})'$ have eigenvalues 1 and l_h^2, $h = 1, \dots, q$.

9.5.5 Confirm that the distances between the rows of $\boldsymbol{FL}$ are the same as the distances between the rows of $\boldsymbol{P}^{\text{row}}$, and that the distances between the rows of $\boldsymbol{GL}$ are the same as the distances between the columns of $\boldsymbol{P}^{\text{col}}$.

9.5.6 The matrix $\boldsymbol{P}^{\text{std}}$ is defined in (9.31). Using the Perron–Frobenius Theorem (Exercise 9.6.1), deduce that $\boldsymbol{P}^{\text{std}}(\boldsymbol{P}^{\text{std}})'$ has one eigenvalue equal to 1, and that the remaining nonzero eigenvalues are greater than 0 and less than 1. Why do the matrices $\boldsymbol{P}^{\text{row}}(\boldsymbol{P}^{\text{col}})'$ and $(\boldsymbol{P}^{\text{col}})'\boldsymbol{P}^{\text{row}}$ have the same nonzero eigenvalues as $\boldsymbol{P}^{\text{std}}(\boldsymbol{P}^{\text{std}})'$?

9.5.7 Using population birth data from singleton live births in NHS hospitals in Wales (2005–2014) (MacFarlane et al., 2019, ch. 3), a random sample of 1000 births were classified according to mode of birth and gestation, as shown in Table 9.6.

Table 9.6 Counts obtained from a random sample of size 1000 taken from table 19 of MacFarlane et al. (2019), classifying births by mode of birth and gestation.

	Preterm	Term	Postterm
Elective caesarian section	2	102	1
Emergency caesarian section	19	121	9
Spontaneous	23	567	8
Instrumental	5	121	22

Source: Adapted from MacFarlane et al. (2019).

Carry out a chi-squared test to investigate whether the mode of birth is independent of gestation. After considering a correspondence analysis, plot the first two columns of $\boldsymbol{F}$ and $\boldsymbol{G}$ to determine similarity between the rows of $\boldsymbol{P}^{\text{row}}$ and the columns of $\boldsymbol{P}^{\text{col}}$, respectively.

9.6.1 (Perron–Frobenius Theorem). Let $\boldsymbol{A}$ be a $p \times p$ symmetric matrix whose elements are all positive $a_{ij} > 0$, $i, j = 1, \dots, p$. Let λ denote the largest eigenvalue with the corresponding eigenvector $\boldsymbol{x}$. Show that $\lambda > 0$, λ has multiplicity 1, and the elements of $\boldsymbol{x}$ cannot have different signs. (Hint: Define another vector $\boldsymbol{y}$ by $y_i = |x_i|$. Then, $\boldsymbol{y}'\boldsymbol{A}\boldsymbol{y} \geq \boldsymbol{x}'\boldsymbol{A}\boldsymbol{x}$.)

9.7.1 The following principal components were obtained from the correlation matrix of six bone measurements on 276 fowls (Wright, 1954). Show that by taking the threshold value $\lambda_0 = 0.70$ we reject variables x_4, x_6, x_3, and x_1 in that order.

	Component numbers					
Measurements	**1**	**2**	**3**	**4**	**5**	**6**
x_1, skull length	0.35	0.53	0.76	−0.05	0.04	0.00
x_2, skull breadth	0.33	0.70	−0.64	0.00	0.00	−0.04
x_3, wing humerus	0.44	−0.19	−0.05	0.53	0.19	0.59
x_4, wing ulna	0.44	−0.25	0.02	0.48	−0.15	−0.63
x_5, leg	0.43	−0.28	−0.06	−0.51	0.67	−0.48
x_6, leg tibia	0.44	−0.22	−0.05	−0.48	−0.70	0.15
Eigenvalues	4.57	0.71	0.41	0.17	0.08	0.06

Show that the six components can be interpreted as overall size, head size versus body size, head shape, wing size versus leg size, leg shape, and wing shape (although these last two interpretations are perhaps dubious). Verify that the successive elimination of variables reduces the number of these features that can be discerned.

9.8.1 Hoerl and Kennard (1970) have proposed the method of *ridge regression* (see also Section 16.2) to improve the accuracy of the parameter estimates in the regression model

$$\boldsymbol{y} = \boldsymbol{X\beta} + \mu\boldsymbol{1} + \boldsymbol{u}, \qquad \boldsymbol{u} \sim N_n(\boldsymbol{0}, \sigma^2\boldsymbol{I}).$$

Suppose that the columns of $\boldsymbol{X}$ have been standardized to have mean $\boldsymbol{0}$ and variance 1. Then, the *ridge estimate* of $\boldsymbol{\beta}$ is defined by

$$\boldsymbol{\beta}^* = (\boldsymbol{X}'\boldsymbol{X} + k\boldsymbol{I})^{-1}\boldsymbol{X}'\boldsymbol{y},$$

where for given $\boldsymbol{X}$, $k \geq 0$ is a (small) *fixed* number.

(a) Show that $\boldsymbol{\beta}^*$ reduces to the ordinary least squares (OLS) estimate $\hat{\boldsymbol{\beta}} = (\boldsymbol{X}'\boldsymbol{X})^{-1}\boldsymbol{X}'\boldsymbol{y}$ when $k = 0$.

(b) Let $\boldsymbol{X}'\boldsymbol{X} = \boldsymbol{GLG}'$ be a spectral decomposition of $\boldsymbol{X}'\boldsymbol{X}$, and let $\boldsymbol{W} = \boldsymbol{XG}$ be the principal component transformation given in (9.40). If $\boldsymbol{\alpha} = \boldsymbol{G}'\boldsymbol{\beta}$ represents the parameter vector for the principal components, show that the ridge estimate $\boldsymbol{\alpha}^*$ of $\boldsymbol{\alpha}$ can be simply related to the OLS estimate $\hat{\boldsymbol{\alpha}}$ by

$$\alpha_j^* = \frac{l_j}{l_j + k}\hat{\alpha}_j, \qquad j = 1, \ldots, p,$$

and hence,

$$\boldsymbol{\beta}^* = \boldsymbol{GDG}'\hat{\boldsymbol{\beta}}, \qquad \text{where} \qquad \boldsymbol{D} = \text{diag}\,(l_j/(l_j + k)).$$

(c) One measure of the accuracy of $\boldsymbol{\beta}^*$ is given by the *trace mean square error*,

$$\phi(k) = \operatorname{tr} \mathsf{E}[\boldsymbol{\beta}^* - \boldsymbol{\beta})(\boldsymbol{\beta}^* - \boldsymbol{\beta})'] = \sum_{i=1}^{p} \mathsf{E}(\beta_i^* - \beta_i)^2.$$

Show that we can write $\phi(k) = \gamma_1(k) + \gamma_2(k)$, where

$$\gamma_1(k) = \sum_{i=1}^{p} \operatorname{var}(\beta_i^*) = \sigma^2 \sum_{j=1}^{p} \frac{l_j}{(l_j + k)^2}$$

represents the sum of the variances of β_i^*, and

$$\gamma_2(k) = \sum_{i=1}^{p} [\mathsf{E}(\beta_i^* - \beta_i)]^2 = k^2 \sum_{j=1}^{p} \frac{\alpha_j^2}{(l_j + k)^2}$$

represents the sum of the squared biases of β_i^*.

(d) Show that the first derivatives of $\gamma_1(k)$ and $\gamma_2(k)$ at $k = 0$ are

$$\gamma_1'(0) = -2\sigma^2 \sum \frac{1}{l_j^2}, \qquad \gamma_2'(0) = 0.$$

Hence, there exist values of $k > 0$ for which $\phi(k) < \phi(0)$, that is for which $\boldsymbol{\beta}^*$ has smaller trace mean square error than $\hat{\boldsymbol{\beta}}$. Note that the increase in accuracy is most pronounced when some of the eigenvalues l_j are near 0, that is when the columns of $\boldsymbol{X}$ are nearly collinear. However, the optimal choice for k depends on the unknown value of $\boldsymbol{\beta} = \boldsymbol{G}\boldsymbol{\alpha}$.

10

Factor Analysis

10.1 Introduction

Factor analysis is a mathematical model that attempts to explain the correlation between a large set of variables in terms of a small number of underlying *factors*. A major assumption of factor analysis is that it is not possible to observe these factors directly; the variables depend on the factors but are also subject to random errors. Such an assumption is particularly well suited to subjects such as psychology where it is not possible to measure exactly the concepts one is interested in (e.g. "intelligence"), and in fact, it is often ambiguous just how to define these concepts.

Factor analysis was originally developed by psychologists. The subject was first put on a respectable statistical footing in the early 1940s by restricting attention to one particular form of factor analysis, which is based on maximum-likelihood estimation. In this chapter, we concentrate on maximum-likelihood factor analysis and also on a second method, principal factor analysis, which is closely related to the technique of principal component analysis.

In order to get a feel for the subject, we first describe a simple example.

Example 10.1.1 Spearman (1904). An important early paper in factor analysis dealt with children's examination performance in Classics (x_1), French (x_2), and English (x_3). It is found that the correlation matrix (computed using 33 sets of results) is given by

$$\begin{bmatrix} 1 & 0.83 & 0.78 \\ & 1 & 0.67 \\ & & 1 \end{bmatrix}.$$

Although this matrix has full rank, its dimensionality can be effectively reduced from $p = 3$ to $p = 1$ by expressing the three variables as follows:

$$x_1 = \lambda_1 f + u_1, \qquad x_2 = \lambda_2 f + u_2, \qquad x_3 = \lambda_3 f + u_3. \tag{10.1}$$

In these equations, f is an underlying "common factor," and λ_1, λ_2, and λ_3 are known as "factor loadings". The terms u_1, u_2, and u_3 represent random disturbance terms. The common factor may be interpreted as "general ability," and u_i will have small variance if x_i is closely related to general ability. The variation in u_i consists of two parts, which we will not try to disentangle in practice. First, this variance represents the extent to

Multivariate Analysis, Second Edition. Kanti V. Mardia, John T. Kent, and Charles C. Taylor.

which an individual's ability at Classics, say, differs from his general ability, and second, it represents the fact that the examination is only an approximate measure of his ability in the subject. □

The model defined in (10.1) will now be generalized to include $k > 1$ common factors.

10.2 The Factor Model

10.2.1 Definition

Let $\boldsymbol{x}(p \times 1)$ be a random vector with mean $\boldsymbol{\mu}$ and covariance matrix $\boldsymbol{\Sigma}$. Then, we say that the k-factor model holds for $\boldsymbol{x}$ if $\boldsymbol{x}$ can be written in the form

$$\boldsymbol{x} = \boldsymbol{\Lambda f} + \boldsymbol{u} + \boldsymbol{\mu}, \tag{10.2}$$

where $\boldsymbol{\Lambda}(p \times k)$ is a matrix of constants, and $\boldsymbol{f}(k \times 1)$ and $\boldsymbol{u}(p \times 1)$ are random vectors. The elements of $\boldsymbol{f}$ are called *common* factors, and the elements of $\boldsymbol{u}$ *specific* or *unique* factors. We suppose that

$$\mathsf{E}(\boldsymbol{f}) = \boldsymbol{0}, \qquad \text{var}(\boldsymbol{F}) = \boldsymbol{I}, \tag{10.3}$$

$$\mathsf{E}(\boldsymbol{u}) = \boldsymbol{0}, \qquad \text{cov}(u_i, u_j) = 0, \quad i \neq j, \tag{10.4}$$

and

$$\text{cov}(\boldsymbol{f}, \boldsymbol{u}) = \boldsymbol{0}. \tag{10.5}$$

Denote the covariance matrix of $\boldsymbol{u}$ by $\text{var}(\boldsymbol{u}) = \boldsymbol{\Psi} = \text{diag}(\psi_{11}, \dots, \psi_{pp})$. Thus, all of the factors are uncorrelated with one another, and further, the common factors are each standardized to have variance 1. It is sometimes convenient to suppose that $\boldsymbol{f}$ and $\boldsymbol{u}$ (and hence $\boldsymbol{x}$) are multivariate normally distributed.

Note that

$$x_i = \sum_{j=1}^{k} \lambda_{ij} f_j + u_i + \mu_i, \quad i = 1, \dots, p,$$

so that

$$\sigma_{ii} = \sum_{j=1}^{k} \lambda_{ij}^2 + \psi_{ii}.$$

Thus, the variance of $\boldsymbol{x}$ can be split into two parts. First,

$$h_i^2 = \sum_{j=1}^{k} \lambda_{ij}^2$$

is called the *communality* and represents the variance of x_i which is shared with the other variables via the common factors. In particular, $\lambda_{ij}^2 = \text{cov}(x_i, f_j)$ represents the extent to which x_i depends on the jth common factor. On the other hand, ψ_{ii} is called the *specific* or *unique variance* and is due to the unique factor u_i; it explains the variability in x_i not shared with the other variables.

The validity of the k-factor model can be expressed in terms of a simple condition on $\boldsymbol{\Sigma}$. Using (10.2)–(10.5), we get

$$\boldsymbol{\Sigma} = \boldsymbol{\Lambda\Lambda}' + \boldsymbol{\Psi}. \tag{10.6}$$

The converse also holds. If $\boldsymbol{\Sigma}$ can be decomposed into the form (10.6), then the k-factor model holds for $\boldsymbol{x}$ (see Exercise 10.2.2). However, $\boldsymbol{f}$ and $\boldsymbol{u}$ are not uniquely determined by $\boldsymbol{x}$. We will discuss this point further in Section 10.7 on factor scores.

10.2.2 Scale Invariance

Rescaling the variables of $\boldsymbol{x}$ is equivalent to letting $\boldsymbol{y} = \boldsymbol{C}\boldsymbol{x}$, where $\boldsymbol{C} = \text{diag}(c_i)$. If the k-factor model holds for $\boldsymbol{x}$ with $\boldsymbol{\Lambda} = \boldsymbol{\Lambda}_x$ and $\boldsymbol{\Psi} = \boldsymbol{\Psi}_x$, then

$$\boldsymbol{y} = \boldsymbol{C}\boldsymbol{\Lambda}_x\boldsymbol{f} + \boldsymbol{C}\boldsymbol{u} + \boldsymbol{C}\boldsymbol{\mu}$$

and

$$\text{var}(\boldsymbol{y}) = \boldsymbol{C}\boldsymbol{\Sigma}\boldsymbol{C} = \boldsymbol{C}\boldsymbol{\Lambda}_x\boldsymbol{\Lambda}_x'\boldsymbol{C} + \boldsymbol{C}\boldsymbol{\Psi}_x\boldsymbol{C}.$$

Thus, the k-factor model also holds for $\boldsymbol{y}$ with factor loading matrix $\boldsymbol{\Lambda}_y = \boldsymbol{C}\boldsymbol{\Lambda}_x$ and specific variances $\boldsymbol{\Psi}_y = \boldsymbol{C}\boldsymbol{\Psi}_x\boldsymbol{C} = \text{diag}(c_i^2\psi_{ii})$.

Note that the factor loading matrix for the scaled variables $\boldsymbol{y}$ is obtained by scaling the factor loading matrix of the original variables (multiply the ith row of $\boldsymbol{\Lambda}_x$ by c_i). A similar comment holds for the specific variances. In other words, factor analysis (unlike principal component analysis) is unaffected by a rescaling of the variables.

10.2.3 Nonuniqueness of Factor Loadings

If the k-factor model (10.2) holds, then it also holds if the factors are rotated; that is, if $\boldsymbol{G}$ is a $(k \times k)$ orthogonal matrix, then $\boldsymbol{x}$ can also be written as

$$\boldsymbol{x} = (\boldsymbol{\Lambda}\boldsymbol{G})(\boldsymbol{G}'\boldsymbol{f}) + \boldsymbol{u} + \boldsymbol{\mu}. \tag{10.7}$$

Since the random vector $\boldsymbol{G}'\boldsymbol{f}$ also satisfies the conditions (10.3) and (10.5), we see that the k-factor model is valid with new factors $\boldsymbol{G}'\boldsymbol{f}$ and new factor loadings $\boldsymbol{\Lambda}\boldsymbol{G}$.

Thus, if (10.7) holds, we can also write $\boldsymbol{\Sigma}$ as $\boldsymbol{\Sigma} = (\boldsymbol{\Lambda}\boldsymbol{G})(\boldsymbol{G}'\boldsymbol{\Lambda}') + \boldsymbol{\Psi}$. In fact, for fixed $\boldsymbol{\Psi}$, this rotation is the only indeterminacy in the decomposition of $\boldsymbol{\Sigma}$ in terms of $\boldsymbol{\Lambda}$ and $\boldsymbol{\Psi}$; that is, if $\boldsymbol{\Sigma} = \boldsymbol{\Lambda}\boldsymbol{\Lambda}' + \boldsymbol{\Psi} = \boldsymbol{\Lambda}^*\boldsymbol{\Lambda}^{*\prime} + \boldsymbol{\Psi}$, then $\boldsymbol{\Lambda} = \boldsymbol{\Lambda}^*\boldsymbol{G}$ for some orthogonal matrix $\boldsymbol{G}$ (Exercise 10.2.3).

This indeterminacy in the definition of factor loadings is usually resolved by rotating the factor loadings to satisfy an arbitrary constraint such as

$$\boldsymbol{\Lambda}'\boldsymbol{\Psi}^{-1}\boldsymbol{\Lambda} \quad \text{is diagonal} \tag{10.8}$$

or

$$\boldsymbol{\Lambda}'\boldsymbol{D}^{-1}\boldsymbol{\Lambda} \quad \text{is diagonal}, \qquad \boldsymbol{D} = \text{diag}(\sigma_{11}, \dots, \sigma_{pp}), \tag{10.9}$$

where, in either case, the diagonal elements are written in decreasing order, say. Both constraints are scale invariant, and, except for possible changes of sign of the columns, $\boldsymbol{\Lambda}$ is then in general completely determined by either constraint (see Exercise 10.2.4). Note that when the number of factors $k = 1$, the constraint is irrelevant. Also, if some of the ψ_{ii} equal 0, then the constraint (10.8) cannot be used; an alternative is given in Exercise 10.2.8.

It is of interest to compare the number of parameters in $\boldsymbol{\Sigma}$ when $\boldsymbol{\Sigma}$ is unconstrained, with the number of free parameters in the factor model. Let s denote the difference. At first sight,

$\boldsymbol{\Lambda}$ and $\boldsymbol{\Psi}$ contain $pk + p$ free parameters. However, (10.8) or (10.9) introduces $k(k-1)/2$ constraints. Since the number of distinct elements of $\boldsymbol{\Sigma}$ is $p(p+1)/2$, we see that

$$\begin{aligned} s &= \frac{1}{2}p(p+1) - \left\{pk + p - \frac{1}{2}k(k-1)\right\} \\ &= \frac{1}{2}(p-k)^2 - \frac{1}{2}(p+k). \end{aligned} \tag{10.10}$$

Usually, it will be the case that $s > 0$. Then, s will represent the extent to which the factor model offers a simpler interpretation for the behavior of $\boldsymbol{x}$ than the alternative assumption that $\text{var}(\boldsymbol{x}) = \boldsymbol{\Sigma}$. If $s \geq 0$ and (10.7) holds, and $\boldsymbol{\Lambda}$ and $\boldsymbol{\Psi}$ are known, then $\boldsymbol{\Sigma}$ can be written in terms of $\boldsymbol{\Lambda}$ and $\boldsymbol{\Psi}$, subject to the constraint (10.8) or (10.9) on $\boldsymbol{\Lambda}$.

10.2.4 Estimation of the Parameters in Factor Analysis

In practice, we observe a data matrix $\boldsymbol{X}$ whose information is summarized by the sample mean $\bar{\boldsymbol{x}}$ and sample covariance matrix $\boldsymbol{S}$. The location parameter is not of interest here, and we will estimate it by $\hat{\boldsymbol{\mu}} = \bar{\boldsymbol{x}}$. The interesting problem is how to estimate $\boldsymbol{\Lambda}$ and $\boldsymbol{\Psi}$ (and hence $\boldsymbol{\Sigma} = \boldsymbol{\Lambda}\boldsymbol{\Lambda}' + \boldsymbol{\Psi}$) from $\boldsymbol{S}$; that is, we wish to find estimates $\hat{\boldsymbol{\Lambda}}$ and $\hat{\boldsymbol{\Psi}}$ satisfying the constraint (10.8) or (10.9), for which the equation

$$\boldsymbol{S} \approx \hat{\boldsymbol{\Lambda}}\hat{\boldsymbol{\Lambda}}' + \hat{\boldsymbol{\Psi}} \tag{10.11}$$

is satisfied, at least approximately. Given an estimate $\hat{\boldsymbol{\Lambda}}$, it is then natural to set

$$\hat{\psi}_{ii} = s_{ii} - \sum_{j=1}^{k} \hat{\lambda}_{ij}^2, \quad i = 1, \ldots, p, \tag{10.12}$$

so that the diagonal equations in (10.11) always hold exactly.

We only consider estimates for which (10.11) is satisfied and $\hat{\psi}_{ii} \geq 0$. Setting $\hat{\boldsymbol{\Sigma}} = \hat{\boldsymbol{\Lambda}}\hat{\boldsymbol{\Lambda}}' + \hat{\boldsymbol{\Psi}}$, we get

$$\hat{\sigma}_{ii} = \sum_{j=1}^{k} \hat{\lambda}_{ij}^2 + \hat{\psi}_{ii},$$

so that (10.12) is equivalent to the condition

$$\hat{\sigma}_{ii} = s_{ii}, \quad i = 1, \ldots, p. \tag{10.13}$$

Three cases can occur in (10.11) depending on the value of s in (10.10). If $s < 0$, then (10.11) contains more parameters than equations. Then, in general, we expect to find an infinity of exact solutions for $\boldsymbol{\Lambda}$ and $\boldsymbol{\Psi}$, and hence, the factor model is *not* well defined.

If $s = 0$, then (10.11) can generally be solved for $\boldsymbol{\Lambda}$ and $\boldsymbol{\Psi}$ exactly (subject to the constraint (10.8) or (10.9) on $\boldsymbol{\Lambda}$). The factor model contains as many parameters as $\boldsymbol{\Sigma}$ and hence offers no simplification of the original assumption that $\text{var}(\boldsymbol{x}) = \boldsymbol{\Sigma}$. However, the change in viewpoint can sometimes be very helpful (see Exercise 10.2.7).

If $s > 0$, as will usually be the case, then there will be more equations than parameters. Thus, it is not possible to solve (10.11) exactly in terms of $\hat{\boldsymbol{\Lambda}}$ and $\hat{\boldsymbol{\Psi}}$, and we must look for approximate solutions. In this case, the factor model offers a simpler explanation for the behavior of $\boldsymbol{x}$ than the full covariance matrix.

10.2.5 Use of the Correlation Matrix R in Estimation

Because the factor model is scale invariant, we only consider estimates of $\mathbf{\Lambda} = \mathbf{\Lambda}_x$ and $\mathbf{\Psi} = \mathbf{\Psi}_x$, which are scale invariant. It is then convenient to consider the scaling separately from the relationships between the variables. Let $\boldsymbol{Y} = \boldsymbol{HXD}_S^{-1/2}$, where $\boldsymbol{D}_S = \text{Diag}(\boldsymbol{S})$, denote the standardized variables, so that

$$\sum_{r=1}^{n} y_{ri} = 0 \quad \text{and} \quad \frac{1}{n}\sum_{r=1}^{n} y_{ri}^2 = 1, \quad i = 1, \dots, p.$$

Then, $\boldsymbol{Y}$ will have estimated factor loading matrix $\hat{\mathbf{\Lambda}}_y = \boldsymbol{D}_S^{-1/2}\hat{\mathbf{\Lambda}}_x$ and estimated specific variances $\hat{\mathbf{\Psi}}_y = \boldsymbol{D}_S^{-1}\mathbf{\Psi}_x$, and (10.11) can be written in terms of the correlation matrix of $\boldsymbol{x}$ as

$$\boldsymbol{R} \approx \hat{\mathbf{\Lambda}}_y\hat{\mathbf{\Lambda}}_y' + \hat{\mathbf{\Psi}}_y. \tag{10.14}$$

Note that (10.12) becomes

$$\hat{\psi}_{yii} = 1 - \sum_{j=1}^{k} \hat{\lambda}_{yij}^2, \quad i = 1, \dots, p, \tag{10.15}$$

so that $\mathbf{\Psi}_y$ is not a parameter of the model any more but a function of $\mathbf{\Lambda}_y$. However, $\boldsymbol{R}$ contains p fewer free parameters than $\boldsymbol{S}$, so that s, the difference between the number of equations and the number of free parameters in (10.14), is still given by (10.10). The p equations for the estimates of the scaling parameters are given by (10.13).

Since in practice it is the relationship between the variables that is of interest rather than their scaling, the data is often summarized by $\boldsymbol{R}$ rather than $\boldsymbol{S}$. The scaling estimates given in (10.13) are then not mentioned explicitly, and the estimated factor loadings and specific variances are presented in terms of the standardized variables.

Example 10.2.1 Let us return to the Spearman examination data with one factor. Since $s = (3-1)^2/2 - (3+1)/2 = 0$, there is an exact solution to (10.14), which is

$$\boldsymbol{R} = \begin{bmatrix} \hat{\lambda}_1^2 + \hat{\psi}_{11} & \hat{\lambda}_1\hat{\lambda}_2 & \hat{\lambda}_1\hat{\lambda}_3 \\ & \hat{\lambda}_2^2 + \hat{\psi}_{22} & \hat{\lambda}_2\hat{\lambda}_3 \\ & & \hat{\lambda}_3^2 + \hat{\psi}_{33} \end{bmatrix}. \tag{10.16}$$

The unique solution (except for the sign of $(\hat{\lambda}_1, \hat{\lambda}_2, \hat{\lambda}_3)$) is given by

$$\begin{aligned} \hat{\lambda}_1^2 &= r_{12}r_{13}/r_{23}, & \hat{\lambda}_2^2 &= r_{12}r_{23}/r_{13}, & \hat{\lambda}_3^2 &= r_{13}r_{23}/r_{12}, \\ \hat{\psi}_{11} &= 1 - \hat{\lambda}_1^2, & \hat{\psi}_{22} &= 1 - \hat{\lambda}_2^2, & \hat{\psi}_{33} &= 1 - \hat{\lambda}_3^2. \end{aligned} \tag{10.17}$$

With the value of $\boldsymbol{R}$ given in Example 10.1.1, we find that

$$\begin{aligned} \hat{\lambda}_1 &= 0.983, & \hat{\lambda}_2 &= 0.844, & \hat{\lambda}_3 &= 0.793, \\ \hat{\psi}_{11} &= 0.034 & \hat{\psi}_{22} &= 0.287 & \hat{\psi}_{33} &= 0.370. \end{aligned}$$

Since $\hat{\psi}_{ii} = 1 - \hat{h}_i^2$, the model explains a higher proportion of the variance of x_1 than of x_2 and x_3. □

For general (3×3) correlation matrices $\boldsymbol{R}$, it is possible for $\hat{\lambda}_1$ to exceed 1 (see Exercise 10.2.6). When this happens, $\hat{\psi}_{11}$ is negative. Hence, we adjust $\hat{\lambda}_1$ to be 1, and we use an approximate solution for the other equations in (10.15). This situation is known as a "Heywood case."

We describe two methods of estimating the parameters of the factor model when $s > 0$. The first is *principal factor analysis* and uses some ideas from principal component analysis. The second method, *maximum-likelihood factor analysis*, is applicable when the data is normally distributed and enables a significance test to be made about the validity of the k-factor model. For further techniques, see Lawley and Maxwell (1971) and Comrey and Lee (1992).

10.3 Principal Factor Analysis

Principal factor analysis is one method of estimating the parameters of the k-factor model when s – given by (10.10) – is positive. Suppose that the data is summarized by the correlation matrix $\boldsymbol{R}$, so that an estimate of $\boldsymbol{\Lambda}$ and $\boldsymbol{\Psi}$ is sought for the standardized variables. (We are implicitly assuming that the variances for the original variables are estimated by $\hat{\sigma}_{ii} = s_{ii}$ in (10.13).)

As a first step, preliminary estimates $\tilde{h}_i^2$ of the communalities $h_i^2, i = 1, \ldots, p$, are made. Two common estimates of the ith communality are

(a) the square of the multiple correlation coefficient of the ith variable with all the other variables or
(b) the largest correlation coefficient between the ith variable and one of the other variables, that is $\max_{j \neq i} |r_{ij}|$.

Note that the estimated communality $\tilde{h}_i^2$ is higher when x_i is highly correlated with the other variables, as we could expect.

The matrix $\boldsymbol{R} - \boldsymbol{\Psi}$ is called the *reduced correlation matrix* because the 1s on the diagonal have been replaced by the estimated communalities $\tilde{h}_i^2 = 1 - \tilde{\psi}_{ii}$. By the spectral decomposition theorem (Theorem A.6.4), it can be written as

$$\boldsymbol{R} - \boldsymbol{\Psi} = \sum_{i=1}^{p} a_i \boldsymbol{\gamma}_{(i)} \boldsymbol{\gamma}'_{(i)}, \tag{10.18}$$

where $a_1 \geq \cdots \geq a_p$ are eigenvalues of $\boldsymbol{R} - \boldsymbol{\Psi}$ with the corresponding orthonormal eigenvectors $\boldsymbol{\gamma}_{(1)}, \ldots, \boldsymbol{\gamma}_{(p)}$. Suppose that the first k eigenvalues of $\boldsymbol{R} - \boldsymbol{\Psi}$ are positive. Then, the ith column of $\boldsymbol{\Lambda}$ is estimated by

$$\hat{\boldsymbol{\lambda}}_{(i)} = a_i^{1/2} \boldsymbol{\gamma}_{(i)}, \quad i = 1, \ldots, k; \tag{10.19}$$

that is, $\hat{\boldsymbol{\lambda}}_{(i)}$ is proportional to the ith eigenvector of the reduced correlation matrix. In matrix form,

$$\hat{\boldsymbol{\Lambda}} = \boldsymbol{\Gamma}_1 \boldsymbol{A}_1^{1/2},$$

where $\boldsymbol{\Gamma}_1 = (\boldsymbol{\gamma}_{(1)}, \ldots, \boldsymbol{\gamma}_{(k)})$ and $\boldsymbol{A}_1 = \text{diag}(a_1, \ldots, a_k)$. Since the eigenvectors are orthogonal, we see that $\hat{\boldsymbol{\Lambda}}'\hat{\boldsymbol{\Lambda}}$ is diagonal, so the constraint (10.9) is satisfied. (Recall that we are working with the standardized variables here, each of whose estimated true variance is 1.)

Finally, revised estimates of the specific variances are given in terms of $\hat{\Lambda}$ by

$$\hat{\psi}_{ii} = 1 - \sum_{j=1}^{k} \hat{\lambda}_{ij}^2, \quad i = 1, \dots, p. \tag{10.20}$$

Then, the principal factor solution is permissible if all the $\hat{\psi}_{ii}$ are nonnegative.

As a motivation for the principal factor method, let us consider what happens when the communalities are known exactly and $\boldsymbol{R}$ equals the true correlation matrix. Then, $\boldsymbol{R} - \boldsymbol{\Psi} = \boldsymbol{\Lambda}\boldsymbol{\Lambda}'$ exactly. The constraint (10.9) implies that the columns of $\boldsymbol{\Lambda}$ are eigenvectors of $\boldsymbol{\Lambda}\boldsymbol{\Lambda}'$, and hence, $\boldsymbol{\Lambda}$ is given by (10.19). Further, the last $(p-k)$ eigenvalues of $\boldsymbol{R} - \boldsymbol{\Psi}$ vanish, $a_{k+1} = \cdots = a_p = 0$. This property can be helpful in practice when $\boldsymbol{R}$ and $\boldsymbol{\Psi}$ are not exactly equal to their theoretical values and k is not known in advance. Hopefully, for some values if k, $a_1, \dots, a_k$ will be "large" and positive, the remaining eigenvalues, $a_{k+1}, \dots, a_p$, will be near zero (some possibly negative). Of course, k must always be small enough, so that $s \geq 0$ in (10.10).

Example 10.3.1 Consider the open/closed-book data of Table 1.2 with correlation matrix

$$\begin{bmatrix} 1 & 0.553 & 0.547 & 0.410 & 0.389 \\ & 1 & 0.610 & 0.485 & 0.437 \\ & & 1 & 0.711 & 0.665 \\ & & & 1 & 0.607 \\ & & & & 1 \end{bmatrix}.$$

If $k > 2$, then $s < 0$, and the factor model is not well defined. The principal factor solutions for $k = 1$ and $k = 2$, where we estimate the ith communality h_i^2 by $\max_j |r_{ij}|$, are given in Table 10.1. The eigenvalues of the reduced correlation matrix are 2.84, 0.38, 0.08, 0.02, and −0.05, suggesting that the two-factor solution fits the data well.

The first factor represents the overall performance, and for $k = 2$, the second factor, which is much less important ($a_2 = 0.38 \ll 2.84 = a_1$), represents a contrast across the range of examinations. Note that even for the two-factor solution, $h_i^2 < 0.75$ for all i, and therefore, a fair proportion of the variance of each variable is left unexplained by the common factors. □

Table 10.1 Principal factor solutions for the open/closed-book data with $k = 1$ and $k = 2$ factors.

	$k = 1$		$k = 2$		
Variable	h_i^2	$\lambda_{(1)}$	h_i^2	$\lambda_{(1)}$	$\lambda_{(2)}$
1	0.417	0.646	0.543	0.646	0.354
2	0.506	0.711	0.597	0.711	0.303
3	0.746	0.864	0.749	0.864	−0.051
4	0.618	0.786	0.680	0.786	−0.249
5	0.551	0.742	0.627	0.742	−0.276

10.4 Maximum-likelihood Factor Analysis

When the data under analysis $\boldsymbol{X}$ is assumed to be normally distributed, then estimates of $\boldsymbol{\Lambda}$ and $\boldsymbol{\Psi}$ can be found by maximizing the likelihood. If $\boldsymbol{\mu}$ is replaced by its maximum-likelihood estimate (m.l.e.) $\hat{\boldsymbol{\mu}} = \bar{\boldsymbol{x}}$, then, from (5.9), the log-likelihood function becomes

$$l = -\frac{1}{2} \log |2\pi\boldsymbol{\Sigma}| - \frac{1}{2} n \operatorname{tr} \boldsymbol{\Sigma}^{-1}\boldsymbol{S}. \tag{10.21}$$

Regarding $\boldsymbol{\Sigma} = \boldsymbol{\Lambda}\boldsymbol{\Lambda}' + \boldsymbol{\Psi}$ as a function of $\boldsymbol{\Lambda}$ and $\boldsymbol{\Psi}$, we can maximize (10.21) with respect to $\boldsymbol{\Lambda}$ and $\boldsymbol{\Psi}$. Let $\hat{\boldsymbol{\Lambda}}$ and $\hat{\boldsymbol{\Psi}}$ denote the resulting m.l.e.s subject to the constraint (10.8) on $\boldsymbol{\Lambda}$.

Note that in writing down the likelihood in (10.21), we have summarized the data using the covariance matrix $\boldsymbol{S}$ rather than the correlation matrix $\boldsymbol{R}$. However, it can be shown that the m.l.e.s are scale invariant (Exercise 10.4.1), and that the m.l.e.s of the scale parameters are given by $\hat{\sigma}_{ii} = s_{ii}$ (see Theorem 10.4.2). Thus, (10.13) is satisfied, and the m.l.e.s for the parameters of the *standardized* variables can be found by replacing $\boldsymbol{S}$ by $\boldsymbol{R}$ in (10.21). However, it is more convenient in our theoretical discussion to work with $\boldsymbol{S}$ rather than $\boldsymbol{R}$.

Consider the function

$$F(\boldsymbol{\Lambda}, \boldsymbol{\Psi}) = F(\boldsymbol{\Lambda}, \boldsymbol{\Psi}; \boldsymbol{S}) = \operatorname{tr} \boldsymbol{\Sigma}^{-1}\boldsymbol{S} - \log |\boldsymbol{\Sigma}^{-1}\boldsymbol{S}| - p, \tag{10.22}$$

where

$$\boldsymbol{\Sigma} = \boldsymbol{\Lambda}\boldsymbol{\Lambda}' + \boldsymbol{\Psi}.$$

This is a linear function of the log likelihood l, and a maximum in l corresponds to a minimum in F. Note that F can be expressed as

$$F = p(a - \log g - 1), \tag{10.23}$$

where a and g are, respectively, the arithmetic and geometric means of the eigenvalues of $\boldsymbol{\Sigma}^{-1}\boldsymbol{S}$.

The minimization of this function $F(\boldsymbol{\Lambda}, \boldsymbol{\Psi})$ can be facilitated by proceeding in two stages. Firstly, we minimize $F(\boldsymbol{\Lambda}, \boldsymbol{\Psi})$ over $\boldsymbol{\Lambda}$ for fixed $\boldsymbol{\Psi}$, and secondly, we minimize over $\boldsymbol{\Psi}$. This approach has the advantage that the first minimization can be carried out analytically although the minimization over $\boldsymbol{\Psi}$ must be done numerically. It was successfully developed by Joreskog (1967).

Theorem 10.4.1 *Let $\boldsymbol{\Psi}$ be fixed, and let $\boldsymbol{S}^* = \boldsymbol{\Psi}^{-1/2}\boldsymbol{S}\boldsymbol{\Psi}^{-1/2}$. Using the spectral decomposition theorem, we write*

$$\boldsymbol{S}^* = \boldsymbol{\Gamma}\boldsymbol{\Theta}\boldsymbol{\Gamma}'.$$

Then, the value of $\boldsymbol{\Lambda}$ satisfying the constraint (10.8), which minimizes $F(\boldsymbol{\Lambda}, \boldsymbol{\Psi})$, occurs when the ith column of $\boldsymbol{\Lambda}^* = \boldsymbol{\Psi}^{-1/2}\boldsymbol{\Lambda}$ is given by $\boldsymbol{\lambda}^*_{(i)} = c_i\boldsymbol{\gamma}_{(i)}$, where $c_i = [\max(\theta_i - 1, 0)]^{1/2}$ for $i = 1, \ldots, k$.

Proof: Since $\boldsymbol{\Lambda}^*\boldsymbol{\Lambda}^{*\prime}(p \times p)$ has rank at most k, it may be written as

$$\boldsymbol{\Lambda}^*\boldsymbol{\Lambda}^{*\prime} = \boldsymbol{G}\boldsymbol{B}\boldsymbol{G}',$$

where $\boldsymbol{B} = \text{diag}(b_1, \ldots, b_k)$ contains (nonnegative) eigenvalues, and $\boldsymbol{G} = (\boldsymbol{g}_{(1)}, \ldots, \boldsymbol{g}_{(k)})$ are standardized eigenvectors. Let

$$\boldsymbol{M} = \boldsymbol{G}'\boldsymbol{\Gamma},$$

so that $\boldsymbol{G}$ may be written in terms of $\boldsymbol{\Gamma}$ as $\boldsymbol{G} = \boldsymbol{\Gamma}\boldsymbol{M}'$. We minimize $F(\boldsymbol{\Lambda}, \boldsymbol{\Psi})$ over possible choices for $\boldsymbol{M}$ and $\boldsymbol{B}$ in that order.

Since $\boldsymbol{G}'\boldsymbol{G} = \boldsymbol{I}_k$, we see that $\boldsymbol{M}\boldsymbol{M}' = \boldsymbol{I}_k$. From (10.6),

$$\begin{aligned} \boldsymbol{\Sigma}^* = \boldsymbol{\Psi}^{-1/2}\boldsymbol{\Sigma}\boldsymbol{\Psi}^{-1/2} &= \boldsymbol{\Lambda}^*\boldsymbol{\Lambda}^{*\prime} + \mathbf{I} \\ &= \boldsymbol{\Gamma}\boldsymbol{M}'\boldsymbol{B}\boldsymbol{M}\boldsymbol{\Gamma}' + \boldsymbol{I} \\ &= \boldsymbol{\Gamma}(\boldsymbol{M}'\boldsymbol{B}\boldsymbol{M} + \boldsymbol{I})\boldsymbol{\Gamma}'. \end{aligned}$$

Using (A.29), we can write

$$\boldsymbol{\Sigma}^{*-1} = \boldsymbol{\Gamma}(\boldsymbol{I} - \boldsymbol{M}'\boldsymbol{B}(\boldsymbol{I} + \boldsymbol{B})^{-1}\boldsymbol{M})\boldsymbol{\Gamma}'.$$

Also,

$$|\boldsymbol{\Sigma}^*| = \prod_{i=1}^{k}(b_i + 1).$$

Thus,

$$\begin{aligned} F(\boldsymbol{\Lambda}, \boldsymbol{\Gamma}) &= C + \sum_i \log(b_i + 1) - \text{tr}\, \boldsymbol{M}'\boldsymbol{B}(\boldsymbol{I} + \boldsymbol{B})^{-1}\boldsymbol{M}\boldsymbol{\Theta} \\ &= C + \sum_i \log(b_i + 1) - \sum_{i,j} \frac{b_i\theta_j}{1 + b_i} m_{ij}^2, \end{aligned}$$

where C depends only on $\boldsymbol{S}$ and $\boldsymbol{\Psi}$. If the eigenvalues $\{b_i\}$ and $\{\theta_j\}$ are written in decreasing order, then, for fixed $\{b_i\}$ and $\{\theta_j\}$, this quantity is minimized over $\boldsymbol{M}$ when $m_{ii} = 1$ for $i = 1, \ldots, k$, and $m_{ij} = 0$ for $i \neq j$. Thus, $\hat{\boldsymbol{M}} = (\boldsymbol{I}_k, \boldsymbol{0})$, and so, $\boldsymbol{G} = \boldsymbol{\Gamma}\hat{\boldsymbol{M}}' = (\boldsymbol{\gamma}_{(1)}, \ldots, \boldsymbol{\gamma}_{(k)})$. Then,

$$F(\boldsymbol{\Lambda}, \boldsymbol{\Psi}) = C + \sum_i \left[\log(b_i + 1) - \frac{b_i\theta_i}{b_i + 1}\right].$$

The minimum of this function over $b_i \geq 0$ occurs when

$$b_i = \max(\theta_i - 1, 0) \tag{10.24}$$

(although the value 0 in fact occurs rarely in practice), and so, the minimizing value of $\boldsymbol{\Lambda}^*$ satisfies

$$\boldsymbol{\Lambda}^*\boldsymbol{\Lambda}^{*\prime} = \sum b_i \boldsymbol{\gamma}_{(i)}\boldsymbol{\gamma}_{(i)}'. \tag{10.25}$$

If the constraint (10.8) is satisfied, then the columns of $\boldsymbol{\Lambda}^*$ are the eigenvectors of $\boldsymbol{\Lambda}^*\boldsymbol{\Lambda}^{*\prime}$, and the theorem follows. □

When $|\boldsymbol{\Psi}| = 0$, the constraint of Exercise 10.2.8 can be used, and special care must be taken in the minimization of $F(\boldsymbol{\Lambda}, \boldsymbol{\Psi})$. As in Example 10.2.1, we term a situation in which $|\hat{\boldsymbol{\Psi}}| = 0$, a Heywood case. Usually, the estimate $\hat{\boldsymbol{\Psi}}$ will be positive definite, although Heywood cases are by no means uncommon (Joreskog, 1967).

An important property of the m.l.e.s $\hat{\boldsymbol{\Lambda}}$ and $\hat{\boldsymbol{\Psi}}$ is that the estimate of the variance of the ith variable,

$$\hat{\sigma}_{ii} = \sum_{j=1}^{k} \hat{\lambda}_{ij}^2 + \hat{\psi}_{ii},$$

equals the sample variance s_{ii} for $i = 1, \ldots, p$ (Joreskog, 1967). We prove this result in the special case when the solution is proper, that is when $\hat{\boldsymbol{\Psi}} > 0$, though it is true in general.

Theorem 10.4.2 *If* $\hat{\boldsymbol{\Psi}} > 0$, *then* $\text{diag}(\hat{\boldsymbol{\Lambda}}\hat{\boldsymbol{\Lambda}}' + \hat{\boldsymbol{\Psi}} - \boldsymbol{S}) = \boldsymbol{0}$.

Proof: Writing $\partial F/\partial \boldsymbol{\Psi}$ as a diagonal matrix and using Section A.9 and the chain rule, it is straightforward to show that

$$\frac{\partial F(\boldsymbol{\Lambda}, \boldsymbol{\Psi})}{\partial \boldsymbol{\Psi}} = \text{diag}(\boldsymbol{\Sigma}^{-1}(\boldsymbol{\Sigma} - \boldsymbol{S})\boldsymbol{\Sigma}^{-1}), \tag{10.26}$$

which equals $\boldsymbol{0}$ at $(\hat{\boldsymbol{\Lambda}}, \hat{\boldsymbol{\Psi}})$ when $\hat{\boldsymbol{\Psi}}$ is proper. From (10.24), it is clear that at the maximum-likelihood solution, $b_i^{1/2}(1 + b_i) = b_i^{1/2}\theta_i$ for all $i = 1, \ldots, p$. Hence,

$$\begin{aligned} \boldsymbol{S}^* \hat{\boldsymbol{\lambda}}_{(i)}^* &= b_i^{1/2}\theta_i \boldsymbol{\gamma}_{(i)}^* \\ &= b_i^{1/2}(1 + b_i)\boldsymbol{\gamma}_{(i)}^* \\ &= (1 + b_i)\hat{\boldsymbol{\lambda}}_{(i)}^*, \quad i = 1, \ldots, p. \end{aligned}$$

Since $\boldsymbol{B} = \hat{\boldsymbol{\Lambda}}^{*\prime}\hat{\boldsymbol{\Lambda}}^*$ and $\hat{\boldsymbol{\Sigma}}^* = \hat{\boldsymbol{\Lambda}}^*\boldsymbol{\Lambda}^{*\prime} + \boldsymbol{I}$, we can write this in matrix form as

$$\boldsymbol{S}^*\hat{\boldsymbol{\Lambda}}^* = \hat{\boldsymbol{\Lambda}}^*(\boldsymbol{I} + \hat{\boldsymbol{\Lambda}}^{*\prime}\hat{\boldsymbol{\Lambda}}^*) = (\boldsymbol{I} + \hat{\boldsymbol{\Lambda}}^*\hat{\boldsymbol{\Lambda}}^{*\prime})^{-1}\hat{\boldsymbol{\Lambda}}^*.$$

Thus,

$$(\hat{\boldsymbol{\Sigma}}^* - \boldsymbol{S}^*)\hat{\boldsymbol{\Lambda}}^* = \boldsymbol{0}.$$

Using this result together with the expansion (A.29) for $\hat{\boldsymbol{\Sigma}}^{*-1}$, we get

$$\begin{aligned} (\hat{\boldsymbol{\Sigma}}^* - \boldsymbol{S}^*)\hat{\boldsymbol{\Sigma}}^{*-1} &= (\hat{\boldsymbol{\Sigma}}^* - \boldsymbol{S}^*)(\boldsymbol{I} - \hat{\boldsymbol{\Lambda}}^*(\boldsymbol{I} + \hat{\boldsymbol{\Lambda}}^{*\prime}\hat{\boldsymbol{\Lambda}}^*)^{-1}\hat{\boldsymbol{\Lambda}}^{*\prime}) \\ &= \hat{\boldsymbol{\Sigma}}^* - \boldsymbol{S}^*. \end{aligned}$$

Then pre- and postmultiplying by $\hat{\boldsymbol{\Psi}}^{1/2}$ and $\hat{\boldsymbol{\Psi}}^{-1/2}$, respectively, gives

$$(\hat{\boldsymbol{\Sigma}} - \boldsymbol{S})\hat{\boldsymbol{\Sigma}}^{-1} = (\hat{\boldsymbol{\Sigma}} - \boldsymbol{S})\hat{\boldsymbol{\Psi}}^{-1}.$$

Using this formula in (10.26), we get

$$\begin{aligned} \frac{\partial F(\hat{\boldsymbol{\Lambda}}, \hat{\boldsymbol{\Psi}})}{\partial \boldsymbol{\Psi}} &= \text{Diag}(\hat{\boldsymbol{\Sigma}}^{-1}(\hat{\boldsymbol{\Sigma}} - \boldsymbol{S})\hat{\boldsymbol{\Sigma}}^{-1}) \\ &= \text{Diag}(\hat{\boldsymbol{\Psi}}^{-1}(\hat{\boldsymbol{\Sigma}} - \boldsymbol{S})\hat{\boldsymbol{\Psi}}^{-1}) \\ &= \boldsymbol{0}, \end{aligned}$$

and hence, $\text{Diag}(\hat{\boldsymbol{\Sigma}} - \boldsymbol{S}) = \boldsymbol{0}$ as required. □

Example 10.4.1 Consider the open/closed-book data of Example 10.3.1. A maximum-likelihood factor analysis for $k = 1$ and $k = 2$ factors is presented in Table 10.2. The value and interpretations of the factor loadings are broadly similar to the results of the principal

Table 10.2 Maximum-likelihood factor solutions for the open/closed-book data with $k = 1$ and $k = 2$ factors.

	$k = 1$		$k = 2$			$k = 2$ (rotated)	
Variable	h_i^2	$\lambda_{(1)}$	h_i^2	$\lambda_{(1)}$	$\lambda_{(2)}$	$\delta_{(1)}$	$\delta_{(2)}$
1	0.359	0.599	0.534	0.628	0.373	0.265	0.681
2	0.445	0.667	0.581	0.695	0.312	0.356	0.674
3	0.842	0.917	0.811	0.899	−0.050	0.740	0.514
4	0.597	0.772	0.648	0.780	−0.201	0.738	0.322
5	0.524	0.724	0.569	0.727	−0.200	0.696	0.290

factor analysis. However, caution must be expressed when comparing the two sets of factor loadings when $k = 2$ because they have been rotated to satisfy different constraints ((10.8) and (10.9), respectively). Note that, unlike principal factor analysis, the factor loadings when $k = 1$ do not exactly equal the first column of factor loadings when $k = 2$. The rotated factors are discussed in Example 10.6.1. □

10.5 Goodness-of-fit Test

One of the main advantages of the maximum-likelihood technique is that it provides a test of the hypothesis H_k that k common factors are sufficient to describe the data against the alternative that $\mathbf{\Sigma}$ has no constraints. We showed in (6.16) that the likelihood ratio statistic λ is given by

$$-2\log\lambda = np(\hat{a} - \log\hat{g} - 1), \tag{10.27}$$

where $\hat{a}$ and $\hat{g}$ are the arithmetic and geometric means of the eigenvalues of $\hat{\mathbf{\Sigma}}^{-1}\mathbf{S}$.

Comparing this with (10.23), we see that

$$-2\log\lambda = nF(\hat{\mathbf{\Lambda}}, \hat{\mathbf{\Psi}}). \tag{10.28}$$

The statistic $-2\log\lambda$ has an asymptotic χ_s^2 distribution under H_k, where s is given by (10.10).

In the trivial case where $k = 0$, H_0 is testing the independence of the observed variables. The m.l.e. of $\mathbf{\Sigma}$ in this case is $\hat{\mathbf{\Sigma}} = \text{Diag}(\mathbf{S})$, and the eigenvalues of $\hat{\mathbf{\Sigma}}^{-1}\mathbf{S}$ are the same as those of $\hat{\mathbf{\Sigma}}^{-1/2}\mathbf{S}\hat{\mathbf{\Sigma}}^{-1/2} = \mathbf{R}$. In this case, (10.27) becomes $-n\log|\mathbf{R}|$, a statistic which has already been discussed in (6.21) and Section 9.4.4.

Bartlett (1954) showed that the chi-squared approximation of (10.27) is improved if n is replaced by

$$n' = n - 1 - \frac{1}{6}(2p + 5) - \frac{2}{3}k. \tag{10.29}$$

The chi-squared approximation to which this leads can probably be trusted if $n \geq p + 50$. Hence, for any specified k, we test H_k with the statistic

$$U = n'F(\hat{\mathbf{\Lambda}}, \hat{\mathbf{\Psi}}), \tag{10.30}$$

where n' is given by (10.29), F is given by (10.22), and $\hat{\boldsymbol{\Lambda}}$ and $\hat{\boldsymbol{\Psi}}$ are the m.l.e.s. When H_k is true, this statistic has an asymptotic chi-squared distribution with

$$s = \frac{1}{2}(p-k)^2 - \frac{1}{2}(p+k) \qquad (10.31)$$

degrees of freedom.

Of course, only rarely is k, the number of common factors, specified in advance. In practice, the problem is usually to decide how many common factors it is worth fitting to the data. To cope with this, a sequential procedure for estimating k is needed. The usual method is to start with a small value of k, say $k = 0$ or $k = 1$, and increase the number of common factors one by one until H_k is not rejected. However, this procedure is open to criticism because the critical values of the test criterion have not been adjusted to allow for the fact that a set of hypotheses is being tested in sequence.

For some data it may happen that the k-factor model is rejected for *all* values of k for which $s \geq 0$ in (10.31). In such cases, we conclude that there is *no* factor model which will fit the data. Of course, if there exists a k for which $s = 0$, then it is usually possible to fit the data *exactly* using the factor model. But no reduction in parameters takes place in this case, so no hypothesis testing can be carried out, and it is up to the user to decide whether this reparameterization is helpful.

Example 10.5.1 Consider the open/closed-book data for which we found the maximum-likelihood factor solutions in Example 10.4.1. Here, $n = 88$. For $k = 1$ factor, the goodness of fit statistic in (10.30) is given by

$$U = 8.65 \sim \chi_5^2.$$

Since $\chi_5^2(0.95) = 11.07$, we accept the one-factor solution as adequate for these data. □

10.6 Rotation of Factors

10.6.1 Interpretation of Factors

The constraint (10.8) or (10.9) on the factor loadings is a mathematical convenience to make the loadings unique; however, it can complicate the problem of interpretation. The interpretation of the factor loadings is the most straightforward if each variable is loaded highly on at most one factor, and if all the factor loadings are either large and positive or near zero, with few intermediate values. The variables are then split into disjoint sets, each of which is associated with one factor, and perhaps, some variables are left over. A factor j can then be interpreted as an average quality over those variables i for which λ_{ij} is large.

The factors $\boldsymbol{f}$ in the factor model (10.2) are mathematical abstractions and do not necessarily have any intuitive meaning. In particular, the factors may be rotated using (10.7) without affecting the validity of the model, and we are free to choose such a rotation to make the factors as intuitively meaningful as possible.

It is considered a disadvantage to choose a rotation subjectively because the factor analyst may try to force the factor loadings to fit his own preconceived pattern. A convenient analytical choice of rotation is given by the varimax method described below.

10.6.2 Varimax Rotation

The varimax method of orthogonal rotation was proposed by Kaiser (1958). Its rationale is to provide axes with a few large loadings and as many near-zero loadings as possible. This is accomplished by an iterative maximization of a quadratic function of the loadings.

Let $\mathbf{\Lambda}$ be a $(p \times k)$ matrix of unrotated loadings, and let $\boldsymbol{G}$ be a $(k \times k)$ orthogonal matrix. The matrix of rotated loadings is

$$\mathbf{\Delta} = \mathbf{\Lambda}\boldsymbol{G};$$

that is, δ_{ij} represents the loading of the ith variable on the jth factor.

The function ϕ that the varimax criterion maximizes is the sum of the variances of the *squared* loadings within each column of the loading matrix, where each row of loadings is normalized by its communality; that is,

$$\phi = \sum_{j=1}^{k}\sum_{i=1}^{p}(d_{ij}^2 - \bar{d}_j)^2 = \sum_{j=1}^{k}\sum_{i=1}^{p} d_{ij}^4 - p\sum_{j=1}^{k}\bar{d}_j^2, \tag{10.32}$$

where

$$d_{ij} = \frac{\delta_{ij}}{h_i} \quad \text{and} \quad \bar{d}_j = \frac{1}{p}\sum_{i=1}^{p} d_{ij}^2.$$

The varimax criterion ϕ is a function of $\boldsymbol{G}$, and the iterative algorithm proposed by Kaiser finds the orthogonal matrix $\boldsymbol{G}$ which maximizes ϕ. See Horst (1965, chapter 18) and Lawley and Maxwell (1971, p. 72).

In the case where $k = 2$, the calculations simplify. Then, $\boldsymbol{G}$ is given by

$$\boldsymbol{G} = \begin{bmatrix} \cos\theta & \sin\theta \\ -\sin\theta & \cos\theta \end{bmatrix}$$

and represents a rotation of the coordinate axes clockwise by an angle θ. Thus,

$$d_{i1} = \left\{\lambda_{i1}\cos\theta - \lambda_{i2}\sin\theta\right\}/h_i, \qquad d_{i2} = \{\lambda_{i1}\sin\theta + \lambda_{i2}\cos\theta\}/h_i.$$

Let

$$G_{a,b} = \sum_{i=1}^{p}\frac{\lambda_{i1}^a\lambda_{i2}^b}{h_i^{a+b}}.$$

Substituting in (10.32) and using

$$4(\cos^4\theta + \sin^4\theta) = 3 + \cos 4\theta, \qquad \sin 2\theta = 2\sin\theta\cos\theta,$$

$$\cos 2\theta = \cos^2\theta - \sin^2\theta,$$

it can be shown that

$$4\phi = (A^2 + B^2)^{1/2}\cos(4\theta - \alpha) + C, \tag{10.33}$$

where

$$A = G_{0,4} + G_{4,0} - 6G_{2,2} - G_{0,2}^2 - G_{2,0}^2 + 2G_{0,2}G_{2,0} + 4G_{1,1}^2, \tag{10.34}$$

$$B = 4(G_{1,3} - G_{3,1} - G_{1,1}G_{0,2} + G_{1,1}G_{2,0}), \tag{10.35}$$

$$C = p\left[3(G_{2,0} + G_{0,2})^2 - (3G_{0,2}^2 + 3G_{2,0}^2 + 2G_{0,2}G_{2,0} + 4G_{1,1}^2)\right], \tag{10.36}$$

and

$$(A^2 + B^2)^{1/2} \cos \alpha = A \qquad (A^2 + B^2)^{1/2} \sin \alpha = B. \tag{10.37}$$

In (10.33), the maximum value of ϕ is obtained when $4\theta = \alpha$. The value of α is obtained from (10.37) using

$$\tan \alpha = B/A, \tag{10.38}$$

and α is uniquely determined from a consideration of the signs of A and B. Note that θ can take four possible values.

In the case of $k > 2$ factors, an iterative solution for the rotation is used. The first and second factors are rotated by an angle determined by the above method. The new first factor is then rotated with the original third factor, and so on, until all the $k(k-1)/2$ pairs of factors have been rotated. This sequence of rotations is called a cycle. These cycles are then repeated until one is completed in which all the angles have achieved some predetermined convergence criterion. Since the effect of each rotation is to increase the value of ϕ, and the values of ϕ are bounded above, the iteration will converge (Kaiser, 1958).

Example 10.6.1 Consider the two-factor maximum-likelihood solutions for the open/closed- book data given in Example 10.4.1. Using (10.33)–(10.38), it is found that $\theta = 38.0°$ for the varimax rotation, and the rotated factors are given in Table 10.2. The rotation can be represented graphically (Figure 10.1) by plotting the factor loadings as $p = 5$ points in $k = 2$ dimensions. The first factor now represents as "open-book" effect, and the second factor a "closed-book" effect, although both factors influence the other exams as well. □

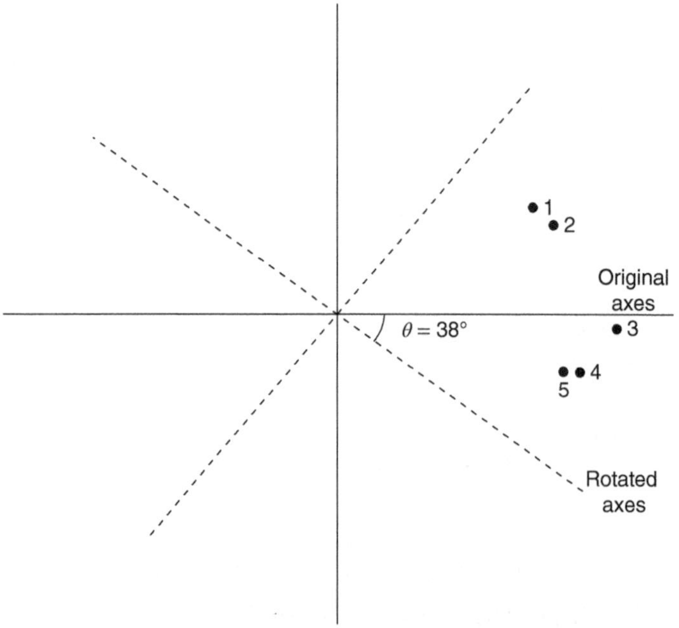

Figure 10.1 Plot of open/closed-book factors before and after varimax rotation.

In this simple example, the overall interpretation is the same whether we use the rotated or the original factors. However, in more complicated situations, the benefits of rotation are much more noticeable.

Example 10.6.2 (Kendall, 1975). In a job interview, 48 applicants were each judged on 15 variables. The variables were

(1)	Form of letter of application	(9)	Experience
(2)	Appearance	(10)	Drive
(3)	Academic ability	(11)	Ambition
(4)	Likeability	(12)	Grasp
(5)	Self-confidence	(13)	Potential
(6)	Lucidity	(14)	Keenness to join
(7)	Honesty	(15)	Suitability
(8)	Salesmanship		

and the correlation matrix between them is given in Table 10.3. Because many of the correlations were quite high, it was felt that the judge might be confusing some of the variables in his own mind. Therefore, a factor analysis was performed to discover the underlying types of variation in the data. A maximum-likelihood factor analysis was carried out, and using the hypothesis test described in Section 10.5, it was found that six factors were needed to explain the data adequately. The original factor loadings are given in Table 10.4, and the factor loadings after the varimax rotation are given in Table 10.5.

Table 10.3 Correlation matrix for the applicant data

	2	3	4	5	6	7	8	9	10	11	12	13	14	15
1	0.24	0.04	0.31	0.09	0.23	−0.11	0.27	0.55	0.35	0.28	0.34	0.37	0.47	0.59
2		0.12	0.38	0.43	0.37	0.35	0.48	0.14	0.34	0.55	0.51	0.51	0.28	0.38
3			0.00	0.00	0.08	−0.03	0.05	0.27	0.09	0.04	0.20	0.29	−0.32	0.14
4				0.30	0.48	0.65	0.35	0.14	0.39	0.35	0.50	0.61	0.69	0.33
5					0.81	0.41	0.82	0.02	0.70	0.84	0.72	0.67	0.48	0.25
6						0.36	0.83	0.15	0.70	0.76	0.88	0.78	0.53	0.42
7							0.23	−0.16	0.28	0.21	0.39	0.42	0.45	0.00
8								0.23	0.81	0.86	0.77	0.73	0.55	0.55
9									0.34	0.20	0.30	0.35	0.21	0.69
10										0.78	0.71	0.79	0.61	0.62
11											0.78	0.77	0.55	0.43
12												0.88	0.55	0.53
13													0.54	0.57
14														0.40

Table 10.4 Maximum-likelihood factor solution of applicant data with $k = 6$ factors, unrotated.

	Factor loadings					
Variable	**1**	**2**	**3**	**4**	**5**	**6**
1	0.361	0.483	−0.378	0.320	0.026	−0.010
2	0.453	0.324	0.267	−0.075	−0.016	−0.338
3	0.024	0.447	0.184	−0.388	−0.278	0.172
4	0.526	0.295	0.365	0.323	0.398	0.087
5	0.840	−0.234	0.144	−0.224	0.025	−0.160
6	0.892	−0.012	−0.019	−0.319	0.287	0.123
7	0.366	−0.036	0.728	0.183	0.300	0.127
8	0.900	−0.043	−0.087	−0.139	−0.057	−0.063
9	0.265	0.589	−0.340	0.150	−0.198	0.135
10	0.911	−0.021	0.004	0.121	−0.342	0.183
11	0.900	0.002	0.000	−0.130	−0.136	−0.384
12	0.872	0.251	0.074	−0.198	0.156	0.016
13	0.857	0.345	0.210	−0.087	−0.043	0.054
14	0.718	−0.009	−0.017	0.614	0.308	−0.057
15	0.556	0.495	−0.269	0.127	−0.219	0.204

Table 10.5 Maximum-likelihood factor solution of applicant data with $k = 6$ factors, varimax rotation

	Factor loadings					
Variable	**1**	**2**	**3**	**4**	**5**	**6**
1	0.112	0.746	0.068	−0.139	0.110	−0.065
2	0.344	0.141	0.256	0.153	0.524	−0.019
3	0.049	0.140	0.000	0.681	0.076	0.014
4	0.238	0.262	0.782	−0.096	0.126	−0.083
5	0.879	−0.091	0.182	−0.073	0.172	0.086
6	0.915	0.142	0.284	0.033	−0.085	−0.219
7	0.198	−0.231	0.834	0.016	0.072	0.104
8	0.870	0.233	0.099	−0.057	0.108	0.099
9	0.061	0.759	−0.038	0.168	0.045	0.031
10	0.756	0.377	0.203	−0.002	−0.026	0.490
11	0.859	0.179	0.052	−0.085	0.442	0.131
12	0.784	0.311	0.352	0.136	0.146	−0.129
13	0.693	0.372	0.424	0.243	0.211	0.090
14	0.400	0.400	0.544	−0.598	0.089	0.104
15	0.341	0.753	0.076	0.159	0.002	0.134

It is very difficult to interpret the unrotated loadings. However, a pattern is slightly clearer from the rotated loadings, The first factor is loaded heavily on variables 5, 6, 8, 10, 11, 12, and 13 and represents perhaps an outward and salesmanlike personality. Factor 2, weighting variables 1, 9, and 15, represents relevant experience, and factor 3, weighting variables 4 and 7, represents likeability. Factor 4 weights variables 3 (academic ability) and 14 (keenness to join), and factor 5 emphasizes appearance (2).

Thus, the judge seems to have measured rather fewer qualities in his candidates than he thought.

A recent paper by Rohe and Zheng (2023) takes a fresh look at Varimax rotation. Their key idea is to give a probabilistic meaning to this rotation by moving away from the assumption of multivariate normality. □

10.7 Factor Scores

So far, our study of the factor model has been concerned with the way in which the observed variables are functions of the (unknown) factors. For example, in Spearman's examination data, we can describe the way in which a child's test scores will depend on his overall intelligence. However, it is also of interest to ask the converse question. Given a particular child's test scores, can we make any statement about his overall intelligence? For the general model we want to know how the factors depend on the observed variables.

One way to approach the problem is to treat the unknown *common factor scores* as parameters to be estimated. Suppose that $\boldsymbol{x}$ is a multivariate normal random vector from the factor model (10.2), and that $\boldsymbol{\Lambda}$ and $\boldsymbol{\Psi}$ and $\boldsymbol{\mu} = \boldsymbol{0}$ are known. Given the vector $\boldsymbol{f}(p \times 1)$ of common factor scores, $\boldsymbol{x}$ is distributed as $N_p(\boldsymbol{\Lambda f}, \boldsymbol{\Psi})$. Thus, the log likelihood of $\boldsymbol{x}$ is given by

$$l(\boldsymbol{x};\boldsymbol{f}) = -\frac{1}{2}(\boldsymbol{x} - \boldsymbol{\Lambda f})'\boldsymbol{\Psi}^{-1}(\boldsymbol{x} - \boldsymbol{\Lambda f}) - \frac{1}{2}\log|2\pi\boldsymbol{\Psi}|. \tag{10.39}$$

Setting the derivative with respect to $\boldsymbol{f}$ equal to $\boldsymbol{0}$ gives

$$\frac{\partial l}{\partial \boldsymbol{f}} = \boldsymbol{\Lambda}'\boldsymbol{\Psi}^{-1}(\boldsymbol{x} - \boldsymbol{\Lambda f}) = \boldsymbol{0},$$

so

$$\hat{\boldsymbol{f}} = (\boldsymbol{\Lambda}'\boldsymbol{\Psi}^{-1}\boldsymbol{\Lambda})^{-1}\boldsymbol{\Lambda}'\boldsymbol{\Psi}^{-1}\boldsymbol{x}. \tag{10.40}$$

The estimate in (10.40) is known as *Bartlett's factor score*. The specific factor scores can then be estimated by $\hat{\boldsymbol{u}} = \boldsymbol{x} - \boldsymbol{\Lambda}\hat{\boldsymbol{f}}$.

Note that (10.39) is the logarithm of the conditional density of $\boldsymbol{x}$, *given* $\boldsymbol{f}$. However, under the factor model, $\boldsymbol{f}$ can be considered as an $N_p(\boldsymbol{0}, \boldsymbol{I})$ random vector, thus giving $\boldsymbol{f}$ a prior distribution. Using this Bayesian approach, the posterior density of $\boldsymbol{f}$ is proportional to

$$\exp\left[-\frac{1}{2}(\boldsymbol{x} - \boldsymbol{\Lambda f})'\boldsymbol{\Psi}^{-1}(\boldsymbol{x} - \boldsymbol{\Lambda f}) - \frac{1}{2}\boldsymbol{f}'\boldsymbol{f}\right],$$

which is a multivariate normal density whose mean

$$\boldsymbol{f}^* = (\boldsymbol{I} + \boldsymbol{\Lambda}'\boldsymbol{\Psi}^{-1}\boldsymbol{\Lambda})^{-1}\boldsymbol{\Lambda}'\boldsymbol{\Psi}^{-1}\boldsymbol{x} \tag{10.41}$$

is the Bayesian estimate of $\boldsymbol{f}$. The estimate in (10.41) is known as *Thompson's factor score*.

Each of these two-factor scores has some favorable properties, and there has been a long controversy as to which is better. For example,

$$\mathrm{E}(\hat{\boldsymbol{f}} \mid \boldsymbol{f}) = \boldsymbol{f}, \qquad \mathrm{E}(\boldsymbol{f}^* \mid \boldsymbol{f}) = (\boldsymbol{I} + \boldsymbol{\Lambda}'\boldsymbol{\Psi}^{-1}\boldsymbol{\Lambda})^{-1}\boldsymbol{\Lambda}'\boldsymbol{\Psi}^{-1}\boldsymbol{\Lambda}\boldsymbol{f}, \tag{10.42}$$

so that Bartlett's factor score is an unbiased estimate of $\boldsymbol{f}$, whereas Thompson's score is biased. However, the average prediction errors are given by

$$\mathrm{E}((\hat{\boldsymbol{f}} - \boldsymbol{f})(\hat{\boldsymbol{f}} - \boldsymbol{f})') = (\boldsymbol{\Lambda}'\boldsymbol{\Psi}^{-1}\boldsymbol{\Lambda})^{-1}, \tag{10.43}$$

$$\mathrm{E}((\boldsymbol{f}^* - \boldsymbol{f})(\boldsymbol{f}^* - \boldsymbol{f})') = (\boldsymbol{I} + \boldsymbol{\Lambda}'\boldsymbol{\Psi}^{-1}\boldsymbol{\Lambda})^{-1}, \tag{10.44}$$

so that Thompson's score is more accurate. Hence, overall Thompson's score seems preferable. If the columns of $\boldsymbol{\Lambda}$ satisfy the constraint (10.8), then, for either factor score, the components of the factor score are uncorrelated with one another. Note that if the eigenvalues of $\boldsymbol{\Lambda}'\boldsymbol{\Psi}^{-1}\boldsymbol{\Lambda}$ are all large, then the prediction errors will be small, and also the two factor scores will be similar to one another.

Of course, in practice, $\boldsymbol{\Lambda}$, $\boldsymbol{\Psi}$, and $\boldsymbol{\mu}$ are not known in advance but are estimated from the same data for which we wish to know the factor scores. It would be theoretically attractive to estimate the factor scores, factor loadings, and specific variances all at the same time from the data, using maximum likelihood. However, there are too many parameters for this to be possible. There are many values of the parameters for which the likelihood becomes infinite; see Exercise 10.7.2.

10.8 Relationships Between Factor Analysis and Principal Component Analysis

Factor analysis, like principal component analysis, is an attempt to explain a set of data in a smaller number of dimensions than one starts with. Because the overall goals are quite similar, it is worth looking at the differences between these two approaches.

Firstly, principal component analysis is merely a transformation of the data. No assumptions are made about the form of the covariance matrix from which the data comes. On the other hand, factor analysis supposes that the data comes from the well-defined model (10.2), where the underlying factors satisfy the assumptions (10.3)–(10.5). If these assumptions are not met, then factor analysis may give spurious results.

Secondly, in principal component analysis, the emphasis is on a transformation from the observed variables to the principal components ($\boldsymbol{y} = \boldsymbol{\Gamma}'\boldsymbol{x}$), whereas in factor analysis, the emphasis is on a transformation *from* the underlying factors *to* the observed variables. Of course, the principal component transformation is invertible ($\boldsymbol{x} = \boldsymbol{\Gamma}\boldsymbol{y}$), and if we have decided to retain the first k components, then $\boldsymbol{x}$ can be approximated by these components,

$$\boldsymbol{x} = \boldsymbol{\Gamma}y = \boldsymbol{\Gamma}_1\boldsymbol{y}_1 + \boldsymbol{\Gamma}_2\boldsymbol{y}_2 \approx \boldsymbol{\Gamma}_1\boldsymbol{y}_1.$$

However, this point of view is less natural than in factor analysis where $\boldsymbol{x}$ can be approximated in terms of the common factors

$$\boldsymbol{x} \approx \boldsymbol{\Lambda}\boldsymbol{f},$$

and the neglected specific factors are explicitly assumed to be "noise."

Note that in Section 10.3, when the specific variances are assumed to be 0, principal factor analysis is equivalent to principal component analysis. Thus, if the factor model holds, and if the specific variances are small, we expect principal component analysis and factor analysis to give similar results. However, if the specific variances are large, they will be absorbed into all the principal components, both retained and rejected, whereas factor analysis makes special provision for them.

10.9 Analysis of Covariance Structures

The factor model can be generalized in the following way (Joreskog, 1970). Let $\boldsymbol{X}(n \times p)$ be a random data matrix with mean

$$\mathsf{E}(\boldsymbol{X}) = \boldsymbol{A\Xi P}, \tag{10.45}$$

where $\boldsymbol{A}(n \times g)$ and $\boldsymbol{P}(h \times p)$ are known matrices, and $\boldsymbol{\Xi}(g \times h)$ is a matrix of parameters. Suppose that each row of $\boldsymbol{X}$ is normally distributed with the same covariance matrix

$$\boldsymbol{\Sigma} = \boldsymbol{B}(\boldsymbol{\Lambda\Phi\Lambda}' + \boldsymbol{\Psi})\boldsymbol{B}' + \boldsymbol{\Theta}. \tag{10.46}$$

Here, $\boldsymbol{B}(p \times q)$, $\boldsymbol{\Lambda}(q \times k)$, the symmetric matrix $\boldsymbol{\Phi} \geq 0$, and the diagonal matrices $\boldsymbol{\Psi} \geq 0$ and $\boldsymbol{\Theta} \geq 0$ are all parameters. In applications, some of the parameters may be fixed or constrained. For example, in factor analysis, all the rows of $\boldsymbol{X}$ have the same mean, and we take $\boldsymbol{B} = \boldsymbol{I}, \boldsymbol{\Phi} = \boldsymbol{I}$, and $\boldsymbol{\Theta} = \boldsymbol{0}$.

The general model represented by (10.45) and (10.46) also includes other models where the means and covariances are structured in terms of parameters to be estimated, for instance multivariate regression, MANOVA, path analysis, and growth curve analysis. For details, see Joreskog (1970) , Joreskog and Sorbom (1977). More general extensions include latent variable analysis and structural equation modeling (e.g. Bollen (1989)).

Exercises and Complements

10.2.1 In the factor model (10.2), prove that $\boldsymbol{\Sigma} = \boldsymbol{\Lambda\Lambda}' + \boldsymbol{\Psi}$.

10.2.2 (a) If $\boldsymbol{x}$ is a random vector whose covariance matrix can be written in the form $\boldsymbol{\Sigma} = \boldsymbol{\Lambda\Lambda}' + \boldsymbol{\Psi}$, show that there exist factors $\boldsymbol{f}, \boldsymbol{u}$ such that the k-factor model holds for $\boldsymbol{x}$. Hint: let $\boldsymbol{y} \sim N_k(\boldsymbol{0}, \boldsymbol{I} + \boldsymbol{\Lambda}'\boldsymbol{\Psi}^{-1}\boldsymbol{\Lambda})$ independently of $\boldsymbol{x}$, and define

$$\begin{pmatrix} \boldsymbol{u} \\ \boldsymbol{f} \end{pmatrix} = \begin{bmatrix} \boldsymbol{I}_p & \boldsymbol{\Lambda} \\ -\boldsymbol{\Lambda}'\boldsymbol{\Psi}^{-1} & \boldsymbol{I}_k \end{bmatrix}^{-1} \begin{pmatrix} \boldsymbol{x} - \boldsymbol{\mu} \\ \boldsymbol{y} \end{pmatrix}.$$

(b) If $\boldsymbol{x}$ is normally distributed, show that $(\boldsymbol{f}, \boldsymbol{u})$ may be assumed to be normally distributed.

(c) Show that $\boldsymbol{f}$ and $\boldsymbol{u}$ are not uniquely determined by $\boldsymbol{x}$.

10.2.3 If $\boldsymbol{\Lambda\Lambda}' = \boldsymbol{\Delta\Delta}'$ ($\boldsymbol{\Lambda}$ and $\boldsymbol{\Delta}$ are $(p \times k)$ matrices), show that $\boldsymbol{\Delta} = \boldsymbol{\Lambda G}$ for some $(k \times k)$ orthogonal matrix $\boldsymbol{G}$. (Hint: use Theorem A.8).

10.2.4 If $\mathbf{\Lambda\Lambda}' = \mathbf{\Delta\Delta}'$, and $\mathbf{\Lambda}'\mathbf{\Psi}^{-1}\mathbf{\Lambda}$ and $\mathbf{\Delta\Psi}^{-1}\mathbf{\Delta}$ are both diagonal matrices with distinct elements written in decreasing order, show that $\lambda_{(i)} = \pm\delta_{(i)}, i = 1, \ldots, k$. (Hint: write

$$\mathbf{\Lambda\Lambda}' = \mathbf{\Psi}^{1/2}\left[\sum_{i=1}^{k} a_i \gamma_{(i)}\gamma'_{(i)}\right]\mathbf{\Psi}^{1/2},$$

where $\gamma_{(i)}$ is the ith standardized eigenvector of $\mathbf{\Psi}^{-1/2}\mathbf{\Lambda\Lambda}'\mathbf{\Psi}^{-1/2}$ with eigenvalue a_i, and show that $\mathbf{\Psi}^{-1/2}\lambda_{(i)} = \pm a_i^{1/2}\gamma_{(i)}$. Now do the same thing for $\mathbf{\Lambda}$.)

10.2.5 If $\boldsymbol{S}(3 \times 3)$ is a sample covariance matrix, solve $\boldsymbol{S} = \hat{\mathbf{\Lambda}}\hat{\mathbf{\Lambda}}' + \hat{\mathbf{\Psi}}$ for $\hat{\mathbf{\Lambda}}$ and $\hat{\mathbf{\Psi}}$. Compare the answer with (10.17) to show that the solution is scale invariant.

10.2.6 Let $\boldsymbol{R}(3 \times 3)$ be a correlation matrix with $r_{12} = r_{13} = 1/3$ and $r_{23} = 1/10$. Check that $\boldsymbol{R}$ is positive definite. Show that $\hat{\lambda}_1 > 1$ in (10.17), and hence that the exact factor solution is unacceptable.

10.2.7 (Structural relationships) Suppose that a scientist has p instruments with which to measure a physical quantity ξ. She knows that each instrument measures a linear function of ξ, but all the measurements are subject to (independent) errors. The scientist wishes to calibrate her instruments (that is, discover what the linear relationship between the different instruments would be if there were no error) and to discover the accuracy of each measurement.

If the scientist takes a sample of measurements for which the ξs can be regarded as realizations of a random variable, then show that her problem can be approached using the 1-factor model (10.2). In particular, if $p = 2$, show that without more information about the error variances, she does not have enough information to solve her problem (see Example 9.2.5).

10.2.8 Suppose that the factor model (10.2) is valid with some of the ψ_{ii} equal to 0. Then, of course, the constraint (10.8) is not appropriate, and we consider an alternative constraint. For simplicity of notation, suppose that $\psi_{11} = \cdots = \psi_{gg} = 0$ and $\psi_{g+1,g+1}, \ldots, \psi_{pp} > 0$ and partition

$$\mathbf{\Sigma} = \begin{bmatrix} \mathbf{\Sigma}_{11} & \mathbf{\Sigma}_{12} \\ \mathbf{\Sigma}_{21} & \mathbf{\Sigma}_{22} \end{bmatrix} \quad \text{and} \quad \mathbf{\Lambda} = \begin{bmatrix} \mathbf{\Lambda}_{11} & \mathbf{\Lambda}_{12} \\ \mathbf{\Lambda}_{21} & \mathbf{\Lambda}_{22} \end{bmatrix}.$$

We will explain the variation in the first g variables exactly in terms of the first g factors and the part of the variation of the last $(p - g)$ variables not explained by the first g variables ($\mathbf{\Sigma}_{22.1} = \mathbf{\Sigma}_{22} - \mathbf{\Sigma}_{22}\mathbf{\Sigma}_{11}^{-1}\mathbf{\Sigma}_{12}$) in terms of the remaining k factors. Let $\mathbf{\Sigma}_{11} = \mathbf{\Gamma\Theta\Gamma}'$ be a spectral decomposition of $\mathbf{\Sigma}_{11}(g \times g)$, where $\mathbf{\Theta}$ is a diagonal matrix of eigenvalues, and the columns of $\mathbf{\Gamma}$ are standardized eigenvectors. Suppose that $\mathbf{\Lambda}^*$ satisfies

$$\mathbf{\Lambda}_{11}^* = \mathbf{\Gamma\Theta}^{1/2}, \qquad \mathbf{\Lambda}_{12}^* = \mathbf{0},$$

$$\mathbf{\Lambda}_{21}^* = \mathbf{\Sigma}_{21}\mathbf{\Gamma\Theta}^{-1/2}, \qquad \mathbf{\Lambda}_{22}^{*\prime}\mathbf{\Psi}_2^{-1}\mathbf{\Lambda}_{22}^* \quad \text{is diagonal},$$

$$\mathbf{\Lambda}_{22}^*\mathbf{\Lambda}_{22}^{*\prime} + \mathbf{\Psi}_2 = \mathbf{\Sigma}_{22.1},$$

where $\Psi_2(p-g) \times (p-g)$ is a diagonal matrix with elements ψ_{ii}, $i > g$. Show that
(a) if Λ^* satisfies these conditions, then

$$\Sigma = \Lambda\Lambda' + \Psi = \Lambda^*\Lambda^{*\prime} + \Psi;$$

(b) there always exists an orthogonal matrix $G(k \times k)$ such that

$$\Lambda = \Lambda^* G;$$

(c) except for the signs and order of the columns, Λ^* is uniquely determined by the above conditions.

10.3.1 Suppose in a principal factor analysis that all the preliminary estimates of the specific variances are given by the same number $\tilde{\psi}_{ii} = c$.
(a) If $c = 0$, show that the columns of $\hat{\Lambda}$ are the first k principal component loading vectors of the correlation matrix R, scaled so that $\hat{\lambda}'_{(i)}\hat{\lambda}_{(i)}$ equals the variance of the ith principal component.
(b) If $c > 0$, show that the columns of $\hat{\Lambda}$ are still given by the principal component loading vectors of R but with a different scaling.
(c) If $k > 1$, and $R = (1-\rho)I + \rho\mathbf{1}\mathbf{1}'$, $0 < \rho < 1$, is the equicorrelation matrix, show that the largest value of c which gives an acceptable principal factor solution is $c = 1 - \rho$.

10.4.1 Let $F(\Lambda, \Psi; S)$ be given by (10.22), and let $\hat{\Lambda}, \hat{\Psi}$ denote the minimizing values of Λ and Ψ. If C is a fixed diagonal matrix with positive elements, show that $\Lambda = C\hat{\Lambda}$ and $\Psi = C\hat{\Psi}C$ minimize the functions $F(\Lambda, \Psi; CSC)$. Hence, deduce that the maximum likelihood estimators are scale invariant.

10.4.2 Explain why the hypothesis test described in Section 10.5 can be constructed with only a knowledge of R (the sample variances s_{ii} not being needed).

10.6.1 (Joreskog, 1967; Maxwell, 1961). A set of $p = 10$ psychological variables were measured on $n = 810$ normal children, with correlations given as follows:

Tests	2	3	4	5	6	7	8	9	10
1	0.345	0.594	0.404	0.579	−0.280	−0.449	−0.188	−0.303	−0.200
2		0.477	0.338	0.230	−0.159	−0.205	−0.120	−0.168	−0.145
3			0.498	0.505	−0.251	−0.377	−0.186	−0.273	−0.154
4				0.389	−0.168	−0.249	−0.173	−0.195	−0.055
5					−0.151	−0.285	−0.129	−0.159	−0.079
6						0.363	0.359	0.227	0.260
7							0.448	0.439	0.511
8								0.429	0.316
9									0.301

A maximum-likelihood factor analysis was carried out, with $k = 4$ factors yielding the following estimate $\hat{\boldsymbol{\Lambda}}$ of the factor loadings (rotated by varimax):

$$\begin{bmatrix} -0.03 & 0.59 & -0.31 & 0.41 \\ -0.04 & 0.09 & -0.12 & 0.59 \\ -0.06 & 0.42 & -0.20 & 0.69 \\ -0.11 & 0.33 & -0.08 & 0.48 \\ -0.06 & 0.76 & -0.07 & 0.24 \\ 0.23 & -0.11 & 0.36 & -0.17 \\ 0.15 & -0.24 & 0.78 & -0.16 \\ 0.93 & -0.04 & 0.37 & -0.06 \\ 0.27 & -0.11 & 0.44 & -0.18 \\ 0.09 & -0.00 & 0.63 & -0.04 \end{bmatrix}.$$

(a) Give interpretations of the four factors using the factor loadings.

(b) The maximum-likelihood factor loadings give a value of $F(\hat{\boldsymbol{\Lambda}}, \hat{\boldsymbol{\Psi}}) = 0.0228$ in (10.22). Using (10.30), carry out a test of the hypothesis that four factors are adequate to describe the data.

10.7.1 Verify formulae (10.42)–(10.44) for the factor scores. (Hint: in (10.44), use (A.29) to show that

$$\begin{aligned} \mathsf{E}(\boldsymbol{f}^*\boldsymbol{f}^{*\prime}) = \mathsf{E}(\boldsymbol{f}^*\boldsymbol{f}') &= (\boldsymbol{I} + \boldsymbol{\Lambda}'\boldsymbol{\Psi}^{-1}\boldsymbol{\Lambda})^{-1}\boldsymbol{\Lambda}'\boldsymbol{\Psi}^{-1}\boldsymbol{\Lambda} \\ &= \boldsymbol{I} - (\boldsymbol{I} + \boldsymbol{\Lambda}'\boldsymbol{\Psi}^{-1}\boldsymbol{\Lambda})^{-1}.) \end{aligned}$$

10.7.2 The following example shows that the specific variances and factor scores cannot be estimated together. For simplicity, let $k = 1$, and suppose that $\boldsymbol{\Lambda} = \lambda(p \times 1)$ and $\boldsymbol{\mu} = \boldsymbol{0}$ are known. Let $\boldsymbol{F} = \boldsymbol{f}(n \times 1)$ denote the unknown factor scores. Then, extending (10.39), the likelihood of the data $\boldsymbol{X}$ can be written as

$$L = |2\pi\boldsymbol{\Psi}|^{-1/2} \exp\left\{-\frac{1}{2}\sum_{i=1}^{p}(\boldsymbol{x}_{(i)} - \lambda_i\boldsymbol{f})'(\boldsymbol{x}_{(i)} - \lambda_i\boldsymbol{f})/\psi_{ii}\right\}.$$

(Note that $\boldsymbol{f}(n \times 1)$ here represents the scores of n individuals on one factor, whereas in (10.39), $\boldsymbol{f}(k \times 1)$ represented the scores on k factors of one individual.) Suppose that $\lambda_1 \neq 0$. Show that for any values of $\psi_{22}, \ldots, \psi_{pp}$ if $\boldsymbol{f} = \lambda_1^{-1}\boldsymbol{x}_{(1)}$ and $\psi_{11} \to 0$, then the likelihood becomes infinite. Hence, m.l.e.s do not exist for this problem.

11

Canonical Correlation Analysis

11.1 Introduction

Canonical correlation analysis involves partitioning a collection of variables into two sets, an $\boldsymbol{x}$-set and a $\boldsymbol{y}$-set. The object is then to find linear combinations $\eta = \boldsymbol{a}'\boldsymbol{x}$ and $\phi = \boldsymbol{b}'\boldsymbol{y}$ such that η and ϕ have the *largest possible* correlation. Such linear combinations can give insight into the relationships between the two sets of variables. Canonical correlation analysis has certain maximal properties similar to those of principal component analysis. However, while principal component analysis considers interrelationships *within* a set of variables, the focus of canonical correlation is on the relationship *between* two groups of variables. One way to view canonical correlation analysis is as an extension of multiple regression. Recall that in multiple regression analysis the variables are partitioned into an $\boldsymbol{x}$-set containing q variables and a y-set containing $p = 1$ variable. The regression solution involves finding the linear combination $\boldsymbol{a}'\boldsymbol{x}$, which is most highly correlated with y.

In canonical correlation analysis, the $\boldsymbol{y}$-set contains $p \geq 1$ variables, and we look for vectors $\boldsymbol{a}$ and $\boldsymbol{b}$ for which the correlation between $\boldsymbol{a}'\boldsymbol{x}$ and $\boldsymbol{b}'\boldsymbol{y}$ is maximized. If $\boldsymbol{x}$ is interpreted as "causing" $\boldsymbol{y}$, then $\boldsymbol{a}'\boldsymbol{x}$ may be called the "best predictor," and $\boldsymbol{b}'\boldsymbol{y}$ the "most predictable criterion." However, there is no assumption of causal asymmetry in the mathematics of canonical correlation analysis; $\boldsymbol{x}$ and $\boldsymbol{y}$ are treated symmetrically.

Example 11.1.1 Consider the open/closed-book data of Table 1.2 and split the five variables into two sets – the closed-book exams (x_1, x_2) and the open-book exams (y_1, y_2, y_3). One possible quantity of interest here is how highly a student's ability on closed-book exams is correlated with his ability on open-book exams.

Alternatively, one might try to use the open-book exam results to predict the closed-book results (or vice versa). This data is explored further in Example 11.2.5. □

Multivariate Analysis, Second Edition. Kanti V. Mardia, John T. Kent, and Charles C. Taylor.

11.2 Mathematical Development

11.2.1 Population Canonical Correlation Analysis

Suppose that $\boldsymbol{x}$ is a q-dimensional random vector, and that $\boldsymbol{y}$ is a p-dimensional random vector. Suppose further that $\boldsymbol{x}$ and $\boldsymbol{y}$ have means $\boldsymbol{\mu}$ and $\boldsymbol{v}$, and that

$$\mathsf{E}\{(\boldsymbol{x}-\boldsymbol{\mu})(\boldsymbol{x}-\boldsymbol{\mu})'\} = \boldsymbol{\Sigma}_{11}, \qquad \mathsf{E}\{(\boldsymbol{y}-\boldsymbol{v})(\boldsymbol{y}-\boldsymbol{v})'\} = \boldsymbol{\Sigma}_{22},$$
$$\mathsf{E}\{(\boldsymbol{x}-\boldsymbol{\mu})(\boldsymbol{y}-\boldsymbol{v})'\} = \boldsymbol{\Sigma}_{12} = \boldsymbol{\Sigma}_{21}'.$$

Now, consider the two linear combinations $\eta = \boldsymbol{a}'\boldsymbol{x}$ and $\phi = \boldsymbol{b}'\boldsymbol{y}$. The correlation between η and ϕ is

$$\rho(\boldsymbol{a},\boldsymbol{b}) = \frac{\boldsymbol{a}'\boldsymbol{\Sigma}_{12}\boldsymbol{b}}{(\boldsymbol{a}'\boldsymbol{\Sigma}_{11}\boldsymbol{a}\,\boldsymbol{b}'\boldsymbol{\Sigma}_{22}\boldsymbol{b})^{1/2}}. \tag{11.1}$$

We use the notation $\rho(\boldsymbol{a},\boldsymbol{b})$ to emphasize the fact that this correlation varies with different values of $\boldsymbol{a}$ and $\boldsymbol{b}$.

Now, we may ask what values of $\boldsymbol{a}$ and $\boldsymbol{b}$ maximize $\rho(\boldsymbol{a},\boldsymbol{b})$. Equivalently, we can solve the problem

$$\max_{\boldsymbol{a},\boldsymbol{b}} \boldsymbol{a}'\boldsymbol{\Sigma}_{12}\boldsymbol{b} \qquad \text{subject to} \quad \boldsymbol{a}'\boldsymbol{\Sigma}_{11}\boldsymbol{a} = \boldsymbol{b}'\boldsymbol{\Sigma}_{22}\boldsymbol{b} = 1, \tag{11.2}$$

because (11.1) does not depend on the scaling of $\boldsymbol{a}$ and $\boldsymbol{b}$.

The solution to this problem is given in Theorem 11.2.1. First, we need some notation.

Suppose that $\boldsymbol{\Sigma}_{11}$ and $\boldsymbol{\Sigma}_{22}$ are nonsingular, and let

$$\boldsymbol{K} = \boldsymbol{\Sigma}_{11}^{-1/2}\boldsymbol{\Sigma}_{12}\boldsymbol{\Sigma}_{22}^{-1/2}. \tag{11.3}$$

Then, set

$$\boldsymbol{N}_1 = \boldsymbol{K}\boldsymbol{K}', \qquad \boldsymbol{N}_2 = \boldsymbol{K}'\boldsymbol{K} \tag{11.4}$$

and

$$\begin{aligned} \boldsymbol{M}_1 &= \boldsymbol{\Sigma}_{11}^{-1/2}\boldsymbol{N}_1\boldsymbol{\Sigma}_{11}^{1/2} = \boldsymbol{\Sigma}_{11}^{-1}\boldsymbol{\Sigma}_{12}\boldsymbol{\Sigma}_{22}^{-1}\boldsymbol{\Sigma}_{21}, \\ \boldsymbol{M}_2 &= \boldsymbol{\Sigma}_{22}^{-1/2}\boldsymbol{N}_2\boldsymbol{\Sigma}_{22}^{1/2} = \boldsymbol{\Sigma}_{22}^{-1}\boldsymbol{\Sigma}_{21}\boldsymbol{\Sigma}_{11}^{-1}\boldsymbol{\Sigma}_{12}. \end{aligned} \tag{11.5}$$

Note that $\boldsymbol{N}_1$ and $\boldsymbol{M}_1$ are $(q \times q)$ matrices, and $\boldsymbol{N}_2$ and $\boldsymbol{M}_2$ are $(p \times p)$ matrices. From Theorem A.6.2, $\boldsymbol{M}_1, \boldsymbol{N}_1, \boldsymbol{M}_2$, and $\boldsymbol{N}_2$ all have the same nonzero eigenvalues. Further, since $\boldsymbol{N}_1 = \boldsymbol{K}\boldsymbol{K}' \geq 0$, all the nonzero eigenvalues are positive. Let k denote the number of nonzero eigenvalues. Then, by (A.41) and (A.42), $k = \operatorname{rank}(\boldsymbol{K}) = \operatorname{rank}(\boldsymbol{\Sigma}_{12})$. For simplicity, suppose that the eigenvalues are all distinct, $\lambda_1 > \cdots > \lambda_k > 0$.

By the singular value decomposition theorem (Theorem A.6.5), $\boldsymbol{K}$ can be written in the form

$$\boldsymbol{K} = (\boldsymbol{\alpha}_1, \ldots, \boldsymbol{\alpha}_k)\boldsymbol{D}(\boldsymbol{\beta}_1, \ldots, \boldsymbol{\beta}_k)', \tag{11.6}$$

where $\boldsymbol{\alpha}_i$ and $\boldsymbol{\beta}_i$ are the standardized eigenvectors of $\boldsymbol{N}_1$ and $\boldsymbol{N}_2$ respectively, for λ_i, and $\boldsymbol{D} = \operatorname{diag}(\lambda_1^{1/2}, \ldots, \lambda_k^{1/2})$. Since the λ_i are distinct, the eigenvectors are unique up to sign. (The signs of the eigenvectors are chosen so that the square roots in $\boldsymbol{D}$ are positive.) Also, since $\boldsymbol{N}_1$ and $\boldsymbol{N}_2$ are symmetric, the eigenvectors are orthogonal. Thus,

$$\boldsymbol{\alpha}_i'\boldsymbol{\alpha}_j = \delta_{ij}, \qquad \boldsymbol{\beta}_i'\boldsymbol{\beta}_j = \delta_{ij}, \tag{11.7}$$

where δ_{ij} is the Kronecker delta, equal to 1 for $i = j$ and 0 otherwise.

Definition *Using the above notation, let*

$$\boldsymbol{a}_i = \boldsymbol{\Sigma}_{11}^{-1/2}\boldsymbol{\alpha}_i, \qquad \boldsymbol{b}_i = \boldsymbol{\Sigma}_{22}^{-1/2}\boldsymbol{\beta}_i, \quad i = 1, \ldots, k. \tag{11.8}$$

Then,

(a) the vectors $\boldsymbol{a}_i$ *and* $\boldsymbol{b}_i$ *are called the* ith canonical correlation vectors *for* $\boldsymbol{x}$ *and* $\boldsymbol{y}$, *respectively;*
(b) the random variables $\eta_i = \boldsymbol{a}_i'\boldsymbol{x}$ *and* $\phi_i = \boldsymbol{b}_i'\boldsymbol{y}$ *are called the* ith canonical correlation variables;
(c) $\rho_i = \lambda_i^{1/2}$ *is called the* ith canonical correlation coefficient.

Note that

$$\begin{aligned} \text{cov}(\eta_i, \eta_j) &= \boldsymbol{a}_i'\boldsymbol{\Sigma}_{11}\boldsymbol{a}_j = \boldsymbol{\alpha}_i'\boldsymbol{\alpha}_j = \delta_{ij}, \\ \text{cov}(\phi_i, \phi_j) &= \boldsymbol{b}_i'\boldsymbol{\Sigma}_{22}\boldsymbol{b}_j = \boldsymbol{\beta}_i'\boldsymbol{\beta}_j = \delta_{ij}. \end{aligned} \tag{11.9}$$

Thus, the ith canonical correlation variables for $\boldsymbol{x}$ are uncorrelated and are standardized to have variance 1; similarly, for the ith canonical correlation variables for $\boldsymbol{y}$. The main properties of canonical correlation analysis are given by the following theorem.

Theorem 11.2.1 *Using the above notation, fix r, with* $1 \le r \le k$, *and let*

$$f_r = \max_{\boldsymbol{a},\boldsymbol{b}} \boldsymbol{a}'\boldsymbol{\Sigma}_{12}\boldsymbol{b}$$

subject to

$$\boldsymbol{a}'\boldsymbol{\Sigma}_{11}\boldsymbol{a} = 1, \qquad \boldsymbol{b}'\boldsymbol{\Sigma}_{22}\boldsymbol{b} = 1, \qquad \boldsymbol{a}_i'\boldsymbol{\Sigma}_{11}\boldsymbol{a} = 0, \quad i = 1, \ldots, r-1. \tag{11.10}$$

Then, the maximum is given by $f_r = \rho_r \ (= \lambda_r^{1/2})$ *and is attained when* $\boldsymbol{a} = \boldsymbol{a}_r, \boldsymbol{b} = \boldsymbol{b}_r$.

Proof: Before giving the proof, note that (11.10) is the largest correlation between linear combinations $\boldsymbol{a}'\boldsymbol{x}$ and $\boldsymbol{b}'\boldsymbol{y}$ subject to the restriction that $\boldsymbol{a}'\boldsymbol{x}$ is uncorrelated with the first $r-1$ canonical correlation variables for $\boldsymbol{x}$. Further, the case $r = 1$ gives the required maximum in (11.2).

Note that the sign of $\boldsymbol{a}'\boldsymbol{\Sigma}_{12}\boldsymbol{b}$ is irrelevant because it can be changed by replacing $\boldsymbol{a}$ by $-\boldsymbol{a}$. Then, it is convenient to solve (11.10) for f_r^2 instead of f_r. The proof splits naturally into three parts.

Step 1. Fix $\boldsymbol{a}$ and maximize over $\boldsymbol{b}$; that is, solve

$$\max_{\boldsymbol{b}} (\boldsymbol{a}'\boldsymbol{\Sigma}_{12}\boldsymbol{b})^2 = \max_{\boldsymbol{b}} \boldsymbol{b}'\boldsymbol{\Sigma}_{21}\boldsymbol{a}\boldsymbol{a}'\boldsymbol{\Sigma}_{12}\boldsymbol{b} \quad \text{subject to} \quad \boldsymbol{b}'\boldsymbol{\Sigma}_{22}\boldsymbol{b} = 1.$$

By Theorem A.2, this maximum is given by the largest (and only nonzero) eigenvalue of the matrix $\boldsymbol{\Sigma}_{22}^{-1}\boldsymbol{\Sigma}_{21}\boldsymbol{a}\boldsymbol{a}'\boldsymbol{\Sigma}_{12}$. By Corollary A.3, this eigenvalue equals

$$\boldsymbol{a}'\boldsymbol{\Sigma}_{12}\boldsymbol{\Sigma}_{22}^{-1}\boldsymbol{\Sigma}_{21}\boldsymbol{a}. \tag{11.11}$$

Step 2. Now, maximize (11.11) over $\boldsymbol{a}$ subject to the constraints in (11.10). Setting $\boldsymbol{\alpha} = \boldsymbol{\Sigma}_{11}^{1/2}\boldsymbol{a}$, the problem becomes

$$\max_{\boldsymbol{\alpha}} \boldsymbol{\alpha}'\boldsymbol{N}_1\boldsymbol{\alpha} \quad \text{subject to} \quad \boldsymbol{\alpha}'\boldsymbol{\alpha} = 1, \quad \boldsymbol{\alpha}_i'\boldsymbol{\alpha} = 0, \quad i = 1, \ldots, r-1, \tag{11.12}$$

where $\boldsymbol{a}_i = \boldsymbol{\Sigma}^{-12}\boldsymbol{\alpha}_i$ is the ith canonical correlation vector. Note that the $\boldsymbol{\alpha}_i$ are the eigenvectors of $\boldsymbol{N}_1$ corresponding to the $(r-1)$ largest eigenvalues of $\boldsymbol{N}_1$. Thus, as in Theorem 9.2.4, the maximum in (11.12) is attained by setting $\boldsymbol{\alpha}$ equal to the eigenvector corresponding to the largest eigenvalue not forbidden to us; that is, take $\boldsymbol{\alpha} = \boldsymbol{\alpha}_r$, or equivalently $\boldsymbol{a} = \boldsymbol{a}_r$. Then,

$$f_r^2 = \boldsymbol{\alpha}_r'\boldsymbol{N}_1\boldsymbol{\alpha}_r = \lambda_r\boldsymbol{\alpha}_r'\boldsymbol{\alpha}_r = \lambda_r.$$

Step 3. Lastly, we show that this maximum is attained when $\boldsymbol{a} = \boldsymbol{a}_r$ and $\boldsymbol{b} = \boldsymbol{b}_r$. From the decomposition (11.6), note that $\boldsymbol{K}\boldsymbol{\beta}_r = \rho_r\boldsymbol{\alpha}_r$. Thus,

$$\boldsymbol{a}_r'\boldsymbol{\Sigma}_{12}\boldsymbol{b}_r = \boldsymbol{\alpha}_r'\boldsymbol{K}\boldsymbol{\beta}_r = \rho_r\boldsymbol{\alpha}_r'\boldsymbol{\alpha}_r = \rho_r.$$ □

As standardized eigenvectors of $\boldsymbol{N}_1$ and $\boldsymbol{N}_2$, respectively, $\boldsymbol{\alpha}_r$ and $\boldsymbol{\beta}_r$ are uniquely defined up to sign. Their signs are usually chosen to make the correlation $\boldsymbol{a}_r'\boldsymbol{\Sigma}_{12}\boldsymbol{b}_r > 0$, although in some situations it is more intuitive to think of a negative correlation.

The above theorem is sometimes stated in a weaker but more symmetric form as follows.

Theorem 11.2.2 *Let*

$$g_r = \max_{\boldsymbol{a},\boldsymbol{b}} \boldsymbol{a}'\boldsymbol{\Sigma}_{12}\boldsymbol{b}$$

subject to

$$\boldsymbol{a}'\boldsymbol{\Sigma}_{11}\boldsymbol{a} = \boldsymbol{b}'\boldsymbol{\Sigma}_{22}\boldsymbol{b} = 1, \qquad \boldsymbol{a}_i'\boldsymbol{\Sigma}_{11}\boldsymbol{a} = \boldsymbol{b}_i'\boldsymbol{\Sigma}_{22}\boldsymbol{b} = 0, \qquad i = 1, \dots, r-1.$$

Then, $g_r = \rho_r$, and the maximum is attained when $\boldsymbol{a} = \boldsymbol{a}_r$ and $\boldsymbol{b} = \boldsymbol{b}_r$.

Proof: Since $\boldsymbol{a}_r$ and $\boldsymbol{b}_r$ satisfy the constraints and since these constraints are more restrictive than the constraints in Theorem 11.2.1, the result follows immediately from Theorem 11.2.1. □

The correlations between the canonical correlation variables are summarized in the following theorem.

Theorem 11.2.3 *Let η_i and ϕ_i be the ith canonical correlation variables, $i = 1, \dots, k$, and let $\boldsymbol{\eta} = (\eta_1, \dots, \eta_k)'$ and $\boldsymbol{\phi} = (\phi_1, \dots, \phi_k)'$. Then,*

$$\text{var}\begin{pmatrix} \boldsymbol{\eta} \\ \boldsymbol{\phi} \end{pmatrix} = \begin{bmatrix} \boldsymbol{I} & \boldsymbol{\Lambda}^{1/2} \\ \boldsymbol{\Lambda}^{1/2} & \boldsymbol{I} \end{bmatrix}. \tag{11.13}$$

Proof: From (11.9), we see that

$$\text{var}(\boldsymbol{\eta}) = \text{var}(\boldsymbol{\phi}) = \boldsymbol{I}.$$

From Theorem 11.2.1, $\text{cov}(\eta_i, \phi_i) = \lambda_i^{1/2}$. Thus, the proof will be complete if we show that $\text{cov}(\eta_i, \phi_j) = 0$ for $i \neq j$. This result follows from the decomposition (11.6) since

$$\text{cov}(\eta_i, \phi_j) = \boldsymbol{a}_i'\boldsymbol{\Sigma}_{12}\boldsymbol{b}_j = \boldsymbol{\alpha}_i'\boldsymbol{K}\boldsymbol{\beta}_j = \lambda_j\boldsymbol{\alpha}_i'\boldsymbol{\alpha}_j = 0.$$ □

Note that (11.13) is also the correlation matrix for $\boldsymbol{\eta}$ and $\boldsymbol{\phi}$.

The next theorem proves an important invariance property.

Theorem 11.2.4 *If $\boldsymbol{x}^* = \boldsymbol{U}'\boldsymbol{x} + \boldsymbol{u}$ and $\boldsymbol{y}^* = \boldsymbol{V}'\boldsymbol{y} + \boldsymbol{v}$, where $\boldsymbol{U}(q \times q)$ and $\boldsymbol{V}(p \times p)$ are nonsingular matrices and $\boldsymbol{u}(q \times 1)$, $\boldsymbol{v}(p \times 1)$ are fixed vectors, then*

(a) *the canonical correlations between* $\boldsymbol{x}^*$ *and* $\boldsymbol{y}^*$ *are the same as those between* $\boldsymbol{x}$ *and* $\boldsymbol{y}$;
(b) *the canonical correlation vectors for* $\boldsymbol{x}^*$ *and* $\boldsymbol{y}^*$ *are given by* $\boldsymbol{a}_i^* = \boldsymbol{U}^{-1}\boldsymbol{a}_i$ *and* $\boldsymbol{b}_i^* = \boldsymbol{V}^{-1}\boldsymbol{b}_i, i = 1, \ldots, k$, *where* $\boldsymbol{a}_i$ *and* $\boldsymbol{b}_i$ *are the canonical correlation vectors for* $\boldsymbol{x}$ *and* $\boldsymbol{y}$.

Proof: The matrix for $\boldsymbol{x}^*$ and $\boldsymbol{y}^*$ corresponding to $\boldsymbol{M}_1$ is

$$\boldsymbol{M}_1^* = (\boldsymbol{U}'\boldsymbol{\Sigma}_{11}\boldsymbol{U})^{-1}\boldsymbol{U}'\boldsymbol{\Sigma}_{12}\boldsymbol{V}(\boldsymbol{V}'\boldsymbol{\Sigma}_{22}\boldsymbol{V})^{-1}\boldsymbol{V}'\boldsymbol{\Sigma}_{21}\boldsymbol{U} = \boldsymbol{U}^{-1}\boldsymbol{M}_1\boldsymbol{U}.$$

Then, $\boldsymbol{M}_1^*$ has the same eigenvalues as $\boldsymbol{M}_1$, and the relationship between the eigenvectors is given by Theorem A.6.2. Note that the standardization,

$$\boldsymbol{a}_i^*(\boldsymbol{U}'\boldsymbol{\Sigma}_{11}\boldsymbol{U})\boldsymbol{a}_i^* = \boldsymbol{a}_i'\boldsymbol{\Sigma}_{11}\boldsymbol{a}_i = 1,$$

remains unaltered. Working with $\boldsymbol{M}_2^*$ gives the eigenvectors $\boldsymbol{b}_i^*$. □

If we put $\boldsymbol{U} = (\text{Diag}\,\boldsymbol{\Sigma}_{11})^{-1/2}$ and $\boldsymbol{V} = (\text{Diag}\,\boldsymbol{\Sigma}_{22})^{-1/2}$ in Theorem 11.2.2, then we obtain $\boldsymbol{M}^* = \boldsymbol{P}_{11}^{-1}\boldsymbol{P}_{12}\boldsymbol{P}_{22}^{-1}\boldsymbol{P}_{21}$, where $\boldsymbol{P}$ is the correlation matrix of $\boldsymbol{x}$ and $\boldsymbol{y}$. Hence, we deduce that canonical correlation analysis using $\boldsymbol{P}$ leads to essentially the same results as canonical correlation analysis applied to $\boldsymbol{\Sigma}$. Recall that principal component analysis did not have this convenient invariance property.

Example 11.2.1 Working with the correlation matrix $\boldsymbol{P}$, consider the situation $q = p = 2$ with all the elements of $\boldsymbol{P}_{12}$ equal; that is, let

$$\boldsymbol{P}_{11} = \begin{bmatrix} 1 & \alpha \\ \alpha & 1 \end{bmatrix}, \qquad \boldsymbol{P}_{22} = \begin{bmatrix} 1 & \gamma \\ \gamma & 1 \end{bmatrix}, \qquad \boldsymbol{P}_{12} = \beta \boldsymbol{J},$$

where $\boldsymbol{J} = \boldsymbol{1}\boldsymbol{1}'$. Then,

$$\boldsymbol{P}_{11}^{-1}\boldsymbol{P}_{12} = \frac{\beta}{1-\alpha^2}\begin{bmatrix} 1 & -\alpha \\ -\alpha & 1 \end{bmatrix}\begin{bmatrix} 1 & 1 \\ 1 & 1 \end{bmatrix} = \beta\boldsymbol{J}/(1+\alpha),$$

if $|\alpha| < 1$. Similarly, for $|\gamma| < 1$,

$$\boldsymbol{P}_{22}^{-1}\boldsymbol{P}_{21} = \beta\boldsymbol{J}/(1+\gamma).$$

Noting that $\boldsymbol{J}^2 = 2\boldsymbol{J}$, we have

$$\boldsymbol{P}_{11}^{-1}\boldsymbol{P}_{12}\boldsymbol{P}_{22}^{-1}\boldsymbol{P}_{21} = \left\{2\beta^2/(1+\alpha)(1+\gamma)\right\}\boldsymbol{J}.$$

Now, the eigenvalues of $\boldsymbol{1}\boldsymbol{1}'$ are 2 and 0. Therefore, the nonzero eigenvalue of the above matrix is

$$\lambda_1 = 4\beta^2/\{(1+\alpha)(1+\gamma)\}.$$

Hence, the first canonical correlation coefficient is

$$\rho_1 = 2\beta/\{(1+\alpha)(1+\gamma)\}^{1/2}.$$

Note that $|\alpha|, |\gamma| < 1$, and therefore, $\rho_1 > \beta$ provided $\beta > 0$. □

Example 11.2.2 In general, if $\boldsymbol{P}_{12}$ has rank 1, then we may write $\boldsymbol{P}_{12} = \boldsymbol{a}\boldsymbol{b}'$. Therefore,

$$\boldsymbol{P}_{11}^{-1}\boldsymbol{P}_{12}\boldsymbol{P}_{22}^{-1}\boldsymbol{P}_{21} = \boldsymbol{P}_{11}^{-1}\boldsymbol{a}\boldsymbol{b}'\boldsymbol{P}_{22}^{-1}\boldsymbol{b}\boldsymbol{a}'.$$

The nonzero eigenvalue of this matrix is

$$\lambda_1 = (\boldsymbol{a}'\boldsymbol{P}_{11}^{-1}\boldsymbol{a})(\boldsymbol{b}'\boldsymbol{P}_{22}^{-1}\boldsymbol{b}).$$

Note that this holds whatever the values of q and p. If $\boldsymbol{a} \propto \mathbf{1}, \boldsymbol{b} \propto \mathbf{1}$, then for $q = p = 2$, we get the result already shown in Example 11.2.1. In general, when $\boldsymbol{a} \propto \mathbf{1}, \boldsymbol{b} \propto \mathbf{1}$, λ_1 is proportional to the product of the sum of the elements of $\boldsymbol{P}_{11}^{-1}$ with the sum of the elements of $\boldsymbol{P}_{22}^{-1}$. □

Note that, in general, $\boldsymbol{M}_1$ and $\boldsymbol{M}_2$ will have k nonzero eigenvalues, where k is the rank of $\boldsymbol{\Sigma}_{12}$. Usually, $k = \min(p, q)$. Hence, in regression, where $p = 1$, there is just one nontrivial canonical correlation vector, and this is indeed the (standardized) least squares regression vector. If the matrices $\boldsymbol{\Sigma}_{11}$ and $\boldsymbol{\Sigma}_{22}$ are not of full rank, then similar results can be developed using generalized inverses (Rao and Mitra, 1971).

11.2.2 Sample Canonical Correlation Analysis

The above development for a population may be followed through for the analysis of sample data. All that is needed is that $\boldsymbol{S}_{ij}$ should be substituted for $\boldsymbol{\Sigma}_{ij}$ wherever it occurs $(i, j = 1, 2)$, l_i for λ_i, r_i for ρ_i, etc. If the data is normal, then $\boldsymbol{S}$ is the maximum-likelihood estimator of $\boldsymbol{\Sigma}$, so the sample canonical correlation values are maximum-likelihood estimators of the corresponding population values, except when there are repeated population eigenvalues.

Example 11.2.3 For the head-length data of Table 6.1, we have the correlation matrix

$$\boldsymbol{R}_{11} = \begin{bmatrix} 1 & 0.7346 \\ 0.7346 & 1 \end{bmatrix}, \qquad \boldsymbol{R}_{22} = \begin{bmatrix} 1 & 0.8392 \\ 0.8392 & 1 \end{bmatrix},$$

$$\boldsymbol{R}_{12} = \boldsymbol{R}_{21}' = \begin{bmatrix} 0.7108 & 0.7040 \\ 0.6932 & 0.7086 \end{bmatrix}.$$

(See Example 5.1.1.) Note that all the elements of $\boldsymbol{R}_{12}$ are about 0.7, so this matrix is nearly of rank 1. This means that the second canonical correlation will be near zero, and the situation approximates that of Example 11.2.1 (see Exercise 11.2.12). In fact, the eigenvalues of $\boldsymbol{M}_1 = \boldsymbol{R}_{11}^{-1}\boldsymbol{R}_{12}\boldsymbol{R}_{22}^{-1}\boldsymbol{R}_{21}$ are 0.6214 and 0.0029. Therefore, the canonical correlation coefficients are $r_1 = 0.7886$ and $r_2 = 0.0537$. As expected, r_2 is close to zero, since $\boldsymbol{R}_{12}$ is almost of rank 1. Note that r_1 exceeds any of the individual correlations between a variable of the first set and a variable of the second set (in particular $0.7886 > 0.7108$). The canonical correlation vectors for the standardized variables are obtained from the eigenvectors of $\boldsymbol{M}_1$ and $\boldsymbol{M}_2$. We have

$$\boldsymbol{a}_1 = \begin{pmatrix} 0.552 \\ 0.522 \end{pmatrix}, \qquad \boldsymbol{b}_1 = \begin{pmatrix} 0.504 \\ 0.538 \end{pmatrix}$$

and

$$\boldsymbol{a}_2 = \begin{pmatrix} 1.366 \\ -1.378 \end{pmatrix}, \qquad \boldsymbol{b}_2 = \begin{pmatrix} 1.769 \\ -1.759 \end{pmatrix}.$$

Hence, the first canonical correlation variables are $\eta_1 = 0.552x_1 + 0.522x_2$ and $\phi_1 = 0.504y_1 + 0.538y_2$. These are approximately the sum of length and breadth of the head size of each brother and may be interpreted as a measure of "girth." These variables are highly correlated between brothers. The second canonical correlation variables η_2 and ϕ_2 seem to be measuring the *difference* between length and breadth. This measure of "shape" would distinguish, for instance, between long thin heads and short fat heads. The head shape of the first and second brothers appears therefore to have little correlation; see also Section 9.6. □

11.2.3 Sampling Properties and Tests

The sampling distributions associated with canonical correlation analysis are very complicated, and we will not go into them in detail here. The interested reader is referred to Kshirsagar (1972, pp. 261–277). We will merely describe briefly an associated significance test.

First, consider the hypothesis $\boldsymbol{\Sigma}_{12} = \mathbf{0}$, which means that the two sets of variables are uncorrelated with one another. From (6.18) (with $q = p_1, p = q_2$), we see that under normality, the likelihood ratio (LR) statistic for testing H_0 is given by

$$\lambda^{2/n} = |\boldsymbol{I} - \boldsymbol{S}_{22}^{-1}\boldsymbol{S}_{21}\boldsymbol{S}_{11}^{-1}\boldsymbol{S}_{12}| = \prod_{i=1}^{k}(1 - r_i^2),$$

which has a Wilks' $\Lambda(p, n-1-q, q)$ distribution. Here, $r_1, \dots, r_k$ are the sample canonical correlation coefficients, and $k = \min(p, q)$. Using Bartlett's approximation (4.43), we see that

$$-\left\{n - \frac{p+q+3}{2}\right\} \log \prod_{i=1}^{k}(1 - r_i^2) \sim \chi^2_{pq},$$

asymptotically for large n.

Bartlett (1939) proposed a similar statistic to test the hypothesis that only s of the population canonical correlation coefficients are nonzero. This test is based on the statistic

$$-\left\{n - \frac{p+q+3}{2}\right\} \log \prod_{i=s+1}^{k}(1 - r_i^2) \sim \chi^2_{(p-s)(q-s)}, \tag{11.14}$$

asymptotically. An alternative large sample test was proposed by Marriott (1952).

Example 11.2.4 Consider the head data of Example 11.2.3. There are $n = 25$ observations, and $q = p = 2$. First, let us test whether or not the head measurements of one brother are independent of those of the other, that is the hypothesis $\rho_1 = \rho_2 = 0$. The LR test for this hypothesis is given in Example 6.3.3, and the null hypothesis is strongly rejected.

Second, to test whether $\rho_2 = 0$, we use (11.14) with $s = 1$. This gives the statistic

$$\left(25 - \frac{7}{2}\right) \log\left(1 - 0.537^2\right) = 0.062,$$

which, when tested against the χ_1^2 distribution, is clearly nonsignificant. Hence, we accept the hypothesis $\rho_2 = 0$ for this data. □

11.2.4 Scoring and Prediction

Let $(\boldsymbol{X}, \boldsymbol{Y})$ be a data matrix of n individuals of $(q+p)$ variables, and let $\boldsymbol{a}_i, \boldsymbol{b}_i$ denote the ith canonical correlation vectors. Then, the n-vectors $\boldsymbol{X}\boldsymbol{a}_i$ and $\boldsymbol{Y}\boldsymbol{b}_i$ denote the *scores* of the n individuals of the ith canonical correlation variables for $\boldsymbol{x}$ and $\boldsymbol{y}$. In terms of the values of the variables for a particular individual, these scores take the form

$$\eta_i = \boldsymbol{a}_i'\boldsymbol{x}, \qquad \phi_i = \boldsymbol{b}_i'\boldsymbol{y}. \tag{11.15}$$

Since the correlations are unaltered by linear transformations, it is sometimes convenient to replace these scores by new scores

$$\eta_i^* = c_1\boldsymbol{a}_i'\boldsymbol{x} + d_1, \qquad \phi_i^* = c_2\boldsymbol{b}_i'\boldsymbol{y} + d_2, \tag{11.16}$$

where $c_1, c_2 > 0$ and d_1, d_2 are real numbers. These scores are most important for the first canonical correlation vectors and can be calculated for each of the n individuals.

Let r_i denote the ith canonical correlation coefficient. If the $\boldsymbol{x}$ and $\boldsymbol{y}$ variables are interpreted as the "predictor" and "predicted" variables, respectively, then the η_i score can be used to predict a value of the ϕ_i score using least squares regression. Since $\boldsymbol{Xa}_i$ and $\boldsymbol{Yb}_i$ each have sample variance 1, the predicted value of ϕ_i, given η_i, is

$$\hat{\phi}_i = r_i(\eta_i - \boldsymbol{a}_i'\overline{\boldsymbol{x}}) + \boldsymbol{b}_i'\overline{\boldsymbol{y}} \tag{11.17}$$

or, equivalently,

$$\hat{\phi}_i^* = \frac{c_2 r_i}{c_1}(\eta_i^* - d_1 - c_1\boldsymbol{a}_i'\overline{\boldsymbol{x}}) + d_2 + c_2\boldsymbol{b}_i'\overline{\boldsymbol{y}}. \tag{11.18}$$

Note that r_i^2 represents the proportion of the variance of ϕ_i, which is "explained" by the regression on $\boldsymbol{X}$. Of course, prediction is most important for the first canonical correlation variable.

Example 11.2.5 Consider again the open/closed-book data discussed in Example 6.1.1. The covariance matrix for this data is given in Example 9.2.3.

The canonical correlations are $r_1 = 0.6631$ and $r_2 = 0.0409$, and the first canonical correlation vectors are

$$\begin{aligned}\eta_1 &= 0.0258\,x_1 + 0.0515\,x_2,\\ \phi_1 &= 0.0819\,y_1 + 0.0080\,y_2 + 0.0035\,y_3.\end{aligned}$$

Thus, the highest correlation between the open- and closed-book exams occurs between an average of x_1 and x_2 weighted on x_2 and an average of y_1, y_2, and y_3 heavily weighted on y_1.

The means of the exam results are

$$38.9546, \quad 50.5909, \quad 50.6023, \quad 46.6818, \quad 42.3068.$$

If we use these linear combinations to predict open-book results from the closed-book results, we get the predictor

$$\hat{\phi}_1 = 0.0171x_1 + 0.0341x_2 + 2.2720.$$

Note that this function essentially predicts the value of y_1 from the values of x_1 and x_2. □

11.3 Qualitative Data and Dummy Variables

Canonical correlation analysis can also be applied to *qualitative* data. Consider a two-way contingency table $\boldsymbol{N}(r \times c)$ in which individuals are classified according to each of two characteristics. The attributes on each characteristic are represented by the r row and c column categories, respectively, and n_{ij} denotes the number of individuals with the ith row and jth column attributes. We wish to explore the relationship between the two characteristics. For example, Table 11.1 gives a (5×5) contingency table of $n = 3497$ individuals comparing their social status with the social status of their fathers.

Now, a contingency table $\boldsymbol{N}$ is not a data matrix since it does not have the property that rows correspond to individuals, while columns represent variables. However, it is possible

Table 11.1 Social mobility contingency table (Glass (1954); see also Goodman (1972)).

		Son's status				
		1	2	3	4	5
	1	50	45	8	18	8
Father's	2	28	174	84	154	55
status	3	11	78	110	223	96
	4	14	150	185	714	447
	5	0	42	72	320	411

to represent the data in an $(n \times (r + c))$ data matrix $\boldsymbol{Z} = (\boldsymbol{X}, \boldsymbol{Y})$, where the columns of $\boldsymbol{X}$ and $\boldsymbol{Y}$ are dummy zero–one variables for the row and column categories, respectively; that is, let

$$x_{ki} = \begin{cases} 1 & \text{if the } k\text{th individual belongs to the } i\text{th row category,} \\ 0 & \text{otherwise,} \end{cases}$$

and

$$y_{kj} = \begin{cases} 1 & \text{if the } k\text{th individual belongs to the } j\text{th column category,} \\ 0 & \text{otherwise,} \end{cases}$$

for $k = 1, \dots, n$, and $i = 1, \dots, r; j = 1, \dots, c$. Note that $\boldsymbol{x}'_{(i)}\boldsymbol{y}_{(j)} = n_{ij}$. Also, the columns of $\boldsymbol{X}$ and the columns of $\boldsymbol{Y}$ each sum to $\boldsymbol{1}_n$.

The purpose of canonical correlation analysis is to find vectors $\boldsymbol{a}(r \times 1)$ and $\boldsymbol{b}(c \times 1)$ such that the variables $\eta = \boldsymbol{a}'\boldsymbol{x}$ and $\phi = \boldsymbol{b}'\boldsymbol{y}$ are maximally correlated. Since $\boldsymbol{x}$ has only one nonzero component, $\eta = a_1, a_2 \dots$, or a_r. Similarly, $\phi = b_1, b_2, \dots$, or b_c. Thus, an individual in the ith row category and jth column category can be associated with his "score" (a_i, b_j). These scores may be plotted on a scattergram – there will be n_{ij} points at (a_i, b_j) – and the correlation represented by these points can be evaluated. Of course, the correlation depends on the values of $\boldsymbol{a}$ and $\boldsymbol{b}$, and if these vectors maximize this correlation, then they give the vector of first canonical correlation loadings. Such scores are known as "canonical scores."

Of course, having calculated the first set of canonical scores, one could ask for a second set that maximizes the correlation among all the sets that are uncorrelated with the first. This would correspond to the second canonical correlation.

Let $\boldsymbol{Z} = (\boldsymbol{X}, \boldsymbol{Y})$ be the data matrix corresponding to a contingency table $\boldsymbol{N}$. Let $f_i = \sum_j n_{ij}$ and $g_j = \sum_i n_{ij}$, so that $\boldsymbol{f}$ and $\mathbf{g}$ denote the marginal and column sums, respectively, for the table. For simplicity, suppose that $f_i > 0, g_j > 0$ for all i and j. Then, a little calculation shows that

$$\begin{aligned} n\boldsymbol{S} &= \boldsymbol{Z}'\boldsymbol{H}\boldsymbol{Z} = \boldsymbol{Z}'\boldsymbol{Z} - n\bar{\boldsymbol{z}}\bar{\boldsymbol{z}}' \\ &= \begin{bmatrix} n\boldsymbol{S}_{11} & n\boldsymbol{S}_{12} \\ n\boldsymbol{S}_{21} & n\boldsymbol{S}_{22} \end{bmatrix} = \begin{bmatrix} \operatorname{diag}(\boldsymbol{f}) - n^{-1}\boldsymbol{f}\boldsymbol{f}' & \boldsymbol{N} - \hat{\boldsymbol{N}} \\ \boldsymbol{N}' - \hat{\boldsymbol{N}}' & \operatorname{diag}(\mathbf{g}) - n^{-1}\mathbf{g}\mathbf{g}' \end{bmatrix}, \end{aligned} \tag{11.19}$$

where $\hat{\boldsymbol{N}} = \boldsymbol{f}\mathbf{g}'$ is the estimated value of $\boldsymbol{N}$ under the assumption that the row and column categories are independent.

Because the columns of $\boldsymbol{X}$ and $\boldsymbol{Y}$ each sum to $\mathbf{1}$, $\boldsymbol{S}_{11}^{-1}$ and $\boldsymbol{S}_{22}^{-1}$ do not exist (see Exercise 11.3.1). One way out of this difficulty is to drop a column of each of $\boldsymbol{X}$ and $\boldsymbol{Y}$, say the first column. Let $\boldsymbol{S}_{ij}^{*}$ denote the component of the covariance matrix obtained by deleting the first row and first column from $\boldsymbol{S}_{ij}$ for $i, j = 1, 2$. Similarly, let $\boldsymbol{f}^{*}$ and $\mathbf{g}^{*}$ denote the vectors obtained by deleting the first components of $\boldsymbol{f}$ and $\mathbf{g}$. Then, it is easy to check that

$$(n\boldsymbol{S}_{11}^{*})^{-1} = \left[\operatorname{diag}(\boldsymbol{f}^{*})\right]^{-1} + f_1^{-1}\mathbf{11}', \qquad (n\boldsymbol{S}_{22}^{*})^{-1} = \left[\operatorname{diag}(\mathbf{g}^{*})\right]^{-1} + g_1^{-1}\mathbf{11}', \tag{11.20}$$

so that $\boldsymbol{M}_1^{*} = \boldsymbol{S}_{11}^{*-1}\boldsymbol{S}_{12}^{*}\boldsymbol{S}_{22}^{*-1}\boldsymbol{S}_{21}^{*}$ and $\boldsymbol{M}_2^{*}$ are straightforward to calculate for the reduced set of variables.

Note that the score associated with an individual who lies in the first row category of $\boldsymbol{N}$ is 0, and similarly for the first column category of $\boldsymbol{N}$.

Example 11.3.1 A canonical correlation analysis on Glass' social mobility data in Table 11.1 yields the following scores (coefficients of the first canonical correlation vectors) for the various social classes:

Father's status:	1	2	3	4	5
	0	3.15	4.12	4.55	4.96
Son's status:	1	2	3	4	5
	0	3.34	4.49	4.87	5.26

The first canonical correlation is $r_1 = 0.504$. Note that for both father's and son's social class, the scores appear in their natural order. Thus, the social class of the father appears to be correlated with the social class of the son. Note that social classes 1 and 2 seem to be more distinct from one another than the other adjacent social classes, both for the son's and the father's status. □

Hypothesis testing based on normal theory is difficult for canonical scores because the data matrix $\boldsymbol{Z}$ is clearly nonnormal. However, it is straightforward to develop tests based on the bootstrap.

11.4 Qualitative and Quantitative Data

The ideas of the last section can be used when the data is a mixture of qualitative and quantitative characteristics. Each quantitative characteristic can of course be represented on one variable, and each qualitative characteristic, with say g attributes, can be represented by dummy zero–one values on $g - 1$ variables.

We illustrate this procedure for an example on academic prediction of Barnett and Lewis (1963) and Lewis (1970). Lewis (1970) gives data collected from 382 university students on the number of A levels taken (a secondary school examination normally obtained for university entrance) and the student's average grade (scored 1–5, with 5 denoting the best score). These represented the "explanatory variables." The dependent variables concerned the student's final degree result at university, which was classified as First, Upper Second, Lower Second, Third, and Pass. Alternatively, some students were classified as 3(4) if they took four years over a three-year course (the final degree results were not available for these students) or as "→" if they left without completing the course.

This information may be represented in terms of the following variables:

x_1 = average A-level grade;
x_2 = 1 if two A levels taken, 0 otherwise;
x_3 = 1 if four A levels taken, 0 otherwise;
y_1 = 1 if degree class II(i) obtained, 0 otherwise;
y_2 = 1 if degree class II(ii) obtained, 0 otherwise;
y_3 = 1 if degree class III obtained, 0 otherwise;
y_4 = 1 if Pass obtained, 0 otherwise;
y_5 = 1 if 3(4) obtained, 0 otherwise;
y_6 = 1 if "→" obtained, 0 otherwise.

Note that a student who obtains a First Class (I) degree would score zero on *all* the $\mathbf{y}$ variables. The data is summarized in Table 11.2. Note that most of these variables are "dummies" in the sense that they take either the value 0 or 1. Hence, the assumptions of normality would be completely unwarranted in this example.

Since the $\mathbf{y}$ variables are in some sense dependent on the $\mathbf{x}$ variables, we may ask what scoring system maximizes the correlation between the *y*s and the *x*s. Lewis (1970) showed that the first canonical correlation vectors are given by $\mathbf{a}_1 = (-1.096, 1.313, 0.661)'$ and $\mathbf{b}_1 = (0.488, 1.877, 2.401, 2.971, 2.527, 3.310)'$. Thus, the scores corresponding to each of the degree results are

I	II(i)	II(ii)	III	Pass	3(4)	→
0	0.488	1.877	2.401	2.971	2.527	3.310

The first canonical correlation coefficient is $r_1 = 0.400$, which means that with the given scores, only $r_1^2 = 0.160$ of the variation in the first canonical correlation variable for $\mathbf{y}$ is explained by variation in A-level scores. However, any other scoring system (such as a "natural" one, $1y_1 + 2y_2 + \cdots + 6y_6$) would explain less than 0.16 of the variance.

The above scores may be interpreted as follows. The scores for I, II(i), II(ii), III, and Pass come out in the "natural" order, but they are not equally spaced. Moreover, the 3(4) group comes between III and Pass, while "→" scores higher than Pass. Note the large gap between II(i) and II(ii).

The canonical correlation vector on the $\mathbf{x}$ variables indicates that the higher one's A-level average, the better the degree one is likely to get. Also, those who take two or four A levels are likely to get poorer degree results than those who take three A levels. In fact, taking two A levels instead of three is equivalent to a drop of about 0.5 in average grade. The fact that three A levels is better than four A levels is somewhat surprising. However, because the value of r_1^2 is not high, perhaps one should not read a great deal into these conclusions. (The test of Section 11.2.3 is not valid here because the population is clearly nonnormal.)

The means for the nine variables are

$$(\overline{\mathbf{x}}', \overline{\mathbf{y}}') = (3.138, 0.173, 0.055, 0.175, 0.306, 0.134, 0.194, 0.058, 0.068).$$

Thus, we can use the A-level performance to predict degree results using

$$\hat{\phi} = -0.439x_1 + 0.525x_2 + 0.264x_3 + 3.199.$$

For a student taking three A levels with an average grade of 3.75, $\hat{\phi} = 1.55$, so we would predict a degree result slightly better than II(ii).

Bartlett (1965) gives various other examples on the use of dummy variables.

Table 11.2 Data from 382 Hull University students (f denotes frequencies).

Class:			**I**		**II(i)**				**II(ii)**				**III**			**Pass**					**3(4)**				→		
$\boldsymbol{y}$: (0, 0, 0, 0, 0, 0)				(1, 0, 0, 0, 0, 0)				(0, 1, 0, 0, 0, 0)				(0, 0, 1, 0, 0, 0)				(0, 0, 0, 1, 0, 0)				(0, 0, 0, 0, 1, 0)				(0, 0, 0, 0, 0, 1)			
x_1	x_2	x_3	f	x_1	x_2	x_3	f	x_1	x_2	x_3	f	x_1	x_2	x_3	f	x_1	x_2	x_3	f	x_1	x_2	x_3	f	x_1	x_2	x_3	f
4.5	0	1	1	4.0	0	1	1	4.25	0	1	1	3.0	0	1	1	3.75	0	1	1	3.25	0	1	1	2.25	0	1	1
3.25	0	1	1	3.25	0	1	1	3.25	0	1	3	2.75	0	1	1	3.25	0	1	1	2.75	0	1	1	3.67	0	0	3
4.67	0	0	6	5.0	0	0	1	3.75	0	1	1	5.0	0	0	1	2.75	0	1	2	2.5	0	1	1	3.0	0	0	3
4.33	0	0	3	4.67	0	0	2	4.67	0	0	1	4.67	0	0	1	2.0	0	1	2	4.0	0	0	4	2.67	0	0	5
4.0	0	0	3	4.33	0	0	10	4.33	0	0	6	4.33	0	0	1	4.67	0	0	2	3.67	0	0	1	2.33	0	0	2
3.33	0	0	3	4.0	0	0	11	4.0	0	0	10	4.0	0	0	2	4.33	0	0	1	3.33	0	0	3	2.0	0	0	1
3.0	0	0	2	3.67	0	0	6	3.67	0	0	7	3.67	0	0	1	4.0	0	0	1	3.0	0	0	1	1.33	0	0	1
2.67	0	0	4	3.33	0	0	12	3.33	0	0	18	3.33	0	0	7	3.67	0	0	3	2.67	0	0	3	1.0	0	0	1
1.0	0	0	1	3.0	0	0	8	3.0	0	0	14	3.0	0	0	7	3.33	0	0	7	2.0	0	0	3	4.0	1	0	1
4.0	1	0	1	2.67	0	0	4	2.67	0	0	15	2.67	0	0	8	3.0	0	0	12	1.67	0	0	1	3.5	1	0	3
				3.33	0	0	3	2.33	0	0	11	2.33	0	0	6	2.67	0	0	7	3.0	1	0	1	3.0	1	0	4
				2.0	0	0	1	2.0	0	0	5	2.0	0	0	3	2.33	0	0	7	2.5	1	0	1	2.5	1	0	1
				1.67	0	0	1	1.67	0	0	1	1.67	0	0	1	2.0	0	0	8	2.0	1	0	1				
				1.33	0	0	1	1.33	0	0	2	4.5	1	0	1	1.67	0	0	3								
				5.0	1	0	1	5.0	1	0	2	4.0	1	0	1	1.33	0	0	1								
				4.5	1	0	1	4.5	1	0	1	3.5	1	0	4	1.0	0	0	1								
				4.0	1	0	1	4.0	1	0	6	2.5	1	0	4	4.5	1	0	1								
				3.5	1	0	1	3.5	1	0	3	2.0	1	0	1	4.0	1	0	1								
				2.0	1	0	1	3.0	1	0	6					3.5	1	0	3								
								2.5	1	0	1					3.0	1	0	4								
								2.0	1	0	1					2.5	1	0	4								
								1.5	1	0	2					2.0	1	0	2								
Total f			25				67				117				51				74				22				26

Source: Lewis (1970).

Exercises and Complements

11.2.1 Show that for fixed $\boldsymbol{b}$,

$$\max_{\boldsymbol{a}} \frac{(\boldsymbol{a}'\boldsymbol{\Sigma}_{12}\boldsymbol{b})^2}{(\boldsymbol{a}'\boldsymbol{\Sigma}_{11}\boldsymbol{a})(\boldsymbol{b}'\boldsymbol{\Sigma}_{22}\boldsymbol{b})} = \frac{\boldsymbol{b}'\boldsymbol{\Sigma}_{21}\boldsymbol{\Sigma}_{11}^{-1}\boldsymbol{\Sigma}_{12}\boldsymbol{b}}{\boldsymbol{b}'\boldsymbol{\Sigma}_{22}\boldsymbol{b}},$$

and that the maximum of this quantity over $\boldsymbol{b}$ is given by the largest eigenvalue of $\boldsymbol{\Sigma}_{11}^{-1}\boldsymbol{\Sigma}_{12}\boldsymbol{\Sigma}_{22}^{-1}\boldsymbol{\Sigma}_{21}$.

11.2.2 Show that $\boldsymbol{M}_1, \boldsymbol{M}_2, \boldsymbol{N}_1$, and $\boldsymbol{N}_2$ of Section 11.2.1 all have the same nonzero eigenvalues, and that these eigenvalues are real and positive.

11.2.3 Show that

$$\boldsymbol{M}_1 = \boldsymbol{I}_p - \boldsymbol{\Sigma}_{11}^{-1}\boldsymbol{\Sigma}_{11.2}, \quad \text{where} \quad \boldsymbol{\Sigma}_{11.2} = \boldsymbol{\Sigma}_{11} - \boldsymbol{\Sigma}_{12}\boldsymbol{\Sigma}_{22}^{-1}\boldsymbol{\Sigma}_{21}.$$

Hence, deduce that the first canonical correlation vector for $\boldsymbol{x}$ is given by the eigenvector corresponding to the smallest eigenvalue of $\boldsymbol{\Sigma}_{11.2}$.

11.2.4 Show that the squared canonical correlations are the roots of the equation

$$|\boldsymbol{\Sigma}_{12}\boldsymbol{\Sigma}_{22}^{-1}\boldsymbol{\Sigma}_{21} - \lambda\boldsymbol{\Sigma}_{11}| = 0,$$

and that the canonical correlation vectors for $\boldsymbol{x}$ satisfy

$$\boldsymbol{\Sigma}_{12}\boldsymbol{\Sigma}_{22}^{-1}\boldsymbol{\Sigma}_{22}\boldsymbol{a}_i = \lambda_i\boldsymbol{\Sigma}_{11}\boldsymbol{a}_i, \quad i = 1, \dots, k.$$

11.2.5 (a) Show that the canonical correlation vectors $\boldsymbol{a}_i$ and $\boldsymbol{b}_i$ are eigenvectors of $\boldsymbol{M}_1$ and $\boldsymbol{M}_2$, respectively.

(b) Write $\boldsymbol{M}_1 = \boldsymbol{L}_1\boldsymbol{L}_2$ and $\boldsymbol{M}_2 = \boldsymbol{L}_2\boldsymbol{L}_1$, where $\boldsymbol{L}_1 = \boldsymbol{\Sigma}_{11}^{-1}\boldsymbol{\Sigma}_{12}$ and $\boldsymbol{L}_2 = \boldsymbol{\Sigma}_{22}^{-1}\boldsymbol{\Sigma}_{21}$. Using Theorem A.6.2, show that the canonical correlation vectors can be expressed in terms of one another as

$$\boldsymbol{b}_i = \lambda_i^{-1/2}\boldsymbol{L}_2\boldsymbol{a}_i, \qquad \boldsymbol{a}_i = \lambda_i^{-1/2}\boldsymbol{L}_1\boldsymbol{b}_i.$$

11.2.6 Let $\boldsymbol{x}(q \times 1)$ and $\boldsymbol{y}(p \times 1)$ be random vectors such that for all $i, k = 1, \dots, q$ and $j, l = 1, \dots, p, i \neq k$ and $j \neq l$,

$$\text{var}(x_i) = 1, \quad \text{var}(y_i) = 1, \quad \text{cov}(x_i, x_k) = \rho, \quad \text{cov}(y_j, y_l) = \rho', \quad \text{cov}(x_i, y_j) = \tau,$$

where ρ, ρ' and τ lie between 0 and 1. Show that the only canonical correlation variables are

$$\left\{\frac{q[(q-1)\rho + 1]}{3}\right\}^{-1/2} \sum_{i=1}^{q} x_i \quad \text{and} \quad \left\{\frac{p[(p-1)\rho' + 1]}{3}\right\}^{-1/2} \sum_{j=1}^{p} y_j.$$

11.2.7 When $p = 1$, show that the value of ρ_1^2 in Theorem 11.2.1 is the squared multiple correlation $\boldsymbol{\Sigma}_{21}\boldsymbol{\Sigma}_{11}^{-1}\boldsymbol{\Sigma}_{12}/\sigma_{22}$.

11.2.8 (a) The residual variance in predicting x_i by its linear regression on $\boldsymbol{b}'\boldsymbol{y}$ is given by

$$\delta_i^2 = \text{var}(x_i) - \text{cov}(x_i, \boldsymbol{b}'\boldsymbol{y})^2/\text{var}(\boldsymbol{b}'\boldsymbol{y}).$$

Show that the vector $\boldsymbol{b}$ that minimizes $\sum_{i=1}^{q} \delta_i^2$ is the eigenvector corresponding to the largest eigenvalue of $\boldsymbol{\Sigma}_{22}^{-1}\boldsymbol{\Sigma}_{21}\boldsymbol{\Sigma}_{12}$.

(b) Show further that the best k linear functions of $y_1, \ldots, y_p$ for predicting $x_1, \ldots, x_q$ in the sense of minimizing the sum of residual variances correspond to the first k eigenvectors of $\boldsymbol{\Sigma}_{22}^{-1}\boldsymbol{\Sigma}_{21}\boldsymbol{\Sigma}_{12}$.

11.2.9 When $q = p = 2$, the squared canonical correlations can be computed explicitly. If

$$\boldsymbol{\Sigma}_{11} = \begin{bmatrix} 1 & \alpha \\ \alpha & 1 \end{bmatrix}, \quad \boldsymbol{\Sigma}_{22} = \begin{bmatrix} 1 & \beta \\ \beta & 1 \end{bmatrix}, \quad \boldsymbol{\Sigma}_{12} = \begin{bmatrix} a & b \\ c & d \end{bmatrix},$$

then show that the eigenvectors of $\boldsymbol{M}_1 = \boldsymbol{\Sigma}_{11}^{-1}\boldsymbol{\Sigma}_{12}\boldsymbol{\Sigma}_{22}^{-1}\boldsymbol{\Sigma}_{21}$ are given by

$$\lambda = \{B \pm (B^2 - 4C)^{1/2}\}/\{2(1-\alpha^2)(1-\beta^2)\},$$

where

$$B = a^2 + b^2 + c^2 + d^2 + 2(ad + bc)\alpha\beta - 2(ac + bd)\alpha - 2(ab + cd)\beta$$

and

$$C = (ad - bc)^2(1 + \alpha^2\beta^2 - \alpha^2 - \beta^2).$$

11.2.10 (Hotelling, 1936) Four examinations in reading speed, reading power, arithmetic speed, and arithmetic power were given to $n = 148$ children. The question of interest is whether reading ability is correlated with arithmetic ability. The correlations are given by

$$\boldsymbol{R}_{11} = \begin{bmatrix} 1.0 & 0.6328 \\ & 1.0 \end{bmatrix}, \quad \boldsymbol{R}_{22} = \begin{bmatrix} 1.0 & 0.4248 \\ & 1.0 \end{bmatrix}, \quad \boldsymbol{R}_{12} = \begin{bmatrix} 0.2412 & 0.0586 \\ -0.0553 & 0.0655 \end{bmatrix}.$$

Using Exercise 11.2.9, verify that the canonical correlations are given by

$$\rho_1 = 0.3945, \qquad \rho_2 = 0.0688.$$

11.2.11 Using (11.14), test whether $\rho_1 = \rho_2 = 0$ for Hotelling's examination data in Exercise 11.2.10. Show that one gets the test statistic

$$25.13 \sim \chi_4^2.$$

Since $\chi_4^2(0.99) = 13.3$, we strongly reject the hypothesis that reading ability and arithmetic ability are independent for these data.

11.2.12 If in Example 11.2.1, $\beta = 0.7$, $\alpha = 0.74$, and $\gamma = 0.84$, show that $\rho_1 = 0.782$, thus approximating the situation in Example 11.2.3, where $r_1 = 0.7886$.

11.2.13 (a) Using the data matrix for the open/closed-book data in Example 10.2.5 and Table 1.2, show that the scores of the first eight individuals on the first canonical correlation variables are as follows:

Subject	1	2	3	4	5	6	7	8
η_1	6.25	5.68	5.73	5.16	4.90	4.54	4.80	5.16
ϕ_1	6.35	7.44	6.67	6.00	6.14	6.71	6.12	6.30

(b) Plot the above eight points on a scattergram.

(c) Repeat the procedure for the second canonical correlation variable and analyze the difference in the correlations. (The second canonical correlation is $r_2 = 0.041$, and the corresponding loading vectors are given by

$$\boldsymbol{a}_2' = (-0.064, 0.076), \qquad \boldsymbol{b}_2' = (-0.091, 0.099, -0.014).)$$

11.2.14 The technique of ridge regression Hoerl and Kennard (1970) has been extended to canonical correlation analysis by Vinod (1976), giving what he calls the "canonical ridge" technique. This technique involves replacing the sample correlation matrix $\boldsymbol{R}$ by

$$\begin{bmatrix} \boldsymbol{R}_{11} + k_1\boldsymbol{I} & \boldsymbol{R}_{12} \\ \boldsymbol{R}_{21} & \boldsymbol{R}_{22} + k_2\boldsymbol{I} \end{bmatrix},$$

where k_1 and k_2 are small nonnegative numbers, and then carrying out a canonical correlation analysis on this new correlation matrix. For data that are nearly collinear (that is, $\boldsymbol{R}_{11}$ and/or $\boldsymbol{R}_{22}$ have eigenvalues near 0), show that small but nonzero values of k_1, k_2 lead to better estimates of the true canonical correlations and canonical correlation vectors than the usual analysis on $\boldsymbol{R}$ provides.

11.2.15 (Lawley, 1959) If $\boldsymbol{S}$ is based on a large number, n, of observations, then the following asymptotic results hold for the k nonzero canonical correlations, provided ρ_i^2 and $\rho_i^2 - \rho_j^2$ are not too close to zero for all $i,j = 1, \ldots, k, i \neq j$:

$$\begin{aligned}
2\rho_i \mathrm{E}(r_i - \rho_i) &= \frac{1-\rho_i^2}{n-1}\left\{p + q - 2 - \rho_i^2 + 2(1-\rho_i^2)\sum_{\substack{s=1\\ s\neq i}}^{k} \frac{\rho_s^2}{\rho_i^2 - \rho_s^2}\right\} + O\left(\frac{1}{n^2}\right), \\
\mathrm{var}(r_i) &= \frac{(1-\rho_i^2)^2}{n-1} + O\left(\frac{1}{n^2}\right), \\
\mathrm{cor}(r_i, r_j) &= \frac{2\rho_i\rho_j(1-\rho_i^2)(1-\rho_j^2)}{(n-1)(\rho_i^2 - \rho_j^2)^2} + O\left(\frac{1}{n^2}\right).
\end{aligned}$$

11.3.1 Show that $\boldsymbol{S}_{11}\boldsymbol{1} = \boldsymbol{0}$ and $\boldsymbol{S}_{22}\boldsymbol{1} = \boldsymbol{0}$ in (11.19). Hence, $\boldsymbol{S}_{11}^{-1}$ and $\boldsymbol{S}_{22}^{-1}$ do not exist.

11.3.2 The following example illustrates that it does not matter which row and column are deleted from a contingency table when constructing canonical scores. Let

$$\boldsymbol{N} = \begin{bmatrix} 4 & 0 \\ 1 & 2 \\ 0 & 3 \end{bmatrix}.$$

(a) Show that deleting the first row and first column from $\boldsymbol{N}$ leads in (11.20) to

$$\boldsymbol{M}_1^* = \frac{1}{15}\begin{bmatrix} 2 & 6 \\ 3 & 9 \end{bmatrix},$$

and hence, the canonical scores for the row categories are proportional to $(0, 2, 3)$.

(b) Similarly, show that deleting the second row and first column from $\boldsymbol{N}$ in (11.20) leads to

$$\boldsymbol{M}_1^* = \frac{1}{15}\begin{bmatrix} 8 & -6 \\ -4 & 3 \end{bmatrix}$$

and hence to canonical scores proportional to $(-2, 0, 1)$. Since these scores are related to the scores in (a) by an additive constant, they are equivalent.

(c) Show that the canonical correlation between the row and column categories equals $(11/15)^{1/2}$.

(d) Note that because there are only two column categories, all scoring functions for column categories are equivalent (as long as they give distinct values to each of the two categories).

12

Discriminant Analysis and Statistical Learning

12.1 Introduction

Consider g populations or groups $\Pi_1, \dots, \Pi_g, g \geq 2$. Suppose that associated with each population Π_j, there is a probability density $f_j(\boldsymbol{x})$ on $\mathbb{R}^p$, so that items belonging to population Π_j have probability density function (p.d.f.) $f_j(\boldsymbol{x})$. (More generally, we could consider the domain of $\boldsymbol{x}$ to be $\Omega_1 \times \cdots \times \Omega_p$, where each Ω_k is either $\mathbb{R}$ or a discrete set, but most of the emphasis will be on the continuous case.)

The task of discriminant analysis is to allocate a *new* item to one of these g groups on the basis of its measurements $\boldsymbol{x}$. In most applications, the $f_j(\boldsymbol{x})$ are unknown, and the required information is contained in a sample data matrix $\boldsymbol{X}(n \times p)$ in which each observation $\boldsymbol{x}_r$, $r = 1, \dots, n$, has an associated group label y_r, with $y_r \in \{1, \dots, g\}$. Of course, the allocation should make as few "mistakes" as possible in a sense to be made precise later. We can rearrange the rows of $\boldsymbol{X}$, so that

$$\boldsymbol{X} = \begin{bmatrix} \boldsymbol{X}_1 \\ \vdots \\ \boldsymbol{X}_g \end{bmatrix} \tag{12.1}$$

in which the $(n_j \times p)$ matrix $\boldsymbol{X}_j$ represents a sample of n_j items with label $y = j$. Note that in this chapter it is the *items* (rows) of $\boldsymbol{X}$ that are grouped into categories, whereas in the last chapter it was the variables (columns) that were grouped.

As an example, consider the samples from three species of iris given in Table 1.3. The object is then to allocate a new iris to one of these species based on the measurements of length and width for the sepals and petals. As another example, suppose that the populations consist of different diseases, and $\boldsymbol{x}$ might measure the symptoms of a patient. Based on previous experience of the symptoms and diseases of many patients, one is trying to diagnose the new patient's disease on the basis of her symptoms.

A *discriminant rule* δ corresponds to a partition of $\mathbb{R}^p$ into disjoint (though not necessarily connected) regions $R_1, \dots, R_g$ (with $\cup R_j = \mathbb{R}^p$). The rule δ is then given by $\delta(\boldsymbol{x}) = j$ if $\boldsymbol{x} \in R_j$ for $j = 1, \dots, g$. Discrimination will be more accurate if Π_j has most of its probability concentrated in R_j for each j. In some applications, it may be useful to permit a new observation to lie in a region of "doubt" (between two or more competing groups) or to "none of the groups," but this refinement will not be pursued here. Historically, this task has been called discrimination but is also known as classification, pattern recognition, or machine learning. We have called this chapter *discriminant analysis and statistical*

Multivariate Analysis, Second Edition. Kanti V. Mardia, John T. Kent, and Charles C. Taylor.

learning to reflect the breadth of recent developments in this topic, as well as connections to artificial intelligence (Mardia, 2023).

Usually, we have no prior information about which population an item may belong to. However, if such information is available, it can easily be incorporated through a Bayesian approach.

Although it is very unrealistic, in Section 12.2, we first examine the case when the p.d.f.s $f_j(\boldsymbol{x})$ are fully known. Section 12.3 addresses the issue of measuring how good the discrimination rule is. This is straightforward in the case that the $f_j(\boldsymbol{x})$ are known, but in practice, it requires estimation. This is an important issue when faced with a choice of methods, since they may not all perform equally.

In Section 12.4, we slightly relax the assumption that the distributions are known by assuming that the populations have a specific distribution but with unknown parameters that must then be estimated using the data $\boldsymbol{X}$ and associated labels $\boldsymbol{y}$. Specifically, we consider the case when the populations are assumed to have a normal distribution. Section 12.5 gives a distributional result to test whether one of the variables can be discarded, which may be useful for variable selection.

Section 12.6 describes Fisher's linear discriminant function. It is motivated by the case of two normal distributions, but in more general settings, it can be viewed as an empirical approach to discriminant analysis that merely looks for a "sensible" rule to discriminate between populations. Alternative nonparametric methods are also available. The simplest of these methods – known as k-nearest neighbor – allocates a new observation $\boldsymbol{x}_0$ according to a majority vote based on the labels of the k nearest neighbors of the observations in $\boldsymbol{X}$ to $\boldsymbol{x}_0$. This simple rule can be modified, so that the vote of a neighbor $\boldsymbol{x}_r$ is weighted according to the distance from $\boldsymbol{x}_r$ to $\boldsymbol{x}_0$. These methods are discussed in Section 12.7.

In Section 12.8, we examine approaches that use recursive partitioning; these also make no parametric assumptions about the data. As the name suggests, this method divides the data into subdivisions. Each partition corresponds to a rule based on an inequality for a one-dimensional function of the attribute data $\boldsymbol{x}$, for example whether $\boldsymbol{a}'\boldsymbol{x} \geq 0$. The choice of partition depends on the purity (which uses the associated labels) of the data in each subdivision. Continuing this process, we can describe δ by a sequence of rules (usually based on components of $\boldsymbol{x}$), and this can be represented as a classification (or decision) tree.

Logistic regression, which is introduced in Section 12.9, also makes few assumptions about the distribution of the data $\boldsymbol{X}$. It seeks to model the *log-odds* as a linear function of the variables. Finally, in Section 12.10, we briefly describe three of the most commonly used neural network families: *multilayer perceptrons* (of which logistic regression represents a simple case), *radial basis functions*, and *support vector machines* (SVMs), which have been widely used for discrimination in the framework described above.

The iris data in Table 1.2 contain 50 measurements for each of $g = 3$ species of iris (*Iris setosa*, *Iris versicolor*, and *Iris virginica*) on $p = 4$ variables (sepal length, sepal width, petal length, and petal width). This chapter makes frequent use of various subsets of the iris data to illustrate different classification methods. The examples sometimes involve just two of the species and often – for ease of graphical presentation – use just two of the four variables. Table 12.1 summarizes the ways in which the data have been used.

Some recent approaches to supervised learning – for example ensemble methods, boosting, and random forests – can be closely connected to classification trees. For a good introduction to such methods – many of which are more algorithmic than model based – as well as a statistical perspective, see, e.g. Hastie et al. (2009). For further details of

Table 12.1 Summary of the examples in this chapter using the iris data, including the choice of species, the choice of variables, and the methods used.

Example	Iris species	Variables	Classification method
12.4.3	*I. setosa, I. versicolor*	Sepal length, width	Linear discrimination
12.4.4	*I. setosa, I. versicolor*	Sepal length, width	Quadratic discrimination
12.4.5	All three	Sepal length, width	Linear/quadratic discrimination
12.5.1	*I. versicolor, I. virginica*	All four	Variable selection
12.7.1	*I. setosa, I. versicolor*	Sepal length, width	Nearest neighbor
12.7.2	All three	Sepal length, width	Kernel classifier
12.8.1	All three	Sepal length, width	Classification tree

connections to AI, deep learning and machine learning, readers should consult the recent works of Krantz et al. (2023), Govindaraju et al. (2023) and Govindaraju and Rao (2013). See also Sell et al. (2023) and the references therein, for approaches to classification in situations of missing or corrupted data — for either the measurement vector $\boldsymbol{x}$ or the class labels.

12.2 Bayes' Discriminant Rule

Consider the situation in which the exact distributions of the populations $\Pi_1, \ldots, \Pi_g$ are known. Of course, this is extremely rare, although it may be possible to estimate the distributions fairly accurately, provided the samples are large enough. In any case, this is a useful framework against which other situations can be compared. Let $\pi_j = \mathsf{P}(y = j), j = 1, \ldots, g$, be the prior probability that an observation belongs to Π_j, and let $f_j(\boldsymbol{x})$ denote the conditional p.d.f. (continuous case) or probability mass function (discrete case) of an observation $\boldsymbol{x}$, given that the observation comes from population Π_j. Given a specific observation $\boldsymbol{x}_0$, the posterior probability of group membership can then be found from Bayes' rule:

$$\mathsf{P}(y = j \mid \boldsymbol{x}_0) = \frac{f(\boldsymbol{x}_0 \mid y = j)\mathsf{P}(y = j)}{f(\boldsymbol{x}_0)} = \frac{f(\boldsymbol{x}_0 \mid y = j)\mathsf{P}(y = j)}{\sum_l f(\boldsymbol{x}_0 \mid y = l)\mathsf{P}(y = l)}$$

$$= \frac{\pi_j f_j(\boldsymbol{x}_0)}{\sum_l \pi_l f_l(\boldsymbol{x}_0)},$$

where $f(\boldsymbol{x}) = \sum_l \pi_l f_l(\boldsymbol{x})$ is the marginal density of $\boldsymbol{x}$. In the case that the p.d.f.s $f_j(\boldsymbol{x})$ are known, we can then define Bayes' discriminant rule by choosing the population that maximizes the posterior probability.

Definition 12.2.1 *If populations* $\Pi_1, \ldots, \Pi_g$ *have prior probabilities* $(\pi_1, \ldots, \pi_g) = \boldsymbol{\pi}'$, *then the* Bayes' discriminant rule *(with respect to* $\boldsymbol{\pi}$*) allocates an observation* $\boldsymbol{x}$ *to the population* Π_j *for which*

$$\pi_j f_j(\boldsymbol{x}) = \max_l \pi_l f_l(\boldsymbol{x}). \tag{12.2}$$

The function $\pi_j f_j(\boldsymbol{x})$, as a function of j, in (12.2) is proportional to the posterior probability of Π_j, given the observed $\boldsymbol{x}$. In the case that all the prior probabilities are equal, we then have the maximum-likelihood (ML) rule defined as follows.

Definition 12.2.2 *The ML discrimination rule for allocating an observation* $\boldsymbol{x}$ *to one of the populations* $\Pi_1, \dots, \Pi_g$ *is to allocate* $\boldsymbol{x}$ *to the population* Π_j *for which the likelihood* $L_j(\boldsymbol{x}) = f_j(\boldsymbol{x})$ *is largest.*

If several likelihoods take the same maximum value, then any one of these may be chosen. This point will not always be repeated in what follows. Further, in the continuous examples we consider, it will usually be the case that

$$\mathsf{P}\left(f_i(\boldsymbol{x}) = f_k(\boldsymbol{x}) \quad \text{for some} \quad i \neq k \mid y = j\right) = 0$$

for all $j = 1, \dots, g$, so that the form of the allocation rule in case of ties has no practical importance.

Example 12.2.1 If x is a 0–1 binary random variable, and if Π_1 is the population with probabilities $(1/2, 1/2)$ and Π_2 is the population with probabilities $(1/4, 3/4)$, then the ML discriminant rule allocates x to Π_1 when $x = 0$ and x to Π_2 when $x = 1$. This is because $f_1(0) = 1/2 > f_2(0) = 1/4$, and $f_2(1) = 3/4 > f_1(1) = 1/2$. □

Example 12.2.2 Suppose that $\boldsymbol{x}$ is a multinomial random vector, which comes *either* from Π_1, with multinomial probabilities $\alpha_1, \dots, \alpha_k$, *or* from Π_2 with multinomial probabilities $\beta_1, \dots, \beta_k$, where $\sum \alpha_i = \sum \beta_i = 1$, and $\sum x_i = n$ is fixed. If $\boldsymbol{x}$ comes from Π_1, its likelihood is

$$\frac{n!}{x_1! \cdots x_k!} \alpha_1^{x_1} \cdots \alpha_k^{x_k}. \tag{12.3}$$

If $\boldsymbol{x}$ comes from Π_2, the likelihood is

$$\frac{n!}{x_1! \cdots x_k!} \beta_1^{x_1} \cdots \beta_k^{x_k}.$$

If λ is the ratio of these likelihoods, then the log-likelihood ratio is

$$\log \lambda = \sum x_i \log \frac{\alpha_i}{\beta_i} = \sum x_i s_i, \tag{12.4}$$

where $s_i = \log(\alpha_i/\beta_i)$. The ML discriminant rule allocates $\boldsymbol{x}$ to Π_1 if $\lambda > 1$, i.e. if $\log \lambda > 0$. (See also Example 12.6.2.) □

12.3 The Error Rate

12.3.1 Probabilities of Misclassification

Let $\boldsymbol{x}$ be a continuous random vector in $\mathbb{R}^p$. For any $\boldsymbol{x} \in \mathbb{R}^p$, we can associate with a given classifier $\delta(\boldsymbol{x})$ a set of indicator variables given by

$$\phi_i(\boldsymbol{x}) = \begin{cases} 1 & \text{if } \delta(\boldsymbol{x}) = i \text{ or, equivalently, if } \boldsymbol{x} \in R_i \\ 0 & \text{otherwise} \end{cases} \tag{12.5}$$

for $i = 1, \dots, g$. Then, the probability of allocating a randomly selected item to group Π_i when it actually belongs to group Π_j is

$$p_{ij} = \int \phi_i(\boldsymbol{x}) f_j(\boldsymbol{x})\, \mathrm{d}\boldsymbol{x}. \tag{12.6}$$

In particular, if an item belongs to population Π_j, the probability of a correct allocation is p_{jj}, and the *error rate* associated with this group is thus

$$\sum_{i \neq j} p_{ij} = 1 - p_{jj}.$$

The performance of the classifier can be summarized by the numbers $p_{11}, \ldots, p_{gg}$. The following definition gives a partial order on the set of classifiers.

Definition 12.3.1 *Say that one classifier δ with probabilities of correct classification $\{p_{jj}\}$ is as good as another classifier δ^* with probabilities $\{p^*_{jj}\}$ if*

$$p_{jj} \geq p^*_{jj} \qquad \text{for all} \quad j = 1, \ldots, g.$$

Say that δ is better *than δ^* if at least one of the inequalities is strict. If δ is a rule for which there is no better rule, say that δ is* admissible.

Equation (12.6) is the (conditional) probability of misclassification, and taking account of the priors, we obtain the overall *expected error rate* as

$$\sum_{j=1}^{g} \pi_j (1 - p_{jj}) = 1 - \sum_{j=1}^{g} \pi_j p_{jj}, \tag{12.7}$$

since the priors sum to 1.

It is clear from (12.7) that in order to minimize the expected error rate, we need to maximize $\sum \pi_j p_{jj}$. So, using (12.6), the task is to maximize

$$\sum_{j=1}^{g} \int \pi_j \phi_j(\boldsymbol{x}) f_j(\boldsymbol{x})\, d\boldsymbol{x}, \quad \text{i.e.} \quad \sum_{j=1}^{g} \pi_j \int_{R_j} f_j(\boldsymbol{x})\, d\boldsymbol{x},$$

over all possible $\phi_j(\boldsymbol{x})$, $j = 1, \ldots, g$, or, equivalently, over all possible nonoverlapping (not necessarily connected) regions $R_1, \ldots, R_g$. It is clear that the maximum occurs when R_i is defined by $R_i = \{\boldsymbol{x};\, i = \arg\max \pi_k f_k(\boldsymbol{x})\}$. Equivalently, we have the rule $\delta(\boldsymbol{x}) = \arg\max \pi_k f_k(\boldsymbol{x})$, which we recognize, from (12.2), as Bayes' discriminant rule. Hence, we note that Bayes' discriminant rule minimizes the overall expected error rate when the priors π_j have been "correctly" specified, i.e. $\pi_j = \mathsf{P}(y = j)$. A more powerful result, which can be viewed as a generalization of the Neyman-Pearson lemma (Rao, 1973, p. 448), states that all Bayes rules are admissible.

Theorem 12.3.2 *All Bayes discriminant rules (including the ML rule) are admissible.*

Proof: Let $\boldsymbol{\tau}$ be a vector of prior probabilities with $\tau_j > 0$ for all $j = 1, \ldots, g$, and let δ^* denote the corresponding Bayes rule with respect to $\boldsymbol{\tau}$. Suppose that there exists a classifier δ which is better. Let $\{p_{jj}\}$ and $\{p^*_{jj}\}$ denote the probabilities of correct classification for this classifier and Bayes rule, respectively. Then, because δ is better than δ^*, and since $\tau_j > 0$ for all j, we have $\sum \tau_j p_{jj} > \sum \tau_j p^*_{jj}$. However, using (12.5) and (12.6),

$$\sum \tau_j p_{jj} = \sum \int \phi_j(\boldsymbol{x}) \tau_j f_j(\boldsymbol{x})\, d\boldsymbol{x}$$

$$
\begin{aligned}
&\leq \sum_j \int \phi_j(\boldsymbol{x}) \max_i \tau_i f_i(\boldsymbol{x}) \mathrm{d}\boldsymbol{x} = \int \left\{ \sum \phi_j(\boldsymbol{x}) \right\} \max_i \tau_i f_i(\boldsymbol{x}) \mathrm{d}\boldsymbol{x} \\
&= \int \max_i \tau_i f_i(\boldsymbol{x}) \mathrm{d}\boldsymbol{x} \\
&= \int \sum \phi_j^*(\boldsymbol{x}) \tau_j f_j(\boldsymbol{x}) \mathrm{d}\boldsymbol{x} \\
&= \sum \tau_j p_{jj}^*,
\end{aligned}
$$

which contradicts the above statement. Hence, the theorem is proved. □

Note that in the above theorem, classifiers can be judged on their g probabilities of correct allocation $p_{11}, \ldots, p_{gg}$. However, when priors exist, then a classifier can also be judged on the basis of a single number (12.7). Using this criterion, *any* two classifiers can be compared, and we have the following result, first given in the case of $g = 2$ groups by Welch (1939).

Theorem 12.3.3 *If populations $\Pi_1, \ldots, \Pi_g$ have prior probabilities $\tau_1, \ldots, \tau_g$, then no discriminant rule has a larger posterior probability of correct allocation than the Bayes rule with respect to this prior.*

Proof: Let δ^* denote the Bayes rule with respect to the prior probabilities $\tau_1, \ldots, \tau_g$ with probabilities of correct allocation $\{p_{jj}^*\}$. Then, using the same argument as in Theorem 12.3.2, it is easily seen that for any other rule δ with probabilities of correct allocation $\{p_{jj}\}$, $\sum \tau_j p_{jj} \leq \sum \tau_j p_{jj}^*$; that is, the posterior probability of correct allocation is at least as large for the Bayes rule as for the other rule. □

More generally, we now suppose that the loss incurred by misclassification depends on the groups. For example, to misclassify a sick patient as healthy may be more serious than to misclassify a healthy patient as sick. Let a set of coefficients $C(i, j)$, $1 \leq i, j \leq g$, define a *loss function* representing the cost or loss incurred if an observation is allocated to Π_i when in fact it comes from Π_j. For this to be a sensible definition, suppose that $C(i, i) = 0$ and $C(i,j) > 0$ for all $i \neq j$. If δ is a classifier with allocation functions given by (12.5), then the *risk function* is defined by

$$
\begin{aligned}
R(\delta,j) &= \mathsf{E}(C(\delta(\boldsymbol{x}), j \mid \Pi_j) \\
&= \sum_i C(i,j) \int \phi_i(\boldsymbol{x}) f_j(\boldsymbol{x}) \, \mathrm{d}\boldsymbol{x} \\
&= \sum_i C(i,j) p_{ij}.
\end{aligned}
$$

The risk function represents the expected loss for a classifier δ, given that the observation comes from Π_j. In particular, if $C(i, j) = 1$ for $i \neq j$, then

$$R(\delta,j) = 1 - p_{jj}$$

represents the misallocation probabilities, as before.

Taking account of the prior probabilities, the *Bayes' risk* can be defined by

$$r(\delta, \boldsymbol{\pi}) = \sum \pi_j R(\delta,j),$$

representing the posterior expected loss. It is clear that when $C(i,j) = 1$ for $i \neq j$, then Bayes' risk is simply the expected error rate.

Define the *loss-adjusted Bayes' rule* in this more general situation as

$$\text{allocate } \boldsymbol{x} \text{ to } \Pi_i \text{ if } \sum_{k\neq i} C(i,k)\,\pi_k f_k(\boldsymbol{x}) = \min_l \sum_{k\neq l} C(l,k)\,\pi_k f_k(\boldsymbol{x}). \tag{12.8}$$

12.3.2 Estimation of Error Rate

The quality of a classifier $\delta(\boldsymbol{x})$ can be judged by its expected error rate, relative to the error rate of Bayes' discriminant rule. However, in practice, the $f_j(\boldsymbol{x})$, $j = 1, \ldots, g$, are unknown (if they were known, then we could always obtain the optimal Bayes' rule), so we need to estimate the p_{ij} in (12.6) corresponding to a proposed δ. Various methods to obtain δ are described in the subsequent sections, but for a given classifier, we can use data to estimate its performance. Given a data matrix $\boldsymbol{X}$ with known population labels as in (12.1), a natural procedure is to simply compute the proportion of observations in class Π_j, which are classified to class Π_i using

$$\hat{p}_{ij} = \frac{\sum_{r:y_r=j} I[\delta(\boldsymbol{x}_r) = i]}{\sum_{r:y_r=j} 1} = \frac{\sum_{r:y_r=j} \phi_i(\boldsymbol{x}_r)}{n_j}. \tag{12.9}$$

Note that the computation of (12.9) is an estimate of a class-specific error rate for any given classifier $\delta(\cdot)$. These quantities can be used to compute an overall error rate. The usefulness of these quantities will critically depend on whether the information in $\boldsymbol{X}$ is "representative" of the population under study, or whether the priors are known.

There are two common ways to estimate p. If we simply resubstitute the data which was used to "learn" δ, then this gives a *resubstitution estimate* which is prone to bias. Typically, $\hat{p}_{ij} < p_{ij}$ for $i \neq j$. However, in order to achieve an unbiased estimate of the p_{ij}, it is important to use data which is independent of that which was used to obtain the classifier $\delta(\boldsymbol{x})$.

One way to obtain an unbiased estimate is to split the original data into two parts, a *train* part and a *test* part, with the training set being used to obtain δ and the testing set being used to obtain the $\hat{p}_{ij}$ and thence an estimate of the expected error rate. Alternatively, if the original data is limited in size (moderate n), then m-fold cross-validation (Lachenbruch and Mickey, 1968) may be used. Here, the data is split into m (approximately equal-sized) parts, and the number of errors for the kth part is obtained from a classifier derived from the remaining $\{1, \ldots, m\}\backslash k$ parts. This is repeated for $k = 1, \ldots, m$, and the results are combined to obtain $\hat{p}_{ij}$. When $m = n$, this is referred to as *leave-one-out cross-validation*, which is a common choice when n is small.

12.3.3 Confusion Matrix

Given a classifier $\delta(\cdot)$ and a (new) set of data that contains both explanatory variables $\boldsymbol{x}$ and a known classes y, the corresponding *confusion matrix* is a $g \times g$ matrix in which entry (i,j) is the number of observations that were actually in class j but predicted to be in class i. If we denote the entries of this table as n_{ij} where

$$n_{ij} = \sum_k I[\{\delta(\boldsymbol{x}_k) = i\} \cap \{y_k = j\}],$$

then the overall error rate corresponding to this $\delta(\cdot)$ for these data is given by $\sum_{i\neq j} n_{ij}/\sum_{i,j} n_{ij}$.
In general, the confusion matrix is not symmetric.

12.4 Discrimination Using the Normal Distribution

12.4.1 Population Discriminant Rules

If the populations are normally distributed, such that Π_j is $N_p(\boldsymbol{\mu}_j, \boldsymbol{\Sigma}_j)$, $j = 1, \dots, g$, then Bayes' rule allocates $\boldsymbol{x}$ to Π_l, where $l \in \{1, \dots, g\}$ is that value of j which maximizes

$$\log\left(\pi_j f_j(\boldsymbol{x})\right) = \log \pi_j - \frac{1}{2}\log|2\pi\boldsymbol{\Sigma}_j| - \frac{1}{2}(\boldsymbol{x} - \boldsymbol{\mu}_j)'\boldsymbol{\Sigma}_j^{-1}(\boldsymbol{x} - \boldsymbol{\mu}_j), \quad j = 1, \dots, g. \tag{12.10}$$

12.4.1.1 Linear Discriminant Analysis

In the special case where all the priors are equal ($\pi_j = 1/g$) and the covariance matrices are equal $\boldsymbol{\Sigma}_j = \boldsymbol{\Sigma}$, $j = 1, \dots, g$, we have the following simplification.

Theorem 12.4.1 *(a) If Π_j is the $N_p(\boldsymbol{\mu}_j, \boldsymbol{\Sigma})$ population, $\pi_j = 1/g$, $j = 1, \dots, g$, and $\boldsymbol{\Sigma} > 0$, then the ML discriminant rule allocates $\boldsymbol{x}$ to Π_l, where $l \in \{1, \dots, g\}$ is that value of j which minimizes the square of the Mahalanobis distance between $\boldsymbol{x}$ and $\boldsymbol{\mu}_j$, i.e. $\Delta^2 = (\boldsymbol{x} - \boldsymbol{\mu}_j)'\boldsymbol{\Sigma}^{-1}(\boldsymbol{x} - \boldsymbol{\mu}_j)$.*

(b) When $g = 2$, the rule allocates $\boldsymbol{x}$ to Π_1 if

$$\boldsymbol{\alpha}'(\boldsymbol{x} - \boldsymbol{\mu}) > 0, \tag{12.11}$$

where $\boldsymbol{\alpha} = \boldsymbol{\Sigma}^{-1}(\boldsymbol{\mu}_1 - \boldsymbol{\mu}_2)$ and $\boldsymbol{\mu} = \frac{1}{2}(\boldsymbol{\mu}_1 + \boldsymbol{\mu}_2)$, and to Π_2 otherwise.

Proof: From (2.49), the jth likelihood is

$$L_j(\boldsymbol{x}) = |2\pi\boldsymbol{\Sigma}|^{-1/2} \exp\left\{-\frac{1}{2}(\boldsymbol{x} - \boldsymbol{\mu}_j)'\boldsymbol{\Sigma}^{-1}(\boldsymbol{x} - \boldsymbol{\mu}_j)\right\}.$$

This is maximized when the exponent is minimized, which proves part (a) of the theorem.

For part (b), note that $L_1(\boldsymbol{x}) > L_2(\boldsymbol{x})$ if and only if

$$(\boldsymbol{x} - \boldsymbol{\mu}_1)'\boldsymbol{\Sigma}^{-1}(\boldsymbol{x} - \boldsymbol{\mu}_1) < (\boldsymbol{x} - \boldsymbol{\mu}_2)'\boldsymbol{\Sigma}^{-1}(\boldsymbol{x} - \boldsymbol{\mu}_2).$$

Canceling and simplifying leads to the condition stated in (12.11). (See Exercise 12.4.1.) □

Example 12.4.1 Singular $\boldsymbol{\Sigma}$ (Rao and Mitra, 1971, p. 204). Consider the situation of the above theorem when $\boldsymbol{\Sigma}$ is singular. Then the ML rule must be modified. Note that Π_1 is concentrated on the hyperplane $\boldsymbol{N}'(\boldsymbol{x} - \boldsymbol{\mu}_1) = \boldsymbol{0}$, where the columns of $\boldsymbol{N}$ span the null space of $\boldsymbol{\Sigma}$, and that Π_2 is concentrated on the hyperplane $\boldsymbol{N}'(\boldsymbol{x} - \boldsymbol{\mu}_2) = \boldsymbol{0}$. (See Section 2.5.4.) If $\boldsymbol{N}'(\boldsymbol{\mu}_1 - \boldsymbol{\mu}_2) \neq \boldsymbol{0}$, then these two hyperplanes are distinct, and discrimination can be carried out with perfect accuracy; namely, if $\boldsymbol{N}'(\boldsymbol{x} - \boldsymbol{\mu}_1) = \boldsymbol{0}$, allocate to Π_1, and if $\boldsymbol{N}'(\boldsymbol{x} - \boldsymbol{\mu}_2) = \boldsymbol{0}$, allocate to Π_2.

The more interesting case occurs when $\boldsymbol{\mu}_1 - \boldsymbol{\mu}_2$ is orthogonal to the null space of $\boldsymbol{\Sigma}$. If $\boldsymbol{N}'(\boldsymbol{\mu}_1 - \boldsymbol{\mu}_2) = \boldsymbol{0}$, then the ML allocation rule is given by allocating $\boldsymbol{x}$ to Π_1 if

$$\boldsymbol{\alpha}'\boldsymbol{x} > \frac{1}{2}\boldsymbol{\alpha}'(\boldsymbol{\mu}_1 + \boldsymbol{\mu}_2),$$

where $\boldsymbol{\alpha} = \boldsymbol{\Sigma}^{-}(\boldsymbol{\mu}_1 - \boldsymbol{\mu}_2)$, and $\boldsymbol{\Sigma}^{-}$ is a g-inverse of $\boldsymbol{\Sigma}$. In the rest of this section, we assume $\boldsymbol{\Sigma} > 0$. □

Note that when there are just $g = 2$ groups, the ML discriminant rule is defined in terms of the *discriminant function*

$$\begin{aligned} h(\boldsymbol{x}) &= \log L_1(\boldsymbol{x}) - \log L_2(\boldsymbol{x}) \\ &= (\boldsymbol{\mu}_1 - \boldsymbol{\mu}_2)'\boldsymbol{\Sigma}^{-1}\{\boldsymbol{x} - \frac{1}{2}(\boldsymbol{\mu}_1 + \boldsymbol{\mu}_2)\}, \end{aligned} \tag{12.12}$$

and the ML rule takes the form

$$\begin{array}{llll} \text{allocate } \boldsymbol{x} \text{ to } \Pi_1 & \text{if} & h(\boldsymbol{x}) > 0, \\ \text{allocate } \boldsymbol{x} \text{ to } \Pi_2 & \text{if} & h(\boldsymbol{x}) < 0. \end{array} \tag{12.13}$$

In particular, note that this discriminant function for two multivariate normal populations with the same covariance matrix is *linear*. Thus, the boundary between the allocation regions in this case is a hyperplane passing through the midpoint of the line segment connecting the two group means, although the hyperplane is not necessarily perpendicular to this line segment. See Figure 12.3 for an example in the sample case.

Formally, the probabilities of misclassification p_{ij} are given by (12.6). In the case of two normal populations $N_p(\boldsymbol{\mu}_1, \boldsymbol{\Sigma})$ and $N_p(\boldsymbol{\mu}_2, \boldsymbol{\Sigma})$, we can obtain p_{12} (the probability of an observation from Π_2 being misallocated) as follows. We first note that, if $\boldsymbol{x}$ comes from Π_2, then $h(\boldsymbol{x}) \sim N(-\Delta^2/2, \Delta^2)$, where

$$\Delta^2 = (\boldsymbol{\mu}_1 - \boldsymbol{\mu}_2)'\boldsymbol{\Sigma}^{-1}(\boldsymbol{\mu}_1 - \boldsymbol{\mu}_2)$$

is the squared Mahalanobis distance between the populations. We then have

$$\begin{aligned} p_{12} &= \mathsf{P}(h(\boldsymbol{x}) > 0 \mid \boldsymbol{x} \text{ belongs to } \Pi_2) \\ &= \Phi(-\mathsf{E}(h)/\sqrt{\text{var}(h)}) = \Phi(-\Delta/2), \end{aligned} \tag{12.14}$$

where $\Phi(\cdot)$ is the standard normal distribution function. Similarly, $p_{21} = \Phi(-\Delta/2)$ also.

12.4.1.2 Quadratic Discriminant Analysis

In the more general case, in which the covariance matrices are not equal, Bayes' rule (12.10) leads to quadratic discriminant functions. Here, the boundaries between regions are curved, and the regions may be disconnected. The rules are based on Eq. (12.10).

Example 12.4.2 Univariate case. Suppose that Π_1 is the $N(\mu_1, \sigma_1^2)$ distribution, and Π_2 is the $N(\mu_2, \sigma_2^2)$ distribution. This situation is illustrated in Figure 12.1 for the case where $\mu_2 > \mu_1$ and $\sigma_1 > \sigma_2$. The likelihood $L_j (j = 1, 2)$ is

$$L_j(x) = (2\pi\sigma_j^2)^{-1/2} \exp\left\{-\frac{1}{2}\left(\frac{x - \mu_j}{\sigma_j}\right)^2\right\}.$$

Note that $L_1(x)$ exceeds $L_2(x)$ if

$$\exp\left\{-\frac{1}{2}\left[\left(\frac{x - \mu_1}{\sigma_1}\right)^2 - \left(\frac{x - \mu_2}{\sigma_2}\right)^2\right]\right\} > \frac{\sigma_1}{\sigma_2}.$$

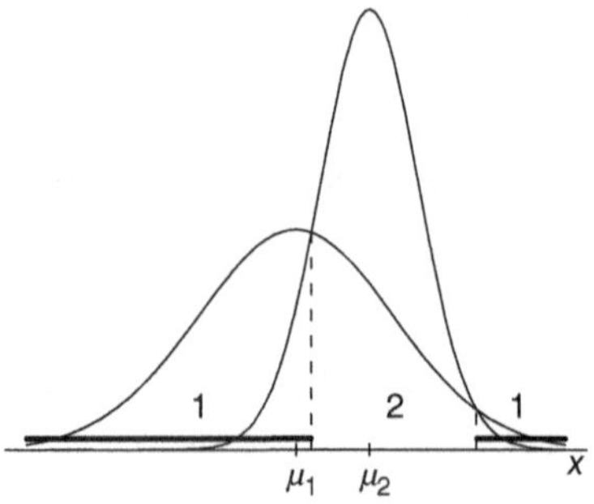

Figure 12.1 Normal likelihoods with unequal means and unequal variances (from Example 12.4.2). The solid line shows the disconnected region of x for which $L_1(x) > L_2(x)$.

On taking logarithms and rearranging, this inequality becomes

$$x^2\left(\frac{1}{\sigma_1^2}-\frac{1}{\sigma_2^2}\right)-2x\left(\frac{\mu_1}{\sigma_1^2}-\frac{\mu_2}{\sigma_2^2}\right)+\left(\frac{\mu_1^2}{\sigma_1^2}-\frac{\mu_2^2}{\sigma_2^2}\right)<2\log\frac{\sigma_2}{\sigma_1}.$$

If $\sigma_1 > \sigma_2$ as in Figure 12.1, then the coefficient of x^2 is negative. Therefore, the set of xs for which this inequality is satisfied forms two distinct regions, one having low values of x, and the other having high values of x (see Exercise 12.4.9 and Anderson and Bahadur (1962)). □

12.4.2 The Sample Discriminant Rules

The *sample ML discriminant rule* is useful when the forms of the distributions of $\Pi_1, \dots, \Pi_g$ are known, but their parameters must be estimated from a data matrix $\boldsymbol{X}(n \times p)$. We suppose that the rows of $\boldsymbol{X}$ are partitioned into g groups, $\boldsymbol{X}' = (\boldsymbol{X}_1', \dots, \boldsymbol{X}_g')$, and that $\boldsymbol{X}_j$ contains the n_j observations from Π_j.

For example, suppose that the groups are assumed to be samples from the multivariate normal distribution with different means and the same covariance matrix. Let $\overline{\boldsymbol{x}}_j$ and $\boldsymbol{S}_j$ denote the sample mean and covariance matrix of the jth group. Unbiased estimates of $\boldsymbol{\mu}_1, \dots, \boldsymbol{\mu}_g$ and $\boldsymbol{\Sigma}$ are given by $\overline{\boldsymbol{x}}_1, \dots, \overline{\boldsymbol{x}}_g$, and $\boldsymbol{S}_u = \sum n_j \boldsymbol{S}_j/(n-g)$. The sample ML discriminant rule is then obtained by inserting these estimates in Theorem 12.4.1. In particular, when g = 2, the sample ML discriminant rule allocates $\boldsymbol{x}$ to Π_1 if and only if

$$\boldsymbol{a}'\left\{\boldsymbol{x}-\frac{1}{2}(\overline{\boldsymbol{x}}_1+\overline{\boldsymbol{x}}_2)\right\}>0, \tag{12.15}$$

where $\boldsymbol{a} = \boldsymbol{S}_u^{-1}(\overline{\boldsymbol{x}}_1 - \overline{\boldsymbol{x}}_2)$.

Example 12.4.3 (Iris data, two species, linear discriminant analysis (LDA)). Consider the $n_1 = n_2 = 50$ observations on two species of iris, *I. setosa* and *I. versicolor*, given in Table 1.3. For simplicity of exposition, we discriminate between them only on the basis of the first two variables, sepal length and sepal width. Then, the sample means and variances for each group are given by

$$\overline{\boldsymbol{x}}_1 = (5.006, 3.428)', \qquad \overline{\boldsymbol{x}}_2 = (5.936, 2.770)',$$
$$\boldsymbol{S}_1 = \begin{bmatrix} 0.1218 & 0.0972 \\ & 0.1408 \end{bmatrix}, \qquad \boldsymbol{S}_2 = \begin{bmatrix} 0.2611 & 0.0835 \\ & 0.0965 \end{bmatrix}.$$

Thus,

$$\begin{aligned} \boldsymbol{a} &= [(50\boldsymbol{S}_1 + 50\boldsymbol{S}_2)/98]^{-1}(\bar{\boldsymbol{x}}_1 - \bar{\boldsymbol{x}}_2) \\ &= \begin{bmatrix} 0.1953 & 0.0922 \\ 0.0922 & 0.1211 \end{bmatrix}^{-1} \begin{pmatrix} -0.930 \\ 0.658 \end{pmatrix} = \begin{pmatrix} -11.436 \\ 14.143 \end{pmatrix}, \end{aligned}$$

and the linear discriminant rule is given by allocating to Π_1 if

$$\begin{aligned} h(\boldsymbol{x}) &= (-11.436, \quad 14.143) \begin{pmatrix} x_1 - \frac{1}{2}(5.006 + 5.936) \\ x_2 - \frac{1}{2}(3.428 + 2.770) \end{pmatrix} \\ &= -11.436x_1 + 14.143x_2 + 18.739 > 0, \end{aligned}$$

and to Π_2 otherwise. A picture of the corresponding allocation regions is given in Figure 12.2 (adapted from Dagnelie (1975, p. 313)).

The estimated ellipses containing 50%, 90%, and 99% of the probability mass within each group have been drawn to give a visual impression of the accuracy of the discrimination. See also Exercise 12.3.3.

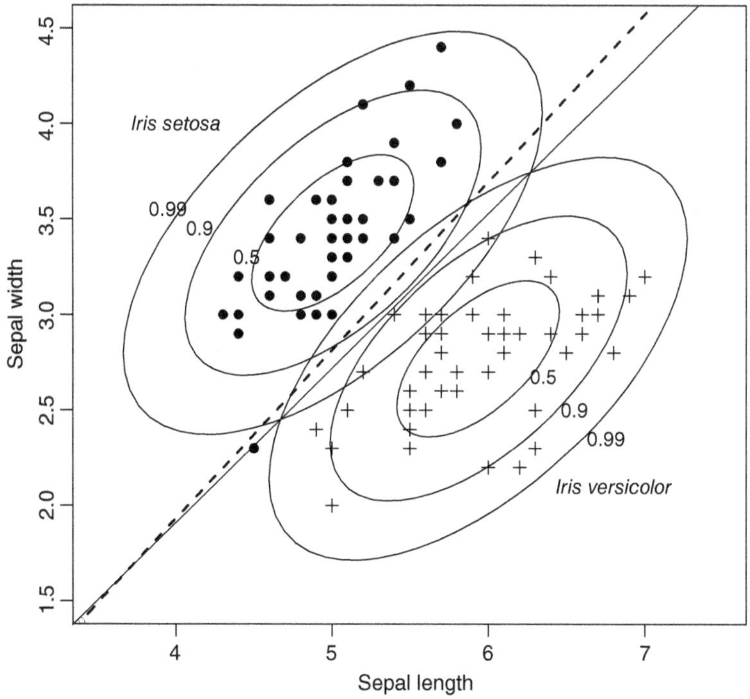

Figure 12.2 Discrimination between two species of iris: a plot of the data points ((•) *I. setosa* and (+) *I. versicolor*) in two dimensions, together with ellipses of the cumulative probability distribution for probabilities 0.5, 0.9, and 0.99, assuming equal covariance matrices. Also shown are the boundaries between the two allocation regions for the linear discriminant function (continuous line) and the quadratic discriminant function (dashed curve).

Note that the sample means themselves have scores $h(\overline{\boldsymbol{x}}_1) = 9.97$ and $h(\overline{\boldsymbol{x}}_2) = -9.97$, so that the boundary line passes through the midpoint of the line segment connecting the two means.

In Section 1.2.4, the question was posed as to how to allocate a new iris from an unknown species to one of the two populations. The illustration there had measurements on the first two variables given by $\boldsymbol{x} = (5.1, 3.2)'$. Since $h(\boldsymbol{x}) = 5.67$, we allocate $\boldsymbol{x}$ to Π_1 (i.e. *I. setosa*). □

If we do not assume that the covariance matrices are equal, then when g = 2, the sample ML rule becomes the quadratic discriminant rule that allocates $\boldsymbol{x}$ to Π_1 when

$$(\boldsymbol{x} - \overline{\boldsymbol{x}}_2)'\boldsymbol{S}_2^{-1}(\boldsymbol{x} - \overline{\boldsymbol{x}}_2) - (\boldsymbol{x} - \overline{\boldsymbol{x}}_1)'\boldsymbol{S}_1^{-1}(\boldsymbol{x} - \overline{\boldsymbol{x}}_1) > \log\frac{|\boldsymbol{S}_1|}{|\boldsymbol{S}_2|}.$$

The rule is thus of the form

$$\sum_{\{i,j\geq 0|i+j\leq 2\}} a_{ij}x_1^i x_2^j = b_{00} + \underbrace{b_{10}x_1 + b_{01}x_2}_{\text{linear terms}} + \underbrace{b_{11}x_1x_2 + b_{20}x_1^2 + b_{02}x_2^2}_{\text{quadratic terms}} > 0 \tag{12.16}$$

for some coefficients a_{ij}, b_{ij}. In this case, the boundary between regions is not linear. The quadratic part is given by $\boldsymbol{x}'(\boldsymbol{S}_2^{-1} - \boldsymbol{S}_1^{-1})\boldsymbol{x}$. The shape of the boundary depends on the eigenvalues of $\boldsymbol{S}_2^{-1} - \boldsymbol{S}_1^{-1}$. If all the eigenvalues are nonzero, then the boundary is an ellipsoid or a hyperboloid. If all the eigenvalues have the same sign, then the boundary is an ellipsoid; if some are positive and some negative, then it is a hyperboloid.

Example 12.4.4 (Iris data, two species, quadratic discriminant analysis). Continuing with the calculations of Example 12.4.3, it is straightforward to obtain the quadratic discriminant rule. We additionally note that

$$\begin{aligned} |\boldsymbol{S}_1| &= 0.00769, & |\boldsymbol{S}_2| &= 0.01823, \\ \boldsymbol{S}_1^{-1} &= \begin{bmatrix} 18.3062 & -12.6402 \\ & 15.8294 \end{bmatrix}, & \boldsymbol{S}_2^{-1} &= \begin{bmatrix} 5.2942 & -4.5799 \\ & 14.3246 \end{bmatrix}. \end{aligned}$$

Thus, we can identify the coefficients of the terms in x_1 and x_2 in (12.16) as

$$\begin{aligned} b_{00} &= -64.2320, & b_{10} &= 59.1403, & b_{01} &= -43.0139, \\ b_{11} &= 16.1208, & b_{20} &= -13.0120, & b_{02} &= -1.5048 \end{aligned},$$

and this gives the boundary as shown in Figure 12.2. We note that this quadratic boundary is very similar to the linear discriminant function of Example 12.4.3; the allocated classes for the data are in fact identical. □

Example 12.4.5 (Iris data, three species, linear and quadratic discriminant analysis). Extend Example 12.4.3 to include the $n_3 = 50$ observations on the third species of iris, *I. virginica*, in Table 1.3, which has sample mean and variance

$$\overline{\boldsymbol{x}}_3 = \begin{pmatrix} 6.588 \\ 2.974 \end{pmatrix}, \qquad \boldsymbol{S}_3 = \begin{bmatrix} 0.3963 & 0.0919 \\ & 0.1019 \end{bmatrix}.$$

In this case, $\boldsymbol{\Sigma}$ is estimated by

$$\boldsymbol{S}_u = (50\boldsymbol{S}_1 + 50\boldsymbol{S}_2 + 50\boldsymbol{S}_3)/147 = \begin{bmatrix} 0.2650 & 0.0927 \\ 0.0927 & 0.1154 \end{bmatrix}.$$

Writing

$$h_{ij}(\boldsymbol{x}) = (\overline{\boldsymbol{x}}_i - \overline{\boldsymbol{x}}_j)'\boldsymbol{S}_u^{-1}\boldsymbol{x} - \frac{1}{2}\overline{\boldsymbol{x}}_i'\boldsymbol{S}_u^{-1}\overline{\boldsymbol{x}}_i + \frac{1}{2}\overline{\boldsymbol{x}}_j'\boldsymbol{S}_u^{-1}\overline{\boldsymbol{x}}_j$$

for $i \neq j$, discrimination is then based on the three linear functions (using an alternative layout of the above):

$$\begin{aligned} h_{12}(\boldsymbol{x}) &= (\overline{\boldsymbol{x}}_1 - \overline{\boldsymbol{x}}_2)'\boldsymbol{S}_u^{-1}\{\boldsymbol{x} - \frac{1}{2}(\overline{\boldsymbol{x}}_1 + \overline{\boldsymbol{x}}_2)\} \\ &= -\ \ 7.657x_1 + 11.856x_2 + 5.153, \\ h_{13}(\boldsymbol{x}) &= -10.219x_1 + 12.147x_2 + 20.361, \\ h_{23}(\boldsymbol{x}) &= -\ \ 2.562x_1 + \ \ 0.291x_2 + 15.208. \end{aligned}$$

Notice that $h_{12}(\boldsymbol{x})$ is not identical to the discriminant function $h(\boldsymbol{x})$ of Example 12.4.1, because we are using a slightly different estimate of $\boldsymbol{\Sigma}$.

Then, the allocation regions are defined by

$$\text{allocate } \boldsymbol{x} \text{ to} \begin{cases} \Pi_1 & h_{12}(\boldsymbol{x}) > 0 \quad \text{and} \quad h_{13}(\boldsymbol{x}) > 0, \\ \Pi_2 & h_{12}(\boldsymbol{x}) < 0 \quad \text{and} \quad h_{23}(\boldsymbol{x}) > 0, \\ \Pi_3 & h_{13}(\boldsymbol{x}) < 0 \quad \text{and} \quad h_{23}(\boldsymbol{x}) < 0. \end{cases}$$

It is easy to see that the discriminant functions are linearly related by

$$h_{12}(\boldsymbol{x}) + h_{23}(\boldsymbol{x}) = h_{13}(\boldsymbol{x})$$

(see Exercise 12.4.7). Thus, the boundary consists of three lines meeting at the point where $h_{12}(\boldsymbol{x}) = h_{23}(\boldsymbol{x}) = h_{13}(\boldsymbol{x}) = 0$.

A picture of the allocation regions for the iris data using the linear discriminant functions is given in panel (a) of Figure 12.3. The boundaries between the three regions form three lines meeting at a point. The boundary between *I. setosa* and *I. versicolor* is very similar to the boundary found in Example 12.4.3. It is more difficult to discriminate accurately between *I. versicolor* and *I. virginica* than it is to discriminate between either of these species and *I. setosa*.

Panel (b) of Figure 12.3 shows the allocation regions for the quadratic discriminant functions. Although there is some similarity with panel (a), the boundaries are now curved, and the allocation regions for each species are now disconnected. In the predictions (based on sepal length and sepal width), the two rules agree in 140 out of the 150 observations. The differences between the two rules are most apparent in the left side of each plot where the allocation regions contain no data. □

In the next example, we can observe a more marked difference between the linear and quadratic discrimination boundaries.

Example 12.4.6 (Mental health data). Smith (1947) collected information on normal individuals and psychotic individuals. Two variables were measured, an unweighted total score (x_1) and a weighted score (x_2). The data are plotted in Figure 12.4. The diagonal line

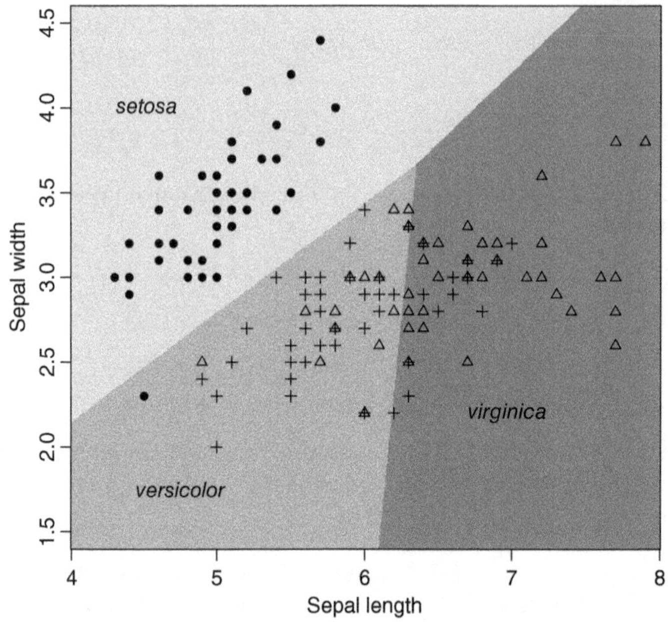

(a)

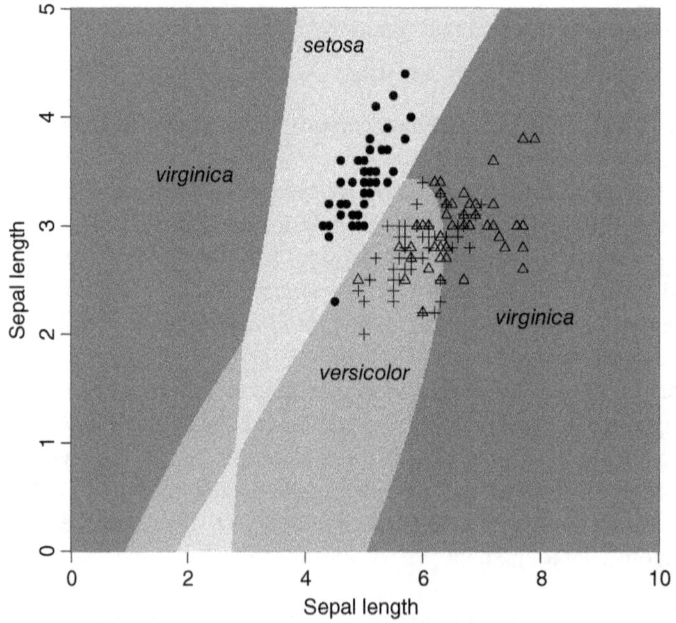

(b)

Figure 12.3 Discrimination between three species of iris using two variables. Both plots show the observations with labels: • (*I. setosa*), + (*I. versicolor*) and Δ (*I. virginica*). (a) linear discrimination allocation regions (b) quadratic discrimination allocation regions.

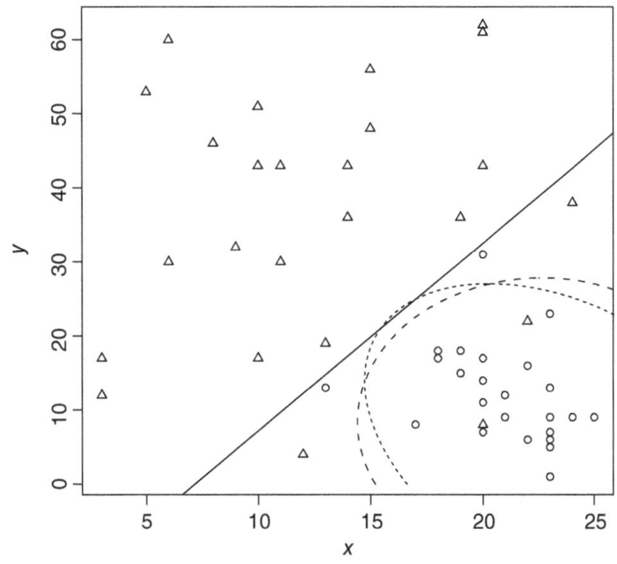

Figure 12.4 (Mental health data). Discrimination between normal individuals (o) and psychotics (Δ) on the basis of two variables *x* and *y*; a plot of the data points, with a linear discriminant boundary (continuous) and quadratic discriminant boundaries using the full covariance matrices (dotted) and setting the off-diagonal entries to zero (dashed).

in this figure represents the boundary given by the sample linear discriminant rule given by (12.15), which assumes equal covariance matrices for the populations. However, it is clear from the figure, and the fact that the sample covariance matrices

$$\boldsymbol{S}_1 = \begin{bmatrix} 6.64 & -5.06 \\ -5.06 & 39.26 \end{bmatrix}, \qquad \boldsymbol{S}_2 = \begin{bmatrix} 5.28 & 13.36 \\ 13.36 & 276.40 \end{bmatrix}$$

are quite different that a quadratic discrimination rule, which is also shown in the figure, may be more appropriate.

The sample correlations are given by −0.31 (normal) and 0.14 (psychotic), which led Smith (1947) to consider a simpler model in which the covariance matrices are allowed to differ but with the off-diagonal terms set to *zero*. This boundary (also plotted) is elliptical (see Exercise 12.4.9) and quite similar to that using the full covariance structure; the rule is perhaps easier to interpret.

The overall resubstitution error for all three rules is the same (4/50), but the confusion matrices corresponding to the quadratic rules are symmetric, unlike the linear discriminant rule that incorrectly allocates four normal individuals to the psychotic class. □

12.4.3 Is Discrimination Worthwhile?

Consider the situation where we are discriminating between g multivariate normal populations with the same covariance matrix, and the parameters are estimated from the data $\boldsymbol{X} = (\boldsymbol{X}_1', \dots, \boldsymbol{X}_g')'$. If all the population means are equal, $\boldsymbol{\mu}_1 = \cdots = \boldsymbol{\mu}_g$, then of course it is pointless to try to obtain a linear discriminant function between the groups on the basis of the data.

However, even if the true means are equal, the sample means $\overline{\boldsymbol{x}}_1, \ldots, \overline{\boldsymbol{x}}_g$ will be different from one another, so an apparently plausible discriminant analysis can be carried out. Thus, to check whether or not the discriminant analysis is worthwhile, it is interesting to test the hypothesis $\boldsymbol{\mu}_1 = \cdots = \boldsymbol{\mu}_g$, given $\boldsymbol{\Sigma}_1 = \cdots = \boldsymbol{\Sigma}_g$. This hypothesis is exactly the one-way multivariate analysis of variance described in Section 6.3.3.1.

Recall that two possible tests of the hypothesis are obtained by partitioning the "total" sum of squares and products (SSP) matrix $\boldsymbol{T} = \boldsymbol{X}'\boldsymbol{H}\boldsymbol{X}$ as $\boldsymbol{T} = \boldsymbol{W} + \boldsymbol{B}$, where $\boldsymbol{W}$ and $\boldsymbol{B}$ are the "within-groups" and "between-groups" SSP matrices, respectively, and $\boldsymbol{H}$ is the centering matrix. The Wilks' Λ test and the greatest root test are given by functions of the eigenvalues of $\boldsymbol{W}^{-1}\boldsymbol{B}$. In particular, if $g = 2$, then $\boldsymbol{W}^{-1}\boldsymbol{B}$ has only one nonzero eigenvalue, and the two tests are the same and are equivalent to the two-sample Hotelling's T^2 test. Then, under the null hypothesis,

$$\left\{\frac{n_1 n_2(n-2)}{n}\right\} \boldsymbol{d}'\boldsymbol{W}^{-1}\boldsymbol{d} \sim T^2(p, n-2) = \{(n-2)p/(n-p-1)\}F_{p,n-p-1},$$

where $\boldsymbol{d} = \overline{\boldsymbol{x}}_1 - \overline{\boldsymbol{x}}_2$, and the null hypothesis is rejected for large values of this statistic.

12.5 Discarding of Variables

Consider the discrimination problem between two multivariate normal populations with means $\boldsymbol{\mu}_1, \boldsymbol{\mu}_2$ and common covariance matrix $\boldsymbol{\Sigma}$. The coefficients of the theoretical ML discriminant function $\boldsymbol{\alpha}'\boldsymbol{x}$ are given by

$$\boldsymbol{\alpha} = \boldsymbol{\Sigma}^{-1}\boldsymbol{\delta}, \quad \text{where} \quad \boldsymbol{\delta} = \boldsymbol{\mu}_1 - \boldsymbol{\mu}_2. \tag{12.17}$$

In practice, of course, the parameters are estimated by $\overline{\boldsymbol{x}}_1, \overline{\boldsymbol{x}}_2$, and $\boldsymbol{S}_u = m^{-1}(n_1\boldsymbol{S}_1 + n_2\boldsymbol{S}_2) = m^{-1}\boldsymbol{W}$, where $m = n_1 + n_2 - 2$. Letting $\boldsymbol{d} = \overline{\boldsymbol{x}}_1 - \overline{\boldsymbol{x}}_2$, the coefficients of the sample ML rule are given by

$$\boldsymbol{a} = m\boldsymbol{W}^{-1}\boldsymbol{d}.$$

Partition $\boldsymbol{\alpha}' = (\boldsymbol{\alpha}_1', \boldsymbol{\alpha}_2')$ and $\boldsymbol{\delta}' = (\boldsymbol{\delta}_1', \boldsymbol{\delta}_2')$, where $\boldsymbol{\alpha}_1$ and $\boldsymbol{\delta}_1$ have k components, and suppose that $\boldsymbol{\alpha}_2 = \boldsymbol{0}$; that is, suppose that the variables $x_{k+1}, \ldots, x_p$ have no discriminating power once the other variables have been taken into account and hence may be safely discarded. Note that the hypothesis $\boldsymbol{\alpha}_2 = \boldsymbol{0}$ is equivalent to $\boldsymbol{\delta}_{2.1} = \boldsymbol{0}$, where $\boldsymbol{\delta}_{2.1} = \boldsymbol{\delta}_2 - \boldsymbol{\Sigma}_{21}\boldsymbol{\Sigma}_{11}^{-1}\boldsymbol{\delta}_1$. (See Exercise 12.5.1.) It is also equivalent to $\Delta_p^2 = \Delta_k^2$, where

$$\Delta_p^2 = \boldsymbol{\delta}'\boldsymbol{\Sigma}^{-1}\boldsymbol{\delta}, \qquad \Delta_k^2 = \boldsymbol{\delta}_1\boldsymbol{\Sigma}_{11}^{-1}\boldsymbol{\delta}_1;$$

that is, the Mahalanobis distance between the populations is the same whether based on the first k components or on all p components.

A test of the hypothesis $H_0 : \boldsymbol{\alpha}_2 = \boldsymbol{0}$ using the sample Mahalanobis distances

$$D_p^2 = m\boldsymbol{d}'\boldsymbol{W}^{-1}\boldsymbol{d} \quad \text{and} \quad D_k^2 = m\boldsymbol{d}_1'\boldsymbol{W}_{11}^{-1}\boldsymbol{d}_1$$

has been proposed by Rao (1973, p. 568). This test uses the statistic

$$\{(m-p+1)/(p-k)\}c^2(D_p^2 - D_k^2)/(m + c^2D_k^2), \tag{12.18}$$

where $c^2 = n_1 n_2/n$. Under the null hypothesis, (12.18) has the $F_{p-k,m-p+1}$ distribution, and we reject H_0 for large values of this statistic. See Theorem 4.6.2.

The most important application of this test occurs with $k = p - 1$, when we are testing the importance of one particular variable once all the other variables have been taken into account. In this case, the statistic in (12.18) can be simplified using the inverse of the total SSP matrix $\boldsymbol{T}^{-1} = (t^{ij})$. Consider the hypothesis $H_0 : \alpha_i = 0$, where i is fixed. Then, (12.18), with D_k now representing the Mahalanobis distance based on all the variables except the ith, equals

$$(m - p + 1)c^2 a_i^2 / \{mt^{ii}(m + c^2 D_p^2)\} \tag{12.19}$$

and has the $F_{1,m-p+1}$ distribution when $\alpha_i = 0$. (See Exercise 12.5.1.)

Of course, the statistic (12.19) is strictly valid only if the variable i is selected ahead of time. However, it is often convenient to look at the value of this statistic on all the variables to see which ones are important.

Example 12.5.1 (Iris data, discarding variables, all $p = 4$ variables). Consider again the two iris species *I. versicolor* and *I. virginica*, this time using all $p = 4$ variables. Here, $n_1 = n_2 = 50, m = 100 - 2 = 98$, and the discriminant function has coefficients

$$\boldsymbol{a}' = (-3.56, -5.58, 6.97, 12.39).$$

It is easily shown that

$$\boldsymbol{T}^{-1} = \begin{bmatrix} 0.092 & -0.044 & -0.075 & 0.054 \\ -0.044 & 0.157 & 0.027 & -0.072 \\ -0.075 & 0.027 & 0.107 & -0.115 \\ 0.054 & -0.072 & -0.115 & 0.221, \end{bmatrix}$$

and that $D_p^2 = 14.22$. Thus, the four F-statistics testing the importance of one variable, given the other three, are given from (12.19) by

Sepal length	Sepal width	Petal length	Petal width
7.368	10.587	24.157	37.092

Since $F_{1,95}(0.99) = 6.91$, all the F-statistics are significant. However, if we proceed in a stepwise manner to remove sepal length (which is only just significant at the 1% level), then the next least significant variable is petal length followed at the next step by sepal width – though all of these three have highly significant F-statistics — so we would have an ordering for removal. □

An alternative approach here is to use the ML linear discriminant function in a forward selection of variables. At each step, we can include the variable that reduces the error rate the most. In this case, for the above example, we include the variables in the following order, with the corresponding error rates for the cumulative set of variables.

Variable	Petal width	Sepal width	Petal length	Sepal length
Error rate	0.06	0.05	0.04	0.03

Of course, these are resubstitution error rates, and one might query whether the improvement at each step makes it worth including each additional explanatory variable. It is interesting to note that the order of variable selection in this forward stagewise process is consistent with the earlier ordering given by the backward removal using the F-statistics. A cross-validation implementation is used in the R function `sepclass` in the library `klaR` (Weihs et al., 2005). Here, a forward selection using cross-validation shows that only petal width is included.

Note that inclusion of irrelevant variables can actually worsen the performance of some classifiers. Selection (or discarding) of variables is also important when it is desired to interpret or understand the classifier, and most methods considered later in the chapter consider this issue. Stepwise variable selection procedures are readily available for logistic discrimination (Section 12.9). Classification trees, which are introduced in Section 12.8, learn splitting rules that select one variable at each split – and so these have a "built-in" mechanism for discarding those variables that are never selected. Variable selection is also important for classifiers based on distance-based measures (Section 12.7).

12.6 Fisher's Linear Discriminant Function

Another approach to the discrimination problem based on a data matrix $\boldsymbol{X}$ can be made without assuming any particular parametric form for the distribution of the populations $\Pi_1, \dots, \Pi_g$ but by merely looking for a simple "sensible" rule to discriminate between them. Fisher's (Fisher, 1936) suggestion was to look for the *linear function* $\boldsymbol{a}'\boldsymbol{x}$ that maximizes the ratio of the between-groups sum of squares to the within-group sum of squares; that is, let

$$\boldsymbol{z} = \boldsymbol{X}\boldsymbol{a} = \begin{pmatrix} \boldsymbol{X}_1\boldsymbol{a} \\ \vdots \\ \boldsymbol{X}_g\boldsymbol{a} \end{pmatrix} = \begin{pmatrix} \boldsymbol{z}_1 \\ \vdots \\ \boldsymbol{z}_g \end{pmatrix}$$

be a linear combination of the columns of $\boldsymbol{X}$. The $\boldsymbol{z}$ has total sum of squares

$$\boldsymbol{z}'\boldsymbol{H}\boldsymbol{z} = \boldsymbol{a}'\boldsymbol{X}'\boldsymbol{H}\boldsymbol{X}\boldsymbol{a} = \boldsymbol{a}'\boldsymbol{T}\boldsymbol{a}, \tag{12.20}$$

which can be partitioned as a sum of within-groups sum of squares,

$$\sum \boldsymbol{z}_j\boldsymbol{H}_j\boldsymbol{z}_j = \sum \boldsymbol{a}'\boldsymbol{X}_j'\boldsymbol{H}_j\boldsymbol{X}_j\boldsymbol{a} = \boldsymbol{a}'\boldsymbol{W}\boldsymbol{a}, \tag{12.21}$$

plus the between-groups sum of squares,

$$\sum n_j(\overline{z}_j - \overline{z})^2 = \sum n_j\{\boldsymbol{a}'(\overline{\boldsymbol{x}}_j - \overline{\boldsymbol{x}})\}^2 = \boldsymbol{a}'\boldsymbol{B}\boldsymbol{a}, \tag{12.22}$$

where $\overline{z}_j$ is the mean of the jth subvector $\boldsymbol{z}_j$ of $\boldsymbol{z}$, and $\boldsymbol{H}_j$ is the $(n_j \times n_j)$ centering matrix.

Fisher's criterion is intuitively attractive because it is easier to tell the groups apart if the between-groups sum of squares for $\boldsymbol{z}$ is large relative to the within-groups sum of squares. The ratio is given by

$$\boldsymbol{a}'\boldsymbol{B}\boldsymbol{a}/\boldsymbol{a}'\boldsymbol{W}\boldsymbol{a}. \tag{12.23}$$

If $\boldsymbol{a}$ is the vector that maximizes (12.23), we call the linear function $\boldsymbol{a}'\boldsymbol{x}$ *Fisher's linear discriminant function* or the *first canonical variate*. Notice that the vector $\boldsymbol{a}$ can be rescaled without affecting the ratio (12.23).

Theorem 12.6.1 *The vector $\boldsymbol{a}$ in Fisher's linear discriminant function is the eigenvector of $\boldsymbol{W}^{-1}\boldsymbol{B}$ corresponding to the largest eigenvalue.*

Proof: This result follows by an application of Theorem A.2. □

Once the linear discriminant function has been calculated, an observation $\boldsymbol{x}$ can be allocated to one of the g populations on the basis of its "discriminant score" $\boldsymbol{a}'\boldsymbol{x}$. The sample means $\overline{\boldsymbol{x}}_i$ have scores $\boldsymbol{a}'\overline{\boldsymbol{x}}_i = \overline{z}_i$. Then, $\boldsymbol{x}$ is allocated to that population whose mean score is closest to $\boldsymbol{a}'\boldsymbol{x}$; that is, allocate $\boldsymbol{x}$ to Π_l if

$$|\boldsymbol{a}'\boldsymbol{x} - \boldsymbol{a}'\overline{\boldsymbol{x}}_l| < |\boldsymbol{a}'\boldsymbol{x} - \boldsymbol{a}'\overline{\boldsymbol{x}}_j| \quad \text{for all} \quad j \neq l. \tag{12.24}$$

Example 12.6.1 For the iris data of Table 1.3, we have mean vectors

$$\overline{\boldsymbol{x}}_1 = \begin{pmatrix} 5.006 \\ 3.428 \\ 1.462 \\ 0.246 \end{pmatrix}, \quad \overline{\boldsymbol{x}}_2 = \begin{pmatrix} 5.936 \\ 2.770 \\ 4.260 \\ 1.326 \end{pmatrix}, \quad \overline{\boldsymbol{x}}_3 = \begin{pmatrix} 6.588 \\ 2.974 \\ 5.552 \\ 2.026 \end{pmatrix}.$$

The SSP matrix between species is

$$\boldsymbol{B} = \begin{bmatrix} 63.21 & -19.95 & 165.25 & 71.28 \\ & 11.34 & -57.24 & -22.93 \\ & & 437.10 & 186.77 \\ & & & 80.41 \end{bmatrix}.$$

The SSP matrix within species is

$$\boldsymbol{W} = \begin{bmatrix} 38.96 & 13.63 & 24.62 & 5.65 \\ & 16.96 & 8.12 & 4.81 \\ & & 27.22 & 6.27 \\ & & & 61.6 \end{bmatrix}.$$

The eigenvector $\boldsymbol{W}^{-1}\boldsymbol{B}$ corresponding to the largest eigenvalue is thus $\boldsymbol{a}' = (0.21, 0.39, -0.55, -0.71)$. We can then compute $|\boldsymbol{a}'\boldsymbol{x} - \boldsymbol{a}'\overline{\boldsymbol{x}}_l|$ for the first observation $\boldsymbol{x}' = (5.9, 3.0, 5.1, 1.8)$ to obtain the values 0.45, 9.89, and 13.84 for $l = 1, 2, 3$, respectively. This would classify this observation to the first class (*I. setosa*), which is correct – and consistent with the allocation when only two variables are used (see Example 12.4.5). Continuing this process for each of the 150 observations we can obtain the (resubstitution) confusion matrix given below. We note that two observations are misclassified (observations 73 and 84).

		Actual		
		I. setosa	*I. versicolor*	*I. virginica*
	I. setosa	50	0	0
Predicted	*I. versicolor*	0	48	0
	I. virginica	0	2	50

As a historical comment, we note that Fisher (1936) derived a genetic-based discriminant function of his own, which misclassified two observations (71 and 84). This discriminant is based on the genetic hypothesis that *I. versicolor* lies between the other two species — half the distance from *I. virginica* as that from *I. setosa*; see Mardia (2023) for more details. □

Fisher's discriminant function is interesting in the special case of $g = 2$ groups. Then, $\boldsymbol{B}$ has rank 1 and can be written as

$$\boldsymbol{B} = \left(\frac{n_1 n_2}{n}\right) \boldsymbol{d}\boldsymbol{d}',$$

where $\boldsymbol{d} = \bar{\boldsymbol{x}}_1 - \bar{\boldsymbol{x}}_2$. Thus, $\boldsymbol{W}^{-1}\boldsymbol{B}$ has only one nonzero eigenvalue which can be found explicitly. This eigenvalue equals

$$\text{tr}\ \boldsymbol{W}^{-1}\boldsymbol{B} = \left(\frac{n_1 n_2}{n}\right) \boldsymbol{d}'\boldsymbol{W}^{-1}\boldsymbol{d},$$

and the corresponding eigenvector is

$$\boldsymbol{a} = \boldsymbol{W}^{-1}\boldsymbol{d}. \tag{12.25}$$

(See Exercise 12.6.1.) Then, the discriminant rule becomes

$$\text{allocate } \boldsymbol{x} \text{ to } \Pi_1 \text{ if } \boldsymbol{d}'\boldsymbol{W}^{-1}\left\{\boldsymbol{x} - \frac{1}{2}(\bar{\boldsymbol{x}}_1 + \bar{\boldsymbol{x}}_2)\right\} > 0 \tag{12.26}$$

and to Π_2 otherwise.

Note that the allocation rule given by (12.26) is exactly the same as the ML sample rule for two groups from the multivariate normal distribution with the same covariance given in (12.15). However, the justifications for this rule are quite different in the two cases. In (12.15), there is an explicit assumption of normality, whereas in (12.26) we have merely sought a "sensible" rule based on a linear function of $\boldsymbol{x}$. Thus, we might hope that this sensible rule will be appropriate where the hypothesis of normality is not exactly satisfied.

For $g \geq 3$ groups, the allocation rule based on Fisher's linear discriminant function and the sample ML rule for multivariate normal populations with the same covariance matrix will not be the same unless the sample means are collinear (although the two rules will be similar if the means are nearly collinear). See Exercise 12.6.3.

In general, $\boldsymbol{W}^{-1}\boldsymbol{B}$ has $\min(p, g-1)$ nonzero eigenvalues. The corresponding eigenvectors define the second, third, and subsequent "canonical variates." (Canonical variates have an important connection with the canonical correlations of Chapter 11, see Exercise 12.6.4.) The first k canonical variates, $k \leq \min(p, g-1)$, are useful when it is hoped to summarize the difference between the groups in k dimensions. See Section 13.5.

Discriminant analysis can also be used as an aid to seriation (putting objects into chronological order), as seen in the following example.

Example 12.6.2 Cox and Brandwood (1959) used discriminant analysis to place in chronological order seven works of Plato – *Republic, Laws, Critias, Philebus, Politicus, Sophist*, and *Timaeus*, where it is known that *Republic* was written before *Laws*, and that the other five works were written in between. However, the order of these five works is not known. The stylistic property on which the statistical analysis is based is the type of the sentence ending. The last five syllables of a sentence are each either long or short, leading to $2^5 = 32$ types of sentence ending. For each work, the percentage of sentences having each type of ending is given in Table 12.2.

It is assumed that sentence style varied systematically over time, and a function measuring the change of style is sought in order to sort the works chronologically. (Note that this problem is not really discriminant analysis as previously defined, because we do not want to allocate these intermediate works to *Republic* or *Laws*.)

Table 12.2 Percentage distribution of sentence endings in seven works of Plato.

Type of ending	Π_0 *Rep.*	Π_1 *Laws*	*Crit.*	*Phil.*	*Pol.*	*Soph.*	*Tim.*
U U U U U	1.1	2.4	3.3	2.5	1.7	2.8	2.4
- U U U U	1.6	3.8	2.0	2.8	2.5	3.6	3.9
U - U U U	1.7	1.9	2.0	2.1	3.1	3.4	6.0
U U - U U	1.9	2.6	1.3	2.6	2.6	2.6	1.8
U U U - U	2.1	3.0	6.7	4.0	3.3	2.4	3.4
U U U U -	2.0	3.8	4.0	4.8	2.9	2.5	3.5
- - U U U	2.1	2.7	3.3	4.3	3.3	3.3	3.4
- U - U U	2.2	1.8	2.0	1.5	2.3	4.0	3.4
- U U - U	2.8	0.6	1.3	0.7	0.4	2.1	1.7
- U U U -	4.6	8.8	6.0	6.5	4.0	2.3	3.3
U - - U U	3.3	3.4	2.7	6.7	5.3	3.3	3.4
U - U - U	2.6	1.0	2.7	0.6	0.9	1.6	2.2
U - U U -	4.6	1.1	2.0	0.7	1.0	3.0	2.7
U U - - U	2.6	1.5	2.7	3.1	3.1	3.0	3.0
U U - U -	4.4	3.0	3.3	1.9	3.0	3.0	2.2
U U U - -	2.5	5.7	6.7	5.4	4.4	5.1	3.9
- - - U U	2.9	4.2	2.7	5.5	6.9	5.2	3.0
- - U - U	3.0	1.4	2.0	0.7	2.7	2.6	3.3
- - U U -	3.4	1.0	0.7	0.4	0.7	2.3	3.3
- U - - U	2.0	2.3	2.0	1.2	3.4	3.7	3.3
- U - U -	6.4	2.4	1.3	2.8	1.8	2.1	3.0
- U U - -	4.2	0.6	4.7	0.7	0.8	3.0	2.8
U U - - -	2.8	2.9	1.3	2.6	4.6	3.4	3.0
U - U - -	4.2	1.2	2.7	1.3	1.0	1.3	3.3
U - - U -	4.8	8.2	5.3	5.3	4.5	4.6	3.0
U - - - U	2.4	1.9	3.3	3.3	2.5	2.5	2.2
U - - - -	3.5	4.1	2.0	3.3	3.8	2.9	2.4
- U - - -	4.0	3.7	4.7	3.3	4.9	3.5	3.0
- - U - -	4.1	2.1	6.0	2.3	2.1	4.1	6.4
- - - U -	4.1	8.8	2.0	9.0	6.8	4.7	3.8
- - - - U	2.0	3.0	3.3	2.9	2.9	2.6	2.2
- - - - -	4.2	5.2	4.0	4.9	7.3	3.4	1.8
Number of sentences	3778	3783	150	958	770	919	762

Source: Cox and Brandwood (1959) / John Wiley & Sons.

Table 12.3 Mean scores and their standard errors from seven works of Plato.

Work	Mean score	Estimated variance	Estimated standard error
Critias	−0.0346	0.003 799	0.0616
Philebus	0.1966	0.000 334 2	0.0183
Politicus	0.1303	0.000 397 3	0.019 93
Sophist	−0.0407	0.000 571 9	0.0239
Timaeus	−0.1170	0.000 721 8	0.0269
Republic	−0.2652	NA	NA
Laws	0.2176	NA	NA

Source: Cox and Brandwood (1959) / John Wiley & Sons.

Suppose that each of the works has a multinomial distribution, and in particular, suppose that the parameters for *Republic* and *Laws* are $\alpha_1, \ldots, \alpha_{32}$ and $\beta_1, \ldots, \beta_{32}$, respectively. Then, from Example 12.2.2, the ML discriminant function between *Republic* and *Laws* (standardized by the number of sentences) is given by

$$h(\boldsymbol{x}) = \sum x_i s_i / \sum x_i,$$

where $s_i = \log(\alpha_i/\beta_i)$, and $x_1, \ldots, x_{32}$ are the number of sentences with each type of ending in a particular work. We do not know the parameters α_i and β_i, $i = 1, \ldots, 32$. However, since the number of sentences in *Republic* and *Laws* is much larger than the other works, we replace α_i and β_i by their sample estimates from Table 12.2.

The scores of each work on this discriminant function are given in Table 12.3. The table also gives their standard errors; the formulae for these are given in Exercise 12.6.5. From these scores it appears that the most likely order is *Republic, Timaeus, Sophist, Critias, Politicus, Philebus, and Laws*. This order is not in accord with the view held by the majority of classical scholars, although there is a minority group who reached a similar ordering by apparently independent arguments. For a discussion of questions related to the statistical significance of this ordering, see Exercise 12.6.5 and Cox and Brandwood (1959). □

In the remainder of this chapter we describe a collection of techniques that make few (or no) distributional assumptions for the data. These are sometimes called nonparametric or distribution-free methods. We will see that, if normality (approximately) holds, then the allocation regions for the nonparametric techniques will be similar to those obtained for discriminant analysis (using linear or quadratic discrimination according to whether the covariance matrices are similar or not). However, these nonparametric methods will usually perform better in those cases where the groups are not normally distributed. For reviews and comparative studies of various discrimination methods, see, e.g. Jain et al. (2000), Lim et al. (2000), Michie et al. (1994), and Demšar (2006).

12.7 Nonparametric Distance-based Methods

12.7.1 Nearest-neighbor Classifier

A k-nearest-neighbor classifier, $\delta(\boldsymbol{x})$, uses a data matrix $\boldsymbol{X}$ (12.1) with the corresponding class labels to classify a new observation $\boldsymbol{x}$ using a majority vote of the classes y_r of the k nearest neighbors $\boldsymbol{x}_r$ to $\boldsymbol{x}$.

Given data $\boldsymbol{x}_1, \ldots \boldsymbol{x}_n \in \mathbb{R}^p$ with associated class labels $y_1, \ldots, y_n$, let $d_r = d(\boldsymbol{x}, \boldsymbol{x}_r)$, $r = 1, \ldots, n$, denote the *distance* (see Section 14.6) from $\boldsymbol{x}_r$ to $\boldsymbol{x}$. We can then reorder the observations so that the distances are sorted with $d_{(1)} < d_{(2)} < \cdots < d_{(n)}$. Denote by $y_{(r)}$ the class label of the observation which has distance $d_{(r)}$. Then, the k-nearest-neighbor classifier $\delta(\boldsymbol{x})$ is the most frequently occurring class label among the $\{y_{(1)}, \ldots y_{(k)}\}$.

The concept of "nearest" requires measurement of distance (see Table 14.6), so clearly the scaling of the data will be important, or one could use Pearson distance or (a class-conditional) Mahalanobis distance. Note also that it is possible to have ties among the distances, and that there may be ties in the majority vote. In practice, ties may be broken using the distances of the k nearest neighbors, resorting to a random allocation, if necessary. If the data are from a continuous distribution, and $k = 1$, then ties will not pose a problem. The case when $k = 1$ is known as the *nearest-neighbor* rule. In many applications, one would choose k (together with the tie-breaking strategy) by cross-validation.

Example 12.7.1 (Iris data, two species, nearest-neighbor classifier). Again, using the first two variables (sepal length and sepal width) for the two species (*I. versicolor* and *I. setosa*) of the iris data, we compute the partition of the sample space using a (1-)nearest-neighbor rule. In this case, the classifier allocates each new observation to the class of the nearest observation. This gives regions corresponding to the union of the Voronoi polygons – as shown in Figure 12.5 – associated with each observation; i.e., the classifier takes the form $\delta(\boldsymbol{x}) = i$ for $\boldsymbol{x} \in R_i$, where

$$R_i = \cup_{r:y_r=i} V_r,$$

and the Voronoi polygon is defined by $\boldsymbol{x} \in V_r$ if $r = \arg\ \min_i d_i$. Note that the boundary between the allocation regions is similar to the straight line boundary in Figure 12.2 for Example 12.4.3. Indeed, based on the first two variables, the predicted classes of the observations are identical for both species (*I. versicolor* and *I. setosa*). □

12.7.2 Large Sample Behavior of the Nearest-neighbor Classifier

In Section 12.3.1, we considered the overall error rates associated with a classifier. To gain more insight for the nearest-neighbor classifier, suppose that the priors and the $f_j(\boldsymbol{x})$ are known and that the training sample is large. For a given point $\boldsymbol{x}$, let $p_j(\boldsymbol{x})$ be the posterior probability that an observation at $\boldsymbol{x}$ belongs to Π_j. This is proportional to $\pi_j f_j(\boldsymbol{x})$. Suppose that $j^*(= j^*(\boldsymbol{x})) = \arg\ \max\ p_j(\boldsymbol{x})$ defines the class corresponding to Bayes' discriminant rule. If an observation $\boldsymbol{x}$ is allocated to Π_j, then the error rate is $1 - p_j(\boldsymbol{x}), j = 1, \ldots, g$; hence the Bayes' error (at $\boldsymbol{x}$) is $1 - p_{j^*}(\boldsymbol{x})$.

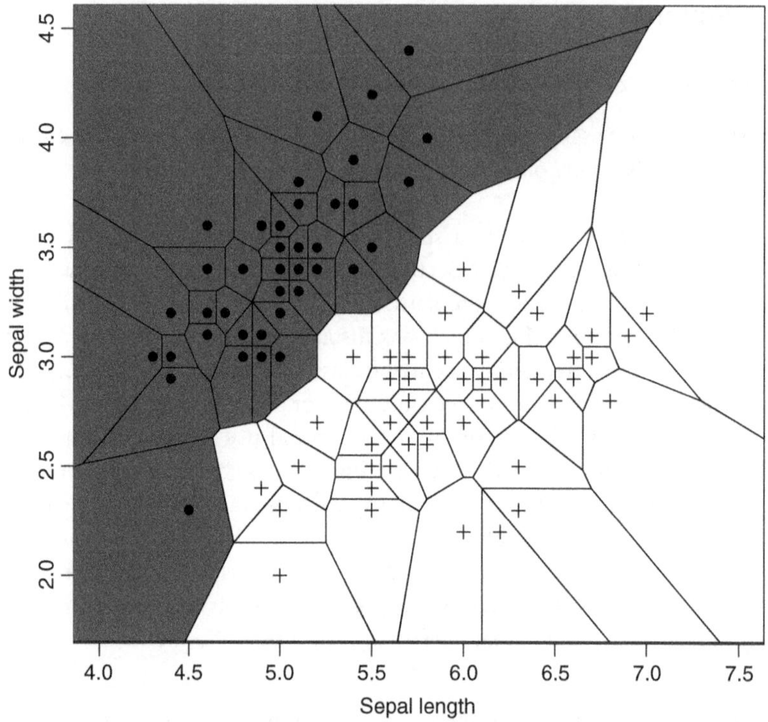

Figure 12.5 Classification regions (polygons obtained using the R library `spatstat` (Baddeley et al., 2015)) for the first two variables of two species of the iris data (*I. versicolor* (+) and *I. setosa* (•)) using the nearest-neighbor classifier.

Given a new observation at $\boldsymbol{x}$, let $\boldsymbol{x}_0$ denote the nearest neighbor. Then, the nearest neighbor error rate at $\boldsymbol{x}$ can be expressed as

$$\sum_{j=1}^{g} \text{P}(\boldsymbol{x}_0 \text{ does not belong to } \Pi_j \mid \boldsymbol{x} \text{ belongs to } \Pi_j)\text{P}(\boldsymbol{x} \text{ belongs to } \Pi_j).$$

Asymptotically, the nearest neighbor $\boldsymbol{x}_0$ will be coincident with $\boldsymbol{x}$, and so, the probability that the nearest neighbor to $\boldsymbol{x}$ does not belong to class Π_j is $1 - p_j(\boldsymbol{x})$. This gives an asymptotic expression of the error rate at $\boldsymbol{x}$ for the nearest-neighbor classifier as

$$\sum_{j=1}^{g} (1 - p_j(\boldsymbol{x}))p_j(\boldsymbol{x}).$$

Hence, the nearest-neighbor classifier error rate will always be at least as large as the Bayes' error rate. This is not surprising, since in Section 12.3.1 we recognized Bayes' error as optimal.

Theorem 12.7.1 *(Cover and Hart, 1967) Asymptotically, the error rate for the nearest-neighbor classifier is less than twice the Bayes' error. Specifically, at a point $\boldsymbol{x}$, we have*

$$\sum_{j=1}^{g} p_j(\boldsymbol{x})(1 - p_j(\boldsymbol{x})) \le 2(1 - p_{j^*}(\boldsymbol{x})) - \frac{g}{g-1}(1 - p_{j^*}(\boldsymbol{x}))^2,$$

where $\delta(\boldsymbol{x}) = j^*$ denotes the classifier corresponding to Bayes' rule.

Proof: We have

$$\sum_{j=1}^{g} p_j(\boldsymbol{x})(1 - p_j(\boldsymbol{x})) = \sum_j p_j(\boldsymbol{x}) - \sum_j p_j(\boldsymbol{x})^2 = 1 - \sum_j p_j(\boldsymbol{x})^2.$$

Using the Cauchy–Schwartz inequality, we have

$$\sum_{j\neq j^*} 1 \sum_{j\neq j^*} p_j(\boldsymbol{x})^2 = (g-1)\sum_{j\neq j^*} p_j(\boldsymbol{x})^2 \geq \left[\sum_{j\neq j^*} 1 \cdot p_j(\boldsymbol{x})\right]^2 = \left[1 - p_{j^*}(\boldsymbol{x})\right]^2.$$

So,

$$\begin{aligned} 1 - \sum_j p_j(\boldsymbol{x})^2 &= 1 - \sum_{j\neq j^*} p_j(\boldsymbol{x})^2 - p_{j^*}(\boldsymbol{x})^2 \\ &\leq 1 - p_{j^*}(\boldsymbol{x})^2 - \frac{\left(1 - p_{j^*}(\boldsymbol{x})\right)^2}{g-1} \\ &= 2(1 - p_{j^*}(\boldsymbol{x})) - \frac{g\left(1 - p_{j^*}(\boldsymbol{x})\right)^2}{g-1}. \end{aligned}$$

□

This theorem can be useful for large samples since it gives an upper bound on the error rate of the nearest-neighbor rule as twice the (optimal) error rate of Bayes' rule.

In general, the choice of k in the k-nearest-neighbor classifier will affect the performance of the classifier, and this can be problematic, particularly when the priors are very unequal; see Hand and Vinciotti (2003). In practice, it is common to choose k small and odd, to help break ties, and many users will resort to cross-validation (see Section 12.3.2) to help with the choice of k. When $k > 1$, the possibility of using weights to downweight points further away from $\boldsymbol{x}$ is taken up in the following section.

12.7.3 Kernel Classifiers

The motivation for kernel classification is twofold. Firstly, we can consider the direct use of Eq. (12.5) in which the unknown $f_j(\boldsymbol{x})$ are estimated by a kernel density estimate

$$\hat{f}_j(\boldsymbol{x}) = \frac{1}{n_j} \sum_{i:y_r=j} K_{\boldsymbol{H}_j}(\boldsymbol{x} - \boldsymbol{x}_r), \tag{12.27}$$

and the $\hat{\pi}_j$ are simply the sample proportions $n_j/n, j = 1, \ldots, g$. In Eq. (12.27), $K_{\boldsymbol{H}_j}(\boldsymbol{x})$ is a kernel function, centered on the origin, with a set of smoothing parameters $\boldsymbol{H}_j, j = 1, \ldots, g$. It is standard to require that $K(\cdot)$ is itself a proper probability density, so that $f_j(\boldsymbol{x})$ integrates to 1 and is nonnegative. For example, we could use

$$K_{\boldsymbol{H}_j}(\boldsymbol{x} - \boldsymbol{x}_r) = \frac{1}{|2\pi\boldsymbol{H}_j|^{1/2}} \exp\left\{-\frac{1}{2}(\boldsymbol{x} - \boldsymbol{x}_r)'\boldsymbol{H}_j^{-1}(\boldsymbol{x} - \boldsymbol{x}_r)\right\}$$

in which a full symmetric positive-definite matrix $\boldsymbol{H}_j$ of smoothing parameters is required. Or, more simply, we could use a multiplicative kernel

$$K_{\boldsymbol{H}}(\boldsymbol{x} - \boldsymbol{x}_r) = \prod_{k=1}^{p} \frac{1}{h_k} K\left(\frac{x_k - x_{rk}}{h_k}\right)$$

for some $h_1, \ldots, h_p$. Density estimation as an end in itself is a well-studied field; see Silverman (1986) and Wand and Jones (1995), but there are fewer theoretical results for the case when it is to be used for classification; see Chacón and Duong (2018), Di Marzio and Taylor (2005), and Hand (1982).

The tricky part in using kernel density estimates for classification is to get the smoothing parameters correct, and this is much more important than the choice of kernel function. In the limiting case, as $\boldsymbol{H} \to \boldsymbol{0}$, it can be seen that the kernel density classifier is essentially equivalent to the nearest-neighbor classifier, provided the kernel function K has infinite support; see Exercise 12.7.3. More generally, we can interpret the kernel classifier as a weighted version of the n-nearest-neighbor classifier, where each observation $\boldsymbol{x}_r$ contributes a weighted vote which is a function of the class-conditional distance (as measured by $\boldsymbol{H}_j$) from $\boldsymbol{x}_0$ to $\boldsymbol{x}_r$. This interpretation is a second motivation for the kernel classifier.

Example 12.7.2 (Iris data, kernel classifier) Consider again the first two variables, sepal length and sepal width, for the three species or classes of iris. For each of the three classes, a kernel density estimate is formed. Using equal priors, the sample space can then be partitioned into three regions (not necessarily connected) corresponding to the three classes. The result is shown in Figure 12.6. Note the similarity to Figure 12.3 for Example 12.4.5 using the linear discriminant functions. In particular, to a first approximation, the boundaries in Figure 12.6 are three straight lines meeting at a point, and the predicted classes (based on sepal length and sepal width) agree on 139 out of 150 observations.

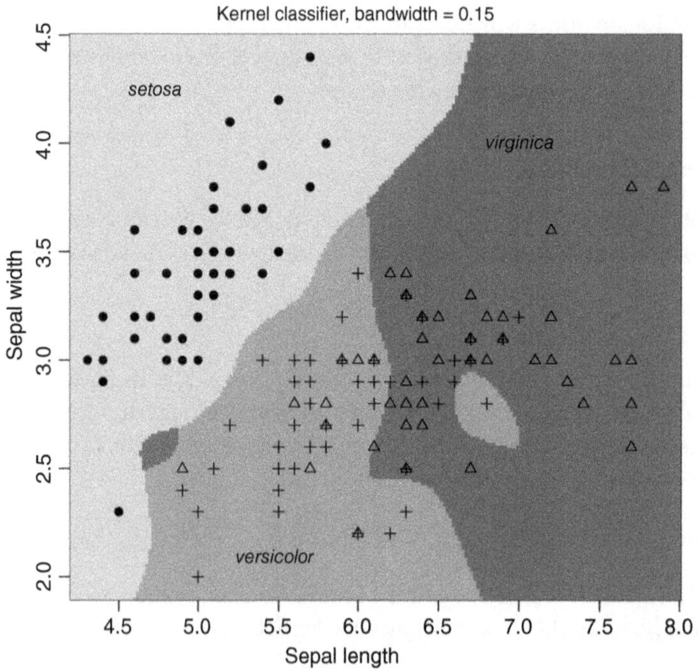

Figure 12.6 Classification regions for the first two variables of the iris data (all three species) using kernel density estimates with bandwidths $h_1 = h_2 = 0.15$. The data are plotted with labels: • (*I. setosa*), + (*I. versicolor*), and Δ (*I. virginica*).

If the plotting region were extended, there might be some numerical instability near the edges of the sample space where all of the estimated densities are close to zero. (Indeed, if the kernel function has finite support, then estimated densities may sometimes be exactly zero.) In this case, we could assign a new region of "none of the groups." □

12.8 Classification Trees

We begin with a brief introduction to classification trees, in which we describe the key components using a small dataset as a running illustrative example.

Consider the illustrative data in Table 12.4, which gives 31 travel expense claims made by employees in a company. The variables are distance traveled, mode of transport, grade of employee, and daily rate claimed. The quantity to be predicted is the outcome of the travel claim. Note that some of the feature variables are categorical, and some are numeric; unlike previous methods of this chapter, classification trees do not require any coding of the categorical variables and (as will be seen later) are not affected by any monotonic transformations of the variables.

Tree-based methods *recursively partition* the feature space – usually into hyper-cuboids – and then allocate each region according to a majority vote of the observations in that region. In this example, before the data is subdivided (so all are in one region), there are 23 claims that were paid, and 8 claims were rejected, so a *naive* classifier would allocate all observations to the class "Pay".

Trees are grown recursively. Initially, all the observations are in one set (at the *root node*). These are split into two or more subsets such that the subsets are more "pure" with respect to the class labels (in this example, Pay/Reject) than the data at the parent node. Each subset is subsequently split into further subsets and so on.

For the data given in Table 12.4, we consider two (among many) possible ways to initially divide the data.

1. Daily rate > 600. This would lead to a contingency table:

	Pay	**Reject**
Daily rate > 600	11	0
Daily rate ≤ 600	12	8

   It can be seen that the first subset is pure (all claims are paid), while the second subset is more impure than the complete dataset, and we would still allocate this region to "Pay".

2. grade $\in$ {Engineer, Jun. Manager} $= G_1$, say. The resulting subsets are then tabulated with respect to the label as:

	Pay	**Reject**
grade $\notin G_1$	17	0
grade $\in G_1$	6	8

Table 12.4 Illustrative data for travel claims.

Distance	Mode	Grade	Daily rate	Outcome
100	Car	Consultant	50	Pay
100	Car	Consultant	250	Pay
100	Car	Consultant	1000	Pay
120	Car	Board	1150	Pay
170	Car	Sen. Manager	100	Pay
220	Car	Board	1450	Pay
270	Car	Engineer	180	Reject
300	Rail	Engineer	180	Reject
400	Rail	Jun. Manager	50	Pay
400	Boat	Engineer	300	Reject
400	Air	Engineer	290	Reject
480	Air	Engineer	400	Reject
700	Air	Engineer	100	Pay
1000	Boat	Sen. Manager	500	Pay
1000	Boat	Sen. Manager	1000	Pay
1200	Car	Sen. Manager	350	Pay
1500	Boat	Jun. Manager	100	Pay
1500	Car	Sen. Manager	300	Pay
1500	Rail	Engineer	300	Reject
1500	Air	Board	900	Pay
1700	Air	Board	1500	Pay
2000	Air	Board	1300	Pay
3000	Air	Jun. Manager	300	Pay
3000	Air	Jun. Manager	400	Pay
3000	Air	Jun. Manager	400	Pay
3000	Air	Jun. Manager	500	Reject
5000	Boat	Sen. Manager	700	Pay
20 000	Air	Board	1400	Pay
20 000	Air	Engineer	200	Reject
20 000	Air	Board	1100	Pay
20 000	Air	Board	1100	Pay

Again, the first subset is pure (all claims are paid), while the second subset is still impure, but we would now allocate this region to "Reject".

An obvious question is: Which of the above two splits is preferred, and are there better ones available? This is considered in Section 12.8.1.

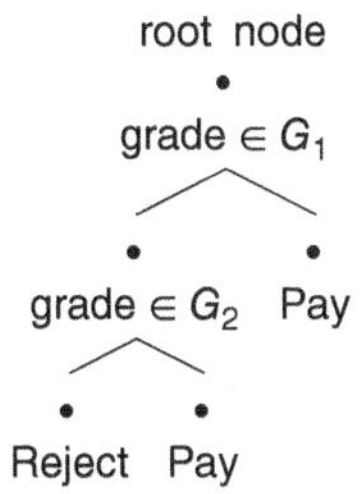

Figure 12.7 A possible classification tree for data in Table 12.4. The branch to the left is used when the condition is TRUE, with sets given by $G_1 = \{\text{Engineer, Jun. Manager}\}$ and $G_2 = \{\text{Engineer}\}$.

If we consider the second option above, then for those observations not in G_1, no further splits are necessary. For those observations in G_1 we could consider a subsequent split: grade $\in \{\text{Engineer}\} = G_2$ (say), which then gives, for this subset:

	Pay	**Reject**
grade $\in G_2$	1	7
grade $\notin G_2$	5	1

and we now have two nearly pure nodes. We could continue to subdivide these two groups (using one of the other variables), but there is then a danger of overfitting. Such overgrown trees will often give poor performance on new data, and we give further details of getting a tree of the right size in Section 12.8.2.

Trees are displayed upside down, with the root node at the top and leaves (also known as terminal nodes) at the bottom. The tree grown for this example is shown in Figure 12.7.

The language used has obvious parallels with physical tree descriptors: for example, there is a *root* as well as *branches* (each branch is a line connecting two nodes), *nodes* (where branches join), and *leaves* (also known as terminal nodes). There are two main components to any algorithm: *splitting criteria* and *pruning* or *stopping* rules. These are now described more fully.

12.8.1 Splitting Criteria

To construct a classification tree (or decision tree) from a set of observations, the set of data is refined into subsets of cases that are increasingly pure with respect to the class labels. Usually, a question is based on one of the variables with two or more mutually exclusive possible answers. The data is then partitioned according to the outcome of the chosen test. The same procedure is then applied recursively to each subset of observations. Such tests or questions are very simple to explain for continuous, discrete, and categorical variables, which makes classification trees very attractive for such "mixed" data.

Most decision tree construction methods are nonbacktracking, *greedy* algorithms; i.e., once a split has been selected to partition the current set of cases, the choice is cast in concrete, so it is important to make good choice. However, stochastic searches, which use Markov Chain Monte Carlo methods and a Bayesian model, are also becoming popular (Chipman et al., 2010; Denison et al., 1998).

Let us suppose that there are N observations in the subset of data that are to be split, with n_{+i} belonging to population i for $i = 1$. For simplicity, we consider a binary split which performs a test on one of the variables $x_1, \dots, x_p$ (for example $x_k > c$ for some $1 \leq k \leq p, c \in \mathbb{R}$), the outcome of which is either T or F. It is straightforward to extend to multiway splits which have more than two outcomes, for example when x_k can take one of 3 categorical values. The split will create a frequency distribution at each branch of the form

Condition	Class 1	2	...	g	
T	n_{11}	n_{12}	...	n_{1g}	n_{1+}
F	n_{21}	n_{22}	...	n_{2g}	n_{2+}
	n_{+1}	n_{+2}	...	n_{+g}	N

If we want to associate a class label j with each condition, then we would let

$$\delta(\boldsymbol{x} \mid \mathsf{T}) = j_{\mathsf{T}} \quad \text{if} \quad n_{1j_{\mathsf{T}}} \geq n_{1i},\ i = 1, \dots, g$$

and

$$\delta(\boldsymbol{x} \mid \mathsf{F}) = j_{\mathsf{F}} \quad \text{if} \quad n_{2j_{\mathsf{F}}} \geq n_{2i},\ i = 1, \dots, g.$$

We note that j_{T} may be the same as j_{F} in practice.

We now describe various splitting criteria, each of which can be "optimized" over the choice of the variable (k) and the choice of the split point (c). The condition used ($x_k < c$) will be applied to each observation, which then creates two subsets of the data at each node. Note that reversing the condition to $x_k > c$ makes no material difference since this will only relabel the branches. There are a variety of splitting criteria that can be used, although some of these are essentially identical, and others are approximately the same (see Exercise 12.8.1). The most commonly used criteria are Gini, entropy, and error rate. The list below also considers two further options.

Error rate. Since our goal is often to minimize the error rate of the classifier, an obvious choice is to select the split to minimize the error rate of the resulting nodes. This is equivalent to choosing the variable k and the split point c so that $n_{1j_{\mathsf{T}}} + n_{2j_{\mathsf{F}}}$ is as large as possible, i.e.

$$\max \ (\text{over all possible splits})\ n_{1j_{\mathsf{T}}} + n_{2j_{\mathsf{F}}}. \tag{12.28}$$

Entropy. A measure of *information* that measures the uncertainty in class membership which is removed by knowledge of the split is given by

$$-\sum_j \frac{n_{j+}}{N} \log \frac{n_{j+}}{N} - \sum_k \frac{n_{+k}}{N} \log \frac{n_{+k}}{N} + \sum_{j,k} \frac{n_{jk}}{N} \log \frac{n_{jk}}{N},$$

which is the average mutual information. Since the $n_{+k}, k = 1, \dots, g$, are given by the distribution of observations to be split, we thus choose the split to maximize the change in entropy measure

$$-\sum_j \frac{n_{j+}}{N} \log \frac{n_{j+}}{N} + \sum_{j,k} \frac{n_{jk}}{N} \log \frac{n_{jk}}{N}. \tag{12.29}$$

Gini index. The **Gini index** chooses the split to maximize

$$\sum_{j,k} n_{jk}\left(1-\frac{n_{jk}}{n_{j+}}\right), \tag{12.30}$$

which is a weighted sum of diversity measures.

Chi-squared statistic. Given a contingency table created by a split (such as the above), the usefulness of the split could be assessed by its association with the class labels. Alternatively, we could choose the split to maximize a chi-squared test statistic (see Exercise 9.5.1), which is calculated on the assumption of independence, i.e.

$$X^2 = N\sum_{j,k}\frac{(n_{jk}-n_{j+}n_{+k}/N)^2}{n_{j+}n_{+k}}, \tag{12.31}$$

which is then maximized over all possible splits. This approach has been incorporated into the algorithm CHAID (Kass, 1980). In practice, one may want to avoid splits that create small expected counts ($n_{j+}n_{+k}/N$) since this will lead to instability.

Deviance. A final criterion is to use the deviance associated with a proposed split, which is based on the difference in the maximized log likelihood when the frequency distributions are modeled by a multinomial distribution. We have a conditional likelihood which is proportional to

$$\prod_{\text{leaves } i}\ \prod_{\text{classes } k} p_{ik}^{n_{ik}}.$$

Before the split, the probabilities p_{ik} are estimated by $n_{+k}n_{i+}/N^2$, whereas after the split, they will be estimated by n_{ik}/N. Then, maximizing **deviance**, which is twice the difference in the log likelihood, is the same as choosing the split to minimize

$$2\sum_{i,k}\left\{n_{ik}\log\frac{n_{+k}n_{i+}}{N^2}-n_{ik}\log\frac{n_{ik}}{N}\right\} = 2\sum_{i,k} n_{ik}\log\frac{n_{+k}n_{i+}}{n_{ik}N}. \tag{12.32}$$

In principle, we can consider splits of the form $x_j \le t$ for each variable x_j, $j = 1, \dots, p$, and each value of t. In practice, we only need to consider at most $n-1$ possible values of t for each j since any split between neighboring points will give exactly the same measure. Thus, the search requires at most $p(n-1)$ evaluations to find the next splitting point.

Example 12.8.1 (Iris data, classification tree). Consider again the first two variables, sepal length and sepal width, for the three species or classes of iris. By considering x_1 and x_2 in turn, we calculate the above measures for each possible t, and these are shown in Figure 12.8 for the first split. Based on these values we select the first split at $x_1 = 5.45$ for the Gini and chi-squared indices and $x_1 = 5.55$ for deviance, error rate, and entropy indices. Variable x_2 is seen to be much less useful. The final partition (using the Gini index as the splitting criterion) is shown in Figure 12.9, and the corresponding tree is shown in Figure 12.10.

Note that these boundaries are quite similar to those obtained by LDA (Figure 12.3 (a) in Example 12.4.5), even though the straight line boundaries there are not constrained to be parallel to either axis. In particular, the predicted classes of the two methods are in agreement for 141 out of 150 observations. □

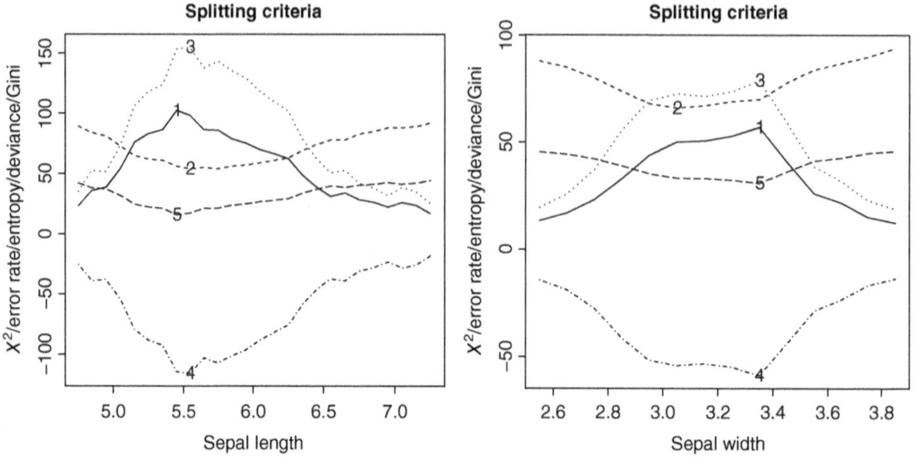

Figure 12.8 Splitting indices (some linearly transformed) of first two variables, using all three species of iris data for **1**: X^2, **2**: error, **3**: entropy, **4**: deviance, and **5**: Gini.

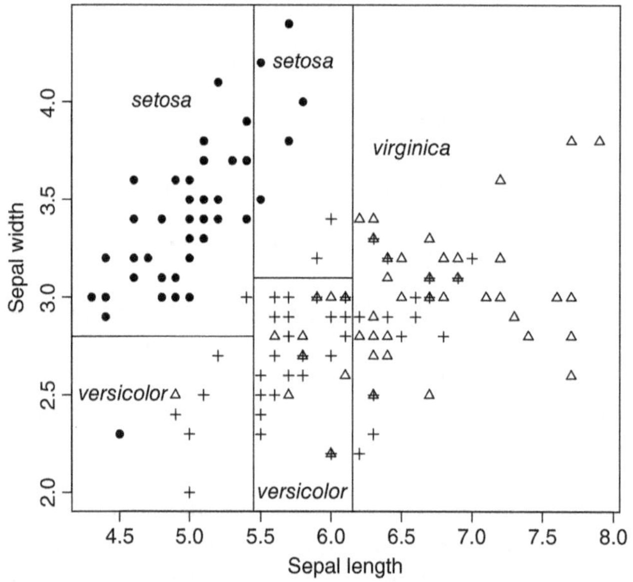

Figure 12.9 Scatter plot of the first two variables for iris data (• *I. setosa*; + *I. versicolor*; and Δ *I. virginica*), with partition regions (labeled as "setosa," "virginica," and "versicolor") as defined by the decision tree in Figure 12.10 which uses the Gini index.

12.8.2 Pruning

Given any of the above measures, we could continue splitting until each node is pure (i.e. only contains observations from one class) at which point no further splits are necessary. If the $\boldsymbol{x}$ are all distinct, then this can usually be achieved.

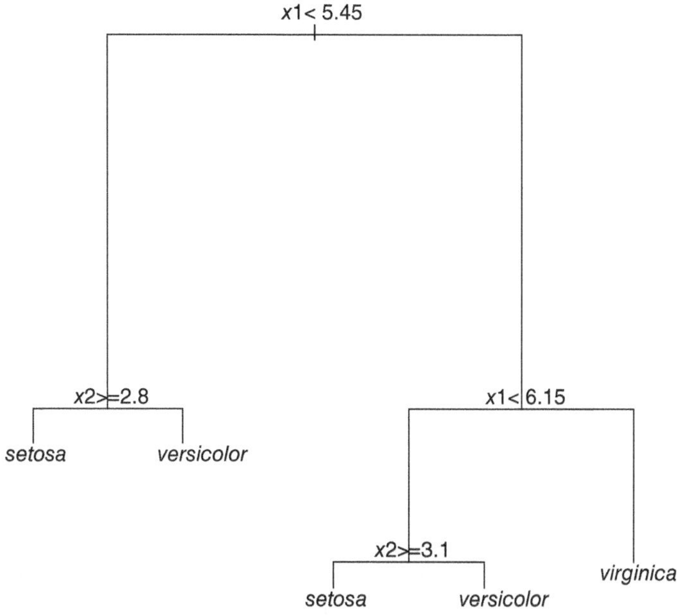

Figure 12.10 Decision tree (created using `rpart` (Therneau and Atkinson, 2019)) for first two variables (sepal length and sepal width) of the iris data.

However, in practice, the resulting tree will give high error rates on new test data. Two strategies can be adopted. Either:

- splitting can stop according to some *stopping rule* (for example, the number of observations is too small, or the increase in impurity too slight to be worth continuing) or
- the tree can be overgrown, and then some branches can be removed retrospectively. This is known as *pruning*.

It is possible that a split can be of very little value, but subsequent splits are of great benefit, so it is very hard to get stopping rules to work well. So, despite their computational savings in that there is no waste, stopping rules are rarely used to get the tree size correct.

Two commonly used methods of pruning are based on a pessimistic estimate of error rates (Quinlan, 1993) and cost-complexity pruning (Breiman et al., 1984).

12.8.2.1 Pessimistic Pruning

Suppose that, at a given node, we observe n misclassified observations out of a total of N. A natural estimate of the error rate at this node is simply n/N, and based on this, the expected number misclassified is simply $\frac{n}{N}N = n$. The pessimistic view is to seek the largest plausible error rate which could have given rise to n misclassified observations. If we let X be the number of misclassified observations at this node, then for given error rate, p, the probability, say β, of observing this many errors or fewer is

$$\beta = \mathrm{P}(X \leq n \mid p) = \sum_{r=0}^{n} \binom{N}{r} p^r (1-p)^{N-r}.$$

Given a specific probability β, we can solve the above equation for p. Given this value of $p(\beta)$, we can thus compute the expected number of errors at this node as $Np(\beta)$.

This calculation can be carried out at any node, and we can thus determine whether a split should be pruned by comparing the expected number of misclassified observations before and after splitting. The algorithm C4.5 which was developed by Quinlan (1993) uses $\beta = 0.25$ as a default value. Increasing β will lead to smaller trees, and decreasing β will lead to larger trees.

Example 12.8.2 Consider a split that results in the following contingency table:

	Class				
Condition	1	2	3	4	
T	3	1	2	10	16
F	1	0	5	1	7
	4	1	7	11	23

At node T, we have $j_{\mathsf{T}} = 4$ with $n = 6$ errors out of $N = 16$ observations, so $p = 0.491$.
At node F, we have $j_{\mathsf{F}} = 3$ with $n = 2$ errors out of $N = 7$ observations, so $p = 0.486$.

Hence, the (pessimistic) expected number of misclassified observations is $16 \times 0.491 + 7 \times 0.486 = 11.258$. Before the split (at the parent node), we have $n = 12$ errors out of $N = 23$ observations, so $p = 0.611$, and so, the expected number misclassified (for the same β) is $23 \times 0.611 = 14.053$. Since $11.258 < 14.053$, we would not prune this split. □

12.8.2.2 Cost Complexity Pruning

This procedure, proposed by Breiman et al. (1984), is based on the relationship between tree size (measured by the number of leaves) and an overall measure of the purity (for example, one of the five splitting criteria mentioned on p. 345) of the partition. Let T be a classification tree used to classify N observations in the training set of data, and denote by $\text{size}(T)$ the number of leaves of T. Suppose that $R(T)$ is a measure of purity of the partition; for example, if there are n misclassified observations, then we could use the error rate $R(T) = n/N$. The *cost complexity* of T, for some parameter α, is given by

$$R_\alpha(T) = R(T) + \alpha \,\text{size}(T),$$

and the objective is, given α, to find a subtree of T to minimize this quantity. If α is small, then there is a low penalty for large trees, and the final tree will be large, whereas a large value of α will lead to a smaller tree. It can be seen that $R_\alpha(T)$ is a linear combination of the error rate and the tree size, where α can be regarded as the cost for each leaf.

Pruning the tree just above one of the interior nodes will replace the subtree (say S) rooted at that node by a new leaf, and the size of the pruned tree will then be $\text{size}(T) - \text{size}(S) + 1$. However, the pruned tree will misclassify more (say k) observations, so the cost complexity of the pruned tree (say T_p) is $R_\alpha(T_p) = R_\alpha(T) + k/N + \alpha(1 - \text{size}(S))$. This will have the same cost complexity as the original tree if

$$\alpha = \frac{k}{N(\text{size}(S) - 1)}. \tag{12.33}$$

So, corresponding to each possible pruning location (often called a "snip"), there is a corresponding value of α that satisfies (12.33). These values of α can be sorted.

Breiman et al. (1984) showed that there is a unique pruned tree that minimizes $R_\alpha(T)$ for any value of α such that all other trees have either a higher cost complexity or the same cost complexity with the pruned tree as a subtree. Thus, there is a minimizing sequence of pruned trees $T_1 \supset T_2 \supset \cdots$, where each tree is obtained by selecting a larger value of α to prune the previous tree. To produce T_{i+1} from T_i, we examine each nonleaf pruned tree of T_i and find the minimum value of α that satisfies (12.33). The subtree corresponding to that value of α is then replaced by a leaf.

The remaining problem is to select α, so that the final pruned tree will perform well on new, unseen data. Breiman et al. (1984) proposed to use m-fold cross-validation to select α.

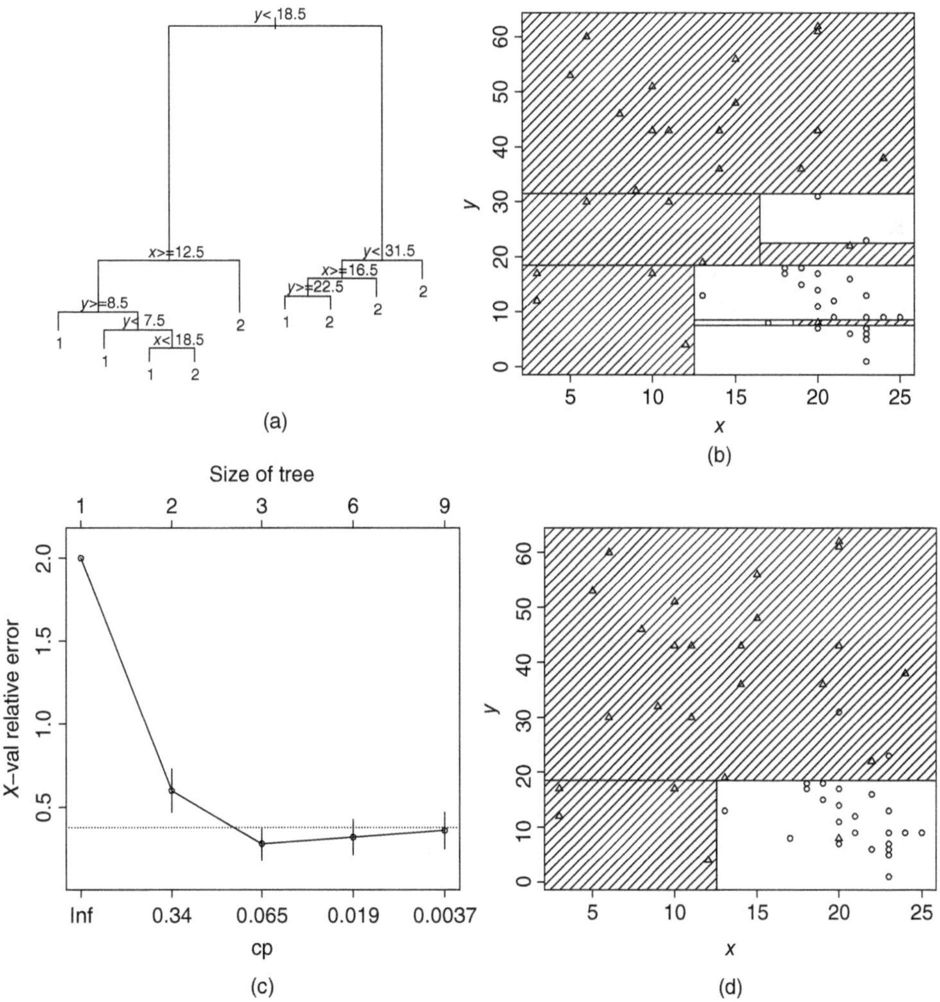

Figure 12.11 (a) Classification tree and the corresponding partition (b) for the mental health data in Example 12.8.3 for normal individuals (o) and psychotics (Δ); (c) cost-complexity pruning plot α/tree size vs. cross-validation estimate of error rate (created using `rpart` (Therneau and Atkinson, 2019)); (d) partition region corresponding to the optimal choice of the pruned tree.

Here, the dataset, say E, is randomly split into m subsamples, say $E_1, \ldots, E_m$, and for each $i = 1, \ldots, m$:

1. a tree is grown using the training dataset $E \backslash E_i$;
2. the error rate R_i is determined using the test set E_i, and so the tree can be optimally pruned to find α_i.

In practice, the α_i can be quite variable, and Breiman et al. (1984) suggest to obtain a final tree using all the data, pruned with a value of α slightly above the minimum of the $\alpha_1, \ldots \alpha_v$.

Example 12.8.3 To illustrate cost complexity pruning, we return to the mental health data of Example 12.4.6 and obtain a classification tree that has no errors (when resubstituting the original data). The tree and the corresponding boundaries are shown in Figure 12.11. The tree, which has been "overtrained," is unlikely to perform well on new data. For each value of α, there exists a corresponding size of tree, and a cross-validation error can be obtained. We can then obtain a plot of tree size vs. cross-validated error, and this can be used to prune the tree using the corresponding value of α. The "cp plot", shown in Figure 12.11, shows the cross-validation (relative) error rate, together with some error bars. This plot suggests that a tree with three leaves is appropriate (two leaves would also be possible since the vertical bar at size = 3 just hits the horizontal dashed line), and the partition corresponding to these two splits ($y < 18.5$, if true followed by $x < 12.5$) is also shown. We note the similarity of this region with that obtained by the quadratic discriminant rules in Example 12.4.6. □

12.9 Logistic Discrimination

12.9.1 Logistic Regression Model

We suppose, initially, that there are $g = 2$ groups, with prior probabilities $\pi_1, \pi_2 = 1 - \pi_1$. For convenience, we label group memberships by $y_i \in \{0, 1\}$ corresponding to Π_1 and Π_2. If the distributions of both groups are multivariate normal with equal covariance, then after incorporating prior probabilities into (12.11), we would allocate $\boldsymbol{x}$ to Π_1 if

$$\log\left(\frac{\pi_1 f_1(\boldsymbol{x})}{\pi_2 f_2(\boldsymbol{x})}\right) = \boldsymbol{\alpha}'(\boldsymbol{x} - \boldsymbol{\mu}) + \log\left(\frac{\pi_1}{\pi_2}\right) > 0.$$

That is, the log of the group-conditional densities is of the form (MacLachlan, 1992)

$$\log\left(\frac{\pi_1 f_1(\boldsymbol{x})}{\pi_2 f_2(\boldsymbol{x})}\right) = \log\left(\frac{\pi_1}{\pi_2}\right) + \beta_0^* + \boldsymbol{\beta}'\boldsymbol{x}. \tag{12.34}$$

Denoting $\beta_0 = \beta_0^* + \log(\pi_1/\pi_2)$, Eq. (12.34) is equivalent to

$$\mathsf{P}(y = 0 \mid \boldsymbol{x}) = \frac{\exp\{\beta_0 + \boldsymbol{\beta}'\boldsymbol{x}\}}{1 + \exp\{\beta_0 + \boldsymbol{\beta}'\boldsymbol{x}\}}. \tag{12.35}$$

If we now relax the assumption of multivariate normality (since normality determines the parameters) and allow $\boldsymbol{\beta}$ ($p \times 1$) and β_0 to be directly *estimated*, then the model given by (12.34) assumes only that the log of the posterior odds is linear in $\boldsymbol{x}$.

12.9.2 Estimation and Inference

Consider data with explanatory variables $\boldsymbol{x}_r$, $r = 1, \dots, n$, and group memberships labeled by $y_r \in \{0, 1\}$, $i = 1, \dots, n$, $n_1 = \sum_r (1 - y_r)$, and $n_2 = n - n_1$. Using (12.35), the likelihood is given by

$$\prod_{r=1}^{n} \mathsf{P}(y_r = 0)^{1-y_r} \mathsf{P}(y_r = 1)^{y_r},$$

and the log likelihood (provided the y_r are independent) can be written as

$$\sum_{r=1}^{n} \left((1 - y_r)(\beta_0 + \boldsymbol{\beta}'\boldsymbol{x}_r) - \log[1 + \exp\{\beta_0 + \boldsymbol{\beta}'\boldsymbol{x}_r\}] \right).$$

Numerical maximization of this will lead to estimates $\hat{\beta}_0 = \hat{\beta}_0^* + \log(n_1/n_2)$ and $\hat{\boldsymbol{\beta}}$. The correct likelihood estimation will depend on the way that the samples have been collected. For example, if the n data values come from a simple random sample of the population, the procedure will differ from the case where the values of n_i are selected in advance; see MacLachlan (1992, pp. 259–266) for more details. In this section, we assume a "mixture sampling" design in which the n_i are not fixed in advance.

As in standard linear regression, each regression coefficient, $\hat{\beta}_0, \hat{\beta}_1, \dots, \hat{\beta}_p$, has a standard error, which can be used to test whether this term should be in the model. Alternatively, likelihood ratio tests (Section 6.1) or Akaike's information criterion (AIC) (Akaike, 1974) can be used to find a parsimonious model.

12.9.3 Interpretation of the Parameter Estimates

Each of the estimated β coefficients is associated with an explanatory variable. We start with the simplest case in which one of the explanatory variables, say x_k, is binary, taking the values 0 and 1. Let $\boldsymbol{x}^{(0)}$ and $\boldsymbol{x}^{(1)}$ denote two vectors which agree in all components, except for x_k, which takes the value i in $\boldsymbol{x}^{(i)}$ for $i = 0, 1$. In this case, from (12.35), we have the *odds ratio* given by

$$\frac{\mathsf{P}(y = 0 \mid \boldsymbol{x} = \boldsymbol{x}^{(1)})}{\mathsf{P}(y = 1 \mid \boldsymbol{x} = \boldsymbol{x}^{(1)})} \bigg/ \frac{\mathsf{P}(y = 0 \mid \boldsymbol{x} = \boldsymbol{x}^{(0)})}{\mathsf{P}(y = 1 \mid \boldsymbol{x} = \boldsymbol{x}^{(0)})} = e^{\beta_k},$$

and the log of the odds ratio, usually called the *log-odds*, is simply β_k. The odds ratio can be considered as the relative risk (of being in group 1) associated with a particular variable. For example, if Π_1(i.e. $y = 0$) denotes the population of those who have lung cancer, and Π_2(i.e. $y = 1$) those who do not, and if x_k is a binary variable in which $x_k = 1$ for a smoker, and $x_k = 0$ for a nonsmoker, then e^{β_k} is the relative risk of having lung cancer for a smoker (relative to a nonsmoker). For example, if $e^{\beta_k} = 3$ then this indicates that the odds of a smoker having lung cancer is three times that of a nonsmoker. This very simple interpretation of the parameter is one of the most appealing features of logistic regression.

Example 12.9.1 On 15 April 1915, the *Titanic* sank in the North Atlantic Ocean after hitting an iceberg during its maiden voyage. There were around 1317 passengers onboard, of whom more than half died. A summary of the data is given in Table 12.5, which shows the proportion of people surviving, classified by sex (male/female), "age" (child/adult), and cabin class (first/second/third/crew). Clearly, the survival rate is related to class and sex as well as age.

Table 12.5 Summary of the Titanic data for the number of survivors/total number in the group, totaled by age, class, and sex. (There were no crew who were children.)

		Male		Female	
Class	**Age**	**Survive**	**%**	**Survive**	**%**
First	Child	5/5	100	1/1	100
	Adult	57/118	33	140/144	97
Second	Child	11/11	100	13/13	100
	Adult	14/168	8	80/93	86
Third	Child	13/48	27	14/31	45
	Adult	75/462	16	76/165	46
Crew	Adult	192/862	22	20/23	87

Table 12.6 Summary of the estimated parameters and standard errors for a logistic regression model fitted to survival data in the sinking of the Titanic.

	Estimate	Std. error	*z* value	Pr(> |*z*|)
Intercept	3.95	0.36	11.03	0.000
Second class	-1.37	0.23	-5.95	0.000
Third class	-2.35	0.23	-10.28	0.000
Male	-2.58	0.17	-15.11	0.000
Age	-0.04	0.01	-5.99	0.000
Sibling and spouse	-0.33	0.10	-3.29	0.001

A more detailed dataset[1] also includes information (for most passengers) on age, fare paid, embarkation point, name, and number of family members onboard. We fitted a logistic regression model to the more detailed dataset. There were 1309 observations, and the following variables were included: cabin class, sex, age, number of siblings/spouses onboard, number of parents/children onboard, fare, and point of embarkation (Cherbourg, Queenstown, and Southampton). Some observations had missing values (mainly for the variable age), and these observations were removed from the analysis, leaving 1048 observations. The final model was reduced to include only the significant variables, with a summary shown in Table 12.6.

From the fitted values, we see that the log-odds ratio associated with the variable `male` is -2.58, which means that the male:female survival odds were only $\exp(-2.58) = 0.076$ (if all other variables were held fixed). Similarly, an increase of age by 10 years would lead to a reduction in the fitted odds of survival by a factor $\exp(10 \times -0.04) = 0.67$. The indicator variables `Second Class` and `Third Class` both have a negative effect; these are measured relative to First Class passengers, who had the best chance of survival. Figure 12.12 shows the fitted probabilities of survival for the main effects (class, age, and sex).

1 https://hbiostat.org/data/repo/titanic3.csv

Table 12.7 Confusion matrix for logistic regression model of Titanic survival data. Predictions obtained by resubstitution probabilities, thresholded at 0.5.

		Actual		
		Died	Survived	Total
Predicted	Died	519	126	647
	Survived	100	301	401
	Total	619	427	1048

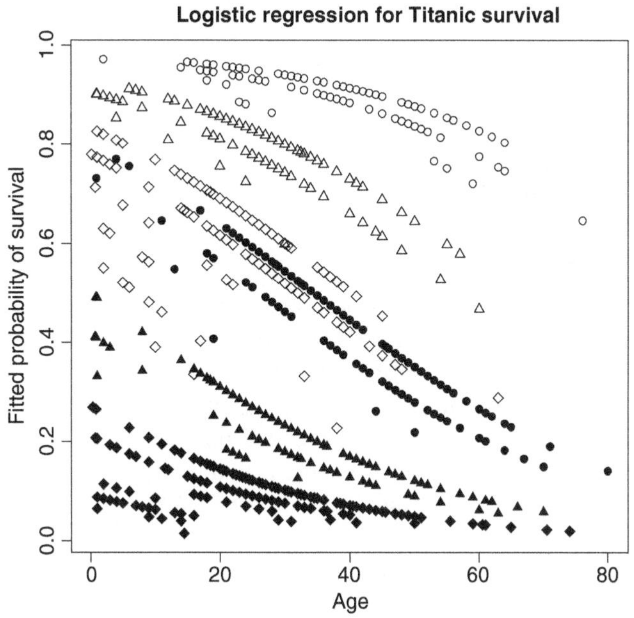

Figure 12.12 Fitted probabilities for survival in the sinking of the Titanic, shown as a function of age (x-axis), sex (filled symbol for male; empty for female), and class (∘ First Class; Δ Second Class; and ◇ Third Class).

The data points lie approximately on *curved* lines because the vertical axis shows probabilities (a log-odds vertical scale would lead to straight lines). However, the points do not follow neat lines due to the sibling and spouse effect (the number of siblings/spouses was in the range [0, 8]); these points are not distinguished in the figure to reduce the complexity of the plotting symbols. The effects of cabin class, sex, and age are all evident in their impact on the fitted probability of survival.

For each passenger, the model can be used to predict the chances of survival, using Eq. (12.35). These probabilities can be thresholded at 0.5 to classify each passenger into one of the two groups (survived/not), and these can be compared to the actual survivals and summarized in a *confusion matrix*, as shown in Table 12.7. Although the classification success rate ($820/1048 = 78\%$) is reasonable, this should be judged while taking account of the fact that the model is being tested on the same data that was used to build it, i.e. a resubstitution estimate. Cross-validation would be more realistic here. □

12.9.4 Extensions

The simplest extension of the model given by (12.35) is to allow the log of the posterior odds to include *nonlinear* terms in the variables. Thus, for example, we could require only that the posterior odds take a form similar to Equation (12.16); this may, or may not, include *interaction* terms. This would extend the logistic regression model in a similar way that the linear discriminant function is extended to a quadratic discriminant.

A more general problem occurs when the number of groups, g, is larger than 2. This is known as multinomial (or polytomous) logistic regression. The response variable is now categorical, and a common approach is to consider a logistic regression model relative to a baseline category (one of the values of y) in which the $g-1$ sets of parameters are estimated simultaneously (Dobson and Barnett, 2018).

12.10 Neural Networks

12.10.1 Motivation

Neural networks include a wide class of approaches to classification. In the simplest (degenerate) version, there are obvious links to logistic discrimination (see Section 12.9) which (in the case of two classes) is itself a generalization of the ML discriminant rule for the multivariate normal distribution. We begin by pointing out some of the connections between neural networks and previously described methods. Recall from Section 12.4.2 that the sample ML discriminant rule (when we assume that the $g \geq 2$ populations are normally distributed with equal covariance matrix and with equal priors) allocates $\boldsymbol{x}$ to $\Pi_l (l \in \{1, \dots, g\})$ if

$$l = \arg\ \min_j (\boldsymbol{x} - \overline{\boldsymbol{x}}_j)' \boldsymbol{S}_u^{-1} (\boldsymbol{x} - \overline{\boldsymbol{x}}_j).$$

This is equivalent to $l = \arg\ \max_j h_j(\boldsymbol{x})$, where

$$h_j(\boldsymbol{x}) = \boldsymbol{w}_j' \boldsymbol{x} + w_{j0}, \tag{12.36}$$

with $\boldsymbol{w}_j' = \overline{\boldsymbol{x}}_j' \boldsymbol{S}_u^{-1}$ and $w_{j0} = -\frac{1}{2} \overline{\boldsymbol{x}}_j' \boldsymbol{S}_u^{-1} \overline{\boldsymbol{x}}_j$. These *discriminant functions* h_j are simple linear combinations of the components of the vector $\boldsymbol{x}$. A more general approach is to allow the $\boldsymbol{w}$ to be arbitrary parameters, and then to estimate them by minimizing the error rate. Neural networks seek to extend the above formulation in a number of ways. For example, if the data are not normally distributed, then by considering the resubstitution error rate, we can obtain alternative $\boldsymbol{w}_j$, w_{j0} which are not based on this assumption, which is the approach used in logistic regression. Note that, if our classifier is based on $\arg\ \max_j h_j(\boldsymbol{x})$, then we can arbitrarily scale all the weights – i.e. multiply all $\boldsymbol{w}$, w_{j0} by the same scaling factor – with no change in the allocation rule. Secondly, we can take more than one linear combination, say of the form

$$h_j(\boldsymbol{x}) = \sum_{k=0}^{M} v_{jk} \psi_k(\boldsymbol{x})$$

in which the ψ_k are known *basis functions*. As a simple example, the formulation in (12.36) can be expressed using $M = p$ basis functions, where $\psi_k(\boldsymbol{x})$ is equal to the kth component of the vector $\boldsymbol{x}$ for $1 \leq k \leq p$ and $\psi_0(\boldsymbol{x}) = 1$. However, this choice does not lead to a new class of functions for $h_j(\boldsymbol{x})$. More generally, the ψ_k can be taken to be nonlinear functions of $\boldsymbol{x}$, typically of the form $\psi_k(\boldsymbol{x}) = \zeta(\boldsymbol{w}_k'\boldsymbol{x} + w_{k0})$.

Logistic discrimination (see Section 12.9) illustrates a motivation for nonlinear ψ. To simplify, we assume that there are $g = 2$ groups, and that the priors satisfy $\pi_1 = \pi_2 = 1/2$. The modeling assumption in logistic regression is that

$$\log\left(\frac{f_1(\boldsymbol{x})}{f_2(\boldsymbol{x})}\right) = \alpha + \boldsymbol{\beta}'\boldsymbol{x},$$

and so,

$$\frac{f_1(\boldsymbol{x})}{f_2(\boldsymbol{x})} = \exp(\alpha + \boldsymbol{\beta}'\boldsymbol{x}).$$

This leads to the posterior probabilities, given $\boldsymbol{x}$, as shown in (12.35):

$$\mathsf{P}(y = 0) = \frac{1}{1 + \exp(-\alpha - \boldsymbol{\beta}\boldsymbol{x})} \quad \text{and} \quad \mathsf{P}(y = 1) = \frac{1}{1 + \exp(\alpha + \boldsymbol{\beta}\boldsymbol{x})},$$

in which we can recognize the *sigmoid* function

$$\zeta(a) = \frac{1}{1 + \mathrm{e}^{-a}}.$$

We thus obtain a generalization of (12.36), which is given by

$$h_j(\boldsymbol{x}) = \sum_{k=1}^{M} w_{jk}\psi_k(\boldsymbol{x}) + w_{j0} \tag{12.37}$$

in which $h(\cdot)$ is no longer interpreted as a probability. Setting $\psi_0(\boldsymbol{x}) \equiv 1$, this can be written in matrix form as

$$\boldsymbol{h}(\boldsymbol{x}) = \boldsymbol{W}\boldsymbol{\psi}(\boldsymbol{x}) \tag{12.38}$$

in which $\boldsymbol{W}$ is a $g \times (M + 1)$ matrix with jth row given by $(w_{j0}, \boldsymbol{w}_j'), j = 1, \ldots, g$, and $\boldsymbol{\psi}$ is a vector of known basis functions $\psi_0, \ldots, \psi_M$.

Sections 12.10.2 and 12.10.3 develop these ideas in two implementations of *feed-forward neural networks*. In both cases, parameters in $\boldsymbol{W}$ and within $\boldsymbol{\Psi}$ are estimated from the data. Multilayer perceptrons are sums of nonlinear functions of different projections of the observations, whereas radial basis functions are weighted sums of radially symmetric functions.

12.10.2 Multilayer Perceptron

The *perceptron* is a function of the form

$$\zeta\left(\sum_{k=0}^{M} w_k\psi_k(\boldsymbol{x})\right) = \zeta(\boldsymbol{w}'\boldsymbol{\psi}) \tag{12.39}$$

in which $\zeta(\cdot)$ is a *threshold* or *activation function*. One choice is $\zeta(a) = \operatorname{sign}(a)$. In (12.39), ζ is kept as a general function as the specific choice will depend on the *layer* of the network. For the purposes of learning the weights (w_k), it is often useful to choose ζ to be differentiable.

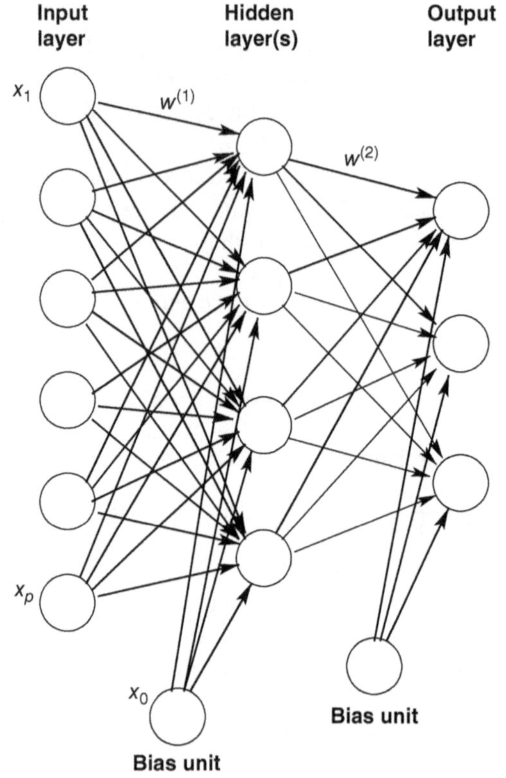

Figure 12.13 Multilayer network with an input layer, one hidden layer, and the output discriminant functions.

A simple example of a multilayer perceptron is shown in Figure 12.13. This is a network with

- an input layer of the p components of the observation $\boldsymbol{x}$ and the intercept term 1 (which is often called the "bias");
- one *hidden layer* of units, the number of which can be chosen adaptively, with another "bias" unit;
- an *output layer* of discriminant functions $h_j, j = 1, \dots, g$.

These units are connected together by weights, and these linear combinations are equivalent to projections of the units. This network is thus a two-layer perceptron. We note that the values taken at the nodes are determined by the data for the input layer and are real-valued quantities at the nodes in the hidden and output layers which are sigmoid (activation) functions of the linear combinations of the values from the previous layer.

The parameters of the model are the weights that connect the units. If there are M hidden units, let $w_{ji}^{(1)}$ denote the weights from the ith input unit (the ith component of $\boldsymbol{x}$) to the jth hidden unit, and $w_{kj}^{(2)}$ denote the weights from the jth hidden unit to the kth output unit

$(k = 1, \ldots, g)$. Then the discriminant function takes the form

$$h_k(\boldsymbol{x}) = \zeta^o\left(\sum_{j=0}^{M} w_{kj}^{(2)} \zeta^h\left(\sum_{i=0}^{p} w_{ji}^{(1)} x_i\right)\right). \tag{12.40}$$

Here, $x_0 \equiv 1$ represents the intercept (bias) term, $\zeta^h(\cdot)$ and $\zeta^o(\cdot)$ the activation functions for the hidden layer and output layer, and $\sum_{i=0}^{p} w_{0i}^{(1)} x_i$ is set to 1 to create the intercept term (x_0) from the hidden layer.

In (12.40), there are $Mp + g(M + 1)$ terms which need to be "estimated" (where p is the number of variables in the data). A commonly used algorithm is backpropagation, which computes the gradient of a loss function (typically the error rate) with respect to the set of weights. Stochastic methods can also be used. See Bishop (1995, 2006) and Ripley (1996) for good introductions.

Example 12.10.1 We return to the mental health data of Example 12.4.6 and consider four multilayer perceptron solutions with one hidden layer with $M \in \{1, 3, 4, 6\}$ hidden units. We use the R library `nnet` (Venables and Ripley, 2002) to estimate the weights; this function uses random starting values, so results may differ slightly. Using the resubstitution method to assess the error rate, we obtain the number of errors (corresponding to each value of M) as 4, 5, 2, and 0, respectively. The decision boundaries for each case are shown in Figure 12.14, along with the linear discriminant boundary, for comparison. It can be seen that the boundary is linear with one hidden unit (with the same number of errors as LDA) and becomes increasingly flexible/erratic with increasing M. Repeat runs of the algorithm will show some randomness in our solution due to a random choice in the initial weights. Apart from the obvious impact of M on the solution, the number of iterations can also be limited to avoid serious overfitting. In the current illustration, this value was set to a large value (1000), so that convergence was reached, and allowed us to focus only on the choice of M. In practice, both M and the number of iterations could be chosen by cross-validation. □

12.10.3 Radial Basis Functions

The basic idea is that the activation of a hidden unit is now determined by the distance from the input vector to a *prototype* vector. Radial basis functions take the form $\psi_j(||\boldsymbol{x} - \boldsymbol{m}_j||)$, where $\psi_j(\cdot)$ is a nonlinear function, and $||\boldsymbol{x} - \boldsymbol{m}_j||$ represents the distance (usually scaled Euclidean distance) from $\boldsymbol{x}$ to the jth prototype vector $\boldsymbol{m}_j, j = 1, \ldots, M$. This gives

$$h_j(\boldsymbol{x}) = w_{j0} + \sum_{k=1}^{M} w_{jk} \psi_k(\boldsymbol{x}), \tag{12.41}$$

where, for example, we could use radial basis functions given by the *Gaussian basis functions*

$$\psi_k(\boldsymbol{x}) = \exp\left(-\frac{||\boldsymbol{x} - \boldsymbol{m}_k||^2}{2s_k^2}\right). \tag{12.42}$$

Here, the $\boldsymbol{m}_k$ represent the center of the basis function (the prototypes), and each basis function is given its own width parameter s_k, whose values are to be determined. As before, the term w_{j0} is called the "bias."

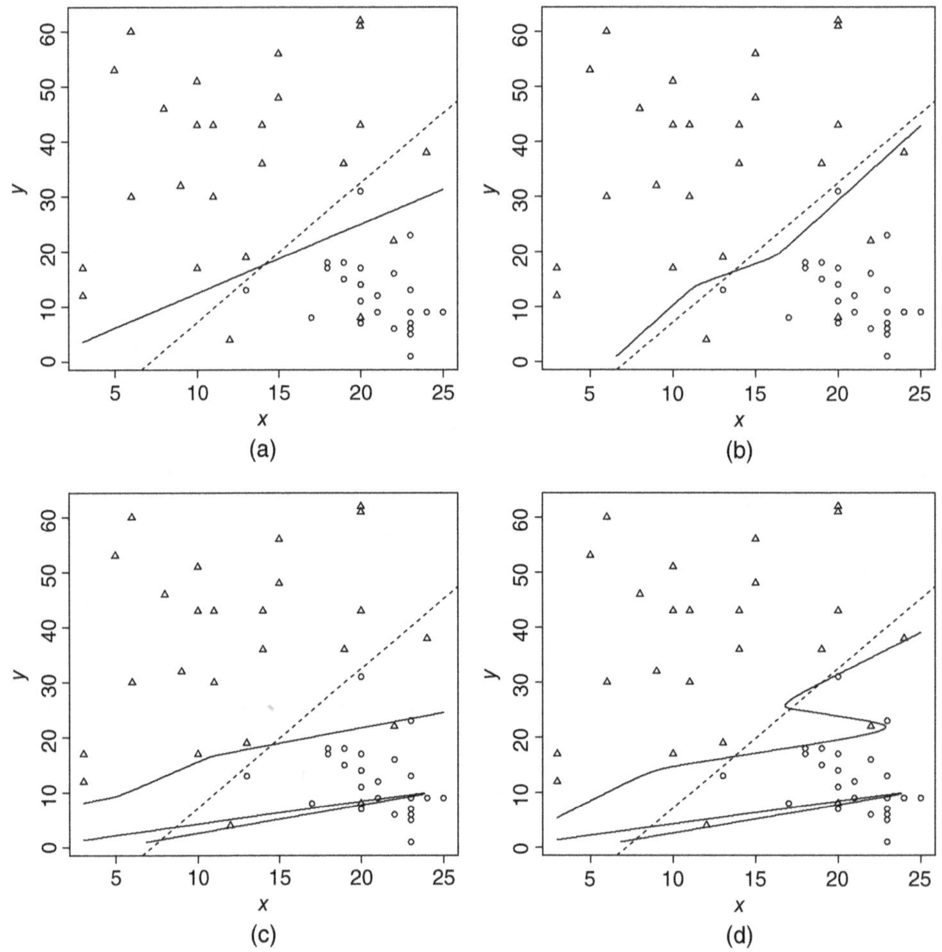

Figure 12.14 Multilayer perceptron boundaries (continuous line) and linear discriminant (dashed line) for mental health data (see Example 12.4.6). The four panels (a)–(d) use one hidden layer, with the number of hidden units given by 1, 3, 4, and 6, respectively. The R package `nnet` (Venables and Ripley, 2002) was used to fit the weights, and the network is then predicted on a fine grid to obtain the boundary.

Usually, estimation of the parameters takes place in two stages. Firstly, the ψ_k (i.e. the $\boldsymbol{m}_k$ and s_k) are determined by "unsupervised learning" techniques (see Chapter 14), and then, the weights can easily be computed by least squares. Noting that Eq. (12.41) is of the same form as (12.38), with $\psi_0(\boldsymbol{x}) \equiv 1$, we thus have the representation

$$\boldsymbol{Y} = \boldsymbol{\Psi}\boldsymbol{W} + \boldsymbol{E} \tag{12.43}$$

in which $\boldsymbol{Y}$ is a $n \times g$ indicator matrix with $y_{rj} = 1$ if observation $\boldsymbol{x}_r \in \Pi_j$, 0 otherwise, and $\boldsymbol{\Psi}$ is a $n \times (M+1)$ matrix with (i, j)th entry given by $\psi_j(\boldsymbol{x}_i)$. Hence, Eq. (12.43) is the same as (7.1), and we can use multivariate regression analysis to estimate the weights in $\boldsymbol{W}$, which is a $(M+1) \times g$ matrix.

Let $\psi_k^o(\boldsymbol{x}), k = 1, \ldots, M$ denote a normalized radial basis function, so that $\psi_k^o(\boldsymbol{x}) = \psi_k(\boldsymbol{x}) / \int \psi_k(\boldsymbol{x})\mathrm{d}\boldsymbol{x}$, and suppose that we can approximate the class conditional densities $f_j(\boldsymbol{x})$ by a weighted sum

$$f_j(\boldsymbol{x}) = \sum_{k=1}^{M} p_{jk}\psi_k^o(\boldsymbol{x}),$$

with $\sum_k p_{jk} = 1, j = 1, \ldots, g$. This gives

$$f(\boldsymbol{x}) = \sum_{j=1}^{g} \pi_j f_j(\boldsymbol{x}) = \sum_{k=1}^{M} \psi_k^o(\boldsymbol{x}) \sum_{j=1}^{g} \pi_j p_{jk},$$

which we can interpret as the sum $\sum_{k=1}^{M} f(\boldsymbol{x} \mid k)\mathrm{P}(k)$ in which $\mathrm{P}(k)$ is the prior for the normalized basis function $\psi_k^o(\boldsymbol{x})$, and $f(\boldsymbol{x} \mid k)$ is the conditional distribution for that function. Then, Bayes' theorem gives

$$\begin{aligned} \mathrm{P}(y = j) &= \frac{\pi_j f_j(\boldsymbol{x})}{\sum_{l=1}^{g} \pi_l f_l(\boldsymbol{x})} \\ &= \frac{\sum_{k=1}^{M} \pi_j p_{jk} \psi_k^o(\boldsymbol{x})}{\sum_{i=1}^{M} f(\boldsymbol{x} \mid i)\mathrm{P}(i)} \\ &= \sum_{k=1}^{M} \left[\frac{p_{jk}\pi_j}{\mathrm{P}(k)}\right] \left[\frac{\psi_k^o(\boldsymbol{x})\mathrm{P}(k)}{\sum_{i=1}^{M} f(\boldsymbol{x} \mid i)\mathrm{P}(i)}\right], \end{aligned}$$

and so, the first square bracket can be interpreted as the weights w_{jk}, and the second square bracket can be interpreted as a multiple of $\psi_k(\boldsymbol{x})$. This gives an interpretation in which the activations of the basis functions are like posterior probabilities of the corresponding features in the input space, and the weights can be interpreted as the posterior probabilities of class membership, given the presence of the features.

12.10.4 Support Vector Machines

Support vector machines (SVMs) tackle the problem of discrimination from a perspective not evidently based on probability theory, by looking for hyperplanes that separate the populations most prominently. To simplify this final section, we suppose that there are two classes with labels $y = 1$ and $y = -1$. A linear discriminant function is of the form

$$h(\boldsymbol{x}) = \boldsymbol{w}'\boldsymbol{x} + w_0,$$

and we predict $y \mid \boldsymbol{x}$ based on the sign of $h(\boldsymbol{x})$. A data set is called *linearly separable* if the two groups can be perfectly separated by a linear discriminant function. In this case there will generally be many solutions for $(w_0, \boldsymbol{w}')$ that give perfect classification. In such cases we would like to find two parallel hyperplanes, say $\boldsymbol{w}'\boldsymbol{x} + w_a$ and $\boldsymbol{w}'\boldsymbol{x} + w_b$, which are as far apart as possible, such that both of them separate the two groups. The preferred discriminant boundary is then the hyperplane which is midway between them – this is called the *maximum-margin* hyperplane. The distance between the two hyperplanes is given by $|w_a - w_b|/||\boldsymbol{w}||$. To maximize this distance, the optimization problem can be shown to solve

$$\min_{(w_0, \boldsymbol{w}')} ||\boldsymbol{w}|| \quad \text{subject to} \quad y_r(\boldsymbol{w}'\boldsymbol{x}_r + w_0) \geq 1, \; r = 1, \ldots, n,$$

and the $\boldsymbol{x}_r$ which are nearest to the hyperplane are called *support vectors*.

This can be extended to classes that are not linearly separable, with a trade-off incorporated (determined by λ below), which balances the desire for a larger margin size with the desire to minimize the resubstitution error. In addition, nonlinear boundaries can be estimated by making use of a radial basis function kernel (12.42), in what is known as the *kernel trick*. This approach transforms the data $\boldsymbol{x}_r$ into a higher dimensional space (for example, to include quadratic terms) and then uses the proximity of the transformed data points together with the class labels to determine appropriate weights. This leads to an optimization problem in which $\alpha_1, \ldots, \alpha_n$ are chosen to maximize

$$\sum_r \alpha_r - \frac{1}{2}\sum_{r,s} y_r \alpha_r k_{rs} \alpha_s y_s \quad \text{subject to} \quad \boldsymbol{\alpha}'\boldsymbol{y} = 0,\ 0 \le \alpha_r \le \frac{1}{2n\lambda},\ r = 1, \ldots, n,$$

where $k_{rs} = k(\boldsymbol{x}_r, \boldsymbol{x}_s)$ denotes the dot product of the transformed points, and λ controls the smoothness of the solution. For example, the function $k(\boldsymbol{x}_r, \boldsymbol{x}_s)$ can be based on a radial basis function (12.42), given by

$$k_{rs} = \exp\left(\frac{-||\boldsymbol{x}_r - \boldsymbol{x}_s||^2}{2s^2}\right),$$

in which case, there are two "tuning parameters" (s and λ) to select. At first sight, it appears that we have made the problem much harder by now requiring to find n weights, rather than the $p+1$ weights in the linear case. However, efficient programming methods exist, and only subsets of the data (the support vectors) are required in the above summations. Note that the radial basis function is based on a distance, and so, the variables should be commensurate, for example by standardizing to unit variance. Moreover, the approach seems to work well with high-dimensional data. For further details, see Cortes and Vapnik (1995).

Exercises and Complements

12.2.1 Suppose, in the univariate case, that there are two classes with p.d.f. given by:

$$f_k(x) = \frac{1}{\sigma_k\sqrt{2\pi}} \exp\left\{-\frac{1}{2}\left(\frac{x-\mu_k}{\sigma_k}\right)^2\right\}, \quad k = 1, 2.$$

(a) Using Bayes' rule, derive the optimal classification rule $\delta(x)$.
(b) Draw a sketch to illustrate your solution.
(c) Simplify your expression in the special case that
 (i) $\mu_1 = \mu_2$ (with $\sigma_1 \neq \sigma_2$)
 (ii) $\sigma_1 = \sigma_2$ (with $(\mu_1 \neq \mu_2)$
 and obtain the expected error rate for case (ii).

12.2.2 Suppose that there are two groups of biased coins in which the probability of tossing a head is p_i for group $i = 1, 2$. A new coin is to be allocated to one of these groups based on the outcome of n tosses. Obtain the maximum likelihood decision rule and the error rate as a function of n, p_1, and p_2.

12.3.1 Suppose that an urn contains some gold dust which weighs X grams, and assume that X is normally distributed with known mean μ and variance σ^2. You are invited to guess the weight of the gold dust in the urn with the promise that, if you guess correctly (to within 1 g), you can keep the contents. Using a suitable cost function, together with Eq. (12.8), obtain your best guess in terms of μ and σ.

12.3.2 Suppose that leave-one-out cross-validation is to be used to estimate the error rate for the sample ML linear discriminant rule in which we use here the *unbiased* estimate of covariance. For each observation $\boldsymbol{x}_r$, we need to compute

$$\exp\left\{-\frac{1}{2}(\boldsymbol{x}_r - \overline{\boldsymbol{x}}_j^{(r)})'\{\boldsymbol{S}_u^{(r)}\}^{-1}(\boldsymbol{x}_r - \overline{\boldsymbol{x}}_j^{(r)})\right\} \quad j = 1, \dots, g,$$

where $\overline{\boldsymbol{x}}_j^{(r)}$ is the sample mean of the jth group, and $\{\boldsymbol{S}_u^{(r)}\}$ is the unbiased pooled estimate of the covariance matrix when the rth observation is omitted. Let $\boldsymbol{S}_{j,u}$ denote the unbiased estimate of the covariance matrix of the jth group.
Show that (Hjort, 1986; MacLachlan, 1992, pp. 342–343)

$$(\boldsymbol{x}_r - \overline{\boldsymbol{x}}_j^{(r)})'\{\boldsymbol{S}_u^{(r)}\}^{-1}(\boldsymbol{x}_r - \overline{\boldsymbol{x}}_j^{(r)}) = \left(\sum_{k=1}^{g} z_{kr} c_{jk,r}\right)(\boldsymbol{x}_r - \overline{\boldsymbol{x}}_j)'\boldsymbol{S}_u^{-1}(\boldsymbol{x}_r - \overline{\boldsymbol{x}}_j),$$

where

$$z_{kr} = \begin{cases} 1 & \text{if } \boldsymbol{x}_r \in \Pi_k, \\ 0 & \text{otherwise} \end{cases},$$

$$c_{jj,r} = \{(n-g-1)n_j^2/(n_j-1)\}/\{(n-g)(n_j-1) - n_j d_{jj,r}\},$$

$$c_{jk,r} = \frac{n-g+1}{n-g}\left[1 + \frac{n_j d_{jk,r}^2}{d_{jj,r}\{(n-g)(n_j-1) - n_j d_{kk,r}\}}\right] \quad k \neq j,$$

and

$$d_{jk,r} = (\boldsymbol{x}_r - \overline{\boldsymbol{x}}_j)'\boldsymbol{S}_u^{-1}(\boldsymbol{x}_r - \overline{\boldsymbol{x}}_k) \quad (1 \le j, k \le g).$$

(Hint: Note that

$$\overline{\boldsymbol{x}}_j^{(r)} = \overline{\boldsymbol{x}}_j - \frac{z_{jr}}{n_j - 1}(\boldsymbol{x}_r - \overline{\boldsymbol{x}}_j).$$

Derive an equivalent expression for

$$\boldsymbol{S}_{k,u}^{(r)} = (n_k - 1 - z_{kr})^{-1} \sum_{\{s: \boldsymbol{x}_s \in \Pi_k, s \neq r\}} (\boldsymbol{x}_s - \overline{\boldsymbol{x}}_k^{(r)})(\boldsymbol{x}_s - \overline{\boldsymbol{x}}_k^{(r)})',$$

and then use the Woodbury formula (A.30).)

12.3.3 If Π_j is the $N_p(\boldsymbol{\mu}_j, \boldsymbol{\Sigma})$ population for $j = 1, 2$, $\boldsymbol{\Sigma} > 0$ and $\boldsymbol{x} \sim N_p(\boldsymbol{\mu}_1, \boldsymbol{\Sigma})$, show that $\boldsymbol{\alpha}'(\boldsymbol{x} - \boldsymbol{\mu}) \sim N_p(-\frac{1}{2}\Delta^2, \Delta^2)$, where

$$\boldsymbol{\mu} = \frac{1}{2}(\boldsymbol{\mu}_1 + \boldsymbol{\mu}_2), \qquad \boldsymbol{\alpha} = \boldsymbol{\Sigma}^{-1}(\boldsymbol{\mu}_1 - \boldsymbol{\mu}_2), \qquad \Delta^2 = (\boldsymbol{\mu}_1 - \boldsymbol{\mu}_2)'\boldsymbol{\Sigma}^{-1}(\boldsymbol{\mu}_1 - \boldsymbol{\mu}_2).$$

12.4.1 If Π_j is the $N_p(\boldsymbol{\mu}_j, \boldsymbol{\Sigma})$ population for $j = 1, 2$ and $\boldsymbol{\Sigma} > 0$, show that the ML discriminant rule is given by

$$\text{allocate } \boldsymbol{x} \text{ to } \Pi_1 \text{ if } \boldsymbol{\alpha}'\{\boldsymbol{x} - \frac{1}{2}(\boldsymbol{\mu}_1 + \boldsymbol{\mu}_2)\} > 0,$$

where $\boldsymbol{\alpha} = \boldsymbol{\Sigma}^{-1}(\boldsymbol{\mu}_1 - \boldsymbol{\mu}_2)$.

12.4.2 Consider three bivariate normal populations with the same covariance matrix in which

$$\mu_1 = \begin{pmatrix} 0 \\ 0 \end{pmatrix}, \quad \mu_2 = \begin{pmatrix} 0 \\ 1 \end{pmatrix}, \quad \mu_3 = \begin{pmatrix} 1 \\ 0 \end{pmatrix}, \quad \Sigma = \begin{bmatrix} 5 & 2 \\ 2 & 1 \end{bmatrix}.$$

(a) Draw the ML allocation regions and show that the three boundary lines meet at the point $(15/2, 7/2)$.

(b) Suppose that the three populations have prior probabilities $1/2$, $1/3$, and $1/6$. Draw the Bayes' allocation regions and find the point where the three boundary lines meet.

12.4.3 The genus *Chaetocnema* (a *genus* of flea beetles) contains two allied species, *Chaetocnema concinna* and *Chaetocnema heikertingeri*, that were long confused with one another. Lubischew (1962) gave the measurements of characteristics for samples of males of the two species, which are shown in Table 12.8.

Table 12.8 Flea-beetle measurements, taken from Lubischew (1962).

Ch. concinna		*Ch. heikertingeri*	
x_1	x_2	x_1	x_2
191	131	186	107
185	134	211	122
200	137	201	114
173	127	242	131
171	118	184	108
160	118	211	118
188	134	217	122
186	129	223	127
174	131	208	125
163	115	199	124
190	143	211	129
174	131	218	126
201	130	203	122
190	133	192	116
182	130	195	123
184	131	211	122
177	127	187	123
178	126	192	109
210	140	223	124

Here, x_1 is the sum of the widths (in micrometers) of the first joints of the first two tarsi ("feet"); x_2 is the corresponding measurement for the second joints.
Source: Data from Lubischew (1962).

Find the sample ML discriminant function for the two species.

Suppose that we have the pair of observations $(x_1, x_2)'$ for a new specimen but do not know to which of the two species the specimen belongs. Calculate the equation of the line (in the (x_1, x_2)-plane) such that, if $(x_1, x_2)'$ lies on one side, the specimen seems more likely to belong to *Ch. concinna*, whereas if $(x_1, x_2)'$ lies on the other side, the specimen seems more likely to belong to *Ch. heikertingeri*. Allocate a new observation (190, 215) to one of the two species. Plot these data together with the line and comment.

12.4.4 Show that the boundary hyperplane for the ML allocation rule between $N_p(\boldsymbol{\mu}_1, \boldsymbol{\Sigma})$ and $N_p(\boldsymbol{\mu}_2, \boldsymbol{\Sigma})$, $\boldsymbol{\Sigma} > 0$, is orthogonal to $\boldsymbol{\delta} = \boldsymbol{\mu}_1 - \boldsymbol{\mu}_2$ if and only if $\boldsymbol{\delta}$ is an eigenvector of $\boldsymbol{\Sigma}$.

12.4.5 (Bartlett, 1965) (a) Consider populations $N_p(\boldsymbol{\mu}_1, \boldsymbol{\Sigma})$ and $N_p(\boldsymbol{\mu}_2, \boldsymbol{\Sigma})$, and let $\boldsymbol{\delta} = \boldsymbol{\mu}_1 - \boldsymbol{\mu}_2$. In the case of biological data, it is sometimes found that the correlation between each pair of variables is approximately the same, so that scaling each variable to have unit variance, we can write $\boldsymbol{\Sigma} = \boldsymbol{E}$, where $\boldsymbol{E} = (1-\rho)\boldsymbol{I} + \rho\boldsymbol{1}\boldsymbol{1}'$ is the equicorrelation matrix. Using (A.35), show that the ML discriminant function is proportional to

$$h(\boldsymbol{x}) = \sum_{i=1}^{p} \delta_i x_i - p\rho\{1 + (p-1)\rho\}^{-1}\bar{\delta} \sum_{i=1}^{p} x_i + \text{const.},$$

where $\bar{\delta} = p^{-1}\sum \delta_i$. Calculate the corresponding allocation regions for the ML rule. Discuss the cases ρ small and ρ near 1. (This discriminant function is very useful in practice, even when the correlations are not exactly equal, as it may help with interpretation.)

(b) Write $h(\boldsymbol{x}) = h_1(\boldsymbol{x}) + h_2(\boldsymbol{x}) + \text{const.}$, where

$$h_1(\boldsymbol{x}) = p\bar{\delta}\{p^{-1} - \rho[1 + (p-1)\rho]^{-1}\} \sum x_i$$

is proportional to the first principal component and

$$h_2(\boldsymbol{x}) = \sum (\delta_i - \bar{\delta})x_i$$

lies in the isotropic $(p-1)$-dimensional subspace of principal components corresponding to the smaller eigenvalue of $\boldsymbol{\Sigma}$. Interpret $h_1(\boldsymbol{x})$ and $h_2(\boldsymbol{x})$ as *size* and *shape* factors, in the sense of Section 9.6.

12.4.6 (Bartlett, 1965) The following problem involves two multivariate normal populations with the *same* means but *different* covariance matrices. In discriminating between monozygotic and dizygotic twins of like sex on the basis of simple physical measurements such as weight and height, the observations recorded are the differences $x_1, \ldots, x_p$ between the corresponding measurements on each set of twins. As either twin might have been measured first, the expected mean differences are automatically zero. Let the covariance matrices for the two types of twins be denoted $\boldsymbol{\Sigma}_1$ and $\boldsymbol{\Sigma}_2$, and assume for simplicity that

$$\boldsymbol{\Sigma}_i = \sigma_i^2\{(1-\rho)\boldsymbol{I} + \rho\boldsymbol{1}\boldsymbol{1}'\}, \qquad i = 1, 2.$$

Under the assumption of multivariate normality, show that the ML discriminant function is proportional to

$$z_1 - \rho\{1 + (p-1)\rho\}^{-1} z_2 + \text{const.},$$

where $z_1 = x_1^2 + \cdots + x_p^2$ and $z_2 = (x_1 + \cdots + x_p)^2$. How would the boundary between the allocation regions be determined, so that the two types of misclassification have equal probability?

12.4.7 Let $h_{12}(\boldsymbol{x}), h_{13}(\boldsymbol{x})$, and $h_{23}(\boldsymbol{x})$ be the three sample discriminant functions used to discriminate between three multivariate normal populations with the same covariance matrix, as in Example 12.4.5. Show that unless $\overline{\boldsymbol{x}}_1, \overline{\boldsymbol{x}}_2$, and $\overline{\boldsymbol{x}}_3$ are collinear, the solutions to the three equations,

$$h_{12}(\boldsymbol{x}) = 0, \qquad h_{13}(\boldsymbol{x}) = 0, \qquad h_{23}(\boldsymbol{x}) = 0,$$

have a point in common. Draw a picture to illustrate what happens if $\overline{\boldsymbol{x}}_1, \overline{\boldsymbol{x}}_2$, and $\overline{\boldsymbol{x}}_3$ are collinear.

12.4.8 Consider two populations, which are distributed as $N_p(\boldsymbol{\mu}_1, \boldsymbol{\Sigma}_1)$ and $N_p(\boldsymbol{\mu}_2, \boldsymbol{\Sigma}_2)$. If $\boldsymbol{\Sigma}_1 \propto \boldsymbol{\Sigma}_2$, then show that the boundary defined by $L_1(\boldsymbol{x}) = L_2(\boldsymbol{x})$ is an ellipsoid and describe its eigenvalues and eigenvectors in terms of the eigenvalues and eigenvectors of $\boldsymbol{\Sigma}_1$ and $\boldsymbol{\Sigma}_2$.

12.4.9 Consider two bivariate normal populations $N_2(\boldsymbol{\mu}, \boldsymbol{\Sigma}_1)$ and $N_2(\boldsymbol{\mu}, \boldsymbol{\Sigma}_2)$, where we suppose that the correlation between the two variables is 0 in each population; that is,

$$\boldsymbol{\Sigma}_1 = \text{diag}(\sigma_1^2, \sigma_2^2), \qquad \boldsymbol{\Sigma}_2 = \text{diag}(\tau_1^2, \tau_2^2).$$

If $\sigma_1^2 > \sigma_2^2$ and $\tau_1^2 > \tau_2^2$, show that the boundary of the ML allocation rule is an ellipse and find its equation.

12.4.10 Why is the discriminant function $h(\boldsymbol{x})$ used in Example 12.4.1 to discriminate between two species of iris different from the discriminant function $h_{12}(\boldsymbol{x})$, which is used in Example 12.4.3 to help discriminate between three species of iris?

12.4.11 It might be thought that a linear combination of two variables would provide a better discriminator if they were correlated than if they were uncorrelated. However, this is not necessarily so, as is seen in Cochran (1962). Let Π_1 and Π_2 be two bivariate normal populations. Suppose that Π_1 is $N_2(\boldsymbol{0}, \boldsymbol{\Sigma})$ and Π_2 is $N_2(\boldsymbol{\mu}, \boldsymbol{\Sigma})$, where $\boldsymbol{\mu} = (\mu_1, \mu_2)'$ and

$$\boldsymbol{\Sigma} = \begin{bmatrix} 1 & \rho \\ \rho & 1 \end{bmatrix}.$$

Show that the Mahalanobis distance between Π_1 and Π_2 is given by

$$\Delta^2 = \boldsymbol{\mu}' \, \boldsymbol{\Sigma} \boldsymbol{\mu} = (\mu_1^2 + \mu_2^2 - 2\rho\mu_1\mu_2)/(1 - \rho^2).$$

Obtain an expression for the Mahalanobis distance, say Δ_0^2, in the case that the variables are uncorrelated.

The correlation will improve discrimination (i.e. reduce the probability of misclassification) if and only if $\Delta^2 > \Delta_0^2$. Show that this happens if and only if

$$\rho\left\{(1+f^2)\rho - 2f\right\} > 0, \qquad \text{where } f = \mu_2/\mu_1,$$

i.e. discrimination is improved unless ρ and f have the same sign and $|\rho| \le 2|f|/\sqrt{1+f^2}$, or ρ and f have opposite signs and $|\rho| \ge 2|f|/\sqrt{1+f^2}$. Sketch a contour plot of the densities of $\Pi_i, i = 1, 2$ to illustrate this result using cases in which $f = \pm 1$.

12.5.1 With a suitable partitioning of $\boldsymbol{\delta}$ and $\boldsymbol{\Sigma}$, let $\boldsymbol{\delta}_{2.1} = \boldsymbol{\delta}_2 - \boldsymbol{\Sigma}_{21}\boldsymbol{\Sigma}_{11}^{-1}\boldsymbol{\delta}_1$ and $\boldsymbol{\alpha} = \boldsymbol{\Sigma}^{-1}\boldsymbol{\delta}$. Show that $\boldsymbol{\delta}_{2.1} = \mathbf{0}$ if and only if $\boldsymbol{\alpha}_2 = \mathbf{0}$ if and only if $\boldsymbol{\delta}'\boldsymbol{\Sigma}^{-1}\boldsymbol{\delta} = \boldsymbol{\delta}_1'\boldsymbol{\Sigma}_{11}^{-1}\boldsymbol{\delta}_1$. (Hint: Using (A.31) and (A.29), show that $\boldsymbol{\delta}'\boldsymbol{\Sigma}^{-1}\boldsymbol{\delta} = \boldsymbol{\delta}_1\boldsymbol{\Sigma}_{11}^{-1}\boldsymbol{\delta}_1 + \boldsymbol{\delta}_{2.1}'\boldsymbol{\Sigma}^{22}\boldsymbol{\delta}_{2.1}$.)

12.5.2 Show that when $k = p - 1$, Eq. (12.18) can be expressed in the form (12.19) (with $i = p$). (Hint: Partition $\boldsymbol{W}^{-1}$, so that

$$\boldsymbol{W}^{22} = (\boldsymbol{W}_{22} - \boldsymbol{W}_{21}\boldsymbol{W}_{11}^{-1}\boldsymbol{W}_{12})^{-1} = w^{pp}.$$

Using (A.25) and (A.22), show that

$$t^{pp} = \frac{|\boldsymbol{T}_{11}|}{|\boldsymbol{T}|} = \frac{|\boldsymbol{W}_{11}|(1 + c^2D_k^2/m)}{|\boldsymbol{W}|(1 + c^2D_p^2/m)} = w^{pp}\frac{1 + c^2D_k^2/m}{1 + c^2D_p^2/m}.$$

Using (A.31) and (A.29), show that

$$\mathbf{d}'\boldsymbol{W}^{-1}\mathbf{d} - \mathbf{d}_1'\boldsymbol{W}_{11}^{-1}\mathbf{d}_1 = \mathbf{d}'\boldsymbol{V}\mathbf{d},$$

where

$$\boldsymbol{V} = w^{pp}\begin{pmatrix}-\boldsymbol{\beta}\\ 1\end{pmatrix}(-\boldsymbol{\beta}', 1) \qquad \text{and} \qquad \boldsymbol{\beta} = \boldsymbol{W}_{11}^{-1}\boldsymbol{W}_{12}.$$

Finally, note that

$$a_p = m(\boldsymbol{W}^{21}\mathbf{d}_1 + \boldsymbol{W}^{22}\mathbf{d}_2) = m\, w^{pp}(\mathbf{d}_2 - \boldsymbol{W}_{21}\boldsymbol{W}_{11}^{-1}\mathbf{d}_1),$$

where $\mathbf{d}_2 = d_p$, and hence,

$$a_p^2 = m^2 w^{pp}\mathbf{d}_{2.1}\boldsymbol{W}^{22}\mathbf{d}_{2.1} = m\, w^{pp}(D_p^2 - D_k^2).)$$

12.5.3 A random sample of 49 old men participating in a study of aging were classified by psychiatric examination into one of two categories: senile or nonsenile (Morrison, 1976, pp. 138–139).

An independently administered adult intelligent test revealed large differences between the two groups in certain standard subsets of the test. The results for these sections of the test are given below.

The group means are as follows:

Subtest	Senile ($N_1 = 37$)	Nonsenile ($N_2 = 12$)
x_1 Information	12.57	8.75
x_2 Similarities	9.57	5.35
x_3 Arithmetic	11.49	8.50
x_4 Picture completion	7.97	4.75

The "within-group" covariance matrix $\boldsymbol{S}_u$ and its inverse are given by

$$\boldsymbol{S}_u = \begin{bmatrix} 11.2553 & 9.4042 & 7.1489 & 3.3830 \\ & 13.5318 & 7.3830 & 2.5532 \\ & & 11.5744 & 2.6170 \\ & & & 5.8085 \end{bmatrix},$$

$$\boldsymbol{S}_u^{-1} = \begin{bmatrix} 0.2591 & -0.1358 & -0.0588 & -0.0647 \\ & 0.1865 & -0.0383 & -0.0144 \\ & & 0.1510 & -0.0170 \\ & & & 0.2112 \end{bmatrix}$$

Calculate the linear discriminant function between the two groups based on the data and investigate the errors of misclassification. Do the subtests "information" and "arithmetic" provide additional discrimination once the other two subtests are taken into account?

12.6.1 In Section 12.4, when $g = 2$, show that the matrix $\boldsymbol{W}^{-1}\boldsymbol{B}$ has only one nonzero eigenvalue, which equals $\{n_1 n_2/n\}\boldsymbol{d}'\boldsymbol{W}\boldsymbol{d}$, and find the corresponding eigenvector.

12.6.2 Show that the following eigenvectors are equivalent (assuming $\boldsymbol{W}$ has rank p):
(a) the eigenvector corresponding to the largest eigenvalue of $\boldsymbol{W}^{-1}\boldsymbol{B}$;
(b) the eigenvector corresponding to the largest eigenvalue of $\boldsymbol{W}^{-1}\boldsymbol{T}$;
(c) the eigenvector corresponding to the smallest eigenvalue of $\boldsymbol{T}^{-1}\boldsymbol{W}$.

12.6.3 (a) When the number of groups $g = 3$, show that the allocation rule based on Fisher's linear discriminant function is different from the ML allocation rule, unless the sample means are collinear. (However, if the means are nearly collinear, the two rules will be very similar.)
(b) Calculate Fisher's discriminant function for the first two variables of the three species of iris and compare the allocation regions with the three given in Example 12.4.5. (The largest eigenvalue of $\boldsymbol{W}^{-1}\boldsymbol{B}$ is 4.17, with eigenvector $(1, -1.29)'$.)

12.6.4 Let $\boldsymbol{X}(n \times p)$ be a data matrix partitioned into g groups. Define a new dummy zero-one data matrix $\boldsymbol{Y}(n \times (g-1))$ by

$$y_{rj} = \begin{cases} 1 & \text{if } \boldsymbol{x}_r \text{ is in the } j \text{ th group,} \\ 0 & \text{otherwise,} \end{cases}$$

for $j = 1, \ldots, g-1$ and $r = 1, \ldots, n$. Let $\boldsymbol{S}$ denote the covariance matrix of $(\boldsymbol{X}, \boldsymbol{Y})$. Show that

$$n\boldsymbol{S}_{11} = \boldsymbol{T} = \text{"total" SSP matrix}$$

and

$$n\boldsymbol{S}_{12}\boldsymbol{S}_{22}^{-1}\boldsymbol{S}_{21} = \boldsymbol{B} = \text{"between-groups" SSP matrix.}$$

Hence, carry out a canonical correlation analysis between $\boldsymbol{X}$ and $\boldsymbol{Y}$ and deduce that the canonical correlation variables for $\boldsymbol{X}$ equal the canonical variates of Section 12.6. (Hint: the canonical correlation variables $\boldsymbol{a}_j'\boldsymbol{x}$ of $\boldsymbol{X}$ are given by $\boldsymbol{T}^{-1}\boldsymbol{B}\boldsymbol{a}_j = \lambda_j\boldsymbol{a}_j$ or equivalently by $\boldsymbol{W}^{-1}\boldsymbol{B}\boldsymbol{a}_j = \{\lambda_j/(1-\lambda_j)\}\boldsymbol{a}_j$.)

12.6.5 In Example 12.6.2, suppose that the $p = 32$ multinomial parameters $\alpha_1, \ldots, \alpha_p$ and $\beta_1, \ldots, \beta_p$ for Plato's works, *Republic* (Π_0), and *Laws* (Π_1), are known exactly. Suppose that the distribution of sentence ending for each of the other works also follows a multinomial distribution, with parameters

$$\gamma_i(\lambda) = \alpha_i^{1-\lambda}\beta_i^{\lambda} \Big/ \sum_{j=1}^{p} \alpha_j^{1-\lambda}\beta_j^{\lambda},$$

where λ is a parameter that is different for each of the five intermediate works.
(a) Show that this family of multinomial distributions varies from $\Pi_0(\lambda = 0)$ to $\Pi_1(\lambda = 1)$ and hence can be used to represent a gradual change in populations from Π_0 to Π_1.
(b) Show that $\phi(\lambda) = \sum \gamma_i(\lambda)s_i$, where $s_i = \log(\alpha_i/\beta_i)$ is a monotonic function of λ, $0 \le \lambda \le 1$. Thus, $\phi(\lambda)$ can be used instead of λ to parameterize this family of distributions.
(c) Suppose that for a particular work containing N sentences, there are x_i endings of type i. Show that the maximum-likelihood estimate (m.l.e.) of $\phi(\lambda)$ is given by

$$\bar{s} = N^{-1} \sum x_i s_i,$$

which is the same as the discriminant score of Example 12.6.2. Thus, the discriminant score estimates a function that gives a measure of the location of the work between Π_0 and Π_1.
(d) Show that the variance of $\bar{s}$ equals

$$V(\bar{s}) = N^{-1}\left\{\sum \gamma_i(\lambda)s_i^2 - \phi(\lambda)^2\right\},$$

and that an unbiased estimate of $V(\bar{s})$ is

$$\hat{V}(\bar{s}) = [N(N-1)]^{-1}\left\{\sum x_i s_i^2 - N\bar{s}^2\right\}.$$

(e) For two works of sizes N' and N'', let the corresponding mean scores and estimated variances be $\bar{s}', \bar{s}''$ and $\hat{V}(\bar{s}')$, $\hat{V}(\bar{s}'')$, respectively. Using the fact that for large N' and N'', $\bar{s}' - \bar{s}''$ will be approximately normally distributed with mean $\phi(\lambda') - \phi(\lambda'')$ and variance $\hat{V}(\bar{s}') + \hat{V}(\bar{s}'')$, show that an approximate significance test of chronological ordering of these two works is given by the statistic

$$\psi = (\bar{s}' - \bar{s}'')/\{\hat{V}(\bar{s}') + \hat{V}(\bar{s}'')\}^{1/2}.$$

If $|\psi|$ is significantly large for an $N(0,1)$ variate, then the observed ordering of the works is significant.

(f) For the two works *Critias* and *Timaeus*, we have $\bar{s}' = -0.0346$, $\bar{s}'' = -0.1170$, $\hat{V}(\bar{s}') = 0.003\,799$, and $\hat{V}(\bar{s}'') = 0.000\,721\,8$. Test the hypothesis $\lambda' = \lambda''$ and hence assess the significance of the ordering given by the discriminant scores.

12.7.1 Suppose that n data points are distributed randomly (uniformly) inside the unit sphere in $\mathbb{R}_d$. Consider the 1-Nearest-Neighbor Classifier of an observation at the origin. Show that the median distance from the origin to the closest data point is

$$\left(1 - 2^{-1/n}\right)^{1/d}.$$

12.7.2 For one-dimensional data $x_1, \ldots, x_n$, consider the kernel density estimate

$$\hat{f}(x) = \frac{1}{nh}\sum_{r=1}^{n} K\left(\frac{x - x_r}{h}\right).$$

By considering a random variable $X \sim f(x)$, show that

$$\mathrm{E}\{\hat{f}(x)\} = \frac{1}{h}\int K\left(\frac{x-y}{h}\right) f(y)\,dy$$

and

$$\mathrm{var}\{\hat{f}(x)\} = \frac{1}{nh^2}\int K\left(\frac{x-y}{h}\right)^2 f(y)\,dy - \left\{\frac{1}{h}\int K\left(\frac{x-y}{h}\right) f(y)\,dy\right\}^2.$$

Making a suitable change of variable, followed by a Taylor series expansion, show that (as $n \to \infty$ and $h \to 0$ with $nh \to \infty$), asymptotically

$$\mathrm{bias}(x) = \frac{h^2}{2} f''(x) \int x^2 K(x)\,dx,$$

$$\mathrm{var}(x) = \frac{f(x)}{nh}\int K^2(x)\,dx.$$

Hence, find an expression for the optimal value of h to minimize the integrated mean squared error (i.e. $\int \{\mathrm{bias}^2(x) + \mathrm{var}(x)\}\,dx$).

Consider the extent to which these calculations apply to the problem of discrimination (as distinct from density estimation). That is, consider the importance of bias and variance (which measure vertical discrepancies) when the goal is to accurately obtain allocation regions (in which discrepancies will be horizontal).

12.7.3 Suppose that $x_{j1}, \ldots, x_{jn_j} \sim f_j(x)$ for $j = 1, 2$. If the two densities are estimated using kernel density estimation in which the kernel has infinite support, show that the ML rule, which determines if $\hat{f}_{h_1}(x) > \hat{f}_{h_2}(x)$, is equivalent to the 1-nearest-neighbor classifier when $h_1 = h_2 \to 0$.

12.8.1 In Example 12.8.1, it was seen that the same splitting variable/value was selected for some of the criteria considered. Is this always true? In this question, we examine the issue in the case of two classes and a single variable, start from the contingency table in which the data has been split to the left (L) and right (R) according to whether $x_1 < c$, say:

	Class		
	1	2	
Condition TRUE	n_{L1}	n_{L2}	n_L
Condition FALSE	n_{R1}	n_{R2}	n_R
	n_1	n_2	N

Obtain the five splitting criteria in terms of the above entries and determine which are equivalent (that is, which are linear functions of another).

12.8.2 Use the pessimistic pruning method (with three possible values: $\alpha = 0.5, 0.25, 0.1$) to determine whether the following split is worthwhile in a decision tree:

	Class		
	A	B	C
Condition TRUE	5	1	0
Condition FALSE	0	1	2

Comment on the way your answers depend on α.

12.8.3 Suppose that data are drawn from two populations with equal priors. The first has a standard exponential distribution ($f_1(x) = e^{-x}$, for $x \geq 0$, and 0 otherwise), and the second represents an exponential distribution with a shift and sign change, with density

$$f_2(x) = \begin{cases} e^{-(4-x)} & \text{if} \quad -\infty < x \leq 4, \\ 0 & \text{otherwise.} \end{cases}$$

Consider the two trees shown below – the left branch is taken if the condition is true. Note that both trees have three splits, but the right tree has depth 2. Work out the expected misclassification rate for each tree. By considering the expected number of splits, or otherwise, say which tree you prefer. Why might the expected number of splits be of interest, from a computational perspective?

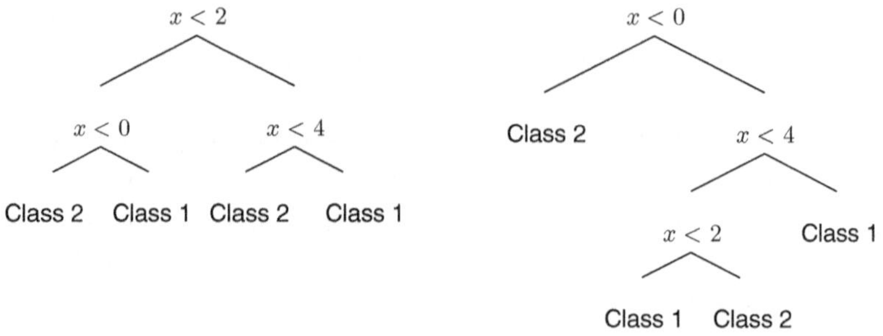

12.9.1 Explain, with reasons, whether the boundary separating the predicted classes in logistic regression is linear or nonlinear.

12.9.2 On 28 January 1986, there was a NASA Space Shuttle disaster when the *Challenger* broke apart about one minute after take-off. The investigation determined that the cause of the accident was due to the failure of an *O*-ring seal in the motor. Before the launch it was already suspected that such failures were more likely to occur at low temperatures, but a flawed scatterplot was inconclusive at the time.

Data are available[2] from the Presidential Commission on the Space Shuttle Challenger Accident (1986, p. 146): they provide the number of *O*-ring failures and the temperature occurring in 23 previous flights. An extraction of these data is given by:

Failures		0	1	0	0	0	0	0	0	1	1	1
Temp		18.9	21.1	20.6	20.0	19.4	22.2	22.8	21.1	13.9	17.2	21.1
Failures	0	0	2	0	0	0	0	0	0	2	0	1
Temp	25.6	19.4	11.7	19.4	23.9	21.1	27.2	24.4	26.1	23.9	24.4	14.4

Plot the variable Temp vs. Failures (the number of *O*-ring failures) to see if there is any relationship. Replot the same data but omitting those cases in which there were no failures (this was what was used in making the original decision as it was thought that these data were not relevant) – and see how/if your conclusions change.

Now, plot the temperature against the binary variable z ($z = 1$ if there were any failures, and $z = 0$ if there were no *O*-ring failures). Fit a logistic regression (i.e. $z \sim$ Temp) to these variables and use it to show that the predicted probability of a failure for a temperature of –0.6 °C (as it was when *Challenger* launched) is 0.9996. Comment on this prediction and its validity.

2 https://raw.githubusercontent.com/egarpor/handy/master/datasets/challenger.txt

12.10.1 Consider a two-layer network of the form

$$h_k(\boldsymbol{x}) = \sigma\left(\sum_{j=1}^{M} w_{kj}^{(2)} \zeta\left(\sum_{i=0}^{p} w_{ji}^{(1)} x_i\right) + w_{k0}^{(2)}\right)$$

in which the nonlinear activation function is given by

$$\sigma(a) = \frac{1}{1+\exp(-a)}.$$

By comparing $\sigma(a)$ with $\tanh(a)$, show that there exists an equivalent network, which computes exactly the same function (12.40) but with hidden activation function given by

$$\tanh(a) = \frac{\exp(a) - \exp(-a)}{\exp(a) + \exp(-a)}.$$

13

Multivariate Analysis of Variance

13.1 Introduction

When there is more than one variable measured per plot in the design of an experiment, the design is analyzed by **m**ultivariate **a**nalysis **o**f **v**ariance techniques (MANOVA techniques for short). Thus, we have the direct multivariate extension of every univariate design, but we confine our main attention to the multivariate one-way classification, which is an extension of the univariate one-way classification.

13.2 Formulation of Multivariate One-way Classification

First, consider a formulation resulting from agricultural experiments. Let there be k treatments that are assigned in a completely random order to some agricultural land. Suppose that there are n_j plots receiving the jth treatment, $j = 1, \dots, k$. Let us assume that $\boldsymbol{x}_{ij}$ is the $(p \times 1)$ yield vector of the ith plot receiving the jth treatment.

In general terminology, the plots are *experimental designs*, the treatments are *conditions*, and the yield is the *response* or *outcome*.

We assume that the $\boldsymbol{x}_{ij}$ are generated from the model

$$\boldsymbol{x}_{ij} = \boldsymbol{\mu} + \boldsymbol{\tau}_j + \boldsymbol{\varepsilon}_{ij}, \qquad i = 1, \dots, n, \quad j = 1, \dots, k, \tag{13.1}$$

where $\boldsymbol{\varepsilon}_{ij}$ = independent $N_p(\mathbf{0}, \boldsymbol{\Sigma})$, $\boldsymbol{\mu}$ = overall effect on the yield vector, and $\boldsymbol{\tau}_j$ = effect due to treatment j.

This design can be viewed as a multisample problem, i.e. we can regard $\boldsymbol{x}_{ij}$, $i = 1, \dots, n_j$, as a random sample from $N_p(\boldsymbol{\mu}_j, \boldsymbol{\Sigma})$, $j = 1, \dots, k$, where

$$\boldsymbol{\mu}_j = \boldsymbol{\mu} + \boldsymbol{\tau}_j, \qquad j = 1, \dots, k.$$

We usually wish to test the hypothesis

$$H_0 : \boldsymbol{\mu}_1 = \cdots = \boldsymbol{\mu}_k, \tag{13.2}$$

which is equivalent to testing that there is no difference between the treatments $\boldsymbol{\tau}_1, \dots, \boldsymbol{\tau}_k$. This problem has already been treated in Section 6.3.3.1 and Example 6.4.1. We now give some further details.

Multivariate Analysis, Second Edition. Kanti V. Mardia, John T. Kent, and Charles C. Taylor.

13.3 The Likelihood Ratio Principle

The Test: To test H_0 against H_1: the $\boldsymbol{\mu}_j$ are unrestricted, we have from Section 6.3.3.1 that the likelihood ratio (LR) criterion is

$$\Lambda = |\boldsymbol{W}|/|\boldsymbol{T}|, \tag{13.3}$$

where

$$\boldsymbol{W} = \sum_{j=1}^{k}\sum_{i=1}^{n_j}(\boldsymbol{x}_{ij} - \overline{\boldsymbol{x}}_j)(\boldsymbol{x}_{ij} - \overline{\boldsymbol{x}}_j)'$$

$$\boldsymbol{T} = \sum_{j=1}^{k}\sum_{i=1}^{n_j}(\boldsymbol{x}_{ij} - \overline{\boldsymbol{x}})(\boldsymbol{x}_{ij} - \overline{\boldsymbol{x}})',$$

with

$$\overline{\boldsymbol{x}}_j = \frac{1}{n_j}\sum_{i=1}^{n_j}\boldsymbol{x}_{ij}, \qquad \overline{\boldsymbol{x}} = \frac{1}{n}\sum_{j=1}^{k}\sum_{i=1}^{n_j}\boldsymbol{x}_{ij}, \qquad n = \sum_{j=1}^{k} n_j.$$

Recall that $\boldsymbol{W}$ and $\boldsymbol{T}$ are, respectively, the "within-samples" and the "total" sum of squares and products (SSP) matrices, respectively.

We can further show that

$$\boldsymbol{T} = \boldsymbol{W} + \boldsymbol{B}, \tag{13.4}$$

where

$$\boldsymbol{B} = \sum_{j=1}^{k} n_j(\overline{\boldsymbol{x}}_j - \overline{\boldsymbol{x}})(\overline{\boldsymbol{x}}_j - \overline{\boldsymbol{x}})'$$

is the "between-samples" SSP matrix. The identity (13.4) is the MANOVA identity.

Under H_0, it is shown in Section 6.3.3.1 that

$$\boldsymbol{W} \sim W_p(\boldsymbol{\Sigma}, n-k), \qquad \boldsymbol{B} \sim W_p(\boldsymbol{\Sigma}, k-1),$$

where $\boldsymbol{W}$ and $\boldsymbol{B}$ are independent. Further, if $n \geq p + k$,

$$\Lambda = |\boldsymbol{W}|/|\boldsymbol{W} + \boldsymbol{B}| \sim \Lambda(p, n-k, k-1),$$

where Λ is a Wilks' lambda variable. We reject H_0 for small values of Λ. H_0 can be tested by forming the MANOVA table as set out in Table 13.1.

For calculations of Λ, the following result is helpful:

$$\Lambda = \prod_{j=1}^{p}(1 + \lambda_j)^{-1},$$

where $\lambda_1, \ldots, \lambda_p$ are the eigenvalues of $\boldsymbol{W}^{-1}\boldsymbol{B}$. This result follows on noting that if $\lambda_1 \ldots, \lambda_p$ are the eigenvalues of $\boldsymbol{W}^{-1}\boldsymbol{B}$, then $(\lambda_i + 1), i = 1, \ldots, p$, are the eigenvalues of $\boldsymbol{W}^{-1}(\boldsymbol{B} + \boldsymbol{W})$. More details are given in Section 4.7. In particular, we need (4.41), namely

$$\frac{(m-p+1)\{1 - \sqrt{\Lambda(p, m, 2)}\}}{p\sqrt{\Lambda(p, m, 2)}} \sim F_{2p, 2(m-p+1)}. \tag{13.5}$$

Table 13.1 Multivariate one-way classification.

Source	d.f.	SSP matrix	Wilks' criterion
Between samples	$k-1$	$\boldsymbol{B} = \sum_{j=1}^{k} n_j(\overline{\boldsymbol{x}}_j - \overline{\boldsymbol{x}})(\overline{\boldsymbol{x}}_j - \overline{\boldsymbol{x}})'$	$\frac{\lvert\boldsymbol{W}\rvert}{\lvert\boldsymbol{W}+\boldsymbol{B}\rvert} \sim \Lambda(p, n-k, k-1)$
Within samples	$n-k$	$\boldsymbol{W} = \boldsymbol{T} - \boldsymbol{B}$	
Total	$n-1$	$\boldsymbol{T} = \sum_{j=1}^{k} \sum_{i=1}^{n_j}(\boldsymbol{x}_{ij} - \overline{\boldsymbol{x}})(\boldsymbol{x}_{ij} - \overline{\boldsymbol{x}})'$	

Example 13.3.1 Consider measurements published by Reeve (1941) on the skulls of $n = 13$ anteaters belonging to the subspecies *chapadensis*, deposited in the British Museum from $k = 3$ different localities. On each skull, $p = 3$ measurements were taken:

$$x_1 = \text{the basal length excluding the premaxilla,}$$
$$x_2 = \text{the occipitonasal length,}$$
$$x_3 = \text{the maximum nasal length.}$$

Table 13.2 shows the logarithms of the original measurements in millimeters and their means. We form the MANOVA table as given in Table 13.3.

We find that

$$\Lambda(3, 10, 2) = |\boldsymbol{W}|/|\boldsymbol{T}| = 0.6014.$$

Table 13.2 Logarithms of multiple measurements on anteater skulls at three localities.

	Minas Graes, Brazil			Matto Grosso, Brazil			Santa Cruz, Bolivia		
	x_1	x_2	x_3	x_1	x_2	x_3	x_1	x_2	x_3
	2.068	2.070	1.580	2.045	2.054	1.580	2.093	2.098	1.653
	2.068	2.074	1.602	2.076	2.088	1.602	2.100	2.106	1.623
	2.090	2.090	1.613	2.090	2.093	1.643	2.104	2.101	1.653
	2.097	2.093	1.613	2.111	2.114	1.643			
	2.117	2.125	1.663						
	2.140	2.146	1.681						
Means	2.097	2.100	1.625	2.080	2.087	1.617	2.099	2.102	1.643

Source: Data from Reeve (1941).

Table 13.3 Matrices in MANOVA table for Reeve's data $\times 10^7$.

Source	d.f.	a_{11}	a_{12}	a_{13}	a_{22}	a_{23}	a_{33}
Between	2	8060	6232	7498	4820	5859	11 844
Within	10	63 423	62 418	76 157	63 528	76 127	109 673
Total	12	71 483	68 651	83 655	68 348	81 986	121 517

We have from (13.5) that

$$\frac{8}{3}(1-\Lambda^{1/2})/\Lambda^{1/2} = 0.722 \sim F_{6,16},$$

which, from Appendix D, Table D.3, is not significant. Therefore, we conclude that there are no significant differences between localities. □

Mardia (1971) gives a modified test, which can be used for moderately nonnormal data.

13.4 Testing Fixed Contrasts

There are problems where the interest is not so much in testing the equality of the means but testing the significance of a fixed *contrast*, i.e.

$$H_0' : a_1\boldsymbol{\mu}_1 + \cdots + a_k\boldsymbol{\mu}_k = \mathbf{0},$$

where $a_1, a_2, \ldots, a_k$ are given constants such that $\sum a_i = 0$. Thus, we wish to test

$$H_0' : a_1\boldsymbol{\mu}_1 + \cdots + a_k\boldsymbol{\mu}_k = \mathbf{0}$$

against

$$H_1\text{: the } \boldsymbol{\mu}_j\text{, are unrestricted.}$$

We show that the LR test leads to the criterion

$$|\boldsymbol{W}|/|\boldsymbol{W}+\boldsymbol{C}| \sim \Lambda(p, n-k, 1), \tag{13.6}$$

where

$$\boldsymbol{C} = \left(\sum_{j=1}^{k} a_j\overline{\boldsymbol{x}}_j\right)\left(\sum_{j=1}^{k} a_j\overline{\boldsymbol{x}}_j'\right) \Bigg/ \left(\sum_{j=1}^{k}\frac{a_j^2}{n_j}\right).$$

The value of max log L under H_1 is the same as in Section 6.3.3.1. Under H_0, let $\boldsymbol{\lambda}$ be a vector of Lagrange multipliers. Then,

$$\log L = -\frac{n}{2}\log|\boldsymbol{\Sigma}| - \frac{1}{2}\sum_{j=1}^{k}\sum_{i=1}^{n_j}(\boldsymbol{x}_{ij}-\boldsymbol{\mu}_j)'\boldsymbol{\Sigma}^{-1}(\boldsymbol{x}_{ij}-\boldsymbol{\mu}_j) + \boldsymbol{\lambda}'(a_1\boldsymbol{\mu}_1 + \cdots + a_k\boldsymbol{\mu}_k). \tag{13.7}$$

On differentiating with respect to $\boldsymbol{\mu}_j$, it is found that

$$\boldsymbol{\mu}_j = \overline{\boldsymbol{x}}_j - n_j^{-1}a_j\boldsymbol{\Sigma}\boldsymbol{\lambda}, \tag{13.8}$$

which on using the constraint

$$\sum_{j=1}^{k} a_j\boldsymbol{\mu}_j = \mathbf{0}$$

leads to

$$\boldsymbol{\lambda} = \boldsymbol{\Sigma}^{-1}\sum_{j=1}^{k} a_j\overline{\boldsymbol{x}}_j \Bigg/ \sum_{j=1}^{k}\frac{a_j^2}{n_j}.$$

On substituting this result in the right-hand side of (13.8), we get $\hat{\boldsymbol{\mu}}_j$. Inserting the $\hat{\boldsymbol{\mu}}_j$ in (13.7) and maximizing over $\boldsymbol{\Sigma}$, it is found using Theorem 5.2.1 that

$$n\hat{\boldsymbol{\Sigma}} = \boldsymbol{W} + \boldsymbol{C}.$$

Hence, we obtain (13.6). Note that H_0' implies $\sum a_j\boldsymbol{\tau}_j = \mathbf{0}$ because $\sum a_j = 0$.

We can describe $\boldsymbol{C}$ as the SSP matrix due to the contrast $\sum a_j\boldsymbol{\mu}_j$. If there is a set of mutually orthogonal contrasts, such matrices will be additive (see Exercise 13.4.1).

13.5 Canonical Variables and A Test of Dimensionality

13.5.1 The Problem

We can regard $\boldsymbol{\mu}_i$ as the coordinates of a point in p-dimensional space for $i = 1, \dots, k$. When the null hypothesis H_0 of (13.2) is true, the vectors $\boldsymbol{\mu}_1, \dots, \boldsymbol{\mu}_k$ are identical. Thus, if r is the dimension of the hyperplane spanned by $\boldsymbol{\mu}_1, \dots, \boldsymbol{\mu}_k$, then H_0 is equivalent to $r = 0$. It can be seen that in any case

$$r \leq \min(p, k-1) = t, \quad \text{say.}$$

If H_0 is rejected, it is of importance to determine the actual dimensionality r, where $r \in \{0, 1, \dots, t\}$. If $r = t$, there is no restriction on the $\boldsymbol{\mu}$s, and $r < t$ occurs if and only if there are exactly $s = t - r$ linearly independent relationships between the k mean vectors.

Note that such a problem does not arise in the univariate case because for $p = 1$, we have $t = 1$, so either $r = 0$ (i.e. H_0) or $r = 1$ (i.e. the alternative).

Thus, we wish to test a new hypothesis

$$H_0 : \text{the } \boldsymbol{\mu}_i \text{ lie in an } r\text{-dimensional hyperplane} \tag{13.9}$$

against

$$H_1 : \text{the } \boldsymbol{\mu}_i \text{ are unrestricted} \tag{13.10}$$

where $i = 1, \dots, k$.

13.5.2 The LR Test ($\boldsymbol{\Sigma}$ Known)

Let us now assume that $\overline{\boldsymbol{x}}_i, i = 1, \dots, k$, are k sample means for samples of sizes $n_1, \dots, n_k$ from $N_p(\boldsymbol{\mu}_i, \boldsymbol{\Sigma})$, respectively. We assume that $\boldsymbol{\Sigma}$ is known. Since $\overline{\boldsymbol{x}}_i \sim N_p(\boldsymbol{\mu}_i, \boldsymbol{\Sigma}/n_i)$, the log-likelihood function is

$$l(\boldsymbol{\mu}_1, \dots, \boldsymbol{\mu}_k) = c - \frac{1}{2}\sum_{i=1}^{k} n_i(\overline{\boldsymbol{x}}_i - \boldsymbol{\mu}_i)'\boldsymbol{\Sigma}^{-1}(\overline{\boldsymbol{x}}_i - \boldsymbol{\mu}_i),$$

where c is a constant. Under H_1, $\hat{\boldsymbol{\mu}}_i = \overline{\boldsymbol{x}}_i$, and we have

$$\max_{H_1} l(\boldsymbol{\mu}_1, \dots, \boldsymbol{\mu}_k) = c. \tag{13.11}$$

Under H_0, the maximum of l is given in the following theorem:

Theorem 13.5.1 *We have*

$$\max_{H_0} l(\boldsymbol{\mu}_1, \dots, \boldsymbol{\mu}_k) = c - \frac{1}{2}(\gamma_{r+1} + \cdots \gamma_p), \tag{13.12}$$

where $\gamma_1 \geq \gamma_2 \geq \cdots \geq \gamma_p$ *are the roots of*

$$|\boldsymbol{B} - \gamma\boldsymbol{\Sigma}| = 0. \tag{13.13}$$

Proof: (Rao, 1973, p. 559) We need to show

$$\min_{H_0} \sum_{i=1}^{k} n_i(\overline{\boldsymbol{x}}_i - \boldsymbol{\mu}_i)'\boldsymbol{\Sigma}^{-1}(\overline{\boldsymbol{x}}_i - \boldsymbol{\mu}_i) = \gamma_{r+1} + \cdots + \gamma_p. \tag{13.14}$$

Let

$$\boldsymbol{y}_i = \boldsymbol{\Sigma}^{-1/2}\overline{\boldsymbol{x}}_i, \qquad \boldsymbol{v}_i = \boldsymbol{\Sigma}^{-1/2}\boldsymbol{\mu}_i. \tag{13.15}$$

Then, H_0 implies that the $\boldsymbol{v}_i$ are confined to an r-dimensional hyperplane. We can determine an r-dimensional hyperplane by a point $\boldsymbol{z}_0$, and r orthonormal direction vectors $\boldsymbol{z}_1, \dots, \boldsymbol{z}_r$, so that the points $\boldsymbol{v}_i$ on this plane can be represented as

$$\boldsymbol{v}_i = \boldsymbol{z}_0 + d_{i1}\boldsymbol{z}_i + \cdots + d_{ir}\boldsymbol{z}_r, \qquad i = 1, \dots, k, \tag{13.16}$$

where the d_{ij} are constants. Hence, (13.14) is equivalent to minimizing

$$\sum_{i=1}^{k} n_i\left(\boldsymbol{y}_i - \boldsymbol{z}_0 - \sum_{j=1}^{r} d_{ij}\boldsymbol{z}_j\right)'\left(\boldsymbol{y}_i - \boldsymbol{z}_0 - \sum_{j=1}^{r} d_{ij}\boldsymbol{z}_j\right), \tag{13.17}$$

subject to $\boldsymbol{z}_j'\boldsymbol{z}_j = 1, \boldsymbol{z}_j'\boldsymbol{z}_l = 0, j \neq l$, for $1 \leq j, l \leq r$. On differentiating with respect to d_{ij} for fixed $\boldsymbol{z}_0, \boldsymbol{z}_1, \dots, \boldsymbol{z}_r$, the minimum of (13.17) is found to occur when

$$d_{ij} = \boldsymbol{y}_i'\boldsymbol{z}_j - \boldsymbol{z}_0'\boldsymbol{z}_j. \tag{13.18}$$

Then, (13.17) reduces to

$$\sum_{i=1}^{k} n_i(\boldsymbol{y}_i - \boldsymbol{z}_0)'(\boldsymbol{y}_i - \boldsymbol{z}_0) - \sum_{i=1}^{k}\sum_{j=1}^{r} n_i[(\boldsymbol{y}_i - \boldsymbol{z}_0)'\boldsymbol{z}_j]^2, \tag{13.19}$$

which is to be minimized with respect to $\boldsymbol{z}_0, \boldsymbol{z}_1, \dots, \boldsymbol{z}_r$. Let $\overline{\boldsymbol{y}} = \sum n_i\boldsymbol{y}_i/n$. Then, Eq. (13.19) can be written as

$$\sum_{i=1}^{k} n_i(\boldsymbol{y}_i - \overline{\boldsymbol{y}})'(\boldsymbol{y}_i - \overline{\boldsymbol{y}}) - \sum_{i=1}^{k}\sum_{j=1}^{r} n_i[(\boldsymbol{y}_i - \overline{\boldsymbol{y}})'\boldsymbol{z}_j]^2 + n(\overline{\boldsymbol{y}} - \boldsymbol{z}_0)'\boldsymbol{F}(\overline{\boldsymbol{y}} - \boldsymbol{z}_0), \tag{13.20}$$

where

$$\boldsymbol{F} = \boldsymbol{I} - \sum_{j=1}^{r} \boldsymbol{z}_j\boldsymbol{z}_j'.$$

Since $\boldsymbol{F}$ is positive semidefinite, we can see that (13.20) is minimized for fixed $\boldsymbol{z}_1, \dots, \boldsymbol{z}_r$, by

$$\boldsymbol{z}_0 = \overline{\boldsymbol{y}}. \tag{13.21}$$

Also, minimizing (13.20) with respect to $\boldsymbol{z}_1, \dots, \boldsymbol{z}_r$ is equivalent to maximizing

$$\sum_{i=1}^{k}\sum_{j=1}^{r} n_i[(\boldsymbol{y}_i - \overline{\boldsymbol{y}})'\boldsymbol{z}_j]^2 = \sum_{j=1}^{r} \boldsymbol{z}_j'\boldsymbol{A}\boldsymbol{z}_j, \text{ say,} \tag{13.22}$$

where

$$\boldsymbol{A} = \sum n_i(\boldsymbol{y}_i - \overline{\boldsymbol{y}})(\boldsymbol{y}_i - \overline{\boldsymbol{y}})'.$$

Using (13.15), it is seen that

$$\boldsymbol{A} = \boldsymbol{\Sigma}^{-1/2}\boldsymbol{B}\boldsymbol{\Sigma}^{-1/2}.$$

Hence, $|\boldsymbol{A} - \gamma \boldsymbol{I}| = 0$ if and only if $|\boldsymbol{B} - \gamma \boldsymbol{\Sigma}| = 0$, so that $\boldsymbol{A}$ and $\boldsymbol{\Sigma}^{-1}\boldsymbol{B}$ have the same eigenvalues, namely $\gamma_1 \geq \cdots \geq \gamma_p$, given by (13.13).

Let $\boldsymbol{g}_1, \ldots, \boldsymbol{g}_p$ be the corresponding standardized eigenvectors of $\boldsymbol{A}$. Then, following the same argument as in Section 9.2.3 item (d), it is easy to see that

$$\max \sum_{j=1}^{r} \boldsymbol{z}_j' \boldsymbol{A} \boldsymbol{z}_j = \gamma_1 + \cdots + \gamma_r, \tag{13.23}$$

and that this maximum is attained for $\boldsymbol{z}_j = \boldsymbol{g}_j, j = 1, \ldots, r$. Further,

$$\sum n_i(\boldsymbol{y}_i - \bar{\boldsymbol{y}})'(\boldsymbol{y}_i - \bar{\boldsymbol{y}}) = \text{tr}\boldsymbol{A} = \gamma_1 + \cdots + \gamma_p. \tag{13.24}$$

Hence, from (13.19) to (13.24), we obtain

$$\text{left-hand side of (13.14)} = \gamma_{r+1} + \cdots + \gamma_p,$$

and the proof of (13.14) is complete. □

Now, the log-LR criterion λ is (13.12) minus (13.11), and thus,

$$-2 \log \lambda = \gamma_{r+1} + \cdots + \gamma_p \tag{13.25}$$

is the test criterion. The hypothesis H_0 is rejected for large values of (13.25).

13.5.3 Asymptotic Distribution of the Likelihood Ratio Criterion

For large values of $n_1, \ldots, n_k$, Eq. (13.25) is distributed as a chi-squared variable with f degrees of freedom (d.f.), where f is the number of restrictions on the pk parameters $\boldsymbol{\mu}_i, i = 1, \ldots, k$, under H_0. It can be seen that an r-dimensional hyperplane is specified by $r + 1$ points, say $\boldsymbol{\mu}_1, \ldots, \boldsymbol{\mu}_{r+1}$. The rest of the points on the hyperplane can be expressed as $a_1\boldsymbol{\mu}_1 + \cdots + a_{r+1}\boldsymbol{\mu}_{r+1}$, where $a_1 + \cdots + a_{r+1} = 1$, say. Thus, the number of unrestricted parameters is $p(r + 1) + (k - r - 1)r$. Consequently,

$$\gamma_{r+1} + \cdots + \gamma_p \sim \chi_f^2, \tag{13.26}$$

asymptotically, where

$$f = pk - p(r + 1) - (k - r - 1)r = (p - r)(k - r - 1).$$

13.5.4 The Estimated Plane

Write

$$\boldsymbol{Z}' = [\boldsymbol{z}_1, \ldots, \boldsymbol{z}_r], \qquad \boldsymbol{v}' = [\boldsymbol{v}_1, \ldots, \boldsymbol{v}_k], \qquad \boldsymbol{Y}' = [\boldsymbol{y}_1, \ldots, \boldsymbol{y}_k].$$

(Note that here a lower case letter $\boldsymbol{v}$ is used to represent a $(k \times p)$ matrix.) From (13.21), $\boldsymbol{z}_0 = \bar{\boldsymbol{y}} = \boldsymbol{\Sigma}^{-1/2}\bar{\boldsymbol{x}}$, so using (13.16) and (13.18), we can write

$$\hat{\boldsymbol{v}}' = \boldsymbol{Z}'\boldsymbol{Z}(\boldsymbol{Y}' - \bar{\boldsymbol{y}}\boldsymbol{1}') + \bar{\boldsymbol{y}}\boldsymbol{1}', \tag{13.27}$$

where $\hat{\boldsymbol{v}}$ is the maximum-likelihood estimate (m.l.e.) of $\boldsymbol{v}$ under H_0.

Let $\boldsymbol{q}_j = \boldsymbol{\Sigma}^{-1/2}\boldsymbol{z}_j$ denote the jth eigenvector of $\boldsymbol{\Sigma}^{-1}\boldsymbol{B} = \boldsymbol{\Sigma}^{-1/2}\boldsymbol{A}\boldsymbol{\Sigma}^{1/2}$, so that

$$\boldsymbol{B}\boldsymbol{q}_j = \gamma_j \boldsymbol{\Sigma}\boldsymbol{q}_j, \qquad \boldsymbol{q}_j'\boldsymbol{\Sigma}\boldsymbol{q}_j = 1, \qquad \boldsymbol{q}_j'\boldsymbol{\Sigma}\boldsymbol{q}_l = 0, \quad j \neq l, \tag{13.28}$$

and write $Q' = [q_1, \dots, q_r]$. Then, since

$$Y' = \Sigma^{-1/2}\overline{X}, \qquad Q' = \Sigma^{-1/2}Z', \qquad \mu' = \Sigma^{1/2}v',$$

we find from (13.27) that

$$\hat{\mu}' = \Sigma Q'Q(\overline{X}' - \overline{x}1') + \overline{x}1', \tag{13.29}$$

where $\hat{\mu}' = [\hat{\mu}_1, \dots, \hat{\mu}_k]$ is the m.l.e. of μ under H_0, and $\overline{X}' = [\overline{x}_1, \dots, \overline{x}_k]$.

Note that (13.29) is of the form $\hat{\mu}_i = D\overline{x}_i + (I - D)\overline{x}$ for $i = 1, \dots, k$, where D is of rank r; that is, $\hat{\mu}_i$ is an estimator of μ_i in an r-dimensional plane passing through $\overline{x}$. It is convenient to represent points x in this plane using the coordinates

$$x^* = (q_1'x, \dots, q_r'x)',$$

which are called *canonical coordinates*. The linear function $y_j = q_j'x$ is called the jth *canonical variable* (or variate), and the vector q_j is called the jth *canonical vector*.

Using $Q\Sigma Q' = I_r$ from (13.28), we observe from (13.29) that

$$Q\hat{\mu}' = Q\overline{X}'.$$

Thus, the canonical coordinates of $\hat{\mu}_i$ are given by

$$(q_1'\overline{x}_i, \dots, q_r'\overline{x}_i)' = \hat{\mu}_i^*, \quad \text{say},$$

for $i = 1, \dots, k$ and are called *canonical means*. Note that the $\hat{\mu}_i^*$ are r-dimensional vectors, whereas the μ_i are p-dimensional vectors.

Suppose for the moment that Q does not depend on $\overline{x}_i - \mu_i$. Since $\overline{x}_i \sim N_p(\mu_i, n_i^{-1}\Sigma)$ and $q_j'\Sigma q_l = 1$ for $j = l$ and $= 0$, for $j \neq l$, it can be seen that

$$\hat{\mu}_i^* \sim N_r(\mu_i^*, n_i^{-1}I),$$

where $\mu_i^* = (q_1'\mu_i, \dots, q_r'\mu_i)'$ is the canonical coordinate representation of the true mean μ_i (under H_0). Hence,

$$n_i(\hat{\mu}_i^* - \mu_i^*)'(\hat{\mu}_i^* - \mu_i^*) \sim \chi_r^2. \tag{13.30}$$

This formula can be used to construct confidence ellipsoids for the μ_i^*. Of course, since Q depends to some extent on $\overline{x}_i - \mu_i$, these confidence regions are only approximate.

13.5.5 The LR Test (Unknown Σ)

Consider the hypotheses H_0 versus H_1 in (13.9) and (13.10), where Σ is now unknown. Unfortunately, the likelihood ratio test (LRT) in this situation is quite complicated. However, an alternative test can be constructed if we replace Σ by the unbiased estimate $W/(n-k)$ and use the test of Section 13.5.2, treating Σ as known. It can be shown that for large n, this test is asymptotically equivalent to the LRT. (See Cox and Hinkley (1974, p. 361).) Then, the test given by (13.26) becomes in this context, asymptotically,

$$(n-k)(\lambda_{r+1} + \cdots + \lambda_p) \sim \chi_f^2, \qquad f = (p-r)(k-r-1), \tag{13.31}$$

where $\lambda_1, \dots, \lambda_p$ are now the roots of

$$|B - \lambda W| = 0. \tag{13.32}$$

Note that $(n-k)\lambda_j \approx \gamma_j$ of the last section because $\boldsymbol{W}/(n-k) \approx \boldsymbol{\Sigma}$. Estimates of the k group means and their r-dimensional plane can be constructed as in Section 13.5.4 and will be explored further in the following section.

Bartlett (1947) suggested the alternative statistic

$$D_r^2 = \left\{n - 1 - \frac{1}{2}(p+k)\right\} \sum_{j=r+1}^{p} \log(1+\lambda_j) \sim \chi_f^2, \tag{13.33}$$

which improves the chi-squared approximation in (13.31), and therefore, (13.33) will be preferred. Note that for large n, (13.33) reduces to (13.31).

We can now perform the tests of dimensionality $r = 0, 1, 2, \ldots, t$ sequentially. First, test $H_0 : r = 0$. If D_0^2 is significant, then test $H_0 : r = 1$, and so on. In general, if $D_0^2, D_1^2, \ldots, D_{r-1}^2$ are significant, but D_r^2 is not significant, then we may infer the dimensionality to be r.

Although we have assumed p roots of (13.32), there will be at most t nonzero roots.

13.5.6 The Estimated Plane (Unknown Σ)

Assume that r is the dimension of the plane spanned by the true group means. As in Section 13.5.4, we can look more deeply into the separation of the groups in this plane (now assuming $\boldsymbol{\Sigma}$ unknown). Let $\boldsymbol{l}_j$ be the eigenvector of $\boldsymbol{W}^{-1}\boldsymbol{B}$ corresponding to λ_j normalized by

$$\boldsymbol{l}_j'[\boldsymbol{W}/(n-k)]\boldsymbol{l}_j = 1. \tag{13.34}$$

Then, the $\boldsymbol{l}_1, \ldots, \boldsymbol{l}_r$ are canonical vectors analogous to $\boldsymbol{q}_1, \ldots, \boldsymbol{q}_r$ of Section 13.5.4 (where we assumed $\boldsymbol{\Sigma}$ was known). As in that section, these can be used to estimate the plane of the true group means and to represent points within this estimated plane.

The projection of a point $\boldsymbol{x}$ onto the estimated plane can be represented in terms of the r-dimensional *canonical coordinates* $(\boldsymbol{l}_1'\boldsymbol{x}, \ldots, \boldsymbol{l}_r'\boldsymbol{x})$. In particular, the *canonical means* of the k groups are $\boldsymbol{m}_i = (\boldsymbol{l}_1'\overline{\boldsymbol{x}}_i, \ldots, \boldsymbol{l}_r'\overline{\boldsymbol{x}}_i)'$ and can be used to study the differences between the groups.

The vector $\boldsymbol{l}_j$ is the *canonical vector* for the jth *canonical variable* $y_j = \boldsymbol{l}_j'\boldsymbol{x}$. Note that, from Section 12.6, the canonical variables are optimal discriminant functions; that is, for the data matrix $\boldsymbol{X}$, the jth canonical variable is that linear function which maximizes the between-group variance relative to within-group variance, subject to the constraint that it is uncorrelated with the preceding canonical variables. Consequently, for any value $r \leq t$, the canonical variables $y_1, \ldots, y_r$ are those linear functions that separate the k sample means as much as possible.

A graph of the canonical means in $r = 1$ or $r = 2$ dimensions can give a useful picture of the data. Note that on such a graph, the estimated variance matrix for the canonical coordinates of an observation coming from any of the k groups is the identity. (The canonical variables are uncorrelated with one another, and their estimated variance is normalized by (13.34) to equal 1.) Thus, the units of each axis on the graph represent one standard deviation for each canonical coordinate of an observation, and the canonical coordinates are uncorrelated with one another. Hence, we can get some idea of the strength of separation between the groups.

Also, we can estimate the accuracy of each of the canonical means on such a graph. From (13.30), a rough $100(1-\alpha)\%$ confidence region for the ith true canonical mean $\boldsymbol{\mu}_i^* = (\boldsymbol{l}_1'\boldsymbol{\mu}_i, \ldots, \boldsymbol{l}_r'\boldsymbol{\mu}_i)'$ is given by the disk of radius $n_i^{-1/2}\chi_r(1-\alpha)$ about the sample canonical mean $\boldsymbol{m}_i = \left(\boldsymbol{l}_1'\overline{\boldsymbol{x}}_i, \ldots, \boldsymbol{l}_r'\overline{\boldsymbol{x}}_i\right)'$, where $\chi_r^2(1-\alpha)$ is the upper α critical point of a χ_r^2 variable. Note that $(\lambda_1 + \cdots + \lambda_r)/(\text{tr}\, \boldsymbol{W}^{-1}\boldsymbol{B}) = (\lambda_1 + \cdots + \lambda_r)/(\lambda_1 + \cdots + \lambda_p)$ represents the proportion of the between-groups variation that is explained by the first r canonical variates.

Canonical analysis is the analogue for *grouped* data of principal component analysis for *ungrouped* data. Further, since an estimate of the inherent variability of the data is given by $\boldsymbol{W}$, canonical coordinates are invariant under changes of scale of the original variables. (This property is not shared by principal component analysis.)

Example 13.5.1 Consider the iris data of Table 1.3 with $p = 4$. Let x_1 = sepal length, x_2 = sepal width, x_3 = petal length, and x_4 = petal width. Here, the three samples are the three species *Iris setosa, Iris versicolor*, and *Iris virginica*. Further, $n_1 = n_2 = n_3 = 50$. It is found that the mean vectors are

$$\overline{\boldsymbol{x}}_1 = \begin{pmatrix} 5.006 \\ 3.428 \\ 1.462 \\ 0.246 \end{pmatrix}, \qquad \overline{\boldsymbol{x}}_2 = \begin{pmatrix} 5.936 \\ 2.770 \\ 4.260 \\ 1.326 \end{pmatrix}, \qquad \overline{\boldsymbol{x}}_3 = \begin{pmatrix} 6.588 \\ 2.974 \\ 5.552 \\ 2.026 \end{pmatrix}.$$

The SSP matrix between species is

$$\boldsymbol{B} = \begin{bmatrix} 63.21 & -19.95 & 165.25 & 71.28 \\ & 11.34 & -57.24 & -22.93 \\ & & 437.10 & 186.77 \\ & & & 80.41 \end{bmatrix}.$$

The SSP matrix within species is

$$\boldsymbol{W} = \begin{bmatrix} 38.96 & 13.63 & 24.62 & 5.65 \\ & 16.96 & 8.12 & 4.81 \\ & & 27.22 & 6.27 \\ & & & 6.16 \end{bmatrix}.$$

It is found that the eigenvalues of $\boldsymbol{W}^{-1}\boldsymbol{B}$ are

$$\lambda_1 = 32.1919, \qquad \lambda_2 = 0.2854,$$

and the corresponding eigenvectors normalized by (13.34) are

$$\begin{aligned} \boldsymbol{l}_1' &= (-0.83, -1.53, 2.20, -2.81), \\ \boldsymbol{l}_2' &= (0.02, 2.16, -0.93, 2.84). \end{aligned}$$

In this case, Eq. (13.33) becomes

$$D_0^2 = 546.12 \sim \chi_8^2,$$

which by Appendix D, Table D.1, is highly significant. Now,

$$D_1^2 = 36.53 \sim \chi_3^2,$$

which is again highly significant. Hence, since $t = 2$, both canonical variables are necessary, and the dimension cannot be reduced. In fact, the canonical means for *I. setosa* are

$$\boldsymbol{m}_1' = (\boldsymbol{l}_1'\overline{\boldsymbol{x}}_1, \boldsymbol{l}_2'\overline{\boldsymbol{x}}_1) = (-5.50, 6.88).$$

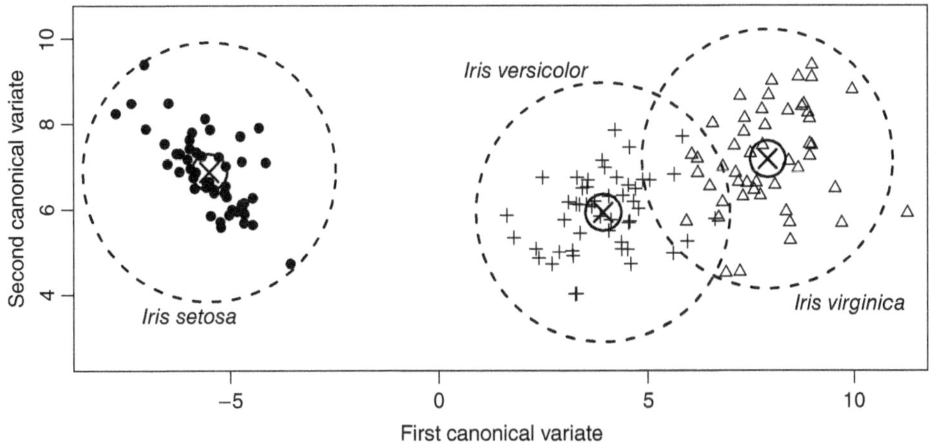

Figure 13.1 Canonical analysis of iris data, with approximate 99% confidence circles (continuous) for the canonical means, and 99% probability contours (dashed). Projected observations are shown with the labels: • (*I. setosa*), + (*I. versicolor*) and Δ (*I. virginica*).

Similarly, for the *I. versicolor* and *I. virginica* varieties, we have, respectively,

$$\boldsymbol{m}_2' = (3.93, 5.93), \qquad \boldsymbol{m}_3' = (7.89, 7.17).$$

Figure 13.1 shows a plot together with the approximate 99% confidence circles of radius $(\chi_2^2(0.99)/50)^{1/2} = (9.21/50)^{1/2} = 0.429$ for the three canonical means. It is quite clear from the figure that the three species are widely different, but *I. versicolor* and *I. virginica* are nearer than *I. setosa*. □

Example 13.5.2 (Delany and Healy, 1966). White-toothed shrews of the genus *Crocidura* occur in the Channel and Scilly Islands of the British Isles and the French mainland. From $p = 10$ measurements on each of $n = 399$ skulls obtained from the $k = 10$ localities, Tresco, Bryher, St Agnes, St Martin's, St Mary's, Sark, Jersey, Alderney, Guernsey, and Cap Gris Nez, the between and within SSP matrices are as given in Exercise 13.5.1. The sample sizes for the data from the localities are, respectively, 144, 16, 12, 7, 90, 25, 6, 26, 53, 20.

The first two canonical vectors normalized by (13.34) are found to be

$$\begin{aligned} \boldsymbol{l}_1' &= (-1.49, 0.38, 2.02, 8.56, -3.94, 0.89, 8.09, 8.80, -0.02, 0.47), \\ \boldsymbol{l}_2' &= (0.42, 1.82, -0.58, -9.77, -6.91, 1.99, 4.01, 7.12, -7.56, 0.34). \end{aligned}$$

The first two canonical variables account for 93.7% of the between-samples variation (see Exercise 13.5.1). The sample canonical means $\boldsymbol{m}_j' = (\boldsymbol{l}_1'\overline{\boldsymbol{x}}_j, \boldsymbol{l}_2'\overline{\boldsymbol{x}}_j), j = 1, \ldots, k$, (centered to have overall weighted mean $\boldsymbol{0}$) and the 99% confidence circles for the true canonical means are shown in Figure 13.2. This suggest that the three populations of *Crocidura russula* (Alderney, Guernsey, and Cap Gris Nez) show more differentiation than the five populations of *Crocidura suaveolens* (Tresco, Bryher, St Agnes, St Mary's, and St Martin's) in the Scilly Isles. The shrews from Sark are distinct from those in the five Scilly Isles and Jersey. Further discussion of this data is given in Exercise 13.5.1 and Chapter 14 on Cluster Analysis.

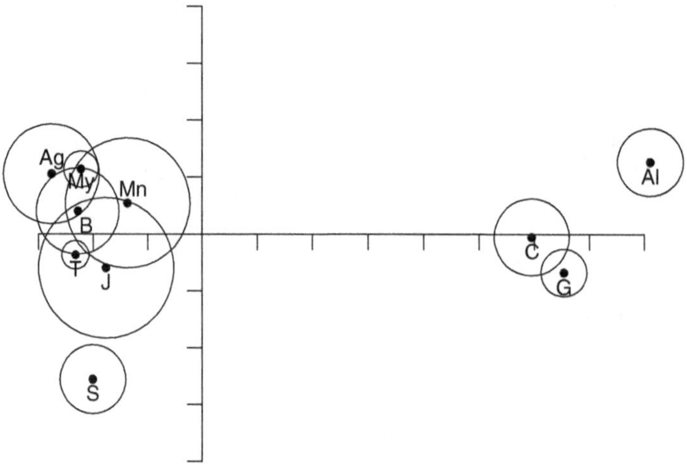

Figure 13.2 Plot of the 10 groups of shrews with respect to the first two canonical variates, with approximate 99% confidence circles for the canonical means. Ag, St Agnes; Al, Alderney; B, Bryher; C, Cap Griz; G, Guernsey; J, Jersey; Mn, St Martin's; My, St Mary's; S, Sark; T, Tresco.

It should be noted that the present analysis differs slightly from the one given by Delany and Healy (1966). In their analysis, the SSP matrix was calculated without using the n_i weights, i.e.

$$\boldsymbol{B} = \sum_{j=1}^{k} (\overline{\boldsymbol{x}}_j - \overline{\boldsymbol{x}})(\overline{\boldsymbol{x}}_j - \overline{\boldsymbol{x}})'.$$

Consequently, the canonical variables obtained differ from those given above, and their diagram corresponding to Figure 13.2 differs mainly in that the canonical means for the Jersey sample become very close to those of Sark. However, the conclusions are broadly the same. □

Example 13.5.3 We consider some image segmentation data in which $n = 2100$ images were drawn randomly from a database of seven outdoor images. The images were hand segmented to create a classification for every pixel. The data set was created by the Vision Group at the University of Massachusetts; it was used in the Statlog project (Michie et al., 1994), and can be obtained from the UCI repository (at the University of California, Irvine).[1] There are seven classes: brickface, sky, foliage, cement, window, path, and grass, with 300 instances per class. The original data set had 19 variables, but here, we omit the variable region-pixel count which is always 9, and those variables that are a linear combination of the others. This leaves 14 explanatory variables:

(i) the column of the center pixel of the region.
(ii) the row of the center pixel of the region.
(iii) the results of a line extraction algorithm that counts how many lines of length 5 (any orientation) with low contrast, less than or equal to 5, go through the region.
(iv) same as above but counts lines of high contrast, greater than 5.

1 https://archive.ics.uci.edu/ml/machine-learning-databases/image/

(v) the contrast of horizontally adjacent pixels in the region (mean). This attribute is used as a vertical edge detector.
(vi) the contrast of horizontally adjacent pixels in the region (standard deviation).
(vii) measures the contrast of vertically adjacent pixels (mean).
(viii) as above with standard deviation.
(ix) the average over the region of the Red value.
(x) the average over the region of the Blue value.
(xi) the average over the region of the Green value.
(xii) nonlinear transformation variable (value mean).
(xiii) nonlinear transformation variable (saturation mean).
(xiv) nonlinear transformation variable (hue mean).

Items (xii)–(xiv) were derived using an algorithm in Foley and Van Dam (1982). It is found that the mean vectors for the seven groups are

$$\bar{\boldsymbol{x}}_1 = \begin{pmatrix} 91.43 \\ 102.56 \\ 0.02 \\ 0.00 \\ 1.13 \\ 1.02 \\ 1.53 \\ 1.47 \\ 14.89 \\ 18.65 \\ 10.51 \\ 18.94 \\ 0.48 \\ -1.34 \end{pmatrix}, \quad \bar{\boldsymbol{x}}_2 = \begin{pmatrix} 129.00 \\ 98.51 \\ 0.02 \\ 0.00 \\ 2.99 \\ 6.06 \\ 2.55 \\ 5.14 \\ 39.96 \\ 55.62 \\ 39.42 \\ 55.62 \\ 0.31 \\ -2.03 \end{pmatrix}, \quad \bar{\boldsymbol{x}}_3 = \begin{pmatrix} 91.97 \\ 111.24 \\ 0.01 \\ 0.01 \\ 2.98 \\ 26.59 \\ 3.44 \\ 29.86 \\ 5.10 \\ 12.66 \\ 6.68 \\ 12.67 \\ 0.77 \\ -2.22 \end{pmatrix}, \quad \bar{\boldsymbol{x}}_4 = \begin{pmatrix} 130.41 \\ 205.29 \\ 0.03 \\ 0.00 \\ 1.59 \\ 1.39 \\ 2.11 \\ 1.96 \\ 12.55 \\ 13.98 \\ 20.28 \\ 20.33 \\ 0.41 \\ 2.23 \end{pmatrix},$$

$$\bar{\boldsymbol{x}}_5 = \begin{pmatrix} 135.85 \\ 186.43 \\ 0.01 \\ 0.01 \\ 2.44 \\ 2.19 \\ 4.95 \\ 13.19 \\ 43.54 \\ 60.50 \\ 42.73 \\ 60.51 \\ 0.30 \\ -2.07 \end{pmatrix}, \quad \bar{\boldsymbol{x}}_6 = \begin{pmatrix} 136.38 \\ 47.04 \\ 0.01 \\ 0.00 \\ 0.88 \\ 0.54 \\ 1.20 \\ 0.81 \\ 106.91 \\ 134.92 \\ 112.44 \\ 134.92 \\ 0.21 \\ -2.30 \end{pmatrix}, \quad \bar{\boldsymbol{x}}_7 = \begin{pmatrix} 159.54 \\ 113.30 \\ 0.01 \\ 0.00 \\ 1.22 \\ 2.18 \\ 1.08 \\ 2.88 \\ 6.70 \\ 13.11 \\ 6.86 \\ 13.14 \\ 0.51 \\ -1.82 \end{pmatrix}.$$

The SSP matrices between groups ($\boldsymbol{B}$) and within groups ($\boldsymbol{W}$) can be computed, and the eigenvalues of $\boldsymbol{W}^{-1}\boldsymbol{B}$ are found to be

$$\lambda_1 = 25.35, \quad \lambda_2 = 15.89, \quad \lambda_3 = 3.32, \quad \lambda_4 = 1.73, \quad \lambda_5 = 0.37, \quad \lambda_6 = 0.12.$$

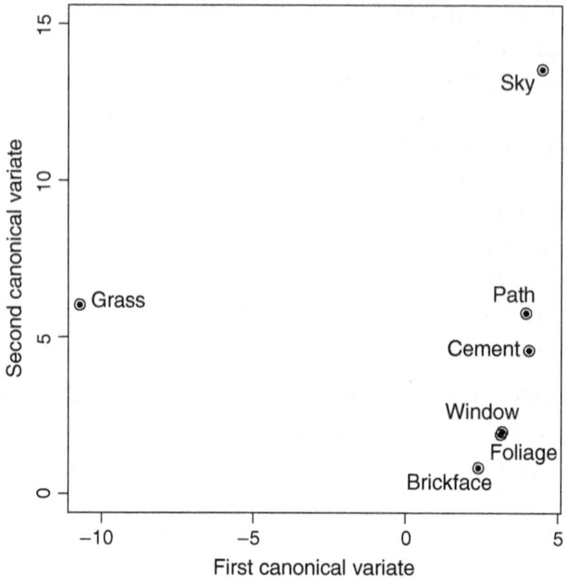

Figure 13.3 Canonical analysis of image segmentation data used in Example 13.5.3.

The first two eigenvalues and the first three eigenvalues account for 88.2% and 95.3% of the variation between samples, respectively. The first three eigenvectors normalized by (13.34) are

$$\begin{aligned}
\boldsymbol{l}_1' &= (-0.1, 0.1, 0.1, 0.1, 0.1, 0.0, 0.2, -0.1, -4.4, -21.3, 4.5, 21.1, 0.5, 3.7),\\
\boldsymbol{l}_2' &= (-0.1, -0.7, 0.1, 0.1, 0.2, 0.0, 0.2, 0.1, 10.0, -7.9, -14.1, 7.7, -0.1, -0.3),\\
\boldsymbol{l}_3' &= (-0.1, -1.4, 0.0, 0.1, 0.1, 0.1, -0.5, 0.2, -12.8, 12.1, 17.5, -17.3, 0.6, -0.1).
\end{aligned}$$

In this example, the values of $D_r^2, r = 0, 1, 2, \ldots, 5$ in (13.33) are

$$18786.6,\ 11953.8,\ 6049.6,\ 2991.6,\ 896.0,\ 244.9,$$

respectively. So, although the first three eigenvalues account for over 95% of the variation, a formal test suggests that six canonical variables are necessary.

Figure 13.3 shows a plot together with the approximate 99% confidence circles of radius $(\chi_2^2(0.99)/300)^{1/2} = (9.21/300)^{1/2} = 0.175$ for a two-dimensional projection of the seven canonical means. It is quite clear from the figure that the groups *window* and *foliage* are more similar, with *grass* being most dissimilar to all other groups. □

13.5.7 Profile Analysis

A related hypothesis to the above dimensionality hypothesis is the *profile hypothesis*, which assumes that the group means lie on a line, and that the direction of this line is given by the vector **1**; that is; the difference between each pair of group means is a vector whose components are all equal.

This hypothesis is of interest when all of the variables measure similar quantities, and it is expected that the difference between the groups will be reflected equally in all of the variables.

If the model is parameterized in terms of the k group means $\boldsymbol{\mu}_1, \dots, \boldsymbol{\mu}_k$, then the profile hypothesis can be written in the form of Section 7.3 as

$$\boldsymbol{C}_1 \boldsymbol{\mu} \boldsymbol{M}_1 = \boldsymbol{0},$$

where $\boldsymbol{\mu}' = (\boldsymbol{\mu}_1, \dots, \boldsymbol{\mu}_k)$, and $\boldsymbol{C}_1((k-1)\times k)$ and $\boldsymbol{M}_1((p \times (p-1))$ are given by

$$\boldsymbol{C}_1 = \begin{bmatrix} 1 & -1 & 0 & \cdots & 0 \\ 0 & 1 & -1 & \cdots & 0 \\ \cdot & \cdot & \cdot & & \cdot \\ \cdot & \cdot & \cdot & & \cdot \\ \cdot & \cdot & \cdot & & \cdot \\ 0 & 0 & 0 & \cdots & -1 \end{bmatrix} \quad \text{and} \quad \boldsymbol{M}_1 = \begin{bmatrix} 1 & 0 & \cdots & 0 \\ -1 & 1 & \cdots & 0 \\ 0 & -1 & \cdots & 0 \\ \cdot & \cdot & & \cdot \\ \cdot & \cdot & & \cdot \\ \cdot & \cdot & & \cdot \\ 0 & 0 & \cdots & -1 \end{bmatrix}.$$

This hypothesis is of interest because $\boldsymbol{M}_1$ is *not* equal to the identity; that is, a relationship between the variables is assumed as well as a relationship between the groups. For further details, see Morrison (1976, pp. 205–216).

13.6 The Union Intersection Approach

For the union intersection approach, we concentrate on the one-way classification given in (13.1). We wish to test the hypothesis $\boldsymbol{\tau}_1 = \cdots = \boldsymbol{\tau}_k$. This multivariate hypothesis is true if and only if all of the univariate hypotheses

$$H_{0\boldsymbol{a},\boldsymbol{c}} : \sum_{i=1}^{k} c_i \boldsymbol{a}' \boldsymbol{\tau}_i = 0$$

are true for all p-vectors $\boldsymbol{a}$ and all k-vector constants $\mathbf{c}$. Following Section 6.3.3.1, it is found that a suitable test criterion is λ_1, the largest eigenvalue of $\boldsymbol{W}^{-1}\boldsymbol{B}$. We reject H_0 for large values of λ_1.

This approach has the advantage that it allows us to construct simultaneous confidence regions for contrasts between the $\boldsymbol{\tau}_i$. The percentage points are given in Appendix D, Table D.7 of $\theta = \lambda_1/(1+\lambda_1)$ for $p = 2$ and selected critical values. We may use these to form $100(1-\alpha)\%$ simultaneous intervals, which include intervals of the form

$$\boldsymbol{a}'(\boldsymbol{\tau}_j - \boldsymbol{\tau}_k) \in \boldsymbol{a}'(\overline{\boldsymbol{x}}_j - \overline{\boldsymbol{x}}_k) \pm \left\{ \frac{\theta_\alpha}{1-\theta_\alpha} \boldsymbol{a}' \boldsymbol{W} \boldsymbol{a} \left(\frac{1}{n_j} + \frac{1}{n_k} \right) \right\}^{1/2},$$

where θ_α denotes the upper critical value of θ. In general, we denote the random variable corresponding to the distribution of the largest eigenvalue of $|\boldsymbol{B} - \theta(\boldsymbol{B}+\boldsymbol{W})| = 0$ by $\theta(p, \nu_1, \nu_2)$, where $\boldsymbol{B} \sim W_p(\boldsymbol{\Sigma}, \nu_2)$ independently of $\boldsymbol{W} \sim W_p(\boldsymbol{\Sigma}, \nu_1)$. Here, ν_1 is the error degrees of freedom, and ν_2 is the hypothesis degrees of freedom. More details are given in Section 4.7. This test is related to the LRT in as much as both are functions of the eigenvalues of $\boldsymbol{W}^{-1}\boldsymbol{B}$.

Example 13.6.1 In Reeve's data given in Example 13.3.1, we find that the nonzero eigenvalues of $\boldsymbol{W}^{-1}\boldsymbol{B}$ are

$$\lambda_1 = 0.3553, \qquad \lambda_2 = 0.2268.$$

(Note rank $(\boldsymbol{W}^{-1}\boldsymbol{B}) = 2$.) From Appendix D, Table D.7, the 1% critical value of $\theta(3, 10, 2) = \theta(2, 9, 3)$ is $\theta_{0.01} = 0.8074$. Since $\lambda_1/(1 + \lambda_1) = 0.2622 < \theta_{0.01}$, we accept the hypothesis that there are no differences, just as in Example 13.3.1.

Even if we look at differences between x_3 for groups 2 and 3 which look plausibly different, by forming a 99% CI with $\boldsymbol{a} = (0, 0, 1)', j = 2, k = 3$, we get an interval

$$\begin{aligned} \tau_{23} - \tau_{33} &\in 1.617 - 1.643 \pm \left\{ \frac{0.010970\theta_{0.01}}{1 - \theta_{0.01}} \left(\frac{1}{4} + \frac{1}{3} \right) \right\}^{1/2} \\ &= -0.026 \pm 0.164 \end{aligned}$$

which contains 0. Because the null hypothesis is accepted, it follows from the construction of the test that *this and all other* contrasts are nonsignificant. □

Example 13.6.2 We now examine Example 13.5.1 by this method. The greatest eigenvalue of $\boldsymbol{W}^{-1}\boldsymbol{B}$ was $\lambda_1 = 32.192$, and so $\theta = \lambda_1/(1 + \lambda_1) = 0.9699 \sim \theta(4, 147, 2) = \theta(2, 145, 4)$. From Appendix D, Table D.7, $\theta_{0.01} = 0.1098$, so we strongly reject the hypothesis of equal mean vectors.

Suppose that we are interested in sepal width (x_2) differences. Choosing $\boldsymbol{a}' = (0, 1, 0, 0)$, the $\pm$ term is always

$$\left\{ \frac{16.96\theta_{0.01}}{1 - \theta_{0.01}} \left(\frac{1}{50} + \frac{1}{50} \right) \right\}^{1/2} = 0.29 \quad \text{as} \quad n_1 = n_2 = n_3 = 50.$$

This leads to the following 99% simultaneous confidence intervals (SCIs) for sepal width differences:

I. virginica and *I. versicolor*	$\tau_{12} - \tau_{22} \in 2.97 - 2.77 \pm 0.29 = 0.20 \pm 0.29;$
I. setosa and *I. versicolor*	$\tau_{32} - \tau_{22} \in 3.43 - 2.77 \pm 0.29 = 0.66 \pm 0.29;$
I. setosa and *I. virginica*	$\tau_{32} - \tau_{12} \in 3.43 - 2.97 \pm 0.29 = 0.46 \pm 0.29.$

Thus, we accept the hypothesis of equal sepal width means at the 1% level for *I. virginica* and *I. versicolor*. It is easily verified that all other confidence intervals on all other variable differences do not contain zero, so apart from this single sepal width difference, the three species are different on all four variables. □

Note that SCIs are helpful because they provide detailed information about each variable and about the differences in means pairwise. However, there is an advantage in canonical analysis because it gives us a global picture of the group differences.

13.7 Two-way Classification

The ANOVA (analysis of variance) has a direct generalization for vector variables, leading to an analysis of a matrix of sums of squares and products, as in the single classification problem. Now, consider a two-way layout. Let us suppose that we have nrc independent observations generated by the model

$$\boldsymbol{x}_{ijk} = \boldsymbol{\mu} + \boldsymbol{\alpha}_i + \boldsymbol{\tau}_j + \boldsymbol{\eta}_{ij} + \boldsymbol{\varepsilon}_{ijk}, \qquad i = 1, \dots, r, \quad j = 1, \dots, c, \quad k = 1, \dots, n,$$

where $\boldsymbol{\alpha}_i$ is the ith row effect, $\boldsymbol{\tau}_j$ is the jth column effect, $\boldsymbol{\eta}_{ij}$ is the interaction effect between the ith row and the jth column, and $\boldsymbol{\varepsilon}_{ijk}$ is the error term which is assumed to be independent $N_p(\mathbf{0}, \boldsymbol{\Sigma})$ for all i, j, k. We require that the number of observations in each (i, j)-cell should be the same, so that the total SSP matrix can be suitably decomposed. We are interested in testing the null hypothesis of equality of the $\boldsymbol{\alpha}_i$, equality of the $\boldsymbol{\tau}_j$, and equality of the $\boldsymbol{\eta}_{ij}$. We partition the SSP matrix in an exactly analogous manner to univariate analysis.

Let $\boldsymbol{T}, \boldsymbol{R}, \boldsymbol{C}$, and $\boldsymbol{E}$ be the total, rows, columns, and errors of SSP matrices, respectively. As in the univariate case, we can show that the following MANOVA identity holds, i.e.

$$\boldsymbol{T} = \boldsymbol{R} + \boldsymbol{C} + \boldsymbol{E}, \tag{13.35}$$

where

$$\begin{aligned}
\boldsymbol{T} &= \sum_{i=1}^{r}\sum_{j=1}^{c}\sum_{k=1}^{n}(\boldsymbol{x}_{ijk} - \overline{\boldsymbol{x}}_{\dots})(\boldsymbol{x}_{ijk} - \overline{\boldsymbol{x}}_{\dots})', \\
\boldsymbol{R} &= cn\sum_{i=1}^{r}(\overline{\boldsymbol{x}}_{i..} - \overline{\boldsymbol{x}}_{\dots}))(\overline{\boldsymbol{x}}_{i..} - \overline{\boldsymbol{x}}_{\dots})', \\
\boldsymbol{C} &= rn\sum_{j=1}^{c}(\overline{\boldsymbol{x}}_{.j.} - \overline{\boldsymbol{x}}_{\dots})(\overline{\boldsymbol{x}}_{.j.} - \overline{\boldsymbol{x}}_{\dots})', \\
\boldsymbol{E} &= \sum_{i=1}^{r}\sum_{j=1}^{c}\sum_{k=1}^{n}(\boldsymbol{x}_{ijk} - \overline{\boldsymbol{x}}_{i..} - \overline{\boldsymbol{x}}_{.j.} + \overline{\boldsymbol{x}}_{\dots})(\boldsymbol{x}_{ijk} - \overline{\boldsymbol{x}}_{i..} - \overline{\boldsymbol{x}}_{.j.} + \overline{\boldsymbol{x}}_{\dots})',
\end{aligned}$$

with

$$\overline{\boldsymbol{x}}_{i..} = \frac{1}{cn}\sum_{j=1}^{c}\sum_{k=1}^{n}\boldsymbol{x}_{ijk}, \qquad \overline{\boldsymbol{x}}_{.j.} = \frac{1}{rn}\sum_{i=1}^{r}\sum_{k=1}^{n}\boldsymbol{x}_{ijk}$$

and

$$\overline{\boldsymbol{x}}_{\dots} = \frac{1}{rcn}\sum_{i=1}^{r}\sum_{j=1}^{c}\sum_{k=1}^{n}\boldsymbol{x}_{ijk}.$$

We may further decompose $\boldsymbol{E}$ into

$$\boldsymbol{E} = \boldsymbol{I} + \boldsymbol{W}, \tag{13.36}$$

where

$$\boldsymbol{I} = n\sum_{i=1}^{r}\sum_{j=1}^{c}(\overline{\boldsymbol{x}}_{ij.} - \overline{\boldsymbol{x}}_{i..} - \overline{\boldsymbol{x}}_{.j.} + \overline{\boldsymbol{x}}_{\dots})(\overline{\boldsymbol{x}}_{ij.} - \overline{\boldsymbol{x}}_{i..} - \overline{\boldsymbol{x}}_{.j.} + \overline{\boldsymbol{x}}_{\dots})'$$

and

$$\boldsymbol{W} = \sum_{i=1}^{r}\sum_{j=1}^{c}\sum_{k=1}^{n}(\boldsymbol{x}_{ijk} - \overline{\boldsymbol{x}}_{ij.})(\boldsymbol{x}_{ijk} - \overline{\boldsymbol{x}}_{ij.})',$$

with

$$\overline{\boldsymbol{x}}_{ij.} = \frac{1}{n}\sum_{k=1}^{n}\boldsymbol{x}_{ijk}.$$

Here, $\boldsymbol{I}$ is the SSP matrix due to interaction, and $\boldsymbol{W}$ is the residual SSP matrix.

13.7.1 Tests for Interactions

Thus, from (13.35) and (13.36), we may partition $\boldsymbol{T}$ into

$$\boldsymbol{T} = \boldsymbol{R} + \boldsymbol{C} + \boldsymbol{I} + \boldsymbol{W}.$$

Clearly under the hypothesis H_0 of all $\boldsymbol{\alpha}_i, \boldsymbol{\tau}_j$, and $\boldsymbol{\eta}_{ij}$ being zero, $\boldsymbol{T}$ must have the $W_p(\boldsymbol{\Sigma}, rcn - 1)$ distribution.

Also, we can write

$$\boldsymbol{W} = \sum_{i=1}^{r}\sum_{j=1}^{c}\boldsymbol{A}_{ij}$$

say, and whether or not H_0 holds, the $\boldsymbol{A}_{ij}$ are independent and identically distributed (i.i.d.) $W_p(\boldsymbol{\Sigma}, n-1)$.

Thus, as in Section 13.3,

$$\boldsymbol{W} \sim W_p(\boldsymbol{\Sigma}, rc(n-1)).$$

In the same spirit as in univariate analysis, whether or not the $\boldsymbol{\alpha}$s and $\boldsymbol{\tau}$s vanish,

$$\boldsymbol{I} \sim W_p(\boldsymbol{\Sigma}, (r-1)(c-1)),$$

if the $\boldsymbol{\eta}_{ij}$ are equal.

It can be shown that the matrices $\boldsymbol{R}$, $\boldsymbol{C}$, $\boldsymbol{I}$, and $\boldsymbol{W}$ are distributed independently of one another, and further, we can show that the LR statistic for testing the quality of the interaction terms is

$$|\boldsymbol{W}|/|\boldsymbol{W} + \boldsymbol{I}| = |\boldsymbol{W}|/|\boldsymbol{E}| \sim \Lambda(p, \nu_1, \nu_2),$$

where

$$\nu_1 = rc(n-1), \qquad \nu_2 = (r-1)(c-1).$$

We reject the hypothesis of no interaction for low values of Λ. We may alternatively look at the largest eigenvalue θ of $\boldsymbol{I}(\boldsymbol{W} + \boldsymbol{I})^{-1}$ as in Section 13.6, which is distributed as $\theta(p, \nu_1, \nu_2)$.

Note that if $n = 1$, i.e. there is one observation per cell, then $\boldsymbol{W}$ has zero d.f., so we cannot make any test for the presence of interaction.

13.7.2 Tests for Main Effects

If the column effects vanish, then $\boldsymbol{C}$ can be shown to have the $W_p(\boldsymbol{\Sigma}, c-1)$ distribution. The LR statistic to test the equality of the $\boldsymbol{\tau}_j$ irrespective of the $\boldsymbol{\alpha}_i$ and $\boldsymbol{\eta}_{ij}$ can be shown to be

$$|\boldsymbol{W}|/|\boldsymbol{W} + \boldsymbol{C}| \sim \Lambda(p, \nu_1, \nu_2),$$

where

$$\nu_1 = rc(n-1), \qquad \nu_2 = c - 1. \tag{13.37}$$

Equality is again rejected for low values of Λ.

Alternatively, we can look at the largest eigenvalue θ of $\mathbf{C}(\boldsymbol{W} + \mathbf{C})^{-1}$, which is distributed as $\theta(p, \nu_1, \nu_2)$, where ν_1 and ν_2 are given by (13.37).

Similarly, to test for the equality of the row effects $\boldsymbol{\alpha}_i$, irrespective of the column effects $\boldsymbol{\tau}_j$, we replace $\boldsymbol{C}$ by $\boldsymbol{R}$ and interchange r and c in the above.

Note that if significant interactions are present, then it does not make much sense to test for row and column effects. One possibility in this situation is to make tests separately on each of the rc row and column categories.

We might alternatively decide to ignore interaction effects completely, either because we have tested for them and shown them to be nonsignificant, or because $n = 1$, or because for various reasons, we may have no $\boldsymbol{\eta}_{ij}$ term in our model. In such cases, we work on the error matrix $\boldsymbol{E}$ instead of $\boldsymbol{W}$, and our test statistic for column effects is

$$|\boldsymbol{E}|/|\boldsymbol{E}+\boldsymbol{C}| \sim \Lambda(p, \nu_1, \nu_2),$$

where

$$\nu_1 = rcn - r - c + 1, \qquad \nu_2 = c - 1. \tag{13.38}$$

We may alternatively look at the largest eigenvalue θ of $\boldsymbol{C}(\boldsymbol{E}+\boldsymbol{C})^{-1}$, which is distributed as $\theta(p, \nu_1, \nu_2)$, with the value of ν_1 and ν_2 given by (13.38).

Example 13.7.1 (Morrison, 1976, p. 190). We wish to compare the weight losses of male and female rats ($r = 2$ sexes) under $c = 3$ drugs, where $n = 4$ rats of each sex are assigned at random to each drug. Weight losses are observed for the first and second weeks ($p = 2$), and the data is given in Table 13.4. We wish to compare the effects of the drugs, the effect of sex, and whether there is any interaction.

We first test for interaction. We construct the MANOVA table (Table 13.5), using the totals in Table 13.4. From Table 13.5, we find that

$$|\boldsymbol{W}| = 4920.75, \qquad |\boldsymbol{W}+\boldsymbol{I}| = 6354.58.$$

Table 13.4 Weight losses (in grams) for the first and second weeks for rats of each sex under drugs A, B, and C.

	Drug			
Sex	**A**	**B**	**C**	**Row sums**
	(5,6)	(7,6)	(21,15)	(33,27)
	(5,4)	(7,7)	(14,11)	(26,22)
Male	(9,9)	(9,12)	(17,12)	(35,33)
	(7,6)	(6,8)	(12,10)	(25,24)
Column sums	(26,25)	(29,33)	(64,48)	(119,106)
	(7,10)	(10,13)	(16,12)	(33,35)
	(6,6)	(8,7)	(14,9)	(28,22)
Female	(9,7)	(7,6)	(14,8)	(30,21)
	(8,10)	(6,9)	(10,5)	(24,24)
Column sums	(30,33)	(31,35)	(54,34)	(115,102)
Treatment sums	(56,58)	(60,68)	(118,82)	
Grand total				(234,208)

Source: Morrison (1976)/McGraw-Hill.

Table 13.5 MANOVA table for the data in Table 13.4.

Source	d.f.	SSP matrix A		
		a_{11}	a_{12}	a_{22}
Sex ($\boldsymbol{R}$)	1	0.667	0.667	0.667
Drugs ($\boldsymbol{C}$)	2	301.0	97.5	36.333
Interaction ($\boldsymbol{I}$)	2	14.333	21.333	32.333
Residual ($\boldsymbol{W}$)	18	94.5	76.5	114.0
Total ($\boldsymbol{T}$)	23	410.5	196.0	183.333

Hence,

$$\Lambda = |\boldsymbol{W}|/|\boldsymbol{W}+\boldsymbol{I}| = 0.7744 \sim \Lambda(2,18,2).$$

From (13.5),

$$(17/2)(1-\sqrt{0.7744})/\sqrt{0.7744} = 1.16 \sim F_{4,34}.$$

This is clearly not significant, so we conclude that there are no interactions and proceed to test for main effects.

First, for drugs, $|\boldsymbol{W}+\boldsymbol{C}| = 29\ 180.83$. Therefore,

$$\Lambda = 4920.75/(29180.83) = 0.1686 \sim \Lambda(2,18,2).$$

From (13.5),

$$(17/2)(1-\sqrt{0.1686})/\sqrt{0.1686} = 12.20 \sim F_{4,34}.$$

This is significant at 0.1%, so we conclude that there are very highly significant differences between drugs.

Finally, for sex, $|\boldsymbol{W}+\boldsymbol{R}| = 4957.75$. Thus,

$$\Lambda = 4920.75/4957.75 = 0.9925 \sim \Lambda(2,18,1).$$

Again from (4.42), the observed value of $F_{2,34}$ is 0.06. This is not significant, so we conclude that there are no significant differences in weight loss between the sexes. □

13.7.3 Simultaneous Confidence Regions

We begin by requiring the interaction parameters $\boldsymbol{\eta}_{ij}$ in the two-way model to vanish, otherwise comparisons among the row and column treatments are meaningless. If so, then, as in Section 13.6, the $100(1-\alpha)\%$ SCIs for linear compounds of the differences of the i_1th and i_2th row effects are given by

$$\boldsymbol{a}'(\boldsymbol{\alpha}_{i_1}-\boldsymbol{\alpha}_{i_2}) \in \boldsymbol{a}'(\overline{\boldsymbol{x}}_{i_1\cdot\cdot}-\overline{\boldsymbol{x}}_{i_2\cdot\cdot}) \pm \left\{\frac{2\theta_\alpha}{cn(1-\theta_\alpha)}\boldsymbol{a}'\boldsymbol{W}\boldsymbol{a}\right\}^{1/2},$$

where θ_α is obtained from Appendix D, Table D.7, with $\nu_1 = rc(n-1)$, $\nu_2 = r-1$, and $\boldsymbol{a}$ is any vector.

Similarly, for the columns,

$$\boldsymbol{a}'(\boldsymbol{\tau}_{j_1} - \boldsymbol{\tau}_{j_2}) \in \boldsymbol{a}'(\overline{\boldsymbol{x}}_{.j_1.} - \overline{\boldsymbol{x}}_{.j_2.}) \pm \left\{ \frac{2\theta_\alpha}{rn(1-\theta_\alpha)} \boldsymbol{a}'\boldsymbol{W}\boldsymbol{a} \right\}^{1/2}, \tag{13.39}$$

where θ_α is again obtained from tables, with r and c interchanged in ν_1 and ν_2 above.

Example 13.7.2 (Morrison, 1976, p. 190). For the drug taking data from Table 13.5,

$$\boldsymbol{W} = \begin{bmatrix} 94.5 & 76.5 \\ 76.5 & 114.0 \end{bmatrix}.$$

From Appendix D, Table D.7, it is found for $p = 2, \nu_1 = 18, \nu_2 = 2$, and $\alpha = 0.01$, then $\theta_\alpha = 0.501$. Taking $\boldsymbol{a}' = (1, 0)$, for the first week, from (13.39), the 99% SCIs for the B–A, C–B, and C–A differences are

$$-4.36 \le \tau_{B1} - \tau_{A1} \le 5.36, \qquad 2.39 \le \tau_{C1} - \tau_{B1} \le 12.11$$

$$2.89 \le \tau_{C1} - \tau_{A1} \le 12.61,$$

where we have used the above value of $\boldsymbol{W}$ and drug means (totals) from Table 13.4. Hence, at the 1% significance level, the drugs A and B are not different with respect to their effects on weight during the first week of the trial. However, the effect of drug C is different from drugs A and B. □

13.7.4 Extension to Higher Designs

By following the same principle of partitioning the total SSP matrix, we are able by analogy with univariate work to analyze many more complex and higher order designs. We refer to Bock (1975) for further work.

Exercises and Complements

13.3.1 (a) Prove the multivariate analysis of variance identity (13.4) after writing

$$\boldsymbol{x}_{ij} - \overline{\boldsymbol{x}} = (\boldsymbol{x}_{ij} - \overline{\boldsymbol{x}}_j) + (\overline{\boldsymbol{x}}_j - \overline{\boldsymbol{x}}).$$

(b) Show from Section 6.3.3.1 that under H_0,

$$\boldsymbol{W} \sim W_p(\boldsymbol{\Sigma}, n-k), \qquad \boldsymbol{B} \sim W_p(\boldsymbol{\Sigma}, k-1).$$

Further, $\boldsymbol{W}$ and $\boldsymbol{B}$ are independent.

13.3.2 (See, for example, Kshirsagar (1972, p. 345)) Under $H_1 : \boldsymbol{\mu}_i \neq \boldsymbol{\mu}_j$ for some $i \neq j$, show that

(a) $\mathsf{E}(\boldsymbol{W}) = (n-k)\boldsymbol{\Sigma}$,

(b) $\mathsf{E}(\boldsymbol{B}) = (k-1)\boldsymbol{\Sigma} + \boldsymbol{\Delta}$,

where

$$\boldsymbol{\Delta} = \sum_{j=1}^{k} n_j(\boldsymbol{\mu}_j - \overline{\boldsymbol{\mu}})(\boldsymbol{\mu}_j - \overline{\boldsymbol{\mu}})', \qquad \overline{\boldsymbol{\mu}} = \frac{1}{n}\sum_{j=1}^{k} n_j\boldsymbol{\mu}_j.$$

13.3.3 The data considered in Example 13.3.1 is a subset of Reeve's data (Reeve, 1941). A fuller data set leads to the following information on skulls at six localities:

Locality	Subspecies	Sample size	Mean vector, $\bar{x}_i'$
Sta. Mata, Columbia	*instabilis*	21	(2.054, 2.066, 1.621)
Minas Geraes, Brazil	*chapadensis*	6	(2.097, 2.100, 1.625)
Matto Grosso, Brazil	*chapadensis*	9	(2.091, 2.095, 1.625)
Sta. Cruz, Bolivia	*chapadensis*	3	(2.099, 2.102, 1.643)
Panama	*chiriquensis*	4	(2.092, 2.110, 1.703)
Mexico	*mexicana*	5	(2.099, 2.107, 1.671)
Total		48	(2.077, 2.086, 1.636)

Show that the "between-groups" SSP matrix is

$$\boldsymbol{B} = \begin{bmatrix} 0.0200211 & 0.0174448 & 0.0130811 \\ & 0.0158517 & 0.0150665 \\ & & 0.0306818 \end{bmatrix}.$$

Assuming the "within-groups" SSP matrix to be

$$\boldsymbol{W} = \begin{bmatrix} 0.0136309 & 0.0127691 & 0.0164379 \\ & 0.0129227 & 0.0171355 \\ & & 0.0361519 \end{bmatrix},$$

show that

$$\lambda_1 = 2.4001, \qquad \lambda_2 = 0.9050, \qquad \lambda_3 = 0.0515.$$

Hence, show that there are differences between the mean vectors of the six groups.

13.3.4 Measurements were taken on the head lengths u_1, head breadths u_2, and weights u_3 of 140 schoolboys of almost the same age belonging to six different schools in an Indian city. The between-schools and total SSP matrices were

$$\boldsymbol{B} = \begin{bmatrix} 752.0 & 214.2 & 521.3 \\ & 151.3 & 401.2 \\ & & 1612.7 \end{bmatrix}$$

and

$$\boldsymbol{T} = \begin{bmatrix} 13561.3 & 1217.9 & 3192.5 \\ & 1650.9 & 4524.8 \\ & & 22622.3 \end{bmatrix}.$$

Show that $|\boldsymbol{T}| = 213629309846$. Obtain $\boldsymbol{W} = \boldsymbol{T} - \boldsymbol{B}$. Hence, or otherwise, show that $|\boldsymbol{W}| = 176005396257$. Consequently, $|\boldsymbol{W}|/|\boldsymbol{T}| = 0.82388$, which is distributed as $\Lambda(3, 134, 5)$. Show that Bartlett's approximation (4.43) gives $26.06 \sim \chi^2_{15}$. Hence, conclude at the 5% level of significance that there are differences between schools.

13.4.1 Let $\boldsymbol{A} = (a_{ij})$ be a known $(k \times r)$ matrix such that $\boldsymbol{A}'\boldsymbol{D}^{-1}\boldsymbol{A} = \boldsymbol{I}_r$, where $\boldsymbol{D} = \text{diag}(n_1, \ldots, n_k)$. Suppose that we wish to test $H_0 : \boldsymbol{A}'\boldsymbol{\mu} = \boldsymbol{0}$, $\boldsymbol{\mu}' = [\boldsymbol{\mu}_1, \ldots, \boldsymbol{\mu}_k]$ against H_1: the $\boldsymbol{\mu}_j$ are unrestricted. Following the method of Section 13.4 show, under H_0, that

$$\hat{\boldsymbol{\mu}}_j = \overline{\boldsymbol{x}}_j - \frac{1}{n_j}\sum_{i=1}^{r} a_{ji}\boldsymbol{\Sigma}\boldsymbol{\lambda}_i,$$

where

$$\boldsymbol{\Sigma}\boldsymbol{\lambda}_i = \sum_{j=1}^{k} a_{ji}\overline{\boldsymbol{x}}_j.$$

Hence, prove that

$$n\hat{\boldsymbol{\Sigma}} = \boldsymbol{W} + \sum_{i=1}^{r}\left(\sum_{j=1}^{k} a_{ji}\overline{\boldsymbol{x}}_j\right)\left(\sum_{j=1}^{k} a_{ji}\overline{\boldsymbol{x}}_j'\right)$$

and derive the LRT.

13.5.1 For the shrew data of Delany and Healy (1966) referred to in Example 13.5.2, we have, from a total sample of 399,

$$\boldsymbol{B} = \begin{bmatrix} 208.02 & 88.56 & 101.39 & 41.16 & 10.06 & 98.35 & 57.46 & 33.01 & 55.12 & 63.99 \\ & 39.25 & 42.14 & 17.20 & 4.16 & 41.90 & 24.35 & 14.16 & 22.66 & 27.32 \\ & & 50.26 & 20.23 & 4.98 & 47.75 & 28.04 & 16.04 & 27.54 & 31.14 \\ & & & 8.76 & 2.14 & 19.33 & 11.66 & 6.41 & 11.03 & 12.88 \\ & & & & 0.78 & 4.73 & 2.73 & 1.45 & 3.05 & 3.13 \\ & & & & & 47.47 & 27.97 & 15.76 & 25.92 & 30.10 \\ & & & & & & 17.48 & 9.47 & 15.11 & 17.68 \\ & & & & & & & 5.44 & 8.58 & 10.06 \\ & & & & & & & & 15.59 & 16.88 \\ & & & & & & & & & 19.87 \end{bmatrix}$$

$$\boldsymbol{W} = \begin{bmatrix} 57.06 & 21.68 & 21.29 & 4.29 & 2.68 & 22.31 & 2.87 & 3.14 & 11.78 & 14.54 \\ & 20.76 & 6.08 & 2.65 & 1.78 & 8.63 & 0.73 & 1.06 & 3.66 & 6.91 \\ & & 13.75 & 0.98 & 0.41 & 7.16 & 1.14 & 1.25 & 6.20 & 5.35 \\ & & & 1.59 & 0.68 & 2.06 & 0.28 & 0.21 & 0.62 & 1.49 \\ & & & & 1.50 & 1.53 & 0.36 & 0.13 & 0.22 & 0.94 \\ & & & & & 14.34 & 1.38 & 1.21 & 4.39 & 5.92 \\ & & & & & & 2.13 & 0.27 & 0.72 & 0.77 \\ & & & & & & & 0.93 & 0.73 & 0.74 \\ & & & & & & & & 4.97 & 3.16 \\ & & & & & & & & & 5.79 \end{bmatrix}$$

The eigenvalues of $\boldsymbol{W}^{-1}\boldsymbol{B}$ are

$$\lambda_1 = 15.98,\ \lambda_2 = 0.99,\ \lambda_3 = 0.48,\ \lambda_4 = 0.36,\quad \lambda_5 = 0.15,$$
$$\lambda_6 = 0.10,\quad \lambda_7 = 0.05,\ \lambda_8 = 0.01,\ \lambda_9 = 0.0006,\ \lambda_{10} = 0,$$

and the first two canonical variables are as given in Example 13.5.2.

Show that $D_0^2 = 1751.2$, $D_1^2 = 652.4$, $D_2^2 = 385.6$, $D_3^2 = 233.5$, $D_4^2 = 114.2$, $D_5^2 = 60.0$, $D_6^2 = 23.0$, $D_7^2 = 4.1$, and $D_8^2 = 0.2$. Hence, show that the dimension of the data should be six if we assume the approximation (13.15) for the distribution of D_r^2 and use a test of size $\alpha = 0.01$. Comment on why this approximation could be inadequate.

13.7.1 For the drug data of Example 13.7.1, obtain the 99% SCIs for the drug differences B–A, C–B, and C–A for the second week. Show that the pairwise differences between the drugs for the second week are not significant at the 1% level. Comment on whether the significance of the original MANOVA could be due to the effect of drug C in the first week.

14

Cluster Analysis and Unsupervised Learning

14.1 Introduction

If, instead of the categorized iris data of Table 1.3, we were presented with the 150 observations in an unclassified manner, then the aim might have been to group the data into homogeneous classes or *clusters*. Such would have been the goal before the species of iris were established. For illustrative purposes, Table 14.1 gives a small mixed subsample using only two of the variables from the iris data, and the aim is to divide it into groups.

This kind of problem appears in any new investigation where one wishes to establish not only the identity but the affinity of new specimens. In general, we can summarize the problem of cluster analysis as follows.

Let $\boldsymbol{x}_1, \dots, \boldsymbol{x}_n$ be measurements of p variables on each of n objects which are believed to be heterogeneous. Then, the aim of cluster analysis is to group these objects into g homogeneous classes where g is also unknown (but usually assumed to be much smaller than n). We call a group "homogeneous" if its members are close to each other, but the members of that group differ considerably from those of another.

The term "clustering" is also referred to as "numerical taxonomy" and "classification" (but note that classification can also be used in the sense of discrimination as in Chapter 12), and it is also known as "unsupervised learning." Other terms used in this context are Q-analysis, typology, pattern recognition (also used in the sense of Chapter 12), and clumping. Good introductions to the topic are given in Everitt et al. (2001) and Jain and Dubes (1988).

The chapter starts by looking at probabilistic models in which the cluster memberships are treated as unknown parameters (Section 14.2) or unobserved random variables (Section 14.3). In an alternative "algorithmic" approach, the purpose of the analysis is purely descriptive with no assumptions about the form of the underlying population. From this point of view, clustering is simply a useful condensation of the data. The algorithmic approach includes partitioning methods (Section 14.4) and hierarchical methods (Section 14.5). All clustering methods depend on a notion of dissimilarity (or its converse, similarity); these concepts are examined more broadly in Section 14.6. Specialized methods for grouped data are given in Section 14.7. Section 14.8 gives a nonparametric approach using mode seeking. The important issue of cluster validation is studied in Section 14.9.

Multivariate Analysis, Second Edition. Kanti V. Mardia, John T. Kent, and Charles C. Taylor.

14.2 Probabilistic Membership Models

Let $f_k(\boldsymbol{x}) = f(\boldsymbol{x}; \boldsymbol{\theta}_k), k = 1, \ldots, g$, denote the p.d.f.s of g groups or populations $\Pi_1, \ldots, \Pi_g$, where the parameters $\boldsymbol{\theta}_k$ are possibly unknown. Consider n independent observations $\boldsymbol{x}_1, \ldots, \boldsymbol{x}_n$, where each $\boldsymbol{x}_r$, $r = 1, \ldots, n$, comes from one of the g populations. Define a label $y_r = k$ if the rth observation comes from Π_k. If the labels were known, these assumptions would be the same as for the standard discrimination problem (Chapter 12). However, in the current setting, the labels are unknown.

There are two ways to treat the unknown vector of group labels $\boldsymbol{y} = (y_r)$. In this section, $\boldsymbol{y}$ is treated as a set of unknown parameters to be estimated. In Section 14.3, $\boldsymbol{y}$ is treated as an unobserved latent random vector.

Thus, suppose that the number of groups g is known, but that the parameters $\boldsymbol{\theta}_k$, $k = 1, \ldots, g$, and $\boldsymbol{y}$ are unknown and need to be estimated. The likelihood function based on the data $\boldsymbol{x}_1, \ldots, \boldsymbol{x}_n$ is

$$L(\boldsymbol{y}; \boldsymbol{\theta}_1, \ldots, \boldsymbol{\theta}_g) = \prod_{k=1}^{g} \prod_{r \in C_k} f(\boldsymbol{x}_r; \boldsymbol{\theta}_k), \tag{14.1}$$

where the sets $C_k = \{r : y_r = k\}$, $k = 1, \ldots, g$, denote the partition of the data into clusters determined by $\boldsymbol{y}$. We can attempt to estimate the parameters by maximum likelihood. Let $\hat{\boldsymbol{y}}, \hat{\boldsymbol{\theta}}_k$ denote the maximum likelihood estimates (m.l.e.s) of $\boldsymbol{y}$ and $\boldsymbol{\theta}_k$, $k = 1, \ldots, g$. Similarly, let $\hat{C}_1, \ldots, \hat{C}_g$ denote the clusters of the estimated partition.

The maximum-likelihood method possesses an important allocation property. Suppose that a particular observation $\boldsymbol{x}_r$, say, has been allocated under the maximum-likelihood estimate to $\hat{C}_k$. Since moving $\boldsymbol{x}_r$ from $\hat{C}_k$ to $\hat{C}_l$ reduces the likelihood, we have

$$L(\hat{\boldsymbol{y}}; \hat{\boldsymbol{\theta}}_1, \ldots, \hat{\boldsymbol{\theta}}_g) f(\boldsymbol{x}_r; \hat{\boldsymbol{\theta}}_l) / f(\boldsymbol{x}_r; \hat{\boldsymbol{\theta}}_k) \leq L(\hat{\boldsymbol{y}}; \hat{\boldsymbol{\theta}}_1, \ldots, \hat{\boldsymbol{\theta}}_g).$$

Thus,

$$f(\boldsymbol{x}; \hat{\boldsymbol{\theta}}_l) \leq f(\boldsymbol{x}; \hat{\boldsymbol{\theta}}_k) \qquad \text{for} \qquad \hat{y}_r = k, \quad l \neq k, \quad l = 1, \ldots, g. \tag{14.2}$$

This is quite a familiar allocation rule in the context of discriminant analysis (Chapter 12).

The normal case

Let $f(\boldsymbol{x}; \boldsymbol{\theta}_k)$ denote the probability density function (p.d.f.) of $N_p(\boldsymbol{\mu}_k, \boldsymbol{\Sigma}_k), k = 1, \ldots, g$. Then, the log likelihood function is

$$l(\boldsymbol{y}; \boldsymbol{\theta}) = \text{const} - \frac{1}{2} \sum_{k=1}^{g} \sum_{r \in C_k} (\boldsymbol{x}_r - \boldsymbol{\mu}_k)' \boldsymbol{\Sigma}_k^{-1} (\boldsymbol{x}_r - \boldsymbol{\mu}_k) - \frac{1}{2} \sum_{k=1}^{g} n_k \log |\boldsymbol{\Sigma}_k|, \tag{14.3}$$

where there are n_k observations in C_k. Hence, for a given $\boldsymbol{y}$, the likelihood is maximized by the ordinary m.l.e. of $\boldsymbol{\mu}$ and $\boldsymbol{\Sigma}$, i.e.

$$\hat{\boldsymbol{\mu}}_k(\boldsymbol{y}) = \overline{\boldsymbol{x}}_k, \qquad \hat{\boldsymbol{\Sigma}}_k(\boldsymbol{y}) = \boldsymbol{S}_k, \tag{14.4}$$

where $\overline{\boldsymbol{x}}_k$ is the mean, and $\boldsymbol{S}_k$ is the covariance of the n_k observations in C_k. On substituting (14.4) in (14.3), we obtain

$$l(\boldsymbol{y}; \hat{\boldsymbol{\theta}}(\boldsymbol{y})) = \text{const} - \frac{1}{2} \sum n_k \log |\boldsymbol{S}_k|.$$

Table 14.1 A subsample of six observations (and two variables) from the iris data.

Observation	1	2	3	4	5	6
Sepal Length	7.0	6.4	6.9	6.3	5.8	7.1
Sepal Width	3.2	3.2	3.1	3.3	2.7	3.0

Hence, the m.l.e. of $\boldsymbol{y}$ is the grouping that minimizes

$$\prod_{k=1}^{g} |\boldsymbol{S}_k|^{n_k}. \tag{14.5}$$

To avoid the degenerate case of infinite likelihood, we assume that there are at least $p+1$ observations assigned to each group, so that $n_k \geq p+1$, which means that $n \geq g(p+1)$.

If, however, the groups are assumed to have identical covariance matrices, $\boldsymbol{\Sigma}_1 = \cdots = \boldsymbol{\Sigma}_g = \boldsymbol{\Sigma}$ (unknown), the same method leads to the grouping that minimizes

$$|\boldsymbol{W}|, \tag{14.6}$$

where

$$\boldsymbol{W} = \sum_{k=1}^{g} \sum_{r \in C_k} (\boldsymbol{x}_r - \overline{\boldsymbol{x}}_k)(\boldsymbol{x}_r - \overline{\boldsymbol{x}}_k)'$$

is the pooled within-groups sums of squares and products (SSP) matrix.

Example 14.2.1 (Iris subsample data) For illustrative purposes, consider a subsample of size $n = 6$ from the iris data in Table 1.3 given by the first three observations from *I. versicolor* and the first three observations from *I. virginica*, in both cases on the first two variables only. The subsample is given in Table 14.1. Assume that the species of iris is unknown. The purpose of the exercise is to investigate the clustering criteria (14.5) and (14.6).

To satisfy $n \geq 3g$, we take $g = 2$, with three observations in each group. For all possible partitions, Table 14.2 gives the values of $|\boldsymbol{S}_1|, |\boldsymbol{S}_2|$, and $|\boldsymbol{W}| = |\boldsymbol{S}_1 + \boldsymbol{S}_2|$. Both ML methods lead to the clusters $(1, 3, 6)$ and $(2, 4, 5)$. A scatter diagram of the points is given in Figure 14.1, and from the figure it can be seen that this clustering is intuitively reasonable. In particular, the points $(1, 3, 6)$ visually form a cluster separate from the other points.

Under the correct grouping $(1, 2, 3)$ and $(4, 5, 6)$, the values of $|\boldsymbol{S}_1| = 0.00013$ and $|\boldsymbol{S}_2| = 0.0147$ are disparate, indicating that the two true groups have different sizes on this very small data set. Further, it is not surprising that any clustering method has difficulty finding the true groups here since that there is considerable overlap between the species *I. versicolor* and *I. virginica* (see Example 12.4.5). □

Before starting clustering, we may wish to look into whether the data are unstructured, that is whether there is only one cluster. Hence, the hypothesis of interest is that the class labels are all equal:

$$H_0 : y_1 = \cdots = y_n,$$

Table 14.2 Clusters for the data in Table 14.1.

Group 1	Group 2	$\lvert S_1 \rvert$	$\lvert S_2 \rvert$	$10^3 \lvert S_1 \rvert \lvert S_2 \rvert$	$\lvert S_1 + S_2 \rvert$
1 2 3	4 5 6	0.000133	0.014700	0.001960	0.020048
1 2 4	3 5 6	0.000133	0.001337	0.000178	0.006557
1 2 5	3 4 6	0.003333	0.000015	0.000050	0.021946
1 2 6	3 4 5	0.000533	0.007837	0.004180	0.018563
1 3 4	2 5 6	0.000237	0.008181	0.001939	0.016779
1 3 5	2 4 6	0.000181	0.000093	0.000017	0.020894
1 3 6	2 4 5	0.000033	0.000448	0.000015	0.001837
1 4 5	2 3 6	0.008181	0.000033	0.000273	0.020104
1 4 6	2 3 5	0.000626	0.003559	0.002228	0.018864
1 5 6	2 3 4	0.003115	0.000059	0.000185	0.012720

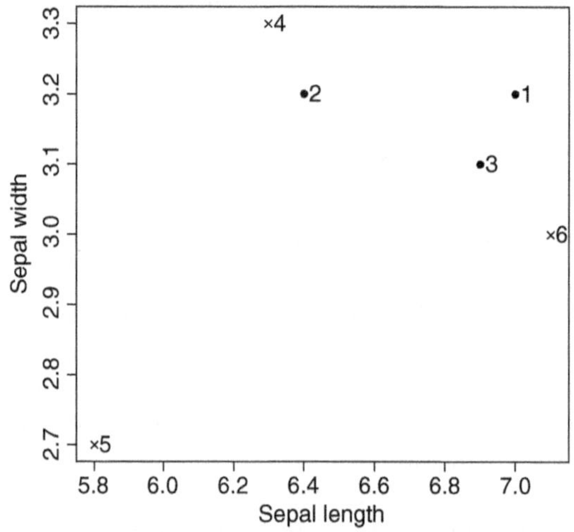

Figure 14.1 Scatter plot for the small subsample of six observations of iris data as shown in Table 14.1. Symbols are used to denote the original groups.

against the alternative that not all the y_r are equal. Assuming normality with equal covariance matrices, we get the statistic

$$-2\log\lambda = n\log\{\max_{y}(|\boldsymbol{T}|/|\boldsymbol{W}|)\}, \tag{14.7}$$

where

$$\boldsymbol{T} = \sum_{r=1}^{n}(\boldsymbol{x}_r - \overline{\boldsymbol{x}})(\boldsymbol{x}_r - \overline{\boldsymbol{x}})'$$

is the total sum of squares and products matrix. Thus, if the number of groups g is unknown, we need to minimize $|\boldsymbol{W}|$ over all permissible partitions of $\boldsymbol{y}$ (those partitions for which $g \leq n - p$, so that rank $(\boldsymbol{W}) = p$. Unfortunately, even for large n, the distribution of (14.7) is not known. (However, some progress has been made in the case $p = 1$; see Hartigan (1978).) A more tractable test using the likelihood for a mixture model is given in (14.16).

It may also be worthwhile to check that the data is not spherically symmetric. One method would be to apply the Rayleigh test of uniformity on the directions $\boldsymbol{z}_i/\|\boldsymbol{z}_i\|$ (Mardia and Jupp, 2000), where $\boldsymbol{z}_i = \boldsymbol{S}^{-1/2}(\boldsymbol{x}_i - \overline{\boldsymbol{x}})$ is the Mahalanobis transformation of Section 1.5.2.

The criterion of minimizing $|\boldsymbol{W}|$ was first put forward on an intuitive basis by Friedman and Rubin (1967), whereas the maximum-likelihood justification was given by Scott and Symonds (1971). An alternative procedure based on minimizing tr $\boldsymbol{W}$ was developed by Edwards and Cavalli-Sforza (1965). This idea can also be given a maximum-likelihood interpretation and this is the criterion that k-means (Section 14.4) uses – see also Exercise 14.2.1.

Remarks (1) Let $\boldsymbol{z}_r = \boldsymbol{A}\boldsymbol{x}_r + \boldsymbol{b}$, where $\boldsymbol{A}$ is a nonsingular matrix. Then, (14.5) and (14.6) can be written as

$$|\boldsymbol{A}|^{-2n}\prod_{k=1}^{g}|\boldsymbol{S}_{k,z}|^{n_k}, \qquad \text{and} \qquad |\boldsymbol{A}|^{-2}|\boldsymbol{W}_z|,$$

respectively, where $\boldsymbol{S}_{k,z}$ and $\boldsymbol{W}_z$ denote $\boldsymbol{S}_k$ and $\boldsymbol{W}$ for the transformed variables $\boldsymbol{z}_r$. Hence, in particular, as far as the minimization of these criteria is concerned, the scale of the variables is immaterial.

(2) The task of minimizing (14.5) or (14.6) is formidable even for a high-performance computer. In practice, to circumvent this problem, a relative minimum is found, with the property that any reassignment of one or two observations results in a larger value, although it may not be an absolute minimum. It is worthwhile examining individually the partitions in the neighborhood of the (estimated) true split.

(3) *Estimation of* g We have assumed g to be preassigned. In practice, if g is allowed to vary, the maximum-likelihood methods will always partition the data into the maximum number of partitions allowed. Thus, g must be chosen by some other method. For equal covariance matrices, Marriott (1971) has suggested taking the correct number of groups to be the value of g for which

$$g^2|\boldsymbol{W}| \tag{14.8}$$

is minimum. The most suitable algorithms for implementing the optimization of the criterion (14.8) are those of Friedman and Rubin (1967) and McRae (1971). See also Hartigan (1975).

In a simulation study, Milligan and Cooper (1985) compared 30 methods for determining g through the use of a "stopping rule." Their results suggest that the best criterion is that of Caliński and Harabasz (1974), which corresponds to choosing g to maximize

$$\frac{\text{tr}\,\boldsymbol{B}}{g-1}\Big/\frac{\text{tr}\,\boldsymbol{W}}{n-g}$$

See also Everitt et al. (2001, pp. 102–105) for further advice on how to choose the number of clusters.

(4) *Criticisms* Algorithms attempting to maximize the likelihood usually require large amounts of computer time and cannot be recommended for use with large data sets. The number of parameters to be estimated increases indefinitely with the sample size, and therefore, the estimates are not "consistent" (Marriott, 1971). Further, even if g is fixed and the p.d.f.s are unimodal, the estimated groups will be the truncated centers of these p.d.f.s,

mixed with the tails of other p.d.f.s. Thus, even for the case of equal covariance matrices, $\boldsymbol{\Sigma}_1 = \cdots = \boldsymbol{\Sigma}_g = \boldsymbol{\Sigma}$, $\boldsymbol{W}$ will not be a consistent estimate of $\boldsymbol{\Sigma}$. Further, in this case, the criterion (14.6) has a tendency to partition the sample into groups of about the same size even when the true clusters are of unequal sizes (Scott and Symonds, 1971). When the modes are near together and the distributions overlap considerably, separation may be impossible even for very large sample sizes. A better approach is to use mixture models, studied in Section 14.3.

(5) For our discussion, we have assumed $\boldsymbol{x}_1, \dots, \boldsymbol{x}_n$ to be independent random variables, but this may not be the case in some situations, e.g. when $\boldsymbol{x}_1, \dots, \boldsymbol{x}_n$ are measurements of the various stages of evolution of a particular creature.

14.3 Parametric Mixture Models

Using the same notation as Section 14.2, suppose that the labels y_r are independent and identically distributed (i.i.d.) random variables with $\mathsf{P}(y_r = k) = p_k, k = 1, \dots, g$, where the elements of the vector of mixture proportions $\boldsymbol{p} = (p_k)$ satisfy $p_k > 0$ and $\sum p_k = 1$. Then, the observations $\boldsymbol{x}_r$ are i.i.d. with p.d.f. given by the mixture density

$$f_{\text{mix}}(\boldsymbol{x}) = \sum_{k=1}^{g} p_k f(\boldsymbol{x}; \boldsymbol{\theta}_k). \tag{14.9}$$

If the parameters p_k and $\boldsymbol{\theta}_k$, $k = 1, \dots, g$ are known, then the data can be partitioned into clusters using the Bayes allocation rule of Section 12.2. This rule allocates each observation $\boldsymbol{x}_r$ to the population for which the contribution to the mixture density is largest. That is, allocate $\boldsymbol{x}_r$ to population k if $k = \arg\max_l p_l f(\boldsymbol{x}_r; \boldsymbol{\theta}_k)$.

Of course, in practice, the parameters will usually be unknown. In this case, they can be estimated by maximum likelihood. One way to compute the m.l.e.s in a mixture model is to use the EM algorithm (MacLachlan and Krishnan, 1997). To understand how the algorithm works, start with the *complete data* log likelihood, assuming that both $\boldsymbol{x}_r$ and y_r are known, $r = 1, \dots, g$, which can be written as

$$l(\boldsymbol{x}_r, y_r,\ r = 1, \dots, n; p_k, \boldsymbol{\theta}_k,\ k = 1, \dots g) = \sum_{r=1}^{n} \sum_{k=1}^{g} I[y_r = k]\{\log p_k + \log f(\boldsymbol{x}_r; \boldsymbol{\theta}_k)\}, \tag{14.10}$$

where $I[\cdot]$ is an indicator function. Since the labels are unobserved, the indicator functions $I[y_r = k]$ should be replaced by their conditional expectations

$$p_{rk} = \mathsf{E}(y_r = k | \boldsymbol{x}_r) = \frac{p_k f(\boldsymbol{x}_r; \boldsymbol{\theta}_k)}{\sum_{l=1}^{g} p_l f(\boldsymbol{x}_r; \boldsymbol{\theta}_l)}, \quad r = 1, \dots, n, \quad k = 1, \dots, g, \tag{14.11}$$

to get the "imputed" log likelihood,

$$l(\boldsymbol{x}_r,\ r = 1, \dots, n; p_k, \boldsymbol{\theta}_k,\ k = 1, \dots g) = \sum_{r=1}^{n} \sum_{k=1}^{g} p_{rk} \{\log p_k + \log f(\boldsymbol{x}_r; \boldsymbol{\theta}_k)\}. \tag{14.12}$$

Note that (14.12) does not depend on the labels y_r.

Given the initial values for the parameters, the EM algorithm updates the parameter estimates in two alternating steps:

(1) (E-step) Given the values for the parameters $p_k, \theta_k,\ k = 1, \dots, g$, update the individual membership probabilities p_{rk} using (14.11).
(2) (M-step) Given the values for the individual membership probabilities p_{rk}, update the estimates of mixture proportions p_k and the parameters $\boldsymbol{\theta}_k$ by maximizing (14.11). In particular, the overall membership probabilities are updated to

$$p_k = \frac{1}{n}\sum_{r=1}^{n} p_{rk}. \tag{14.13}$$

For a multivariate normal mixture, $\boldsymbol{\theta}_k$ stands for $\boldsymbol{\mu}_k$ and $\boldsymbol{\Sigma}_k$, and the M-step yields a weighted sample mean and variance for each population,

$$\boldsymbol{\mu}_k = (np_k)^{-1}\sum_{r=1}^{n} p_{rk}\boldsymbol{x}_r, \tag{14.14}$$

$$\boldsymbol{\Sigma}_k = (np_k)^{-1}\sum_{r=1}^{n} p_{rk}(\boldsymbol{x}_r - \boldsymbol{\mu}_k)(\boldsymbol{x}_r - \boldsymbol{\mu}_k)'. \tag{14.15}$$

The EM algorithm alternates between these two steps and is iterated until convergence. It can be shown that in general each step of the algorithm increases the mixture log likelihood based on (14.9), and that the algorithm converges to a *local* maximum of the likelihood. However, whether this local maximum is global (or even sensible) depends critically on the starting values. In practice, it is recommended to use multiple starting points. More sophisticated strategies have also been proposed. See Karlis and Xekalaki (2003) and Biernacki et al. (2003) for some comparative studies based on finite mixture models as well as several refinements to the basic EM algorithm presented here. In particular, for a normal mixture model in which the covariance matrices are assumed to be isotropic and equal to one another, the limiting case of the EM algorithm when $\boldsymbol{\Sigma} \to \boldsymbol{0}$ is equivalent to the k-means algorithm (Section 14.4).

The EM algorithm is a likelihood-based method of inference, and it is possible to carry out asymptotic hypothesis tests. Let $\lambda = L_g/L_{g'}$, where λ is the ratio of the likelihood of g groups against that of g' groups ($g < g'$). From a Monte Carlo investigation, Wolfe (1971) recommends the approximation

$$-\frac{2}{n}\left(n - 1 - p - \frac{1}{2}g'\right)\log\lambda \sim \chi^2_\nu, \qquad \nu = 2p(g' - g) \tag{14.16}$$

for the distribution of λ under the null hypothesis that there are g groups.

Another way to choose between models is to use the Bayes information criterion (BIC) defined by

$$\text{BIC} = 2\hat{l} - \nu\log n \tag{14.17}$$

(Fraley and Raftery, 2002), where the number of estimated parameters ν depends on the number of groups and the number of constraints on the covariance matrices. The R package `mclust` (Scrucca et al., 2016) implements this procedure, using a version of agglomerative hierarchical clustering (e.g. using Ward's criterion; see Eq. (14.23) in Section 14.5) to initialize the EM algorithm. It also allows a variety of options for the covariance matrices in the mixture model to reduce the number of parameters. These options include information about the shape of each covariance matrix (e.g. isotropic, diagonal, and unconstrained) and the homogeneity of covariance matrices for different groups (e.g. equal and unequal). Overall, the methodology in `mclust` offers a useful balance between flexibility and tractability for many practical problems.

Note that in spite of the term "Bayes", the BIC criterion does not involve any assumptions about the prior distribution of the mean or covariance parameters. At the cost of more computation, it is possible to take a more fully Bayesian approach to parametric mixture modeling by including prior distributions for all the parameters; see, e.g. Scott and Symonds (1971).

14.4 Partitioning Methods

Given a set of n observations and a given (known) number of clusters g, partitioning methods seek to allocate the observations to clusters such that some clustering criterion is optimized. Intuitively, we expect that most criteria will assess the within-cluster variability compared with the between-cluster variability. Thus, most criteria depend on

$$\boldsymbol{W} = \sum_{k=1}^{g} \sum_{\boldsymbol{x}_r \in \Pi_k} (\boldsymbol{x}_r - \overline{\boldsymbol{x}}_k)(\boldsymbol{x}_r - \overline{\boldsymbol{x}}_k)',$$

the pooled within-groups sums of squares and products (SSP) matrix, and

$$\boldsymbol{T} = \sum_{r=1}^{n} (\boldsymbol{x}_r - \overline{\boldsymbol{x}})(\boldsymbol{x}_r - \overline{\boldsymbol{x}})',$$

the total sum of squares and products matrix. Examples of clustering criteria include

(A) Minimization of tr $\boldsymbol{W}$. Given that $\boldsymbol{W} = \boldsymbol{W}_1 + \cdots + \boldsymbol{W}_g$, this is equivalent to minimizing the total within-group sum of squares about the g means $\overline{\boldsymbol{x}}_k, k = 1, \ldots, g$. Such methods tend to produce clusters that are spherical in shape. Since the criterion is not invariant to scale, some form of standardization should first be applied.

(B) Minimization of $|\boldsymbol{W}|$. This criterion is invariant to nonsingular transformations of the data. Although it does not impose a spherical structure on the data, it does assume that all clusters have the same shape; see Everitt et al. (2001, pp. 93–94) for further details.

(C) Maximization of tr $(\boldsymbol{W}^{-1}\boldsymbol{B})$, where $\boldsymbol{B} = \boldsymbol{T} - \boldsymbol{W}$ is the "between-groups" SSP matrix (6.23). This is equivalent to minimizing the sum of squares under the Mahalanobis metric and is also invariant to nonsingular transformations of the data.

Although the criteria are easily stated, the problem is a computationally demanding combinatorial one in that there are

$$N(n, g) = \frac{1}{g!} \sum_{i=1}^{g} (-1)^{g-i} \binom{g}{i} i^n$$

ways of allocating n observations into g groups (approximately $g^n/g!$) (e.g. Abramowitz and Stegun, 1964, Section 24.1.4). So, for example, $N(50, 5) = 7.40 \times 10^{32}$ and $N(100, 5) = 6.57 \times 10^{67}$. Thus, even for moderate values of n, suboptimal solutions are required. We now describe some popular algorithms.

k-Means

The k-means algorithm aims to minimize criterion (A) above. Given g, in its simplest form it alternates two steps:

(1) Assign each point $\boldsymbol{x}_r, r = 1, \ldots, n$, to the mean $\overline{\boldsymbol{x}}_j$ which is closest (for example, using Euclidean distance);
(2) Recalculate $\overline{\boldsymbol{x}}_j, j = 1, \ldots, g$, based on the group membership defined in step 1.

The process eventually terminates when the allocation does not change any of the cluster memberships. Although it is clear that at each step of the algorithm the within-group sum of squares decreases, it is also the case that the final solution is probably just a local minimum, which depends, in part, on the starting point of the algorithm described above. This version of k-means is called hmeans (Späth, 1980) and also has the additional shortcoming that there is no guarantee that the final solution actually has g clusters (i.e. some clusters may be empty).

k-Medoids

The above method is usually applied to Euclidean distances, perhaps after an initial scaling, so that each variable has unit variance. However, Euclidean distance can be sensitive to outliers that produce very large distances. Hence, it is sometimes of interest to consider a non-Euclidean distance that downweights large distances. For a general distance function D (see Section 14.6), let $D(\boldsymbol{x}_r, \boldsymbol{x}_s), 1 \leq r, s \leq n$, denote the distances between the observations. In this case, we may seek a partition to solve

$$\min_{\substack{c_1,\ldots,c_g,\\ \Pi_1,\ldots,\Pi_g}} \sum_{k=1}^{g} \sum_{\{r:\boldsymbol{x}_r \in \Pi_k\}} D(\boldsymbol{x}_r, \boldsymbol{c}_k)$$

over cluster centers $\boldsymbol{c}_1, \ldots, \boldsymbol{c}_g$. The k-medoids algorithm aims to minimize this criterion by alternating two steps:

(1) Given a set of cluster centers $\boldsymbol{c}_1, \ldots \boldsymbol{c}_g$, allocate each observation to the closest center, i.e. assign $\boldsymbol{x}_r$ to Π_j if
$$j = \arg\min_{1 \leq k \leq g} D(\boldsymbol{x}_r, \boldsymbol{c}_k).$$
(2) Given partitions $\Pi_k, k = 1, \ldots, g$, find the new cluster center $\boldsymbol{c}_k$ such that
$$\boldsymbol{c}_k = \arg\min_{\boldsymbol{c}} \sum_{\{r:\boldsymbol{x}_r \in \Pi_k\}} D(\boldsymbol{x}_r, \boldsymbol{c})$$
in which the search space for $\boldsymbol{c}$ is often restricted to the discrete set $\boldsymbol{x}_r \in \Pi_k$.

As with the k-means algorithm, the solution may be a local rather than global minimum and will depend on the starting point.

Fuzzy k-Means

The above methods assign each observation to exactly one group. The basic idea behind fuzzy clustering is that each object can belong to all clusters with different degrees of membership. There are many similarities to mixture models; see Section 14.3. That is, we allow each observation $\boldsymbol{x}_r$ to belong partly to all the clusters rather than wholly to one cluster.

Let y_{rk} represent the membership proportion of the rth observation in cluster k, where

$$\sum_{k=1}^{g} y_{rk} = 1, \quad 1 \le r \le n, \qquad y_{rk} \ge 0 \quad r = 1, \dots, n, \ k = 1, \dots, g.$$

The objective, for given value of a "fuzziness" parameter $\gamma > 0$, is to minimize the weighted sum of Euclidean distances

$$\sum_{r=1}^{n}\sum_{k=1}^{g} y_{rk}^{\gamma} \|\boldsymbol{x}_r - \mathbf{c}_k\|,$$

where $\mathbf{c}_k$ is the "centroid" of cluster k

$$\mathbf{c}_k = \frac{\sum_{r=1}^{n} y_{rk}^{\gamma}\boldsymbol{x}_r}{\sum_{r=1}^{n} y_{rk}^{\gamma}}, \qquad k = 1, \dots, g. \tag{14.18}$$

The basic algorithm (Bezdek, 1981) is given by alternating the two steps

(1) Update the cluster centers $\mathbf{c}_k$ using (14.18).
(2) Update the membership function according to

$$y_{rk}^{-1} = \sum_{j=1}^{g} \left(\frac{\|\boldsymbol{x}_r - \mathbf{c}_k\|}{\|\boldsymbol{x}_r - \mathbf{c}_j\|} \right)^{2/(\gamma-1)}.$$

Example 14.4.1 (Iris area data, k-means, and `mclust` clustering). For the iris data in Table 1.3, define two natural measures of area, *sepal area* and *petal area*, by

$$x_1 = \text{sepal length} \times \text{sepal width}, \quad x_2 = \text{petal length} \times \text{petal width},$$

respectively. This choice of variables has been made because the resulting clusters are heterogeneous in shape. Panel (a) of Figure 14.2 shows the true species for each observation. *I. setosa* forms a long thin cluster near the bottom, well separated from the other two. *I. versicolor* and *I. viriginica* partially overlap, with *I. versicolor* to the lower left of *I. virginica*.

Next, we treat the species membership as unknown and investigate whether various algorithms can correctly pick out the true clusters. The clusters found by the k-means algorithm for $g = 3$ and $g = 4$ clusters are shown in panels (b) and (c) of Figure 14.2. The $g = 3$ clusters are reasonably close to the true species clusters but not perfect. Finally, the result from `mclust` is also shown in panel (d). For `mclust`, the number of clusters and the choice of covariance matrix constraint are selected by BIC. Here, the number of clusters (3) is correctly selected, and the covariance matrices are selected as unconstrained. To evaluate the solutions, we can compare the cluster memberships (which are arbitrarily numbered) with the actual class labels. For the three solutions in Figure 14.2, this gives the tables shown on the next page

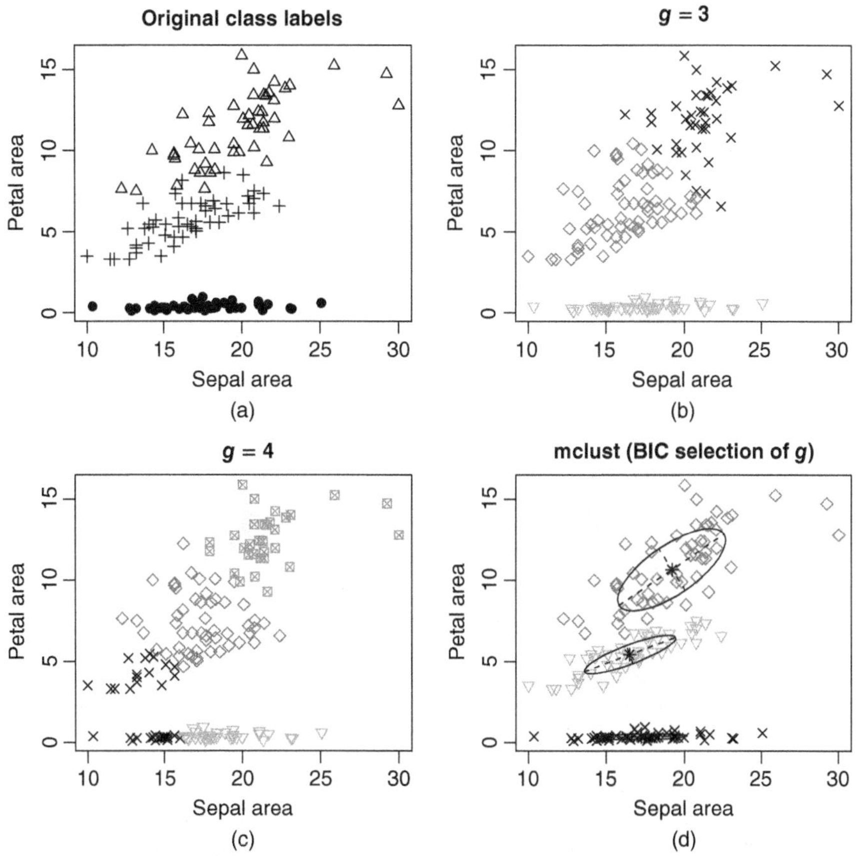

Figure 14.2 Application of the k-means algorithm, and `mclust` to the iris area data in Example 14.4.1. (a) Original labels for the three species (• = *I. setosa*, + = *I. versicolor*, and △ = *I. virginica*). (b) and (c) Labeling after application of the k-means algorithm for $g = 3, 4$ clusters. (d) The result from `mclust`. Symbols are used to differentiate the cluster memberships, which are arbitrarily ordered.

	Allocated cluster number									
	`mclust`			k-means ($g = 3$)			k-means ($g = 4$)			
	1	2	3	1	2	3	1	2	3	4
I. setosa	50	0	0	50	0	0	32	0	18	0
I. versicolor	0	44	6	0	46	4	0	34	16	0
I. virginica	0	0	50	0	16	34	0	19	0	31

It can be seen that the clusters found by `mclust` are closer to the true species clusters. Further ways to evaluate a cluster solution are considered in Section 14.9.

For comparative purposes, we apply the fuzzy k-means algorithm to the same data. Now, we hold $g = 3$ fixed and try different values of γ. The output is in the form of mixtures

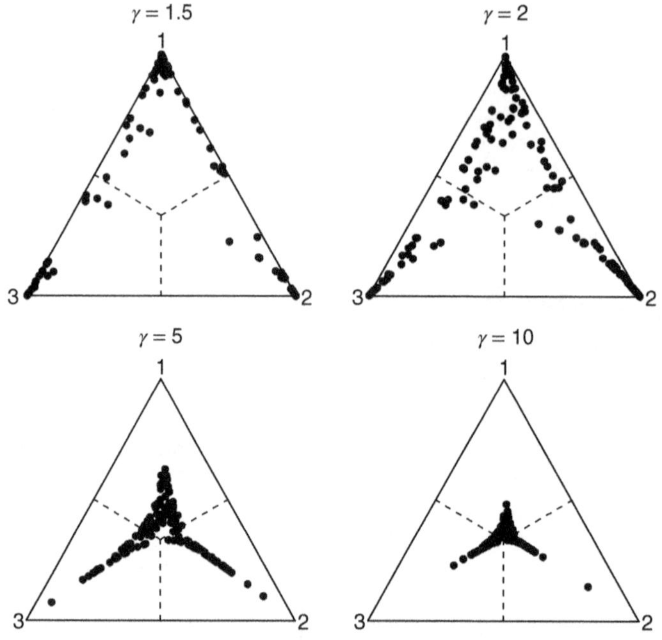

Figure 14.3 Application of the fuzzy k-means algorithm to the iris area data in Example 14.4.1. Each panel shows a ternary plot of the iris data for various fuzziness parameter values γ and $g = 3$ clusters. The dashed lines could be applied to partition the data set into the three clusters.

(y_{r1}, y_{r2}, y_{r3}), $r = 1, \ldots, n$, and these can be plotted in a *ternary* plot as shown in Figure 14.3. As γ decreases toward 1, the algorithm tends to the k-means algorithm. As γ increases, the membership proportions become fuzzier, and all the memberships converge to the point $(1/3, 1/3, 1/3)$. To finally cluster the data, one could subsequently apply a majority rule to the mixture proportions in order to make an allocation, with the possibility of adding a "doubt" category in case that the mixtures are approximately equal. For example, if the solution for $\gamma = 1.5$ is partitioned using the dashed lines shown in Figure 14.3, the cluster labels can then be compared to the true classes, as shown below. It can be seen that this solution is very similar to that given by k-means with $g = 3$, which is to be expected as γ is close to 1.

	Fuzzy k-means ($\gamma = 1.5$)		
		Cluster	
	1	2	3
I. setosa	0	50	0
I. versicolor	47	0	3
I. virginica	16	0	34

□

The methods of this section are mainly looking for compact spherical clusters. However, in some cases, a cluster can have a different shape, e.g. a curve in two or higher dimensions. Mardia et al. (2022) discuss a method based on principal nested spheres. Their method can be applied to data in Euclidean space and other manifolds such as spheres and tori.

14.5 Hierarchical Methods

The clustering methods in Section 14.4 can be described as *optimization partitioning techniques* since the clusters are formed by optimizing a clustering criterion. We now consider *hierarchical methods* in which the clusterings into g and g + 1 groups, respectively, have the property that (i) they have $g - 1$ identical groups and (ii) the remaining single group in the g-groups clustering is a union of two groups in the $(g + 1)$-groups clustering.

By very nature of these techniques, once an object is allocated to a group, it cannot be reallocated as groups are combined, unlike the optimization techniques of Section 14.4. The end product of these methods is a tree diagram (dendrogram); see Section 14.5.2.

These techniques operate on a matrix $\boldsymbol{D} = (d_{rs})$ of distances between the points $\boldsymbol{x}_1, \dots, \boldsymbol{x}_n$ rather than directly on the points themselves. The possible choices for the distance matrix will be discussed in Section 14.6.

In hierarchical methods, clustering can be obtained through two types of algorithm: *agglomerative* algorithms start with each observation belonging to its own cluster, and items are then successively pooled together into larger sets; *divisive* algorithms start with all observations belonging to a single cluster, which is then recursively divided into subsets.

To apply a divisive procedure it is generally necessary to consider all possible splits. At the first stage, there are $2^{n-1} - 1$ possibilities, which is a prohibitively large number of choices for even moderate n. One of the few tractable algorithms that computes a divisive hierarchical clustering is DIANA (Kaufman and Rousseeuw, 1990, ch. 6), which recursively divides the cluster with the largest diameter (largest within-cluster distance) by selecting the most isolated observation within that cluster. This cluster is then split by allocating observations to this new center.

On the other hand, agglomerative methods are much less computationally demanding and much more widely used. We focus on agglomerative algorithms here.

14.5.1 Agglomerative Algorithms

Agglomerative algorithms work as follows:

(a) Start with n clusters $C_1, \dots C_n$, each containing one observation, namely $C_i = \{i\}$, $i = 1, \dots, n$.
(b) At each step of the algorithm, join the two clusters that are closest together.
(c) After $n - 1$ steps, all the points form a single cluster, and the algorithm stops.

Agglomerative algorithms differ in the way they decide to combine clusters. Starting from a distance matrix $\boldsymbol{D} = (d_{rs})$ between the points, one general approach is to define a corresponding distance between two clusters. Some common choices include *single linkage, complete linkage, average linkage, median linkage*, and *Ward's criterion*, for which the distances between clusters are given by:

$$d^{\mathrm{SL}}(C_1, C_2) = \min\, d_{rs} \; : \; r \in C_1,\; s \in C_2, \tag{14.19}$$

$$d^{\mathrm{CL}}(C_1, C_2) = \max\, d_{rs} \; : \; r \in C_1,\; s \in C_2, \tag{14.20}$$

$$d^{AL}(C_1, C_2) = \text{ave } d_{rs} : r \in C_1,\ s \in C_2, \tag{14.21}$$

$$d_{ij}^{\text{med}} = \text{median } d_{rs} : r \in C_1,\ s \in C_2, \tag{14.22}$$

$$d_{ij}^{\text{Ward}} = \frac{n_i n_j}{n_i + n_j} \|\mathbf{c}_i - \mathbf{c}_j\|. \tag{14.23}$$

In the final line, $\mathbf{c}_i$ is the arithmetic mean of a cluster C_i of size n_i, and $\| \cdot \|$ denotes the Euclidean norm (Ward, 1963). Note that (14.23) uses the data points themselves, and not just the interpoint distances.

Given a threshold value d_0, say, the algorithm can be stopped when all the clusters are at least a distance d_0 apart. The result of the algorithm is then a split of the data into a number of clusters, which can be denoted $C_1^*, \ldots, C_g^*$, say. The threshold is often chosen by eye, with the aim of ensuring that the within-cluster distances are "small," and the between-cluster distances are "large."

Example 14.5.1 (Single linkage clustering for the shrew data). Consider the shrew data in Example 13.5.2. This is a multisample data set consisting of 299 skulls found at $n = 10$ localities. The distances between the localities are taken to be the Mahalanobis distances between the *means* at each of the 10 localities, with the covariance matrix estimated from the within-groups SSP matrix (see Section 6.3.3). The distance matrix is given in Table 14.3 (Delany and Healy, 1966; Gower and Ross, 1969).

One way to compute single linkage clustering is to list the distances from smallest to largest. Then, go through the distances d_{ij} in turn. If i and j are already in the same cluster, then nothing changes. If they are in different clusters, then the two clusters are joined. This

Table 14.3 Mahalanobis distances between 10 island races of white-toothed shrews.

	The Scilly Islands					The Channel Islands				France
	1	2	3	4	5	6	7	8	9	10
1 Tresco	0									
2 Bryher	1.80	0								
3 St Agnes	2.39	2.56	0							
4 St Martin's	2.33	2.93	3.20	0						
5 St Mary's	1.76	2.11	1.53	2.64	0					
6 Sark	3.02	4.09	4.09	4.60	3.90	0				
7 Jersey	3.31	4.47	4.02	3.44	3.45	3.36	0			
8 Alderney	10.66	10.72	11.09	9.85	10.39	10.87	10.27	0		
9 Guernsey	8.91	9.03	9.62	8.21	9.02	9.03	8.95	3.11	0	
10 Cap Gris-Nez	8.56	8.65	9.10	8.17	8.55	8.72	9.00	3.70	2.85	0

Source: From Delany and Healy (1966); Gower and Ross (1969).

Table 14.4 Single linkage procedure for shrew data.

	Distances	Clusters
Order	(ordered)	
1	$d_{35} = 1.53$	(1), (2), (3,5), (4), (6), (7), (8), (9), (10)
2	$d_{15} = 1.76$	(1, 3, 5), (2), (4), (6), (7), (8), (9), (10)
3	$d_{12} = 1.80$	(1, 2, 3, 5), (4), (6), (7), (8), (9), (10)
4	$d_{25} = 2.11$	(1, 2, 3, 5), (4), (6), (7), (8), (9), (10)[a]
5	$d_{14} = 2.33$	(1, 2, 3, 4, 5), (6), (7), (8), (9), (10)
6	$d_{13} = 2.39$	(1, 2, 3, 4, 5), (6), (7), (8), (9), (10)[a]
7	$d_{23} = 2.56$	(1, 2, 3, 4, 5), (6), (7), (8), (9), (10)[a]
8	$d_{45} = 2.64$	(1, 2, 3, 4, 5), (6), (7), (8), (9), (10)[a]
9	$d_{9,10} = 2.85$	(1, 2, 3, 4, 5), (6), (7), (8), (9, 10)
10	$d_{24} = 2.93$	(1, 2, 3, 4, 5), (6), (7), (8), (9, 10)[a]
11	$d_{16} = 3.02$	(1, 2, 3, 4, 5, 6), (7), (8), (9, 10)
12	$d_{89} = 3.11$	(1, 2, 3, 4, 5, 6), (7), (8, 9, 10)
13	$d_{34} = 3.20$	(1, 2, 3, 4, 5, 6), (7), (8, 9, 10)[a]
14	$d_{17} = 3.31$	(1, 2, 3, 4, 5, 6, 7), (8, 9, 10)
15	$d_{67} = 3.36$	(1, 2, 3, 4, 5, 6, 7), (8, 9, 10)[a]
16	$d_{47} = 3.44$	(1, 2, 3, 4, 5, 6, 7), (8, 9, 10)[a]
17	$d_{57} = 3.45$	(1, 2, 3, 4, 5, 6, 7), (8, 9, 10)[a]
⋮	⋮	⋮
45	$d_{38} = 11.09$	(1, 2, 3, 4, 5, 6, 7, 8, 9, 10)

a) No new clusters formed.

procedure is illustrated in Table 14.4. In the first step, (3,5) becomes a cluster. At the second step, locality 1 is joined to this cluster. At steps 4, 6, 7, and 8, no new clusters result. By the 17th step, we have clusters at the level $d_{57} = 3.45$,

$$(1, 2, 3, 4, 5, 6, 7), \qquad (8, 9, 10).$$

Of course, if we carry on further, we get a single cluster. At $d_0 = 3.45$, we infer that there are two species, one containing (1)–(7) and the other (8)–(10). In fact, the species containing shrews from Alderney (8), Guernsey (9), and Cap Gris-Nez (10) is named *Crocidura russula*, and the other is named *C. suaveolens*. Some of these facts were only appreciated in 1958; see Delany and Healy (1966). □

14.5.1.1 Dendrogram

A dendrogram is a graphical display of a hierarchical clustering. The diagram places the objects on the x-axis and the distances on the y-axis. Moving upward, the branching of the tree gives the order of the $n - 1$ links; the first fork represents the first link, the second fork the second link, and so on until all join together at the trunk. We illustrate this definition by an example.

Example 14.5.2 (Dendrogram for the single linkage cluster analysis of the shrew data). For the shrew data in Example 14.5.1, the single linkage cluster analysis is given in Table 14.4. The corresponding dendrogram is given in Figure 14.4. The first link joins $(3, 5)$ = (St Agnes, St Mary's), etc. The level of each horizontal line shows the distance at which a link is formed (e.g. d_{35} is the level of the first link). The order of objects along the x-axis has been chosen to ensure that the branches do not cross. The dendrogram indicates that the distances between animals from the Scilly Isles are small, suggesting the origin from a common stock. The distance between shrews from Sark (6) and Jersey (7) is greater than between any pair of Scilly Island populations. The three populations of *Crocidura russula* (8, 9, 10) do not form as close a cluster as those of the Scilly Islands.

The value of the threshold $d_0 = 3.45$ is chosen by eye, without any probabilistic justification. For the complete dendrogram, take $d_0 = \infty$. Obviously, for a given d_0, the clusters can be read off from a dendrogram; e.g. for $d_0 = 3.70$, we obtain from Figure 14.4 the two clusters (1, 2, 3, 4, 5, 6, 7) and (8, 9, 10). Conversely, if we know g in advance, it is possible to choose a suitable threshold to produce this number of clusters. □

The dendrogram can be used to construct a new distance matrix between the objects. For two objects i and j, let u_{ij} be the *smallest* threshold d_0 for which i and j lie in the same cluster. The horizontal lines in the dendrogram are called *nodes*. Then, u_{ij} can be found from the dendrogram as the smallest node which is linked to both i and j. For the single linkage

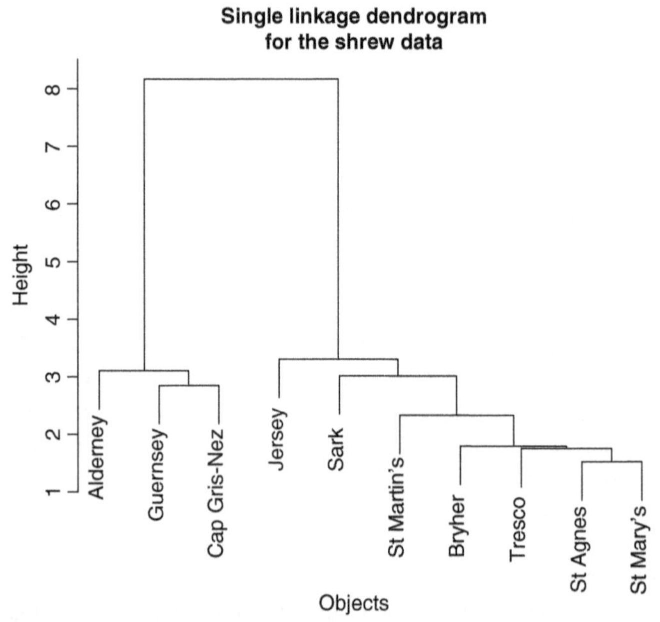

Figure 14.4 Dendrogram for shrew group means (single linkage) in Example 14.5.2.

dendrogram (although not in general for other dendrograms), the u_{ij} satisfy the *ultrametric inequality*

$$u_{ij} \leq \max(u_{ik}, u_{kj}) \qquad \text{for all} \quad i, j, k \tag{14.24}$$

(see Exercises 14.5.4 and 14.5.5).

14.5.2 Minimum Spanning Tree and Single Linkage

Start with a set of points $\boldsymbol{x}_1, \ldots, \boldsymbol{x}_n$, and suppose that some of the pairs of points $\boldsymbol{x}_r, \boldsymbol{x}_s$, say, $r \neq s$, are connected by edges. A *tree* is defined as a set of edges such that every point is connected to at least one other point by an edge, and there are no closed circuits (Chapter 8) of edges joining a point to itself. The tree property ensures that there is a unique path between $\boldsymbol{x}_r$ and $\boldsymbol{x}_s$ for all $1 \leq r \neq s \leq n$. If each edge is given a *length* equal to the distance between the points, the length of the tree is defined to be the sum of the lengths of all the edges. A *minimum spanning tree* (MST) is then defined to be the tree with minimum length.

Gower and Ross (1969) describe algorithms for computing the MST. A simple iterative algorithm begins with two sets of edges, A and B, say, where A is the empty set, and B contains all the edges in the tree. Each step of the iteration moves the smallest d_{rs} from B to A subject to the constraint that the move does not produce a circuit in A. It is easy to see that the process terminates when A contains $n - 1$ edges.

The MST has a number of useful features, the most obvious being that it highlights close neighbors when appropriately plotted. More importantly, it contains the same information as a single linkage cluster analysis. For simplicity, assume that there are no tied distances. Each horizontal bar in the dendrogram of a single linkage cluster analysis corresponds to an edge in the MST between the two closest points in the two clusters being joined.

Example 14.5.3 (MST for the shrew data). Consider again the shrew data, whose distances are given in Example 14.5.1. The minimal spanning tree is shown in Figure 14.5. The locations of the vertices are the first two canonical variates (as in 13.2), so the Euclidean distances in the figure approximate the Mahalanobis distances in Table 14.3.

The horizontal bars in the single linkage dendrogram (Figure 14.4) represent the joining of two clusters, and the distances of the closest points in the two clusters correspond to the

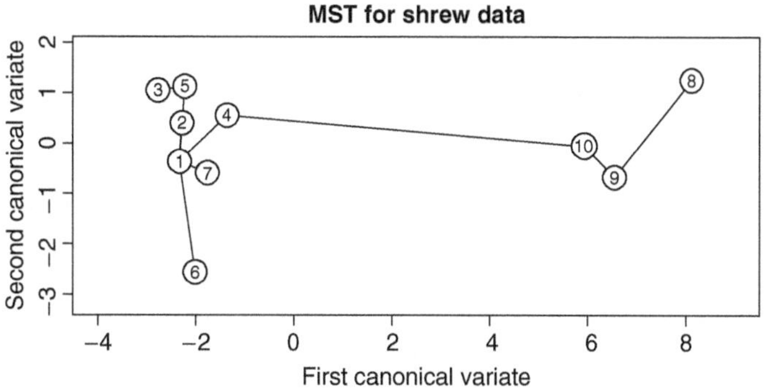

Figure 14.5 Minimum spanning tree for the shrew data whose distances are given in Table 14.3.

edges (3,5), (1,5), (1,2), (1,4), (9,10), (1,6), (8,9), (1,7), and (3,8), respectively. These are the edges appearing in the MST, which is a projection of the data. □

14.5.3 Properties of Different Agglomerative Algorithms

It is possible to make some general points about the agglomerative clustering methods in this section.

(1) *Chaining.* The single linkage (nearest-neighbor) method is the simplest to apply. However, it often leads to elongated clusters shaped like "rods" or "chains," partly because the links once made cannot be broken and partly because two clusters can be joined if there is at least one small distance between their members. In particular, single linkage will often give unsatisfactory results in the case that there are intermediates (as a result of noise) between clusters.

(2) *Compact cluster.* In contrast, the complete linkage method will tend to produce compact clusters since at threshold d_0, we have

$$\max_{r,s \in C_i} d_{rs} \leq d_0$$

for each cluster C_i. Spherical clusters are also more likely to occur with average linkage, median linkage, and Ward's minimum variance method. Further, the maximum-likelihood methods of Sections 14.2–14.3 tend to lead to spherical or elliptical clusters.

Example 14.5.4 (Complete linkage for the shrew data). For the shrew data, the smallest distance is $d_{35} = 1.53$, so that these two localities are the first to be joined into a cluster. The next smallest distance is $d_{15} = 1.76$, but we cannot add locality 1 to this cluster because $d_{13} = 2.39$ is too large. The next cluster is $(1, 2)$ since $d_{12} = 1.80$. Following this procedure, we find the dendrogram as given in Figure 14.6. Note that there is no chaining as in the single linkage case (see Figure 14.4). At $d_0 = 4$, we have groups $(1, 2, 3, 4, 5), (6, 7)$, and $(8, 9, 10)$, but at each stage the groups are *not* chained as seen in Figure 14.6. □

(2) *Monotonicity.* It can be seen that the clusters produced by single linkage are unchanged under any monotone transformation of the dissimilarities d_{ij}. The same property holds for complete linkage. However, this is not true for the other methods. Jardine and Sibson (1971) argue that this is a very important property.

(3) *Ties.* If there are ties in the smallest distance, then for the single linkage method, an arbitrary choice will not make a difference. However, for the other methods, different clusters (and different dendrograms) will sometimes arise depending on how the tie is broken for different selections.

(4) *Outliers.* These will often be apparent with the single linkage method as isolated clusters. However, they may be absorbed in larger clusters by other methods, which favor clusters of more equal size; hence, the outliers will be less obvious.

(5) The relation of the single linkage method to the Minimum Spanning Tree is fully described by Gower and Ross (1969). Jardine and Sibson (1971) present a set of axioms for hierarchical clustering – in which the single linkage method is the only one that is fully consistent. However, the chaining characteristic of the single linkage method makes it less attractive in practice.

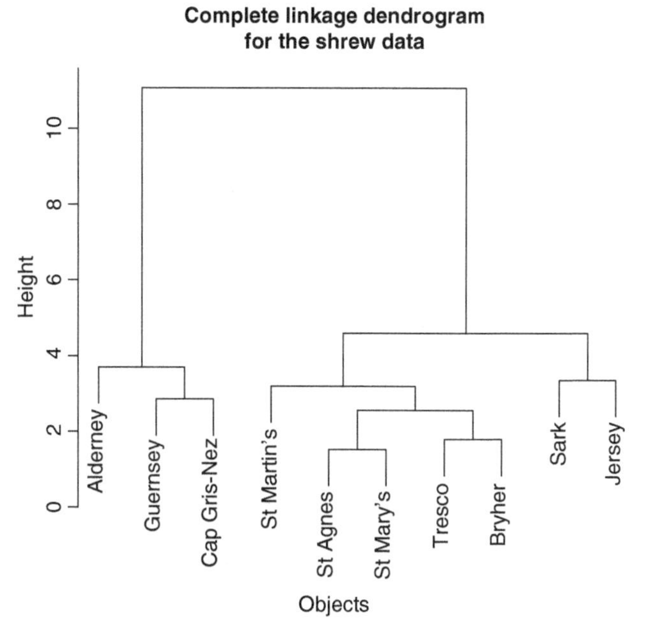

Figure 14.6 Dendrogram for the shrew data (complete linkage).

14.6 Distances and Similarities

14.6.1 Distances

Definition 14.6.1 *Let P and Q be two points where these may represent measurements on* $\boldsymbol{x}$ *and* $\boldsymbol{y}$ *on two objects. A real-valued function* $d(P, Q)$ *is a* distance function *if it has the following properties:*

(I) symmetry, $d(P, Q) = d(Q, P)$;
(II) nonnegativity, $d(P, Q) \geq 0$;
(III) identification mark $d(P, P) = 0$.

For many distance functions, the following properties also hold:

(IV) definiteness, $d(P, Q) = 0$ if and only if $P = Q$;
(V) triangle inequality, $d(P, Q) \leq d(P, R) + d(R, Q)$.

If (I)–(V) hold, d is called a metric.

For some purposes, it is sufficient to consider distance functions satisfying only (I)–(III), but we only consider distances for which (I)–(V) are satisfied unless otherwise mentioned. Note that (I) need not always be true; in sociology perhaps, subject P's opinion of subject Q measured by $d(P, Q)$ may differ from subject Q's opinion of subject P, $d(Q, P)$. Further, (V) is not satisfied for some distances (see Exercise 14.6.6). On the other hand, the distance may satisfy some stronger condition than (V), such as the ultrametric inequality in Equation (14.24).

One would expect $d(P, Q)$ to increase as "dissimilarity" or "divergence" between P and Q increases. Thus, $d(P, Q)$ is also described as a coefficient of dissimilarity even when it does not satisfy the metric properties (IV) and (V).

Quantitative Data

Euclidean distance. Let $\boldsymbol{X}$ be an $(n \times p)$ data matrix with rows $\boldsymbol{x}_1', \ldots, \boldsymbol{x}_n'$. Then, the Euclidean distance between the points $\boldsymbol{x}_r$ and $\boldsymbol{x}_s$ is d_{rs}, where

$$d_{rs}^2 = \sum_{k=1}^{p} (x_{rk} - x_{sk})^2 = ||\boldsymbol{x}_r - \boldsymbol{x}_s||^2. \tag{14.25}$$

The distance function satisfies properties (I)–(V). It also satisfies the following properties:

(a) *Positive semidefinite property.* Let $\boldsymbol{A} = (-d_{rs}^2/2)$. Then, $\boldsymbol{HAH}$ is p.s.d., where $\boldsymbol{H} = \boldsymbol{I} - n^{-1}\boldsymbol{11}'$ is the centering matrix. (For a proof of this result, see Section 15.2. In fact, Theorem 15.2.2 also gives the converse result.) This property will be of importance when we examine similarity coefficients.
(b) The distances d_{rs} are invariant under orthogonal transformations of the $\boldsymbol{x}$ data.
(c) We have the cosine law

$$d_{rs}^2 = b_{rr} + b_{ss} - 2b_{rs}, \tag{14.26}$$

where $b_{rs} = (\boldsymbol{x}_r - \overline{\boldsymbol{x}})'(\boldsymbol{x}_s - \overline{\boldsymbol{x}})$ is the centered inner product between $\boldsymbol{x}_r$ and $\boldsymbol{x}_s$. Another useful identity for calculation purposes is given by

$$\sum_{r=1}^{n}\sum_{s=1}^{n} d_{rs}^2 = 2n \sum_{r=1}^{n} b_{rr}. \tag{14.27}$$

Karl Pearson distance. When the variables are not commensurable, it is desirable to standardize (14.25); that is, we can use

$$d_{rs}^2 = \sum_{k=1}^{p} \frac{(x_{rk} - x_{sk})^2}{s_k^2}, \tag{14.28}$$

where s_k^2 is the variance of the kth variable. We call such a standardized distance the "Karl Pearson distance." This distance is then invariant under changes of scale. Another way is to replace s_k by the range

$$R_k = \max_{r,s} |x_{rk} - x_{sk}|. \tag{14.29}$$

For cluster analysis, one usually uses (14.28) except when the difference in scale between two variables is intrinsic, when one uses Euclidean distance.

Mahalanobis distance We define squared Mahalanobis distance between points $\boldsymbol{x}_r$ and $\boldsymbol{x}_s$ as

$$D_{rs}^2 = (\boldsymbol{x}_r - \boldsymbol{x}_s)'\hat{\boldsymbol{\Sigma}}^{-1}(\boldsymbol{x}_r - \boldsymbol{x}_s). \tag{14.30}$$

Its variants and properties have already been discussed in Section 1.6.

Qualitative Data (Multinomial Populations)

Consider a classification of individuals into p categories. For each $r = 1, \ldots, g$, let $\boldsymbol{x}_r' = (x_{r1}, \ldots, x_{rp})$ denote the observed proportions from a population of size n_r lying in each of these categories. For example, $\boldsymbol{x}_r$ might denote the proportions of people having blood group types A, B, AB, and O in each of g countries. Of course,

$$\sum_{i=1}^{p} x_{ri} = 1, \qquad r = 1, \ldots, g.$$

We now consider several choices of a distance measure between these populations.

Euclidean distance The points $\boldsymbol{x}_r$ lie on the hyperplane $x_1 + \cdots + x_p = 1$ in the positive orthant of $\mathbb{R}^p$. An obvious measure of distance between these points is Euclidean distance, given by

$$d_{rs}^2 = \sum_{i=1}^{p} (x_{ri} - x_{si})^2.$$

This distance may be suitable if the proportions are merely measured quantities for which no model of stochastic variation is being considered. However, if the proportions are thought of as random vectors, then a Mahalanobis-like distance is more appropriate.

A Mahalanobis-like distance Suppose that for each $r = 1, \ldots, g$, $\boldsymbol{x}_r$ represents the proportions based on a sample n_r from a multinomial distribution with parameter $\boldsymbol{a} = (a_1, \ldots, a_p)'$ (the same parameter for each r). Then, $\boldsymbol{x}_r$ has mean $\boldsymbol{a}$ and covariance matrix $\boldsymbol{\Sigma}_r = n_r^{-1}\boldsymbol{\Sigma}$, where $\boldsymbol{\Sigma} = (\sigma_{ij})$ is given by

$$\sigma_{ij} = \begin{cases} a_i(1 - a_i), & i = j, \\ -a_i a_j, & i \neq j; \end{cases} \tag{14.31}$$

that is, $\boldsymbol{\Sigma} = \text{diag}\,(\boldsymbol{a}) - \boldsymbol{a}\boldsymbol{a}'$. Since $\boldsymbol{x}_r$ lies on a hyperplane, $\boldsymbol{\Sigma}$ is singular. However, it is easily checked that a g-inverse of $\boldsymbol{\Sigma}$ is given by

$$\boldsymbol{\Sigma}^- = \text{diag}\,(a_1^{-1}, \ldots, a_p^{-1}). \tag{14.32}$$

Thus (see Example 12.4.1), we can define a (generalized) squared Mahalanobis distance between $\boldsymbol{x}_r$ and $\boldsymbol{x}_s$ as

$$\frac{n_r n_s}{n_r + n_s} \sum_{i=1}^{p} \frac{(x_{ri} - x_{si})^2}{a_i}. \tag{14.33}$$

Remarks (1) Unfortunately, there are two problems with this approach. First, in practice, we wish to compare multinomial populations with *different* parameters (and hence different covariance matrices). Thus, the sum in (14.33) only represents an "approximate" Mahalanobis distance, and $\boldsymbol{a}$ must be thought of as an "average" parameter for the populations. To reduce the effect of the differences between populations on sample size, we will drop the factor $n_r n_s/(n_r + n_s)$ in Equation (14.33).

(2) The second problem involves the estimate of the "average" parameter $\boldsymbol{a}$. A common procedure is to estimate $\boldsymbol{a}$ pairwise using

$$\hat{a}_i(r, s) = \frac{1}{2}(x_{ri} + x_{si})$$

(so $\hat{\boldsymbol{a}}(r,s)$ depends on r and s). Then, (14.33) becomes the Mahalanobis-like distance

$$D^2_{1;rs} = 2\sum_{i=1}^{p} \frac{(x_{ri} - x_{si})^2}{x_{ri} + x_{si}}. \tag{14.34}$$

This form was suggested by Mahalanobis (Bhattacharyya, 1946) and rediscovered by Sanghvi (1957).

(3) Other possibilities for estimating $\boldsymbol{a}$ include a global average of all the proportions ($\hat{a}_1 = g^{-1}(x_{1i} + \cdots + x_{gi})$) and/or weighting each proportion x_{ri} by its sample size n_r (Balakrishnan and Sanghvi, 1968).

(4) The distance in (14.34) is obtained using a pooled estimate of the parameter $\boldsymbol{a}$. An alternative procedure is to use a pooled estimate of $\boldsymbol{\Sigma}$, based on the sample covariance matrices, thus giving yet another Mahalanobis-like distance. See Exercise 14.6.1 and Balakrishnan and Sanghvi (1968).

Bhattacharyya distance. Let $\boldsymbol{v}_r = (x_{r1}^{1/2}, \ldots, x_{rp}^{1/2})'$, $r = 1, \ldots, g$, so that the vectors $\boldsymbol{v}_r$ are points on the unit sphere in $\mathbb{R}^p$ with center at the origin. The cosine of the angle B_{rs}, say, between $\boldsymbol{v}_r$ and $\boldsymbol{v}_s$, is

$$\cos B_{rs} = \sum_{i=1}^{p} v_{ri} v_{si} = \sum_{i=1}^{p} (x_{ri} x_{si})^{1/2}, \tag{14.35}$$

so that the angle B_{rs} is the great circle distance between $\boldsymbol{v}_r$ and $\boldsymbol{v}_s$. The squared Euclidean distance of the chord between $\boldsymbol{v}_r$ and $\boldsymbol{v}_s$ is given by

$$D^2_{2;rs} = \sum_{i=1}^{p} (x_{ri}^{1/2} - x_{si}^{1/2})^2. \tag{14.36}$$

We call $D_{2;rs}$ the *Bhattacharyya distance* (although sometimes the angle B_{rs} is given this name). The two measures are connected by

$$D^2_{2;rs} = 4\sin^2 \frac{1}{2} B_{rs}.$$

Remarks (1) If $\boldsymbol{x}_r$ and $\boldsymbol{x}_s$ come from multinomial populations with the *same* parameter $\boldsymbol{a}$, then $D^2_{1;rs}$ and $4D^2_{2;rs}$ are asymptotically the same for large n_r, n_s. See Exercise 14.6.2.

(2) Bhattacharyya distance can be interpreted as an asymptotic Mahalanobis distance. From Example 2.7.1, we see that $\boldsymbol{v}_r \sim N_p(\boldsymbol{b}, (4n_r)^{-1}\boldsymbol{\Sigma})$ for large n_r, where $b_i = a_i^{1/2}$, $i = 1, \ldots, p$, and $\boldsymbol{\Sigma} = \boldsymbol{I} - \boldsymbol{b}\boldsymbol{b}'$. Although $\boldsymbol{\Sigma}$ is singular, it is easy to see that a g-inverse of $\boldsymbol{\Sigma}$ is given by $\boldsymbol{\Sigma}^- = \boldsymbol{I}$. Thus, if $\boldsymbol{v}_r$ and $\boldsymbol{v}_s$ are proportions from multinomial distributions with the same parameter, the asymptotic squared Mahalanobis distance between them is given by

$$4n_r n_s (n_r + n_s)^{-1} D^2_{2;rs}.$$

(3) Note that with both D_1^2 and D_2^2, differences between very small proportions are given more weight than differences between intermediate or large proportions. However, with Euclidean distance, all such differences are weighted equally.

(4) Practical studies (Sanghvi and Balakrishnan, 1972) have shown that D_1^2 and D_2^2 are very similar to one another, and in practice, it hardly matters which one is chosen.

Table 14.5 Relative frequencies of blood groups A_1, A_2, B, and O for four populations.

Blood groups	Populations Eskimo	Bantu	English	Korean
A_1	0.2914	0.1034	0.2090	0.2208
A_2	0.0000	0.0866	0.0696	0.0000
B	0.0316	0.1200	0.0612	0.2069
O	0.6770	0.6900	0.6602	0.5723

Source: Data from Cavalli-Sforza and Edwards (1967).

Bhattacharyya distance is perhaps preferable because there is no need to estimate unknown parameters and because it has a simple geometric interpretation.

Example 14.6.1 Table 14.5 gives the relative gene frequencies for blood-group systems with types A_1, A_2, B, and O for large samples from four human populations: (1) Eskimo, (2) Bantu, (3) English, and (4) Korean. The object is to assess the affinities between the populations.

The (Bhattacharyya) distance matrix with elements B_{rs} (14.35) is found to be

	Bantu	English	Korean
Eskimo	23.27	16.40	16.87
Bantu		9.84	20.44
English			19.61

Use of the complete linkage clustering method suggests the two clusters Bantu-English and Eskimo-Korean. Cavalli-Sforza and Edwards (1967) came to this conclusion by a maximum-likelihood method. However, the single linkage method, which one might think would be appropriate in this situation, does not support this conclusion. □

Multiclassification Case

If there are t classifications instead of a single one; e.g. if there are gene frequencies for each of the main blood-group systems of classification, then we can sum the distances using

$$D^2 = \sum_{k=1}^{t} D^2_{f,k},$$

where $D^2_{f,k}, f = 1, 2$, is either the Mahalanobis distance or the Bhattacharyya distance between two populations on the kth classification.

The process of summing distances in this way is meaningful when the t types of classifications are independent.

A list of various distances is given in Table 14.6.

Table 14.6 Distances between two points $\boldsymbol{x}_r$ and $\boldsymbol{x}_s$.

1	Euclidean distance: $\left\{\sum_{k=1}^{p} w_k(x_{rk} - x_{sk})^2\right\}^{1/2}$. (a) Unstandardized, $w_k = 1$. (b) Standardized by SD, $w_k = 1/s_k^2$ (Karl Pearson distance). (c) Standardized by range $w_k = 1/R_k^2$.
2	Mahalanobis distance: $\{(\boldsymbol{x}_r - \boldsymbol{x}_s)'\boldsymbol{\Sigma}^{-1}(\boldsymbol{x}_r - \boldsymbol{x}_s)\}^{1/2}$, with $\boldsymbol{\Sigma}$ any transforming positive definite matrix.
3	City-block metric (Manhattan distance): $\sum_{k=1}^{p} w_k\lvert x_{rk} - x_{sk}\rvert$. Mean character difference $w_k = 1/p$.
4	Minkowski metric: $\left\{\sum_{k=1}^{p} w_k\lvert x_{rk} - x_{sk}\rvert^{\lambda}\right\}^{1/\lambda}, \lambda \geq 1$.
5	Canberra metric: $\sum_{k=1}^{p} \dfrac{\lvert x_{rk} - x_{sk}\rvert}{x_{rk} + x_{sk}}$. (Scaling does not depend on the whole range of the variable.)
6	Bhattacharyya distance (for proportions): $\left\{\sum_{r=1}^{p}\left(x_{rk}^{1/2} - x_{sk}^{1/2}\right)^2\right\}^{1/2}$.
7	Distances between groups (see Section 14.7): (a) Karl Pearson dissimilarity coefficient: $\left\{\frac{1}{p}\sum_{k=1}^{p}\dfrac{(\overline{x}_{rk} - \overline{x}_{sk})^2}{(s_{rk}^2/n_r) + (s_{sk}^2/n_s)}\right\}^{1/2}$, where n_j = size of the jth sample, $j = r, s$ and $\overline{x}_{jk}, s_{jk}^2$ = mean and variance of kth variable for the jth sample. (b) Mahalanobis distance: $\{(\overline{\boldsymbol{x}}_r - \overline{\boldsymbol{x}}_s)'\hat{\boldsymbol{\Sigma}}^{-1}(\overline{\boldsymbol{x}}_r - \overline{\boldsymbol{x}}_s)\}^{1/2}$.

14.6.2 Similarity Coefficients

So far, we have concentrated on measures of distance or dissimilarity, but there are situations as in taxonomy where it is often common to use measure of similarity between points A and B.

Definition 14.6.2 *A reasonable measure of* similarity, $s(A, B)$, *should have the following properties:*

(i) $s(A, B) = s(B, A)$,
(ii) $s(A, B) > 0$,
(iii) $s(A, B)$ *increases as the similarity between A and B increases.*

Because greater similarity means less dissimilarity, similarity coefficients can be used in any of the hierarchical techniques of Section 14.4 simply by changing the sign of all the inequalities.

We now consider some examples.

Qualitative Variables

Let the presence or absence of p attributes on two objects P and Q be denoted $(x_1, \dots, x_p)$ and $(y_1, \dots, y_p)$, where $x_i = 1$ or 0 depending on whether the ith attribute is present or absent for object P.

Set

$$\begin{aligned} a &= \sum x_i y_i, & b &= \sum (1 - x_i) y_i, \\ c &= \sum x_i (1 - y_i), & d &= \sum (1 - x_i)(1 - y_i); \end{aligned} \tag{14.37}$$

that is, a, b, c, and d are the frequencies of $(x_i, y_i) = (1,1)$, $(0,1)$, $(1,0)$, and $(0,0)$, respectively.

The simplest measure of similarity between P and Q is

$$s_1(P, Q) = \frac{a}{p}. \tag{14.38}$$

An alternative is the *simple matching coefficient* (Sokal and Michener, 1958) defined as

$$s_2(P, Q) = \frac{a + d}{p}, \tag{14.39}$$

which satisfies $s_2(P, P) = 1$.

It is not clear in practice whether to use s_1, s_2 or some other association coefficient. In s_2, all matched pairs of variables are equally weighted, whereas in s_1, negative matches are excluded. In both s_1 and s_2, every attribute is given equal weight, but in some applications, it might be preferable to use a differential weighting of attributes. For a discussion of these problems, see Jardine and Sibson (1971, ch. 4) and Everitt et al. (2001, pp. 35–39).

Let $\boldsymbol{X}$ be the data matrix containing the presence/absence information on p attributes for n objects. The matrix $\boldsymbol{X}$ is called an *incidence matrix* since $x_{rs} = 1$ or 0. In view of (14.37), the matrix of similarities based on (14.38) is simply

$$\boldsymbol{S}_1 = \frac{(\boldsymbol{XX}')}{p}, \qquad \boldsymbol{S}_1 = (s_{rs}^{(1)}), \tag{14.40}$$

whereas the matrix of similarities based on (14.39) is

$$\boldsymbol{S}_2 = \frac{\boldsymbol{XX}' + (\boldsymbol{J} - \boldsymbol{X})(\boldsymbol{J} - \boldsymbol{X})'}{p}, \tag{14.41}$$

where $\boldsymbol{J} = \boldsymbol{11}'$, $\boldsymbol{S}_2 = (s_{rs}^{(2)})$. Note that the diagonal elements of $\boldsymbol{S}_2$ are 1. Clearly, (14.40) and (14.41) satisfy $\boldsymbol{S}_1 \geq 0$ and $\boldsymbol{S}_2 \geq 0$.

Example 14.6.2 Consider the (6×5) data matrix of Example 9.5.2 in which $x_{rj} = 1$ if the rth grave contains the jth variety of pottery and 0 otherwise. The aim here is to see which graves have similar varieties of pottery to one another.

It is found that $5\boldsymbol{S}_2$ for $A, \dots, F$ is

$$\begin{array}{c c} & \begin{matrix} A & B & C & D & E & F \end{matrix} \\ \begin{matrix} A \\ B \\ C \\ D \\ E \\ F \end{matrix} & \begin{bmatrix} 5 & 0 & 3 & 5 & 1 & 3 \\ 0 & 5 & 2 & 0 & 4 & 2 \\ 3 & 2 & 5 & 3 & 1 & 3 \\ 5 & 0 & 3 & 5 & 1 & 3 \\ 1 & 4 & 1 & 1 & 5 & 3 \\ 3 & 2 & 3 & 3 & 3 & 5 \end{bmatrix} \end{array}.$$

We can now use any clustering method of Section 14.4 to see if there are any clusters. For example, using single linkage, A and D are grouped together first (which is not surprising since they are identical), and then B and E. The next link joins all the graves into a single group. □

Mixed Variables

If there are qualitative as well as quantitative variables, Gower (1971a) has proposed the following similarity coefficient between the rth and sth points:

$$s_{rs}^{(3)} = 1 - \frac{1}{p}\sum_{k=1}^{p} w_k |x_{rk} - x_{sk}|, \tag{14.42}$$

where $w_k = 1$ if k is qualitative, and $w_k = 1/R_k$ if k is quantitative, where R_k is the range of the kth variable.

It can be shown that the matrix $(s_{rs}^{(3)})$ is positive semidefinite, but if R_k is replaced by the sample standard deviation s_k, this may not be so (see Exercise 14.6.3).

In some applications, the data consists of an $(n \times n)$ matrix $\boldsymbol{D}$ consisting of the distances (or similarities) between the points, rather than an $(n \times p)$ data matrix $\boldsymbol{X}$. In this situation, the choice of distance has already been made, and one can immediately apply any of the hierarchical techniques of Section 14.5. Examples of such data are quite common in the context of multidimensional scaling (see Chapter 15).

14.7 Grouped Data

In the case in which each observation belongs to a known group, it may seem as if cluster analysis has been made redundant. However in this situation, which is known as the *multisample* case, the aim is to group the samples into those that are more similar. The problem can be better understood through a specific situation in zoology already described in Example 13.5.2. It is known that the white-toothed shrews of the known genus *Crocidura* occur in the Channel and Scilly Islands of the British Isles and the French mainland. A large number of observations from 10 localities in the Channel and Scilly Islands were obtained to examine the belief that there may be two species of *Crocidura*. The localities were geographically close, but it is assumed that only one subspecies was present in any one place. The problem here is to group "samples" rather than individual observations. In general, the problem can be summarized as follows.

Multisample problem Let $\boldsymbol{x}_{jr}, r = 1, \ldots, n_j$, be the observations in the jth (random) sample, $j = 1, \ldots, m$. The aim of cluster analysis is to group the m samples into g homogeneous classes where g is unknown, $g \leq m$.

It would be possible to assume that the m groups are each normally distributed with a common covariance matrix. In this case, we could use the model $\overline{\boldsymbol{x}}_j \sim N_p(\boldsymbol{\mu}_j, n^{-1}\boldsymbol{\Sigma})$, independently for $j = 1, \ldots, m$, with a consistent estimate of $\boldsymbol{\Sigma}$ given by

$$\hat{\boldsymbol{\Sigma}} = \boldsymbol{W}/(n - m), \tag{14.43}$$

where, in the notation of Chapter 13, $\boldsymbol{W} = \boldsymbol{W}_1 + \cdots + \boldsymbol{W}_m$ is the within-samples SSP matrix and $n = n_1 + \cdots + n_m$. Letting C_k denote the set of $\overline{\boldsymbol{x}}_j$ assigned to the kth group with

$$n_k^* = \sum_{\{j:\overline{\boldsymbol{x}}_j \in C_k\}} n_j, \qquad \overline{\overline{\boldsymbol{x}}}_k = \sum_{\{j:\overline{\boldsymbol{x}}_j \in C_k\}} \frac{n_j \overline{\boldsymbol{x}}_j}{n_k^*}$$

so that $\overline{\overline{\boldsymbol{x}}}_k$ is the weighted mean vector of the means in the kth grouping, then our objective is to choose the C_k to maximize

$$b_g^2 = \sum_{k=1}^{g} n_k^* (\overline{\overline{\boldsymbol{x}}}_k - \overline{\boldsymbol{x}})' \hat{\boldsymbol{\Sigma}}^{-1} (\overline{\overline{\boldsymbol{x}}}_k - \overline{\boldsymbol{x}}), \tag{14.44}$$

with $\overline{\boldsymbol{x}}$ denoting the mean vector of the pooled sample.

In practice, it may be rather unrealistic to suppose that all of the groups have the same covariance, with the result that a pooled estimate leads to unwanted clusters of groups. An alternative strategy is to use the two-sample Hotelling T^2 test (see Section 6.3.3.1) to compare each pair of groups. The test that produces the largest p-value then suggests which two groups to "merge," and the procedure can continue with the merged group replacing its two components. This process can be repeated, either until there is only one group or until the largest p-value is smaller than some user-specified threshold.

Example 14.7.1 In a study into the possibility of using forensic data to determine how easy it is to distinguish different glass types, various measurements were taken on 214 samples of glass. The types of glass belonged to one of six groups (with sample sizes):

- building windows – float processed (70)
- building windows – nonfloat processed (76)
- vehicle windows – float processed (17)
- containers (13)
- tableware (9)
- headlamps (29)

and on each glass sample, nine real-valued measurements were taken: Refractive index Sodium, Magnesium, Aluminum, Silicon, Potassium, Calcium, Barium, Iron (the chemical measurements were expressed as a weight percent in the corresponding oxide). The data are available on the UCI machine learning repository[1] , and the background to the original problem is given in Evett and Spiehler (1987).

At the first step, we compare all pairs of groups, with the largest p-value observed in the test between the float-processed windows (vehicle and building). After merging of these groups, the next largest p-value was observed between the container and the tableware groups. Using a "distance" between clusters which is $-\log(p\text{-value})$, it is possible to represent the potential clustering as a dendrogram, and this is shown in Figure 14.7. It can be seen that the window glass is clearly distinct from the nonwindow glass – the final two groups had a p-value very close to zero. □

1 http://archive.ics.uci.edu/ml/machine-learning-databases/glass/

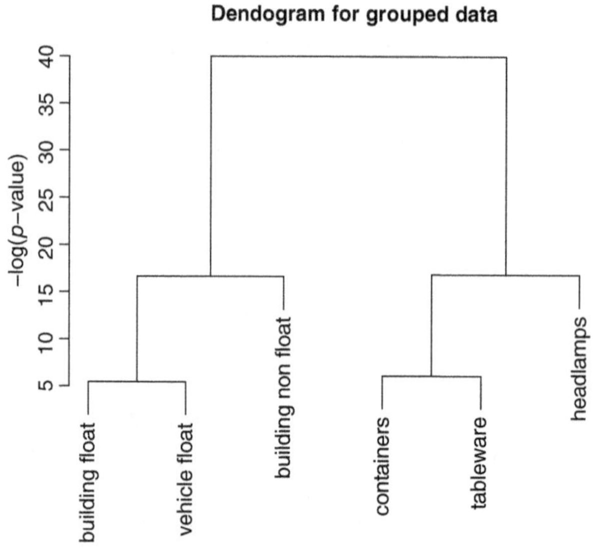

Figure 14.7 Hierarchical clusters for grouped glass data. Groups are sequentially grouped according to a p-value from Hotelling's T^2 statistic, with a distance between groups plotted as $-\log P(X > T^2_{\text{obs}})$, where X has a Hotelling T^2 distribution with parameters determined by the relevant sample sizes.

14.8 Mode Seeking

Wishart's method (Wishart, 1969) searches for "dense points" where k or more observations are located within a distance R. Starting with small values of R, the method looks at a hypersphere of radius R around each observation and counts the number of other observations within this hypersphere. If the number of observations is at least k, then the center is called a *dense point*. The parameter R is gradually increased, so that more and more points become dense, until all the points lie within a single hypersphere.

The concept of a dense point can be made more precise by considering the kernel density estimate based on all the data, i.e.

$$\hat{f}(\boldsymbol{x};\boldsymbol{H}) = \frac{1}{n}\sum_{r=1}^{n} K_{\boldsymbol{H}}(\boldsymbol{x}-\boldsymbol{x}_r), \tag{14.45}$$

where $K_{\boldsymbol{H}}(\boldsymbol{x})$ is a kernel function which depends on a matrix $\boldsymbol{H}$ of smoothing parameters (see Section 12.7.3). For example, we could use

$$K_{\boldsymbol{H}}(\boldsymbol{x}) = \begin{cases} 1/v_p(R) & \text{if } \|\boldsymbol{x}\| < R, \\ 0 & \text{otherwise,} \end{cases}$$

where $v_p(R)$ is the volume of a p-dimensional hypersphere of radius R (in which R takes the role of the smoothing parameter). In this case, Wishart's method is equivalent to finding all points $\boldsymbol{x}$ for which $\hat{f}(\boldsymbol{x};\boldsymbol{H})$ is greater than some threshold.

More generally, we may seek to find all points $\boldsymbol{x}$ for which $\hat{f}(\boldsymbol{x};\boldsymbol{H})$ is a local maximum. If we use a smooth kernel function K, then there will be a finite number of such points. Clearly, the number and location of these critical points will depend on the set of smoothing parameters $\boldsymbol{H}$. In the limiting case (as $\boldsymbol{H} \to \boldsymbol{0}$), there will be n points, located at each

$\boldsymbol{x}_r$, $r = 1, \ldots, n$. Conversely, as $\boldsymbol{H}$ becomes large, there will be only one mode, located at $\overline{\boldsymbol{x}}$. In the one-dimensional case ($p = 1$), Silverman (1981) proved that, if K is a normal kernel, then the number of modes is a nonincreasing function of the smoothing parameter h, although this is not generally the case when $p > 1$.

For a given smoothing parameter matrix $\boldsymbol{H}$, we can define the cluster centers as those $\boldsymbol{x}$ for which $\hat{f}(\boldsymbol{x}; \boldsymbol{H})$ is a local maximum. Moreover, using standard hill-climbing methods, we can assign each $\boldsymbol{x}_r$, $r = 1, \ldots, n$ to a cluster center by simply moving uphill from $\hat{f}(\boldsymbol{x}_r; \boldsymbol{H})$ until a mode is reached. In this way, each observation is assigned to a cluster.

Such methods are quite computationally demanding, but the main difficulty lies in knowing which smoothing parameter to use.

Example 14.8.1 (Iris area data, mode-based clustering) We again use the iris area data of Example 14.4.1. Given a smoothing parameter matrix $\boldsymbol{H} = h\boldsymbol{I}_2$ and a normal kernel, we compute $\hat{f}(\boldsymbol{x}; h\boldsymbol{I}_2)$ for various values of h. These estimates can be visualized through contour plots as shown in Figure 14.8. Using numerical methods, we can locate all points where

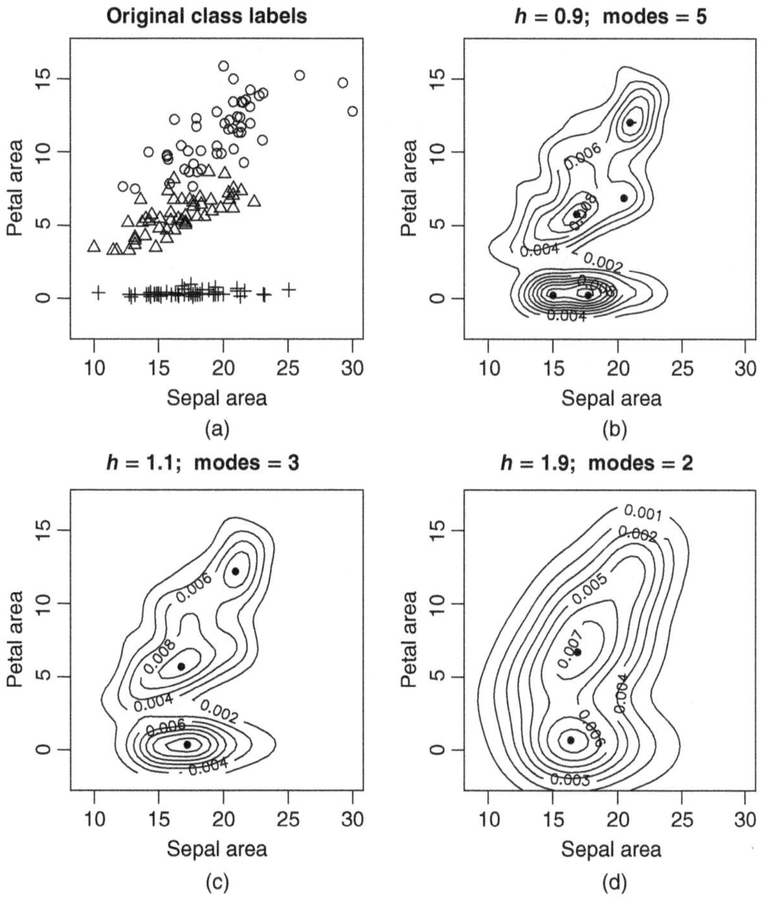

Figure 14.8 Contour plots of the kernel density estimate for the transformed iris data. (a) Original labels for the three species. Other panels (b–d) show contours of the kernel density estimate (of the pooled data) for various smoothing parameters h, with the location of the modes also identified.

Table 14.7 Comparison of cluster allocations based on hill -climbing and mode identification of kernel density estimate corresponding to $h = 1.1$ (see lower left plot of Figure 14.8) with the true classification.

		Allocated to cluster		
		1	2	3
	I. setosa	50	0	0
Actual class	*I. versicolor*	0	50	0
	I. virginica	0	19	31

Note that the three modes are labeled arbitrarily, and that three outlying observations were allocated to the nearest mode.

$\hat{f}(\boldsymbol{x})$ is a local maximum, and these have been added to the plots. As expected, the number of modes, and hence the number of identified clusters, is dependent on h. Using the three mode solution ($h = 1.1$), we use standard hill climbing to allocate each observation to one of the three modes. This allocation is then compared with the true clusters, as shown in Table 14.7. Small modes for which the local maximum was less than $0.1\times$ the global maximum have been excluded. The allocations in Table 14.7 can be compared to the allocations in Example 14.4.1. In particular, they are similar to the allocations from the k-means cluster solution with $g = 3$ groups. □

14.9 Measures of Agreement

Deciding whether the results of a clustering algorithm are meaningful is by no means straightforward. Given the nature of the problem, the "true" subgroups are unknown, so there is no natural goodness-of-fit measure as there is in the case of regression or discrimination problems. It is, of course, possible to simulate data from specified models and then to compare the results of an algorithm with the known group memberships, but there are countless models, and finding one which is appropriate for a given real data set is problematic.

Several *internal* measures have been proposed, which measure the *distortion* of a cluster analysis. Clusters will usually have a representation – for example, the cluster mean – and so we can compare the similarity/dissimilarity between observations and the similarity/dissimilarity of the cluster representation of those observations. In the review by Cormack (1971), there is a list of distortion measures that have been proposed. For example, if d_{rs} denotes the dissimilarity between observations $\boldsymbol{x}_r$ and $\boldsymbol{x}_s$, and d^*_{rs} denotes the dissimilarity of the cluster representations of these observations, then Guttman (1968) proposes that

$$\frac{\sum_{r,s} d_{rs} d^*_{rs}}{\left(\sum_{r,s} d^2_{rs} \sum_{r,s} d^{*2}_{rs}\right)^{1/2}}$$

as a distortion measure that can be used as an internal measure of cluster validity. This can be thought of as a measure of classifiability which is scale-free.

Rousseeuw (1987) proposed a measure, designed for graphical analysis, known as the *silhouette* value. Start with a clustering with clusters C_l, $l = 1, \dots, g$. If an observation $\boldsymbol{x}_r$ has been allocated to cluster C_k, let

$$\overline{d}(r,k) = \frac{1}{n_k} \sum_{\boldsymbol{x}_s \in C_k} d_{rs}$$

be the average distance from observation $\boldsymbol{x}_r$ to observations in cluster C_k. The silhouette value for this observation is then defined by

$$h(r) = \frac{b(r) - a(r)}{\max\{a(r), b(r)\}},$$

where $a(r) = \overline{d}(r,k)$ and $b(r) = \min_{l \neq k} \overline{d}(r,l)$. This measure compares the similarity of $\boldsymbol{x}_r$ to its assigned cluster to the similarity with its second nearest cluster.

Suppose that a clustering has been defined for various choices of g (e.g. using k-means clustering). The silhouette coefficient (Kaufman and Rousseeuw, 1990) allows for a comparison between these choices; it is defined by $\overline{h}(g)$, the average of $h(r), r = 1, \dots, n$, when using the clustering into g clusters. Then, the "optimal" choice of g is defined to be the value of g maximizing $\overline{h}(g)$.

A simulation study that compares the recovery capacity of the various indices has been carried out by Milligan (1981). In this study, one of the most reliable measures was found to be the Gamma index, as proposed by Goodman and Kruskal (1954). This is given by

$$(S_1 - S_2)/(S_1 + S_2),$$

where

$$S_1 = \sum_{r<s} I[C(\boldsymbol{x}_r) \neq C(\boldsymbol{x}_s)] I[d_{rs} > D],$$
$$S_2 = \sum_{r<s} I[C(\boldsymbol{x}_r) \neq C(\boldsymbol{x}_s)] I[d_{rs} < D],$$

$C(\boldsymbol{x})$ denotes the cluster label associated with observation $\boldsymbol{x}$, and $D = \max_k D_k$, with $D_k = \max_{\{\boldsymbol{x}_r, \boldsymbol{x}_s \in C_k\}} d_{rs}$.

External measures have been proposed to summarize the agreement between two cluster partitions. These could be partitions that have resulted from distinct clustering algorithms or – in the case of simulated data – a comparison between the "true" classes, and the output of a cluster analysis. There are many agreement measures that have been suggested; see Meilă (2007), Wagner and Wagner (2007) for some reviews. These measures can be grouped into those which:

- count pairs of observations that the partitions agree are in the same cluster (or not);
- use measures based on the concept of overlapping sets;
- use measures of mutual information.

An example of a measure that uses the first criterion is the Rand index (Rand, 1971). Here, a count is made of the number of pairs of elements that are in the same set in the first partition and in the same set in the second partition (say a) and the number of pairs of elements that are in different sets in the first partition and in different sets in the second partition (say b). Then, the Rand index is given by

$$I_R = \frac{a+b}{\binom{n}{2}}. \tag{14.46}$$

A different approach is taken by Tibshirani and Walther (2005) who develop a validation measure based on a train-test split of the data which is assessed using "prediction strength." This is also used to choose the number of clusters.

Example 14.9.1 Information on some development indicators is available from the `World DataBank` (World Bank Group, 2014), and the following variables were selected for each country:

- Birth rate, crude (per 1,000 people)
- Death rate, crude (per 1,000 people)
- Life expectancy at birth, total (years)
- GDP per capita (constant 2005 US$)
- GDP growth (annual %)

After removing the missing data, there were 182 countries for which full information was available in 2001. A log transformation of GDP per capita was used to symmetrize this variable, and then all variables were standardized before a cluster analysis. Using hierarchical clustering with complete linkage, we cut the dendrogram so as to obtain 11 clusters. Using the resulting cluster means, we plot the dendrogram of these 11 clusters in Figure 14.9.

It may be noted that the richer countries dominate cluster 3, clusters 2 and 7 (which are close to each other) contain only countries in the African continent, and cluster 5 almost entirely countries from the former communist blocks. A map that allocates a shading to each country based on cluster membership (only for clusters 1– 7 and 9) is shown in Figure 14.10.

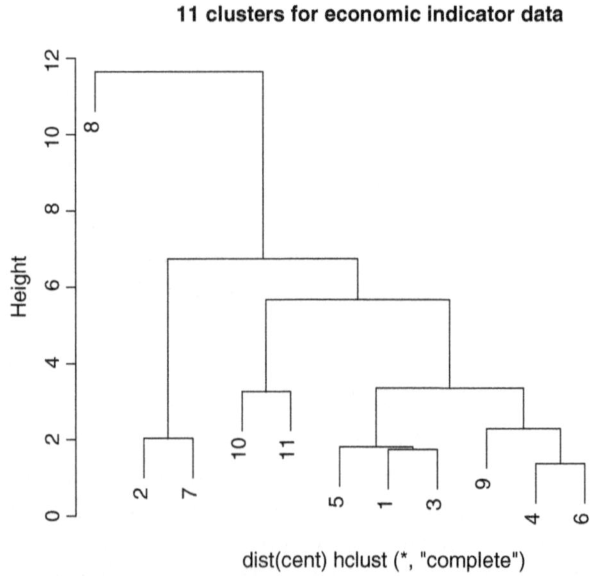

Figure 14.9 Clusters of countries based on five indicators, after selecting complete linkage, and 11 clusters from the full dendrogram. A map showing the members of the first seven clusters is shown in Figure 14.10.

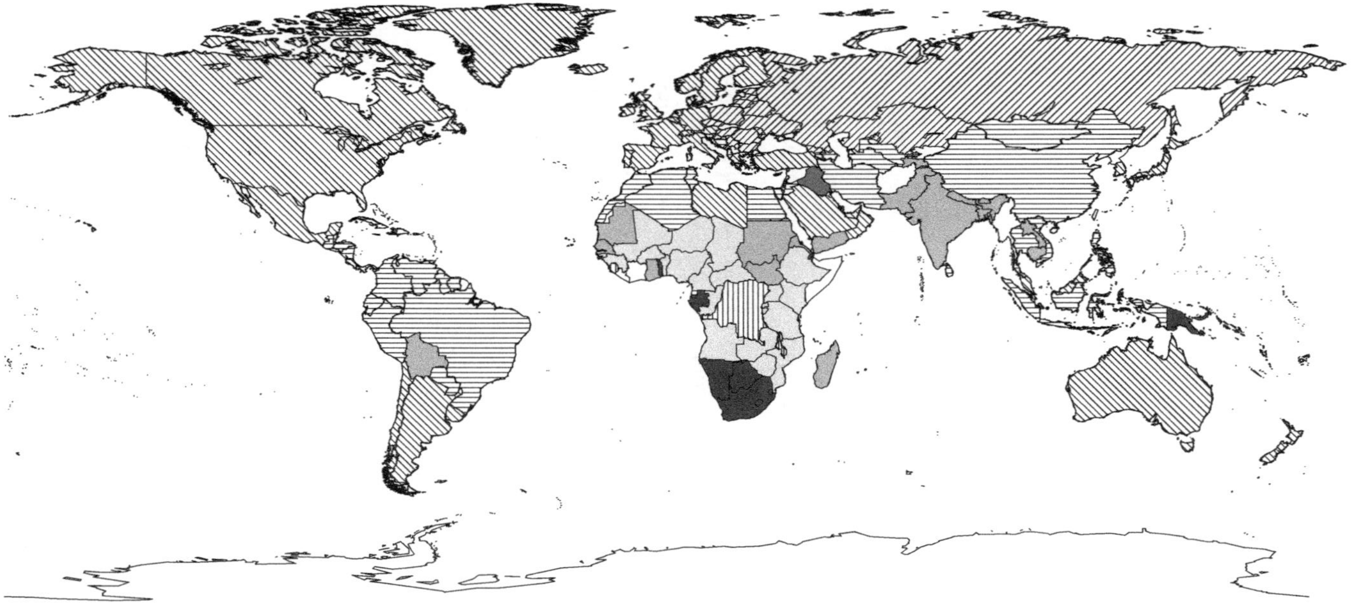

Figure 14.10 World map for the 11 clusters of the dendogram of Figure 14.9. The members of 8 clusters are indicated by shading: (1) horizontal lines, (2) light gray, (3) angled lines (\\), (4) mid-gray, (5) angled lines (//), (6) very dark gray, (7) vertical lines, and (9) dark gray. Those clusters not marked are: (8) Equatorial Guinea; (10) Kosovo; (11) Liberia. The remaining countries (unshaded) were omitted from the analysis due to missing data.

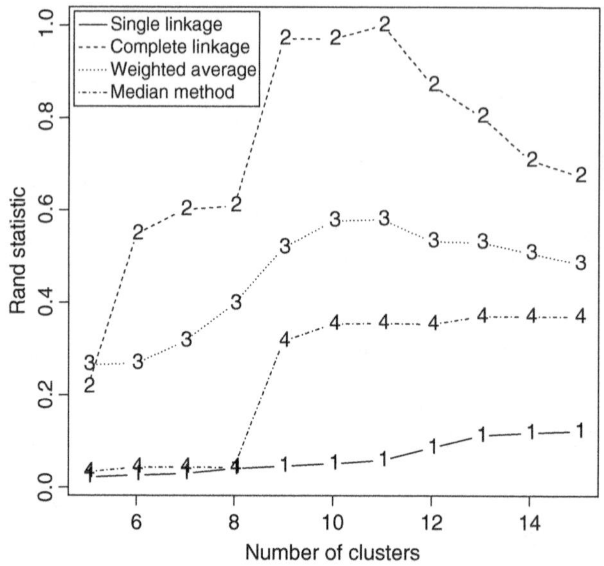

Figure 14.11 Rand index showing the agreement between the complete linkage cluster solution found in Figure 14.9, with various other clustering methods and cluster size solutions.

In this example, we have some idea of whether the clustering "makes sense," but we can carry out a more formal (though still somewhat *ad hoc*) validation by comparing this 11-cluster solution with those obtained by other clustering methods. Figure 14.11 plots the Rand statistic (see Eq. (14.46)) that measures an agreement between this 11-cluster solution and the solution obtained by the following clustering methods for various numbers of clusters: single linkage, complete linkage, average linkage, and median method (see Section 14.5.1). It can be seen that the average linkage method tends to give more similar clusters for this data set, with very different clusters given by the single linkage method. The somewhat inconclusive nature of this example illustrates the difficulties in validating a cluster analysis, when the true clusters are unknown.

Exercises and Complements

14.2.1 (Scott and Symonds, 1971) Let us suppose $\boldsymbol{x}_1, \dots, \boldsymbol{x}_n$ to be a mixed sample from the $N_p(\boldsymbol{\mu}_k, \boldsymbol{\Sigma}_k), k = 1, \dots, g,$ populations where the $\boldsymbol{\mu}_k$ are unknown. If $\boldsymbol{\Sigma}_1 = \cdots = \boldsymbol{\Sigma}_g = \boldsymbol{\Sigma}$ where $\boldsymbol{\Sigma}$ is known, show that the ML partition $\hat{\gamma}$ minimizes $\operatorname{tr}(\boldsymbol{W}\boldsymbol{\Sigma}^{-1})$. Show that it is equivalent to maximizing the weighted between-groups sum of squares

$$\sum_{k=1}^{g} n_k(\overline{\boldsymbol{x}}_k - \overline{\boldsymbol{x}})'\boldsymbol{\Sigma}^{-1}(\overline{\boldsymbol{x}}_k - \overline{\boldsymbol{x}}),$$

where $\overline{\boldsymbol{x}}_k$ is the mean for a cluster C_k, $k = 1, \dots, g$.

14.2.2 Let us suppose $\boldsymbol{x}_1, \dots, \boldsymbol{x}_n$ to be a mixed sample from the $N_p(\boldsymbol{\mu}_k, \boldsymbol{\Sigma}_k), k = 1, \dots, g,$ populations where $\boldsymbol{\mu}_k$ and $\boldsymbol{\Sigma}_k$ are unknown. Further, let $\boldsymbol{y}_{k1}, \dots, \boldsymbol{y}_{km_k}, k = 1, \dots, g,$

be previous samples of independent observations known to come from $N_p(\boldsymbol{\mu}_k, \boldsymbol{\Sigma}_k)$, $k = 1, \ldots, g$, respectively. Show that the m.l.e.s of $\boldsymbol{\mu}_k, \boldsymbol{\Sigma}_k$ and the partition $\boldsymbol{\gamma}$ of the $\boldsymbol{x}$s are given by

$$\hat{\boldsymbol{\mu}}_k = \frac{n_k\bar{\boldsymbol{x}}_k + m_k\bar{\boldsymbol{y}}_k}{m_k + n_k}, \qquad \hat{\boldsymbol{\Sigma}}_k = \frac{\boldsymbol{W}_{kx} + \boldsymbol{W}_{ky} + \boldsymbol{W}_{kxy}}{m_k + n_k},$$

and $\hat{\boldsymbol{\gamma}}$ minimizes

$$\prod_{k=1}^{g} |\hat{\boldsymbol{\Sigma}}_k|^{m_k+n_k},$$

where

$$\boldsymbol{W}_{kx} = \sum_{r\in C_k} (\boldsymbol{x}_r - \bar{\boldsymbol{x}})(\boldsymbol{x}_r - \bar{\boldsymbol{x}})', \qquad \boldsymbol{W}_{ky} = \sum_{r\in C_k} (\boldsymbol{y}_{kr} - \bar{\boldsymbol{y}}_k)(\boldsymbol{y}_{kr} - \bar{\boldsymbol{y}}_k)',$$

and

$$\boldsymbol{W}_{kxy} = \frac{m_k n_k}{m_k + n_k}(\bar{\boldsymbol{x}}_k - \bar{\boldsymbol{y}}_k)(\bar{\boldsymbol{x}}_k - \bar{\boldsymbol{y}}_k)'.$$

14.2.3 (Day, 1969) Let $\boldsymbol{x}_1, \ldots, \boldsymbol{x}_n$ be drawn from the mixture

$$f(\boldsymbol{x}) = p_1 f_1(\boldsymbol{x}) + p_2 f_2(\boldsymbol{x}),$$

where f_i is the p.d.f. of $N_p(\boldsymbol{\mu}_i, \boldsymbol{\Sigma})$, $i = 1, 2$, and $p_1 + p_2 = 1$.
(a) Show that the mean $\boldsymbol{m}$ and the covariance matrix $\boldsymbol{V}$ of the mixture are given by

$$\boldsymbol{m} = p_1\boldsymbol{\mu}_1 + p_2\boldsymbol{\mu}_2, \qquad \boldsymbol{V} = \boldsymbol{\Sigma} + p_1p_2(\boldsymbol{\mu}_1 - \boldsymbol{\mu}_2)(\boldsymbol{\mu}_1 - \boldsymbol{\mu}_2)'.$$

Show that the m.l.e.s of $\boldsymbol{m}$ and $\boldsymbol{V}$ are given by

$$\hat{\boldsymbol{m}} = \bar{\boldsymbol{x}} \qquad \text{and} \qquad \hat{\boldsymbol{V}} = \boldsymbol{S}.$$

Let

$$\boldsymbol{a} = \boldsymbol{\Sigma}^{-1}(\boldsymbol{\mu}_2 - \boldsymbol{\mu}_1), \qquad b = \frac{1}{2}\left(\boldsymbol{\mu}_1'\boldsymbol{\Sigma}^{-1}\boldsymbol{\mu}_1 - \boldsymbol{\mu}_2'\boldsymbol{\Sigma}^{-1}\boldsymbol{\mu}_2\right) + \log(p_2/p_1).$$

Show that

$$\hat{\boldsymbol{a}} = \hat{\boldsymbol{V}}^{-1}(\hat{\boldsymbol{\mu}}_1 - \hat{\boldsymbol{\mu}}_2)/\{1 - \hat{p}_1\hat{p}_2(\hat{\boldsymbol{\mu}}_1 - \hat{\boldsymbol{\mu}}_2)'\hat{\boldsymbol{V}}^{-1}(\hat{\boldsymbol{\mu}}_1 - \hat{\boldsymbol{\mu}}_2)\}, \tag{*}$$

$$\hat{b} = -\frac{1}{2}\hat{\boldsymbol{a}}'(\hat{\boldsymbol{\mu}}_1 + \hat{\boldsymbol{\mu}}_2) + \log(\hat{p}_2/\hat{p}_1). \tag{**}$$

(b) If $\mathsf{P}(k \mid \boldsymbol{x}_j)$ is the probability that observation $\boldsymbol{x}_j$ arises from the component k, then the m.l.e.s of $\mathsf{P}(k \mid \boldsymbol{x}_j)$ are

$$\hat{\mathsf{P}}(1 \mid \boldsymbol{x}_j) = \hat{p}_1 e_{1j}/\{\hat{p}_1 e_{1j} + \hat{p}_2 e_{2j}\},$$
$$\hat{\mathsf{P}}(2 \mid \boldsymbol{x}_j) = 1 - \hat{P}(1 \mid \boldsymbol{x}_j),$$

where

$$e_{ij} = \exp\left\{-\frac{1}{2}(\boldsymbol{x}_j - \hat{\boldsymbol{\mu}}_i)'\hat{\boldsymbol{\Sigma}}^{-1}(\boldsymbol{x}_j - \hat{\boldsymbol{\mu}}_i)\right\}.$$

Hence, show that the maximum-likelihood equations can be written as

$$\frac{\hat{p}_2}{\hat{p}_1} = \sum_{j=1}^{n}\hat{\mathsf{P}}(2 \mid \boldsymbol{x}_j) \Big/ \sum_{j=1}^{n}\hat{\mathsf{P}}(1 \mid \boldsymbol{x}_j),$$

$$\hat{\boldsymbol{\mu}}_i = \left\{ \sum_j \boldsymbol{x}_j \hat{\mathrm{P}}(i \mid \boldsymbol{x}_j) \right\} \Big/ \sum_j \hat{\mathrm{P}}(i \mid \boldsymbol{x}_j), \qquad i = 1, 2,$$

$$\hat{\boldsymbol{\Sigma}} = \frac{1}{n} \sum_j \left\{ (\boldsymbol{x}_j - \hat{\boldsymbol{\mu}}_1)(\boldsymbol{x}_j - \hat{\boldsymbol{\mu}}_1)'\hat{\mathrm{P}}(1 \mid \boldsymbol{x}_j) + (\boldsymbol{x}_j - \hat{\boldsymbol{\mu}}_2)(\boldsymbol{x}_j - \hat{\boldsymbol{\mu}}_2)'\hat{\mathrm{P}}(2 \mid \boldsymbol{x}_j) \right\}.$$

Hence, deduce that $\hat{p}_1$, $\hat{\boldsymbol{\mu}}_1$, and $\hat{\boldsymbol{\mu}}_2$ are some functions of the $\boldsymbol{x}$s and of $\hat{\boldsymbol{a}}$ and $\hat{b}$, whereas $\hat{\boldsymbol{V}}$ is a function of the $\boldsymbol{x}$s only. Hence, conclude that (*) and (**) form a set of equations of the form

$$\hat{\boldsymbol{a}} = \phi_1(\hat{\boldsymbol{a}}, \hat{b}; \boldsymbol{x}_1, \dots, \boldsymbol{x}_n), \qquad \hat{b} = \phi_2(\hat{\boldsymbol{a}}, \hat{b}; \boldsymbol{x}_1, \dots, \boldsymbol{x}_n),$$

which can be solved numerically by iteration.

14.3.1 Verify the maximum-likelihood estimates in Eqs. (14.14) and (14.15).

14.3.2 Consider a special case of a multivariate normal mixture model in which the covariance matrices $\boldsymbol{\Sigma}_k$ of the g components are all assumed to have a common value $\boldsymbol{\Sigma}$. Derive the EM equations for maximizing the likelihood function under such a model. (Hint: see MacLachlan and Krishnan (1997, sec. 2.7).)

14.3.3 Derive the E- and M-step solutions in the case that the data are modeled by a mixture of Poisson distributions. (Hint: see MacLachlan and Krishnan (1997, sec. 2.5).)

14.5.1 (Lance and Williams, 1967) Let $\boldsymbol{D} = (d_{rs})$ be a distance matrix. Show that the distance between group t and a group (rs) formed by the merger of groups r and s can be written as

$$d_{t(rs)} = \alpha d_{rt} + \alpha d_{st} + \gamma |d_{rt} - d_{st}|,$$

with

(i) $\alpha = 1/2, \gamma = -1/2$ for single linkage,
(ii) $\alpha = 1/2, \gamma = 1/2$ for complete linkage,
(iii) $\alpha = 1/2, \gamma = 0$ for average linkage.

14.5.2 Show that the squared Euclidean distance matrix (d_{rs}^2) for the iris data of Table 14.1 is

$$\begin{bmatrix} 0 & 0.36 & 0.02 & 0.50 & 1.69 & 0.05 \\ & 0 & 0.26 & 0.02 & 0.61 & 0.53 \\ & & 0 & 0.40 & 1.37 & 0.05 \\ & & & 0 & 0.61 & 0.73 \\ & & & & 0 & 1.78 \\ & & & & & 0 \end{bmatrix}.$$

Draw the dendrograms obtained on applying single and complete linkage clustering methods. Show that (i) at threshold 0.30, the single linkage method leads to clusters (1, 3, 6), (2, 4), and (5), and (ii) at threshold 0.75, the complete linkage method leads to clusters (1,2,3,4,6), (5). From the scatter diagram of the points in Example 14.2.1, compare and contrast these results with the clusters (1, 3, 6) and

(2, 4, 5) obtained in Example 14.2.1. Show that up to tie breaking, the minimum spanning tree contains the edges (1,3), (2,4), (3,6), (2,3), and (4,5).

14.5.3 Suppose that two bivariate normal populations A_1, A_2 have means

$$\begin{pmatrix} 0 \\ 0 \end{pmatrix} \quad \text{and} \quad \begin{pmatrix} 3 \\ 2 \end{pmatrix},$$

and each has dispersion matrix

$$\begin{bmatrix} 1 & 0 \\ 0 & 1 \end{bmatrix}.$$

The following observations were drawn at random from A_1 and A_2, the first six from A_1 and the last three from A_2:

$$\begin{pmatrix} 1.14 \\ 1.01 \end{pmatrix}, \begin{pmatrix} 0.98 \\ -0.94 \end{pmatrix}, \begin{pmatrix} 0.49 \\ 1.02 \end{pmatrix}, \begin{pmatrix} -0.31 \\ 0.76 \end{pmatrix}, \begin{pmatrix} 1.78 \\ 0.69 \end{pmatrix}, \begin{pmatrix} 0.40 \\ -0.43 \end{pmatrix},$$

$$\begin{pmatrix} 2.42 \\ 1.37 \end{pmatrix}, \begin{pmatrix} 2.55 \\ 1.06 \end{pmatrix}, \begin{pmatrix} 2.97 \\ 2.37 \end{pmatrix}.$$

Using single linkage cluster analysis, and the Euclidean distance matrix (unsquared distances shown below) for these nine observations, divide the observations into two groups and comment on the result.

	2	3	4	5	6	7	8	9
1	1.96	0.65	1.47	0.72	1.62	1.33	1.41	2.28
2		2.02	2.13	1.82	0.77	2.72	2.54	3.86
3			0.84	1.33	1.45	1.96	2.06	2.82
4				2.09	1.39	2.80	2.88	3.65
5					1.78	0.93	0.85	2.06
6						2.71	2.62	3.80
7							0.34	1.14
8								1.38

14.5.4 (Ultrametric distances) If d is a distance that satisfies the ultrametric inequality

$$d(P, Q) \leq \max\{d(P, R), d(Q, R)\},$$

for all points P, Q, and R, then d is called an *ultrametric* distance.

(a) For three points P, Q, and R, let $a = d(P, Q)$, $b = d(P, R)$, and $c = d(Q, R)$ and suppose that $a \leq b \leq c$. Show that $b = c$, i.e. the two larger distances are equal.
(b) Using (a), show that d satisfies the triangle inequality.

14.5.5 Let $\boldsymbol{D} = (d_{rs})$ be a distance matrix, and let $\boldsymbol{U} = (u_{rs})$ denote the distance matrix obtained from the corresponding single linkage dendrogram; that is, u_{rs} is the smallest threshold d_0 for which objects r and s lie in the same cluster.

(a) Show that $\boldsymbol{U}$ is an ultrametric distance matrix; that is,

$$u_{rs} \leq \max(u_{rt}, u_{st}) \qquad \text{for all} \quad r, s, t.$$

(b) Show that $\boldsymbol{D} = \boldsymbol{U}$ if and only if $\boldsymbol{D}$ is ultrametric.

(c) Consider the matrix of distances between five objects:

$$\begin{bmatrix} 0 & 4 & 1 & 4 & 3 \\ & 0 & 4 & 2 & 4 \\ & & 0 & 4 & 3 \\ & & & 0 & 4 \\ & & & & 0 \end{bmatrix}.$$

Draw the dendrogram for the single linkage method. Verify that $\boldsymbol{D} = \boldsymbol{U}$, and hence, that $\boldsymbol{D}$ is ultrametric.

14.5.6 Show that, up to the scale of the vertical axis, the single link dendrogram is invariant to a nonlinear monotone transformation of the dissimilarities.

14.5.7 Suppose that the squared distance between two clusters of objects, say A and B, is denoted by $d_{AB} = |\boldsymbol{m}_A - \boldsymbol{m}_B|^2$, where $\boldsymbol{m}_j$ is the mean of objects in cluster $j \in \{A, B\}$. Show that the distance between a cluster k and another cluster formed by joining i and j is

$$d_{i+j,k} = \frac{n_i}{n_i + n_j} d_{ik} + \frac{n_j}{n_i + n_j} d_{jk} - \frac{n_i n_j}{(n_i + n_j)^2} d_{ij},$$

where n_i denotes the number of points in group i.

14.6.1 (Balakrishnan and Sanghvi, 1968) Let $x_{r1}, \ldots, x_{rp}$ be proportions based on n_r observations for $r = 1, \ldots, g$. Let

$$\hat{\sigma}_{ij}^{(r)} = \begin{cases} x_{ri}(1 - x_{ri}), & i = j, \\ -x_{ri}x_{rj}, & i \neq j. \end{cases}$$

Define

$$\hat{\boldsymbol{\Sigma}} = \sum_{r=1}^{g} n_r \hat{\boldsymbol{\Sigma}}^{(r)} \Big/ \sum_{r=1}^{g} n_r.$$

Why might $(\boldsymbol{x}_r - \boldsymbol{x}_s)'\hat{\boldsymbol{\Sigma}}^{-1}(\boldsymbol{x}_r - \boldsymbol{x}_s)$ be a suitable measure of distance when it is suspected that the $\boldsymbol{\Sigma}^{(r)}$ are not all equal. Is $\hat{\boldsymbol{\Sigma}}$ singular?

14.6.2 Suppose that $\boldsymbol{x}$ and $\boldsymbol{y}$ are proportions based on samples of sizes n_1 and n_2 from multinomial distributions with the same parameter $\boldsymbol{a}$. Let $m = \min(n_1, n_2)$. Using (14.34) and (14.36), show that the ratio between the Mahalanobis distance and the Bhattacharyya distance is given by

$$\frac{D_1^2}{D_2^2} = 2\sum_{i=1}^{p} \frac{(x_i^{1/2} + y_i^{1/2})^2}{x_i + y_i}.$$

Let $\boldsymbol{b} = (a_1^{1/2}, \ldots, a_p^{1/2})'$. Using the fact that $x_i^{1/2} - b_i$ and $y_i^{1/2} - b_i$ have order $O_p(m^{-1/2})$, show that $D_1^2/D_2^2 - 4$ has order $O_p(m^{-1})$, and hence, D_1^2 is asymptotically equivalent to $4D_2^2$ as $m \to \infty$. (The notation, $O_p(\cdot)$, order in probability, is defined in (2.75).)

14.6.3 (Gower, 1971a) (a) For quantitative observations with values $x_1, \ldots, x_n$, consider the similarity coefficient $s_{ij} = 1 - \{|x_i - x_j|/R\}$, where R is the range. Show that we can assume $R = 1$ by rescaling, and $1 = x_1 \geq x_2 \geq \cdots \geq x_{n-1} \geq x_n = 0$ by permuting the rows of the data matrix and shifting the origin, so that $x_n = 0$. Thus, the similarity matrix $\boldsymbol{S}_n = (s_{ij})$ is

$$\begin{bmatrix} 1 & 1-(x_1-x_2) & 1-(x_1-x_3) & \cdots & 1-(x_1-x_n) \\ & 1 & 1-(x_2-x_3) & \cdots & 1-(x_2-x_n) \\ & & 1 & \cdots & 1-(x_3-x_n) \\ & & & \ddots & \vdots \\ & & & & 1 \end{bmatrix}.$$

By a series of elementary transformations, show that for $1 \leq p \leq n$, the principal $p \times p$ leading minor is

$$\Delta_{pp} = 2^{p-1}\left[1 - \frac{1}{2}(x_1 - x_p)\right]\prod_{i=1}^{p-1}(x_i - x_{i+1}).$$

Since $x_1 - x_p \leq x_1 - x_n = 1$ and $x_i - x_{i+1} \geq 0$, show that $\Delta_{pp} \geq 0$ for all p, and therefore, $\boldsymbol{S}_n \geq 0$. Extend this result to t quantitative variates, noting that the average of t p.s.d. matrices is again p.s.d. Hence, deduce that $(s_{ij}^{(3)})$ (see (14.42)) is p.s.d.
(b) Suppose that $\boldsymbol{S}_n^*$ is defined by $s_{ij}^* = 1 - \{|x_i - x_j|/T\}$, where $T > R$. Show that again $\Delta_{pp} \geq 0$, and so, $\boldsymbol{S}_n^*$ is p.s.d.
(c) If R is replaced by the standard deviation s, then we may have $s < R$, so $1 - (x_1 - x_p)/2$ need not be positive, and therefore, Δ_{pp} need not be positive. Hence, $\boldsymbol{S}_n$ obtained from $s_{ij} = 1 - \{|x_i - x_j|/s\}$ need not be p.s.d.

14.6.4 Obtain $\boldsymbol{S}_1$ (14.40) for the data in Example 14.6.2 and apply the single linkage method to obtain clusters. Why is this an unsatisfactory approach?

14.6.5 (Kendall, 1971) Let $\boldsymbol{X}$ be a $(n \times p)$ matrix of proportions; that is, x_{ik} is a fraction between 0 and 1 representing the fraction of artifacts of type k which are found on individual i (so $\sum_{k=1}^p x_{ik} = 1$). Define a similarity measure

$$s_{ij} = \sum_{k=1}^{p} \min(x_{ik}, x_{jk}).$$

If $\sum_{k=1}^p x_{ik} = 1$, show that

$$s_{ij} = 1 - \frac{1}{2}\sum_{k=1}^{p}|x_{ik} - x_{jk}|,$$

and that (s_{ij}) need not be p.s.d.

14.6.6 Show that the measure of divergence between two p.d.f.s f_1 and f_2 given by

$$\int (f_1 - f_2) \log\left(\frac{f_1}{f_2}\right) \mathrm{d}x$$

does not satisfy the triangular inequality.

14.6.7 Let $\boldsymbol{x}_i \sim N_p(\boldsymbol{\mu}_i, \boldsymbol{\Sigma}), i = 1, 2$. Show that the squared Bhattacharyya distance

$$\int \left(f_1^{1/2} - f_2^{1/2}\right)^2 \mathrm{d}\boldsymbol{x} = 2 - 2\int (f_1 f_2)^{1/2} \mathrm{d}\boldsymbol{x}$$

between the points is a function of the squared Mahalanobis distance,

$$(\boldsymbol{\mu}_1 - \boldsymbol{\mu}_2)'\boldsymbol{\Sigma}^{-1}(\boldsymbol{\mu}_1 - \boldsymbol{\mu}_2).$$

14.7.1 (a) Using the notation in Section 14.7, substitute

$$\overline{\boldsymbol{x}}_j - \overline{\overline{\boldsymbol{x}}}_k = \frac{1}{n_k^*} \sum_{\overline{\boldsymbol{x}}_r \in C_k} n_r(\overline{\boldsymbol{x}}_j - \overline{\boldsymbol{x}}_r)$$

in

$$w_g^2 = \sum_{k=1}^{g} \sum_{\overline{\boldsymbol{x}}_j \in C_k} n_j(\overline{\boldsymbol{x}}_j - \overline{\overline{\boldsymbol{x}}}_k)'\boldsymbol{\Sigma}^{-1}(\overline{\boldsymbol{x}}_j - \overline{\overline{\boldsymbol{x}}}_k),$$

and then, using

$$(\overline{\boldsymbol{x}}_j - \overline{\boldsymbol{x}}_r)'\boldsymbol{\Sigma}^{-1}(\overline{\boldsymbol{x}}_j - \overline{\boldsymbol{x}}_s) = \frac{1}{2}(D_{jr}^2 + D_{js}^2 - D_{rs}^2), \qquad j, r, s \in C_k,$$

where $D_{rs} = (\overline{\boldsymbol{x}}_r - \overline{\boldsymbol{x}}_s)'\boldsymbol{\Sigma}^{-1}(\overline{\boldsymbol{x}}_r - \overline{\boldsymbol{x}}_s)$, prove that

$$w_g^2 = \frac{1}{2} \sum_{k=1}^{g} \frac{1}{n_k^*} \sum_{\overline{\boldsymbol{x}}_i, \overline{\boldsymbol{x}}_j \in C_k} n_i n_j D_{ij}^2.$$

(b) Using

$$\overline{\boldsymbol{x}}_k - \overline{\overline{\boldsymbol{x}}} = \frac{1}{n_k^*} \sum_{\overline{\boldsymbol{x}}_j \in C_k} n_j(\overline{\boldsymbol{x}}_j - \overline{\boldsymbol{x}})$$

in (14.44), prove that

$$b_g^2 = \sum_{k=1}^{g} \frac{1}{n_k^*} \sum_{C_k} n_i n_j B_{ij}$$

where $B_{ij} = (\overline{\boldsymbol{x}}_i - \overline{\boldsymbol{x}})'\boldsymbol{\Sigma}^{-1}(\overline{\boldsymbol{x}}_j - \overline{\boldsymbol{x}})$ and the second summation is extended over all i, j such that $\overline{\boldsymbol{x}}_i, \overline{\boldsymbol{x}}_j \in C_k$. For $g = 2$, use $n_2^*(\overline{\overline{\boldsymbol{x}}}_2 - \overline{\boldsymbol{x}}) = -n_1^*(\overline{\overline{\boldsymbol{x}}}_1 - \overline{\boldsymbol{x}})$ to show that

$$b_g^2 = \left\{\frac{n_1^* + n_2^*}{n_1^* n_2^*}\right\} \sum_{j, j' \in C_1} n_j n_{j'} B_{jj'}.$$

14.7.2 By considering the likelihood for groups of clusters in Section 14.7, show that maximizing the likelihood implies the minimization of

$$w_g^2 = \sum_{k=1}^{g} \sum_{\{j : \overline{\boldsymbol{x}}_j \in C_k\}} n_j(\overline{\boldsymbol{x}}_j - \overline{\overline{\boldsymbol{x}}}_k)'\hat{\boldsymbol{\Sigma}}^{-1}(\overline{\boldsymbol{x}}_j - \overline{\overline{\boldsymbol{x}}}_k)$$

over the possible sets C_k, and show that this is an equivalent criterion to maximizing (14.44).

15

Multidimensional Scaling

15.1 Introduction

Multidimensional scaling (MDS) is concerned with the problem of constructing a configuration of n points in Euclidean space using the information about the distances between the n objects. The interpoint distances themselves may be subject to error. Note that this technique differs from those described in earlier chapters in an important way. Previously, the data points were directly observable as n points in p-space, but here, we can only observe a function of the data points, namely the $n(n-1)/2$ distances.

A simple test example is given in Table 15.1, where we are given the road distances (not the "shortest distances") between towns, and the aim is to construct a geographical map of Britain based on this information. Since these road distances equal the true distances subject to small perturbations, we expect that any sensible MDS method will produce a configuration that is "close" to the true map of these towns.

However, the distances need not be based on Euclidean distances and can represent many types of dissimilarities between objects. Also, in some cases, we start not with dissimilarities but with *similarities* between objects. (We have already given a general account of different types of distances and similarities in Section 14.6.) An example of similarity between two Morse code signals could be the percentage of people who think that the Morse code sequences corresponding to the pair of characters are identical after hearing them in rapid succession. Table 15.2 gives such data for characters consisting of the 10 numbers in Morse code. These "similarities" can then be used to plot the signals in two-dimensional space. The purpose of this plot is to observe which signals were "like," i.e. near, and which were "unlike," i.e. far from each other, and also to observe the general interrelationship between the signals.

Definition 15.1.1 *An $(n \times n)$ matrix $\boldsymbol{D}$ is called a distance matrix if it is symmetric and*

$$d_{rr} = 0, \qquad d_{rs} \geq 0, \qquad r \neq s.$$

Starting with a distance matrix $\boldsymbol{D}$, the object of MDS is to find points $P_1, \ldots, P_n$ in k dimensions such that if $\hat{d}_{rs}$ denotes the Euclidean distance between P_r and P_s, then $\hat{\boldsymbol{D}}$ is "similar" in

Multivariate Analysis, Second Edition. Kanti V. Mardia, John T. Kent, and Charles C. Taylor.

Table 15.1 Road distances in miles between 12 British towns.

	1	2	3	4	5	6	7	8	9	10	11
2	244										
3	218	350									
4	284	77	369								
5	197	167	347	242							
6	312	444	94	463	441						
7	215	221	150	236	279	245					
8	469	583	251	598	598	169	380				
9	166	242	116	257	269	210	55	349			
10	212	53	298	72	170	392	168	531	190		
11	253	325	57	340	359	143	117	264	91	273	
12	270	168	284	164	277	378	143	514	173	111	256

1, Aberystwyth; 2, Brighton; 3, Carlisle; 4, Dover; 5, Exeter; 6, Glasgow; 7, Hull; 8, Inverness; 9, Leeds; 10, London; 11, Newcastle; 12, Norwich.

Table 15.2 Percentage of times that the pairs of Morse code signals for two numbers were declared to be the same by 598 subjects (Rothkopf 1957, the reference contains entries for 26 letters as well).

	1	2	3	4	5	6	7	8	9	0
1	84									
2	62	89								
3	16	59	86							
4	6	23	38	89						
5	12	8	27	56	90					
6	12	14	33	34	30	86				
7	20	25	17	24	18	65	85			
8	37	25	16	13	10	22	65	88		
9	57	28	9	7	5	8	31	58	91	
0	52	18	9	7	5	18	15	39	79	94

Source: Data from Rothkopf (1957).

some sense to $\boldsymbol{D}$. The points P_r are unknown, and usually, the dimension k is also unknown. In practice, one usually limits the dimension to $k = 1, 2$, or 3 in order to facilitate the visualization and interpretation of the solution.

Nature of the solution It is important to realize that the configuration produced by any MDS method is indeterminate with respect to translation, rotation, and reflection. In the two-dimensional case of road distances (Table 15.1), the whole configuration of points can be "shifted" from one place in the plane to another, and the whole configuration can be "rotated" or "reflected."

In general, if $P_1, \dots, P_n$ with coordinates $\boldsymbol{x}_i' = (x_{i1}, \dots, x_{ip}), i = 1, \dots, n$, represents an MDS solution in p dimensions, then

$$\boldsymbol{y}_i = \boldsymbol{A}\boldsymbol{x}_i + \boldsymbol{b}, \qquad i = 1, \dots, n$$

is also a solution, where $\boldsymbol{A}$ is an orthogonal matrix, and $\boldsymbol{b}$ is any vector.

Types of solution Methods of solution using only the rank order of the distances

$$d_{r_1 s_1} < d_{r_2 s_2} < \cdots < d_{r_m s_m}, \qquad m = n(n-1)/2,$$

are termed *nonmetric methods of MDS*. Here, $(r_1, s_1), \dots, (r_m, s_m)$ denotes all pairs of subscripts of r and s, $r < s$.

The rank orders are invariant under monotone increasing transformations f of the d_{rs}, i.e.

$$d_{r_1 s_1} < d_{r_2 s_2} < \cdots \Longleftrightarrow f(d_{r_1 s_1}) < f(d_{r_2 s_2}) < \cdots$$

Therefore, the configurations that arise from the nonmetric scaling are indeterminate not only with respect to translation, rotation, and reflection but also with respect to uniform expansion or contraction.

Solutions that try to obtain the points directly from the given distances are called *metric methods*. These methods derive P_r such that, in some sense, the new distances $\hat{d}_{rs}$ between points P_r and P_s are as close to the original as possible.

In general, the purpose of MDS is to provide a "picture" that can be used to give a meaningful interpretation of the data. Hopefully, the picture will convey useful information about the relationships between the objects. Note that this chapter differs from most of the earlier chapters in that no probabilistic framework is set up; the technique is purely data-analytic.

One important use of MDS is seriation, and this is described in Section 15.5. The aim here is to order a set of objects chronologically on the basis of dissimilarities or similarities between them. Suppose that the points in the MDS configuration in $k = 2$ dimensions lie nearly on a smooth curve. This property then suggests that the differences in the data are in fact one-dimensional, and the ordering of the points along this curve can then be used to seriate the data; see Kendall (1971).

15.2 Classical Solution

15.2.1 Some Theoretical Results

Definition 15.2.1 *A distance matrix* $\boldsymbol{D}$ *is called* Euclidean *if there exists a configuration of points in some Euclidean space whose interpoint distances are given by* $\boldsymbol{D}$; *that is, if for some* p, *there exist points* $\boldsymbol{x}_1, \dots, \boldsymbol{x}_n \in \mathbb{R}^p$ *such that*

$$d_{rs}^2 = (\boldsymbol{x}_r - \boldsymbol{x}_s)'(\boldsymbol{x}_r - \boldsymbol{x}_s). \tag{15.1}$$

The following theorem enables us to tell whether $\boldsymbol{D}$ is Euclidean, and if so, how to find a corresponding configuration of points. First, we need some notation. For any distance matrix $\boldsymbol{D}$, let

$$\boldsymbol{A} = (a_{rs}), \qquad a_{rs} = -\frac{1}{2}d_{rs}^2, \tag{15.2}$$

and set

$$\boldsymbol{B} = \boldsymbol{HAH}, \tag{15.3}$$

where $\boldsymbol{H} = \boldsymbol{I} - n^{-1}\boldsymbol{11}'$ is the $(n \times n)$ centering matrix.

Theorem 15.2.2 ***(Torgerson–Gower)*** *Let* $\mathbf{D}$ *be a distance matrix, and define* $\mathbf{B}$ *by (15.3). Then,* $\mathbf{D}$ *is Euclidean if and only if* $\mathbf{B} \geq 0$*. In particular, the following results hold:*

(a) If $\mathbf{D}$ *is the matrix of Euclidean interpoint distances for a configuration* $\boldsymbol{Z} = (\boldsymbol{z}_1, \ldots, \boldsymbol{z}_n)'$*, then*

$$b_{rs} = (\boldsymbol{z}_r - \overline{\boldsymbol{z}})'(\boldsymbol{z}_s - \overline{\boldsymbol{z}}), \qquad r, s = 1, \ldots, n. \tag{15.4}$$

In matrix form, (15.4) becomes $\mathbf{B} = (\boldsymbol{HZ})(\boldsymbol{HZ})'$*, so* $\mathbf{B} \geq 0$*. Note that* $\mathbf{B}$ *can be interpreted as the "centered inner product matrix" for the configuration* $\boldsymbol{Z}$*.*

(b) Conversely, if $\mathbf{B} \geq 0$ *of rank p, then a configuration corresponding to* $\mathbf{B}$ *can be constructed as follows. Let* $\mathbf{B} = \boldsymbol{\Gamma\Lambda\Gamma}'$ *be the spectral decomposition of* $\mathbf{B}$*, and* $\lambda_1 \geq \cdots \geq \lambda_p > 0$ *denote the positive eigenvalues. Define an* $n \times p$ *configuration* $\boldsymbol{X}$ *to have columns*

$$\boldsymbol{x}_{(i)} = \lambda_i^{1/2} \boldsymbol{\gamma}_{(i)}, \qquad i = 1, \ldots, p. \tag{15.5}$$

Then, the points P_r *in* $\mathbb{R}^p$ *with coordinates* $\boldsymbol{x}_r = (x_{r1}, \ldots, x_{rp})'$ *(so,* $\boldsymbol{x}_r'$ *is the rth row of* $\boldsymbol{X}$*) have interpoint distances given by* $\mathbf{D}$*. Further, this configuration has center of gravity* $\overline{\boldsymbol{x}} = \mathbf{0}$*, and* $\mathbf{B}$ *represents the inner product matrix for this configuration.*

Proof: We first prove (a). Suppose that

$$d_{rs}^2 = -2a_{rs} = (\boldsymbol{z}_r - \boldsymbol{z}_s)'(\boldsymbol{z}_r - \boldsymbol{z}_s). \tag{15.6}$$

We can write

$$\boldsymbol{B} = \boldsymbol{HAH} = \boldsymbol{A} - n^{-1}\boldsymbol{AJ} - n^{-1}\boldsymbol{JA} + n^{-2}\boldsymbol{JAJ},$$

where $\boldsymbol{J} = \mathbf{11}'$. Now,

$$\frac{1}{n}\boldsymbol{AJ} = \begin{bmatrix} \bar{a}_{1\cdot} & \cdots & \bar{a}_{1\cdot} \\ \vdots & \ddots & \vdots \\ \bar{a}_{n\cdot} & \cdots & \bar{a}_{n\cdot} \end{bmatrix}, \quad \frac{1}{n}\boldsymbol{JA} = \begin{bmatrix} \bar{a}_{\cdot 1} & \cdots & \bar{a}_{\cdot n} \\ \vdots & \ddots & \vdots \\ \bar{a}_{\cdot 1} & \cdots & \bar{a}_{\cdot n} \end{bmatrix}, \quad \frac{1}{n^2}\boldsymbol{JAJ} = \begin{bmatrix} \bar{a}_{\cdot\cdot} & \cdots & \bar{a}_{\cdot\cdot} \\ \vdots & \ddots & \vdots \\ \bar{a}_{\cdot\cdot} & \cdots & \bar{a}_{\cdot\cdot} \end{bmatrix},$$

where

$$\bar{a}_{r\cdot} = \frac{1}{n}\sum_{s=1}^{n} a_{rs}, \qquad \bar{a}_{\cdot s} = \frac{1}{n}\sum_{r=1}^{n} a_{rs}, \qquad \bar{a}_{\cdot\cdot} = \frac{1}{n^2}\sum_{r,s=1}^{n} a_{rs}. \tag{15.7}$$

Thus,

$$b_{rs} = a_{rs} - \bar{a}_{r\cdot} - \bar{a}_{\cdot s} + \bar{a}_{\cdot\cdot}. \tag{15.8}$$

After substituting for a_{rs} from (15.6) and using (15.7), this formula simplifies to

$$b_{rs} = (\boldsymbol{z}_r - \overline{\boldsymbol{z}})'(\boldsymbol{z}_s - \overline{\boldsymbol{z}}). \tag{15.9}$$

(See Exercise 15.2.1 for further details.) Thus, (a) is proved.

Conversely, to prove (b), suppose that $\boldsymbol{B} \geq 0$ and consider the configuration given in the theorem. Let $\boldsymbol{\Lambda}_1 = \text{diag}(\lambda_1, \ldots, \lambda_p)$, and $\boldsymbol{\Gamma}_1 = [\boldsymbol{\gamma}_{(1)}, \ldots, \boldsymbol{\gamma}_{(p)}]$ contain the eigenvalues and standardized eigenvectors for the positive eigenvalues. Then, by the spectral decomposition theorem (Remark 4 after Theorem A.5),

$$\boldsymbol{B} = \boldsymbol{\Gamma}\boldsymbol{\Lambda}\boldsymbol{\Gamma}' = \boldsymbol{\Gamma}_1\boldsymbol{\Lambda}_1\boldsymbol{\Gamma}_1' = \boldsymbol{X}\boldsymbol{X}';$$

that is, $b_{rs} = \boldsymbol{x}_r'\boldsymbol{x}_s$, so $\boldsymbol{B}$ represents the inner product matrix for this configuration.

We must now show that $\boldsymbol{D}$ represents the matrix of interpoint distances for this configuration. Using (15.8) to write $\boldsymbol{B}$ in terms of $\boldsymbol{A}$, we get

$$\begin{aligned}(\boldsymbol{x}_r - \boldsymbol{x}_s)'(\boldsymbol{x}_r - \boldsymbol{x}_s) &= \boldsymbol{x}_r'\boldsymbol{x}_r - 2\boldsymbol{x}_r'\boldsymbol{x}_s + \boldsymbol{x}_s'\boldsymbol{x}_s \\ &= b_{rr} - 2b_{rs} + b_{ss} \\ &= (a_{rr} - \overline{a}_{r\cdot} - \overline{a}_{\cdot r} + \overline{a}_{\cdot\cdot}) - 2(a_{rs} - \overline{a}_{r\cdot} - \overline{a}_{\cdot s} + \overline{a}_{\cdot\cdot}) \\ &\quad + (a_{ss} - \overline{a}_{s\cdot} - \overline{a}_{\cdot s} + \overline{a}_{\cdot\cdot}) \\ &= a_{rr} - 2a_{rs} + a_{ss} \\ &= -2a_{rs} \\ &= d_{rs}^2\end{aligned}$$

because $a_{rr} = -d_{rr}^2/2 = 0$ and $-2a_{rs} = d_{rs}^2$. Note $\overline{a}_{r\cdot} = \overline{a}_{\cdot r}$ since $\boldsymbol{D}$ is symmetric.

Finally, note that $\boldsymbol{B}\boldsymbol{1} = \boldsymbol{H}\boldsymbol{A}\boldsymbol{H}\boldsymbol{1} = \boldsymbol{0}$, so that $\boldsymbol{1}$ is an eigenvector of $\boldsymbol{B}$ corresponding to the eigenvalue 0. Thus, $\boldsymbol{1}$ is orthogonal to all the columns of $\boldsymbol{X}$, i.e. $\boldsymbol{x}_{(i)}'\boldsymbol{1} = 0, i = 1, \ldots, p$. Hence,

$$n\overline{\boldsymbol{x}} = \sum_{r=1}^{n} \boldsymbol{x}_r = \boldsymbol{X}'\boldsymbol{1} = (\boldsymbol{x}_{(1)}'\boldsymbol{1}, \ldots, \boldsymbol{x}_{(p)}'\boldsymbol{1})' = \boldsymbol{0},$$

so that the center of this configuration lies at the origin. □

Remarks (1) The matrix $\boldsymbol{X}$ can be visualized as follows:

$$\begin{array}{lcccccc} & & & & & & \text{Rows} \\ \text{Points} & \left\{\begin{array}{l} P_1 \\ P_2 \\ \vdots \\ P_n \end{array}\right. & \begin{array}{c} x_{11} \\ x_{21} \\ \vdots \\ x_{n1} \end{array} & \begin{array}{c} x_{12} \\ x_{22} \\ \vdots \\ x_{n2} \end{array} & \begin{array}{c} \cdots \\ \cdots \\ \\ \cdots \end{array} & \begin{array}{c} x_{1p} \\ x_{2p} \\ \vdots \\ x_{np} \end{array} & \begin{array}{c} \boldsymbol{x}_1' \\ \boldsymbol{x}_2' \\ \vdots \\ \boldsymbol{x}_n' \end{array} \\ \text{Columns} & & \boldsymbol{x}_{(1)} & \boldsymbol{x}_{(2)} & \cdots & \boldsymbol{x}_{(p)} & \\ \text{Eigenvectors} & & \boldsymbol{\gamma}_{(1)} & \boldsymbol{\gamma}_{(2)} & \cdots & \boldsymbol{\gamma}_{(p)} & \\ \text{Eigenvalues} & & \lambda_1 & \lambda_2 & \cdots & \lambda_p & \end{array}$$

where $\boldsymbol{x}_{(i)} = \lambda_i^{1/2}\boldsymbol{\gamma}_{(i)}$. In short, the rth *row* of $\boldsymbol{X}$ contains the coordinates of the rth point, whereas the ith *column* of $\boldsymbol{X}$ contains the component corresponding to the ith eigenvector. The center of gravity of $\boldsymbol{X}$ lies at the origin,

$$\overline{\boldsymbol{x}} = (\overline{x}_1, \overline{x}_2, \ldots, \overline{x}_p)' = \boldsymbol{0}.$$

(2) Geometrically, if $\boldsymbol{B}$ is the centered inner product matrix for a configuration $\boldsymbol{Z}$, then $b_{rr}^{1/2}$ equals the distance between $\boldsymbol{z}_r$ and $\overline{\boldsymbol{z}}$, and $b_{rs}/(b_{rr}b_{ss})^{1/2}$ equals the cosine of the angle subtended at $\overline{\boldsymbol{z}}$ between $\boldsymbol{z}_r$ and $\boldsymbol{z}_s$.

(3) Note that $\boldsymbol{1}$ is an eigenvector of $\boldsymbol{B}$ whether $\boldsymbol{D}$ is Euclidean or not.

(4) The theorem does not hold if $\boldsymbol{B}$ has negative eigenvalues. The reason can be found in (15.5) because it is impossible to take the square root of a negative number in real arithmetic.

(5) *History.* This result was first proved by Schoenberg (1935) and Young and Householder (1938). Its use as a basis for MDS was put forward by Torgerson (1958), and the ideas were substantially amplified by Gower (1966).

15.2.2 An Algorithm for the Classical MDS Solution

Suppose that we are given a distance matrix $\boldsymbol{D}$ which we hope can approximately represent the interpoint distances of a configuration in a Euclidean space of low dimension k (usually $k = 1, 2,$ or 3). The matrix $\boldsymbol{D}$ may or may not be Euclidean; however, even if $\boldsymbol{D}$ is Euclidean, the dimension of the space in which it can be represented will usually be too large to be of practical interest.

One possible choice of configuration in k dimensions is suggested by Theorem 15.2.2. *Choose the configuration in $\mathbb{R}^k$ whose coordinates are determined by the first k eigenvectors of $\boldsymbol{B}$.* If the first k eigenvalues of $\boldsymbol{B}$ are "large" and positive and the other eigenvalues are near 0 (positive or negative), then hopefully, the interpoint distances of this configuration will closely approximate $\boldsymbol{D}$.

This configuration is called the *classical solution to the MDS problem in k dimensions.* It is a metric solution, and its optimality properties are discussed in Section 15.4. For computational purposes, we summarize the calculations involved:

(a) From $\boldsymbol{D}$ construct the matrices

$$\boldsymbol{A} = \left(-\frac{1}{2}d_{rs}^2\right), \qquad \boldsymbol{B} = \boldsymbol{HAH}. \tag{15.10}$$

(b) Let $\boldsymbol{B} = \boldsymbol{\Gamma\Lambda\Gamma}'$ denote the spectral decomposition of $\boldsymbol{B}$. Partition

$$\boldsymbol{\Lambda} = \begin{bmatrix} \boldsymbol{\Lambda}_1 & \boldsymbol{0} \\ \boldsymbol{0} & \boldsymbol{\Lambda}_2 \end{bmatrix}, \quad \boldsymbol{\Gamma} = \begin{bmatrix} \boldsymbol{\Gamma}_1 & \boldsymbol{\Gamma}_2 \end{bmatrix}, \tag{15.11}$$

where $\boldsymbol{\Lambda}_1 (k \times k)$ contains the "large" eigenvalues, and $\boldsymbol{\Lambda}_2$ of size $(n-k) \times (n-k)$ contains the remaining eigenvalues. Similarly, partition $\boldsymbol{\Gamma}$, where $\boldsymbol{\Gamma}_1 (n \times k)$ contains the standardized eigenvectors for $\boldsymbol{\Lambda}_1$. Define a configuration

$$\hat{\boldsymbol{X}} = \boldsymbol{\Gamma}_1\boldsymbol{\Lambda}_1^{1/2}, \qquad \text{i.e. } \hat{\boldsymbol{x}}_{(i)} = \lambda_i^{1/2}\boldsymbol{\gamma}_{(i)}, \quad i = 1, \ldots, k. \tag{15.12}$$

(We are supposing here that the first k eigenvalues are all positive.) Then $\hat{\boldsymbol{X}}$ represents the classical solution to the MDS problem in k dimensions. The circumflex or "hat" on $\boldsymbol{X}$ indicates that it is an estimated configuration based on the data in $\boldsymbol{D}$.

(d) Let $\hat{\boldsymbol{D}}$ denote the Euclidean distance matrix between the rows of the MDS solution $\hat{\boldsymbol{X}}$, i.e.

$$\hat{d}_{rs}^2 = (\hat{\boldsymbol{x}}_r - \hat{\boldsymbol{x}}_s)'(\hat{\boldsymbol{x}}_r - \hat{\boldsymbol{x}}_s), \quad r, s = 1, \ldots, n.$$

Then, we "expect" the distances $\hat{d}_{rs}$ of the MDS solution to be close to the original distances d_{rs}.

(e) By the Torgerson–Gower Theorem 15.2.2, the MDS solution will be exact, $\hat{\boldsymbol{D}} = \boldsymbol{D}$, if all the eigenvalues in $\boldsymbol{\Lambda}_2$ in (15.11) are exactly equal to 0. If the ignored eigenvalues in $\boldsymbol{\Lambda}_2$ are not exactly 0 (the most important situation in practice), the MDS reconstruction will not be exact; see Section 15.4.

(f) Finally, the configuration $\hat{\boldsymbol{X}}$ can be plotted as n points in $\mathbb{R}^k$ to get a visual interpretation of the data. The coordinates of the points P_r are the *rows* of $\hat{\boldsymbol{X}}$,

$$\hat{\boldsymbol{x}}_r' = (\hat{x}_{r1}, \ldots, \hat{x}_{rk}), \quad r = 1, \ldots, n.$$

Example 15.2.1 To illustrate the algorithm, consider a (7×7) distance matrix

$$\boldsymbol{D} = \begin{bmatrix} 0 & 1 & \sqrt{3} & 2 & \sqrt{3} & 1 & 1 \\ & 0 & 1 & \sqrt{3} & 2 & \sqrt{3} & 1 \\ & & 0 & 1 & \sqrt{3} & 2 & 1 \\ & & & 0 & 1 & \sqrt{3} & 1 \\ & & & & 0 & 1 & 1 \\ & & & & & 0 & 1 \\ & & & & & & 0 \end{bmatrix}.$$

Constructing the matrix $\boldsymbol{A}$ from (15.2), it is found that

$$\bar{a}_{r.} = -\frac{13}{14}, \quad r = 1, \ldots, 6, \qquad \bar{a}_{7.} = -\frac{3}{7}$$

$$\bar{a}_{r.} = \bar{a}_{.r}, \qquad \bar{a}_{..} = -\frac{6}{7}.$$

Hence, from (15.8), the matrix $\boldsymbol{B}$ is given by

$$\boldsymbol{B} = \frac{1}{2}\begin{bmatrix} 2 & 1 & -1 & -2 & -1 & 1 & 0 \\ & 2 & 1 & -1 & -2 & -1 & 0 \\ & & 2 & 1 & -1 & -2 & 0 \\ & & & 2 & 1 & -1 & 0 \\ & & & & 2 & 1 & 0 \\ & & & & & 2 & 0 \\ & & & & & & 0 \end{bmatrix}.$$

The columns of $\boldsymbol{B}$ are linearly dependent. It can be seen that

$$\boldsymbol{b}_{(3)} = \boldsymbol{b}_{(2)} - \boldsymbol{b}_{(1)}, \quad \boldsymbol{b}_{(4)} = -\boldsymbol{b}_{(1)}, \quad \boldsymbol{b}_{(5)} = -\boldsymbol{b}_{(2)}, \quad \boldsymbol{b}_{(6)} = \boldsymbol{b}_{(1)} - \boldsymbol{b}_{(2)}, \quad \boldsymbol{b}_{(7)} = \boldsymbol{0}.$$

Hence, the rank of the matrix $\boldsymbol{B}$ is at most 2. From the leading (2×2) matrix, it is clear that the rank is 2. Thus, a configuration exactly fitting the distance matrix can be constructed in $k = 2$ dimensions.

The eigenvalues of $\boldsymbol{B}$ are found to be

$$\lambda_1 = 3, \qquad \lambda_2 = 3, \qquad \lambda_3 = \cdots = \lambda_7 = 0.$$

The configuration can be constructed using any two orthogonal vectors for the eigenspace corresponding to $\lambda = 3$, such as

$$\boldsymbol{x}_{(1)}' = (a, a, 0, -a, -a, 0, 0), \qquad a = \frac{1}{2}\sqrt{3},$$

$$\boldsymbol{x}_{(2)}' = (b, -b, -2b, -b, b, 2b, 0), \qquad b = \frac{1}{2}.$$

Then, the coordinates of the seven points are

$$\begin{array}{ccccccc} \text{A} & \text{B} & \text{C} & \text{D} & \text{E} & \text{F} & \text{G} \\ (\sqrt{3}/2, 1/2) & (\sqrt{3}/2, -1/2) & (0, -1) & (-\sqrt{3}/2, -1/2) & (-\sqrt{3}/2, 1/2) & (0, 1) & (0, 0) \end{array}.$$

The center of gravity of these points is of course $(0, 0)$, and it can be verified that the distance matrix for these points is $\boldsymbol{D}$. In fact, A–F are vertices of a hexagon with each side of length 1, and the line FC is the y-axis. Its center is G. (Indeed, $\boldsymbol{D}$ was constructed with the help of these points.) Since $\boldsymbol{X}$ reproduces the distances in $\boldsymbol{D}$ exactly, we write it without a hat. A similar configuration based on a non-Euclidean distance is described in Exercise 15.2.7. □

Example 15.2.2 The simplest possible example is a 2×2 distance matrix

$$\boldsymbol{D} = \begin{bmatrix} 0 & a \\ a & 0 \end{bmatrix},$$

where $a > 0$. The matrix $\boldsymbol{B}$ in (15.10) is given by

$$\boldsymbol{B} = \frac{1}{4} \begin{bmatrix} a^2 & -a^2 \\ -a^2 & a^2 \end{bmatrix}.$$

The eigenvalues of $\boldsymbol{B}$ are $\lambda_1 = \frac{1}{2}a^2$ and $\lambda_2 = 0$ with the corresponding standardized eigenvectors given by the columns of

$$\boldsymbol{\Gamma} = \frac{1}{\sqrt{2}} \begin{bmatrix} 1 & -1 \\ 1 & 1 \end{bmatrix}.$$

Hence, there is an exact one-dimensional MDS solution

$$\boldsymbol{X} = \begin{bmatrix} \frac{1}{2}a \\ -\frac{1}{2}a \end{bmatrix}.$$

□

15.2.3 Similarities

In some situations we start not with distances between n objects but with similarities. Recall that an $(n \times n)$ matrix $\boldsymbol{C}$ is called a similarity matrix if $c_{rs} = c_{sr}$ and if

$$c_{rs} \le c_{rr} \quad \text{for all } r, s. \tag{15.13}$$

Examples of possible similarity matrices are given in Section 14.6.

To use the techniques of the preceding sections, it is necessary to transform the similarities to distances. A useful transformation is the following.

Definition 15.2.3 *The standard transformation from a similarity matrix* $\boldsymbol{C}$ *to a distance matrix* $\boldsymbol{D}$ *is defined by*

$$d_{rs} = (c_{rr} - 2c_{rs} + c_{ss})^{1/2}. \tag{15.14}$$

Note that if (15.13) holds, then the quantity under the square root in (15.14) must be non-negative, and that $d_{rr} = 0$. Hence, $\boldsymbol{D}$ is a distance matrix.

Many of the similarity matrices in Section 14.6 were positive semi-definite (p.s.d.) This property is attractive because the resulting distance matrix, using the standard transformation, is Euclidean.

Theorem 15.2.4 *If $\boldsymbol{C} \geq 0$, then the distance matrix $\boldsymbol{D}$ defined by the standard transformation (15.14) is Euclidean, with centered inner product matrix $\boldsymbol{B} = \boldsymbol{HCH}$.*

Proof: First, note that since $\boldsymbol{C} \geq 0$,

$$d_{rs}^2 = c_{rr} - 2c_{rs} + c_{ss} = \boldsymbol{x}'\boldsymbol{Cx} \geq 0,$$

where $\boldsymbol{x}$ is a vector with +1 in the rth place and −1 in the sth place, for $r \neq s$. Thus, the standard transformation is well defined, and $\boldsymbol{D}$ is a distance matrix.

Let $\boldsymbol{A}$ and $\boldsymbol{B}$ be defined by (15.2) and (15.3). Since $\boldsymbol{HCH}$ is also p.s.d., it is sufficient to prove that $\boldsymbol{B} = \boldsymbol{HCH}$ in order to conclude that $\boldsymbol{D}$ is Euclidean with centered inner product matrix $\boldsymbol{HCH}$.

Now, $\boldsymbol{B} = \boldsymbol{HAH}$ can be written elementwise using (15.8). Substituting for $a_{rs} = -d_{rs}^2/2$ using (15.14) gives

$$\begin{aligned}
-2b_{rs} &= d_{rs}^2 - \frac{1}{n}\sum_{i=1}^{n} d_{ri}^2 - \frac{1}{n}\sum_{j=1}^{n} d_{js}^2 + \frac{1}{n^2}\sum_{i,j=1}^{n} d_{ij}^2 \\
&= c_{rr} - 2c_{rs} + c_{ss} - \frac{1}{n}\sum_{i=1}^{n}(c_{rr} - 2c_{ri} + c_{ii}) - \frac{1}{n}\sum_{j=1}^{n}(c_{jj} - 2c_{js} + c_{ss}) \\
&\quad + \frac{1}{n^2}\sum_{i,j=1}^{n}(c_{ii} - 2c_{ij} + c_{jj}) \\
&= -2c_{rs} + 2\bar{c}_{r\cdot} + 2\bar{c}_{\cdot s} - 2\bar{c}_{\cdot\cdot}.
\end{aligned}$$

Hence,

$$b_{rs} = c_{rs} - \bar{c}_{r\cdot} - \bar{c}_{\cdot s} + \bar{c}_{\cdot\cdot}$$

or, in matrix form, $\boldsymbol{B} = \boldsymbol{HCH}$. Thus, the theorem is proved. □

Example 15.2.3 We now consider the Morse code data given in Table 15.2 and described in Section 15.1. The data is presented as a similarity matrix $\boldsymbol{C} = (c_{rs})$. Using the standard transformation from similarities to distances, take

$$d_{rs} = (c_{rr} + c_{ss} - 2c_{rs})^{1/2}$$

to define $\boldsymbol{D}$. We obtain the eigenvectors and eigenvalues of $\boldsymbol{HCH}$ in accordance with Theorem 15.2.4. It is found that

$$\lambda_1 = 187.4, \quad \lambda_2 = 121.0, \quad \lambda_3 = 95.4, \quad \lambda_4 = 55.4, \quad \lambda_5 = 46.6,$$
$$\lambda_6 = 31.5, \quad \lambda_7 = 9.6, \quad \lambda_8 = 4.5, \quad \lambda_9 = 0.0, \quad \lambda_{10} = -4.1.$$

The first two eigenvectors appropriately normalized are

$$(-4.2, -0.3, 3.7, 5.6, 5.4, 3.8, 0.9, -3.0, -6.2, -5.7),$$
$$(-3.2, -5.8, -4.3, -0.6, 0.0, 4.0, 5.5, 3.6, 0.6, 0.2).$$

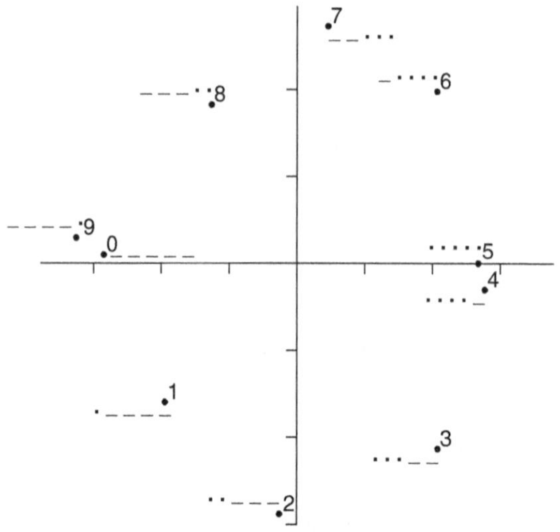

Figure 15.1 Classical solution for Morse code data in Table 15.2.

However, the first two principal coordinates account for only $100(\lambda_1 + \lambda_2)/\sum|\lambda_i|$ percent = 56% of the total configuration. The points P_r are plotted in Figure 15.1. It can be seen that the x_1-axis measures the increasing number of dots, whereas the x_2-axis measures the heterogeneity of the signal. If we regard the OP_rs as vectors, then angles between consecutive vectors are about 45° except between the vectors from 0 and 9, and 4 and 5. The small separation between these latter points might be expected because the change in just the last character may not make much impact on the untrained ear. Thus, the configuration brings out the main features of the general interrelationship between these signals. The points are roughly on a circle in the order 0, 1, … , 9. □

15.3 Duality Between Principal Coordinate Analysis and Principal Component Analysis

So far in this chapter, we have treated the $(n \times n)$ matrix $\boldsymbol{D}$ of distances between n objects as the starting point for our analysis. However, in many situations, we start with a data matrix $\boldsymbol{X}(n \times p)$ and must make a choice of distance function.

Several possibilities for a distance function are discussed in Section 14.6. The simplest choice is, of course, Euclidean distance. In this case, there is a close connection between the work in Section 15.2 and principal component analysis.

Let $\boldsymbol{X}(n \times p)$ be a data matrix and $\lambda_1 \geq \cdots \geq \lambda_p$ the eigenvalues of $n\boldsymbol{S} = \boldsymbol{X}'\boldsymbol{H}\boldsymbol{X}$, where $\boldsymbol{S}$ is the sample covariance matrix. For simplicity, we suppose that the eigenvalues are all nonzero and distinct. Then, $\lambda_1, \dots, \lambda_p$ are also the nonzero eigenvalues of $\boldsymbol{B} = \boldsymbol{H}\boldsymbol{X}\boldsymbol{X}'\boldsymbol{H}$. Note that the rows of $\boldsymbol{H}\boldsymbol{X}$ are just the centered rows of $\boldsymbol{X}$, so that $\boldsymbol{B}$ represents the centered inner product matrix,

$$b_{rs} = (\boldsymbol{x}_r - \overline{\boldsymbol{x}})'(\boldsymbol{x}_s - \overline{\boldsymbol{x}}).$$

Definition 15.3.1 *Let $\boldsymbol{v}_{(i)}$ be the ith eigenvector of* $\boldsymbol{B}$,

$$\boldsymbol{B}\boldsymbol{v}_{(i)} = \lambda_i \boldsymbol{v}_{(i)},$$

normalized by $\boldsymbol{v}'_{(i)}\boldsymbol{v}_{(i)} = \lambda_i, i = 1, \ldots, p$. For fixed k $(1 \le k \le p)$, the rows of $\boldsymbol{V}_k = [\boldsymbol{v}_{(1)}, \ldots, \boldsymbol{v}_{(k)}]$ are called principal coordinates of $\boldsymbol{X}$ in k dimensions.

Thus, from Theorem 15.2.2, if $\boldsymbol{D}$ is the Euclidean distance matrix between the rows of $\boldsymbol{X}$, then the k-dimensional classical solution to the MDS problem is given by the principal coordinates of $\boldsymbol{X}$ in k dimensions. Principal coordinates are closely linked to principal components as shown by the following result.

Theorem 15.3.2 *The principal coordinates of $\boldsymbol{X}$ in k dimensions are given by the centered* scores *of the n objects on the first k principal components.*

Proof: Let $\boldsymbol{\gamma}_{(i)}$ denote the ith principal component loading vector, standardized so that $\boldsymbol{\gamma}'_{(i)}\boldsymbol{\gamma}_{(i)} = 1$, and so by the spectral decomposition theorem (Theorem A.5),

$$\boldsymbol{X}'\boldsymbol{H}\boldsymbol{X} = \boldsymbol{\Gamma}\boldsymbol{\Lambda}\boldsymbol{\Gamma}',$$

where $\boldsymbol{\Gamma} = [\boldsymbol{\gamma}_{(1)}, \ldots, \boldsymbol{\gamma}_{(p)}]$ and $\boldsymbol{\Lambda} = \text{diag}(\lambda_1, \ldots, \lambda_p)$. By the singular value decomposition theorem (Theorem A.8), we can choose the signs of $\boldsymbol{\gamma}_{(i)}$ and $\boldsymbol{v}_{(i)}$, so that $\boldsymbol{H}\boldsymbol{X}$ can be written in terms of these eigenvectors as

$$\boldsymbol{H}\boldsymbol{X} = \boldsymbol{V}\boldsymbol{\Gamma}',$$

where $\boldsymbol{V} = \boldsymbol{V}_p = [\boldsymbol{v}_{(1)}, \ldots, \boldsymbol{v}_{(p)}]$.

The scores of the n rows of $\boldsymbol{H}\boldsymbol{X}$ on the ith principal component are given by the n elements of $\boldsymbol{H}\boldsymbol{X}\boldsymbol{\gamma}_{(i)}$. Thus, writing $\boldsymbol{\Gamma}_k = [\boldsymbol{\gamma}_{(1)}, \ldots, \boldsymbol{\gamma}_{(k)}]$, the scores on the first k principal components are given by

$$\boldsymbol{H}\boldsymbol{X}\boldsymbol{\Gamma}_k = \boldsymbol{V}\boldsymbol{\Gamma}'\boldsymbol{\Gamma}_k = \boldsymbol{V}[\boldsymbol{I}_k, \boldsymbol{0}]' = \boldsymbol{V}_k,$$

since the columns of $\boldsymbol{\Gamma}$ are orthogonal to one another. Hence, the theorem is proved. □

Since the columns of $\boldsymbol{\Gamma}_k$ are orthogonal to one another, $\boldsymbol{\Gamma}'_k\boldsymbol{\Gamma}_k = \boldsymbol{I}_k$, we see that $\boldsymbol{V}_k = \boldsymbol{X}\boldsymbol{\Gamma}_k$ represents a projection of $\boldsymbol{X}$ onto a k-dimensional subspace of $\mathbb{R}^p$. The projection onto principal coordinates is optimal out of all k-dimensional projections because it is closest to the original p-dimensional configuration. (See Theorem 15.4.1.)

This result is dual to the result in principal component analysis in that the sum of the variances of the first k principal components is larger than the sum of the variances of any other k uncorrelated linear combinations of the columns of $\boldsymbol{X}$. (See Exercise 9.2.5.)

15.4 Optimal Properties of the Classical Solution and Goodness of Fit

Given a distance matrix $\boldsymbol{D}$, the object of MDS is to find a configuration $\hat{\boldsymbol{X}}$ in a low-dimensional Euclidean space $\mathbb{R}^k$ whose interpoint distances, $\hat{d}^2_{rs} = (\hat{\boldsymbol{x}}_r - \hat{\boldsymbol{x}}_s)'(\hat{\boldsymbol{x}}_r - \hat{\boldsymbol{x}}_s)$, say, closely match $\boldsymbol{D}$. The circumflex or "hat" indicates that the interpoint distances $\hat{\boldsymbol{D}}$ for the configuration $\hat{X}$ are "fitted" to the original distances D. Similarly, let $\hat{B}$ denote the fitted centered inner product matrix.

Now, let X be a configuration of n points in $\mathbb{R}^p$ and $L = [L_1, L_2]$ be a $(p \times p)$ orthogonal matrix where L_1 is $(p \times k)$. Then, XL_1 represents a projection of the configuration X onto the k-dimensional subspace of $\mathbb{R}^p$ spanned by the columns of L_1. We can think of $\hat{X} = XL_1$ as a "fitted' configuration in k dimensions.

Since the columns of L are orthogonal, the distances between the rows of X are the same as the distances between the rows of XL,

$$d_{rs}^2 = \sum_{i=1}^{p} (x_{ri} - x_{si})^2 = \sum_{i=1}^{p} (x_r' l_{(i)} - x_s' l_{(i)})^2. \tag{15.15}$$

If we denote the distances between the rows of XL_1 by $\hat{D}$, then

$$\hat{d}_{rs}^2 = \sum_{i=1}^{k} (x_r' l_{(i)} - x_s' l_{(i)})^2. \tag{15.16}$$

Thus, $\hat{d}_{rs} \leq d_{rs}$; that is, projecting a configuration reduces the interpoint distances. Hence, a measure of the discrepancy between the original configuration X and the projected configuration $\hat{X}$ is given by

$$\phi = \sum_{r,s=1}^{n} (d_{rs}^2 - \hat{d}_{rs}^2). \tag{15.17}$$

Then, the classical solution to the MDS problem in k dimensions has the following optimal property:

Theorem 15.4.1 *Let D be a Euclidean distance matrix corresponding to configuration X in $\mathbb{R}^p$, and fix k $(1 \leq k \leq p)$. Then, among all projections XL_1 of X onto k-dimensional subspaces of $\mathbb{R}^p$, the quantity (15.17) is minimized when X is projected onto its principal coordinates in k dimensions.*

Proof: Using (15.15) and (15.16), we see that

$$\begin{aligned} \phi &= \sum_{r,s=1}^{n} \sum_{i=k+1}^{p} (x_r' l_{(i)} - x_s' l_{(i)})^2 \\ &= \operatorname{tr} L_2' \left\{ \sum_{r,s=1}^{n} (x_r - x_s)(x_r - x_s)' \right\} L_2 \\ &= 2n^2 \operatorname{tr} L_2' S L_2, \end{aligned}$$

where S is the sample variance matrix of X, since

$$\begin{aligned} \sum_{r,s=1}^{n} (x_r - x_s)(x_r - x_s)' &= 2n \sum_{r=1}^{n} (x_r - \bar{x})(x_r - \bar{x})' - 2 \sum_{r=1}^{n} (x_r - \bar{x}) \sum_{s=1}^{n} (x_s - \bar{x})' \\ &= 2n^2 S. \end{aligned}$$

Letting $\lambda_1 \geq \cdots \geq \lambda_p$ denote the eigenvalues of nS with standardized eigenvectors $\Gamma = [\gamma_{(1)}, \ldots, \gamma_{(p)}]$, we can write

$$\phi = 2n \operatorname{tr} F_2' \Lambda F_2,$$

where $F_2 = \Gamma' L_2$ is a column orthonormal matrix ($F_2' F_2 = I_{p-k}$). Using Exercise 9.2.7, we see that ϕ is minimized when $F_2 = (0, I_{p-k})'$, that is when $L_2 = [\gamma_{(1)}, \dots, \gamma_{(p)}]$. Thus, the columns of L_1 span the space of the first k eigenvectors of nS, and so XL_1 represents the principal coordinates of X in k dimensions. Note that for this principal coordinate projection,

$$\phi = 2n(\lambda_{k+1} + \cdots + \lambda_p). \qquad \square$$

When D is not necessarily Euclidean, it is more convenient to work with the matrix $B = HAH$. If $\hat{X}$ is a fitted configuration with centered inner product matrix $\hat{B}$, then a measure of the discrepancy between B and $\hat{B}$ is given by

$$\psi = \sum_{r,s=1}^{n} (b_{rs} - \hat{b}_{rs})^2 = \text{tr}\,(B - \hat{B})^2, \tag{15.18}$$

(Mardia, 1978). For this measure also, we can prove that the classical solution to the MDS problem is optimal.

Theorem 15.4.2 *If D is a distance matrix (not necessarily Euclidean), then for fixed k, (15.18) is minimized over all configurations $\hat{X}$ in k dimensions when $\hat{X}$ is the classical solution to the MDS problem.*

Proof: Let $\lambda_1 \geq \cdots \geq \lambda_n$ denote the eigenvalues of B, some of which might be negative, and Γ the corresponding standardized eigenvectors. For simplicity, suppose that $\lambda_k > 0$ (the situation $\lambda_k < 0$ is discussed in Exercise 15.4.2). Let $\hat{X}$ be an $n \times k$ configuration with corresponding matrix $\hat{B}$. Let $\hat{\lambda}_1 \geq \cdots \geq \hat{\lambda}_n \geq 0$ denote the eigenvalues of $\hat{B}$. By the spectral decomposition theorem (Theorem A.5), we can write the symmetric matrix $\Gamma' \hat{B} \Gamma$ as

$$\Gamma' \hat{B} \Gamma = G \hat{\Lambda} G',$$

where G is orthogonal. Then,

$$\psi = \text{tr}\,(B - \hat{B})^2 = \text{tr}\,\Gamma'(B - \hat{B})\Gamma\Gamma'(B - \hat{B})\Gamma = \text{tr}\,(\Lambda - G\hat{\Lambda}G')(\Lambda - G\hat{\Lambda}G').$$

We see that for fixed $\hat{\Lambda}$ (see Exercise 15.4.2), ψ is minimized when $G = I$, so that

$$\psi = \sum_{i=1}^{n} (\lambda_i - \hat{\lambda}_i)^2.$$

Since $\hat{X}$ lies in $\mathbb{R}^k$, $B = H\hat{X}\hat{X}'H$ will have at most k nonzero eigenvalues, which must be nonnegative. Thus, it is easy to see that ψ is minimized when

$$\hat{\lambda}_i = \begin{cases} \lambda_i, & i = 1, \dots, k, \\ 0, & i = k+1, \dots, n. \end{cases}$$

Hence, $\hat{B} = \Gamma_1 \Lambda_1 \Gamma_1'$, where $\Gamma_1 = [\gamma_{(1)}, \dots, \gamma_{(k)}]$ and $\Lambda_1 = \text{diag}(\lambda_1, \dots, \lambda_k)$, so that $\hat{X}$ can be taken to equal $\Gamma_1 \Lambda_1^{1/2}$, the classical solution to the MDS problem in k dimensions. Note that the minimum value of ψ is given by

$$\psi = \lambda_{k+1}^2 + \cdots + \lambda_p^2. \qquad \square$$

The above two theorems suggest possible *agreement measures* for the "proportion of a distance matrix $\boldsymbol{D}$ explained" by the k-dimensional classical MDS solutions. Supposing $\lambda_k > 0$, these measures are

$$\alpha_{1,k} = \left(\sum_{i=1}^{k} \lambda_i \Big/ \sum_{i=1}^{n} |\lambda_i| \right) \times 100\% \tag{15.19}$$

and

$$\alpha_{2,k} = \left(\sum_{i=1}^{k} \lambda_i^2 \Big/ \sum_{i=1}^{n} \lambda_i^2 \right) \times 100\%,$$

(Mardia, 1978). We need to use absolute values in (15.19) because some of the smaller eigenvalues might be negative.

Example 15.4.1 We now consider the example of constructing a map of Britain from the road distances between 12 towns (Table 15.1). From this data, it is found that

$$\lambda_1 = 394\,473, \qquad \lambda_2 = 63\,634, \qquad \lambda_3 = 13\,544, \qquad \lambda_4 = 10\,245,$$

$$\lambda_5 = 2465, \qquad \lambda_6 = 1450, \qquad \lambda_7 = 501, \qquad \lambda_8 = 0,$$

$$\lambda_9 = -17, \qquad \lambda_{10} = -214, \qquad \lambda_{11} = -1141, \qquad \lambda_{12} = -7063.$$

We note that the last four eigenvalues are negative, but they are small in relation to $\lambda_1, \ldots, \lambda_4$. We know from Theorem 15.2.2 that some negative values are expected because the distance metric is not Euclidean.

The percentage variation explained by the first two eigenvectors is

$$\alpha_{1,2} = 92.6\% \qquad \text{and} \qquad \alpha_{2,2} = 99.8\%.$$

The first two eigenvectors, standardized so that $\boldsymbol{x}'_{(i)}\boldsymbol{x}_{(i)} = \lambda_i$, are

$$(45, 203, -138, 212, 189, -234, -8, -382, -32, 153, -120, 112),$$

$$(140, -18, 31, -76, 140, 31, -50, -26, -5, -27, -34, -106).$$

Since the MDS solution is invariant under rotations and translations, the coordinates have been superimposed on the true map in Figure 15.2 by Procrustes rotation and scaling (see Example 14.7.1). We find that the two eigenvectors closely reproduce the true map. □

Next, we look more closely at the differences between the data distances d_{rs} and the fitted distances $\hat{d}_{rs}$ for the classical k-dimensional MDS solution $\hat{\boldsymbol{X}}$. Starting from the spectral decompositions

$$\boldsymbol{B} = \boldsymbol{\Gamma}\boldsymbol{\Lambda}\boldsymbol{\Gamma}', \quad \hat{\boldsymbol{B}} = \boldsymbol{\Gamma}_1\boldsymbol{\Lambda}_1\boldsymbol{\Gamma}_1',$$

it follows that

$$b_{rs} = \sum_{i=1}^{n} \lambda_i \gamma_{ri} \gamma_{si}, \quad \hat{b}_{rs} = \sum_{i=1}^{k} \lambda_i \gamma_{ri} \gamma_{si}.$$

Further,

$$d_{rs}^2 = b_{rr} + b_{ss} - 2b_{rs}, \quad \hat{d}_{rs}^2 = \hat{b}_{rr} + \hat{b}_{ss} - 2\hat{b}_{rs}.$$

Figure 15.2 MDS solutions for the road data in Table 15.1, •, original points; △, classical solution; ▽, Shepard–Kruskal solution.

Setting

$$g_{rs}^{(i)} = \lambda_i(\gamma_{ri} - \gamma_{si})^2, \qquad (15.20)$$

it follows that

$$d_{rs}^2 = \sum_{i=1}^{n} g_{rs}^{(i)}, \quad \hat{d}_{rs}^2 = \sum_{i=1}^{k} g_{rs}^{(i)},$$

so that

$$d_{rs}^2 - \hat{d}_{rs}^2 = \sum_{i=k+1}^{n} g_{rs}^{(i)}. \qquad (15.21)$$

In particular, a positive eigenvalue λ_i will contribute a positive term to the difference (15.21), whereas a negative eigenvalue λ_i will contribute a negative term to the difference (15.21).

Here is a numerical example to illustrate these effects (see also Mardia and Riley (2021)).

Example 15.4.2 We look at the journey times between a selection of five rail stations in Yorkshire (UK) to investigate how negative eigenvalues in $\boldsymbol{B}$ can affect the MDS solution.

There are two rail lines between Leeds and York: a fast line with direct trains and a slow line that stops at various intermediate stations including Headingley, Horsforth, and Harrogate.

Here, the "journey time" is defined as the time taken to reach the destination station for a passenger who begins a journey at the starting station at 12:00 noon. For example, consider a passenger beginning a journey at Leeds station at 12:00. If the next train for York leaves at 12:08 and arrives in York at 12:31, then the journey time is 31 min (8 min waiting in Leeds plus 23 min on the train). The times here are taken from a standard weekday timetable.

Table 15.3 gives the observed and fitted distances between all pairs of stations, where the observed distance between two stations $S1$ and $S2$ is defined as the smaller of two times: the journey time from $S1$ to $S2$ and the journey time from $S2$ to $S1$. Further, the observed distance between a station and itself is taken to be 0. The fitted distances come from the two-dimensional MDS solution.

The eigenvalues of $\boldsymbol{B}$ are

$$\lambda_1 = 3210, \quad \lambda_2 = 1439, \quad \lambda_3 = 61, \quad \lambda_4 = 0, \quad \lambda_5 = -964,$$

Table 15.3 Observed and fitted journey time distance matrices for Example 15.4.2.

	$\boldsymbol{D}$					$\hat{\boldsymbol{D}}$				
	1	2	3	4	5	1	2	3	4	5
1: Leeds	0	23	23	53	31	0.0	38.4	35.2	53.3	45.5
2: Headingley	23	0	11	34	71	38.4	0.0	3.9	42.3	70.8
3: Horsforth	23	11	0	34	67	35.2	3.9	0.0	40.0	67.0
4: Harrogate	53	34	34	0	44	53.3	42.3	40.0	0.0	52.1
5: York	31	71	67	44	0	45.5	70.8	67.0	52.1	0.0

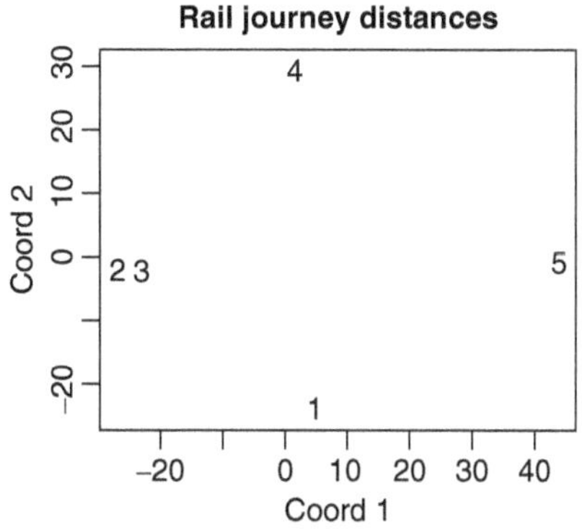

Figure 15.3 Two-dimensional MDS solution for Example 15.4.2. Train journey times between five rail stations in Yorkshire. 1, Leeds; 2, Headingley; 3, Horsforth; 4, Harrogate; 5, York.

and the corresponding eigenvectors in $\boldsymbol{\Gamma}$ are

$$\boldsymbol{\Gamma} = \begin{bmatrix} 0.08 & 0.63 & -0.06 & -0.45 & 0.63 \\ -0.48 & 0.06 & -0.66 & -0.45 & -0.38 \\ -0.41 & 0.06 & 0.75 & -0.45 & -0.26 \\ 0.03 & -0.77 & -0.04 & -0.45 & 0.45 \\ 0.77 & 0.02 & 0.00 & -0.45 & -0.45 \end{bmatrix},$$

where the rows correspond to the stations, 1, 2, 3, 4, and 5. Since $\boldsymbol{B}$ has a negative eigenvalue, $\boldsymbol{D}$ is not a Euclidean distance matrix.

The first two eigenvalues are considerably larger than the rest in absolute value, suggesting that the two-dimensional MDS solution should be a good representation; see Figure 15.3. The stations lie roughly on a circle (not surprising since there are two lines between Leeds and York). Also, Headingley and Horsforth are close together, and Leeds is further from Harrogate than from York in terms of the distance, though geographically Harrogate is nearer to Leeds than to York.

Of the ignored eigenvalues, $\lambda_3 = 61$ is positive, $\lambda_4 = 0$ always appears with eigenvector $\mathbf{1}_n$, and $\lambda_5 = -964$ is negative. The purpose of this example is to explore the distortion in the MDS solution due to λ_3 and λ_5.

In the MDS solution for journey time distances, the Euclidean distance between Headingley and Horsforth is 3.9, which is smaller than the observed journey time distance 11. On the other hand, in the MDS solution, the Euclidean distance between Leeds and York is 45.5, which is larger than the observed journey time distance 31. We can explain this behavior using the spectral decomposition of $\boldsymbol{B}$ and (15.21).

The eigenvector entries for selected stations and eigenvalues $i = 3$ and $i = 5$ are given by:

Station	$i = 3$	$i = 5$
2: Headingley	−0.66	−0.38
3: Horsforth	0.75	−0.26
Absolute difference	**1.41**	0.12
1: Leeds	−0.06	0.63
5: York	0.00	−0.45
Absolute difference	0.06	**1.07**

Hence, the difference $d_{23} - \hat{d}_{23}$ for Headingley and Horsforth is dominated by the eigenvector $i = 3$ (with positive eigenvalue, 61), whereas the difference $d_{15} - \hat{d}_{15}$ for Leeds and York is dominated by the eigenvector $i = 5$ (with negative eigenvalue, −964). In fact, the numerical values of the terms (15.20) are

$$g_{23}^{(3)} = 120.1, \quad g_{23}^{(4)} = 0, \quad g_{23}^{(5)} = -13.9$$

and

$$g_{15}^{(3)} = 0.2, \quad g_{15}^{(4)} = 0, \quad g_{15}^{(5)} = -1108.3,$$

so the dominating contributions $g_{23}^{(3)}$ and $g_{15}^{(5)}$ in (15.21) are clearly seen. This discussion explains why in the MDS solution, the Euclidean distance between Headingley and Horsforth is smaller than the observed journey time distance, whereas the Euclidean distance between Leeds and York is larger than the observed journey time distance. □

15.5 Seriation

15.5.1 Description

MDS can be used to pick out one-dimensional structure in a data set; that is, we expect the data to be parameterized by a single axis. The most common example is seriation, where we want to ascertain the chronological ordering of the data. Note that although MDS can be used to order the data in time, the *direction* of time must be determined independently.

Example 15.5.1 Consider the archaeological problem of Example 9.5.2 where the similarity between graves is measured by the number of types of pottery they have in common. Using the similarity matrix $\boldsymbol{S}_2$ of Example 14.6.2 and the transformation of Theorem 15.2.4 (see Exercise 15.2.6), it is found that

$$\lambda_1 = 1.75, \qquad \lambda_2 = 0.59, \qquad \lambda_3 = 0.35, \qquad \lambda_4 = 0.05, \qquad \lambda_5 = \lambda_6 = 0,$$

with coordinates in two dimensions

$$(-0.60, 0.77, -0.19, -0.60, 0.64, -0.01)$$

and

$$(-0.15, 0.20, 0.60, -0.15, -0.35, -0.14).$$

See Figure 15.4. The coordinates in one dimension suggest the order (A,D), C, F, E, B, which is similar to but not identical with the ordering given by correspondence analysis (Example 9.5.2). □

It is often a good idea to plot the data in more than one dimension to see if the data is in fact one-dimensional. For example, the artificial data in the above example does not particularly

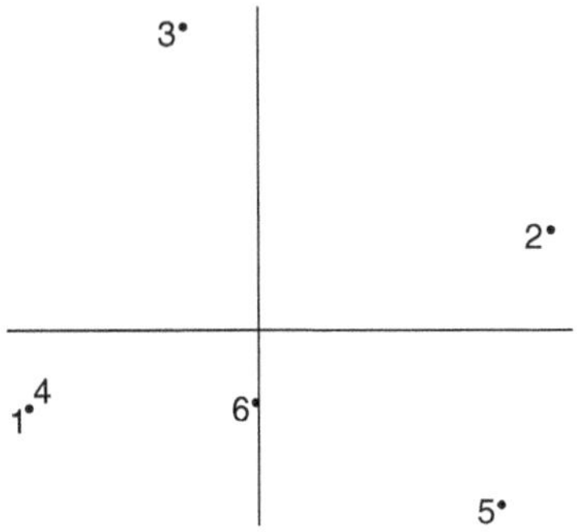

Figure 15.4 Classical MDS solution in two dimensions using similarity matrix S_2 for grave data in Example 15.5.1.

seem to lie in one dimension. However, even when the data is truly one-dimensional, it need not lie on the axis of the first dimension but can sometimes lie on a curve as shown in the following section.

15.5.2 Horseshoe Effect

In some situations we can accurately measure the distance between two objects when they are close together but not when they are far apart. Distances that are "moderate" and those that are "large" appear to be the same. For example, consider the archaeology example described above. Graves that are close together in time will have some varieties of pottery in common, but those that are separated by more than a certain gap of time will have *no* varieties in common.

This merging of all "large" distances tends to pull the farthest objects closer together and has been called the "horseshoe effect" by Kendall (1971). This effect can be observed in the following artificial example.

Example 15.5.2 (Kendall, 1971) Consider the (51×51) similarity matrix $\boldsymbol{C}$ defined by

$$c_{rr} = 9, \qquad c_{rs} = \begin{cases} k & \text{if } 25 - 3k \le |r - s| \le 27 - 3k, \quad k = 1, \dots, 8, \\ 0 & \text{if } |r - s| \ge 25. \end{cases}$$

Using the standard transformation from similarities to distances leads to eight negative eigenvalues (varying from −0.09 to −2.07) and 43 nonnegative eigenvalues,

$$126.09, 65.94, 18.17, 7.82, 7.61, 7.38, 7.02, 5.28, 3.44, \dots, 0.10, 0.$$

A plot of the configuration in two dimensions is given in Figure 15.5. The furthest points are pulled together, so that the configuration looks roughly like part of the circumference of a circle. Note that while the ordering is clear from the figure, it is not ascertainable from the one-dimensional classical solution. □

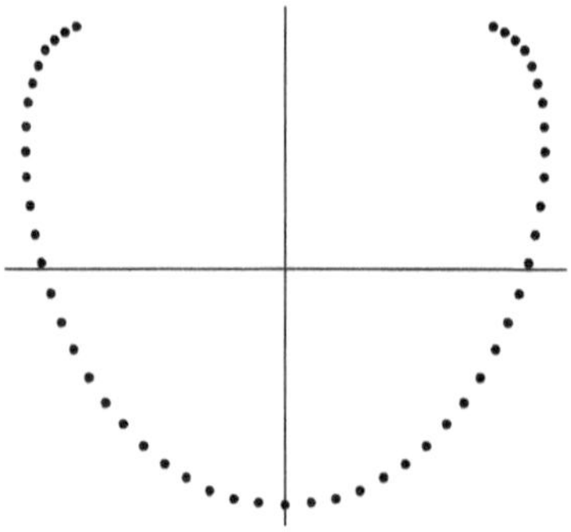

Figure 15.5 Two-dimensional representation of Kendall's similarity matrix for Example 15.5.2.

15.6 Nonmetric Methods

Implicit in the preceding sections is the assumption that there is a "true" configuration in k dimensions with interpoint distances δ_{rs}. We wish to reconstruct this configuration using an observed distance matrix $\boldsymbol{D}$ whose elements are of the form

$$d_{rs} = \delta_{rs} + e_{rs}.$$

Here, the e_{rs} represent errors of measurement plus distortion errors arising because the distances do not exactly correspond to a configuration in $\mathbb{R}^k$.

However, in some situations it is more realistic to hypothesize a less rigid relationship between d_{rs} and δ_{rs}; namely, suppose that

$$d_{rs} = f(\delta_{rs} + e_{rs}),$$

where f is an unknown monotone increasing function. For this "model," the only information we can use to reconstruct the δ_{rs} is the *rank order* of the d_{rs}. For example, for the road map data, we could try to reconstruct the map of Britain using the information

the shortest journey is that from Brighton to London;
the next shortest journey is that from Hull to Leeds;
...
the longest journey is that from Dover to Inverness.

In this nonmetric approach, $\boldsymbol{D}$ is not thought of as a "distance" matrix but as a "dissimilarity" matrix. In fact, the nonmetric approach is often most appropriate when the data is presented as a similarity matrix. For in this situation the transformation from similarities to distances is somewhat arbitrary, and perhaps, the strongest statement one should make is that greater similarity implies less dissimilarity.

An algorithm to construct a configuration based on the rank order information has been developed by Shepard (1962a,b) and Kruskal (1964).

Shepard–Kruskal Algorithm

(a) Given a dissimilarity matrix $\boldsymbol{D}$, order the off-diagonal elements so that

$$d_{r_1 s_1} \leq \cdots \leq d_{r_m s_m}, \qquad m = n(n-1)/2, \tag{15.22}$$

where $(r_1, s_1), \ldots, (r_m, s_m)$ denote all pairs of unequal subscripts, $r_i < s_i$. Say that numbers d^*_{rs} are monotonically related to the d_{rs} (and write $d^*_{rs} \overset{\text{mon}}{\sim} d_{rs}$) if

$$d_{rs} < d_{uv} \Rightarrow d^*_{rs} \leq d^*_{uv} \qquad \text{for all} \quad r < s \quad \text{and} \quad u < v. \tag{15.23}$$

(b) Let $\hat{\boldsymbol{X}}(n \times k)$ be a configuration on $\mathbb{R}^k$ with interpoint distances $\hat{d}_{rs}$. Define the (squared) *stress* of $\hat{\boldsymbol{X}}$ by

$$S^2(\hat{\boldsymbol{X}}) = \min \sum_{r<s} (d^*_{rs} - \hat{d}_{rs})^2 \Bigg/ \sum_{r<s} \hat{d}^2_{rs}, \tag{15.24}$$

where the minimum is taken over d^*_{rs} such that $d^*_{rs} \overset{\text{mon}}{\sim} d_{rs}$. The d^*_{rs} that minimizes (15.24) represent the *least-squares monotone regression* of $\hat{d}_{rs}$ on d_{rs}. Thus, Eq. (15.24) represents the extent to which the rank order of the $\hat{d}_{rs}$ disagrees with the rank order of the d_{rs}. If the rank orders match exactly (which is very rare in practice), then $S(\hat{\boldsymbol{X}}) = 0$. The presence of the denominator in (15.24) standardizes the stress and makes it invariant under transformations of the sort $\boldsymbol{y}_r = c\boldsymbol{x}_r$, $r = 1, \ldots, n$, $c \neq 0$. The stress is also invariant under transformations of the form $\boldsymbol{y}_r = \boldsymbol{A}\boldsymbol{x}_r + \boldsymbol{b}$ when $\boldsymbol{A}$ is orthogonal.

(c) For each dimension k, the configuration $\hat{\boldsymbol{X}}$ which has the smallest stress is called the *best fitting configuration in k dimensions*. Let

$$S_k = \min_{\hat{\boldsymbol{X}}(n \times k)} S(\hat{\boldsymbol{X}})$$

denote the minimal stress.

(d) To choose the correct dimension, calculate $S_1, S_2, \ldots$ until the value becomes low. Say, for example, S_k is low for $k = k_0$. Since S_k is a decreasing function of k, $k = k_0$ is the "right dimension." A rule of thumb is provided by Kruskal (1964) to judge the tolerability of S_k; $S_k \geq 20\%$, poor; $S_k = 10\%$, fair; $S_k \leq 5\%$, good; $S_k = 0$, perfect.

Remarks (1) The "best configuration" starting from an arbitrary initial configuration can be obtained using the method of steepest descent to find the local minimum. The initial configuration can be taken as the classical solution. Unfortunately, there is no ways of distinguishing in practice between a local minimum and the global minimum.

(2) The Shepard–Kruskal solution is invariant under rotation, translation, and uniform expansion or contraction of the best-fitting configuration.

(3) The Shepard–Kruskal solution is nonmetric since it utilizes only the rank orders (15.22). However, we still need a sufficiently objective numerical measure of distance to determine the rank order of the d_{rs}.

(4) *Similarities.* The nonmetric method works just as well with similarities as with dissimilarities. One simply changes the direction of the inequalities.

(5) *Missing values.* The Shepard–Kruskal method is easily adapted to the situation where there are missing values. One simply omits the missing dissimilarities in the ordering (15.22) and deletes the corresponding terms from the numerator and denominator of (15.24). As long as not too many values are missing, the method still seems to work well.

(6) *Treatment of ties.* The constraint given by (15.23) is called the *primary treatment of ties* (PTT). If $d_{rs} = d_{uv}$, then no constraint is made on d^*_{rs} and d^*_{uv}. An alternative constraint, called the *secondary treatment of ties* (STT), is given by

$$d_{rs} \leq d_{uv} \Rightarrow d^*_{rs} \leq d^*_{uv},$$

which has the property that $d_{rs} = d_{uv} \Rightarrow d^*_{rs} = d^*_{uv}$. However, one must be cautious when using STT on data with a large number of ties. The use of STT on such data such as Example 14.5.2 leads to the horseshoe effect; see Kendall (1971).

(7) *Comparison of methods.* The computation is simpler for the classical method than it is for the nonmetric method. It is not known how robust the classical method is to monotone transformations of the distance function; however, both methods seem to give similar answers when applied to well-known examples in the field. Figure 15.2 gives the two solutions for the road data. For the Shepard–Kruskal solution for the Morse code data, see Shepard (1963).

(8) The Shepard–Kruskal solution is implemented in an iterative algorithm in the R function `isoMDS` in the MASS library (Venables and Ripley, 2002). An alternative nonlinear mapping uses a stress suggested by Sammon (1969):

$$\frac{1}{\sum_{i<j} d_{ij}} \sum_{i<j} \frac{(\hat{d}_{ij} - d_{ij})^2}{d_{ij}},$$

which is minimized over k-dimensional configurations which have distances $\hat{d}_{ij}$.

15.7 Goodness of Fit Measure: Procrustes Rotation

We now describe a *goodness of fit* measure (Gower, 1971b; Green, 1952) used to compare two configurations. Let $\boldsymbol{X}$ be the $(n \times p)$ matrix of the coordinates on n points obtained from $\boldsymbol{D}$ by one technique. Suppose that $\boldsymbol{Y}$ is the $(n \times q)$ matrix of coordinates of another set of points obtained by another technique or using another measure of distance. Let $q \leq p$. By adding columns of zeros to $\boldsymbol{Y}$, we may also assume $\boldsymbol{Y}$ to be $(n \times p)$.

The measure of goodness of fit adopted is obtained by moving the points $\boldsymbol{y}_r$, relative to the points $\boldsymbol{x}_r$, until the "residual" sum of squares

$$\sum_{r=1}^{n} (\boldsymbol{x}_r - \boldsymbol{y}_r)'(\boldsymbol{x}_r - \boldsymbol{y}_r)$$

is minimal. We can move $\boldsymbol{y}_r$ relative to $\boldsymbol{x}_r$ through rotation, reflection, and translation, i.e. by

$$\boldsymbol{A}'\boldsymbol{y}_r + \boldsymbol{b}, \qquad r = 1, \dots, n, \tag{15.25}$$

where $\boldsymbol{A}'$ is a $(p \times p)$ orthogonal matrix. Hence, we wish to solve

$$R^2 = \min_{\boldsymbol{A},\boldsymbol{b}} \sum_{r=1}^{n} (\boldsymbol{x}_r - \boldsymbol{A}'\boldsymbol{y}_r - \boldsymbol{b})'(\boldsymbol{x}_r - \boldsymbol{A}'\boldsymbol{y}_r - \boldsymbol{b}) \tag{15.26}$$

for $\boldsymbol{A}$ and $\boldsymbol{b}$. Note that $\boldsymbol{A}$ and $\boldsymbol{b}$ are found by least squares. Their values are given in the following theorem.

Theorem 15.7.1 *Let $\boldsymbol{X}(n \times p)$ and $\boldsymbol{Y}(n \times p)$ be two configurations of n points, for convenience centered at the origin, so $\overline{\boldsymbol{x}} = \overline{\boldsymbol{y}} = \boldsymbol{0}$. Let $\boldsymbol{Z} = \boldsymbol{Y}'\boldsymbol{X}$, and using the singular value decomposition theorem (Theorem A.8), write*

$$\boldsymbol{Z} = \boldsymbol{V\Gamma U}', \tag{15.27}$$

where $\boldsymbol{V}$ and $\boldsymbol{U}$ are orthogonal $(p \times p)$ matrices, and $\boldsymbol{\Gamma}$ is a diagonal matrix of nonnegative elements. Then, the minimizing values of $\boldsymbol{A}$ and $\boldsymbol{b}$ in (15.26) are given by

$$\hat{\boldsymbol{b}} = \boldsymbol{0}, \qquad \hat{\boldsymbol{A}} = \boldsymbol{VU}',$$

and further

$$R^2 = \operatorname{tr} \boldsymbol{XX}' + \operatorname{tr} \boldsymbol{YY}' - 2 \operatorname{tr} \boldsymbol{\Gamma}. \tag{15.28}$$

Proof: On differentiating with respect to $\boldsymbol{b}$, we have

$$\hat{\boldsymbol{b}} = \overline{\boldsymbol{x}} - \boldsymbol{A}'\overline{\boldsymbol{y}}$$

where $\overline{\boldsymbol{y}} = \sum \boldsymbol{y}_r/n, \overline{\boldsymbol{x}} = \sum \boldsymbol{x}_r/n$. Since both configurations are centered, $\hat{\boldsymbol{b}} = \boldsymbol{0}$. Then, we can rewrite (15.26) as

$$R^2 = \min_{\boldsymbol{A}} \operatorname{tr} (\boldsymbol{X} - \boldsymbol{YA})(\boldsymbol{X} - \boldsymbol{YA})' = \operatorname{tr} \boldsymbol{XX}' + \operatorname{tr} \boldsymbol{YY}' - 2 \max_{\boldsymbol{A}} \boldsymbol{X}'\boldsymbol{YA}. \tag{15.29}$$

The constraints on $\boldsymbol{A}$ are $\boldsymbol{AA}' = \boldsymbol{I}$, i.e. $\boldsymbol{a}_i'\boldsymbol{a}_i = 1, \boldsymbol{a}_i'\boldsymbol{a}_j = 0, i \neq j$, where $\boldsymbol{a}_i'$ is the ith row of $\boldsymbol{A}$. Hence, there are $p(p+1)/2$ constraints.

Let $\Lambda/2$ be a $(p \times p)$ symmetric matrix of Lagrange multipliers for these constraints. The aim is to maximize

$$\operatorname{tr} \{\boldsymbol{Z}'\boldsymbol{A} - \frac{1}{2}\Lambda(\boldsymbol{AA}' - \boldsymbol{I})\}, \tag{15.30}$$

where $\boldsymbol{Z}' = \boldsymbol{X}'\boldsymbol{Y}$. By direct differentiation, it can be shown that

$$\frac{\partial}{\partial \boldsymbol{A}} \operatorname{tr} (\boldsymbol{Z}'\boldsymbol{A}) = \boldsymbol{Z}, \qquad \frac{\partial}{\partial \boldsymbol{A}} \operatorname{tr} (\boldsymbol{\Lambda AA}') = 2\boldsymbol{\Lambda A}.$$

Hence, on differentiating (15.30) and equating the derivatives to zero, we find that $\boldsymbol{A}$ must satisfy

$$\boldsymbol{Z} = \boldsymbol{\Lambda A}. \tag{15.31}$$

Write $\boldsymbol{Z}$ using (15.27). Noting that Λ is symmetric and that $\boldsymbol{A}$ is to be orthogonal, we get, from (15.31),

$$\Lambda^2 = \boldsymbol{ZA}'\boldsymbol{AZ} = \boldsymbol{ZZ}' = (\boldsymbol{V\Gamma U}')(\boldsymbol{U\Gamma V}').$$

Thus, we can take $\boldsymbol{\Lambda} = \boldsymbol{V\Gamma V}'$. Substituting this value of Λ in (15.31), we see that

$$\hat{\boldsymbol{A}} = \boldsymbol{VU}' \tag{15.32}$$

is a solution of (15.31). Note that $\hat{\boldsymbol{A}}$ is orthogonal. Using this value of $\hat{\boldsymbol{A}}$ in (15.29) gives (15.28).

Finally, to verify that $\hat{\boldsymbol{A}}$ maximizes (15.30) (and is not just a stationary point), we must differentiate (15.30) with respect to $\boldsymbol{A}$ a second time. For this purpose, it is convenient to write $\boldsymbol{A}$ as a vector $\boldsymbol{a} = (\boldsymbol{a}'_{(1)}, \dots, \boldsymbol{a}'_{(p)})'$. Then, (15.30) is a quadratic function of the elements of $\boldsymbol{a}$, and the second derivative of (15.30) with respect to $\boldsymbol{a}$ can be expressed as the matrix $-\boldsymbol{I}_p \otimes \boldsymbol{\Lambda}$. Since $\boldsymbol{\Lambda} = \boldsymbol{V}\boldsymbol{\Gamma}\boldsymbol{V}'$, and the diagonal elements of $\boldsymbol{\Gamma}$ are nonnegative, we see that the second derivative matrix is negative semidefinite. Hence, $\hat{\boldsymbol{A}}$ maximizes (15.30). □

We have assumed that the column means of $\boldsymbol{X}$ and $\boldsymbol{Y}$ are zero. Then, the "best" rotation of $\boldsymbol{Y}$ relative to $\boldsymbol{X}$ is $\boldsymbol{Y}\hat{\boldsymbol{A}}$, where $\hat{\boldsymbol{A}}$ is given by (15.32), and $\hat{\boldsymbol{A}}$ is called the *Procrustes rotation* of $\boldsymbol{Y}$ relative to $\boldsymbol{X}$. Noting from (15.27) that $\boldsymbol{X}'\boldsymbol{Y}\boldsymbol{Y}'\boldsymbol{X} = \boldsymbol{U}\boldsymbol{\Gamma}^2\boldsymbol{U}'$, we can rewrite (15.29) as

$$R^2 = \operatorname{tr} \boldsymbol{X}\boldsymbol{X}' + \operatorname{tr} \boldsymbol{Y}\boldsymbol{Y}' - 2 \operatorname{tr} (\boldsymbol{X}'\boldsymbol{Y}\boldsymbol{Y}'\boldsymbol{X})^{1/2}.$$

It can be seen that (15.29) is zero if and only if the $\boldsymbol{y}_r$ can be rotated to the $\boldsymbol{x}_r$ exactly.

Scale factor. If the scales of the two configurations are different, then the transformation (15.25) should be of the form

$$c\boldsymbol{A}'\boldsymbol{y}_r + \boldsymbol{b},$$

where $c > 0$. Following the above procedure, it can be seen that

$$\hat{c} = (\operatorname{tr} \boldsymbol{\Gamma})/(\operatorname{tr} \boldsymbol{Y}\boldsymbol{Y}'), \tag{15.33}$$

and the other estimates remain as before. This transformation is called the *Procrustes rotation with scaling* of $\boldsymbol{Y}$ relative to $\boldsymbol{X}$. Then, the new minimum residual sum of squares is given by (assuming that the column means of $\boldsymbol{X}$ and $\boldsymbol{Y}$ are zero)

$$R^2 = \operatorname{tr} (\boldsymbol{X}\boldsymbol{X}') - \hat{c}^2 \operatorname{tr} (\boldsymbol{Y}\boldsymbol{Y}'), \tag{15.34}$$

where $\hat{c}$ is given by (15.33). Note that this procedure is not symmetrical with respect to $\boldsymbol{X}$ and $\boldsymbol{Y}$.

Symmetry can be obtained by selecting scaling, so that

$$\operatorname{tr} (\boldsymbol{X}\boldsymbol{X}') = c^2 \operatorname{tr} (\boldsymbol{Y}\boldsymbol{Y}').$$

There are many other variations of Procrustes matching. In similarity shape analysis, one uses Procrustes matching between two configurations where the rotation is restricted to $|\boldsymbol{A}| = 1$. Further, when the scale term c is set to $c = 1$, the matching is called shape-size analysis or simply matching of two *forms* (see, for example, Dryden and Mardia (2016)). For an excellent review and further development of this topic, see Sibson (1978) and Dryden and Mardia (2016, ch. 4), respectively.

Example 15.7.1 The actual Ordnance Survey coordinates of the 12 towns in Table 15.1 are

	1	2	3	4	5	6	7	8	9	10	11	12
E	258	532	340	632	292	259	510	267	430	530	425	623
N	282	104	556	142	92	666	428	845	434	181	565	309

Treating these quantities as planar coordinates $\boldsymbol{X}(12 \times 2)$ (the curvature of the Earth has little effect), and using the first two eigenvectors from the classical MDS solution for the data given in Example 15.4.1 as $\boldsymbol{Y}$, it is found that

$$\boldsymbol{X}'\boldsymbol{Y} = \begin{bmatrix} -180919.9 & -91485.2 \\ 496606.3 & -24845.2 \end{bmatrix}, \quad \boldsymbol{U} = \begin{bmatrix} -0.344919 & 0.938633 \\ 0.938633 & 0.344919 \end{bmatrix},$$

$$\boldsymbol{\Gamma} = \begin{bmatrix} 528597.65 & 0 \\ 0 & 94452.02 \end{bmatrix}, \quad \boldsymbol{V} = \begin{bmatrix} 0.999879 & 0.015578 \\ 0.015578 & -0.999879 \end{bmatrix}.$$

This leads to the transforming of the $\boldsymbol{y}$s to match the $\boldsymbol{x}$s by

$$\boldsymbol{y}_r^* = c\boldsymbol{A}'\boldsymbol{y}_r + \boldsymbol{b},$$

where

$$c = 1.360052, \ \boldsymbol{A} = \boldsymbol{V}\boldsymbol{U}' = \begin{bmatrix} -0.330255 & 0.943892 \\ -0.943892 & -0.330255 \end{bmatrix}, \ \boldsymbol{b} = \begin{pmatrix} 424.825 \\ 383.577 \end{pmatrix}$$

The transformation has been used on $\boldsymbol{y}_r$ to obtain $\boldsymbol{y}_r^*$, and these $\boldsymbol{y}_r^*$ are plotted in Figure 15.2 together with the $\boldsymbol{x}_r$. We have, for $\boldsymbol{X}$ transformed to have column means equal to zero,

$$\operatorname{tr}\boldsymbol{X}\boldsymbol{X}' = 855332, \qquad \operatorname{tr}\boldsymbol{Y}\boldsymbol{Y}' = 458107.$$

Hence from (15.34), the residual is $R^2 = 7952$.

The Shepard–Kruskal solution has a stress of 1.26%. Using this solution as $\boldsymbol{Y}$ and the Ordnance Survey coordinates again as $\boldsymbol{X}$ leads to a residual of $R^2 = 7326$. Hence, the Shepard–Kruskal solution fits slightly better for these data. Of course, Figure 15.2 shows little difference between the two solutions.

If we are given two distance matrices $\boldsymbol{D}_1$ and $\boldsymbol{D}_2$ but *not* the corresponding points, (15.28) cannot be computed without some method to find the correspondence between the points. The first two terms are expressible in terms of $\boldsymbol{D}_1$ and $\boldsymbol{D}_2$ but not $\operatorname{tr}\boldsymbol{\Gamma}$. □

15.8 Multisample Problem and Canonical Variates

Consider the case of g p-variate populations with means $\boldsymbol{\mu}_r, r = 1, \ldots, g$, and common covariance matrix $\boldsymbol{\Sigma}$. If we are given a data matrix $\boldsymbol{Y}$ representing samples of size n_r from the rth group, $r = 1, \ldots, g$, let $\overline{\boldsymbol{y}}_r$ denote the rth sample mean and estimate the common covariance matrix $\boldsymbol{\Sigma}$ by $\boldsymbol{W}/\nu$, where $\boldsymbol{W}$ is the within-groups sum of squares and products (SSP) matrix with ν degrees of freedom.

Assume that the overall (unweighted) mean $\overline{\boldsymbol{y}}$ is

$$\overline{\boldsymbol{y}} = \sum_{r=1}^{g} \overline{\boldsymbol{y}}_r = \boldsymbol{0}.$$

We will work with the Mahalanobis distances

$$d_{rs}^2 = \nu(\overline{\boldsymbol{y}}_r - \overline{\boldsymbol{y}}_s)'\boldsymbol{W}^{-1}(\overline{\boldsymbol{y}}_r - \overline{\boldsymbol{y}}_s).$$

It is easily checked that if $\boldsymbol{B}$ is defined by (15.3), then

$$\boldsymbol{B} = \nu\overline{\boldsymbol{Y}}\boldsymbol{W}^{-1}\overline{\boldsymbol{Y}}',$$

where $\overline{\boldsymbol{Y}}' = [\overline{\boldsymbol{y}}_1, \ldots, \overline{\boldsymbol{y}}_g]$.

Thus, $\boldsymbol{B} \geq 0$, and so $\boldsymbol{D}$ is Euclidean. Let $\boldsymbol{X}$ be the configuration for $\boldsymbol{B}$ defined in Theorem 15.2.2, and fix k, $1 \leq k \leq p$. Then, the first k columns of $\boldsymbol{X}$ can be regarded as the coordinates of points representing the g means in k dimensions ($k \leq p$). This configuration has the optimal property that it is the "best" representation in k dimensions.

Note that $\overline{\boldsymbol{Y}}'\overline{\boldsymbol{Y}}$ is the (unweighted) between-groups SSP matrix. Let $\boldsymbol{l}_i$ denote the ith canonical vector of Section 13.5 using this unweighted between-groups SSP matrix; that is, define $\boldsymbol{l}_i$ by

$$\nu \boldsymbol{W}^{-1}\overline{\boldsymbol{Y}}'\overline{\boldsymbol{Y}}\boldsymbol{l}_i = \lambda_i \boldsymbol{l}_i, \qquad \nu^{-1}\boldsymbol{l}_1'\boldsymbol{W}\boldsymbol{l}_i = 1, \quad i = 1, \ldots, \min(p, g),$$

where λ_i is the ith eigenvalue of $\nu \boldsymbol{W}^{-1}\overline{\boldsymbol{Y}}'\overline{\boldsymbol{Y}}$, which is the same as the ith eigenvalue of $\boldsymbol{B}$. Then, the scores of the g groups on the ith canonical coordinate are given by

$$\overline{\boldsymbol{Y}}\boldsymbol{l}_i.$$

Since $\boldsymbol{B}\overline{\boldsymbol{Y}}\boldsymbol{l}_i = \lambda_i \overline{\boldsymbol{Y}}'\boldsymbol{l}_i$ and $\boldsymbol{l}_i'\overline{\boldsymbol{Y}}'\overline{\boldsymbol{Y}}\boldsymbol{l}_i = \lambda_i$, we see from (15.5) that

$$\overline{\boldsymbol{Y}}\boldsymbol{l}_i = \boldsymbol{x}_{(i)},$$

so that $\overline{\boldsymbol{Y}}\boldsymbol{l}_i$ is also the ith eigenvector of $\boldsymbol{B}$. Thus, the canonical means in k dimensions, that is, the scores of the first k canonical variates in the g groups, are the same as the coordinates given by Theorem 15.2.2.

Exercises and Complements

15.2.1 Using (15.6), show that (15.8) can be written in the form (15.9).

15.2.2 In the notation of Theorem 15.2.2, show that

(a) $b_{rr} = \bar{a}_{..} - 2\bar{a}_{r.}, \quad b_{rs} = a_{rs} - \bar{a}_{r.} - \bar{a}_{.s} + \bar{a}_{..}, \quad r \neq s;$

(b) $\boldsymbol{B} = \sum_{i=1}^{p} \boldsymbol{x}_{(i)}\boldsymbol{x}_{(i)}'$;

(c) $\sum_{r=1}^{n} \lambda_r = \sum_{r=1}^{n} b_{rr} = \frac{1}{2n} \sum_{r,s=1}^{n} d_{rs}^2$.

15.2.3 (Gower, 1968) Let $\boldsymbol{D} = (d_{rs})$ be an $(n \times n)$ Euclidean distance matrix with configuration $\boldsymbol{X} = (\boldsymbol{x}_1, \ldots, \boldsymbol{x}_n)'$ in p-dimensional principal coordinates, given by Theorem 15.2.2. Suppose that we wish to add an additional point to the configuration using distances $d_{r,n+1}$, $r = 1, \ldots, n$ (which we know to be Euclidean), allowing for a $(p+1)$th dimension. If the first n points are represented by $(x_{r1}, \ldots, x_{rp}, 0)'$, $r = 1, \ldots, n$, then show that the $(n+1)$th point is given by

$$\boldsymbol{x}_{n+1} = (x_{n+1,1}, \ldots, x_{n+1,p}, x_{n+1,p+1})' = (\boldsymbol{x}', y)', \quad \text{say}$$

where

$$\boldsymbol{x} = \frac{1}{2}\boldsymbol{\Lambda}^{-1}\boldsymbol{X}'\boldsymbol{f}, \qquad \boldsymbol{f} = (f_1, \ldots, f_n)', \qquad f_r = b_{rr} - d_{r,n+1}^2$$

and

$$y^2 = \frac{1}{n}\sum_{r=1}^{n} d_{r,n+1}^2 - \frac{1}{n}\sum_{r=1}^{n} b_{rr} - \boldsymbol{x}'\boldsymbol{x}.$$

Hence, $\boldsymbol{x}$ is uniquely determined, but y is determined only in value, not in sign. Give the reason. (Hint: substitute $f_r = 2\boldsymbol{x}_r'\boldsymbol{x}_{n+1} - \boldsymbol{x}_{n+1}'\boldsymbol{x}_{n+1}$ for f_r in terms of $\boldsymbol{x}_1, \dots, \boldsymbol{x}_{n+1}$ to verify the formula for $\boldsymbol{x}$.)

15.2.4 If $\boldsymbol{C} \geq 0$, then show that $c_{ii} + c_{jj} - 2c_{ij} > 0$. Show that the distance d_{ij} defined by $d_{ij}^2 = c_{ii} + c_{jj} - 2c_{ij}$ satisfies the triangle inequality.

15.2.5 For the Bhattacharyya distance matrix $\boldsymbol{D}$ given in Example 14.6.1, the eigenvalues of $\boldsymbol{B}$ are

$$\lambda_1 = 318.97, \qquad \lambda_2 = 167.72, \qquad \lambda_3 = 11.11, \qquad \lambda_4 = 0.$$

Hence, $\boldsymbol{D}$ is Euclidean, and the two-dimensional representation accounts for $\alpha_{1,2} = 98\%$ of the variation.
Show that the principal coordinates in two dimensions for Eskimo, Bantu, English, and Korean are, respectively,

$$(9.69, 7.29), \qquad (-11.39, -2.51), \qquad (-6.00, 4.57), \qquad (7.70, -9.34).$$

Plot these points and comment on the conclusions drawn in Example 14.6.1.

15.2.6 For the grave data (Example 14.6.2) using the similarity matrix $\boldsymbol{S}_2$, show that the distance matrix given by standard transformation is

$$\boldsymbol{D} = \begin{bmatrix} 0 & \sqrt{10} & 2 & 0 & \sqrt{8} & 2 \\ & 0 & \sqrt{6} & \sqrt{10} & \sqrt{2} & \sqrt{6} \\ & & 0 & 2 & \sqrt{8} & 2 \\ & & & 0 & \sqrt{8} & 2 \\ & & & & 0 & 2 \\ & & & & & 0 \end{bmatrix}.$$

15.2.7 Suppose that $1, 2, \dots, 7$ are regions (enclosed by unbroken lines) in a country arranged as in Figure 15.6. Let the distance matrix be constructed by counting the minimum number of boundaries crossed to pass from region i to region j. Show that the distance matrix is given by

$$\begin{bmatrix} 0 & 1 & 2 & 2 & 2 & 1 & 1 \\ & 0 & 1 & 2 & 2 & 2 & 1 \\ & & 0 & 1 & 2 & 2 & 1 \\ & & & 0 & 1 & 2 & 1 \\ & & & & 0 & 1 & 1 \\ & & & & & 0 & 1 \\ & & & & & & 0 \end{bmatrix}.$$

Show that the distances constructed in this way obey the triangle inequality $d_{ik} \leq d_{ij} + d_{jk}$. By showing that the eigenvalues of the matrix $\boldsymbol{B}$ are

$$\lambda_1 = \lambda_2 = \frac{7}{2}, \qquad \lambda_3 = \lambda_4 = \frac{1}{2}, \qquad \lambda_5 = 0, \qquad \lambda_6 = -\frac{1}{7}, \qquad \lambda_7 = -1,$$

deduce that this metric is non-Euclidean.

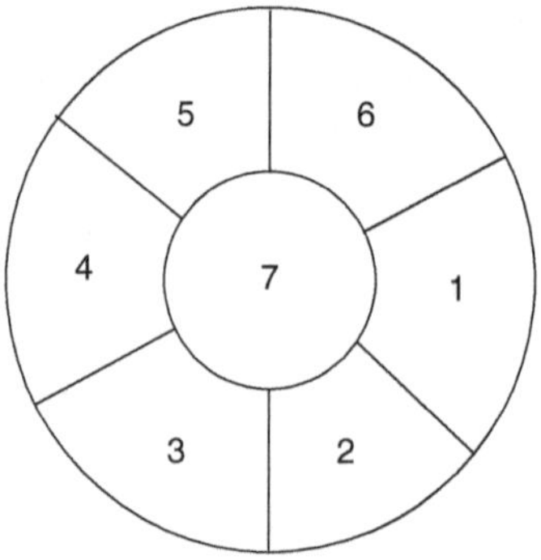

Figure 15.6 Seven regions in a country for Exercise 15.2.7.

Since $\lambda_1 = \lambda_2$, select any two orthogonal eigenvectors corresponding to λ_1 and λ_2, and by plotting the seven points so obtained, show that the original map is reconstructed. As in Example 15.2.1, the points are vertices of a hexagon with center at the origin.

15.2.8 (Lingoes, 1970; Mardia, 1978). Let $\boldsymbol{D}$ be a distance matrix. Show that for some real number a, there exists a Euclidean configuration in $p \leq n-2$ dimensions with interpoint distances d^*_{rs} satisfying

$$d^{*2}_{rs} = d^2_{rs} - 2a, \qquad r \neq s; \qquad d^*_{rr} = 0.$$

Thus, d^{*2}_{rs} is a linear function of d^2_{rs}, so the configuration preserves the rank order of the distances. (Hint: show that the matrix $\boldsymbol{D}^*$ leads to $\boldsymbol{A}^*$ and $\boldsymbol{B}^*$ given by

$$\boldsymbol{A}^* = \left(-\frac{1}{2}d^{*2}_{rs}\right) = \boldsymbol{A} - a(\boldsymbol{I} - \boldsymbol{J}), \qquad \boldsymbol{B}^* = \boldsymbol{H}\boldsymbol{A}^*\boldsymbol{H} = \boldsymbol{B} - a\boldsymbol{H}.$$

If $\boldsymbol{B}$ has eigenvalues $\lambda_1 \geq \cdots \geq \lambda_u > 0 \geq \lambda'_1 \geq \cdots \geq \lambda'_v$, then $\boldsymbol{B}^*$ has eigenvalues $\lambda_r - a$, $r = 1, \ldots, u$; 0; and $\lambda'_r - a$, $r = 1, \ldots, v$. Then the choice $a = \lambda'_v$ makes $\boldsymbol{B}^* \geq 0$ of rank at most $n-2$.)

15.2.9 Let $\boldsymbol{D} \neq \boldsymbol{0}$ be a nontrivial $n \times n$ distance matrix. Define $\boldsymbol{A}$ and $\boldsymbol{B}$ by (15.10), and let $\lambda_1 > \cdots > \lambda_n$ denote the eigenvalues of $\boldsymbol{B}$. Show that $\lambda_1 > 0$. Hence, it is always possible to construct an MDS solution in one dimension.
Hint: Since $\boldsymbol{H} = \boldsymbol{I} - \frac{1}{n}\boldsymbol{1}\boldsymbol{1}^T$, note that

$$\sum \lambda_i = \text{tr}\,\boldsymbol{B} = \text{tr}\,(\boldsymbol{HAH}) = \text{tr}\,(\boldsymbol{HA}) = \text{tr}\,(\boldsymbol{A}) - \frac{1}{n}\boldsymbol{1}'\boldsymbol{A}\boldsymbol{1} = 0 + \frac{1}{n}\sum d^2_{rs} > 0.$$

Since the sum of eigenvalues is positive, the largest eigenvalue must be positive.

15.2.10 (Mardia and Riley, 2021). This exercise extends the result of Example 15.2.2 from a 2×2 distance matrix to a 3×3 distance matrix

$$\boldsymbol{D} = \begin{bmatrix} 0 & a & b \\ a & 0 & c \\ b & c & 0 \end{bmatrix},$$

where $a, b, c > 0$ satisfy the triangle inequalities, $a + b > c$, $a + c > b$, and $b + c > a$. The purpose is to show how the exact MDS solution can be obtained from the algorithm in Section 15.2.2.

(a) Compute $\boldsymbol{B}$ in (15.10) and show that

$$\boldsymbol{B} = \frac{1}{18}\begin{bmatrix} 4a^2 + 4b^2 - 2c^2 & -5a^2 + b^2 + c^2 & a^2 - 5b^2 + c^2 \\ -5a^2 + b^2 + c^2 & 4a^2 - 2b^2 + 4c^2 & a^2 + b^2 - 5c^2 \\ a^2 - 5b^2 + c^2 & a^2 + b^2 - 5c^2 & -2a^2 + 4b^2 + 4c^2 \end{bmatrix}.$$

(b) To get the eigenvalues and eigenvectors of $\boldsymbol{B}$, it is helpful to rotate $\boldsymbol{B}$ using a version of the Helmert matrix

$$\boldsymbol{R} = \begin{bmatrix} -\frac{1}{\sqrt{2}} & \frac{1}{\sqrt{6}} & \frac{1}{\sqrt{3}} \\ 0 & -\frac{2}{\sqrt{6}} & \frac{1}{\sqrt{3}} \\ \frac{1}{\sqrt{2}} & \frac{1}{\sqrt{6}} & \frac{1}{\sqrt{3}} \end{bmatrix}$$

(see (A.32) for the same matrix with a different ordering of the rows and columns). Set $\boldsymbol{R}'\boldsymbol{B}\boldsymbol{R} = \boldsymbol{\Sigma}$, say. Show that the elements of $\boldsymbol{\Sigma}$ are given by

$$\sigma_{11} = \frac{b^2}{2}, \quad \sigma_{12} = \frac{c^2 - a^2}{2\sqrt{3}}, \quad \sigma_{22} = \frac{2a^2 - b^2 + 2c^2}{6},$$

and $\sigma_{3i} = \sigma_{i3} = 0$, $i = 1, 2, 3$, so that $\boldsymbol{\Sigma}$ has a 2×2 symmetric matrix nested within a 3×3 singular matrix. Let

$$\Delta = \sqrt{a^4 + b^4 + c^4 - a^2b^2 - a^2c^2 - b^2c^2}.$$

The representation

$$2(a^4 + b^4 + c^4 - a^2b^2 - a^2c^2 - b^2c^2) = (a^2 - b^2)^2 + (a^2 - c^2)^2 + (b^2 - c^2)^2$$

as a sum of squares shows that $\Delta \geq 0$ is real. Using Exercise 9.2.1, show that if $a \neq c$, then the eigenvalues of $\boldsymbol{\Sigma}$ are

$$\lambda_1 = \frac{1}{6}(a^2 + b^2 + c^2 + 2\Delta) \geq \lambda_2 = \frac{1}{6}(a^2 + b^2 + c^2 - 2\Delta) > \lambda_3 = 0, \quad (15.35)$$

and the corresponding standardized eigenvectors of Σ are the columns of

$$\boldsymbol{Q} = \begin{bmatrix} u & v & 0 \\ v & -u & 0 \\ 0 & 0 & 1 \end{bmatrix},$$

where

$$u = (c^2 - a^2)/(\kappa\sqrt{3}), \quad v = (a^2 - 2b^2 + c^2 + 2\Delta)/3\kappa,$$

and where

$$\kappa = \frac{2}{3}\sqrt{2\Delta^2 + \Delta(a^2 - 2b^2 + c^2)}.$$

(c) Hence, when $a \neq c$, deduce that the exact MDS solution is given by $\boldsymbol{X} = \boldsymbol{RQ\Lambda}^{1/2}$ with columns

$$\boldsymbol{x}_{(1)} = \left(\frac{2\lambda_1}{27\kappa^2}\right)^{1/2} \begin{pmatrix} 2a^2 - b^2 - c^2 + \Delta \\ -a^2 + 2b^2 - c^2 - 2\Delta \\ -a^2 - b^2 + 2c^2 + \Delta \end{pmatrix}, \quad \boldsymbol{x}_{(2)} = \left(\frac{2\lambda_2}{9\kappa^2}\right)^{1/2} \begin{pmatrix} b^2 - c^2 - \Delta \\ -a^2 + c^2 \\ a^2 - b^2 + \Delta \end{pmatrix},$$

and $\boldsymbol{x}_{(3)} = \boldsymbol{0}$. Of course, the final column $\boldsymbol{x}_{(3)}$ can be deleted without affecting the distances between the rows of $\boldsymbol{X}$. Hence, the points of the triangle have coordinates in two dimensions given by the rows of $[\boldsymbol{x}_{(1)}, \boldsymbol{x}_{(2)}]$.

15.2.11 Using the same notation as Exercise 15.2.10, consider now the case of an isosceles triangle with $a = c$. The formula for the eigenvectors in part (c) is not always valid in this case. However, the calculations are simpler since $\boldsymbol{\Sigma}$ in part (b) is now diagonal. Note that the eigenvectors of $\boldsymbol{\Sigma}$ are given by $\boldsymbol{Q} = \boldsymbol{I}$, the identity matrix, with eigenvalues

$$\lambda_1 = b^2/2, \quad \lambda_2 = (4a^2 - b^2)/6, \quad \lambda_3 = 0,$$

though with this ordering, the eigenvalues are not necessarily in decreasing order here. In particular, show that $\lambda_1 > \lambda_2 > 0$ if $2a > b > a$ and $0 < \lambda_1 < \lambda_2$ if $b < a$. Deduce that the two nonzero columns of the exact MDS solution $\boldsymbol{X} = \boldsymbol{RQ\Lambda}^{1/2}$ are given by

$$\boldsymbol{x}_{(1)} = \begin{pmatrix} -b/2 \\ 0 \\ b/2 \end{pmatrix}, \quad \boldsymbol{x}_{(2)} = \begin{pmatrix} e \\ -2e \\ e \end{pmatrix},$$

where $e = \sqrt{(4a^2 - b^2)/6}$. Hence, the coordinates of the three points are $(-b/2, e)$, $(0, -2e)$, and $(b/2, e)$. For the equilateral case $a = b$, note that $e = 1/\sqrt{3}$.

15.4.1 Let $\boldsymbol{l}_1, \ldots, \boldsymbol{l}_p$ be orthonormal vectors in $\mathbb{R}^q (p \leq q)$ and $\boldsymbol{z}_r$, $r = 1, \ldots, n$ be points in $\mathbb{R}^q$. Let H_r denote the foot of the perpendicular of $\boldsymbol{z}_r$ on the subspace spanned by $\boldsymbol{l}_1, \ldots, \boldsymbol{l}_p$. Show that with respect to the new coordinate system with axes $\boldsymbol{l}_1, \ldots, \boldsymbol{l}_p$, the coordinates of H_r are $(\boldsymbol{l}_1'\boldsymbol{z}_r, \ldots, \boldsymbol{l}_p'\boldsymbol{z}_r)'$. What modification must be made if the $\boldsymbol{l}_i$ are orthogonal but not orthonormal?

15.4.2 Let $\boldsymbol{\Lambda} = \text{diag}(\lambda_1, \ldots, \lambda_p)$, where $\lambda_1 \geq \cdots \geq \lambda_p$ are real numbers, and $\hat{\boldsymbol{\Lambda}} = \text{diag}(\hat{\lambda}_1, \ldots, \hat{\lambda}_p)$, where $\hat{\lambda}_1 \geq \cdots \geq \hat{\lambda}_p \geq 0$ are nonnegative numbers. Show that minimizing

$$\text{tr}\,(\boldsymbol{\Lambda} - \boldsymbol{G}\hat{\boldsymbol{\Lambda}}\boldsymbol{G}')^2$$

over orthogonal matrices $\boldsymbol{G}$ is equivalent to maximizing

$$\text{tr}\,(\boldsymbol{\Lambda G}\hat{\boldsymbol{\Lambda}}\boldsymbol{G}') = \sum_{i,j=1}^{p} \lambda_i \hat{\lambda}_j g_{ij}^2 = \sum_{i=1}^{p} \lambda_i h_i = \phi(\boldsymbol{h}) \qquad \text{, say,}$$

where

$$h_i = \sum_{j=1}^{p} \hat{\lambda}_j g_{ij}^2 \geq 0 \quad \text{and} \quad \sum_{i=1}^{p} h_i = \sum_{j=1}^{p} \hat{\lambda}_j.$$

Show that $\phi(\boldsymbol{h})$ is maximized over such vectors $\boldsymbol{h}$ when $h_i = \hat{\lambda}_i$ for $i = 1, \ldots, p$; that is, when $\boldsymbol{G} = \boldsymbol{I}$.

16

High-dimensional Data

16.1 Introduction

The starting point for much multivariate regression analysis is an $n \times q$ data matrix $\boldsymbol{X}$ with n observations and q variables. Many methods in earlier chapters are designed for the case where q is small or moderate relative to the sample size n. In contrast, this chapter focuses on the high-dimensional case of large q, sometimes known as "fat data" (in which the data matrix can be wider than it is tall). An extreme example is genomic data for which there can be over 3 billion variables for just one individual. In contrast to Chapter 7, the matrix $\boldsymbol{X}$ here does not include the intercept term, which must be included separately as in (16.1).

High-dimensional data covers two possibilities. If $n \le q$, then the sample covariance matrix is singular. Methods that require the inverse of a covariance matrix to be estimated (as in Eqs. (6.4), (7.5), Section 12.4.2, etc.) cannot even be defined. A less extreme situation occurs when $n > q$ but n is not much bigger than q. In this case, the sample covariance matrix is mathematically nonsingular but is often numerically unstable. As a result, statistical methods depending on the inverse sample covariance matrix tend to be very noisy and uninformative.

Another issue arising in high dimensions is known as "the curse of dimensionality." One interpretation of the curse of dimensionality is that high-dimensional Euclidean space is very sparsely populated by data, unless the sample size is huge. We can illustrate this problem by a simulation in which we generate a data matrix $\boldsymbol{X}$ containing n observations from a multivariate normal distribution with q independent components. We compute (i) the average of the nearest neighbor Euclidean distance from each observation and (ii) the mean and the standard deviation of the distances from each observation to the remaining observations, averaging the ratio over all observations. The results are shown in Figure 16.1 for various pairs (q, n). It can be seen that the observations become increasingly spread out for large values of q, and that the relative variability in distances rapidly diminishes with q.

One way to address the problems of high-dimensional data is through *regularization*. Regularization includes ideas such as variable selection, dimension reduction, shrinkage, and smoothing. However, there is a price to be paid for regularization. Conventional multivariate methods such as Hotelling's T^2 statistic and Fisher's linear discriminant rule are

Multivariate Analysis, Second Edition. Kanti V. Mardia, John T. Kent, and Charles C. Taylor.

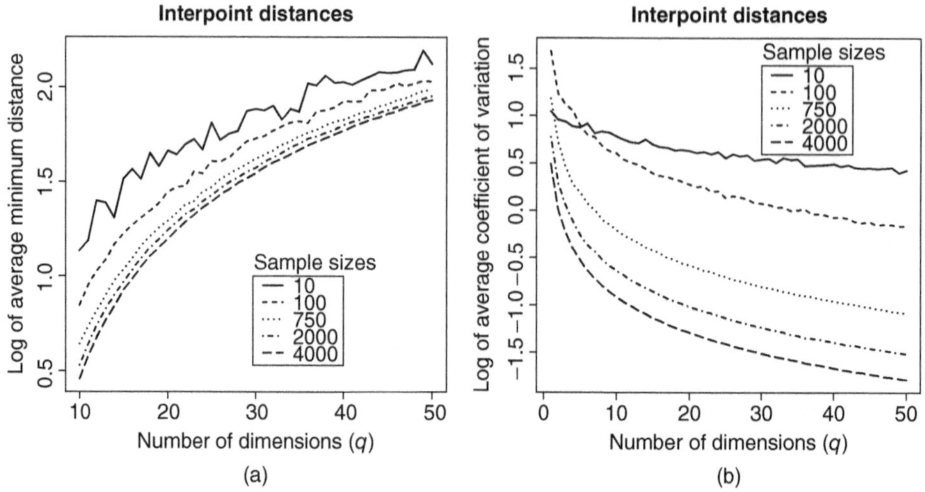

Figure 16.1 (a) Log of average nearest neighbor distances for various dimensions q and different sample sizes taken from a multivariate normal distribution with i.i.d. elements; (b) The coefficient of variation in interpoint distances for each observation, averaged over all observations (log scale).

equivariant under all linear transformations of the data. Regularized methods are equivariant under a smaller set of transformations of the data. After centering the data to have sample mean 0, and scaling the data so each variable has sample variance 1, regularized statistical methods either have no transformations (e.g. variable selection) or are equivariant only under orthogonal transformations. The prime example of a regularization method is principal component analysis, which is equivariant under orthogonal transformations. The concept of equivariance is explained in more detail in Section 16.2.

This chapter focuses on regression methods, where $\boldsymbol{X}$ is the matrix of regressor variables, and where the response is either a one-dimensional vector $\boldsymbol{y}$ ($n \times 1$) (multiple regression) or an $n \times p$ matrix $\boldsymbol{Y}$ (multivariate regression). Variable selection methods tend to be most useful when variables in $\boldsymbol{X}$ have limited correlation with one another and only a few have an important impact on $\boldsymbol{y}$. Ideas based on dimension reduction, shrinkage, and smoothing through linear combinations tend to be more useful when the variables in $\boldsymbol{X}$ are highly correlated with one another.

The rest of this chapter is organized as follows. In Section 16.2, we use shrinkage to derive some variance reduction methods that, although biased, can have the added advantage of improved interpretability as well as lower mean-squared error. In Section 16.3, we examine the use of principal components to reduce the dimensionality before carrying out regression. In Section 16.4, we introduce partial least squares, which considers projections of the data based on the covariances between $\boldsymbol{y}$ and $\boldsymbol{X}$. In Section 16.5, we conclude the chapter with a brief introduction to functional data, which is essentially an infinite-dimensional setting with smoothness penalties.

16.2 Shrinkage Methods in Regression

16.2.1 The Multiple Linear Regression Model

The starting point for much of this chapter is the linear regression model (7.35) with a univariate response variable, also known as multiple linear regression:

$$\boldsymbol{y} = \mu \boldsymbol{1}_n + \boldsymbol{X}\boldsymbol{\beta} + \boldsymbol{u}, \qquad \boldsymbol{u} \sim N_n(\boldsymbol{0}, \sigma^2 \boldsymbol{I}), \tag{16.1}$$

where $\boldsymbol{X}$ is an $n \times q$ matrix. There are two ways to think about this regression model:

(a) *Fixed regressors*. The data matrix $\boldsymbol{X}$ is treated as nonrandom.
(b) *Random regressors*. The rows of $\boldsymbol{X}$ are treated as independent and identically distributed (i.i.d.) random vectors from an $N_p(\boldsymbol{0}, \Sigma_{\boldsymbol{xx}})$ distribution, and $\boldsymbol{X}$ is independent of $\boldsymbol{u}$.

Many discussions of regression treat the regressors as fixed, either working in setting (a) or working in setting (b) and conditioning on $\boldsymbol{X}$. Treating the regressors as fixed is the approach taken in much of the development here. However, especially for the theoretical development of partial least squares, it is helpful to assume random regressors.

In general, the intercept term is not very interesting. Given an estimate $\hat{\boldsymbol{\beta}}$, the intercept can be estimated by $\mu = \bar{y} - \bar{\boldsymbol{x}}'\hat{\boldsymbol{\beta}}$. Hence, it is not considered further here.

The ordinary least squares (OLS) estimate of $\boldsymbol{\beta}$ is the value that minimizes the residual sum of squares

$$(\boldsymbol{y} - \mu \boldsymbol{1}_n - \boldsymbol{X}\boldsymbol{\beta})'(\boldsymbol{y} - \mu \boldsymbol{1}_n - \boldsymbol{X}\boldsymbol{\beta}) \tag{16.2}$$

and, provided $\boldsymbol{S}_{xx}$ is nonsingular, is given by

$$\hat{\boldsymbol{\beta}}^{\text{OLS}} = \boldsymbol{S}_{xx}^{-1} \boldsymbol{s}_{xy}, \tag{16.3}$$

involving the sample covariance matrix of $\boldsymbol{X}$ and the sample covariance vector between $\boldsymbol{X}$ and $\boldsymbol{y}$. As noted in Section 7.2, the OLS estimate is also the maximum likelihood estimate (MLE) under normality.

The OLS criterion (16.2) can be thought of as an affine equivariant criterion in the sense that the resulting estimator is equivariant under all linear transformations of $\boldsymbol{X}$. That is, if $\boldsymbol{X}$ is replaced by $\boldsymbol{XA}$, where $\boldsymbol{A}$ is $q \times q$ nonsingular, then $\boldsymbol{\beta}$ is replaced by $\boldsymbol{A}^{-1}\boldsymbol{\beta}$ in the population model (16.1), and the estimator $\hat{\boldsymbol{\beta}}^{\text{OLS}}$ is replaced by

$$(\boldsymbol{A}'\boldsymbol{S}_{\boldsymbol{xx}}\boldsymbol{A})^{-1}\boldsymbol{A}'\boldsymbol{S}_{\boldsymbol{xy}} = \boldsymbol{A}^{-1}\hat{\boldsymbol{\beta}}^{\text{OLS}}.$$

Several of the methods of this chapter involve replacing the full $n \times q$ matrix $\boldsymbol{X}$ of regressors by a smaller $n \times k$ matrix $\boldsymbol{XG}_1$, where $\boldsymbol{G}_1$ is a known $q \times k$ matrix $k < q$. Then, the regression estimator takes the form

$$\hat{\boldsymbol{\beta}} = \boldsymbol{G}_1(\boldsymbol{G}_1'\boldsymbol{S}_{xx}\boldsymbol{G}_1)^{-1}\boldsymbol{G}_1'\boldsymbol{s}_{xy}. \tag{16.4}$$

For numerical stability, $\boldsymbol{G}_1$ is often assumed to have orthonormal columns.

16.2.2 Ridge Regression

When $q > n$, $\boldsymbol{S}_{xx}$ will not be invertible, so the OLS estimate of $\boldsymbol{\beta}$ in (16.3) does not exist. In addition, as described in Section 7.7.1 when there is high collinearity between the columns of $\boldsymbol{X}$, the inverse of $\boldsymbol{S}_{xx}$ may exist but will be unstable. In both of these situations, *ridge regression* may be considered. One indicator of instability in the inverse of $\boldsymbol{S}_{xx}$ is large absolute values of the components of $\hat{\boldsymbol{\beta}}^{\text{OLS}}$. Ridge regression tries to prevent such values by minimizing a sum of squares that penalizes large values of $\boldsymbol{\beta}$. Ridge regression has already been encountered in Exercises 9.8.1 and 11.2.14.

Let $\lambda > 0$ be a given tuning parameter. The ridge regression estimate is defined to be the value of $\boldsymbol{\beta}$ that minimizes the penalized sum of squares:

$$\sum_{r=1}^{n}(y_r - \mu - \boldsymbol{x}_r'\boldsymbol{\beta})^2 + \lambda\sum_{j=1}^{q}\beta_j^2, \tag{16.5}$$

where $\boldsymbol{x}_r'$ is the rth row of $\boldsymbol{X}$. The value of λ in Eq. (16.5) determines the amount of penalty associated with large components of $\hat{\boldsymbol{\beta}}$. As $\lambda \to 0$, the solution for $\hat{\boldsymbol{\beta}}^{\text{Ridge}}$ tends to the ordinary least squares solution, whereas as $\lambda \to \infty$, the solution tends to $\boldsymbol{0}$. In general, the value of $\boldsymbol{\beta}$ that minimizes (16.5) is given by

$$\hat{\boldsymbol{\beta}}_{\lambda}^{\text{Ridge}} = \boldsymbol{S}_{xx}(\lambda)^{-1}\boldsymbol{s}_{xy}, \qquad \text{where} \quad \boldsymbol{S}_{xx}(\lambda) = \boldsymbol{S}_{xx} + \frac{\lambda}{n}\boldsymbol{I}_q. \tag{16.6}$$

In the special case $\boldsymbol{S}_{xx} = \boldsymbol{I}$, the ridge regression solution reduces to

$$\hat{\boldsymbol{\beta}}_{\lambda}^{\text{Ridge}} = \frac{\boldsymbol{s}_{xy}}{1+\lambda/n} = \frac{\hat{\boldsymbol{\beta}}^{\text{OLS}}}{1+\lambda/n},$$

where $\hat{\boldsymbol{\beta}}^{\text{OLS}}$ is the ordinary least squares solution (16.3). Thus, the effect of λ is to *shrink* $\hat{\boldsymbol{\beta}}^{\text{OLS}}$ toward $\boldsymbol{0}$.

Unless $\boldsymbol{S}_{xx}$ is singular, $\hat{\boldsymbol{\beta}}^{\text{OLS}}$ is an unbiased estimate of $\boldsymbol{\beta}$, so we can immediately conclude that (provided $\lambda > 0$) $\hat{\boldsymbol{\beta}}_{\lambda}^{\text{Ridge}}$ is biased. However, the ridge regression estimator is preferable to the OLS estimator if it has smaller mean-squared error. In order to check this, we compute the bias and variance given $\boldsymbol{X}$ as

$$\mathsf{E}(\hat{\boldsymbol{\beta}}_{\lambda}^{\text{Ridge}} - \boldsymbol{\beta}) = \left[\boldsymbol{S}_{xx}(\lambda)^{-1}\boldsymbol{S}_{xx} - \boldsymbol{I}\right]\boldsymbol{\beta} \tag{16.7}$$

and

$$\mathsf{V}(\hat{\boldsymbol{\beta}}_{\lambda}^{\text{Ridge}}) = \sigma^2\boldsymbol{S}_{xx}(\lambda)^{-1}\boldsymbol{S}_{xx}\boldsymbol{S}_{xx}(\lambda)^{-1}. \tag{16.8}$$

In the random $\boldsymbol{X}$ model, these quantities should be averaged over the marginal distribution of $\boldsymbol{X}$.

Example 16.2.1 We illustrate ridge regression using simulated data. Taking $\boldsymbol{X} = (\boldsymbol{1}, \boldsymbol{Z})$ with $\boldsymbol{Z}$ a standard multivariate normal, we generate $N = 50$ data sets of size $n = 100$ from model (16.1), with $\boldsymbol{X}$ of dimension $n \times q$, and $\boldsymbol{\beta} = \boldsymbol{1}$. For each data set, we compute $\hat{\boldsymbol{\beta}}_{\lambda}^{\text{Ridge}}$ according to (16.6), for a range of values of λ, and then compute the average bias-squared:

$$(\hat{\boldsymbol{\beta}}_{\lambda}^{\text{Ridge}} - \boldsymbol{\beta})'(\hat{\boldsymbol{\beta}}_{\lambda}^{\text{Ridge}} - \boldsymbol{\beta})/(Nq)$$

and the average variance (using the sample mean of the diagonal elements of the covariance matrix of the estimated $\hat{\boldsymbol{\beta}}_{\lambda}^{\text{Ridge}}$). We can then plot the bias-squared and variance as a function

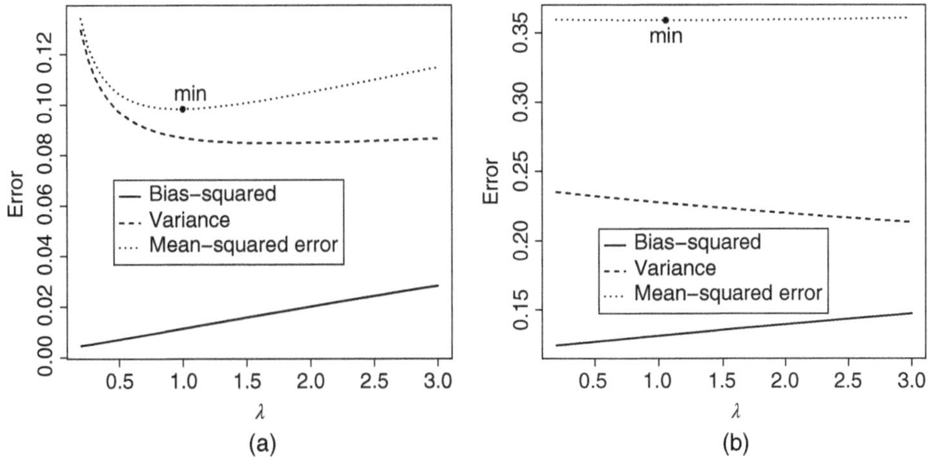

Figure 16.2 Bias-squared and variance for ridge regression estimates of β, with minimum MSE, as a function of the penalty λ. Averages were taken over 50 samples with $n = 100$ observations in each sample. (a) $q = 99$, (b) $q = 150$.

of λ, with the mean-squared error given by bias-squared + variance. Results for $q = 99$ and $q = 150(> n)$ are shown in Figure 16.2. We can clearly see that the bias increases, while the variance decreases, with λ. We also note that in panel (a) when $q < n$, the biased estimator has a smaller mean-squared error than the OLS (unbiased) solution. □

Standardization As outlined above, ridge regression can be applied directly to any set of data. However, the penalty is intuitively reasonable only if the variables in $\boldsymbol{X}$ have variances of a similar magnitude; otherwise, the amount of shrinkage will be higher for variables with smaller variances. Further, if the variables in $\boldsymbol{X}$ are measured in different units, the relative size of their variances is arbitrary. To address these issues, it is common practice to standardize each of the variables in $\boldsymbol{X}$ to have unit variance (though they may still be correlated). In addition, it rarely makes sense to shrink the intercept term of a regression model, so the intercept term is treated separately.

Hoerl and Kennard (1970) proposed an iterative method for choosing λ, but this does not always converge. An alternative is to simply use leave-one-out cross-validation (see 12.3.2), which chooses λ to minimize

$$\sum_r \left(y_r - \hat{y}_r^{(-r)}\right)^2 \tag{16.9}$$

in which $\hat{y}_r^{(-r)}$ is the predicted response at $\boldsymbol{x}_r$ after $\boldsymbol{\beta}$ is estimated by $\hat{\boldsymbol{\beta}}_{\lambda,-r}^{\text{Ridge}}$, the ridge regression estimator using all the data except the rth observation. Efficient computational methods for this quantity can be found in Meijer and Goeman (2013).

16.2.3 Least Absolute Selection and Shrinkage Operator (LASSO)

Ridge regression provides good predicted values ($\hat{\boldsymbol{y}}$), but the coefficients ($\hat{\boldsymbol{\beta}}_{\lambda}^{\text{Ridge}}$) are not so straightforward to interpret when q is large. Moreover, since none of the coefficients will shrink to zero (unless $\lambda = \infty$ in which case they are all zero), there is no variable selection. The LASSO, which was first proposed by Santosa and Symes (1986) but discovered independently and popularized by Tibshirani (1996), combines some of the advantages of shrinking with variable selection. A good overview is given in Hastie et al. (2015), and a Bayesian interpretation in Park and Casella (2008).

Rather than penalizing the squared L_2 norm of $\boldsymbol{\beta}$, $||\boldsymbol{\beta}||_2^2 = \sum \beta_j^2$, as in ridge regression, the LASSO penalizes the L_1 norm, $||\boldsymbol{\beta}||_1 = \sum_{j=1}^{q} |\beta_j|$. That is, the LASSO estimates $\boldsymbol{\beta}$ by minimizing the criterion

$$\sum_{r=1}^{n} (y_r - \mu - \boldsymbol{x}_r'\boldsymbol{\beta})^2 + \lambda \sum_{j=1}^{q} |\beta_j|, \tag{16.10}$$

where $\lambda > 0$ is a tuning parameter. Let $\hat{\boldsymbol{\beta}}^{\text{LASSO}}$ denote the corresponding estimator. Minimizing (16.10) is equivalent to finding $\boldsymbol{\beta}$ to solve the optimization problem

$$\min_{\boldsymbol{\beta}} (\boldsymbol{y} - \mu \boldsymbol{1}_n - \boldsymbol{X}\boldsymbol{\beta})'(\boldsymbol{y} - \mu \boldsymbol{1}_n - \boldsymbol{X}\boldsymbol{\beta}) \qquad \text{such that} \quad \sum_{j=1}^{q} |\beta_j| \le t$$

for some t (which has a one-to-one correspondence with λ). As $\lambda \to 0$ (or $t \to \infty$), the ordinary least squares solution is obtained. Conversely, if λ is large enough (or t is small enough), all the components of $\hat{\boldsymbol{\beta}}^{\text{LASSO}}$ will vanish, $\hat{\boldsymbol{\beta}}^{\text{LASSO}} = \boldsymbol{0}$.

As in the case of ridge regression, it is generally advised to standardize the explanatory variables. Unlike ridge regression, there is no explicit solution for $\hat{\boldsymbol{\beta}}^{\text{LASSO}}$. However, we can easily see that, if $||\hat{\boldsymbol{\beta}}^{\text{OLS}}||_1 \le t$, then $\hat{\boldsymbol{\beta}}^{\text{LASSO}} = \hat{\boldsymbol{\beta}}^{\text{OLS}}$. In general, as t decreases, there will be more and more zero entries in $\hat{\boldsymbol{\beta}}^{\text{LASSO}}$, but this relationship is *not* monotonic. Computation of the solution is a quadratic programming problem.

The LARS algorithm (Efron et al., 2004) was proposed to solve this problem in which the special structure is exploited to efficiently obtain a solution for *all* values of λ. The output of this algorithm is given by a set of threshold values and the corresponding solutions for $\boldsymbol{\beta}^{\text{OLS}}$, that is, a sequence $\{\lambda_k, \hat{\boldsymbol{\beta}}^{(k)}, k = 0, \dots, K\}$ at which a component of $\boldsymbol{\beta}$ changes from nonzero to zero as λ increases. This output contains all the information required, since the $\hat{\boldsymbol{\beta}}^{\text{LASSO}}$ solutions for any value of λ can be found by interpolation between the relevant λ_k. See Efron and Hastie (2016, Algorithm 16.3) for more details.

Example 16.2.2 Figure 16.3 illustrates some results from our implementation of the LARS algorithm using some simulated data. Here, we generate $n = 100$ observations in which the response $y_r \sim N(1 + \boldsymbol{x}_r'\boldsymbol{\beta}, 1)$, $r = 1, \dots, n$, with $\boldsymbol{x}_r$ simulated from a standard multivariate normal with $q = 50$ and $\boldsymbol{\beta}' = (1, 1, \dots, 1)$. We begin by centering and normalizing each column of $\boldsymbol{X}$ (which will have little effect in this case) as well as centering $\boldsymbol{y}$. The vertical axis shows how the coefficients β_j change as λ changes, with the horizontal axis showing the relative change. So, each piecewise linear line shows the relationship between $\beta_j(\lambda)$ and $c(\lambda)$, where $c(\lambda) = |\boldsymbol{\beta}(\lambda)|/\max_\lambda |\boldsymbol{\beta}(\lambda)|$ and $|\boldsymbol{\beta}| = \sum |\beta_j|$, which is a monotone function of λ. The left end of the horizontal axis corresponds to $\lambda = \infty$ with all coefficients equal to zero, and the right end corresponds to $\lambda = 0$ with the OLS solution for $\boldsymbol{\beta}$ – these values can be seen to be around $\boldsymbol{\beta} = \boldsymbol{1}$. The numbers at the top of the panel indicate the number of nonzero coefficients, $\hat{\beta}_j$ (corresponding to the vertical lines), for a selection of the values of λ. □

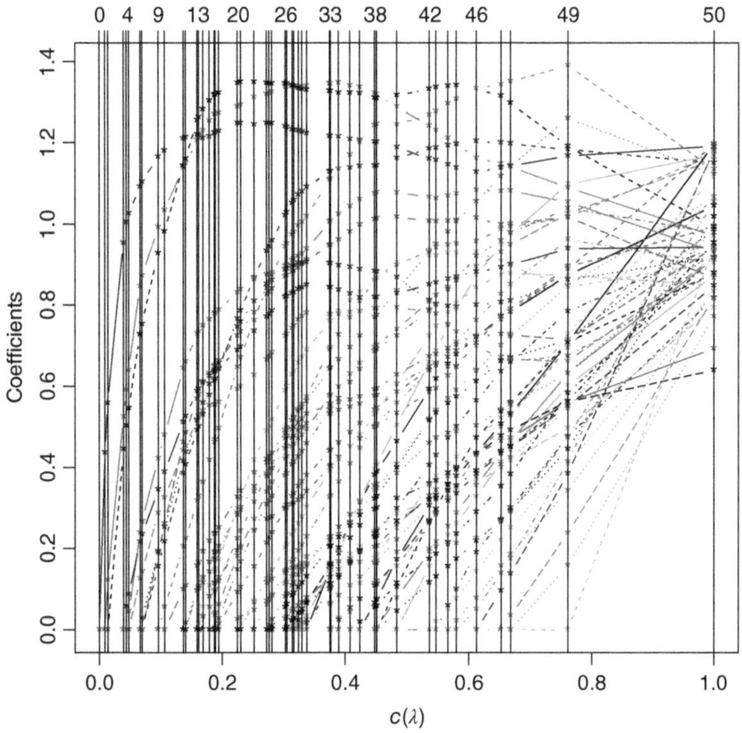

Figure 16.3 Results from using LARS with simulated data with uncorrelated explanatory variables. The vertical axis shows estimates of the β coefficients as λ changes corresponding to $c(\lambda) = |\boldsymbol{\beta}(\lambda)|/\max_{\lambda} |\boldsymbol{\beta}(\lambda)|$.

As can be seen in the above example, the $\hat{\beta}_j$ are not necessarily monotonic functions of λ, and so, it is possible for the path of some coefficients to pass through zero. This is more likely to occur in the case of high correlation among the explanatory variables.

LARS is implemented in the R libraries `glmnet` (Hastie and Efron, 2022) and `lars` (Hastie and Efron, 2013).

16.3 Principal Component Regression

Principal component regression is covered in Section 9.8 as a technique to reduce the dimensionality of the explanatory variables in a linear regression. Here, we note that the method works just as well for high-dimensional data as for low-dimensional data.

Let

$$\boldsymbol{S}_{xx} = \boldsymbol{GLG}'$$

be the spectral decomposition of $\boldsymbol{S}_{xx}$, and let k denote the number of principal components to use in the regression. Let $r = \text{rank}(\boldsymbol{S}_{xx})$ denote the number of nonzero eigenvalues, satisfying $r \leq \min(n-1, p)$. Provided $k \leq r$, principal component regression is straightforward to perform.

Partition $G = [G_1, G_2]$, where G_1 is a $q \times k$ matrix. Then, regressing y on the first k principal components of X leads to the estimator

$$\hat{\beta}^{PC} = G_1(G_1' S_{xx} G_1)^{-1} G_1' s_{xy} = G_1 L_1^{-1} G_1' s_{xy}, \tag{16.11}$$

where L_1 is a $k \times k$ diagonal matrix containing the first k eigenvalues.

Example 16.3.1 We illustrate principal component regression using a data set of $n = 28$ NIR spectra of PET yarns, measured at $q = 268$ wavelengths, with density (y) as a response variable (Swierenga et al., 1999). These data are available in the `pls` R library (Liland et al., 2021). After a standard principal component analysis, Figure 16.4 shows a scree plot, indicating that the first two components explain 98.6% of the variation, and at most it is worth including up to six components (which explain 99.97%). Note that the analysis so far has

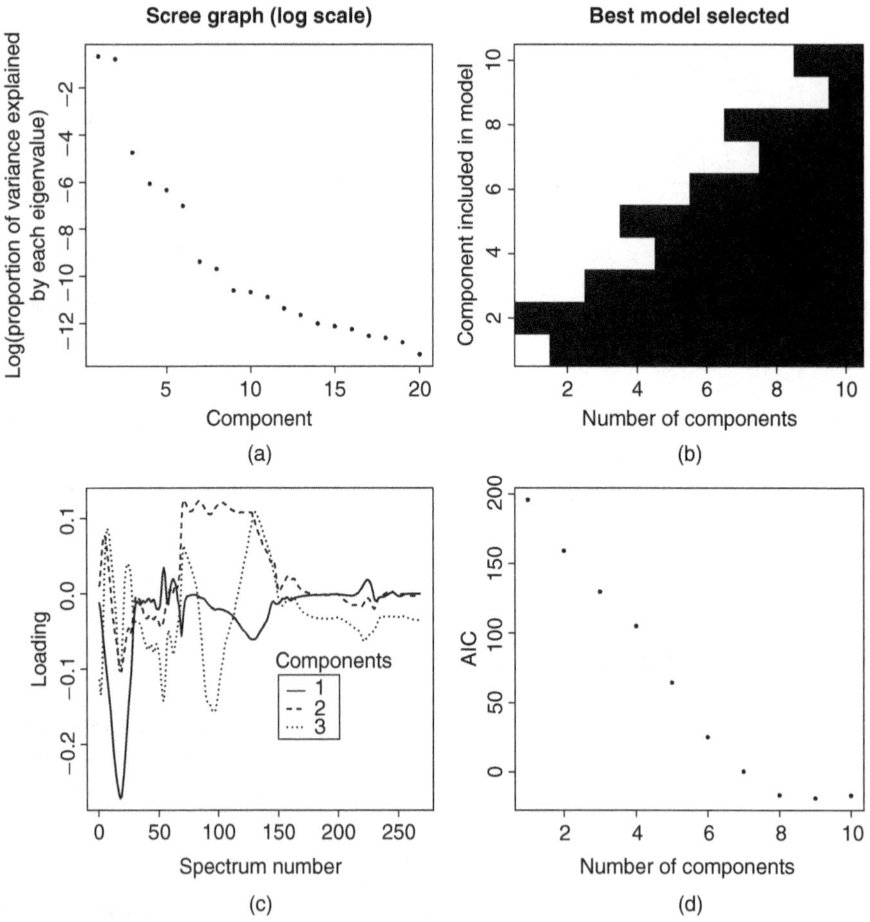

Figure 16.4 (a) Log of scree plot for yarn data; (b) Best subset selection for each model size, among first 10 components, using the R library/function `leaps` (Lumley, 2020) for variable selection in regression; (c) Loadings of first three components; (d) AIC vs. model size.

simply ignored $\boldsymbol{y}$. We consider using the first 10 scores in a regression model for $\boldsymbol{y}$ and carry out a best subset selection for $1 \leq k \leq 10$. The variables selected are shown in Figure 16.4(b) in which a black square in the (k, l)th position indicates that variable l is included in the best model of size k. In this example, the variables chosen are consistent with a stepwise procedure. We can see that the second component is the most important one; it has a correlation of 0.96 with the response variable, whereas the first component has a correlation of only 0.23. □

In this example, we can see an obvious limitation in principal component regression since the construction of the principal component loading vectors does not use any information about the response variable. This limitation is also evident when considering how many components are needed. Although the scree analysis indicated that two components might suffice, using AIC indicated that nine components would be best. That is, instead of a strong relation between just a few principal components and the response variable, there seems to be a weak relation between many principal components and the response variable.

Section 16.4 describes the method of partial least squares. This method also uses linear combinations of the explanatory variables. However, in this case, the construction of the linear combinations also depends on the response variable.

16.4 Partial Least Squares Regression

16.4.1 Overview

Partial least squares (PLS) is a regularized method of estimation for linear regression developed by Wold (1966). The method was originally developed for multiple linear regression and was subsequently extended to multivariate linear regression. It has become very popular because it is an effective way to deal with high collinearity in the explanatory variables. Abdi (2010) and Cook (2022) provide extensive reviews.

The theory behind PLS is easier to present in the population case using random vectors than in the sample case using data matrices. Thus, consider a multivariate normal random vector $(\boldsymbol{x}', \boldsymbol{y}')'$, where $\boldsymbol{x}$ is q-dimensional and $\boldsymbol{y}$ is p-dimensional. It is desired to model the dependence of $\boldsymbol{y}$ on $\boldsymbol{x}$ through a linear regression model. Although it is common practice in regression to condition on the $\boldsymbol{x}$ values, it is helpful here to treat $\boldsymbol{x}$ as random and to consider the joint covariance matrix for $\boldsymbol{x}$ and $\boldsymbol{y}$. The case $p = 1$ corresponds to multiple linear regression, and the case $p > 1$ corresponds to multivariate linear regression. It is common to assume that all the $\boldsymbol{x}$ and $\boldsymbol{y}$ variables have been scaled to have unit variance, so that all the variables have equal importance in the analysis.

The starting point for PLS is the construction of an orthogonal matrix called the *PLS loading matrix*. This matrix is analogous to the loading matrix in principal component analysis given by the eigenvectors of the covariance matrix of $\boldsymbol{x}$. The difference is that in PCA the loading matrix depends only on the covariance matrix for $\boldsymbol{x}$; in PLS, the loading matrix also depends on the covariances between $\boldsymbol{x}$ and $\boldsymbol{y}$. The algorithm to compute the loading matrix is simplest to describe if $p = 1$ (Section 16.4.2). Modifications to the algorithm needed to deal with the case $p > 1$ are described in Section 16.4.3. A *predictor envelope model* assumes that only a limited number of columns of the PLS loading matrix are needed to describe the dependence of $\boldsymbol{y}$ on $\boldsymbol{x}$ in the population (Section 16.4.4). The PLS estimator of the regression coefficient for a data set assumed to come from a predictor envelope model is given in Section 16.4.5. More general envelope models are described in Section 16.4.6.

Throughout this section, let $\boldsymbol{\Sigma}$ denote the joint covariance matrix for $\boldsymbol{x}$ and $\boldsymbol{y}$, partitioned as

$$\boldsymbol{\Sigma} = \begin{bmatrix} \boldsymbol{\Sigma}_{xx} & \boldsymbol{\Sigma}_{xy} \\ \boldsymbol{\Sigma}_{yx} & \boldsymbol{\Sigma}_{yy} \end{bmatrix}, \tag{16.12}$$

where $\boldsymbol{\Sigma}_{xx}$ is a $q \times q$ matrix, $\boldsymbol{\Sigma}_{xy} = \boldsymbol{\Sigma}'_{yx}$ is a $q \times p$ matrix, and $\boldsymbol{\Sigma}_{yy}$ is a $p \times p$ matrix. If $p = 1$, then there is some simplification: the upper-right entry, written as $\boldsymbol{\sigma}_{xy}$, is a vector, and the lower-right entry (σ_{yy}) is a scalar.

16.4.2 The PLS1 Algorithm to Construct the PLS Loading Matrix for $p = 1$ Response Variable

This section describes how to construct the PLS loading matrix when there is just $p = 1$ response variable y. The following algorithm, known as the PLS1 algorithm, produces a $q \times q$ orthogonal matrix $\boldsymbol{\Phi}$ partitioned into two parts, $\boldsymbol{\Phi} = [\boldsymbol{\Phi}_1, \boldsymbol{\Phi}_2]$, where $\boldsymbol{\Phi}_1$ is $q \times q_1$ and $\boldsymbol{\Phi}_2$ is $q \times (q - q_1)$. The value of q_1 is determined during the algorithm. In brief, $\boldsymbol{\Phi}_1$ is created by iteratively maximizing a covariance and constructing a residual covariance matrix. To describe the algorithm in more detail, let $\nu \geq 1$ index the iterations and initialize the algorithm by setting $\nu = 1$ and $\boldsymbol{\Sigma}^{(1)} = \boldsymbol{\Sigma}$. Then, proceed as follows for $\nu = 1, 2, \ldots$

PSL1 algorithm

(a) Let $\boldsymbol{\Sigma}^{(\nu)}$ denote the joint covariance matrix at iteration ν, partitioned as in (16.12) with the four parts inheriting the superscript index (ν). If $\boldsymbol{\sigma}_{xy}^{(\nu)} \neq \boldsymbol{0}$, set $\boldsymbol{\phi}_{(\nu)}$, the νth column of $\boldsymbol{\Phi}$, to be

$$\boldsymbol{\phi}_{(\nu)} = \boldsymbol{\sigma}_{xy}^{(\nu)} / ||\boldsymbol{\sigma}_{xy}^{(\nu)}||. \tag{16.13}$$

This choice can be justified as follows. Assume that $\boldsymbol{x}$ and y are treated as having joint covariance matrix $\boldsymbol{\Sigma}^{(\nu)}$, and let $\boldsymbol{\phi}$ be a q-dimensional standardized coefficient vector (so $\boldsymbol{\phi}'\boldsymbol{\phi} = 1$). The covariance between the linear combination $\boldsymbol{\phi}'\boldsymbol{x}$ and y is $\boldsymbol{\phi}'\boldsymbol{\sigma}_{xy}^{(\nu)}$. This covariance is maximized when $\boldsymbol{\phi}$ is given by (16.13).

If $\boldsymbol{\sigma}_{xy}^{(\nu)} = \boldsymbol{0}$, then stop the algorithm, let $q_1 = \nu - 1$, and set

$$\boldsymbol{\Phi}_1 = \left[\boldsymbol{\phi}_{(1)}, \ldots, \boldsymbol{\phi}_{(q_1)}\right]. \tag{16.14}$$

See below for further discussion.

(b) If the algorithm has not stopped in step (a), let $\boldsymbol{e}_0^{(\nu)} = (\boldsymbol{\phi}'_{(\nu)}, \boldsymbol{0}'_p)'$ be a $(q + p)$-dimensional vector (here, $p = 1$, so $\boldsymbol{0}_p = 0$) whose first q elements are given by $\boldsymbol{\phi}_{(\nu)}$ and define the residual joint covariance matrix

$$\boldsymbol{\Sigma}^{(\nu+1)} = \boldsymbol{\Sigma}^{(\nu)} - (\boldsymbol{\Sigma}^{(\nu)}\boldsymbol{e}_0^{(\nu)})(\boldsymbol{\Sigma}^{(\nu)}\boldsymbol{e}_0^{(\nu)})' / \boldsymbol{e}_0^{(\nu)\prime}\boldsymbol{\Sigma}^{(\nu)}\boldsymbol{e}_0^{(\nu)}. \tag{16.15}$$

That is, if $(\boldsymbol{x}', y)'$ is treated as having joint covariance matrix $\boldsymbol{\Sigma}^{(\nu)}$, then $\boldsymbol{\Sigma}^{(\nu+1)}$ is the conditional covariance matrix of $(\boldsymbol{x}', y)'$, given the linear combination $\boldsymbol{e}_0^{(\nu)\prime}(\boldsymbol{x}', y)' = \boldsymbol{\phi}'_{(\nu)}\boldsymbol{x}$. See Exercise 16.4.1. Note that (ν) is a label for an iteration when used as a superscript, but is a column index when used as a subscript.

(c) Increment v and go back to step (a).

It can be shown that the vectors $\boldsymbol{\phi}_{(j)}$ are orthogonal to one another, $j = 1, \ldots, q_1$ (see Exercise 16.4.1), so that $\boldsymbol{\Phi}_1$ is a column orthonormal matrix. If $q_1 < q$, it is helpful to let $\boldsymbol{\Phi}_2$ be an arbitrary complementary column orthonormal matrix, so that $\boldsymbol{\Phi}_2'\boldsymbol{\Phi}_2 = \boldsymbol{I}_{q-q_1}$ and $\boldsymbol{\Phi}_2'\boldsymbol{\Phi}_1 = \mathbf{0}$. Then,

$$\boldsymbol{\Phi} = [\boldsymbol{\Phi}_1, \boldsymbol{\Phi}_2] \tag{16.16}$$

is an orthogonal matrix.

If $\boldsymbol{\sigma}_{xy}^{(v)} \neq \mathbf{0}$ for all $v = 1, \ldots, q$, the algorithm runs for q iterations. In this case, $\boldsymbol{\Phi} = \boldsymbol{\Phi}_1$, and $\boldsymbol{\Phi}_2$ is empty.

Here is a rationale for the PLS1 algorithm. The first column of $\boldsymbol{\Phi}_1$ describes the "direct" effect of $\boldsymbol{x}$ on y. The loading vector is proportional to the covariance vector $\boldsymbol{\sigma}_{xy}$ between $\boldsymbol{x}$ and y. Subsequent columns of $\boldsymbol{\Phi}_1$ describe "indirect" effects of $\boldsymbol{x}$ on y. For example, although $\boldsymbol{\phi}_{(2)}'\boldsymbol{x}$ is uncorrelated with y, it is correlated with $\boldsymbol{\phi}_{(1)}'\boldsymbol{x}$. Thus, $\boldsymbol{\phi}_{(2)}'\boldsymbol{x}$ affects y indirectly through its effect on $\boldsymbol{\phi}_{(1)}'\boldsymbol{x}$. Finally, the linear combinations defined by $\boldsymbol{\Phi}_2'\boldsymbol{x}$ are uncorrelated with both $\boldsymbol{\Phi}_1'\boldsymbol{x}$ and y.

The PLS1 algorithm identifies important structure in $\boldsymbol{\Sigma}$. Hence, it is helpful to make some formal definitions.

Definition 16.4.1 *The value of q_1 in the PLS1 algorithm is called the* PLS rank *of $\boldsymbol{\Sigma}$. The orthogonal matrix $\boldsymbol{\Phi} = [\boldsymbol{\Phi}_1, \boldsymbol{\Phi}_2]$ is called the* PLS loading matrix, *and its columns are called* PLS loading vectors. *The loading vectors are mainly of interest for $\boldsymbol{\Phi}_1$. The elements of $\boldsymbol{\Phi}_1'\boldsymbol{x}$ are called the* relevant components, *and the elements of $\boldsymbol{\Phi}_2'\boldsymbol{x}$ are called the* irrelevant components.

Example 16.4.1 Let $q = 3$, and suppose that x_1, x_2, x_3, y all have variance 1. Consider two correlation parameters $0 < \rho, \delta < 1$. Suppose that the only nonzero correlations are $\text{cor}(x_1, y) = \delta$, so x_1 has a direct effect on y, and $\text{cor}(x_1, x_2) = \rho$, so x_2 has an indirect effect on y. Assume $\rho^2 + \delta^2 < 1$ to ensure that $\boldsymbol{\Sigma}$ is positive definite. The application of the PLS1 algorithm to the joint covariance matrix of $\boldsymbol{x}$ and y yields the following results:

$$\boldsymbol{\Sigma} = \boldsymbol{\Sigma}^{(1)} = \begin{bmatrix} 1 & \rho & 0 & \delta \\ \rho & 1 & 0 & 0 \\ 0 & 0 & 1 & 0 \\ \delta & 0 & 0 & 1 \end{bmatrix}, \quad \text{so} \quad \boldsymbol{e}_0^{(1)} = \begin{pmatrix} 1 \\ 0 \\ 0 \\ 0 \end{pmatrix}, \qquad \boldsymbol{\Sigma}^{(1)}\boldsymbol{e}_0^{(1)} = \begin{pmatrix} 1 \\ \rho \\ 0 \\ \delta \end{pmatrix},$$

and $\boldsymbol{e}_0^{(1)\prime}\boldsymbol{\Sigma}^{(1)}\boldsymbol{e}_0^{(1)} = 1$. Hence,

$$\boldsymbol{\Sigma}^{(2)} = \begin{bmatrix} 1 & \rho & 0 & \delta \\ \rho & 1 & 0 & 0 \\ 0 & 0 & 1 & 0 \\ \delta & 0 & 0 & 1 \end{bmatrix} - \begin{bmatrix} 1 & \rho & 0 & \delta \\ \rho & \rho^2 & 0 & \delta\rho \\ 0 & 0 & 0 & 0 \\ \delta & \delta\rho & 0 & \delta^2 \end{bmatrix}$$

$$= \begin{bmatrix} 0 & 0 & 0 & 0 \\ 0 & 1-\rho^2 & 0 & -\delta\rho \\ 0 & 0 & 1 & 0 \\ 0 & -\delta\rho & 0 & 1-\delta^2 \end{bmatrix}, \quad \text{so} \quad \boldsymbol{e}_0^{(2)} = \begin{pmatrix} 0 \\ -1 \\ 0 \\ 0 \end{pmatrix}, \quad \boldsymbol{\Sigma}^{(2)}\boldsymbol{e}_0^{(2)} = \begin{pmatrix} 0 \\ -(1-\rho^2) \\ 0 \\ \delta\rho \end{pmatrix},$$

and $\boldsymbol{e}_0^{(2)\prime}\boldsymbol{\Sigma}^{(2)}\boldsymbol{e}_0^{(2)} = 1-\rho^2$. Hence,

$$\boldsymbol{\Sigma}^{(3)} = \begin{bmatrix} 0 & 0 & 0 & 0 \\ 0 & 1-\rho^2 & 0 & -\delta\rho \\ 0 & 0 & 1 & 0 \\ 0 & -\delta\rho & 0 & 1-\delta^2 \end{bmatrix} - \begin{bmatrix} 0 & 0 & 0 & 0 \\ 0 & (1-\rho^2)^2 & 0 & -(1-\rho^2)\delta\rho \\ 0 & 0 & 0 & 0 \\ 0 & -(1-\rho^2)\delta\rho & 0 & (\delta\rho)^2 \end{bmatrix} /(1-\rho^2)$$

$$= \begin{bmatrix} 0 & 0 & 0 & 0 \\ 0 & 0 & 0 & 0 \\ 0 & 0 & 1 & 0 \\ 0 & 0 & 0 & c \end{bmatrix}, \quad \text{where} \quad c = (1-\delta^2-\rho^2)/(1-\rho^2).$$

Note that the first three elements of the final column of $\boldsymbol{\Sigma}^{(3)}$ vanish, so $\boldsymbol{\sigma}_{xy}^{(3)} = \mathbf{0}$, and the algorithm stops after $p_1 = 2$ iterations. The vectors $\boldsymbol{\phi}_{(j)}$ are given by the first three elements of $\boldsymbol{e}_0^{(j)}, j = 1, 2$. The third column of $\boldsymbol{\Phi}$ is allowed to be any vector orthogonal to the first two columns. In this case, up to sign, there is only one choice. Putting the columns together yields

$$\boldsymbol{\Phi} = \begin{bmatrix} 1 & 0 & 0 \\ 0 & -1 & 0 \\ 0 & 0 & 1 \end{bmatrix}.$$

The first loading vector $\boldsymbol{\phi}_{(1)} = (1, 0, 0)'$ picks out the direct effect of x_1 on y. The second loading vector $\boldsymbol{\phi}_{(2)} = (0, -1, 0)'$ picks out the indirect effect of x_2 on y. The linear combinations $\boldsymbol{\phi}_{(1)}'\boldsymbol{x} = x_1$ and $\boldsymbol{\phi}_{(2)}'\boldsymbol{x} = -x_2$ are the relevant components in this example. Finally, since x_3 is uncorrelated with all the other variables, the third loading vector $\boldsymbol{\phi}_{(3)} = (0, 0, 1)'$ defines the irrelevant component $\boldsymbol{\phi}_{(3)}'\boldsymbol{x} = x_3$.

Note that the linear combinations in this example are concentrated on single variables. In general, each of the relevant and irrelevant components will involve linear combinations of all the variables.

Finally, it can be shown that the residual variance of y after regressing on the first ν relevant components is $\sigma_{yy}^{(\nu+1)}$, for $\nu = 1, 2$. This quantity gets smaller as ν increases. In particular, the residual variance after regressing on all the relevant components is $\sigma_{yy}^{(3)} = (1-\delta^2-\rho^2)/(1-\rho^2)$. This quantity is the same as the conditional variance of y, given $\boldsymbol{x}$ appearing in the population version of OLS regression, which in turn is the same as one minus the squared multiple correlation coefficient between y and $\boldsymbol{x}$, and the same as one minus the single nonzero squared canonical correlation between y and $\boldsymbol{x}$. □

16.4.3 The PLS2 Algorithm to Construct the PLS Loading Matrix for $p > 1$ Response Variables

Extending the last section, let the response $\boldsymbol{y}$ now be a p-dimensional random vector instead of a scalar random variable. Most of the notation and theory carries over with little change. However, the construction of the PLS loading matrix needs more care. An algorithm called the PLS2 algorithm has been designed to deal with the case $p > 1$. It is essentially the same as the PLS1 algorithm of Section 16.4.2, except for the following change to step (a).

PLS2 algorithm (same as PLS1 with a modified step (a))

(a) If $\boldsymbol{\Sigma}_{xy}^{(\nu)} \neq \mathbf{0}$, express it using the singular value decomposition as

$$\boldsymbol{\Sigma}_{xy}^{(\nu)} = \boldsymbol{U}^{(\nu)}\boldsymbol{L}^{(\nu)}\boldsymbol{V}^{(\nu)\prime}. \tag{16.17}$$

Set $\boldsymbol{\phi}_{(\nu)}$, the νth column of $\boldsymbol{\Phi}$, to be equal to the first column of the matrix of left singular vectors, i.e.

$$\boldsymbol{\phi}_{(\nu)} = \boldsymbol{u}_{(1)}^{(\nu)}. \tag{16.18}$$

Note again that the superscript (ν) on the vector $u_{(1)}^{(\nu)}$ is an index labeling the iteration, whereas the subscript (ν) on the vector $\boldsymbol{\phi}_{(\nu)}$ indicates the νth column of the matrix $\boldsymbol{\Phi}$.

The justification for this choice is similar to justification in the PLS1 algorithm. Treating $\boldsymbol{x}$ and $\boldsymbol{y}$ as having cross-covariance matrix $\boldsymbol{\Sigma}_{xy}^{(\nu)}$, the covariance between a linear combination $\boldsymbol{\phi}'\boldsymbol{x}$ and a linear combination $\boldsymbol{\gamma}'\boldsymbol{y}$ is maximized when $\boldsymbol{\phi} = \boldsymbol{u}_{(1)}^{(\nu)}$ and $\boldsymbol{\gamma} = \boldsymbol{v}_{(1)}^{(\nu)}$. The choice for $\boldsymbol{\gamma}$ is not of direct interest here; the choice for $\boldsymbol{\phi}$ defines the νth column of $\boldsymbol{\Phi}$. As before, if $\boldsymbol{\Sigma}_{xy}^{(\nu)} = \mathbf{0}$, then stop the algorithm, set $q_1 = \nu - 1$, and define $\boldsymbol{\Phi}_1$ by (16.14).

16.4.4 The Predictor Envelope Model

Assumptions about the PLS rank provide a way to impose structure on a joint covariance matrix.

Definition 16.4.2 *Let a q-dimensional random vector $\boldsymbol{x}$ and a p-dimensional random vector $\boldsymbol{y}$ follow a joint multivariate normal distribution with a positive definite covariance matrix $\boldsymbol{\Sigma}$. Fix an index q_1, $1 \leq q_1 \leq q$. Say that $\boldsymbol{\Sigma}$ satisfies the* predictor envelope model *of order q_1 if the PLS rank of $\boldsymbol{\Sigma}$ is constrained to equal q_1.*

Envelope models were developed by R D Cook and his colleagues; see, e.g. Cook (2022) and the references therein. If $\boldsymbol{\Phi}$ is the PLS loading matrix partitioned as in (16.12), then under a predictor envelope model of order q_1, three key properties hold

(a) $\boldsymbol{y}$ and the relevant components of $\boldsymbol{x}$ have nonzero covariance, $\text{cov}(\boldsymbol{\Phi}_1'\boldsymbol{x}, \boldsymbol{y}) \neq \mathbf{0}$;
(b) $\boldsymbol{y}$ and the irrelevant components of $\boldsymbol{x}$ are uncorrelated, $\text{cov}(\boldsymbol{\Phi}_2'\boldsymbol{x}, \boldsymbol{y}) = \mathbf{0}$;
(c) the relevant and irrelevant components of $\boldsymbol{x}$ are uncorrelated, $\text{cov}(\boldsymbol{\Phi}_1'\boldsymbol{x}, \boldsymbol{\Phi}_2'\boldsymbol{x}) = \mathbf{0}$.

That is,

$$\text{var}\begin{pmatrix} \boldsymbol{\Phi}_1'\boldsymbol{x} \\ \boldsymbol{\Phi}_2'\boldsymbol{x} \\ \boldsymbol{y} \end{pmatrix} = \begin{bmatrix} * & & \\ 0 & * & \\ * & 0 & * \end{bmatrix}, \tag{16.19}$$

where a "*" indicates a nonzero block, and just the lower triangle has been shown.

Note that the predictor envelope model involves regularization because it requires orthogonality between different coefficient vectors. Hence, the model is not equivariant under all linear transformations of $\boldsymbol{x}$.

16.4.5 PLS Regression

Now consider data matrices $\boldsymbol{X}(n \times q)$ and $\boldsymbol{Y}(n \times p)$, with sample covariance matrix $\boldsymbol{S}$, partitioned as in (16.12). Suppose that all the variables have been standardized to have sample variance 1. The PLS rank of $\boldsymbol{S}$ is usually $\min(n-1, q)$ since usually $\operatorname{rank}(\boldsymbol{S}_{xx}) = \min(n-1, q)$, and there is no special structure for $\boldsymbol{S}_{xy}$. Let $\boldsymbol{F}$ denote the PLS loading matrix of $\boldsymbol{S}$. Thus, $\boldsymbol{S}$ and $\boldsymbol{F}$ are the sample counterparts of $\boldsymbol{\Sigma}$ and $\boldsymbol{\Phi}$ in the population case. Although $\boldsymbol{F}$ can be partitioned into two parts, $\boldsymbol{F} = [\boldsymbol{F}_1, \boldsymbol{F}_2]$, as in (16.16), it is generally the case – when running the algorithm for real data – that $\boldsymbol{F}_2$ is empty and $\boldsymbol{F} = \boldsymbol{F}_1$ has q columns. To develop suitable PLS regression estimators, let $\boldsymbol{F}^{(k)}$ denote the $q \times k$ matrix given by the first k columns of $\boldsymbol{F}$, $1 \le k \le \min(n-1, q)$.

Choose an index k. It is desired to regress the response matrix $\boldsymbol{Y}$ on the predictor matrix $\boldsymbol{X}$. PLS regression involves replacing the q-dimensional predictors in $\boldsymbol{X}$ by the k-dimensional predictors in $\boldsymbol{X}\boldsymbol{F}^{(k)}$ to give the estimated regression coefficient as a $q \times p$ matrix

$$\hat{\boldsymbol{\beta}}^{\mathrm{PLS},k} = \boldsymbol{F}^{(k)}(\boldsymbol{F}^{(k)\prime}\boldsymbol{S}_{xx}\boldsymbol{F}^{(k)})^{-1}\boldsymbol{F}^{(k)\prime}\boldsymbol{S}_{xy}. \tag{16.20}$$

If $p = 1$, the cross-covariance matrix $\boldsymbol{S}_{xy}$ becomes the cross-covariance vector $\boldsymbol{s}_{xy}$, and (16.20) provides an alternative to the OLS regression estimator in (16.3).

PLS regression is well suited to high-dimensional problems when q is large. It even makes sense when $n \le q$ in which case $\boldsymbol{S}_{xx}$ is a singular matrix. Provided $k \le \min(n-1, q)$, $\boldsymbol{F}^{(k)\prime}\boldsymbol{S}_{xx}\boldsymbol{F}^{(k)}$ will be nonsingular.

The PLS estimator of the population regression coefficient $\boldsymbol{\beta} = \boldsymbol{\Sigma}_{xx}^{-1}\boldsymbol{\Sigma}_{xy}$ can be viewed in two ways:

(a) a regularized estimator where $\boldsymbol{\Sigma}$ is assumed to be a positive definite covariance matrix with no further restrictions or
(b) a "plug-in" estimator where $\boldsymbol{\Sigma}$ is assumed to be a positive definite covariance matrix satisfying the predictor envelope model of order $q_1 = k$.

In case (b), the term "plug-in" means that the population PLS loading matrix $\boldsymbol{\Phi}$ has been replaced by the sample PLS loading matrix $\boldsymbol{F}$, and the population covariance matrix $\boldsymbol{\Sigma}$ has been replaced by the sample covariance matrix $\boldsymbol{S}$. The PLS estimator is a root-n consistent estimator of $\boldsymbol{\beta}$ in this case, but it is not the maximum likelihood estimator (e.g. Cook et al., 2013; Helland, 1992).

Example 16.4.2 We illustrate the multivariate PLS methods using data on sensory and physicochemical data of olive oils (Massart et al., 1998), which is included in the `pls` package in R (Mevik and Wehrens, 2007). There are $n = 16$ olive oil samples. The first five oils are Greek, the next five are Italian, and the last six are Spanish. The response ($\boldsymbol{Y}$) has $p = 6$ sensory variables: Scores for attributes "yellow" (1), "green" (2), "brown" (3), "glossy" (4), "transp" (5), and "syrup" (6), and there are $q = 5$ chemical explanatory variables: measurements of acidity, peroxide, K232, K270, and DK. We begin by centering all the data. The explanatory variables are measured on different scales, so these are also standardized, so that each column standard deviation is 1. The response variables are scores and measured on a common scale, so these are only mean centered and not standardized. All the plots are shown on the original data scales.

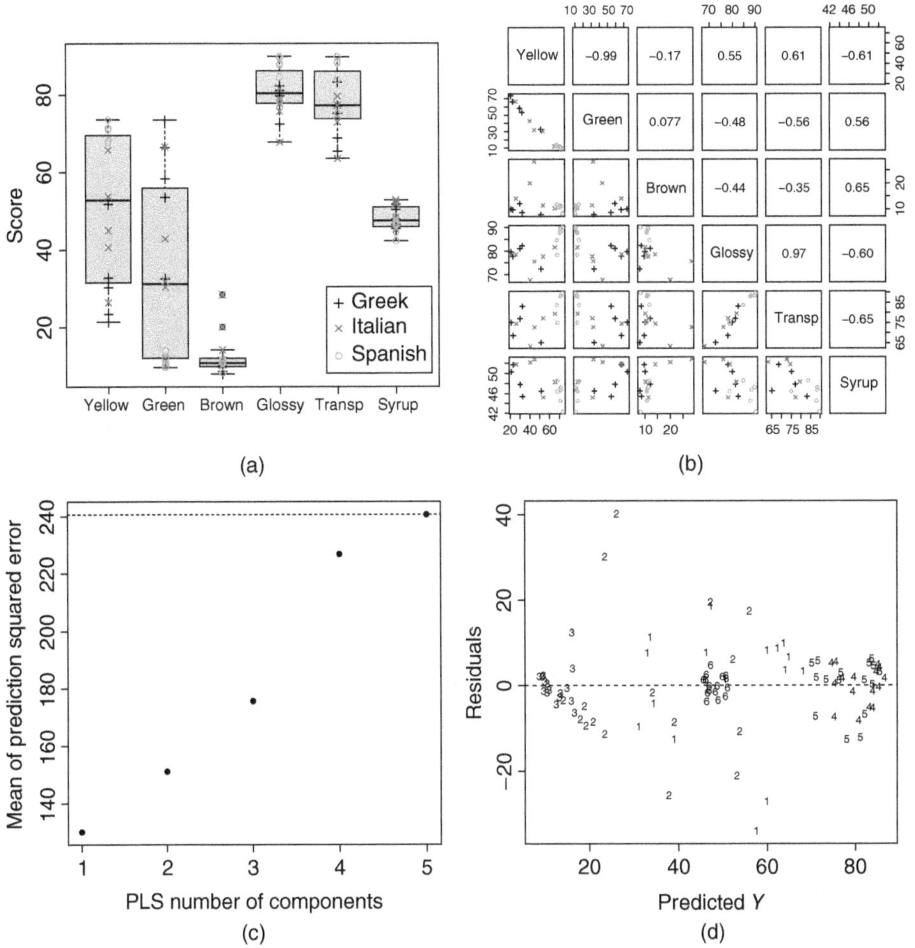

Figure 16.5 (a) Boxplots of scores for six sensory variables ($\boldsymbol{Y}$); (b) Pairwise scatterplots and sample correlations; (c) Mean prediction squared error, using leave-one-out cross-validation for various number of components (k), with a horizontal line showing the mean-squared prediction error for multivariate ordinary least squares regression; (d) Residual plot for $k = 1$ component, with the plotting number corresponding to the column of $\boldsymbol{Y}$.

Figure 16.5 shows a boxplot of the response variable (scores) and a pairwise scatterplot, which show very strong negative correlation between "yellow" and "green," with the Spanish oils having the highest yellow score and lowest green score. To choose a suitable number of components, we can use leave-one-out cross-validation, using $n - 1$ observations with PLS to predict the omitted observation. The mean-squared prediction error can then be minimized. In general, one could use a different value of k for each of the response variables, but here we just consider an average, so the same value of k is used for all responses. In this example, $k = 1$ is optimal (see Figure 16.5) – which yields a somewhat smaller cross-validated mean-squared prediction error than OLS. Note also that when there is no cross-validation and all components are used, the predicted values for PLS and OLS are the same. The residual plot using $k = 1$ shows that the "green" (2) scores are hardest to predict, which might have been anticipated from the larger interquartile range in the boxplot.

The PLS loading matrix $\boldsymbol{F}$ is given by

Component	1	2	3	4	5
acidity	0.42	0.61	−0.37	0.03	0.56
peroxide	0.37	−0.59	0.08	0.61	0.37
K232	0.48	−0.29	0.32	−0.74	0.21
K270	0.59	0.34	0.39	0.26	−0.57
DK	0.33	−0.30	−0.78	−0.12	−0.43
C_j^2	1278.8	30.0	8.9	0.1	0.2

The first column is an average of the $q = 5$ (standardized) $\boldsymbol{X}$ variables, and the subsequent columns represent contrasts between the variables. For each PLS component, the last row of the above table contains the sum of squared covariances over the $p = 6$ $\boldsymbol{Y}$ variables, given by

$$C_j^2 = \sum_{l=1}^{p} \text{cov}^2(\boldsymbol{X}\boldsymbol{f}_{(j)}, \boldsymbol{y}_{(l)}).$$

Only the first PLS component has been included in the PLS solution in Figure 16.5. The relatively small values of C_j^2, $j \geq 2$ help to explain why there is no benefit in including the other components. Note that (as in this example), the C_j^2 are not necessarily monotone decreasing in j. □

16.4.6 Joint Envelope Models

Envelope models can be defined for the *response* variables as well as the *predictor* variables. Here, we give a brief overview. First, it is necessary to extend some terminology. The PLS rank q_1 and the $q \times q$ PLS loading matrix $\boldsymbol{\Phi}$ in Sections 16.4.2 and 16.4.3 can be described more completely as the PLS rank *for* $\boldsymbol{x}$ and the PLS loading matrix *for* $\boldsymbol{x}$. They depend on the joint covariance matrix $\boldsymbol{\Sigma}$ through $\boldsymbol{\Sigma}_{xx}$ and $\boldsymbol{\Sigma}_{xy}$. Conversely, by swapping the roles of $\boldsymbol{x}$ and $\boldsymbol{y}$, it is possible to focus on the response variables in $\boldsymbol{y}$ to define the PLS loading matrix *for* $\boldsymbol{y}$, denoted by a $p \times p$ orthogonal matrix $\boldsymbol{\Gamma}$, say, and the corresponding PLS rank *for* $\boldsymbol{y}$, denoted by p_1, say. These depend on $\boldsymbol{\Sigma}$ through $\boldsymbol{\Sigma}_{yy}$ and $\boldsymbol{\Sigma}_{yx}$.

Definition 16.4.3 *Let a q-dimensional random vector* $\boldsymbol{x}$ *and a p-dimensional random vector* $\boldsymbol{y}$ *follow a joint multivariate normal distribution with a positive definite covariance matrix* $\boldsymbol{\Sigma}$. *Fix two indices* q_1, $1 \leq q_1 \leq q$ *and* p_1, $1 \leq p_1 \leq p$. *Say that* $\boldsymbol{\Sigma}$ *satisfies the* joint envelope model *of order* q_1 *for* $\boldsymbol{x}$ *and order* p_1 *for* $\boldsymbol{y}$ *if the PLS rank for* $\boldsymbol{x}$ *is constrained to equal* q_1 *and the PLS rank for* $\boldsymbol{y}$ *is constrained to equal* p_1.

The joint envelope model was developed by Cook and Zhang (2015). If $p_1 = p$, it reduces to the predictor envelope model of Section 16.4.4. If the loading matrices are partitioned as

$$\boldsymbol{\Phi} = [\boldsymbol{\Phi}_1, \ \boldsymbol{\Phi}_2] \quad \text{and} \quad \boldsymbol{\Gamma} = [\boldsymbol{\Gamma}_1, \ \boldsymbol{\Gamma}_2],$$

where $\mathbf{\Phi}_1$ is $q \times q_1$ and $\mathbf{\Gamma}_1$ is $p \times p_1$, then under the joint envelope model,

$$\text{var}\begin{pmatrix} \mathbf{\Phi}_1'\boldsymbol{x} \\ \mathbf{\Phi}_2'\boldsymbol{x} \\ \mathbf{\Gamma}_1'\boldsymbol{y} \\ \mathbf{\Gamma}_2'\boldsymbol{y} \end{pmatrix} = \begin{bmatrix} * & & & \\ 0 & * & & \\ * & 0 & * & \\ 0 & 0 & 0 & * \end{bmatrix}, \tag{16.21}$$

where a $*$ indicates that a block is nonzero, and only the lower diagonal of the covariance matrix is given. Thus, $\mathbf{\Phi}_1'\boldsymbol{x}$ is correlated with $\mathbf{\Gamma}_1'\boldsymbol{y}$, but otherwise the blocks are uncorrelated. The first sets of variables $\mathbf{\Phi}_1'\boldsymbol{x}$ and $\mathbf{\Gamma}_1'\boldsymbol{y}$ are called the *relevant components* of $\boldsymbol{x}$ and $\boldsymbol{y}$, respectively. Similarly, the second sets of variables $\mathbf{\Phi}_2'\boldsymbol{x}$ and $\mathbf{\Gamma}_2'\boldsymbol{y}$ are called the *irrelevant components* of $\boldsymbol{x}$ and $\boldsymbol{y}$, respectively. Thus, the relevant components of $\boldsymbol{x}$ and $\boldsymbol{y}$ are correlated with one another, but the irrelevant components of $\boldsymbol{x}$ and $\boldsymbol{y}$ are uncorrelated with one another and with the relevant components.

Bookstein et al. (2002) and Bookstein (2018) noted that if $q_1 = p_1$, then in the population case, $\mathbf{\Phi}_1$ and $\mathbf{\Gamma}_1$ can be read off immediately using all the left and right singular vectors in the singular value decomposition of $\mathbf{\Sigma}_{xy}$. In the sample case, the simultaneous extraction of $q_1 = p_1$ left and right singular vectors provides a way to estimate $\mathbf{\Phi}_1$ and $\mathbf{\Gamma}_1$. el Bouhaddani et al. (2018) have developed a *probabilistic PLS* model, which can be viewed as a joint envelope model with $q_1 = p_1$ plus additional independence and isotropy assumptions.

Finally, the analysis of the joint envelope model can be contrasted with canonical correlation analysis (CCA) of Chapter 11. Since CCA allows more general transformations of $\boldsymbol{x}$ and $\boldsymbol{y}$ than the joint envelope model (namely, arbitrary nonsingular linear transformations are allowed rather than just orthogonal transformations), it is able to achieve a simpler representation of the joint distribution. For example, consider a joint envelope model of orders q_1 and p_1 for $\boldsymbol{x}$ and $\boldsymbol{y}$, respectively, and suppose that the matrix of singular values $\boldsymbol{D}$ in the CCA decomposition has rank k in (11.6). Then, it is necessarily the case that $k \leq \min(q_1, p_1)$. A particular case is given by multiple linear regression ($p = 1$). It is entirely possible for the PLS rank to exceed 1. However, there is just one singular value in the CCA decomposition.

Hence, at first sight, it might seem that for data sets with substantial correlation within $\boldsymbol{x}$ and within $\boldsymbol{y}$, the CCA approach is able to produce a simpler description through a low rank approximation to $\mathbf{\Lambda}$. However, for high-dimensional data with low or moderate sample size, there is a large amount of sampling variation in $\boldsymbol{S}$, and the envelope and PLS approximations are generally more useful and numerically stable.

16.5 Functional Data

In functional data analysis (FDA), the underlying *item* or *data object* is a smooth curve $x(t)$, a function of a continuous variable t. For convenience, we think of t as time, but other applications are also possible. Observations are made on each function at a discrete set of times leading to data

$$z_{rj} = x_r(t_{rj}) + \varepsilon_{rj} = x_{rj} + \varepsilon_{rj}, \qquad r = 1, \ldots, n, \quad j \in \{1, \ldots, p_r\}, \tag{16.22}$$

where observations are made for the rth item $x_r(t)$ at time points $t_{r1}, t_{r2}, \dots, t_{rp_r}$, not necessarily equally spaced, and where x_{rj} denotes the exact function values. Typically, the errors ε_{rj} are assumed to be i.i.d. $N(0, \sigma^2)$ random variables. In the functional setting, we may be interested in identifying structural differences and similarities among the $x_r(\cdot)$ functions, and whether differences can be explained by covariates. Sometimes, interest may lie in the derivatives of the x_r.

Although the measurements are taken at a finite number of discrete points, the nature of functional data modeling is essentially infinite dimensional. However, since $x_r(t)$ is smooth, we expect the exact function values x_{rj} to be similar for each item r at nearby times. It is this assumption of smoothness that mitigates against the curse of dimensionality.

It is helpful to distinguish several possibilities within the framework of (16.22).

- Accuracy of the measurements. The measurements can be either exact ($\sigma^2 = 0$) or noisy ($\sigma^2 > 0$). In the former case, $z_{rj} = x_{rj}$, so the notation simplifies.
- Density of data. The spacing between the measurement times can range from widely spaced to finely spaced. The functional aspect of the data becomes more important as the spacing between times becomes finer, and in the limit, we effectively observe the complete functions $x_r(t)$. On the other hand, if the measurement times are widely spaced, then temporal continuity is not very relevant, and the analysis is closer to a standard multivariate analysis.
- Spacing of the measurement times. The measurement times of the data can be either *equally spaced* or *unequally spaced* for each individual.
- Commonality of measurement times. Measurements for different individuals can be made at common times (so the times $t_{rj} = t_j$ do not depend on r) or at a set of times unique to each individual.

Example 16.5.1 A typical example of functional data is shown in Figure 16.6. This shows growth curves (Tuddenham and Snyder, 1954) for 54 girls and 39 boys in their childhood years. The points show the times of the measurements; lines are drawn to connect these points (and so are not part of the data). For this example, the data – which have a measurement error with a standard deviation of about 3 mm (Ramsay and Silverman, 2002, p. 1) – are quite finely spaced, unequally spaced, and measured at common ages with 31 measurements for each individual. □

Useful books focusing on functions of time include Ferraty and Vieu (2006), Horváth and Kokoszka (2012), Ramsay and Silverman (2002), and Wood (2017). For an extension to functions of a spatial variable, see, e.g. Kent and Mardia (2022). In the brief overview of the subject given here, we focus on functional principal component analysis and functional linear models.

16.5.1 Functional Principal Component Analysis

We first recall that Chapter 9 described principal components for multivariate data. In the population case, a covariance matrix $\boldsymbol{\Sigma}$ is given an expansion

$$\boldsymbol{\Sigma} = \sum_{k=1}^{p} \lambda_k \boldsymbol{\gamma}_{(k)} \boldsymbol{\gamma}'_{(k)} \tag{16.23}$$

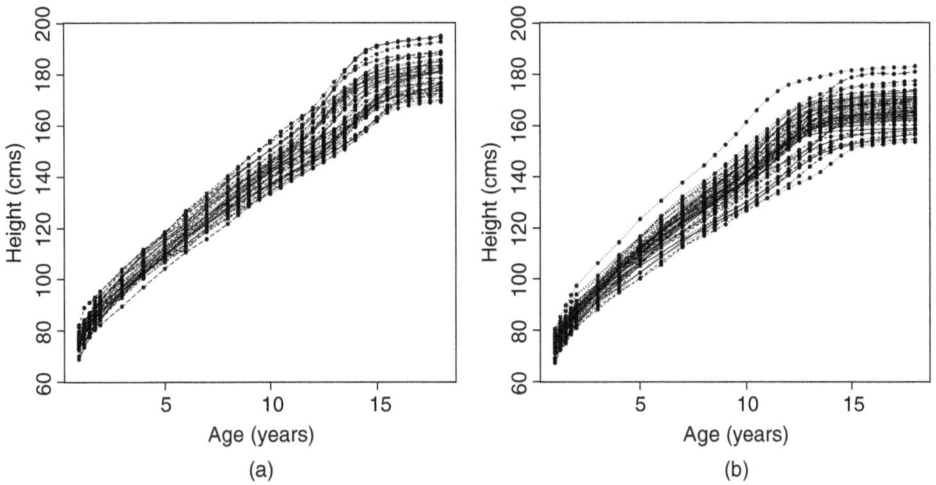

Figure 16.6 Height measurements for 39 boys (a) and 54 girls (b). Points show times of measurements, and lines connect the points Source: Adapted from Ramsay and Silverman (2002)/Springer Nature.

in terms of eigenvalues and eigenvectors. If a p-dimensional random vector $\boldsymbol{x}$ has mean $\boldsymbol{\mu}$ and population covariance matrix $\boldsymbol{\Sigma}$, then the linear combinations,

$$y_k = \boldsymbol{\gamma}'_{(k)}(\boldsymbol{x} - \boldsymbol{\mu}), \quad k = 1, \ldots, p,$$

define the principal components of $\boldsymbol{x}$.

The construction of functional principal components is very similar; the main change is that the summations are replaced by integrals. Let $\{x(t),\ 0 \le t \le T\}$ be a random process on a finite time interval $[0, T]$, with mean function $\mu(t)$ and covariance function

$$G(s,t) = \text{cov}\{x(s), x(t)\}, \qquad 0 \le s, t \le T, \tag{16.24}$$

where it is assumed here that $\mu(t)$ and $G(s,t)$ are continuous functions. Mercer's Theorem (Mercer 1909) is the analogue for a continuous covariance function on a compact interval of the spectral decomposition theorem for a covariance matrix. The representation of $G(s,t)$ by Mercer's Theorem is analogous to (16.23) and takes the form

$$G(s,t) = \sum_{k=1}^{\infty} \lambda_k \phi_k(s)\phi_k(t),$$

where the eigenvalues are positive, $\lambda_k > 0$, the eigenfunctions are orthonormal,

$$\int_0^T \phi_j(t)\phi_k(t)\mathrm{d}t = \begin{cases} 1 & j = k, \\ 0 & j \neq k, \end{cases}$$

and the sum now runs from $k = 1$ to ∞. The population principal components are defined by

$$y_k = \int_0^T \phi_k(t)(x(t) - \mu(t))\mathrm{d}t, \qquad k = 1, 2, \ldots$$

The representation of the process $x(t)$ in terms of its principal components is known as the *Karhunen–Loève expansion*,

$$x(t) = \mu(t) + \sum_{k=1}^{\infty} y_k \phi_k(t), \tag{16.25}$$

where the y_k are uncorrelated random variables with $\mathsf{E}y_k = 0$ and $\mathsf{E}y_k^2 = \lambda_k$.

A sample version of the Karhunen–Loève representation can be developed for data $\{z_{rj}\}$ which are available at equally spaced times t_j, $j = 1, \ldots, p$ between 0 and T. Let $\boldsymbol{S}$ denote the $p \times p$ sample covariance matrix with eigenvalues l_k and eigenvectors $\mathbf{g}_{(k)}$. Define estimated eigenfunctions $\hat{\phi}_k(t)$ at the data times by

$$\hat{\phi}_k(t_j) = g_{jk}, \tag{16.26}$$

with piecewise linear interpolation between. Similarly, estimate the mean function $\mu(t)$ by $\overline{x}(t)$, where $\overline{x}(t_j) = \overline{x}_j$ equals the sample mean of the data at each data time, with piecewise linear interpolation between. Then, the unobserved data functions $x_r(t)$ can be approximated by

$$\hat{x}_r(t) = \overline{x}(t) + \sum_{k=1}^{K} y_{rk} \hat{\phi}_k(t), \tag{16.27}$$

where the y_{rk} are the principal component scores (Section 9.2.2), and K is the number of "important" principal components. The higher order principal components will tend to be dominated by the noise terms ε_{rj} in (16.22).

An alternative approach is to replace the eigenfunctions in (16.25) by another, simpler, set of *basis functions* to represent $x(t)$. The basis functions are generally chosen to be smooth and computationally convenient and lead to the representation

$$x(t) = \sum_{k=1}^{K} c_k \phi_k(t) = \boldsymbol{\phi}(t)\mathbf{c}, \tag{16.28}$$

where $\boldsymbol{\phi}(t) = [\phi_1(t), \ldots, \phi_K(t)]$ is a finite-dimensional *basis system* of continuous functions on $[0, T]$.

The most natural basis functions are based on trigonometric functions, giving a Fourier basis. In this case, the $\phi(t)$ are given by

$$\{\cos(k\omega t), k = 0, 1, \ldots\} \quad \text{and} \quad \{\sin(k\omega t),\ k = 1, 2, , \ldots\},$$

where $\omega = 2\pi/T$ determines the period of the first pair of sine and cosine terms. This basis choice is very computationally efficient but inevitably leads to periodic representations. This is fine if the data are periodic but can be problematic when they are not – as in Example 16.5.1.

Another commonly used basis system is derived from *splines*. A spline of *order* $m \geq 1$ is a piecewise polynomial of degree $m - 1$ which is joined at a set of *knots* in such a way that the spline and its derivatives up to order $m - 2$ are continuous at the knots. See De Boor (2001) for a comprehensive introduction to splines. A vector space of spline functions is determined by the choice of m and the location of the knots. A convenient basis is given by the set of B-splines. Given a collections of knots $\{u_k\}$, the B-splines are defined recursively in the order m by

$$B_{k,1}(t) = \begin{cases} 1 & \text{if} \quad u_k \leq t < u_{k+1}, \\ 0 & \text{otherwise,} \end{cases}$$

$$B_{k,m+1}(t) = \frac{t - u_k}{u_{k+m} - u_j} B_{k,m}(t) + \frac{u_{k+m+1} - t}{u_{k+m+1} - u_{k+1}} B_{k+1,m}(t).$$

Note that $B_{k,m}(t)$ depends on the knots $u_k, \ldots, u_{k+m}$ (Exercise 16.5.1). The functions $B_{k,1}$ are piecewise constant, the functions $B_{k,2}$ are piecewise linear, and so on. In particular, for $m = 2$, the splines are called *linear splines*, and for $m = 4$, the splines are called *cubic splines*. For the remainder of the discussion here, we assume that $m \geq 2$ is even.

Consider a time interval $[0, T]$ and a set of knots $0 = u_1 < \cdots < u_q = T$. If m is even, augment the list of knots with $m/2$ extra *control knots* smaller than 0 and $m/2$ extra control knots larger than T. Then, with the index k for the knots running from $-m/2$ to $q + m/2$, the q B-splines $B_{k-m/2,m}$, $k = 1, \ldots, q$, define a vector space of functions on $[0, T]$ of dimension q. If the control knots are allowed to tend to $-\infty$ and $+\infty$, respectively, then the limiting B-splines become *natural B-splines*; that is, their derivatives of order greater than or equal to $m/2$ vanish at 0 and T. These natural B-splines can be used for the functions $\phi_k(t)$ in the representation (16.28).

One way to motivate the use of splines is through a penalized least squares problem. Consider the case of p noisy observations $\{z_j\}_1^p$ of a single random function $x(t)$ at times $0 = t_1 < \cdots < t_p = T$. Suppose that it is desired to estimate $x(t)$ by minimizing the penalized least squares criterion

$$\sum_{j=1}^{p} ((z_j - x(t_j))^2 + \lambda R(x), \tag{16.29}$$

where the "roughness penalty",

$$R(x) = R_{m/2}(x) = \int_0^T [x^{(m/2)}(t)]^2 \, dt, \tag{16.30}$$

is the integral of the squared derivative of x of order $m/2$. In particular, for $m = 4$, $R_2(x) = \int \{x''(t)\}^2 dt$ is the integrated squared second derivative of x. Here, $\lambda \geq 0$ is a *tuning* parameter. The smoothing spline theorem of Reinsch (1967) states that the solution to this optimization problem is a natural spline of order m with p knots are located at the data times. When $\lambda \to 0$, the solution becomes an interpolating spline, and when $\lambda \to \infty$, the solution tends to the least squares solution of a regression model with polynomial drift of degree $m/2 - 1$.

In practice, the most important choice for m is $m = 4$, corresponding to cubic splines. The reason is that this is the smallest (even) choice of m for which the human eye cannot easily see the joint in a spline at the knots. At the same time, cubic splines provide a very flexible class of functions to model data. For noisy data, the tuning parameter λ can be estimated by cross-validation (Craven and Wahba, 1979).

Using the representation (16.28) for a cubic smoothing spline, the penalized least squares criterion (16.29) becomes

$$(\boldsymbol{z} - \boldsymbol{\Phi c})'(\boldsymbol{z} - \boldsymbol{\Phi c}) + \lambda \boldsymbol{c}' \boldsymbol{R}_2 \boldsymbol{c},$$

where $\boldsymbol{R}_2$ is a matrix, with the (j, k)th entry given by

$$\int \frac{d^2\phi_j(t)}{dt^2} \frac{d^2\phi_k(t)}{dt^2} \, dt. \tag{16.31}$$

Minimizing this criterion leads to the penalized least squares estimate of $\boldsymbol{c}$,

$$\hat{\boldsymbol{c}} = (\boldsymbol{\Phi}'\boldsymbol{\Phi} + \lambda \boldsymbol{R}_2)^{-1} \boldsymbol{\Phi}' \boldsymbol{z}. \tag{16.32}$$

Example 16.5.2 We use the data shown in Figure 16.6 to estimate $\mu(t)$ (by a simple point-wise average) and the $\phi_k(t)$ for both the girls' and boys' growth curves. A comparison of $\mu(t)$ in Figure 16.7 shows very little difference in heights (boys tend to be a little taller, with a reversal around the age of 12) until around age 14, when the gap widens. The first four eigenvalues account for nearly 99% of the variability in the data (for both sexes), and the corresponding eigenfunctions $\phi_k(t)$ are also shown in Figure 16.7.

Both sample-based eigenfunctions and B-splines can be used to fit functions $x_r(t)$ for each individual. When using eigenfunctions, the residuals arise due to truncating the number of terms, whereas when using B-splines, the residuals are due to an imposition of smoothness. If K eigenfunctions are used, then there will usually be a corresponding value of λ such that the two approaches will have the same sum of squared residuals. If K is chosen appropriately, then the use of B-splines sacrifices unbiasedness for a smaller variance.

Using $K = 4$ (which was selected as the first four eigenvalues account for almost 99% of the variability), the residual sum of squares is 517.5, summing over data times and individuals, using eigenfunctions to fit the boys' growth data, and 778.8 for the girls' growth data. The corresponding residual sums of squares using B-splines are 476.4 (boys) and 746.1 (girls), which are obtained with $\lambda = 0.79$ and $\lambda = 1.00$, respectively. These choices for λ were made by leave-one-out cross-validation. Figure 16.7 shows the mean $\pm$2SD intervals of the residuals over the children at each age. Comparing the solutions, it can be seen that the residual sum of squares is less for the B-spline solution (particularly for the boys), the mean of the solution based on the eigenfunctions is zero, and the (symmetric) intervals for the residuals are typically wider than for the B-spline solution. So, overall, the B-spline solution seems preferable in this case for this example. □

16.5.2 Functional Linear Regression Models

The previous methodology can be considered primarily as an exploratory data analysis. We now examine predictive relationships, which can be viewed as a generalization of linear models. There are three different cases that can be considered:

- the response is a real-valued scalar, and the explanatory variables are functions;
- the response is a function, and the explanatory variables are real-valued;
- both the response and explanatory variables are functions.

Here, we consider only the first case, where the values of response variable are denoted by w_r, $r = 1, \dots, n$, and the regressor functions by $x_r(t)$, $r = 1, \dots, n$. Suppose that these functions have been observed without error at the same times t_j, $j = 1, \dots, p$, on $[0, T]$, and that they have been summarized by the first K sample principal components, as in (16.27), with scores y_{rk}. A natural linear regression model is given by

$$w_r = \beta_0 + \sum_{k=1}^{K} \beta_k y_{rk} + \eta_r, \qquad r = 1, 2, \dots, n, \tag{16.33}$$

where the η_r are i.i.d. error terms.

Let $\hat{\phi}_k(t)$ denote the fitted eigenfunctions from (16.26). It is also possible to describe the dependence of w_r on $x_r(t)$ using the *functional regression expression*

$$\beta(t) = \sum_{k=1}^{K} \beta_k \hat{\phi}_k(t), \tag{16.34}$$

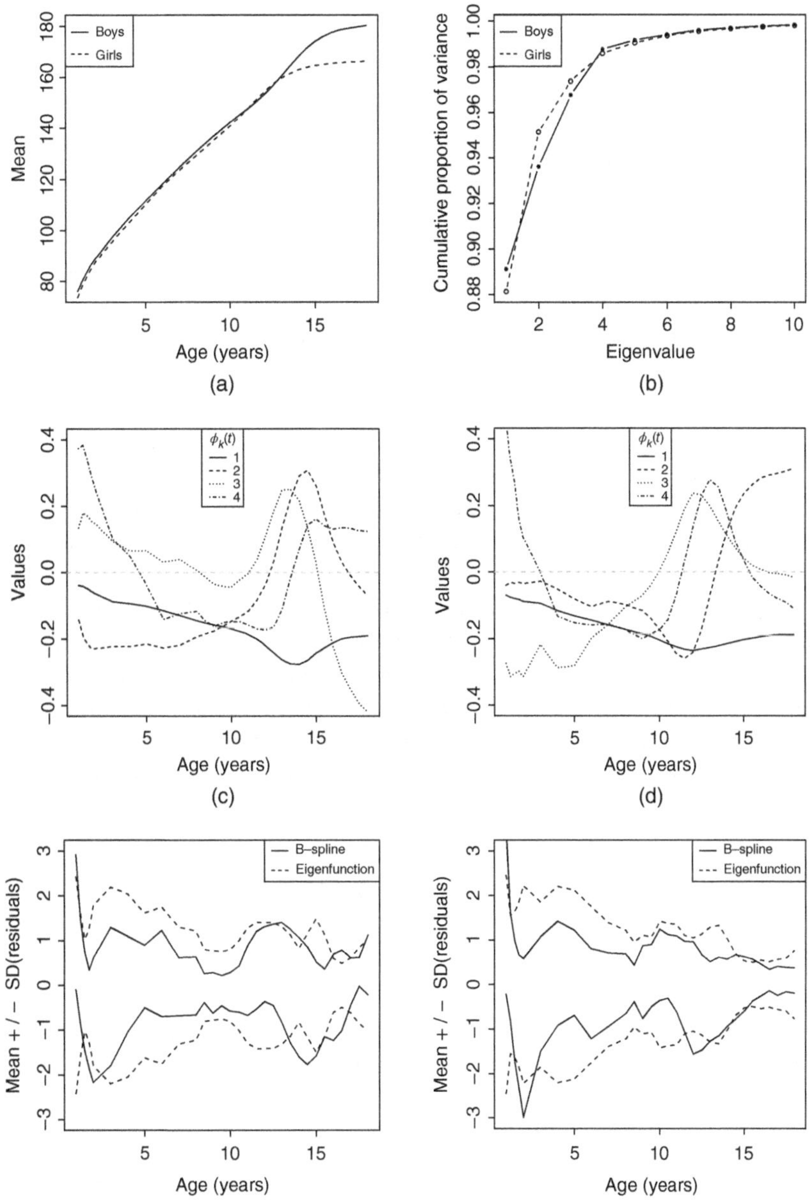

Figure 16.7 Growth Data. (a) Estimated $\mu(t)$ (Eq. (16.25); (b) Cumulative proportion of variance explained by first 10 eigenvalues (right) for boys (continuous) and girls (dashed); estimated $\phi_k(t)$ for $k = 1, 2, 3, 4$ for boys (c) and girls (d); empirically computed mean $\pm$ 2 SD of the residuals for the eigenfunction solution (dashed) and the B-spline solution (continuous), respectively, for boys (e) and girls (f).

since

$$\sum_{k=1}^{K} \beta_k y_{rk} \approx \int_0^T \beta(t)(x_r(t) - \overline{x}(t))\mathrm{d}t. \tag{16.35}$$

Equation (16.35) would be an exact equality if the fitted eigenfunctions were exactly orthonormal on $[0, T]$. (The fitted eigenfunctions are piecewise linear functions defined through an orthonormal set of eigenvectors.)

Example 16.5.3 We continue to use the data of Example 16.5.2, but we now pool the growth data from the girls and boys. We compute a pooled mean growth curve and use principal component scores y_{rk} corresponding to the largest four eigenvalues in the sample Karhunen–Loève representation in (16.25). The response w_r is a binary variable (with 1 for girls and 0 for boys), and the regression model (16.33) can be used to predict the sex of the child, depending on whether the predicted value of w_r is greater or less than 1/2.

The coefficients and standard errors of $\boldsymbol{\beta}$ are shown in Table 16.1. The table indicates that the fourth term is not significant and can be dropped from the equation. The refitted $\beta_0, \ldots, \beta_3$ are unchanged, and the standard errors are reduced by a small amount.

These predictions from the model with $K = 3$ components can be compared to the actual sex, leading to the (resubstitution) confusion matrix given below. The estimated regression function $\hat{\beta}(t) = \sum_{k=1}^{K} \hat{\beta}_k \hat{\phi}_k(t)$ is shown in Figure 16.8. It can be seen that this function has a maximum near time $t = 12$ years and a minimum at time $t = 18$ years. Hence, for an individual whose growth curve $x_r(t)$ is larger than average at time $t = 12$ and smaller than average at time $t = 18$ will tend to yield a prediction of "girl." This interpretation reflects the fact that teenage girls typically have a growth spurt at an earlier time than boys.

Confusion Matrix:

		Actual Boy	Actual Girl
Predicted	Boy	38	2
	Girl	1	52

In passing we note that a stepwise regression on the *original variables* leads to a model with six terms, which results in a slightly worse performance with four resubstitution errors in the corresponding confusion matrix. □

Table 16.1 Estimated regression coefficients $\boldsymbol{\beta}$ using (16.33), with regressors y_{rk} obtained from (16.24).

	Estimate	Std error
β_0	0.5806	0.0244
β_1	0.0062	0.0007
β_2	0.0277	0.0018
β_3	0.0125	0.0040
β_4	−0.0033	0.0055

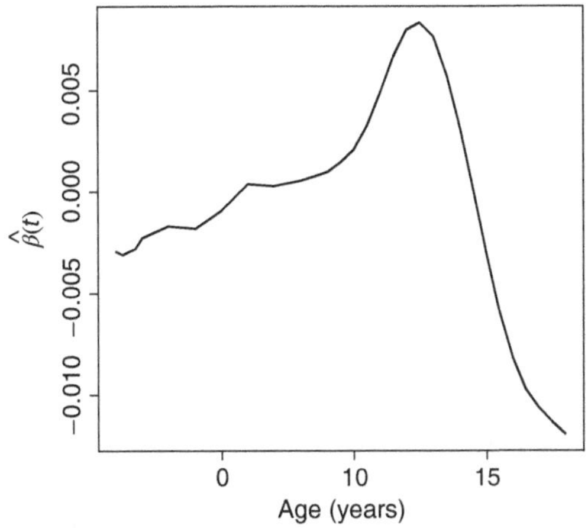

Figure 16.8 Growth Data. The estimated regression function $\hat{\beta}(t)$ obtained using the first three principal components in (16.33).

Exercises and Complements

16.2.1 In the regression model (16.1), show that, if each regression coefficient has a prior distribution with expected value zero, and variance σ^2/λ, then the mean of the posterior distribution is the same as the ridge regression estimate in (16.6).

16.2.2 Show that the ridge regression estimate in (16.6) is the same as the OLS solution for an artificial data set in which the original data is augmented with "phoney data" consisting of p rows in which the jth row has all variables (including y) equal to zero, except for the jth element, which equals $\sqrt{\lambda}$.

16.2.3 Starting from Eq. (16.6), show that

$$\boldsymbol{X}\hat{\boldsymbol{\beta}}_\lambda^{\text{Ridge}} = \boldsymbol{U}\boldsymbol{L}(\boldsymbol{L}^2 + \lambda \boldsymbol{I})^{-1}\boldsymbol{L}\boldsymbol{U}'\boldsymbol{y},$$

where the SVD of $\boldsymbol{X}$ is given by $\boldsymbol{X} = \boldsymbol{U}\boldsymbol{L}\boldsymbol{V}'$.
Show that this can be further written as

$$\boldsymbol{X}\hat{\boldsymbol{\beta}}_\lambda^{\text{Ridge}} = \sum_{j=1}^{p} \boldsymbol{u}_{(j)} \frac{l_j^2}{l_j^2 + \lambda} \boldsymbol{u}'_{(j)} \boldsymbol{y},$$

where $\boldsymbol{u}_{(j)}$ is the jth column of $\boldsymbol{U}$, and $\boldsymbol{L} = \text{diag}(l_1, \ldots, l_p)$.
Hence show that, as $\lambda \to \infty$, the expectation of $\hat{\boldsymbol{\beta}}_\lambda^{\text{Ridge}}$ tends to $\boldsymbol{0}$.

16.2.4 Show that the mean-squared error of the OLS solution exceeds the mean-squared error of the ridge regression solution by

$$\lambda(\boldsymbol{X}'\boldsymbol{X} + \lambda\boldsymbol{I})^{-1}\left\{2\sigma^2\boldsymbol{I} + \lambda\sigma^2(\boldsymbol{X}'\boldsymbol{X})^{-1} - \lambda\boldsymbol{\beta}\boldsymbol{\beta}'\right\}(\boldsymbol{X}'\boldsymbol{X} + \lambda\boldsymbol{I})^{-1}.$$

Using the result that (Farebrother, 1976) $c\boldsymbol{A} - \boldsymbol{b}\boldsymbol{b}'$ is positive definite if and only if $\boldsymbol{b}'\boldsymbol{A}^{-1}\boldsymbol{b} > c$, when $\boldsymbol{A}$ is positive definite, $\boldsymbol{b} \neq \boldsymbol{0}$, and $c > 0$, show that this difference in mean-squared error is positive if $2\sigma^2(\boldsymbol{\beta}'\boldsymbol{\beta})^{-1} > \lambda$. (See also Exercise 9.8.1.)

16.2.5 Show that there exists a value of t (which depends on λ) such that the value of $\boldsymbol{\beta}$ which minimizes

$$\sum_{r=1}^{n}(y_r - \boldsymbol{x}'_r\boldsymbol{\beta})^2 + \lambda\sum_{j=1}^{p}|\beta_j|$$

is equal to

$$\arg\min_{\boldsymbol{\beta}}(\boldsymbol{y} - \boldsymbol{X}\boldsymbol{\beta})'(\boldsymbol{y} - \boldsymbol{X}\boldsymbol{\beta}) \qquad \text{such that} \quad \sum_{j=1}^{p}|\beta_j| \le t.$$

16.3.1 Show that the estimate of $\hat{\boldsymbol{\beta}}^{\text{PC}}$ using the first k principal components of $\boldsymbol{X}$ takes the form given in (16.11).

16.3.2 Carry out a forward stepwise regression for the yarn data set (Example 16.3.1), which is available in the R library `pls` (Liland et al., 2021). Compare the model that has seven variables with the model that regresses y on the same number of principal components (scores). Explain which model you prefer and why.

16.4.1 The purpose of this exercise is to confirm that the PLS loading vectors, i.e. the columns of $\boldsymbol{\Phi}_1$ in the PLS1 algorithm of Section 16.4.2, are orthogonal to one another. The easiest way is to proceed by induction using the induction hypothesis that $\boldsymbol{\Sigma}_{xx}^{(\nu)}\boldsymbol{\phi}_{(j)} = \boldsymbol{0}$ and $\boldsymbol{\sigma}_{xy}^{(\nu)\prime}\boldsymbol{\phi}_{(j)} = 0$ for $j = 1, \dots, \nu - 1$. This hypothesis clearly holds for index $\nu = 1$ since there are no indices j to consider.

(a) Suppose that the induction hypothesis holds for index ν, where $1 \le \nu \le q_1 - 1$. Using (16.13) and (16.15), show that $\boldsymbol{\phi}'_{(\nu)}\boldsymbol{\phi}_{(j)} = 0$ for $j = 1, \dots, \nu - 1$, $\boldsymbol{\Sigma}_{xx}^{(\nu+1)}\boldsymbol{\phi}_{(\nu)} = \boldsymbol{0}$, and $\boldsymbol{\sigma}_{xy}^{(\nu+1)\prime}\boldsymbol{\phi}_{(\nu)} = 0$.

(b) Hence, conclude that the induction hypothesis holds for index $\nu + 1$.

(c) Putting these results together, deduce that $\boldsymbol{\Phi}_1$ is a column orthonormal matrix.

(d) Using a similar argument show that the PLS loading vectors are orthogonal to one another for the PLS2 algorithm in Section 16.4.3.

16.4.2 What is the maximum number of steps (l) that can be used in partial least squares regression for the case of (a) univariate response, and (b) multivariate response? Explain, or show, why – when this value is used – the fitted values are the same as those obtained from the ordinary least squares solution.

16.5.1 Given a collection of knots $u_1 < u_2 < \cdots < u_n$, show (by induction) that the B-spline $B_{k,m}(t)$ has support $[u_k, u_{k+m}]$, and that in this interval $B_{k,m}(t) \ge 0$. For which value of t is $B_{k.m}(t)$ a maximum for $k = 2, 3, 4$?

16.5.2 Using the representation of $x(t)$ in (16.28) as a linear combination of basis functions $\phi_j(t), j = 1, \dots, K$, show that the roughness, as defined by (16.30), leads to $\boldsymbol{R}_2$ described by (16.31).

A

Matrix Algebra

A.1 Introduction

This appendix gives (i) a summary of the basic definitions and results in matrix algebra with comments and (ii) details of those results and proofs that are used in this book but normally not treated in undergraduate Mathematics courses. It is designed as a convenient source of reference to be used in the rest of the book. A geometrical interpretation of some of the results is also given. If the reader is unfamiliar with any of the results not proved here, he or she should consult a text such as Graybill (1969, especially pp. 4–52, 163–196, and 222–235) and (Rao, 1973, pp. 1–78). For the computational aspects of matrix operations, see, for example, Wilkinson (1965). Matrix algebra is standard tool in multivariate analysis but historically it took some time for this to be accepted; see Mardia (2023).

Definition A.1.1 *A matrix* $\boldsymbol{A}$ *is a rectangular array of numbers. If* $\boldsymbol{A}$ *has* n *rows and* p *columns, we say it is of order* $n \times p$. *For example,* n *observations on* p *random variables are arranged in this way.*

Notation 1 *We write matrix* $\boldsymbol{A}$ *of order* $n \times p$ *as*

$$\boldsymbol{A} = \begin{bmatrix} a_{11} & a_{12} & \cdots & a_{1p} \\ a_{21} & a_{22} & \cdots & a_{2p} \\ \vdots & \vdots & & \vdots \\ a_{n1} & a_{n2} & \cdots & a_{np} \end{bmatrix} = \left(a_{ij}\right), \tag{A.1}$$

where a_{ij} *is the element in row* i *and column* j *of the matrix* $\boldsymbol{A}$, $i = 1, \ldots, n; j = 1, \ldots, p$. *Sometimes, we write* $(\boldsymbol{A})_{ij}$ *for* a_{ij}.

We may write the matrix $\boldsymbol{A}$ as $\boldsymbol{A}(n \times p)$ to emphasize the row and column order. In general, matrices are represented by boldface upper case letters throughout the book, e.g. $\boldsymbol{A}, \boldsymbol{B}, \boldsymbol{A}, \boldsymbol{Y}, \boldsymbol{Z}$. Their elements are represented by small letters with subscripts.

Multivariate Analysis, Second Edition. Kanti V. Mardia, John T. Kent, and Charles C. Taylor.

Definition A.1.2 *The* transpose *of a matrix* $\boldsymbol{A}$ *is formed by interchanging the rows and columns:*

$$\boldsymbol{A}' = \begin{bmatrix} a_{11} & a_{21} & \cdots & a_{n1} \\ a_{12} & a_{22} & \cdots & a_{n2} \\ \vdots & \vdots & & \vdots \\ a_{1p} & a_{2p} & \cdots & a_{np} \end{bmatrix}.$$

Definition A.1.3 *A matrix with column-order 1 is called a* column vector. *Thus,*

$$\boldsymbol{a} = \begin{pmatrix} a_1 \\ a_2 \\ \vdots \\ a_n \end{pmatrix}$$

is a column vector with n components.

Note that, in general, we use round brackets for vectors and square brackets for matrices.

In general, boldface lower case letters represent column vectors. Row vectors are written as column vectors transposed, i.e.

$$\boldsymbol{a}' = (a_1, \dots, a_n).$$

Notation 2 *We write the columns of the matrix* $\boldsymbol{A}$ *as* $\boldsymbol{a}_{(1)}, \boldsymbol{a}_{(2)}, \dots, \boldsymbol{a}_{(p)}$, *and the rows (if written as column vectors) as* $\boldsymbol{a}_1, \boldsymbol{a}_2, \dots, \boldsymbol{a}_n$, *so that*

$$\boldsymbol{A} = [\boldsymbol{a}_{(1)}, \boldsymbol{a}_{(2)}, \dots, \boldsymbol{a}_{(p)}] = \begin{bmatrix} \boldsymbol{a}_1' \\ \boldsymbol{a}_2' \\ \vdots \\ \boldsymbol{a}_n' \end{bmatrix}, \tag{A.2}$$

where

$$\boldsymbol{a}_{(j)} = \begin{pmatrix} a_{1j} \\ \vdots \\ a_{nj} \end{pmatrix}, \qquad \boldsymbol{a}_i = \begin{pmatrix} a_{i1} \\ \vdots \\ a_{ip} \end{pmatrix}.$$

Definition A.1.4 *A matrix written in terms of its submatrices is called a* partitioned matrix.

Notation 3 *Let* $\boldsymbol{A}_{11}$, $\boldsymbol{A}_{12}$, $\boldsymbol{A}_{21}$, *and* $\boldsymbol{A}_{22}$ *be submatrices of* $\boldsymbol{A}$ *such that* $\boldsymbol{A}_{11}(r \times s)$ *has elements* $a_{ij}, i = 1, \dots, r; j = 1, \dots, s$, *and so on. Then, we write*

$$\boldsymbol{A}(n \times p) = \begin{bmatrix} \boldsymbol{A}_{11}(r \times s) & \boldsymbol{A}_{12}(r \times (p-s)) \\ \boldsymbol{A}_{21}((n-r) \times s) & \boldsymbol{A}_{22}((n-r) \times (p-s)) \end{bmatrix}.$$

Obviously, this notation can be extended to contain further partitions of $\boldsymbol{A}_{11}, \boldsymbol{A}_{12}$, *etc.*

A list of some important types of particular matrices is given in Table A.1. Another list that depends on the next section appears in Table A.3.

As shown in Table A.1, a square matrix $\boldsymbol{A}(p \times p)$ is *diagonal* if $a_{ij} = 0$ for all $i \neq j$. There are two convenient ways to construct diagonal matrices. If $\boldsymbol{a} = (a_1, \dots, a_p)'$ is any vector,

Table A.1 Particular matrices and types of matrices (List 1). For List 2, see Table A.3.

	Name	Definition	Notation	Trivial examples
1	Scalar	$p = n = 1$	a, b	(1)
2a	Column vector	$p = 1$	$\boldsymbol{a}, \boldsymbol{b}$	$\begin{pmatrix} 1 \\ 2 \end{pmatrix}$
2b	Unit vector	$(1, \ldots, 1)'$	$\mathbf{1}$ or $\mathbf{1}_p$	$\begin{pmatrix} 1 \\ 1 \end{pmatrix}$
3	Rectangular	$n \times p$	$\boldsymbol{A}(n \times p)$	
4	Square	$p = n$	$\boldsymbol{A}(p \times p)$	$\begin{bmatrix} 1 & 3 \\ 4 & 5 \end{bmatrix}$
4a	Diagonal	$p = n, a_{ij} = 0, i \neq j$	$\text{diag}(a_{ii})$	$\begin{bmatrix} 2 & 0 \\ 0 & 1 \end{bmatrix}$
4b	Identity	$\text{diag}(\mathbf{1})$ (or $\text{diag}(p)$)	$\boldsymbol{I}$ or $\boldsymbol{I}_p$	$\begin{bmatrix} 1 & 0 \\ 0 & 1 \end{bmatrix}$
4c	Symmetric	$a_{ij} = a_{ji}$		$\begin{bmatrix} 3 & 2 \\ 2 & 5 \end{bmatrix}$
4d	Skew symmetric	$a_{ij} = -a_{ji}$		$\begin{bmatrix} 0 & 2 \\ -2 & 0 \end{bmatrix}$
4e	Unit matrix	$p = n, a_{ij} = 1$	$\boldsymbol{J}_p = \mathbf{11}'$	$\begin{bmatrix} 1 & 1 \\ 1 & 1 \end{bmatrix}$
4f	Triangular matrix (upper)	$a_{ij} = 0$ below the diagonal	$\boldsymbol{\Delta}'$	$\begin{bmatrix} 1 & 0 & 0 \\ 2 & 2 & 0 \\ 3 & 2 & 5 \end{bmatrix}$
	Triangular matrix (lower)	$a_{ij} = 0$ above the diagonal	$\boldsymbol{\Delta}$	
5	Asymmetric	$a_{ij} \neq a_{ji}$		$\begin{bmatrix} 1 & 1 \\ 2 & 3 \end{bmatrix}$
6	Null	$a_{ij} = 0$	$\mathbf{0}$	$\begin{bmatrix} 0 & 0 & 0 \\ 0 & 0 & 0 \end{bmatrix}$
7	Permutation	$a_{ij} \in \{0,1\}$ $\Sigma_i a_{ij} = \Sigma_j a_{ij} = 1$		$\begin{bmatrix} 0 & 1 \\ 1 & 0 \end{bmatrix}$

and $\boldsymbol{B}(p \times p)$ is any square matrix, then

$$\text{diag}(\boldsymbol{a}) = \text{diag}(a_i) = \text{diag}(a_1, \dots, a_p) = \begin{bmatrix} a_1 & \cdots & 0 \\ \vdots & \ddots & \vdots \\ 0 & \cdots & a_p \end{bmatrix}$$

and

$$\text{Diag}(\boldsymbol{B}) = \begin{bmatrix} b_{11} & \cdots & 0 \\ \vdots & \ddots & \vdots \\ 0 & \cdots & b_{pp} \end{bmatrix},$$

each defines a diagonal matrix, with the notations *diag* and *Diag*, respectively.

A.2 Matrix Operations

Table A.2 gives a summary of various important matrix operations. We deal with some of these in detail, assuming the definitions in the table.

A.2.1 Transpose

The transpose satisfies the simple properties

$$(\boldsymbol{A}')' = \boldsymbol{A}, \qquad (\boldsymbol{A}+\boldsymbol{B})' = \boldsymbol{A}'+\boldsymbol{B}', \qquad (\boldsymbol{AB})' = \boldsymbol{B}'\boldsymbol{A}'. \tag{A.3}$$

Table A.2 Basic matrix operations.

Operation	Restrictions	Definitions	Remarks
Addition	$\boldsymbol{A}, \boldsymbol{B}$ of same order	$\boldsymbol{A}+\boldsymbol{B} = (a_{ij}+b_{ij})$	
Subtraction	$\boldsymbol{A}, \boldsymbol{B}$ of same order	$\boldsymbol{A}-\boldsymbol{B} = (a_{ij}-b_{ij})$	
Scalar multiplication		$c\boldsymbol{A} = (ca_{ij})$	
Inner product	$\boldsymbol{a}, \boldsymbol{b}$ of same order	$\boldsymbol{a}'\boldsymbol{b} = \sum a_i b_i$	
Outer product	$\boldsymbol{a}, \boldsymbol{b}$ of same order	$\boldsymbol{ab}'$	
Multiplication	Number of columns of $\boldsymbol{A}$ equals number of rows of $\boldsymbol{B}$	$\boldsymbol{AB} = (\boldsymbol{a}_i'\boldsymbol{b}_{(j)})$	$\boldsymbol{AB} \neq \boldsymbol{BA}$
Transpose		$\boldsymbol{A}' = (\boldsymbol{a}_1, \dots, \boldsymbol{a}_n)$	Section A.2.1
Trace	$\boldsymbol{A}$ square	$\text{tr}\,\boldsymbol{A} = \sum a_{ii}$	Section A.2.2
Determinant	$\boldsymbol{A}$ square	$\lvert\boldsymbol{A}\rvert$	Section A.2.3
Inverse	$\boldsymbol{A}$ square, $\lvert\boldsymbol{A}\rvert \neq 0$	$\boldsymbol{AA}^{-1} = \boldsymbol{A}^{-1}\boldsymbol{A} = \boldsymbol{I}$	$(\boldsymbol{A}+\boldsymbol{B})^{-1} \neq \boldsymbol{A}^{-1}+\boldsymbol{B}^{-1}$ Section A.2.4
g-inverse ($\boldsymbol{A}^-$)	$\boldsymbol{A}(n \times p)$	$\boldsymbol{AA}^-\boldsymbol{A} = \boldsymbol{A}$	Section A.8

For partitioned $\boldsymbol{A}$,

$$\boldsymbol{A}' = \begin{bmatrix} \boldsymbol{A}'_{11} & \boldsymbol{A}'_{21} \\ \boldsymbol{A}'_{12} & \boldsymbol{A}'_{22} \end{bmatrix}.$$

If $\boldsymbol{A}$ is a symmetric matrix, $a_{ij} = a_{ji}$, then

$$\boldsymbol{A}' = \boldsymbol{A}.$$

A.2.2 Trace

The trace function, $\text{tr}\,\boldsymbol{A} = \sum a_{ii}$, satisfies the following properties for matrices $\boldsymbol{A}(p \times p), \boldsymbol{B}(p \times p), \boldsymbol{C}(p \times n), \boldsymbol{D}(n \times p)$, and scalar α:

$$\text{tr}\,\alpha = \alpha \qquad \text{tr}\,(\boldsymbol{A} \pm \boldsymbol{B}) = \text{tr}\,\boldsymbol{A} \pm \text{tr}\,\boldsymbol{B}, \qquad \text{tr}\,\alpha\boldsymbol{A} = \alpha\text{tr}\,\boldsymbol{A}, \tag{A.4}$$

$$\text{tr}\,\boldsymbol{CD} = \text{tr}\,\boldsymbol{DC} = \sum_{i,j} c_{ij} d_{ji}, \tag{A.5}$$

$$\sum \boldsymbol{x}_i' \boldsymbol{A} \boldsymbol{x}_i = \text{tr}\,(\boldsymbol{AT}), \qquad \text{where} \quad \boldsymbol{T} = \sum \boldsymbol{x}_i \boldsymbol{x}_i'. \tag{A.6}$$

To prove this last property, note that since $\sum \boldsymbol{x}_i' \boldsymbol{A} \boldsymbol{x}_i$ is a scalar, the left-hand side of (A.6) is

$$\begin{aligned} \text{tr} \sum \boldsymbol{x}_i' \boldsymbol{A} \boldsymbol{x}_i &= \sum \text{tr}\,\boldsymbol{x}_i' \boldsymbol{A} \boldsymbol{x}_i \quad \text{by (A.4)} \\ &= \sum \text{tr}\,\boldsymbol{A} \boldsymbol{x}_i \boldsymbol{x}_i' \quad \text{by (A.5)} \\ &= \text{tr}\,\boldsymbol{A} \sum \boldsymbol{x}_i \boldsymbol{x}_i' \quad \text{by (A.4)}. \end{aligned}$$

As a special case of (A.5), note that

$$\text{tr}\,\boldsymbol{CC}' = \text{tr}\,\boldsymbol{C}'\boldsymbol{C} = \sum c_{ij}^2. \tag{A.7}$$

A.2.3 Determinants and Cofactors

Definition A.2.1 *The determinant of a square matrix* $\boldsymbol{A}$ *is defined as*

$$|\boldsymbol{A}| = \sum (-1)^{|\tau|} a_{1\tau(1)} \cdots a_{p\tau(p)}, \tag{A.8}$$

where the summation is taken over all permutations τ *of* $(1, 2, \ldots, p)$, *and* $|\tau|$ *equals* $+1$ *or* -1, *depending on whether* τ *can be written as the product of an even or odd number of transpositions.*

For $p = 2$,

$$|\boldsymbol{A}| = a_{11}a_{22} - a_{12}a_{21}. \tag{A.9}$$

Definition A.2.2 *The* cofactor *of* a_{ij} *is defined by* $(-1)^{i+j}$ *times the* minor *of* a_{ij}, *where the* minor *of* a_{ij} *is the value of the determinant obtained after deleting the ith row and the jth column of* $\boldsymbol{A}$.

We denote the cofactor of a_{ij} by A_{ij}. Thus, for $p = 3$,

$$A_{11} = \begin{vmatrix} a_{22} & a_{23} \\ a_{32} & a_{33} \end{vmatrix}, \qquad A_{12} = -\begin{vmatrix} a_{21} & a_{23} \\ a_{31} & a_{33} \end{vmatrix}, \qquad A_{13} = \begin{vmatrix} a_{21} & a_{22} \\ a_{31} & a_{32} \end{vmatrix}. \tag{A.10}$$

Definition A.2.3 *A square matrix is* nonsingular *if* $|\boldsymbol{A}| \neq 0$*; otherwise it is* singular.

We have the following results:

(I) $$|\boldsymbol{A}| = \sum_{j=1}^{p} a_{ij}A_{ij} = \sum_{i=1}^{p} a_{ij}A_{ij}, \quad \text{any } i,j, \tag{A.11}$$

$$\text{but} \quad \sum_{k=1}^{p} a_{ik}A_{jk} = 0, \quad i \neq j. \tag{A.12}$$

(II) If $\boldsymbol{A}$ is triangular or diagonal, $$|\boldsymbol{A}| = \prod a_{ii}. \tag{A.13}$$

(III) $$|c\boldsymbol{A}| = c^p|\boldsymbol{A}|. \tag{A.14}$$

(IV) $$|\boldsymbol{AB}| = |\boldsymbol{A}||\boldsymbol{B}|. \tag{A.15}$$

(V) For square submatrices $\boldsymbol{A}(p \times p)$ and $\boldsymbol{B}(q \times q)$,

$$\begin{vmatrix} \boldsymbol{A} & \boldsymbol{C} \\ \boldsymbol{0} & \boldsymbol{B} \end{vmatrix} = |\boldsymbol{A}||\boldsymbol{B}|. \tag{A.16}$$

(VI)

$$\begin{vmatrix} \boldsymbol{A}_{11} & \boldsymbol{A}_{12} \\ \boldsymbol{A}_{21} & \boldsymbol{A}_{22} \end{vmatrix} = |\boldsymbol{A}_{11}||\boldsymbol{A}_{22} - \boldsymbol{A}_{21}\boldsymbol{A}_{11}^{-1}\boldsymbol{A}_{12}|$$
$$= |\boldsymbol{A}_{22}||\boldsymbol{A}_{11} - \boldsymbol{A}_{12}\boldsymbol{A}_{22}^{-1}\boldsymbol{A}_{21}|, \tag{A.17}$$

$$\begin{vmatrix} \boldsymbol{A} & \boldsymbol{a} \\ \boldsymbol{a}' & b \end{vmatrix} = |\boldsymbol{A}|(b - \boldsymbol{a}'\boldsymbol{A}^{-1}\boldsymbol{a}). \tag{A.18}$$

(VII) For $\boldsymbol{B}(p \times n)$ and $\boldsymbol{C}(n \times p)$, and nonsingular $\boldsymbol{A}(p \times p)$,

$$|\boldsymbol{A} + \boldsymbol{BC}| = |\boldsymbol{A}||\boldsymbol{I}_p + \boldsymbol{A}^{-1}\boldsymbol{BC}| = |\boldsymbol{A}||\boldsymbol{I}_n + \boldsymbol{CA}^{-1}\boldsymbol{B}|, \tag{A.19}$$

$$|\boldsymbol{A} + \boldsymbol{ab}'| = |\boldsymbol{A}|(1 + \boldsymbol{b}'\boldsymbol{A}^{-1}\boldsymbol{a}). \tag{A.20}$$

Remarks

(1) Properties (I)–(III) follow easily from the definition (A.8). As an application of (I), from (A.9)–(A.11), we have, for $p = 3$,

$$|\boldsymbol{A}| = a_{11}(a_{22}a_{33} - a_{23}a_{32}) - a_{12}(a_{21}a_{33} - a_{23}a_{31}) + a_{13}(a_{21}a_{32} - a_{31}a_{22}).$$

(2) To prove (V), note that the only permutations giving nonzero terms in the summation (A.8) are those taking $\{1, \ldots, p\}$ to $\{1, \ldots, p\}$ and $\{p+1, \ldots, p+q\}$ to $\{p+1, \ldots, p+q\}$.

(3) To prove (VI), simplify $\boldsymbol{BAB}'$ and then take its determinant where

$$\boldsymbol{B} = \begin{bmatrix} \boldsymbol{I} & -\boldsymbol{A}_{12}\boldsymbol{A}_{22}^{-1} \\ \boldsymbol{0} & \boldsymbol{I} \end{bmatrix}.$$

From (VI), we deduce, after putting $\boldsymbol{A}_{11} = \boldsymbol{A}, \boldsymbol{A}_{12} = \boldsymbol{b}$, etc.,

$$\begin{vmatrix} \boldsymbol{A} & \boldsymbol{b} \\ \boldsymbol{c}' & d \end{vmatrix} = |\boldsymbol{A}|(d - \boldsymbol{c}'\boldsymbol{A}^{-1}\boldsymbol{b}). \tag{A.21}$$

(4) To prove the second part of (VII), simplify

$$\begin{vmatrix} I_p & -A^{-1}B \\ C & I_n \end{vmatrix}$$

using (VI). As special cases of (VII), we see that, for nonsingular $\boldsymbol{A}$,

$$|A + bc'| = |A|(1 + c'A^{-1}b), \tag{A.22}$$

and that, for $\boldsymbol{B}(p \times n)$ and $\boldsymbol{C}(n \times p)$,

$$|I_p + BC| = |I_n + CB|. \tag{A.23}$$

In practice, we can simplify determinants using the property that the value of a determinant is unaltered if a linear combination of some of the columns (rows) is added to another column (row).

(5) Determinants are usually evaluated on computers as follows. $\boldsymbol{A}$ is decomposed into upper and lower triangular matrices $\boldsymbol{A} = \boldsymbol{LU}$. If $\boldsymbol{A} > 0$, then the Cholesky decomposition is used (i.e. $\boldsymbol{U} = \boldsymbol{L}'$, so $\boldsymbol{A} = \boldsymbol{LL}'$). Otherwise, the Crout decomposition is used where the diagonal elements of $\boldsymbol{U}$ are 1s.

A.2.4 Inverse

Definition A.2.4 *As already defined in Table A.1, the* inverse *of $\boldsymbol{A}$ is the unique matrix $\boldsymbol{A}^{-1}$ satisfying*

$$AA^{-1} = A^{-1}A = I. \tag{A.24}$$

The inverse exists if and only if $\boldsymbol{A}$ is nonsingular, that is if and only if $|\boldsymbol{A}| \neq 0$.

We write the (i,j)th element of $\boldsymbol{A}^{-1}$ by a^{ij}. For partitioned $\boldsymbol{A}$, we write

$$A^{-1} = \begin{bmatrix} A^{11} & A^{12} \\ A^{21} & A^{22} \end{bmatrix}.$$

The following properties hold:

(I) $$A^{-1} = \frac{1}{|A|}(A_{ij})'. \tag{A.25}$$

(II) $$(cA)^{-1} = c^{-1}A^{-1}. \tag{A.26}$$

(III) $$(AB)^{-1} = B^{-1}A^{-1}. \tag{A.27}$$

(IV) The unique solution of $\boldsymbol{Ax} = \boldsymbol{b}$ is $\boldsymbol{x} = \boldsymbol{A}^{-1}\boldsymbol{b}$. (A.28)

(V) If all the necessary inverses exist, then for $\boldsymbol{A}(n \times p)$, $\boldsymbol{B}(p \times n)$, $\boldsymbol{C}(n \times n)$, and $\boldsymbol{D}(n \times p)$,

$$(A + BCD)^{-1} = A^{-1} - A^{-1}B(C^{-1} + DA^{-1}B)^{-1}DA^{-1}, \tag{A.29}$$

which is known as the Woodbury Formula, and

$$(A + ab')^{-1} = A^{-1} - A^{-1}ab'A^{-1}/(1 + b'A^{-1}a). \tag{A.30}$$

(VI) If all the necessary inverses exist, then for partitioned $\boldsymbol{A}$, the elements of $\boldsymbol{A}^{-1}$ are

$$\begin{aligned} \boldsymbol{A}^{11} &= (\boldsymbol{A}_{11} - \boldsymbol{A}_{12}\boldsymbol{A}_{22}^{-1}\boldsymbol{A}_{21})^{-1}, & \boldsymbol{A}^{22} &= (\boldsymbol{A}_{22} - \boldsymbol{A}_{21}\boldsymbol{A}_{11}^{-1}\boldsymbol{A}_{12})^{-1}, \\ \boldsymbol{A}^{12} &= -\boldsymbol{A}^{11}\boldsymbol{A}_{12}\boldsymbol{A}_{22}^{-1}, & \boldsymbol{A}^{21} &= -\boldsymbol{A}_{22}^{-1}\boldsymbol{A}_{21}\boldsymbol{A}^{11}. \end{aligned} \tag{A.31}$$

Alternatively, $\boldsymbol{A}^{12}$ and $\boldsymbol{A}^{21}$ can be defined by

$$\boldsymbol{A}^{12} = -\boldsymbol{A}_{11}^{-1}\boldsymbol{A}_{12}\boldsymbol{A}^{22}, \qquad \boldsymbol{A}^{21} = -\boldsymbol{A}^{22}\boldsymbol{A}_{21}\boldsymbol{A}_{11}^{-1}.$$

(VII) For symmetrical matrices $\boldsymbol{A}$ and $\boldsymbol{D}$, we have, if all the necessary inverses exist,

$$\begin{bmatrix} \boldsymbol{A} & \boldsymbol{B} \\ \boldsymbol{B}' & \boldsymbol{D} \end{bmatrix}^{-1} = \begin{bmatrix} \boldsymbol{A}^{-1} & \boldsymbol{0} \\ \boldsymbol{0} & \boldsymbol{0} \end{bmatrix} + \begin{bmatrix} -\boldsymbol{E} \\ \boldsymbol{I} \end{bmatrix} [\boldsymbol{D} - \boldsymbol{B}'\boldsymbol{A}^{-1}\boldsymbol{B}]^{-1}[-\boldsymbol{E} \quad \boldsymbol{I}],$$

where $\boldsymbol{E} = \boldsymbol{A}^{-1}\boldsymbol{B}$.

Remarks

(1) The result (I) follows on using (A.11) and (A.12). As a simple application, note that, for $p = 2$, we have

$$\boldsymbol{A}^{-1} = \frac{1}{a_{11}a_{22} - a_{12}a_{21}} \begin{bmatrix} a_{22} & -a_{12} \\ -a_{21} & a_{11} \end{bmatrix}.$$

(2) Formulae (II)–(VI) can be verified by checking that the product of the matrix and its inverse reduces to the identity matrix; e.g., to verify (III), we proceed

$$(\boldsymbol{AB})^{-1}(\boldsymbol{AB}) = \boldsymbol{B}^{-1}\boldsymbol{A}^{-1}(\boldsymbol{AB}) = \boldsymbol{B}^{-1}\boldsymbol{IB} = \boldsymbol{I}.$$

(3) We have assumed $\boldsymbol{A}$ to be a square matrix with $|\boldsymbol{A}| \neq 0$ in defining $\boldsymbol{A}^{-1}$. For $\boldsymbol{A}(n \times p)$, a generalized inverse (g-inverse) is defined in Section A.8.
(4) In computer algorithms for evaluating $\boldsymbol{A}^{-1}$, the following methods are commonly used. If $\boldsymbol{A}$ is symmetric, the Cholesky decomposition is used, namely decomposing $\boldsymbol{A}$ into the form $\boldsymbol{LL}'$ where $\boldsymbol{L}$ is lower triangular and then using $\boldsymbol{A}^{-1} = (\boldsymbol{L}^{-1})'\boldsymbol{L}^{-1}$. For nonsymmetric matrices, Crout's method is used, which is a modification of Gaussian elimination.

A.2.5 Kronecker Products

Definition A.2.5 *Let $\boldsymbol{A} = (a_{ij})$ be an $(m \times n)$ matrix and $\boldsymbol{B} = (b_{kl})$ be a $(p \times q)$ matrix. Then, the* Kronecker product *of $\boldsymbol{A}$ and $\boldsymbol{B}$ is defined as*

$$\begin{bmatrix} a_{11}\boldsymbol{B} & a_{12}\boldsymbol{B} & \cdots & a_{1n}\boldsymbol{B} \\ a_{21}\boldsymbol{B} & a_{22}\boldsymbol{B} & \cdots & a_{2n}\boldsymbol{B} \\ \vdots & \vdots & & \vdots \\ a_{m1}\boldsymbol{B} & a_{m2}\boldsymbol{B} & \cdots & a_{mn}\boldsymbol{B} \end{bmatrix},$$

which is an $(mp \times nq)$ matrix. It is denoted by $\boldsymbol{A} \otimes \boldsymbol{B}$.

Definition A.2.6 *If $\boldsymbol{X}$ is an $(n \times p)$ matrix, let $\boldsymbol{X}^V$ denote the np-vector obtained by "vectorizing" $\boldsymbol{X}$, that is by stacking the columns of $\boldsymbol{X}$ on top of one another, so that*

$$\boldsymbol{X}^V = \begin{pmatrix} \boldsymbol{x}_{(1)} \\ \boldsymbol{x}_{(2)} \\ \vdots \\ \boldsymbol{x}_{(p)} \end{pmatrix}.$$

From these definitions, the elementary properties given below easily follow:

(I) $\alpha(\boldsymbol{A} \otimes \boldsymbol{B}) = (\alpha \boldsymbol{A}) \otimes \boldsymbol{B} = \boldsymbol{A} \otimes (\alpha \boldsymbol{B})$ for all scalar α and hence can be written without ambiguity as $\alpha \boldsymbol{A} \otimes \boldsymbol{B}$.
(II) $\boldsymbol{A} \otimes (\boldsymbol{B} \otimes \boldsymbol{C}) = (\boldsymbol{A} \otimes \boldsymbol{B}) \otimes \boldsymbol{C}$. Hence, this can be written as $\boldsymbol{A} \otimes \boldsymbol{B} \otimes \boldsymbol{C}$.
(III) $(\boldsymbol{A} \otimes \boldsymbol{B})' = \boldsymbol{A}' \otimes \boldsymbol{B}'$.
(IV) $(\boldsymbol{A} \otimes \boldsymbol{B})(\boldsymbol{F} \otimes \boldsymbol{G}) = (\boldsymbol{AF}) \otimes (\boldsymbol{BG})$. Here, the parentheses are necessary.
(V) $(\boldsymbol{A} \otimes \boldsymbol{B})^{-1} = \boldsymbol{A}^{-1} \otimes \boldsymbol{B}^{-1}$ for nonsingular $\boldsymbol{A}$ and $\boldsymbol{B}$.
(VI) $(\boldsymbol{A} + \boldsymbol{B}) \otimes \boldsymbol{C} = \boldsymbol{A} \otimes \boldsymbol{C} + \boldsymbol{B} \otimes \boldsymbol{C}$.
(VII) $\boldsymbol{A} \otimes (\boldsymbol{B} + \boldsymbol{C}) = \boldsymbol{A} \otimes \boldsymbol{B} + \boldsymbol{A} \otimes \boldsymbol{C}$.
(VIII) $(\boldsymbol{AXB})^V = (\boldsymbol{B}' \otimes \boldsymbol{A})\boldsymbol{X}^V$.
(IX) $\operatorname{tr}(\boldsymbol{A} \otimes \boldsymbol{B}) = (\operatorname{tr} \boldsymbol{A})(\operatorname{tr} \boldsymbol{B})$.
(X) $|\boldsymbol{A} \otimes \boldsymbol{B}| = |\boldsymbol{A}|^p |\boldsymbol{B}|^n$ for square matrices $\boldsymbol{A}(n \times n)$and $\boldsymbol{B}(p \times p)$.
(XI) $\boldsymbol{A} \otimes \boldsymbol{B} \neq \boldsymbol{B} \otimes \boldsymbol{A}$ in general.

A.3 Further Particular Matrices and Types of Matrices

Table A.3 gives another list of some important types of matrices. Here, $\boldsymbol{J} = \boldsymbol{1}\boldsymbol{1}'$ is the outer product of $\boldsymbol{1}$ with itself. We consider a few in more detail.

A.3.1 Orthogonal Matrices

A square matrix $\boldsymbol{A}(n \times n)$ is *orthogonal* if $\boldsymbol{AA}' = \boldsymbol{I}$. The following properties hold:

(I) $\boldsymbol{A}^{-1} = \boldsymbol{A}'$.
(II) $\boldsymbol{A}'\boldsymbol{A} = \boldsymbol{I}$.
(III) $|\boldsymbol{A}| = \pm 1$.
(IV) $\boldsymbol{a}_i'\boldsymbol{a}_j = 0, i \neq j; \boldsymbol{a}_i'\boldsymbol{a}_i = 1$. $\boldsymbol{a}_{(i)}'\boldsymbol{a}_{(j)} = 0, i \neq j; \boldsymbol{a}_{(i)}'\boldsymbol{a}_{(i)} = 1$.
(V) $\boldsymbol{C} = \boldsymbol{AB}$ is orthogonal if $\boldsymbol{A}$ and $\boldsymbol{B}$ are orthogonal.

Remarks

(1) All of these properties follow easily from the definition $\boldsymbol{AA}' = \boldsymbol{I}$. Result (IV) states that the sum of squares of the elements in each row (column) is unity, whereas the sum of the cross-products of the elements in any two rows (columns) is zero.
(2) The *Helmert matrix* is a particular orthogonal matrix whose columns are defined by

$$\begin{aligned} \boldsymbol{a}_{(1)}' &= (n^{-1/2}, \dots, n^{-1/2}), \\ \boldsymbol{a}_{(j)}' &= (\underbrace{d_j, \dots, d_j}_{j-1 \text{ times}}, -(j-1)d_j, 0, \dots, 0), \qquad j = 2, \dots, n, \end{aligned} \tag{A.32}$$

where

$$d_j = \{j(j-1)\}^{-1/2}. \tag{A.33}$$

(3) Orthogonal matrices can be used to represent a change of basis, or rotation. See Section A.5.

Table A.3 Particular types of matrices (List 2).

Name	Definition	Examples	Section
Nonsingular	$\|A\| \neq 0$	$\begin{bmatrix} 1 & 2 \\ 0 & 1 \end{bmatrix}$	A.2.3
Singular	$\|A\| = 0$	$\begin{bmatrix} 1 & 2 \\ 1 & 2 \end{bmatrix}$	A.2.3
Orthogonal	$\boldsymbol{AA}' = \boldsymbol{A}'\boldsymbol{A} = \boldsymbol{I}$	$\begin{bmatrix} \cos\theta & -\sin\theta \\ \sin\theta & \cos\theta \end{bmatrix}$	A.3.1
Equicorrelation	$\boldsymbol{E} = (1-\rho)\boldsymbol{I} + \rho\boldsymbol{J}$	$\begin{bmatrix} 1 & \rho \\ \rho & 1 \end{bmatrix}$	A.3.2
Idempotent	$\boldsymbol{A}^2 = \boldsymbol{A}$	$\frac{1}{2}\begin{bmatrix} 1 & -1 \\ -1 & 1 \end{bmatrix}$	
Centering matrix, $\boldsymbol{H}_n$	$\boldsymbol{H}_n = \boldsymbol{I}_n - n^{-1}\boldsymbol{J}_n$		A.3.3
Positive definite (p.d.)	$\boldsymbol{x}'\boldsymbol{Ax} > 0$ for all $\boldsymbol{x} \neq \boldsymbol{0}$	$x_1^2 + x_2^2$	A.7
Positive semidefinite (p.s.d.)	$\boldsymbol{x}'\boldsymbol{Ax} \geq 0$ for all $\boldsymbol{x} \neq \boldsymbol{0}$	$(x_1 - x_2)^2$	A.7

A.3.2 Equicorrelation Matrix

Consider the $(p \times p)$ matrix defined by

$$\boldsymbol{E} = (1-\rho)\boldsymbol{I} + \rho\boldsymbol{J}, \tag{A.34}$$

where ρ is any real number, and $\boldsymbol{J} = \boldsymbol{11}'$. Then, $e_{ii} = 1, e_{ij} = \rho$, for $i \neq j$. For statistical purposes, this matrix is most useful for $-(p-1)^{-1} < \rho < 1$, when it is called the *equicorrelation* matrix.

Direct verification shows that, provided $\rho \neq 1, -(p-1)^{-1}$, $\boldsymbol{E}^{-1}$ exists and is given by

$$\boldsymbol{E}^{-1} = (1-\rho)^{-1}[\boldsymbol{I} - \rho\{1 + (p-1)\rho\}^{-1}\boldsymbol{J}]. \tag{A.35}$$

Its determinant is given by

$$|\boldsymbol{E}| = (1-\rho)^{p-1}\{1 + \rho(p-1)\}. \tag{A.36}$$

This formula is most easily verified using the eigenvalues given in Remark (6) of Section A.6.

A.3.3 Centering Matrix

The $(n \times n)$ centering matrix is defined by $\boldsymbol{H} = \boldsymbol{H}_n = \boldsymbol{I} - n^{-1}\boldsymbol{J}$, where $\boldsymbol{J} = \boldsymbol{11}'$.Then:

(I) $\boldsymbol{H}' = \boldsymbol{H},\ \boldsymbol{H}^2 = \boldsymbol{H}$.
(II) $\boldsymbol{H1} = \boldsymbol{0},\ \boldsymbol{HJ} = \boldsymbol{JH} = \boldsymbol{0}$.
(III) $\boldsymbol{Hx} = \boldsymbol{x} - \overline{x}\boldsymbol{1}$, where $\overline{x} = n^{-1}\sum x_i$.
(IV) $\boldsymbol{x}'\boldsymbol{Hx} = \sum (x_i - \overline{x})^2$.

Remarks

(1) Property (I) states that $\boldsymbol{H}$ is symmetric and idempotent.
(2) Property (III) is most important in data analysis. The ith element of $\boldsymbol{Hx}$ is $x_i - \bar{x}$. Therefore, premultiplying a column vector by $\boldsymbol{H}$ has the effect of reexpressing the elements of the vector as *deviations from the mean*. Similarly, premultiplying a matrix by $\boldsymbol{H}$ reexpresses each element of the matrix as a *deviation from its column mean*, i.e. $\boldsymbol{HX}$ has its (i,j)th element $x_{ij} - \bar{x}_j$, where $\bar{x}_j$ is the mean of the jth column of $\boldsymbol{X}$. This "centering" property explains the nomenclature for $\boldsymbol{H}$.

A.4 Vector Spaces, Rank, and Linear Equations

A.4.1 Vector Spaces

The set of vectors in $\mathbb{R}^n$ satisfies the following properties. For all $\boldsymbol{x}, \boldsymbol{y} \in \mathbb{R}^n$ and all $\lambda, \mu \in \mathbb{R}$,

1. $\lambda(\boldsymbol{x} + \boldsymbol{y}) = \lambda\boldsymbol{x} + \lambda\boldsymbol{y}$,
2. $(\lambda + \mu)\boldsymbol{x} = \lambda\boldsymbol{x} + \mu\boldsymbol{x}$,
3. $(\lambda\mu)\boldsymbol{x} = \lambda(\mu\boldsymbol{x})$,
4. $1\boldsymbol{x} = \boldsymbol{x}$.

Thus, $\mathbb{R}^n$ can be considered as a *vector space* over the real numbers $\mathbb{R}$.

Definition A.4.1 *If W is a subset of $\mathbb{R}^n$ such that for all $\boldsymbol{x}, \boldsymbol{y} \in W$ and $\lambda \in \mathbb{R}$*

$$\lambda(\boldsymbol{x} + \boldsymbol{y}) \in W,$$

then W is called a vector subspace *of $\mathbb{R}^n$.*

Two simple examples of subspaces of $\mathbb{R}^n$ are $\{\boldsymbol{0}\}$ and $\mathbb{R}^n$ itself.

Definition A.4.2 *Vectors $\boldsymbol{x}_1, \ldots, \boldsymbol{x}_k$ are called* linearly dependent *if there exist numbers $\lambda_1, \ldots, \lambda_k$, not all zero, such that*

$$\lambda_1\boldsymbol{x}_1 + \cdots + \lambda_k\boldsymbol{x}_k = \boldsymbol{0}.$$

Otherwise, the k vectors are linearly independent.

Definition A.4.3 *Let W be a subspace of $\mathbb{R}^n$. Then, a* basis *of W is a maximal linearly independent set of vectors.*

The following properties hold for a basis of W:

(I) Every basis of W contains the same (finite) number of elements. This number is called the *dimension* of W and denoted $\dim W$. In particular, $\dim \mathbb{R}^n = n$.
(II) If $\boldsymbol{x}_1, \ldots, \boldsymbol{x}_k$ is a basis for W, then every element $\boldsymbol{x}$ in W can be expressed as a linear combination of $\boldsymbol{x}_1, \ldots, \boldsymbol{x}_k$; that is, $\boldsymbol{x} = \lambda_1\boldsymbol{x}_1 + \cdots + \lambda_k\boldsymbol{x}_k$ for some numbers $\lambda_1, \ldots, \lambda_k$.

Definition A.4.4 *The* inner *(or* scalar *or* dot*)* product *between two vectors* $\mathbf{x}, \mathbf{y} \in \mathbb{R}^n$ *is defined by*

$$\mathbf{x} \cdot \mathbf{y} = \mathbf{x}'\mathbf{y} = \sum_{i=1}^{n} x_i y_i.$$

The vectors $\mathbf{x}$ *and* $\mathbf{y}$ *are called* orthogonal *if* $\mathbf{x} \cdot \mathbf{y} = 0$.

Definition A.4.5 *The* norm *of a vector* $\mathbf{x} \in \mathbb{R}^n$ *is given by*

$$||\mathbf{x}|| = (\mathbf{x} \cdot \mathbf{x})^{1/2} = \left(\sum x_i^2\right)^{1/2}.$$

Then, the distance *between two vectors* $\mathbf{x}$ *and* $\mathbf{y}$ *is given by*

$$||\mathbf{x} - \mathbf{y}||.$$

Definition A.4.6 *A* basis $\mathbf{x}_1, \ldots, \mathbf{x}_k$ *of a subspace W of* $\mathbb{R}^n$ *is called* orthonormal *if all the elements have norm 1 and are orthogonal to one another; that is, if*

$$\mathbf{x}_i'\mathbf{x}_j = \begin{cases} 1, & i = j, \\ 0, & i \neq j. \end{cases}$$

In particular, if $\mathbf{A}(n \times n)$ is an orthogonal matrix, then the columns of $\mathbf{A}$ form an orthonormal basis of $\mathbb{R}^n$.

A.4.2 Rank

Definition A.4.7 *The* rank *of a matrix* $\mathbf{A}(n \times p)$ *is defined as the maximum number of linearly independent rows (columns) in* $\mathbf{A}$.

We denote it by $r(\mathbf{A})$ or $\text{rank}(\mathbf{A})$.

The following properties hold:

(I)
$$0 \leq r(\mathbf{A}) \leq \min(n, p). \tag{A.37}$$

(II)
$$r(\mathbf{A}) = r(\mathbf{A}'). \tag{A.38}$$

(III)
$$r(\mathbf{A} + \mathbf{B}) \leq r(\mathbf{A}) + r(\mathbf{B}). \tag{A.39}$$

(IV)
$$r(\mathbf{AB}) \leq \min\{r(\mathbf{A}), r(\mathbf{B})\}. \tag{A.40}$$

(V)
$$r(\mathbf{A}'\mathbf{A}) = r(\mathbf{AA}') = r(\mathbf{A}). \tag{A.41}$$

(VI)
$$\text{If } \mathbf{B}(n \times n) \text{ and } \mathbf{C}(p \times p) \text{ are nonsingular, then } r(\mathbf{BAC}) = r(\mathbf{A}). \tag{A.42}$$

(VII)
$$\text{If } n = p, \text{ then } r(\mathbf{A}) = p \text{ if and only if } \mathbf{A} \text{ is nonsingular.} \tag{A.43}$$

Table A.4 gives the ranks of some particular matrices.

Table A.4 Rank of some matrices.

Matrix	Rank
Nonsingular $\boldsymbol{A}(p \times p)$	p
$\text{diag}(a_i)$	Number of nonzero a_i
$\boldsymbol{H}_n$	$n-1$
Idempotent $\boldsymbol{A}$	$\text{tr}\,\boldsymbol{A}$
$\boldsymbol{CAB}$, nonsingular $\boldsymbol{B}, \boldsymbol{C}$	$r(\boldsymbol{A})$

Remarks

(1) Another definition of $r(\boldsymbol{A})$ is $r(\boldsymbol{A})$ = the largest order of those (square) submatrices that have nonvanishing determinants.

(2) If we define $M(\boldsymbol{A})$ as the vector subspace in $\mathbb{R}^n$ spanned by the columns of $\boldsymbol{A}$, then $r(\boldsymbol{A}) = \dim M(\boldsymbol{A})$, and we may choose linearly independent columns of $\boldsymbol{A}$ as a basis for $M(\boldsymbol{A})$. Note that for any p-vector $\boldsymbol{x}$, $\boldsymbol{Ax} = x_1\boldsymbol{a}_{(1)} + \cdots x_p\boldsymbol{a}_{(p)}$ is a linear combination of the columns of $\boldsymbol{A}$, and hence, $\boldsymbol{Ax}$ lies in $M(\boldsymbol{A})$.

(3) Define the *null space* of $\boldsymbol{A}(n \times p)$ by

$$N(\boldsymbol{A}) = \{\boldsymbol{x} \in \mathbb{R}^p : \boldsymbol{Ax} = \boldsymbol{0}\}.$$

Then, $N(\boldsymbol{A})$ is a vector subspace of $\mathbb{R}^n$ of dimension k, say. Let $\boldsymbol{e}_1, \ldots, \boldsymbol{e}_p$ be a basis of $\mathbb{R}^p$ for which $\boldsymbol{e}_1, \ldots, \boldsymbol{e}_k$ are a basis of $N(\boldsymbol{A})$. Then, $\boldsymbol{Ae}_{k+1}, \ldots, \boldsymbol{Ae}_p$ form a maximally linearly independent set of vectors in $M(\boldsymbol{A})$ and hence are a basis for $M(\boldsymbol{A})$. Thus, we get the important result

$$\dim N(\boldsymbol{A}) + \dim M(\boldsymbol{A}) = p. \tag{A.44}$$

(4) To prove (V), note that if $\boldsymbol{Ax} = \boldsymbol{0}$, then $\boldsymbol{A}'\boldsymbol{Ax} = \boldsymbol{0}$; conversely, if $\boldsymbol{A}'\boldsymbol{Ax} = \boldsymbol{0}$, then $\boldsymbol{x}'\boldsymbol{A}'\boldsymbol{Ax} = ||\boldsymbol{Ax}||^2 = 0$, and so, $\boldsymbol{Ax} = \boldsymbol{0}$. Thus, $N(\boldsymbol{A}) = N(\boldsymbol{A}'\boldsymbol{A})$. Since $\boldsymbol{A}$ and $\boldsymbol{A}'\boldsymbol{A}$ each have p columns, we see from (A.44) that $\dim M(\boldsymbol{A}) = \dim M(\boldsymbol{A}'\boldsymbol{A})$, so that $r(\boldsymbol{A}) = r(\boldsymbol{A}'\boldsymbol{A})$.

(5) If $\boldsymbol{A}$ is symmetric, its rank equals the number of nonzero eigenvalues of $\boldsymbol{A}$. For general $\boldsymbol{A}(n \times p)$, the rank is given by the number of nonzero eigenvalues of $\boldsymbol{A}'\boldsymbol{A}$. See Section A.6.

A.4.3 Linear Equations

For the n linear equations

$$x_1\boldsymbol{a}_{(1)} + \cdots x_p\boldsymbol{a}_{(p)} = \boldsymbol{b} \tag{A.45}$$

or

$$\boldsymbol{Ax} = \boldsymbol{b} \tag{A.46}$$

with the coefficient matrix $\boldsymbol{A}(n \times p)$, we note the following results:

(I) If $n = p$, and $\boldsymbol{A}$ is nonsingular, the unique solution is

$$\boldsymbol{x} = \boldsymbol{A}^{-1}\boldsymbol{b} = \frac{1}{|\boldsymbol{A}|}[A_{ij}]'\boldsymbol{b}. \tag{A.47}$$

(II) The equation is *consistent* (i.e. admits at least one solution) if and only if

$$r(\boldsymbol{A}) = r([\boldsymbol{A}, \boldsymbol{b}]). \tag{A.48}$$

(III) For $\boldsymbol{b} = \boldsymbol{0}$, there exists a nontrivial solution (i.e. $\boldsymbol{x} \neq \boldsymbol{0}$) if and only if $r(\boldsymbol{A}) < p$.

(IV) The equation

$$\boldsymbol{A}'\boldsymbol{A}\boldsymbol{x} = \boldsymbol{A}'\boldsymbol{b} \tag{A.49}$$

is always consistent.

Remarks

(1) To prove (II), note that the vector $\boldsymbol{Ax}$ is a linear combination of the columns of $\boldsymbol{A}$. Thus, the equation $\boldsymbol{Ax} = \boldsymbol{b}$ has a solution if and only if $\boldsymbol{b}$ can be expressed as a linear combination of the columns of $\boldsymbol{A}$.
(2) The proof of (III) is immediate from the definition of rank.
(3) To prove (IV), note that $M(\boldsymbol{A}'\boldsymbol{A}) \subseteq M(\boldsymbol{A}')$ because $\boldsymbol{A}'\boldsymbol{A}$ is a matrix whose columns are linear combinations of the columns of $\boldsymbol{A}'$. From Remark (4) of Section A.4.2, we see that $\dim M(\boldsymbol{A}'\boldsymbol{A}) = \dim M(\boldsymbol{A}) = \dim M(\boldsymbol{A}')$, and hence, $M(\boldsymbol{A}'\boldsymbol{A}) = M(\boldsymbol{A}')$. Thus, $\boldsymbol{A}'\boldsymbol{b} \in M(\boldsymbol{A}'\boldsymbol{A})$, and so, $r(\boldsymbol{A}'\boldsymbol{A}) = r([\boldsymbol{A}'\boldsymbol{A}, \boldsymbol{A}'\boldsymbol{b}])$.

A.5 Linear Transformations

Definition A.5.1 *The transformation from $\boldsymbol{x}(p \times 1)$ to $\boldsymbol{y}(n \times 1)$ given by*

$$\boldsymbol{y} = \boldsymbol{Ax} + \boldsymbol{b}, \tag{A.50}$$

where $\boldsymbol{A}$ is an $(n \times p)$ matrix, is called a linear transformation. *For $n = p$, the transformation is called* nonsingular *if $\boldsymbol{A}$ is nonsingular, and in this case the inverse transformation is*

$$\boldsymbol{x} = \boldsymbol{A}^{-1}(\boldsymbol{y} - \boldsymbol{b}).$$

An orthogonal *transformation is defined by*

$$\boldsymbol{y} = \boldsymbol{Ax}, \tag{A.51}$$

where $\boldsymbol{A}$ is an orthogonal matrix. Geometrically, an orthogonal matrix represents a rotation of the coordinate axes. See Section A.10.

A.6 Eigenvalues and Eigenvectors

A.6.1 General Results

If $\boldsymbol{A}(p \times p)$ is any square matrix, then

$$q(\lambda) = |\boldsymbol{A} - \lambda \boldsymbol{I}| \tag{A.52}$$

is a pth order polynomial in λ. The p roots of $q(\lambda)$, $\lambda_1, \ldots, \lambda_p$, possibly complex numbers, are called *eigenvalues* of $\boldsymbol{A}$. Some of the λ_i will be equal if $q(\lambda)$ has multiple roots.

For each $i = 1, \ldots, p$, $|\boldsymbol{A} - \lambda_i \boldsymbol{I}| = 0$, so $\boldsymbol{A} - \lambda_i \boldsymbol{I}$ is singular. Hence, there exists a nonzero vector $\boldsymbol{\gamma}$ satisfying

$$\boldsymbol{A\gamma} = \lambda_i \boldsymbol{\gamma}. \tag{A.53}$$

Any vector satisfying (A.53) is called a *(right) eigenvector* of $\boldsymbol{A}$ for the eigenvalue λ_i. If λ_i is complex, then $\boldsymbol{\gamma}$ may have complex entries. An eigenvector $\boldsymbol{\gamma}$ with real entries is called *standardized* if

$$\boldsymbol{\gamma}'\boldsymbol{\gamma} = 1. \tag{A.54}$$

If $\boldsymbol{x}$ and $\boldsymbol{y}$ are eigenvectors for λ_i and $\alpha \in \mathbb{R}$, then $\boldsymbol{x} + \boldsymbol{y}$ and $\alpha\boldsymbol{x}$ are also eigenvectors for λ_i. Thus, the set of all eigenvectors for λ_i forms a subspace, which is called the *eigenspace* of $\boldsymbol{A}$ for λ_i.

Since the coefficient of λ^p in $q(\lambda)$ is $(-1)^p$, we can write $q(\lambda)$ in terms of its roots as

$$q(\lambda) = \prod_{i=1}^{p} (\lambda_i - \lambda). \tag{A.55}$$

Setting $\lambda = 0$ in (A.52) and (A.55) gives

$$|\boldsymbol{A}| = \prod \lambda_i; \tag{A.56}$$

that is, $|\boldsymbol{A}|$ is the product of the eigenvalues of $\boldsymbol{A}$. Similarly, matching the coefficient of λ in (A.52) and (A.55) gives

$$\sum a_{ii} = \operatorname{tr} \boldsymbol{A} = \sum \lambda_i; \tag{A.57}$$

that is, $\operatorname{tr} \boldsymbol{A}$ is the sum of the eigenvalues of $\boldsymbol{A}$.

Let $\boldsymbol{C}(p \times p)$ be a nonsingular matrix. Then,

$$|\boldsymbol{A} - \lambda\boldsymbol{I}| = |\boldsymbol{C}|\ |\boldsymbol{A} - \lambda\boldsymbol{C}^{-1}\boldsymbol{C}|\ |\boldsymbol{C}^{-1}| = |\boldsymbol{C}\boldsymbol{A}\boldsymbol{C}^{-1} - \lambda\boldsymbol{I}|. \tag{A.58}$$

Thus, $\boldsymbol{A}$ and $\boldsymbol{C}\boldsymbol{A}\boldsymbol{C}^{-1}$ have the same eigenvalues. Further, if $\boldsymbol{\gamma}$ is an eigenvector of $\boldsymbol{A}$ for λ_i, then $\boldsymbol{C}\boldsymbol{A}\boldsymbol{C}^{-1}(\boldsymbol{C}\boldsymbol{\gamma}) = \lambda_i\boldsymbol{C}\boldsymbol{\gamma}$, so that

$$\boldsymbol{\nu} = \boldsymbol{C}\boldsymbol{\gamma}$$

is an eigenvector of $\boldsymbol{C}\boldsymbol{A}\boldsymbol{C}^{-1}$ for λ_i.

Let $\alpha \in \mathbb{R}$. Then, $|\boldsymbol{A} + \alpha\boldsymbol{I} - \lambda\boldsymbol{I}| = |\boldsymbol{A} - (\lambda - \alpha)\boldsymbol{I}|$, so that $\boldsymbol{A} + \alpha\boldsymbol{I}$ has eigenvalues $\lambda_i + \alpha$. Further, if $\boldsymbol{A}\boldsymbol{\gamma} = \lambda_i\boldsymbol{\gamma}$, then $(\boldsymbol{A} + \alpha\boldsymbol{I})\boldsymbol{\gamma} = (\lambda_i + \alpha)\boldsymbol{\gamma}$, so that $\boldsymbol{A}$ and $\boldsymbol{A} + \alpha\boldsymbol{I}$ have the same eigenvectors.

Bounds on the dimension of the eigenspace of $\boldsymbol{A}$ for λ_i are given by the following theorem.

Theorem A.6.1 *Let λ_1 denote any particular eigenvalue of $\boldsymbol{A}(p \times p)$, with eigenspace H of dimension r. If k denotes the multiplicity of λ_1 in $q(\lambda)$, then $1 \le r \le k$.*

Proof: Since λ_1 is an eigenvalue, there is at least one nontrivial eigenvector, so $r \ge 1$.

Now, let $\boldsymbol{e}_1, \dots, \boldsymbol{e}_r$ be an orthonormal basis of H and extend it so that $\boldsymbol{e}_1, \dots, \boldsymbol{e}_r, \boldsymbol{f}_1, \dots, \boldsymbol{f}_{p-r}$ is an orthonormal basis of $\mathbb{R}^p$. Write $\boldsymbol{E} = [\boldsymbol{e}_1, \dots, \boldsymbol{e}_r], \boldsymbol{F} = [\boldsymbol{f}_1, \dots, \boldsymbol{f}_{p-r}]$. Then, $[\boldsymbol{E}, \boldsymbol{F}]$ is an orthogonal matrix, so that $\boldsymbol{I}_p = [\boldsymbol{E}, \boldsymbol{F}][\boldsymbol{E}, \boldsymbol{F}]' = \boldsymbol{E}\boldsymbol{E}' + \boldsymbol{F}\boldsymbol{F}'$ and $|[\boldsymbol{E}, \boldsymbol{F}]| = 1$. Also, $\boldsymbol{E}'\boldsymbol{A}\boldsymbol{E} = \lambda_1\boldsymbol{E}'\boldsymbol{E} = \lambda_1\boldsymbol{I}_r, \boldsymbol{F}'\boldsymbol{F} = \boldsymbol{I}_{p-r}$ and $\boldsymbol{F}'\boldsymbol{A}\boldsymbol{E} = \lambda_1\boldsymbol{F}'\boldsymbol{E} = \boldsymbol{0}$. Thus,

$$\begin{aligned} q(\lambda) &= |\boldsymbol{A} - \lambda\boldsymbol{I}| = |[\boldsymbol{E}, \boldsymbol{F}]'|\ |\boldsymbol{A} - \lambda\boldsymbol{I}|\ |[\boldsymbol{E}, \boldsymbol{F}]| \\ &= |[\boldsymbol{E}, \boldsymbol{F}]'(\boldsymbol{A}\boldsymbol{E}\boldsymbol{E}' + \boldsymbol{A}\boldsymbol{F}\boldsymbol{F}' - \lambda\boldsymbol{E}\boldsymbol{E}' - \lambda\boldsymbol{F}\boldsymbol{F}')[\boldsymbol{E}, \boldsymbol{F}]| \end{aligned}$$

$$= \begin{vmatrix} (\lambda_1 - \lambda)\boldsymbol{I}_r & \boldsymbol{E}'\boldsymbol{A}\boldsymbol{F} \\ \boldsymbol{0} & \boldsymbol{F}'\boldsymbol{A}\boldsymbol{F} - \lambda\boldsymbol{I}_{p-r} \end{vmatrix}$$
$$= (\lambda_1 - \lambda)^r q_1(\lambda), \quad \text{say},$$

using (A.16). Thus, the multiplicity of λ_1 as a root of $q(\lambda)$ is at least r. □

Remarks

(1) If $\boldsymbol{A}$ is symmetric, then $r = k$; see Section A.6.2. However, if $\boldsymbol{A}$ is not symmetric, it is possible that $r < k$. For example,

$$\boldsymbol{A} = \begin{bmatrix} 0 & 1 \\ 0 & 0 \end{bmatrix}$$

has eigenvalue 0 with multiplicity 2; however, the corresponding eigenspace which is generated by $(1, 0)'$ only has dimension 1.

(2) If $r = 1$, then the eigenspace for λ_1 has dimension 1, and the standardized eigenvector for λ_1 is unique (up to sign).

Now, let $\boldsymbol{A}(n \times p)$ and $\boldsymbol{B}(p \times n)$ be any two matrices, and suppose that $n \geq p$. Then, from (A.17)

$$\begin{vmatrix} -\lambda\boldsymbol{I}_n & -\boldsymbol{A} \\ \boldsymbol{B} & \boldsymbol{I}_p \end{vmatrix} = (-\lambda)^{n-p}|\boldsymbol{B}\boldsymbol{A} - \lambda\boldsymbol{I}_p| = |\boldsymbol{A}\boldsymbol{B} - \lambda\boldsymbol{I}_n|. \tag{A.59}$$

Hence, the eigenvalues of $\boldsymbol{AB}$ equal the p eigenvalues of $\boldsymbol{BA}$, together with the eigenvalue 0, $n - p$ times. The following theorem describes the relationship between the eigenvectors.

Theorem A.6.2 *For $\boldsymbol{A}(n \times p)$ and $\boldsymbol{B}(p \times n)$, the nonzero eigenvalues of $\boldsymbol{AB}$ and $\boldsymbol{BA}$ are the same and have the same multiplicity. If $\boldsymbol{x}$ is a nontrivial eigenvector of $\boldsymbol{AB}$ for an eigenvalue $\lambda \neq 0$, then $\boldsymbol{y} = \boldsymbol{Bx}$ is a nontrivial eigenvector of $\boldsymbol{BA}$.*

Proof: The first part follows from (A.59). For the second part substituting $\boldsymbol{y} = \boldsymbol{Bx}$ in the equation, $\boldsymbol{B}(\boldsymbol{ABx}) = \lambda\boldsymbol{Bx}$ gives $\boldsymbol{BAy} = \lambda\boldsymbol{y}$. The vector $\boldsymbol{x}$ is nontrivial if $\boldsymbol{x} \neq \boldsymbol{0}$. Since $\boldsymbol{Ay} = \boldsymbol{ABx} = \lambda\boldsymbol{x} \neq \boldsymbol{0}$, it follows that $\boldsymbol{y} \neq \boldsymbol{0}$ also. □

Corollary A.6.3 *For $\boldsymbol{A}(n \times p)$, $\mathbf{B}(q \times n)$, $\boldsymbol{a}(p \times 1)$, and $\mathbf{b}(q \times 1)$, the matrix $\boldsymbol{A}\boldsymbol{a}\mathbf{b}'\mathbf{B}$ has rank at most 1. The nonzero eigenvalue, if present, equals $\mathbf{b}'\mathbf{B}\boldsymbol{A}\boldsymbol{a}$, with eigenvector $\boldsymbol{A}\boldsymbol{a}$.*

Proof: The nonzero eigenvalue of $\boldsymbol{A}\boldsymbol{a}\boldsymbol{b}'\boldsymbol{B}$ equals that of $\boldsymbol{b}'\boldsymbol{B}\boldsymbol{A}\boldsymbol{a}$, which is a scalar, and hence is its own eigenvalue. The fact that $\boldsymbol{A}\boldsymbol{a}$ is a corresponding eigenvector is easily checked. □

A.6.2 Symmetric Matrices

If $\boldsymbol{A}$ is symmetric, it is possible to give more detailed information about its eigenvalues and eigenvectors.

Theorem A.6.4 *All the eigenvalues of a symmetric matrix $\boldsymbol{A}(p \times p)$ are real.*

Proof: If possible, let

$$\gamma = x + iy, \qquad \lambda = a + ib, \qquad \gamma \neq \mathbf{0}. \tag{A.60}$$

From (A.53), after equating the real and imaginary parts, we have

$$Ax = ax - by, \qquad Ay = bx + ay.$$

On premultiplying by y' and x', respectively, and subtracting, we obtain $b = 0$. Hence, from (A.60), λ is real. □

In the above discussion, we can choose $y = \mathbf{0}$, so we can assume γ to be real.

Theorem A.6.5 (Spectral decomposition theorem or Jordan decomposition theorem). *Any symmetric matrix $A(p \times p)$ can be written as*

$$A = \Gamma\Lambda\Gamma' = \sum \lambda_i \gamma_{(i)}\gamma'_{(i)}, \tag{A.61}$$

where Λ is a diagonal matrix of eigenvalues of A, and Γ is an orthogonal matrix whose columns are standardized eigenvectors.

Proof: Suppose that we can find orthonormal vectors $\gamma_{(1)}, \dots, \gamma_{(p)}$ such that $A\gamma_{(i)} = \lambda_i\gamma_{(i)}$ for some numbers λ_i. Then,

$$\gamma'_{(i)}A\gamma_{(j)} = \lambda_j\gamma'_{(i)}\gamma_{(j)} = \begin{cases} \lambda_i, & i = j, \\ 0, & i \neq j, \end{cases}$$

or in matrix form

$$\Gamma'A\Gamma = \Lambda. \tag{A.62}$$

Pre- and postmultiplying by Γ and Γ' gives (A.61). From (A.58), A and Λ have the same eigenvalues, so the elements of Λ are exactly the eigenvalues of A with the same multiplicities.

Thus, we must find an orthonormal basis of eigenvectors. Note that if $\lambda_i \neq \lambda_j$ are distinct eigenvalues with eigenvectors x and y, respectively, then $\lambda_i x'y = x'Ay = y'Ax = \lambda_j y'x$, so that $y'x = 0$. *Hence, for a symmetric matrix, eigenvectors corresponding to distinct eigenvalues are orthogonal to one another.*

Suppose that there are k distinct eigenvalues of A with the corresponding eigenspaces $H_1, \dots, H_k$ of dimensions $r_1, \dots, r_k$. Let

$$r = \sum_{j=1}^{k} r_j.$$

Since distinct eigenspaces are orthogonal, there exists an orthonormal set of vectors $e_1, \dots, e_r$ such that the vectors labeled

$$\sum_{i=1}^{j-1} r_i + 1, \dots, \sum_{i=1}^{j} r_i$$

form a basis for H_j. From Theorem A.6.1, r_j is less than or equal to the multiplicity of the corresponding eigenvalue. Hence, by reordering the eigenvalues λ_i if necessary, we may suppose that

$$Ae_i = \lambda_i e_i, \qquad i = 1, \dots, r,$$

and $r \leq p$. (If all p eigenvalues are distinct, then we know from Theorem A.6.1 that $r = p$.)

If $r = p$, set $\gamma_{(i)} = \mathbf{e}_i$, and the proof follows. We show that the situation $r < p$ leads to a contradiction and therefore cannot arise.

Without loss of generality, we may suppose that all of the eigenvalues of $\mathbf{A}$ are strictly positive. (If not, we can replace $\mathbf{A}$ by $\mathbf{A} + \alpha\mathbf{I}$ for a suitable α, because $\mathbf{A}$ and $\mathbf{A} + \alpha\mathbf{I}$ have the same eigenvectors.) Set

$$\mathbf{B} = \mathbf{A} - \sum_{i=1}^{r} \lambda_i \mathbf{e}_i \mathbf{e}_i'.$$

Then,

$$\operatorname{tr} \mathbf{B} = \operatorname{tr} \mathbf{A} - \sum_{i=1}^{r} \lambda_i (\mathbf{e}_i' \mathbf{e}_i) = \sum_{i=r+1}^{p} \lambda_i > 0,$$

since $r < p$. Thus, $\mathbf{B}$ has at least one nonzero eigenvalue, say θ. Let $\mathbf{x} \neq \mathbf{0}$ be a corresponding eigenvector. Then, for $1 \leq j \leq r$,

$$\theta \mathbf{e}_j' \mathbf{x} = \mathbf{e}_j' \mathbf{B} \mathbf{x} = \left\{ \lambda_j \mathbf{e}_j' - \sum_{i=1}^{r} \lambda_i (\mathbf{e}_j' \mathbf{e}_i) \mathbf{e}_i' \right\} \mathbf{x} = 0,$$

so that $\mathbf{x}$ is orthogonal to $\mathbf{e}_j, j = 1, \ldots, r$. Therefore,

$$\theta \mathbf{x} = \mathbf{B} \mathbf{x} = \left(\mathbf{A} - \sum \lambda_i \mathbf{e}_i \mathbf{e}_i' \right) \mathbf{x} = \mathbf{A} \mathbf{x} - \sum \lambda_i (\mathbf{e}_i' \mathbf{x}) \mathbf{e}_i = \mathbf{A} \mathbf{x},$$

so that $\mathbf{x}$ is an eigenvector of $\mathbf{A}$ also. Thus, $\theta = \lambda_i$ for some i, and $\mathbf{x}$ is a linear combination of some of the $\mathbf{e}_i$, which contradicts the orthogonality between $\mathbf{x}$ and the $\mathbf{e}_i$. □

Corollary A.6.6 *If $\mathbf{A}$ is a nonsingular symmetric matrix, then for any integer n,*

$$\mathbf{\Lambda}^n = \operatorname{diag}(\lambda_i^n) \qquad \textit{and} \qquad \mathbf{A}^n = \mathbf{\Gamma} \mathbf{\Lambda}^n \mathbf{\Gamma}'. \tag{A.63}$$

If all the eigenvalues of $\mathbf{A}$ are positive, then we can define the rational powers

$$\mathbf{A}^{r/s} = \mathbf{\Gamma} \mathbf{\Lambda}^{r/s} \mathbf{\Gamma}', \qquad \textit{where} \qquad \mathbf{\Lambda}^{r/s} = \operatorname{diag}(\lambda_i^{r/s}), \tag{A.64}$$

for integers $s > 0$ and r. If some of the eigenvalues of $\mathbf{A}$ are zero, then (A.63) and (A.64) hold if the exponents are restricted to be nonnegative.

Proof: Since

$$\mathbf{A}^2 = (\mathbf{\Gamma \Lambda \Gamma}')^2 = \mathbf{\Gamma \Lambda \Gamma}' \mathbf{\Gamma \Lambda \Gamma}' = \mathbf{\Gamma \Lambda}^2 \mathbf{\Gamma}'$$

and

$$\mathbf{A}^{-1} = \mathbf{\Gamma \Lambda}^{-1} \mathbf{\Gamma}', \qquad \mathbf{\Lambda}^{-1} = \operatorname{diag}(\lambda_i^{-1}),$$

we see that (A.63) can be easily proved by induction. To check that rational powers make sense, note that

$$(\mathbf{A}^{r/s})^s = \mathbf{\Gamma \Lambda}^{r/s} \mathbf{\Gamma}' \cdots \mathbf{\Gamma \Lambda}^{r/s} \mathbf{\Gamma}' = \mathbf{\Gamma \Lambda}^r \mathbf{\Gamma}' = \mathbf{A}^r.$$ □

Motivated by (A.64), we can define powers of $\mathbf{A}$ for real-valued exponents.

Important special cases of (A.64) are

$$\mathbf{A}^{1/2} = \mathbf{\Gamma \Lambda}^{1/2} \mathbf{\Gamma}', \qquad \mathbf{\Lambda}^{1/2} = \operatorname{diag}(\lambda_i^{1/2}) \tag{A.65}$$

when $\lambda_i \geq 0$ for all i, and

$$\boldsymbol{A}^{-1/2} = \boldsymbol{\Gamma}\boldsymbol{\Lambda}^{-1/2}\boldsymbol{\Gamma}', \qquad \boldsymbol{\Lambda}^{-1/2} = \text{diag}(\lambda_i^{-1/2}) \tag{A.66}$$

when $\lambda_i > 0$ for all i. The decomposition (A.65) is called the *symmetric square root decomposition* of $\boldsymbol{A}$.

Corollary A.6.7 *The rank of* $\boldsymbol{A}$ *equals the number of nonzero eigenvalues.*

Proof: By (A.42), $r(\boldsymbol{A}) = r(\boldsymbol{\Lambda})$, whose rank is easily seen to equal the number of nonzero diagonal elements. □

Remarks

(1) Theorem A.6.5 shows that a symmetric matrix $\boldsymbol{A}$ is uniquely determined by its eigenvalues and eigenvectors, or more specifically by its distinct eigenvalues and corresponding eigenspaces.

(2) Since $\boldsymbol{A}^{1/2}$ has the same eigenvectors as $\boldsymbol{A}$ and has eigenvalues that are given functions of the eigenvalues of $\boldsymbol{A}$, we see that the p.s.d. symmetric square root is uniquely defined by (A.66).

(3) If the λ_i are all distinct and written in decreasing order say, then $\boldsymbol{\Gamma}$ is uniquely determined, up to the signs of its columns.

(4) If $\lambda_{k+1} = \cdots = \lambda_p = 0$, then (A.61) can be written more compactly as

$$\boldsymbol{A} = \boldsymbol{\Gamma}_1\boldsymbol{\Lambda}_1\boldsymbol{\Gamma}_1' = \sum_{k=1}^{k} \lambda_i \boldsymbol{\gamma}_{(i)}\boldsymbol{\gamma}_{(i)}',$$

where $\boldsymbol{\Lambda}_1 = \text{diag}(\lambda_1, \ldots, \lambda_k)$ and $\boldsymbol{\Gamma}_1 = [\boldsymbol{\gamma}_{(1)}, \ldots, \boldsymbol{\gamma}_{(k)}]$.

(5) A symmetric matrix $\boldsymbol{A}$ has rank 1 if and only if

$$\boldsymbol{A} = \boldsymbol{x}\boldsymbol{x}'$$

for some $\boldsymbol{x} \neq \boldsymbol{0}$. Then, the nonzero eigenvalue of $\boldsymbol{A}$ is given by

$$\text{tr}\boldsymbol{A} = \text{tr}\boldsymbol{x}\boldsymbol{x}' = \boldsymbol{x}'\boldsymbol{x},$$

and the corresponding eigenspace is generated by $\boldsymbol{x}$.

(6) Since $\boldsymbol{J} = \boldsymbol{1}\boldsymbol{1}'$ has rank 1 with eigenvalue p and the corresponding eigenvector $\boldsymbol{1}$, we see that the equicorrelation matrix $\boldsymbol{E} = (1-\rho)\boldsymbol{I} + \rho\boldsymbol{J}$ has eigenvalues $\lambda_1 = 1 + (p-1)\rho$ and $\lambda_2 = \cdots = \lambda_p = 1 - \rho$ and the same eigenvectors as $\boldsymbol{J}$. For the eigenvectors $\boldsymbol{\gamma}_{(2)}, \ldots, \boldsymbol{\gamma}_{(p)}$, we can select any standardized set of vectors orthogonal to $\boldsymbol{1}$ and each other. A possible choice for $\boldsymbol{\Gamma}$ is the transpose of the Helmert matrix of Section A.3.1. Multiplying the eigenvalues together yields the formula for $|\boldsymbol{E}|$ given in (A.36).

(7) If $\boldsymbol{A}$ is symmetric and idempotent (that is $\boldsymbol{A} = \boldsymbol{A}'$ and $\boldsymbol{A}^2 = \boldsymbol{A}$), then $\lambda_i = 0$ or 1 for all i, because $\boldsymbol{A} = \boldsymbol{A}^2$ implies $\boldsymbol{\Lambda} = \boldsymbol{\Lambda}^2$.

(8) If $\boldsymbol{A}$ is symmetric and idempotent, then $r(\boldsymbol{A}) = \text{tr}\boldsymbol{A}$. This result follows easily from (A.57) and Corollary A.6.7.

(9) As an example, consider

$$\boldsymbol{A} = \begin{bmatrix} 1 & \rho \\ \rho & 1 \end{bmatrix}. \tag{A.67}$$

The eigenvalues of $\boldsymbol{A}$ from (A.52) are the solutions of

$$\begin{vmatrix} 1-\lambda & \rho \\ \rho & 1-\lambda \end{vmatrix} = 0,$$

namely, $\lambda_1 = 1+\rho$ and $\lambda_2 = 1-\rho$. Thus,

$$\boldsymbol{\Lambda} = \mathrm{diag}(1+\rho, 1-\rho). \tag{A.68}$$

For $\rho \neq 0$, the eigenvector corresponding to $\lambda_1 = 1+\rho$ from (A.53) is

$$\begin{bmatrix} 1 & \rho \\ \rho & 1 \end{bmatrix} \begin{pmatrix} x_1 \\ x_2 \end{pmatrix} = (1+\rho) \begin{pmatrix} x_1 \\ x_2 \end{pmatrix},$$

which leads to $x_1 = x_2$; therefore, the first standardized eigenvector is

$$\boldsymbol{\gamma}_{(1)} = \begin{pmatrix} 1/\sqrt{2} \\ 1/\sqrt{2} \end{pmatrix}.$$

Similarly, the eigenvector corresponding to $\lambda_2 = 1-\rho$ is

$$\boldsymbol{\gamma}_{(2)} = \begin{pmatrix} 1/\sqrt{2} \\ -1/\sqrt{2} \end{pmatrix}.$$

Hence,

$$\boldsymbol{\Gamma} = \begin{bmatrix} 1/\sqrt{2} & 1/\sqrt{2} \\ 1/\sqrt{2} & -1/\sqrt{2} \end{bmatrix}. \tag{A.69}$$

If $\rho = 0$, then $\boldsymbol{A} = \boldsymbol{I}$, and any orthonormal basis will do.

(10) Formula (A.65) suggests a method for calculating the symmetric *square root* of a matrix. For example, for the matrix in (A.67) with $\rho^2 < 1$, we find on using $\boldsymbol{\Lambda}$ and $\boldsymbol{\Gamma}$ from (A.62) and (A.65) that

$$\boldsymbol{A}^{1/2} = \boldsymbol{\Gamma}\boldsymbol{\Lambda}^{1/2}\boldsymbol{\Gamma} = \begin{bmatrix} a & b \\ b & a \end{bmatrix},$$

where

$$2a = (1+\rho)^{1/2} + (1-\rho)^{1/2}, \qquad 2b = (1+\rho)^{1/2} - (1-\rho)^{1/2}.$$

(11) The following methods are commonly used to calculate eigenvalues and eigenvectors on computers. For symmetric matrices, the Householder reduction to tridiagonal form (i.e. $a_{ij} = 0$ for $i \geq j+2$ and $i \leq j-2$) is used followed by the QL algorithm. For non-symmetric matrices, reduction to upper Hessenberg form (i.e. $a_{ij} = 0$ for $i \geq j+2$) is used followed by the QR algorithm.

(12) For general matrices $\boldsymbol{A}(n \times p)$, we can use the spectral decomposition theorem to derive the following result.

Theorem A.6.8 (Singular value decomposition (*SVD*) theorem) *If $\boldsymbol{A}$ is an $(n \times p)$ matrix of rank r, then $\boldsymbol{A}$ can be written as*

$$\boldsymbol{A} = \boldsymbol{U}\boldsymbol{L}\boldsymbol{V}', \tag{A.70}$$

where $\boldsymbol{U}(n \times r)$ and $\boldsymbol{V}(p \times r)$ are column orthonormal matrices ($\boldsymbol{U}'\boldsymbol{U} = \boldsymbol{V}'\boldsymbol{V} = \boldsymbol{I}_r$), and $\boldsymbol{L}$ is a diagonal matrix with positive elements.

Proof: Since $\boldsymbol{A}'\boldsymbol{A}$ is a symmetric matrix which also has rank r, we can use the spectral decomposition theorem to write

$$\boldsymbol{A}'\boldsymbol{A} = \boldsymbol{V}\boldsymbol{\Lambda}\boldsymbol{V}', \tag{A.71}$$

where $\boldsymbol{V}(p \times r)$ is a column orthonormal matrix of eigenvectors of $\boldsymbol{A}'\boldsymbol{A}$, and $\boldsymbol{\Lambda} = \text{diag}(\lambda_1, \dots, \lambda_r)$ contains the nonzero eigenvalues. Note that all the λ_i are positive because $\lambda_i = \boldsymbol{v}'_{(i)}\boldsymbol{A}'\boldsymbol{A}\boldsymbol{v}_{(i)} = \|\boldsymbol{A}\boldsymbol{v}_{(i)}\|^2 > 0$. Let

$$l_i = \lambda_i^{1/2}, \qquad i = 1, \dots, r, \tag{A.72}$$

and set $\boldsymbol{L} = \text{diag}(l_1, \dots, l_r)$. Define $\boldsymbol{U}(n \times r)$ by

$$\boldsymbol{u}_{(i)} = l_i^{-1}\boldsymbol{A}\boldsymbol{v}_{(i)}, \qquad i = 1, \dots, r. \tag{A.73}$$

Then,

$$\boldsymbol{u}'_{(j)}\boldsymbol{u}_{(i)} = l_i^{-1}l_j^{-1}\boldsymbol{v}'_{(j)}\boldsymbol{A}'\boldsymbol{A}\boldsymbol{v}_{(i)} = \lambda_i l_i^{-1}l_j^{-1}\boldsymbol{v}'_{(j)}\boldsymbol{v}_{(i)} = \begin{cases} 1, & i = j, \\ 0, & i \neq j. \end{cases}$$

Thus, $\boldsymbol{U}$ is also a column orthonormal matrix.

Any p-vector $\boldsymbol{x}$ can be written as $\boldsymbol{x} = \sum \alpha_i \boldsymbol{v}_{(i)} + \boldsymbol{y}$, where $\boldsymbol{y} \in N(\boldsymbol{A})$, the null space of $\boldsymbol{A}$. Note that $N(\boldsymbol{A}) = N(\boldsymbol{A}'\boldsymbol{A})$ is the eigenspace of $\boldsymbol{A}'\boldsymbol{A}$ for the eigenvalue 0, so that $\boldsymbol{y}$ is orthogonal to the eigenvectors $\boldsymbol{v}_{(i)}$. Let $\boldsymbol{e}_i$ denote the r-vector with 1 in the ith place and 0 elsewhere. Then,

$$\begin{aligned} \boldsymbol{U}\boldsymbol{L}\boldsymbol{V}'\boldsymbol{x} &= \sum \alpha_i \boldsymbol{U}\boldsymbol{L}\boldsymbol{e}_i + \boldsymbol{0} \\ &= \sum \alpha_i l_i \boldsymbol{u}_{(i)} + \boldsymbol{0} \\ &= \sum \alpha_i \boldsymbol{A}\boldsymbol{v}_{(i)} + \boldsymbol{A}\boldsymbol{y} = \boldsymbol{A}\boldsymbol{x}. \end{aligned}$$

Since this formula holds for all $\boldsymbol{x}$ it follows that $\boldsymbol{U}\boldsymbol{L}\boldsymbol{V}' = \boldsymbol{A}$. □

Note that the columns of $\boldsymbol{U}$ are eigenvectors of $\boldsymbol{A}\boldsymbol{A}'$, and the columns of $\boldsymbol{V}$ are eigenvectors of $\boldsymbol{A}'\boldsymbol{A}$. Also, from Theorem A.6.2, the eigenvalues of $\boldsymbol{A}\boldsymbol{A}'$ and $\boldsymbol{A}'\boldsymbol{A}$ are the same.

A.7 Quadratic Forms and Definiteness

Definition A.7.1 *A* quadratic form *in the vector* $\boldsymbol{x}$ *is a function of the form*

$$Q(\boldsymbol{x}) \equiv \boldsymbol{x}'\boldsymbol{A}\boldsymbol{x} = \sum_{i=1}^{p}\sum_{j=1}^{p} a_{ij}x_i x_j, \tag{A.74}$$

where $\boldsymbol{A}$ *is a symmetric matrix; that is*

$$Q(\boldsymbol{x}) = a_{11}x_1^2 + \cdots + a_{pp}x_p^2 + 2a_{12}x_1x_2 + \cdots + 2a_{p-1,p}x_{p-1}x_p.$$

Clearly, $Q(\boldsymbol{0}) = 0$.

Definition A.7.2 *(1)* $Q(\boldsymbol{x})$ *is called a* positive definite (p.d.) quadratic form *if* $Q(\boldsymbol{x}) > 0$ *for all* $\boldsymbol{x} \neq \boldsymbol{0}$.

(2) $Q(\boldsymbol{x})$ *is called a* positive semidefinite (p.s.d.) quadratic form *if* $Q(\boldsymbol{x}) \geq 0$ *for all* $\boldsymbol{x} \neq \boldsymbol{0}$.

(3) A symmetric matrix $\boldsymbol{A}$ *is called p.d. (p.s.d.) if* $Q(\boldsymbol{x})$ *is p.d. (p.s.d.), and we write* $\boldsymbol{A} > 0$ *or* $\boldsymbol{A} \geq 0$ *for* $\boldsymbol{A}$ *p.d. or p.s.d., respectively.*

Negative definite and negative semidefinite quadratic forms are similarly defined. For $p = 2$, $Q(\boldsymbol{x}) = x_1^2 + x_2^2$ is p.d., while $Q(\boldsymbol{x}) = (x_1 - x_2)^2$ is p.s.d.

Canonical form Any quadratic form can be converted into a weighted sum of squares without cross-product terms with the help of the following theorem.

Theorem A.7.3 *For any symmetric matrix $\boldsymbol{A}$, there exists an orthogonal transformation*

$$\boldsymbol{y} = \boldsymbol{\Gamma}'\boldsymbol{x}, \tag{A.75}$$

such that

$$\boldsymbol{x}'\boldsymbol{A}\boldsymbol{x} = \sum \lambda_i y_i^2. \tag{A.76}$$

Proof: Consider the spectral decomposition theorem in Theorem A.6.5:

$$\boldsymbol{A} = \boldsymbol{\Gamma\Lambda\Gamma}'. \tag{A.77}$$

From (A.75),

$$\boldsymbol{x}'\boldsymbol{A}\boldsymbol{x} = \boldsymbol{y}'\boldsymbol{\Gamma}'\boldsymbol{A}\boldsymbol{\Gamma}\boldsymbol{y} = \boldsymbol{y}'\boldsymbol{\Gamma}'\boldsymbol{\Gamma\Lambda\Gamma}'\boldsymbol{\Gamma}\boldsymbol{y} = \boldsymbol{y}'\boldsymbol{\Lambda}\boldsymbol{y}.$$

Hence, (A.76) follows. □

It is important to recall that $\boldsymbol{\Gamma}$ has as its columns the eigenvectors of $\boldsymbol{A}$ and that $\lambda_1, \ldots, \lambda_p$ are the eigenvalues of $\boldsymbol{A}$. Using this theorem, we can deduce the following results for a matrix $\boldsymbol{A} > 0$.

Theorem A.7.4 *If $\boldsymbol{A} > 0$, then $\lambda_i > 0$ for $i = 1, \ldots, p$. If $\boldsymbol{A} \geq 0$, then $\lambda_i \geq 0$.*

Proof: If $\boldsymbol{A} > 0$, we have, for all $\boldsymbol{x} \neq \boldsymbol{0}$,

$$0 < \boldsymbol{x}'\boldsymbol{A}\boldsymbol{x} = \lambda_1 y_1^2 + \cdots + \lambda_p y_p^2.$$

From (A.75), $\boldsymbol{x} \neq \boldsymbol{0}$ implies $\boldsymbol{y} \neq \boldsymbol{0}$. Choosing $y_1 = 1, y_2 = \cdots = y_p = 0$, we deduce that $\lambda_1 > 0$. Similarly $\lambda_i > 0$ for all i. If $\boldsymbol{A} \geq 0$, the above inequalities are weak. □

Corollary A.7.5 *If $\boldsymbol{A} > 0$, then $\boldsymbol{A}$ is nonsingular, and $|\boldsymbol{A}| > 0$.*

Proof: Use the determinant of (A.77), with $\lambda_i > 0$. □

Corollary A.7.6 *If $\boldsymbol{A} > 0$, then $\boldsymbol{A}^{-1} > 0$.*

Proof: From (A.76), we have

$$\boldsymbol{x}'\boldsymbol{A}^{-1}\boldsymbol{x} = \sum y_i^2/\lambda_i. \tag{A.78}$$

□

Corollary A.7.7 (Symmetric decomposition) *Any symmetric matrix $\boldsymbol{A} \geq 0$ can be written uniquely as*

$$\boldsymbol{A} = \boldsymbol{B}^2, \tag{A.79}$$

where $\boldsymbol{B}$ is a p.s.d. symmetric matrix.

Proof: Take $\boldsymbol{B} = \boldsymbol{\Gamma\Lambda}^{1/2}\boldsymbol{\Gamma}'$ from (A.66). Note that positive square roots of the eigenvalues must be used to ensure that $\boldsymbol{B}$ is p.s.d. □

Theorem A.7.8 *If $\boldsymbol{A} \geq 0$ is a $(p \times p)$ matrix, then for any $(p \times n)$ matrix $\boldsymbol{C}$, $\boldsymbol{C}'\boldsymbol{A}\boldsymbol{C} \geq 0$. If $\boldsymbol{A} > 0$, and $\boldsymbol{C}$ is nonsingular (so $p = n$), then $\boldsymbol{C}'\boldsymbol{A}\boldsymbol{C} > 0$.*

Proof: If $\boldsymbol{A} \geq 0$, then for any n-vector $\boldsymbol{x} \neq \boldsymbol{0}$,

$$\boldsymbol{x}'\boldsymbol{C}'\boldsymbol{A}\boldsymbol{C}\boldsymbol{x} = (\boldsymbol{C}\boldsymbol{x})'\boldsymbol{A}(\boldsymbol{C}\boldsymbol{x}) \geq 0, \qquad \text{so} \qquad \boldsymbol{C}'\boldsymbol{A}\boldsymbol{C} \geq 0.$$

If $\boldsymbol{A} > 0$, and $\boldsymbol{C}$ is nonsingular, then $\boldsymbol{C}\boldsymbol{x} \neq \boldsymbol{0}$, so $(\boldsymbol{C}\boldsymbol{x})'\boldsymbol{A}(\boldsymbol{C}\boldsymbol{x}) > 0$, and hence, $\boldsymbol{C}'\boldsymbol{A}\boldsymbol{C} > 0$. □

Corollary A.7.9 *If $\boldsymbol{A} \geq 0$ and $\boldsymbol{B} > 0$ are $(p \times p)$ matrices, then all of the nonzero eigenvalues of $\boldsymbol{B}^{-1}\boldsymbol{A}$ are positive.*

Proof: Since $\boldsymbol{B} > 0$, $\boldsymbol{B}^{-1/2}$ exists, and by Theorem A.6.2, $\boldsymbol{B}^{-1/2}\boldsymbol{A}\boldsymbol{B}^{-1/2}$, $\boldsymbol{B}^{-1}\boldsymbol{A}$, and $\boldsymbol{A}\boldsymbol{B}^{-1}$ have the same eigenvalues. By Theorem A.7.8, $\boldsymbol{B}^{-1/2}\boldsymbol{A}\boldsymbol{B}^{-1/2} \geq 0$, so all of the nonzero eigenvalues are positive. □

Remarks

(1) There are other forms of interest:
   a) *Linear form.* $\boldsymbol{a}'\boldsymbol{x} = a_1x_1 + \cdots + a_px_p$. Generally called a linear combination.
   b) *Bilinear form.* $\boldsymbol{x}'\boldsymbol{A}\boldsymbol{y} = \sum\sum a_{ij}x_iy_j$.
(2) We have noted in Corollary A.7.4 that $|\boldsymbol{A}| > 0$ for $\boldsymbol{A} > 0$. In fact, $|\boldsymbol{A}_{11}| > 0$ for all partitions of $\boldsymbol{A}$, where $\boldsymbol{A}_{11}$ has dimension $(p_1 \times p_1)$ say. The proof follows on considering $\boldsymbol{x}'\boldsymbol{A}\boldsymbol{x} > 0$ for all $\boldsymbol{x}$ with $x_{p_1+1} = \cdots = x_p = 0$. The converse is also true.
(3) For

$$\boldsymbol{\Sigma} = \begin{bmatrix} 1 & \rho \\ \rho & 1 \end{bmatrix}, \qquad \rho^2 < 1$$

the transformation (A.75) is given by (A.69)

$$y_1 = (x_1 + x_2)/\sqrt{2}, \qquad y_2 = (x_1 - x_2)/\sqrt{2}.$$

Thus, from (A.76) and (A.78),

$$\boldsymbol{x}'\boldsymbol{\Sigma}\boldsymbol{x} = x_1^2 + 2\rho x_1x_2 + x_2^2 = (1+\rho)y_1^2 + (1-\rho)y_2^2,$$

$$\boldsymbol{x}'\boldsymbol{\Sigma}^{-1}\boldsymbol{x} = \frac{1}{1-\rho^2}(x_1^2 - 2\rho x_1x_2 + x_2^2) = \frac{y_1^2}{1+\rho} + \frac{y_2^2}{1-\rho}.$$

A geometrical interpretation of these results will be found in Section A.10.4.
(4) Note that the centering matrix $\boldsymbol{H} \geq 0$ because $\boldsymbol{x}'\boldsymbol{H}\boldsymbol{x} = \sum(x_i - \bar{x})^2 \geq 0$.
(5) For any matrix $\boldsymbol{A}$, $\boldsymbol{A}\boldsymbol{A}' \geq 0$ and $\boldsymbol{A}'\boldsymbol{A} \geq 0$. Further, $r(\boldsymbol{A}\boldsymbol{A}') = r(\boldsymbol{A}'\boldsymbol{A}) = r(\boldsymbol{A})$.

A.8 Generalized Inverse

We now consider a method of defining an inverse for any matrix.

Definition A.8.1 *For a matrix* $\boldsymbol{A}(n \times p)$, $\boldsymbol{A}^-$ *is called a* g-inverse *(or* generalized inverse*) if*

$$\boldsymbol{A}\boldsymbol{A}^-\boldsymbol{A} = \boldsymbol{A}. \tag{A.80}$$

A g-inverse always exists although in general it is not unique.

Methods of construction

(1) Using the SVD theorem (Theorem A.6.8) for $\boldsymbol{A}(n \times p)$, write $\boldsymbol{A} = \boldsymbol{U}\boldsymbol{L}\boldsymbol{V}'$. Then, it is easily checked that

$$\boldsymbol{A}^- = \boldsymbol{V}\boldsymbol{L}^{-1}\boldsymbol{U}' \tag{A.81}$$

defines a g-inverse.

(2) If $r(\boldsymbol{A}) = r$, rearrange the rows and columns of $\boldsymbol{A}(n \times p)$ and partition $\boldsymbol{A}$, so that $\boldsymbol{A}_{11}$ is an $(r \times r)$ nonsingular matrix. Then, it can be verified that

$$\boldsymbol{A}^- = \begin{bmatrix} \boldsymbol{A}_{11}^{-1} & \boldsymbol{0} \\ \boldsymbol{0} & \boldsymbol{0} \end{bmatrix} \tag{A.82}$$

is a g-inverse.
The result follows on noting that there exist $\boldsymbol{B}$ and $\boldsymbol{C}$ such that

$$\boldsymbol{A}_{12} = \boldsymbol{A}_{11}\boldsymbol{B}, \quad \boldsymbol{A}_{21} = \boldsymbol{C}\boldsymbol{A}_{11}, \quad \text{and} \quad \boldsymbol{A}_{22} = \boldsymbol{C}\boldsymbol{A}_{11}\boldsymbol{B}.$$

(3) If $\boldsymbol{A}(p \times p)$ is nonsingular, then $\boldsymbol{A}^- = \boldsymbol{A}^{-1}$ is uniquely defined.

(4) If $\boldsymbol{A}(p \times p)$ is symmetric of rank r, then using Remark (4) after Theorem A.6.5, $\boldsymbol{A}$ can be written as $\boldsymbol{A} = \boldsymbol{\Gamma}_1\boldsymbol{\Lambda}_1\boldsymbol{\Gamma}_1'$, where $\boldsymbol{\Gamma}_1$ is a column orthonormal matrix of eigenvectors corresponding to the nonzero eigenvalues $\boldsymbol{\Lambda}_1 = \text{diag}(\lambda_1, \dots, \lambda_r)$ of $\boldsymbol{A}$. Then, it is easily checked that

$$\boldsymbol{A}^- = \boldsymbol{\Gamma}_1\boldsymbol{\Lambda}_1^{-1}\boldsymbol{\Gamma}_1' \tag{A.83}$$

is a g-inverse.

Applications

(1) *Linear equations.* A particular solution of the consistent equations

$$\boldsymbol{A}\boldsymbol{x} = \boldsymbol{b} \tag{A.84}$$

is

$$\boldsymbol{x} = \boldsymbol{A}^-\boldsymbol{b}. \tag{A.85}$$

Proof: From (A.80),

$$\boldsymbol{A}\boldsymbol{A}^-\boldsymbol{A}\boldsymbol{x} = \boldsymbol{A}\boldsymbol{x} \Rightarrow \boldsymbol{A}(\boldsymbol{A}^-\boldsymbol{b}) = \boldsymbol{b},$$

which when compared with (A.84) leads to (A.85). □

It can be shown that a general solution of a consistent equation is

$$\boldsymbol{x} = \boldsymbol{A}^-\boldsymbol{b} + (\boldsymbol{I} - \boldsymbol{G})\boldsymbol{z},$$

where $\boldsymbol{z}$ is arbitrary, and $\boldsymbol{G} = \boldsymbol{A}^-\boldsymbol{A}$. For $\boldsymbol{b} = \boldsymbol{0}$, a general solution is $(\boldsymbol{I} - \boldsymbol{G})\boldsymbol{z}$.

(2) *Quadratic forms.* Let $\boldsymbol{A}(p \times p)$ be a symmetric matrix of rank $r \leq p$. Then, there exists an orthogonal transformation such that for $\boldsymbol{x}$ restricted to $M(\boldsymbol{A})$, the subspace spanned by the columns of $\boldsymbol{A}$, $\boldsymbol{x}'\boldsymbol{A}^{-}\boldsymbol{x}$ can be written as

$$\boldsymbol{x}'\boldsymbol{A}^{-}\boldsymbol{x} = \sum u_i^2/\lambda_i, \tag{A.86}$$

where $\lambda_1, \ldots, \lambda_r$ are the nonzero eigenvalues of $\boldsymbol{A}$.

Proof: First note that if $\boldsymbol{x}$ lies in $M(\boldsymbol{A})$, we can write $\boldsymbol{x} = \boldsymbol{A}\boldsymbol{y}$ for some $\boldsymbol{y}$, so that

$$\boldsymbol{x}'\boldsymbol{A}^{-}\boldsymbol{x} = \boldsymbol{y}'\boldsymbol{A}\boldsymbol{A}^{-}\boldsymbol{A}\boldsymbol{y} = \boldsymbol{y}'\boldsymbol{A}\boldsymbol{y}$$

does not depend on the particular g-inverse chosen. From the spectral decomposition of $\boldsymbol{A}$, we see that $M(\boldsymbol{A})$ is spanned by the eigenvectors of $\boldsymbol{A}$ corresponding to nonzero eigenvalues, say $[\boldsymbol{\gamma}_{(1)}, \ldots, \boldsymbol{\gamma}_{(r)}] = \boldsymbol{\Gamma}_1$. Then, if $\boldsymbol{x} \in M(\boldsymbol{A})$, it can be written as $\boldsymbol{x} = \boldsymbol{\Gamma}_1\boldsymbol{u}$ for some r-vector $\boldsymbol{u}$. Defining $\boldsymbol{A}^{-}$ by (A.83), we see that (A.86) follows. □

Remarks

(1) For the equicorrelation matrix $\boldsymbol{E}$, if $1 + (p-1)\rho = 0$, then $(1-\rho)^{-1}\boldsymbol{I}$ is a g-inverse of $\boldsymbol{E}$.
(2) Under the following conditions, $\boldsymbol{A}^{-}$ is defined uniquely:

$$\boldsymbol{A}\boldsymbol{A}^{-}\boldsymbol{A} = \boldsymbol{A}, \qquad \boldsymbol{A}\boldsymbol{A}^{-} \text{ and } \boldsymbol{A}^{-}\boldsymbol{A} \text{ symmetric}, \qquad \boldsymbol{A}^{-}\boldsymbol{A}\boldsymbol{A}^{-} = \boldsymbol{A}^{-}.$$

(3) For $\boldsymbol{A} \geq 0$, the unique choice in (2) for $\boldsymbol{A}^{-}$ is normally computed using the Cholesky decomposition (see Remark (4), Section A.2.4).

A.9 Matrix Differentiation and Maximization Problems

Let us define the derivative of $f(\boldsymbol{X})$ with respect to $\boldsymbol{X}(n \times p)$ as the matrix

$$\frac{\partial f(\boldsymbol{X})}{\partial \boldsymbol{X}} = \left(\frac{\partial f(\boldsymbol{X})}{\partial x_{ij}}\right).$$

We have the following results:

(I)
$$\frac{\partial \boldsymbol{a}'\boldsymbol{x}}{\partial \boldsymbol{x}} = \boldsymbol{a}. \tag{A.87}$$

(II)
$$\frac{\partial \boldsymbol{x}'\boldsymbol{x}}{\partial \boldsymbol{x}} = 2\boldsymbol{x}, \quad \frac{\partial \boldsymbol{x}'\boldsymbol{A}\boldsymbol{x}}{\partial \boldsymbol{x}} = (\boldsymbol{A} + \boldsymbol{A}')\boldsymbol{x}, \quad \frac{\partial \boldsymbol{x}'\boldsymbol{A}\boldsymbol{y}}{\partial \boldsymbol{x}} = \boldsymbol{A}\boldsymbol{y}. \tag{A.88}$$

(III)
$$\frac{\partial |\boldsymbol{X}|}{\partial x_{ij}} = X_{ij} \quad \text{if all elements of } \boldsymbol{X}(n \times n) \text{ are distinct},$$
$$= \begin{cases} X_{ii}, & i = j \\ 2X_{ij}, & i \neq j \end{cases} \quad \text{if } \boldsymbol{X} \text{ is symmetric}, \tag{A.89}$$

where X_{ij} is the (i, j)th cofactor of $\boldsymbol{X}$.

(IV)
$$\frac{\partial \operatorname{tr} \boldsymbol{X}\boldsymbol{Y}}{\partial \boldsymbol{X}} = \boldsymbol{Y}' \quad \text{if all elements of } \boldsymbol{X}(n \times p) \text{ are distinct},$$
$$= \boldsymbol{Y} + \boldsymbol{Y}' - \operatorname{Diag}(\boldsymbol{Y}) \quad \text{if } \boldsymbol{X}(n \times n) \text{ is symmetric}. \tag{A.90}$$

(V)
$$\frac{\partial \boldsymbol{X}^{-1}}{\partial x_{ij}} = -\boldsymbol{X}^{-1}\boldsymbol{J}_{ij}\boldsymbol{X}^{-1} \quad \text{if all elements of } \boldsymbol{X}(n \times n) \text{ are distinct,}$$
$$= \begin{cases} -\boldsymbol{X}^{-1}\boldsymbol{J}_{ij}\boldsymbol{X}^{-1}, & i = j \\ -\boldsymbol{X}^{-1}(\boldsymbol{J}_{ij} + \boldsymbol{J}_{ji})\boldsymbol{X}^{-1}, & i \neq j \end{cases} \quad \text{if } \boldsymbol{X} \text{ is symmetric,} \tag{A.91}$$

where $\boldsymbol{J}_{ij}$ denotes a matrix with a 1 in the (i,j)th place and zeros elsewhere. Note that (A.91) gives the derivative of a matrix valued function with respect to a one-dimensional variable $\boldsymbol{x}_{ij}$.

We now consider some applications of these results to some stationary value problems.

Theorem A.9.1 *The vector* $\boldsymbol{x}$ *which minimizes*

$$f(\boldsymbol{x}) = (\boldsymbol{y} - \boldsymbol{A}\boldsymbol{x})'(\boldsymbol{y} - \boldsymbol{A}\boldsymbol{x})$$

is given by

$$\boldsymbol{A}'\boldsymbol{A}\boldsymbol{x} = \boldsymbol{A}'\boldsymbol{y}. \tag{A.92}$$

Proof: Differentiate $f(\boldsymbol{x})$ and set the derivative equal to $\boldsymbol{0}$. Note that (A.92) is always a consistent set of equations from (A.49). Further, the second derivative matrix $2\boldsymbol{A}'\boldsymbol{A} \geq 0$, so that the solution to (A.92) is always a minimum. □

Theorem A.9.2 *Let* $\boldsymbol{A}$ *and* $\boldsymbol{B}$ *be two symmetric matrices. Suppose that* $\boldsymbol{B} > 0$. *Then, the maximum (minimum) of* $\boldsymbol{x}'\boldsymbol{A}\boldsymbol{x}$, *given*

$$\boldsymbol{x}'\boldsymbol{B}\boldsymbol{x} = 1, \tag{A.93}$$

is attained when $\boldsymbol{x}$ *is the eigenvector of* $\boldsymbol{B}^{-1}\boldsymbol{A}$ *corresponding to the largest (smallest) eigenvalue of* $\boldsymbol{B}^{-1}\boldsymbol{A}$. *Thus, if* λ_1 *and* λ_p *are the largest and smallest eigenvalues of* $\boldsymbol{B}^{-1}\boldsymbol{A}$, *then, subject to the constraint (A.93),*

$$\max_{\boldsymbol{x}} \boldsymbol{x}'\boldsymbol{A}\boldsymbol{x} = \lambda_1, \qquad \min_{\boldsymbol{x}} \boldsymbol{x}'\boldsymbol{A}\boldsymbol{x} = \lambda_p. \tag{A.94}$$

Proof: Let $\boldsymbol{B}^{1/2}$ denote the symmetric square root of $\boldsymbol{B}$, and let $\boldsymbol{y} = \boldsymbol{B}^{1/2}\boldsymbol{x}$. Then, the maximum of $\boldsymbol{x}'\boldsymbol{A}\boldsymbol{x}$ subject to (A.93) can be written as

$$\max_{\boldsymbol{y}} \boldsymbol{y}'\boldsymbol{B}^{-1/2}\boldsymbol{A}\boldsymbol{B}^{-1/2}\boldsymbol{y} \qquad \text{subject to} \qquad \boldsymbol{y}'\boldsymbol{y} = 1. \tag{A.95}$$

Let $\boldsymbol{\Gamma\Lambda\Gamma}'$ be a spectral decomposition of the symmetric matrix $\boldsymbol{B}^{-1/2}\boldsymbol{A}\boldsymbol{B}^{-1/2}$. Let $\boldsymbol{z} = \boldsymbol{\Gamma}'\boldsymbol{y}$. Then, $\boldsymbol{z}'\boldsymbol{z} = \boldsymbol{y}'\boldsymbol{\Gamma\Gamma}'\boldsymbol{y} = \boldsymbol{y}'\boldsymbol{y}$, so that (A.95) can be written as

$$\max_{\boldsymbol{z}} \boldsymbol{z}'\boldsymbol{\Lambda}\boldsymbol{z} = \max_{\boldsymbol{z}} \sum \lambda_i z_i^2 \qquad \text{subject to} \quad \boldsymbol{z}'\boldsymbol{z} = 1. \tag{A.96}$$

If the eigenvalues are written in descending order, then (A.96) satisfies

$$\max \sum \lambda_i z_i^2 \leq \lambda_1 \max \sum z_i^2 = \lambda_1.$$

Further, this bound is attained for $\boldsymbol{z} = (1, 0, \dots, 0)'$, that is for $\boldsymbol{y} = \boldsymbol{\gamma}_{(1)}$ and $\boldsymbol{x} = \boldsymbol{B}^{-1/2}\boldsymbol{\gamma}_{(1)}$. By Theorem A.6.2, $\boldsymbol{B}^{-1}\boldsymbol{A}$ and $\boldsymbol{B}^{-1/2}\boldsymbol{A}\boldsymbol{B}^{-1/2}$ have the same eigenvalues, and $\boldsymbol{x} = \boldsymbol{B}^{-1/2}\boldsymbol{\gamma}_{(1)}$ is an eigenvector of $\boldsymbol{B}^{-1}\boldsymbol{A}$ corresponding to λ_1. Thus, the theorem is proved for maximization.

The same technique can be applied to prove the minimization result. □

Corollary A.9.3 *If* $R(\boldsymbol{x}) = \boldsymbol{x}'\boldsymbol{A}\boldsymbol{x}/\boldsymbol{x}'\boldsymbol{B}\boldsymbol{x}$, *then for* $\boldsymbol{x} \neq \boldsymbol{0}$,

$$\lambda_p \leq R(\boldsymbol{x}) \leq \lambda_1. \tag{A.97}$$

Proof: Since $R(\boldsymbol{x})$ is invariant under changes of scale of $\boldsymbol{x}$, we can regard the problem as maximizing (minimizing) $\boldsymbol{x}'\boldsymbol{A}\boldsymbol{x}$, given (A.93). □

Corollary A.9.4 *The maximum of* $\boldsymbol{a}'\boldsymbol{x}$ *subject to (A.93) is*

$$(\boldsymbol{a}'\boldsymbol{B}^{-1}\boldsymbol{a})^{1/2}. \tag{A.98}$$

Further,

$$\max_{\boldsymbol{x}} \{(\boldsymbol{a}'\boldsymbol{x})^2/(\boldsymbol{x}'\boldsymbol{B}\boldsymbol{x})\} = \boldsymbol{a}'\boldsymbol{B}^{-1}\boldsymbol{a}, \tag{A.99}$$

and the maximum is attained at $\boldsymbol{x} = \boldsymbol{B}^{-1}\boldsymbol{a}/(\boldsymbol{a}'\boldsymbol{B}^{-1}\boldsymbol{a})^{1/2}$.

Proof: Apply Theorem A.9.2 with $\boldsymbol{x}'\boldsymbol{A}\boldsymbol{x} = (\boldsymbol{a}'\boldsymbol{x})^2 = \boldsymbol{x}'(\boldsymbol{a}\boldsymbol{a}')\boldsymbol{x}$. □

Remarks

(1) A direct method is sometimes instructive. Consider the problem of maximizing the squared distance from the origin

$$x^2 + y^2$$

of a point (x, y) on the ellipse (where we suppose $a \neq b$)

$$\frac{x^2}{a^2} + \frac{y^2}{b^2} = 1. \tag{A.100}$$

When y^2 is eliminated, the problem reduces to finding the maximum of

$$x^2 + b^2(1 - x^2/a^2), \qquad x \in [-a, a].$$

Setting the derivative equal to 0 yields the stationary point $x = 0$ which, from (A.100), gives $y = \pm b$. Also, at the endpoints of the interval ($x = \pm a$), we get $y = 0$. Hence,

$$\max (x^2 + y^2) = \max (a^2, b^2).$$

This solution is not as elegant as the proof of Theorem A.9.2 and does not generalize neatly to more complicated quadratic forms.

(2) The results (A.87) and (A.88) follow by direct substitution; e.g.,

$$\frac{\partial}{\partial x_1}\boldsymbol{a}'\boldsymbol{x} = \frac{\partial}{\partial x_1}(a_1x_1 + \cdots + a_px_p) = a_1$$

proves (A.87). For (A.89), use (A.11).

A.10 Geometrical Ideas

A.10.1 n-dimensional Geometry

Let $\boldsymbol{e}_i$ denote the vector in $\mathbb{R}^n$ with 1 in the ith place and zeros elsewhere, so that $\boldsymbol{e}_1, \ldots, \boldsymbol{e}_n$ form an orthonormal basis of $\mathbb{R}^n$. In terms of this basis, vectors can be represented as $\boldsymbol{x} = \sum x_i\boldsymbol{e}_i$, and $\boldsymbol{e}_i$ is called the ith *coordinate axis*. A point $\boldsymbol{a}$ in $\mathbb{R}^n$ is represented in terms of these coordinates by $x_1 = a_1, \ldots, x_n = a_n$. The point $\boldsymbol{a}$ can also be interpreted as a directed

Table A.5 Basic concepts in n-dimensional geometry.

Concept	Description $\left(\|\boldsymbol{x}\| = \left(\sum x_i^2\right)^{1/2}\right)$
Point $\boldsymbol{a}$	$x_1 = a_1, \ldots, x_n = a_n$
Distance between $\boldsymbol{a}$ and $\boldsymbol{b}$	$\|\boldsymbol{a} - \boldsymbol{b}\| = \left\{\sum (a_i - b_i)^2\right\}^{1/2}$
Line passing through $\boldsymbol{a}, \boldsymbol{b}$	$\boldsymbol{x} = \lambda \boldsymbol{a} + (1 - \lambda)\boldsymbol{b}$ is the equation
Line passing through $\boldsymbol{0}, \boldsymbol{a}$	$\boldsymbol{x} = \lambda \boldsymbol{a}$
Angle between lines from $\boldsymbol{0}$ to $\boldsymbol{a}$ and $\boldsymbol{0}$ to $\boldsymbol{b}$	θ where $\cos\theta = \boldsymbol{a}'\boldsymbol{b}/\{\|\boldsymbol{a}\|\ \|\boldsymbol{b}\|\}$, where $0 \le \theta \le \pi$
Direction cosine vector of a line from $\boldsymbol{0}$ to $\boldsymbol{a}$	$(\cos\gamma_1, \ldots, \cos\gamma_n)$, $\cos\gamma_i = a_i/\|\boldsymbol{a}\|$; γ_i = angle between line and ith axis
Plane P	$\boldsymbol{a}'\boldsymbol{x} = c$ is general equation
Plane through $\boldsymbol{b}_1, \ldots, \boldsymbol{b}_k$	$\boldsymbol{x} = \sum \lambda_i \boldsymbol{b}_i, \quad \sum \lambda_i = 1$
Plane through $\boldsymbol{0}, \boldsymbol{b}_1, \ldots, \boldsymbol{b}_k$	$\boldsymbol{x} = \sum \lambda_i \boldsymbol{b}_i$
Hypersphere with center $\boldsymbol{a}$ and radius r	$(\boldsymbol{x} - \boldsymbol{a})'(\boldsymbol{x} - \boldsymbol{a}) = r^2$
Ellipsoid	$(\boldsymbol{x} - \boldsymbol{a})'\boldsymbol{A}^{-1}(\boldsymbol{x} - \boldsymbol{a}) = c^2 \quad \boldsymbol{A} > 0$

line segment from $\boldsymbol{0}$ to $\boldsymbol{a}$. Some generalizations of various basic concepts of two- and three-dimensional Euclidean geometry are summarized in Table A.5.

A.10.2 Orthogonal Transformations

Let $\boldsymbol{\Gamma}$ be an orthogonal matrix. Then, $\boldsymbol{\Gamma e}_i = \boldsymbol{\gamma}_{(i)}, i = 1, \ldots, n$, also form an orthonormal basis, and points $\boldsymbol{x}$ can be represented in terms of this new basis as

$$\boldsymbol{x} = \sum x_i \boldsymbol{e}_i = \sum y_i \boldsymbol{\gamma}_{(i)},$$

where $y_i = \boldsymbol{\gamma}'_{(i)}\boldsymbol{x}$ are new coordinates. If $\boldsymbol{x}^{(1)}$ and $\boldsymbol{x}^{(2)}$ are two points with new coordinates $\boldsymbol{y}^{(1)}$ and $\boldsymbol{y}^{(2)}$, note that

$$\begin{aligned}(\boldsymbol{y}^{(1)} - \boldsymbol{y}^{(2)})'(\boldsymbol{y}^{(1)} - \boldsymbol{y}^{(2)}) &= (\boldsymbol{x}^{(1)} - \boldsymbol{x}^{(2)})'\boldsymbol{\Gamma\Gamma}'(\boldsymbol{x}^{(1)} - \boldsymbol{x}^{(2)}) \\ &= (\boldsymbol{x}^{(1)} - \boldsymbol{x}^{(2)})'(\boldsymbol{x}^{(1)} - \boldsymbol{x}^{(2)}),\end{aligned}$$

so that orthogonal transformations preserve distances. An orthogonal transformation represents a rotation of the coordinate axes (plus a reflection if $|\boldsymbol{\Gamma}| = -1$). When $n = 2$ and $|\boldsymbol{\Gamma}| = 1$, $\boldsymbol{\Gamma}$ can be represented as

$$\begin{bmatrix}\cos\theta & -\sin\theta \\ \sin\theta & \cos\theta\end{bmatrix}$$

and represents a rotation of the coordinates axes counterclockwise through an angle θ.

A.10.3 Projections

Consider a point $\boldsymbol{a}$, in n dimensions (see Figure A.1). Its projection onto a plane P (or onto a line) through the origin is the point $\hat{\boldsymbol{a}}$ at the foot of the perpendicular from $\boldsymbol{a}$ to P. The vector $\hat{\boldsymbol{a}}$ is called the *orthogonal projection* of the vector $\boldsymbol{a}$ onto the plane.

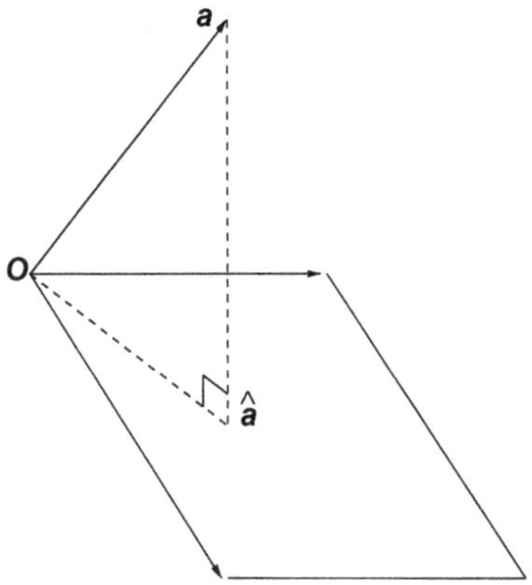

Figure A.1 $\hat{\boldsymbol{a}}$ is the projection of $\boldsymbol{a}$ onto the plane P.

Let the plane P pass through points $\mathbf{0}, \boldsymbol{b}_1, \ldots, \boldsymbol{b}_k$, so that its equation, from Table A.5, is

$$\boldsymbol{x} = \sum \lambda_i \boldsymbol{b}_i, \qquad \boldsymbol{B} = [\boldsymbol{b}_1, \ldots, \boldsymbol{b}_k].$$

Suppose that $\operatorname{rank}(\boldsymbol{B}) = k$, so that the plane is a k-dimensional subspace. The point $\hat{\boldsymbol{a}}$ is defined by $\boldsymbol{x} = \sum \hat{\lambda}_i \boldsymbol{b}_i$, where $\hat{\lambda}_1, \ldots, \hat{\lambda}_k$ minimize

$$\left\| \boldsymbol{a} - \sum \lambda_i \boldsymbol{b}_i \right\|$$

since $\hat{\boldsymbol{a}}$ is the point on the plane closest to $\boldsymbol{a}$. Using Theorem A.9.9, we deduce the following result.

Theorem A.10.1 *The point $\hat{\boldsymbol{a}}$ is given by*

$$\hat{\boldsymbol{a}} = \boldsymbol{B}(\boldsymbol{B}'\boldsymbol{B})^{-1}\boldsymbol{B}'\boldsymbol{a}. \tag{A.101}$$

Note that $\boldsymbol{B}(\boldsymbol{B}'\boldsymbol{B})^{-1}\boldsymbol{B}'$ is a symmetric idempotent matrix, and can be interpreted as a projection from $\mathbb{R}^n$ to the plane P.

A.10.4 Ellipsoids

Let $\boldsymbol{A} > 0$. Then,

$$(\boldsymbol{x} - \boldsymbol{\alpha})'\boldsymbol{A}^{-1}(\boldsymbol{x} - \boldsymbol{\alpha}) = c^2 \tag{A.102}$$

represents an ellipsoid in n dimensions. We note that the center of the ellipsoid is at $\boldsymbol{x} = \boldsymbol{\alpha}$. On shifting the center to $\boldsymbol{x} = \mathbf{0}$, the equation becomes

$$\boldsymbol{x}'\boldsymbol{A}^{-1}\boldsymbol{x} = c^2. \tag{A.103}$$

Definition A.10.2 *Let $\boldsymbol{x}$ be a point on the ellipsoid defined by (A.102), and let $f(\boldsymbol{x}) = \|\boldsymbol{x} - \boldsymbol{\alpha}\|^2$ denote the squared distance between $\boldsymbol{\alpha}$ and $\boldsymbol{x}$. A line from $\boldsymbol{\alpha}$ to $\boldsymbol{x}$ for which $\boldsymbol{x}$ is a stationary point of $f(\boldsymbol{x})$ is called a principal axis of the ellipsoid. The distance $\|\boldsymbol{x} - \boldsymbol{\alpha}\|$ is called the length of the principal semiaxis.*

Theorem A.10.3 *Let $\lambda_1, \ldots, \lambda_n$ be the eigenvalues of* $\boldsymbol{A}$ *satisfying* $\lambda_1 > \lambda_2 > \cdots > \lambda_n > 0$. *Suppose that* $\boldsymbol{\gamma}_{(1)}, \ldots, \boldsymbol{\gamma}_{(n)}$ *are the corresponding eigenvectors. For the ellipsoids (A.102) and (A.103), we have*

(1) The direction cosine vector of the ith principal axis is $\boldsymbol{\gamma}_{(i)}$.

(2) The length of the ith principal semiaxis is $c\lambda_i^{1/2}$.

Proof: It is sufficient to prove the result for (A.103). The problem reduces to finding the stationary points of $f(\boldsymbol{x}) = \boldsymbol{x}'\boldsymbol{x}$ subject to $\boldsymbol{x}$ lying on the ellipsoid $\boldsymbol{x}'\boldsymbol{A}^{-1}\boldsymbol{x} = c^2$. The derivative of $\boldsymbol{x}'\boldsymbol{A}^{-1}\boldsymbol{x}$ is $2\boldsymbol{A}^{-1}\boldsymbol{x}$. Thus, a point $\boldsymbol{y}$ represents a direction tangent to the ellipsoid at $\boldsymbol{x}$ if $2\boldsymbol{y}'\boldsymbol{A}^{-1}\boldsymbol{x} = 0$.

The derivative of $f(\boldsymbol{x})$ is $2\boldsymbol{x}$, so the directional derivative of $f(\boldsymbol{x})$ in the direction $\boldsymbol{y}$ is $2\boldsymbol{y}'\boldsymbol{x}$. Then, $\boldsymbol{x}$ is a stationary point if and only if for all points $\boldsymbol{y}$ representing tangent directions to the ellipsoid at $\boldsymbol{x}$, we have $2\boldsymbol{y}'\boldsymbol{x} = 0$, that is if

$$\boldsymbol{y}'\boldsymbol{A}^{-1}\boldsymbol{x} = 0 \Rightarrow \boldsymbol{y}'\boldsymbol{x} = 0.$$

This condition is satisfied if and only if $\boldsymbol{A}^{-1}\boldsymbol{x}$ is proportional to $\boldsymbol{x}$, that is if and only if $\boldsymbol{x}$ is an eigenvector of $\boldsymbol{A}^{-1}$.

Setting $\boldsymbol{x} = \beta\boldsymbol{\gamma}_{(i)}$ in (A.103) gives $\beta^2/\lambda_i = c^2$, so $\beta = c\lambda_i^{1/2}$. Thus, the theorem is proved. □

If we rotate the coordinate axes with the transformation $\boldsymbol{y} = \boldsymbol{\Gamma}'\boldsymbol{x}$, we find that (A.103) reduces to

$$\sum y_i^2/\lambda_i = c^2.$$

Figure A.2 gives a pictorial representation.

With $\boldsymbol{A} = \boldsymbol{I}$, Eq. (A.103) reduces to a hypersphere with $\lambda_1 = \cdots = \lambda_n = 1$, so that the λs are not distinct, and the above theorem fails; all unit vectors $\boldsymbol{\gamma}$ define stationary points.

In general, if $\lambda_i = \lambda_{i+1}$, the section of the ellipsoid is *circular* in the plane generated by $\boldsymbol{\gamma}_{(i)}, \boldsymbol{\gamma}_{(i+1)}$. Although we can construct two perpendicular axes for the common root, their positions on the circle are not unique. Similarly, if $\boldsymbol{A}$ equals the equicorrelation matrix, there are $p-1$ isotropic principal axes corresponding to the last $p-1$ eigenvalues.

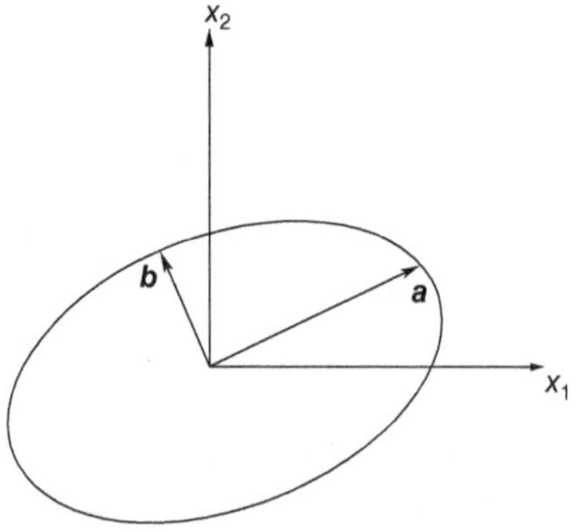

Figure A.2 Ellipsoid $\boldsymbol{x}'\boldsymbol{A}^{-1}\boldsymbol{x} = 1$ in the plane. Vectors given by $\boldsymbol{a} = \lambda_1^{1/2}\boldsymbol{\gamma}_{(1)}$ and $\boldsymbol{b} = \lambda_2^{1/2}\boldsymbol{\gamma}_{(2)}$.

B

Univariate Statistics

B.1 Introduction

In this appendix, we summarize the univariate distributions needed in the text, together with some of their simple properties. Note that we do not distinguish notationally between a random variable x and the argument of its probability density function (p.d.f.) $f(x)$.

B.2 Normal Distribution

If a random variable x has p.d.f.

$$f(x) = (2\pi)^{-1/2}e^{-x^2/2}, \qquad -\infty < x < \infty, \tag{B.1}$$

then x is said to have a standardized *normal distribution*, and we write $x \sim N(0,1)$. Its characteristic function (c.f.) is given by

$$\phi(t) = \mathsf{E}(e^{itx}) = e^{-t^2/2}. \tag{B.2}$$

Using the formula

$$\mu_r = \mathsf{E}(x^r) = (-1)^r \frac{d^r\phi(t)}{dt^r}\bigg|_{t=0}$$

for the rth moment of x, where r is a positive integer, it is found that

$$\mu_1 = \mu_3 = 0, \qquad \mu_2 = 1, \qquad \mu_4 = 3, \tag{B.3}$$

and, more generally (for $i > 2$), that

$$\mu_i = \begin{cases} 0, & i \text{ odd}, \\ (i-1)(i-3)\cdots 1, & i \text{ even}. \end{cases}$$

The p.d.f. of $N(\mu, \sigma^2)$ is given by

$$(2\pi\sigma^2)^{-1/2} \exp\left\{-\frac{1}{2}\frac{(x-\mu)^2}{\sigma^2}\right\}, \qquad -\infty < x < \infty,$$

where $\sigma^2 > 0, -\infty < \mu < \infty$. Its characteristic function is given by

$$\exp\left(i\mu t - \frac{1}{2}t^2\sigma^2\right).$$

Multivariate Analysis, Second Edition. Kanti V. Mardia, John T. Kent, and Charles C. Taylor.

B.3 Chi-squared Distribution

Let $x_1, \dots, x_p$ be independent $N(0, 1)$ variables, and set

$$y = x_1^2 + \cdots + x_p^2. \tag{B.4}$$

Then, y is said to have a *chi-squared distribution* with p degrees of freedom, and we write $y \sim \chi_p^2$.

The density of y is given by

$$f(y) = \left\{2^{p/2}\Gamma\left(\frac{p}{2}\right)\right\}^{-1} y^{p/2-1}e^{-y/2}, \qquad 0 < y < \infty. \tag{B.5}$$

Here, $\Gamma(x)$ is the gamma function defined by

$$\Gamma(x) = \int_0^\infty u^{x-1}e^{-u}du, \qquad x > 0, \tag{B.6}$$

which satisfies the relations $\Gamma(x + 1) = x\Gamma(x)$ for $x > 0$ and $\Gamma(n + 1) = n!$ for integer $n \geq 0$.

Note that (B.5) is a density for any real $p > 0$; p need not be an integer. Using (B.6), the moments of y are easily calculated as

$$\mathsf{E}(y^r) = 2^r\Gamma\left(\frac{p}{2} + r\right) \Big/ \Gamma\left(\frac{p}{2}\right), \tag{B.7}$$

for any real number r for which $r > -\frac{1}{2}p$. Using $r = 1,2$, we get

$$\mathsf{E}(y) = p, \qquad \text{var}(y) = 2p. \tag{B.8}$$

Also, if $p > 2$, taking $r = -1$ in (B.7), we get

$$\mathsf{E}(1/y) = 1/(p - 2). \tag{B.9}$$

B.4 *F* and Beta Variables

Let $u \sim \chi_p^2$ independently of $v \sim \chi_q^2$, and set

$$x = \frac{u/p}{v/q}. \tag{B.10}$$

Then, x is said to have an F distribution with degrees of freedom p and q. We write $x \sim F_{p,q}$.

Using the same notation, set

$$y = u/(u + v). \tag{B.11}$$

Then, y is said to have a *beta distribution* (sometimes called the *beta type I distribution*) with parameters $\frac{1}{2}p$ and $\frac{1}{2}q$. We write $x \sim \text{B}(\frac{1}{2}p, \frac{1}{2}q)$. Its density is given by

$$f(y) = \frac{1}{\text{B}\left(\frac{p}{2}, \frac{q}{2}\right)} y^{p/2-1}(1 - y)^{q/2-1}, \qquad 0 < y < 1, \tag{B.12}$$

where

$$\text{B}(\alpha, \beta) = \frac{\Gamma(\alpha)\Gamma(\beta)}{\Gamma(\alpha + \beta)} \quad (\alpha, \beta > 0) \tag{B.13}$$

is a normalization constant, and $\Gamma(\cdot)$ is defined in (B.6).

The *beta type II distribution* is proportional to the F distribution and is defined in the above notation by

$$w = px/q = u/v.$$

The F and beta type I distributions are related by the one-to-one transformation

$$y = px/(q + px) \tag{B.14}$$

or, equivalently,

$$x = qy/\{p(1 - y)\}.$$

The following theorem shows an important relationship between the beta and chi-squared distributions.

Theorem B.4.1 *Let $u \sim \chi_p^2$ independently of $v \sim \chi_q^2$. Then, $z = u + v \sim \chi_{p+q}^2$ independently of $y = u/(u + v) \sim B(\frac{1}{2}p, \frac{1}{2}q)$.*

Proof: Transform the variables (u, v) to (z, y), where $u = zy$ and $v = z(1 - y)$. The p.d.f. of (u, v) is known from (B.5), and simplifying yields the desired result. □

B.5 *t* Distribution

Let $x \sim N(0,1)$ independently of $u \sim \chi_p^2$. Then,

$$t = x/(u/p)^{1/2} \tag{B.15}$$

is said to have a *t distribution* with p degrees of freedom. We write $t \sim t_p$. Its p.d.f. is given by

$$\left[\Gamma\left(\frac{p}{2} + \frac{1}{2}\right) \Big/ \left\{(p\pi)^{1/2}\Gamma\left(\frac{p}{2}\right)\right\}\right] (1 + t^2/p)^{-(p+1)/2}, \qquad -\infty < t < \infty$$

From (B.4) and (B.10), note that if $t \sim t_p$, then

$$t^2 \sim F_{1,p}. \tag{B.16}$$

The mean and variance of the t distribution can be calculated from (B.3) and (B.9) to be

$$\mathsf{E}(t) = 0, \qquad \text{var}(t) = p/(p - 2), \tag{B.17}$$

provided $p > 1$ and $p > 2$, respectively.

B.6 Poisson Distribution

If a random variable x has probability mass function

$$p(x) = \frac{\exp(-\lambda)\lambda^x}{x!}, \qquad x = 0, 1, 2 \ldots, \qquad \lambda > 0,$$

then x is said to have a *Poisson distribution* with parameter λ. Its characteristic function is given by

$$\phi(t) = \exp[\lambda(e^{it} - 1)].$$

Hence, the expected value and variance of the distribution are both equal to λ.

C

R Commands and Data

C.1 Basic R Commands Related to Matrices

Those who want to learn multivariate analysis by example are referred to Everitt and Hothorn (2011) and Husson et al. (2017), which are dedicated to the use of R. In the latter case, there is also an accompanying R library. A more general guide to matrix algebra using R is given by Vinod (2011). Here, we give some introductory commands for those who already have some familiarity with R, focusing on matrix operations.

Note that data, say $\boldsymbol{X}$, may be read into R as a data frame (with n rows and p columns) or as a matrix. The dimensions (n and p) of a data frame or matrix can be found using `dim(X)`. Alternatively, an $n \times p$ matrix X may be input using, for example,

```
X=matrix(1:10,nrow=5,ncol=2)
```

where the first argument (`1:10`) is a vector containing the elements (x_{ij}) of the matrix, which are placed (by default) column by column, and `nrow` and `ncol` (one of which can be omitted if all the elements are given) are n and p, respectively. If the number of elements in the first argument does not fill the matrix (specified by `nrow` and `ncol`), then they are recycled. For example, `matrix(1:2,nrow=5,ncol=2)` is valid. If a set of data is read from a file, it will usually be stored as a data frame, but many matrix operations require matrices. One can "convert" using `as.matrix`.

Table C.1 gives some basic R commands and may be compared with Table A.2 and the sections mentioned therein. When matrices are added, subtracted, or multiplied, they need to be "conformable" – that is, the dimensions need to appropriately match. However, it is also possible in R to add scalars or vectors to matrices. In this case, the vector (or scalar) is recycled, and the elements of the matrix are used column-wise. Note that, by default, R computes the unbiased estimates of the covariance matrix $\boldsymbol{S}_u$ in the command `var` or `cov`. Note also that matrices (of the same size) *can* be multiplied together – or raised to a power – elementwise, which is rarely used in the methods of this book.

The `diag` command has three different usages, depending on the argument. These are summarized in Table C.2.

For a square matrix, say $\boldsymbol{A}$, the eigenvalues and eigenvectors (see Section A.6) are given by

```
eigen(A)
```

Multivariate Analysis, Second Edition. Kanti V. Mardia, John T. Kent, and Charles C. Taylor.

Table C.1 R commands for some basic matrix operations.

Operation	Notation	R command/syntax
Mean	$\bar{\boldsymbol{x}}$	`colMeans(X)` or `apply(X,2,mean)`
Variance	$\boldsymbol{S}$ – see (1.7)	`var(X)*(n-1)/n` or:
		`xm=colMeans(X); t(X)%*%X/n-xm%*%t(xm)`
Subtraction	$\boldsymbol{A}-\boldsymbol{B}$	`A-B`
Scalar multiplication	$s\boldsymbol{A}$	`s*A`
Transpose	$\boldsymbol{A}'$	`t(A)`
Inner product	$\boldsymbol{a}'\boldsymbol{b}$	`t(a)%*%b`
Outer product	$\boldsymbol{a}\boldsymbol{b}'$	`a%*%t(b)` or `outer(a,b,"*")` or `a%o%b`
Multiplication	$\boldsymbol{AB}$	`A%*%B`
Trace	$\text{tr}\boldsymbol{A}$	`sum(diag(A))`
Determinant	$\lvert\boldsymbol{A}\rvert$	`det(A)`
Inverse	$\boldsymbol{A}^{-1}$	`solve(A)`
g-inverse	$\boldsymbol{A}^{-}$	`ginv(A)` (requires: library(MASS))
Maximum	$\max \boldsymbol{A}$	`max(A)` (where $\max A = \max_{i,j} a_{ij}$)
Position[a)]	$\arg_{i,j} \max \boldsymbol{A}$	`which(A==max(A),arr.ind=TRUE)`
Kronecker	$\boldsymbol{A} \otimes \boldsymbol{B}$	`kronecker(A,B)`
Vectorize	$\boldsymbol{A}^V$	`c(A)`

a) The position is a pair of indices indicating the location of the maximum value.

The eigenvalues are sorted (largest to smallest) with the jth column of `eigen(A)$vectors` corresponding to the jth eigenvalue (given by `eigen(A)$values[j]`). Remember that the sign of each vector is arbitrary.

If $\boldsymbol{A}$ is an $n \times p$ matrix, then the singular value decomposition (SVD) (see Theorem A.6.5), given by $\boldsymbol{A} = \boldsymbol{ULV}'$, is computed by

```
svd(A)
```

The output has three components:

`$u` – the matrix $\boldsymbol{U}$
`$d` – the diagonal elements of $\boldsymbol{L}$. That is, in R: `L=diag(svd(A)$d)`
`$v` – the matrix $\boldsymbol{V}$.

In the case that $\boldsymbol{A}$ is symmetric, then $\boldsymbol{U} = \boldsymbol{V}$ (Theorem A.6.5).

C.2 R Libraries and Commands Used in Exercises and Figures

The website github.com/charlesctaylor/MVAdata-rcode contains the R commands used for most of the examples and figures. These are not intended to illustrate efficient coding, but to provide the ingredients for learning by example. There are many R libraries that implement

Table C.2 Using the `diag` command in R.

Input	Command	Output returned by R
p (positive integer)	`diag(p)`	$\boldsymbol{I}_p$
$\boldsymbol{A}$ (square matrix)	`diag(A)`	$(a_{11}, a_{22}, \ldots, a_{pp})'$
$\boldsymbol{a}$ (vector)	`diag(a)`	$\begin{bmatrix} a_1 & \cdots & 0 \\ \vdots & \ddots & 0 \\ 0 & \cdots & a_p \end{bmatrix}$

the methods very succinctly, but most of our code is more "raw," so that the matrix algebra is not hidden and follows the equations in the text.

However, at times, we used R libraries for ease as well as for checking and other background details. The full list of libraries we have used is: `blighty` (Lucy, 2012), `e1071` (Meyer et al., 2021), `fda` (Ramsay et al., 2020), `fields` (Nychka et al., 2017), `graph` (Gentleman et al., 2020), `gRim` (Højsgaard et al., 2012), `ica` (Helwig, 2018), `igraph` (Csardi and Nepusz, 2006), `KernSmooth` (Wand, 2020), `klaR` (Weihs et al., 2005), `lars` (Hastie and Efron, 2013), `leaps` (Lumley, 2020), `mapdata` (Brownrigg, 2018), `MASS` (Venables and Ripley, 2002), `mclust` (Scrucca et al., 2016), `nnet` (Venables and Ripley, 2002), `moments` (Komsta and Novomestky, 2015), `pls` (Liland et al., 2021), `Pursuit` (Ossani and Cirillo, 2021), `rootWishart` (Turgeon, 2018), `rpart` (Therneau and Atkinson, 2019), `rrcov` (Todorov, 2021), `shapes` (Dryden, 2021), and `spatstat` (Baddeley et al., 2015).

C.3 Data Availability

Some of the data sets used in the book are available in R. Table C.3 gives a list of the data sets (stored as data frames) and, where relevant, the package in R where it can be found.

Other data sets are available on the web site github.com/charlesctaylor/MVAdata-rcode, with a list given in Table C.4.

Table C.3 R data sets used and associated library (see index for usage).

Data set	R library/package
iris	(available on starting R)
growth	`fda`
oliveoil	`pls`
yarn	`pls`

These can be loaded with, for example, `data(growth, package=fda)` or `library(fda);data(growth)`.

Table C.4 The following datasets are available on the website github.com/charlesctaylor/MVAdata-rcode (see the Index for use in this book).

File name	Brief description
accident.dat	Car accident data
anteater.tab	Ant-eater skull measurements
applicants.dat	Applicants for job interview
bodyfat.txt	Body circumference measurements
challenger.txt	Challenger Space Shuttle failure
copen.dat	Copenhagen housing data
cork.dta	Cork deposits
frets.dat	Head length and breadth for fathers and sons
gene.csv	Relative frequencies of blood groups
glass.data	Forensic data on glass types
morse.dat	Percentage of time morse code signals confused
obcb.data	Open-book, closed-book exam data
pitpropC.tab	Pitprops from pine trees (correlation matrix)
rats.tab	Rat weight losses (male and female)
segmentation2.data	Image segmentation classification
shoulders.dat	Distance between shoulders of Cladocera species
shrewmeans.dat	Means of variables for shrews, according to location
titanic3v.csv	Survival data for the Titanic disaster
townlocs.dat	Latitude and longitude of some towns in Great Britain
towns.dat	Road distances between above towns
turtles.dat	Shell dimensions of turtles (male and female)
worldbank.csv	Country summary data from the World Bank

D

Tables

This table gives the critical values $\chi^2_\nu(1-\alpha)$ for various values of α and degrees of freedom ν, as indicated by the figure to the right, plotted in the case $\nu = 3$.

If X is a variable distributed as χ^2 with ν degrees of freedom, α is the probability that $X \geq \chi^2_\nu(1-\alpha)$.

For $\nu > 100$, $\sqrt{2X}$ is approximately normally distributed with mean $\sqrt{2\nu - 1}$ and unit variance.

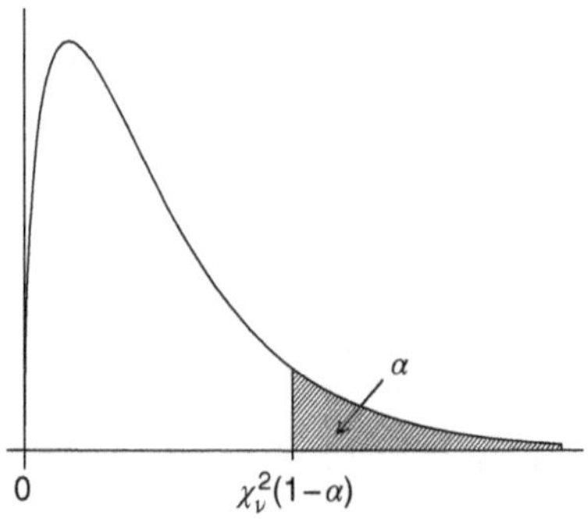

Table D.1 Upper critical values of the χ^2_ν distribution.

	α						
ν	0.100	0.050	0.025	0.01	0.005	0.001	0.0005
1	2.706	3.841	5.024	6.635	7.879	10.828	12.116
2	4.605	5.991	7.378	9.210	10.597	13.816	15.202
3	6.251	7.815	9.348	11.345	12.838	16.266	17.730
4	7.779	9.488	11.143	13.277	14.860	18.467	19.997
5	9.236	11.070	12.833	15.086	16.750	20.515	22.105
6	10.645	12.592	14.449	16.812	18.548	22.458	24.103
7	12.017	14.067	16.013	18.475	20.278	24.322	26.018
8	13.362	15.507	17.535	20.090	21.955	26.124	27.868
9	14.684	16.919	19.023	21.666	23.589	27.877	29.666
10	15.987	18.307	20.483	23.209	25.188	29.588	31.420
11	17.275	19.675	21.920	24.725	26.757	31.264	33.137
12	18.549	21.026	23.337	26.217	28.300	32.909	34.821
13	19.812	22.362	24.736	27.688	29.819	34.528	36.478
14	21.064	23.685	26.119	29.141	31.319	36.123	38.109
15	22.307	24.996	27.488	30.578	32.801	37.697	39.719
16	23.542	26.296	28.845	32.000	34.267	39.252	41.308
17	24.769	27.587	30.191	33.409	35.718	40.790	42.879
18	25.989	28.869	31.526	34.805	37.156	42.312	44.434
19	27.204	30.144	32.852	36.191	38.582	43.820	45.973
20	28.412	31.410	34.170	37.566	39.997	45.315	47.498
25	34.382	37.652	40.646	44.314	46.928	52.620	54.947
30	40.256	43.773	46.979	50.892	53.672	59.703	62.162
40	51.805	55.758	59.342	63.691	66.766	73.402	76.095
50	63.167	67.505	71.420	76.154	79.490	86.661	89.561
80	96.578	101.879	106.629	112.329	116.321	124.839	128.261

This table gives the critical values $t_\nu(1-\alpha)$ for various values of α and degrees of freedom ν, as indicated by the figure to the right.

The lower percentage points are given by symmetry as $-t_\nu(1-\alpha)$, and the probability that $|t| \geq t_\nu(1-\alpha)$ is 2α.

The limiting distribution of t as $\nu \to \infty$ is the normal distribution with zero mean and unit variance.

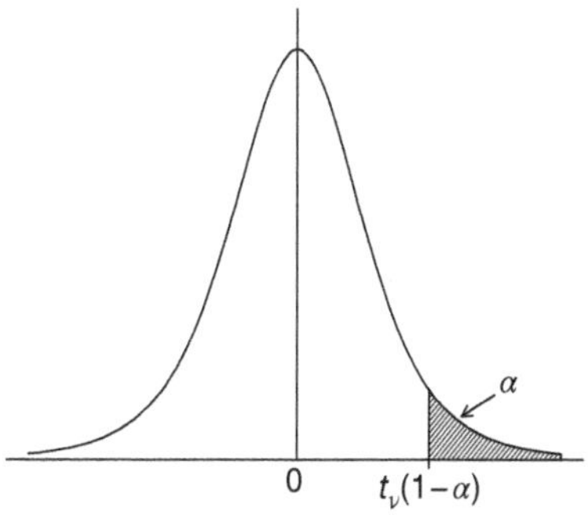

Table D.2 Upper critical values of the t_ν distribution.

ν	α = 0.100	0.050	0.025	0.010	0.005	0.001	0.0005
1	3.078	6.314	12.706	31.821	63.657	318.309	636.619
2	1.886	2.920	4.303	6.965	9.925	22.327	31.599
3	1.638	2.353	3.182	4.541	5.841	10.215	12.924
4	1.533	2.132	2.776	3.747	4.604	7.173	8.610
5	1.476	2.015	2.571	3.365	4.032	5.893	6.869
6	1.440	1.943	2.447	3.143	3.707	5.208	5.959
7	1.415	1.895	2.365	2.998	3.499	4.785	5.408
8	1.397	1.860	2.306	2.896	3.355	4.501	5.041
9	1.383	1.833	2.262	2.821	3.250	4.297	4.781
10	1.372	1.812	2.228	2.764	3.169	4.144	4.587
11	1.363	1.796	2.201	2.718	3.106	4.025	4.437
12	1.356	1.782	2.179	2.681	3.055	3.930	4.318
13	1.350	1.771	2.160	2.650	3.012	3.852	4.221
14	1.345	1.761	2.145	2.624	2.977	3.787	4.140
15	1.341	1.753	2.131	2.602	2.947	3.733	4.073
16	1.337	1.746	2.120	2.583	2.921	3.686	4.015
18	1.330	1.734	2.101	2.552	2.878	3.610	3.922
21	1.323	1.721	2.080	2.518	2.831	3.527	3.819
25	1.316	1.708	2.060	2.485	2.787	3.450	3.725
30	1.310	1.697	2.042	2.457	2.750	3.385	3.646
40	1.303	1.684	2.021	2.423	2.704	3.307	3.551
50	1.299	1.676	2.009	2.403	2.678	3.261	3.496
70	1.294	1.667	1.994	2.381	2.648	3.211	3.435
100	1.290	1.660	1.984	2.364	2.626	3.174	3.390
∞	1.282	1.645	1.960	2.326	2.576	3.090	3.291

This table gives the upper critical values $F_{\nu_1,\nu_2}(1-\alpha)$ for $\alpha = 0.10$ and degrees of freedom ν_1, ν_2, as indicated by the figure to the right.

The lower percentage points, that is the values $F_{\nu_1,\nu_2}(\alpha)$ such that the probability that $F \le F_{\nu_1,\nu_2}(\alpha)$ is equal to α, may be found using the formula

$$F_{\nu_1,\nu_2}(\alpha) = 1/F_{\nu_2,\nu_1}(1-\alpha)$$

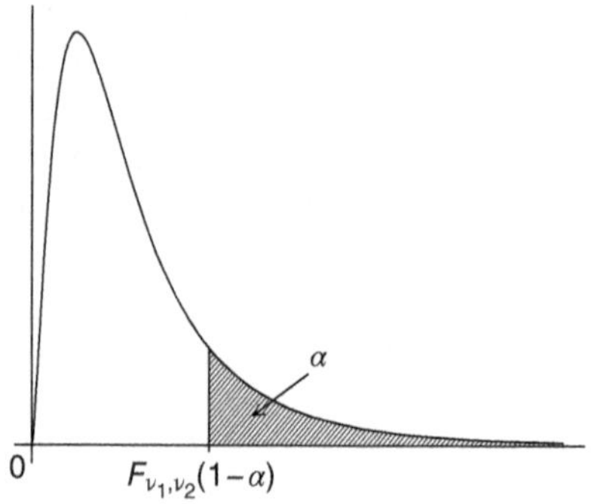

Table D.3 Upper critical values of the F_{ν_1,ν_2} distribution ($\alpha = 0.10$).

	ν_1								
ν_2	1	2	3	4	5	6	12	24	∞
2	8.526	9.000	9.162	9.243	9.293	9.326	9.408	9.450	9.491
3	5.538	5.462	5.391	5.343	5.309	5.285	5.216	5.176	5.134
4	4.545	4.325	4.191	4.107	4.051	4.010	3.896	3.831	3.761
5	4.060	3.780	3.619	3.520	3.453	3.405	3.268	3.191	3.105
6	3.776	3.463	3.289	3.181	3.108	3.055	2.905	2.818	2.722
7	3.589	3.257	3.074	2.961	2.883	2.827	2.668	2.575	2.471
8	3.458	3.113	2.924	2.806	2.726	2.668	2.502	2.404	2.293
9	3.360	3.006	2.813	2.693	2.611	2.551	2.379	2.277	2.159
10	3.285	2.924	2.728	2.605	2.522	2.461	2.284	2.178	2.055
11	3.225	2.860	2.660	2.536	2.451	2.389	2.209	2.100	1.972
12	3.177	2.807	2.606	2.480	2.394	2.331	2.147	2.036	1.904
13	3.136	2.763	2.560	2.434	2.347	2.283	2.097	1.983	1.846
14	3.102	2.726	2.522	2.395	2.307	2.243	2.054	1.938	1.797
15	3.073	2.695	2.490	2.361	2.273	2.208	2.017	1.899	1.755
16	3.048	2.668	2.462	2.333	2.244	2.178	1.985	1.866	1.718
17	3.026	2.645	2.437	2.308	2.218	2.152	1.958	1.836	1.686
18	3.007	2.624	2.416	2.286	2.196	2.130	1.933	1.810	1.657
19	2.990	2.606	2.397	2.266	2.176	2.109	1.912	1.787	1.631
20	2.975	2.589	2.380	2.249	2.158	2.091	1.892	1.767	1.607
25	2.918	2.528	2.317	2.184	2.092	2.024	1.820	1.689	1.518
30	2.881	2.489	2.276	2.142	2.049	1.980	1.773	1.638	1.456
40	2.835	2.440	2.226	2.091	1.997	1.927	1.715	1.574	1.377
50	2.809	2.412	2.197	2.061	1.966	1.895	1.680	1.536	1.327
100	2.756	2.356	2.139	2.002	1.906	1.834	1.612	1.460	1.214
∞	2.706	2.303	2.084	1.945	1.847	1.774	1.546	1.383	1.000

This table gives the upper critical values $F_{\nu_1,\nu_2}(1-\alpha)$ for $\alpha = 0.05$ and degrees of freedom ν_1, ν_2, as indicated by the figure to the right.

The lower percentage points, that is the values $F_{\nu_1,\nu_2}(\alpha)$ such that the probability that $F \le F_{\nu_1,\nu_2}(\alpha)$ is equal to α, may be found using the formula

$$F_{\nu_1,\nu_2}(\alpha) = 1/F_{\nu_2,\nu_1}(1-\alpha)$$

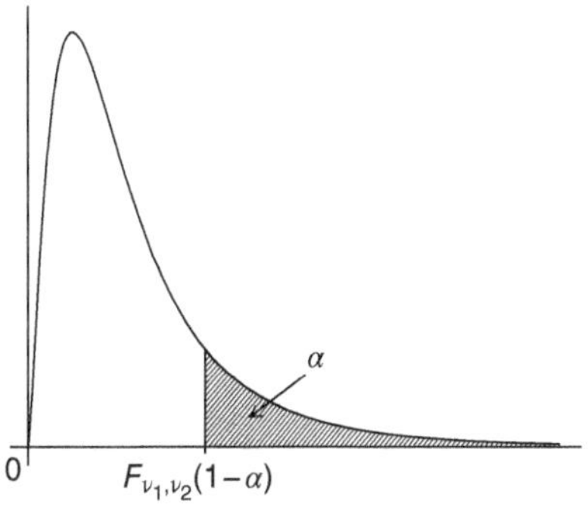

Table D.4 Upper critical values of the F_{ν_1,ν_2} distribution ($\alpha = 0.05$).

	ν_1								
ν_2	1	2	3	4	5	6	12	24	∞
2	18.513	19.000	19.164	19.247	19.296	19.330	19.413	19.454	19.496
3	10.128	9.552	9.277	9.117	9.013	8.941	8.745	8.639	8.526
4	7.709	6.944	6.591	6.388	6.256	6.163	5.912	5.774	5.628
5	6.608	5.786	5.409	5.192	5.050	4.950	4.678	4.527	4.365
6	5.987	5.143	4.757	4.534	4.387	4.284	4.000	3.841	3.669
7	5.591	4.737	4.347	4.120	3.972	3.866	3.575	3.410	3.230
8	5.318	4.459	4.066	3.838	3.687	3.581	3.284	3.115	2.928
9	5.117	4.256	3.863	3.633	3.482	3.374	3.073	2.900	2.707
10	4.965	4.103	3.708	3.478	3.326	3.217	2.913	2.737	2.538
11	4.844	3.982	3.587	3.357	3.204	3.095	2.788	2.609	2.404
12	4.747	3.885	3.490	3.259	3.106	2.996	2.687	2.505	2.296
13	4.667	3.806	3.411	3.179	3.025	2.915	2.604	2.420	2.206
14	4.600	3.739	3.344	3.112	2.958	2.848	2.534	2.349	2.131
15	4.543	3.682	3.287	3.056	2.901	2.790	2.475	2.288	2.066
16	4.494	3.634	3.239	3.007	2.852	2.741	2.425	2.235	2.010
17	4.451	3.592	3.197	2.965	2.810	2.699	2.381	2.190	1.960
18	4.414	3.555	3.160	2.928	2.773	2.661	2.342	2.150	1.917
19	4.381	3.522	3.127	2.895	2.740	2.628	2.308	2.114	1.878
20	4.351	3.493	3.098	2.866	2.711	2.599	2.278	2.082	1.843
25	4.242	3.385	2.991	2.759	2.603	2.490	2.165	1.964	1.711
30	4.171	3.316	2.922	2.690	2.534	2.421	2.092	1.887	1.622
40	4.085	3.232	2.839	2.606	2.449	2.336	2.003	1.793	1.509
50	4.034	3.183	2.790	2.557	2.400	2.286	1.952	1.737	1.438
100	3.936	3.087	2.696	2.463	2.305	2.191	1.850	1.627	1.283
∞	3.841	2.996	2.605	2.372	2.214	2.099	1.752	1.517	1.000

This table gives the upper critical values $F_{\nu_1,\nu_2}(1-\alpha)$ for $\alpha = 0.025$ and degrees of freedom ν_1, ν_2, as indicated by the figure to the right.

The lower percentage points, that is the values $F_{\nu_1,\nu_2}(\alpha)$ such that the probability that $F \le F_{\nu_1,\nu_2}(\alpha)$ is equal to α, may be found using the formula

$$F_{\nu_1,\nu_2}(\alpha) = 1/F_{\nu_2,\nu_1}(1-\alpha)$$

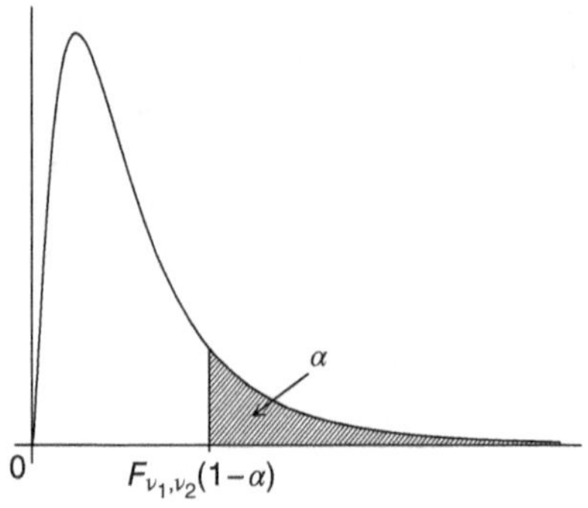

Table D.5 Upper critical values of the F_{ν_1,ν_2} distribution ($\alpha = 0.025$).

	ν_1								
ν_2	1	2	3	4	5	6	12	24	∞
2	38.506	39.000	39.165	39.248	39.298	39.331	39.415	39.456	39.498
3	17.443	16.044	15.439	15.101	14.885	14.735	14.337	14.124	13.902
4	12.218	10.649	9.979	9.605	9.364	9.197	8.751	8.511	8.257
5	10.007	8.434	7.764	7.388	7.146	6.978	6.525	6.278	6.015
6	8.813	7.260	6.599	6.227	5.988	5.820	5.366	5.117	4.849
7	8.073	6.542	5.890	5.523	5.285	5.119	4.666	4.415	4.142
8	7.571	6.059	5.416	5.053	4.817	4.652	4.200	3.947	3.670
9	7.209	5.715	5.078	4.718	4.484	4.320	3.868	3.614	3.333
10	6.937	5.456	4.826	4.468	4.236	4.072	3.621	3.365	3.080
11	6.724	5.256	4.630	4.275	4.044	3.881	3.430	3.173	2.883
12	6.554	5.096	4.474	4.121	3.891	3.728	3.277	3.019	2.725
13	6.414	4.965	4.347	3.996	3.767	3.604	3.153	2.893	2.595
14	6.298	4.857	4.242	3.892	3.663	3.501	3.050	2.789	2.487
15	6.200	4.765	4.153	3.804	3.576	3.415	2.963	2.701	2.395
16	6.115	4.687	4.077	3.729	3.502	3.341	2.889	2.625	2.316
17	6.042	4.619	4.011	3.665	3.438	3.277	2.825	2.560	2.247
18	5.978	4.560	3.954	3.608	3.382	3.221	2.769	2.503	2.187
19	5.922	4.508	3.903	3.559	3.333	3.172	2.720	2.452	2.133
20	5.871	4.461	3.859	3.515	3.289	3.128	2.676	2.408	2.085
25	5.686	4.291	3.694	3.353	3.129	2.969	2.515	2.242	1.906
30	5.568	4.182	3.589	3.250	3.026	2.867	2.412	2.136	1.787
40	5.424	4.051	3.463	3.126	2.904	2.744	2.288	2.007	1.637
50	5.340	3.975	3.390	3.054	2.833	2.674	2.216	1.931	1.545
100	5.179	3.828	3.250	2.917	2.696	2.537	2.077	1.784	1.347
∞	5.024	3.689	3.116	2.786	2.567	2.408	1.945	1.640	1.000

This table gives the upper critical values $F_{\nu_1,\nu_2}(1-\alpha)$ for $\alpha = 0.01$ and degrees of freedom ν_1, ν_2, as indicated by the figure to the right.

The lower percentage points, that is the values $F_{\nu_1,\nu_2}(\alpha)$ such that the probability that $F \le F_{\nu_1,\nu_2}(\alpha)$ is equal to α, may be found using the formula

$$F_{\nu_1,\nu_2}(\alpha) = 1/F_{\nu_2,\nu_1}(1-\alpha)$$

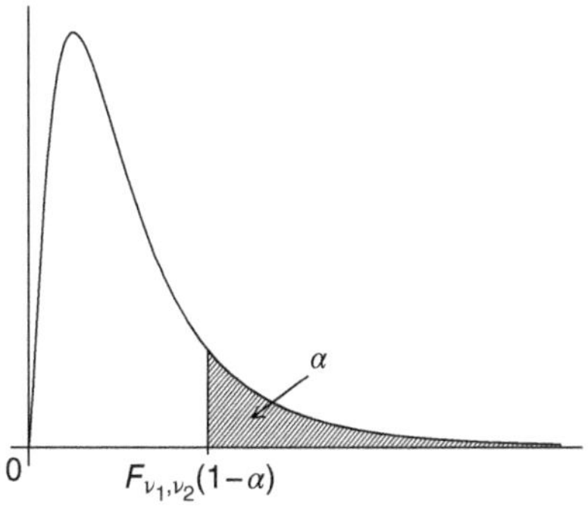

Table D.6 Upper critical values of the F_{ν_1,ν_2} distribution ($\alpha = 0.01$).

	ν_1								
ν_2	1	2	3	4	5	6	12	24	∞
2	98.503	99.000	99.166	99.249	99.299	99.333	99.416	99.458	99.499
3	34.116	30.817	29.457	28.710	28.237	27.911	27.052	26.598	26.125
4	21.198	18.000	16.694	15.977	15.522	15.207	14.374	13.929	13.463
5	16.258	13.274	12.060	11.392	10.967	10.672	9.888	9.466	9.020
6	13.745	10.925	9.780	9.148	8.746	8.466	7.718	7.313	6.880
7	12.246	9.547	8.451	7.847	7.460	7.191	6.469	6.074	5.650
8	11.259	8.649	7.591	7.006	6.632	6.371	5.667	5.279	4.859
9	10.561	8.022	6.992	6.422	6.057	5.802	5.111	4.729	4.311
10	10.044	7.559	6.552	5.994	5.636	5.386	4.706	4.327	3.909
11	9.646	7.206	6.217	5.668	5.316	5.069	4.397	4.021	3.602
12	9.330	6.927	5.953	5.412	5.064	4.821	4.155	3.780	3.361
13	9.074	6.701	5.739	5.205	4.862	4.620	3.960	3.587	3.165
14	8.862	6.515	5.564	5.035	4.695	4.456	3.800	3.427	3.004
15	8.683	6.359	5.417	4.893	4.556	4.318	3.666	3.294	2.868
16	8.531	6.226	5.292	4.773	4.437	4.202	3.553	3.181	2.753
17	8.400	6.112	5.185	4.669	4.336	4.102	3.455	3.084	2.653
18	8.285	6.013	5.092	4.579	4.248	4.015	3.371	2.999	2.566
19	8.185	5.926	5.010	4.500	4.171	3.939	3.297	2.925	2.489
20	8.096	5.849	4.938	4.431	4.103	3.871	3.231	2.859	2.421
25	7.770	5.568	4.675	4.177	3.855	3.627	2.993	2.620	2.169
30	7.562	5.390	4.510	4.018	3.699	3.473	2.843	2.469	2.006
40	7.314	5.179	4.313	3.828	3.514	3.291	2.665	2.288	1.805
50	7.171	5.057	4.199	3.720	3.408	3.186	2.562	2.183	1.683
100	6.895	4.824	3.984	3.513	3.206	2.988	2.368	1.983	1.427
∞	6.635	4.605	3.782	3.319	3.017	2.802	2.185	1.791	1.000

Table D.7 Using the R function doubleWishart (Turgeon, 2018) with $p = 2$, these are the upper critical values θ_α of $\theta(p, \nu_1, \nu_2)$, the largest root of $|\boldsymbol{B} - \theta(\boldsymbol{W} + \boldsymbol{B})| = 0$, with ν_1 = hypothesis degrees of freedom and ν_2 = error degrees of freedom.

		Degrees of freedom ν_2									
ν_1	$1-\alpha$	2	4	6	8	10	12	14	16	18	20
	0.90	0.724	0.824	0.870	0.897	0.914	0.926	0.936	0.943	0.949	0.953
6	0.95	0.795	0.871	0.905	0.925	0.938	0.947	0.954	0.959	0.963	0.966
	0.99	0.895	0.935	0.953	0.963	0.969	0.974	0.977	0.980	0.982	0.983
	0.90	0.563	0.687	0.753	0.796	0.826	0.848	0.865	0.878	0.889	0.899
9	0.95	0.638	0.745	0.800	0.835	0.860	0.878	0.892	0.903	0.912	0.919
	0.99	0.763	0.836	0.874	0.897	0.912	0.924	0.933	0.940	0.945	0.950
	0.90	0.457	0.583	0.658	0.709	0.746	0.774	0.797	0.815	0.831	0.844
12	0.95	0.528	0.642	0.708	0.753	0.785	0.810	0.829	0.845	0.858	0.869
	0.99	0.654	0.743	0.793	0.826	0.850	0.867	0.881	0.893	0.902	0.910
	0.90	0.384	0.505	0.581	0.636	0.677	0.710	0.736	0.758	0.776	0.792
15	0.95	0.448	0.562	0.632	0.681	0.718	0.747	0.771	0.790	0.807	0.821
	0.99	0.569	0.664	0.721	0.760	0.789	0.812	0.830	0.845	0.857	0.868
	0.90	0.330	0.444	0.520	0.575	0.619	0.653	0.682	0.706	0.727	0.745
18	0.95	0.388	0.498	0.569	0.621	0.660	0.692	0.718	0.740	0.759	0.775
	0.99	0.501	0.598	0.658	0.701	0.734	0.760	0.782	0.799	0.814	0.827
	0.90	0.290	0.397	0.470	0.525	0.569	0.605	0.635	0.661	0.683	0.702
21	0.95	0.343	0.447	0.517	0.569	0.610	0.644	0.671	0.695	0.716	0.733
	0.99	0.448	0.543	0.604	0.650	0.685	0.713	0.737	0.756	0.773	0.788
	0.90	0.258	0.358	0.428	0.482	0.526	0.562	0.593	0.620	0.643	0.664
24	0.95	0.307	0.405	0.473	0.525	0.566	0.601	0.630	0.655	0.676	0.695
	0.99	0.404	0.496	0.558	0.604	0.641	0.671	0.696	0.717	0.735	0.751
	0.90	0.232	0.326	0.393	0.446	0.489	0.525	0.556	0.584	0.607	0.629
27	0.95	0.277	0.370	0.436	0.487	0.528	0.563	0.592	0.618	0.640	0.660
	0.99	0.368	0.457	0.518	0.564	0.601	0.632	0.658	0.681	0.700	0.718
	0.90	0.211	0.299	0.363	0.414	0.456	0.492	0.524	0.551	0.575	0.597
30	0.95	0.253	0.341	0.404	0.454	0.494	0.529	0.559	0.585	0.608	0.628
	0.99	0.338	0.423	0.483	0.529	0.566	0.598	0.624	0.648	0.668	0.686

Table D.7 (Continued)

		Degrees of freedom ν_2									
ν_1	$1-\alpha$	2	4	6	8	10	12	14	16	18	20
	0.90	0.194	0.276	0.337	0.387	0.428	0.463	0.494	0.522	0.546	0.568
33	0.95	0.232	0.316	0.376	0.424	0.465	0.499	0.529	0.555	0.578	0.599
	0.99	0.312	0.394	0.452	0.497	0.535	0.566	0.593	0.617	0.638	0.657
	0.90	0.179	0.257	0.315	0.363	0.403	0.438	0.468	0.495	0.520	0.542
36	0.95	0.215	0.294	0.352	0.399	0.438	0.472	0.502	0.528	0.551	0.572
	0.99	0.290	0.368	0.424	0.469	0.506	0.538	0.565	0.589	0.611	0.630
	0.90	0.166	0.240	0.295	0.341	0.380	0.414	0.444	0.471	0.496	0.517
39	0.95	0.200	0.275	0.331	0.376	0.415	0.448	0.477	0.503	0.527	0.548
	0.99	0.271	0.346	0.400	0.444	0.481	0.512	0.539	0.564	0.585	0.604
	0.90	0.155	0.225	0.278	0.322	0.360	0.394	0.423	0.450	0.473	0.495
42	0.95	0.187	0.258	0.312	0.356	0.393	0.426	0.455	0.481	0.504	0.525
	0.99	0.254	0.326	0.379	0.421	0.457	0.488	0.516	0.540	0.562	0.581
	0.90	0.146	0.212	0.263	0.305	0.342	0.375	0.404	0.430	0.453	0.475
45	0.95	0.176	0.243	0.295	0.338	0.374	0.406	0.435	0.460	0.483	0.504
	0.99	0.239	0.308	0.359	0.401	0.436	0.467	0.494	0.518	0.540	0.559
	0.90	0.137	0.200	0.249	0.290	0.326	0.357	0.386	0.411	0.435	0.456
48	0.95	0.166	0.230	0.280	0.321	0.357	0.388	0.416	0.441	0.464	0.485
	0.99	0.226	0.292	0.342	0.382	0.417	0.447	0.474	0.498	0.519	0.539
	0.90	0.130	0.190	0.237	0.276	0.311	0.342	0.369	0.395	0.418	0.439
51	0.95	0.157	0.218	0.266	0.306	0.341	0.371	0.399	0.423	0.446	0.467
	0.99	0.214	0.278	0.326	0.365	0.399	0.429	0.455	0.479	0.501	0.520
	0.90	0.123	0.180	0.225	0.264	0.297	0.327	0.354	0.379	0.402	0.423
54	0.95	0.148	0.208	0.254	0.292	0.326	0.356	0.383	0.407	0.430	0.450
	0.99	0.203	0.264	0.311	0.350	0.383	0.412	0.438	0.462	0.483	0.502
	0.90	0.117	0.172	0.215	0.252	0.285	0.314	0.341	0.365	0.387	0.408
57	0.95	0.141	0.198	0.242	0.280	0.313	0.342	0.368	0.392	0.414	0.434
	0.99	0.194	0.253	0.298	0.335	0.368	0.396	0.422	0.445	0.467	0.486
	0.90	0.111	0.164	0.206	0.242	0.273	0.302	0.328	0.351	0.373	0.394
60	0.95	0.135	0.189	0.232	0.268	0.300	0.329	0.355	0.378	0.400	0.420
	0.99	0.185	0.242	0.285	0.322	0.354	0.382	0.407	0.430	0.451	0.470

(continued)

Table D.7 (Continued)

		Degrees of freedom ν_2									
ν_1	$1-\alpha$	2	4	6	8	10	12	14	16	18	20
	0.90	0.096	0.142	0.180	0.212	0.241	0.267	0.291	0.313	0.334	0.353
70	0.95	0.116	0.165	0.203	0.236	0.265	0.292	0.316	0.338	0.358	0.378
	0.99	0.161	0.211	0.251	0.285	0.314	0.340	0.364	0.386	0.406	0.425
	0.90	0.084	0.126	0.159	0.189	0.215	0.240	0.262	0.283	0.302	0.320
80	0.95	0.102	0.146	0.181	0.211	0.238	0.262	0.284	0.305	0.325	0.343
	0.99	0.142	0.188	0.224	0.255	0.282	0.307	0.329	0.350	0.369	0.387
	0.90	0.068	0.102	0.130	0.155	0.178	0.198	0.218	0.236	0.253	0.270
100	0.95	0.083	0.119	0.148	0.173	0.196	0.218	0.237	0.256	0.273	0.289
	0.99	0.115	0.153	0.184	0.211	0.235	0.256	0.276	0.295	0.312	0.329
	0.90	0.057	0.086	0.110	0.131	0.151	0.169	0.187	0.203	0.218	0.233
120	0.95	0.069	0.100	0.125	0.147	0.167	0.186	0.204	0.220	0.236	0.250
	0.99	0.097	0.130	0.156	0.180	0.201	0.220	0.238	0.255	0.271	0.286
	0.90	0.049	0.074	0.095	0.114	0.131	0.148	0.163	0.178	0.191	0.205
140	0.95	0.060	0.086	0.108	0.128	0.146	0.163	0.178	0.193	0.207	0.221
	0.99	0.083	0.112	0.136	0.156	0.175	0.193	0.209	0.224	0.239	0.252
	0.90	0.043	0.065	0.084	0.101	0.116	0.131	0.145	0.158	0.171	0.183
160	0.95	0.052	0.076	0.096	0.113	0.129	0.144	0.158	0.172	0.185	0.197
	0.99	0.073	0.099	0.120	0.139	0.156	0.171	0.186	0.200	0.213	0.226

References and Author Index

Abdi, H. (2010). Partial least squares regression and projection on latent structure regression (PLS Regression). *Wiley Interdiscip. Rev. Comput. Stat.* 2, 97–106. (p. 457)

Abramowitz, M. and I. A. Stegun (1964). *Handbook of Mathematical Functions with Formulas, Graphs, and Mathematical Tables* (ninth Dover printing, tenth GPO printing ed.). New York: Dover. (p. 386)

Akaike, H. (1974). A new look at the statistical model identification. *IEEE Trans. Autom. Control* 19, 716–723. (pp. 194 and 333)

Alashwali, F. and J. T. Kent (2016). The use of a common location measure in the invariant coordinate selection and projection pursuit. *J. Multivariate Anal.* 152, 145–161. (p. 245)

Anderson, T. W. (1958). *An Introduction to Multivariate Statistical Analysis.* New York: Wiley. (pp. xxiii, 79, 91, 92, 130, 138, and 161)

Anderson, E. (1960). A semigraphical method for the analysis of complex problems. *Technometrics* 2, 387–391. (p. 18)

Anderson, T. W. (1963). Asymptotic theory for principal component analysis. *Ann. Math. Stat.* 34, 122–148. (pp. 222, 223, and 230)

Anderson, T. W. (2003). *An Introduction to Multivariate Statistical Analysis* (3rd ed.). Wiley Series in Probability and Mathematical Statistics. Hoboken, NJ: Wiley. (p. 56)

Anderson, T. W. and R. R. Bahadur (1962). Classification into two multivariate normal distributions with different covariance matrices. *Ann. Math. Stat.* 33, 420–431. (p. 306)

Anderson, T. W., S. P. Das Gupta, and G. P. H. Styan (1972). *A Bibliography of Multivariate Analysis*. Edinburgh: Oliver and Boyd. (p. xxii)

Andrews, D. F. (1972). Plots of high dimensional data. *Biometrics* 28, 125–136. (p. 18)

Azzalini, A. (2005). The skew-normal distribution and related multivariate families. *Scand. J. Stat.* 32(2), 159–188. (p. 51)

Azzalini, A. and A. Dalla Valle (1996). The multivariate skew-normal distribution. *Biometrika* 83, 715–726. (pp. 50 and 51)

Baddeley, A., E. Rubak, and R. Turner (2015). *Spatial Point Patterns: Methodology and Applications with R*. London: Chapman and Hall/CRC Press. (pp. 320 and 511)

Balakrishnan, V. and L. D. Sanghvi (1968). Distance between populations on the basis of attribute data. *Biometrics* 24, 859–865. (pp. 400 and 416)

Ball, G. H. and D. J. Hall (1970). Some implications of inter-active graphic computer systems for data analysis and statistics. *Technometrics* 12, 17–31. (p. 18)

Barnett, V. D. (1976). The ordering of multivariate data. *J. R. Stat. Soc. A* 139, 318–355. (p. 114)

Multivariate Analysis, Second Edition. Kanti V. Mardia, John T. Kent, and Charles C. Taylor.

Barnett, V. D. and T. Lewis (1963). A study of the relation between GCE and degree results. *J. R. Stat. Soc. A* 126, 187–216. (p. 290)

Bartlett, M. S. (1939). The standard errors of discriminant function coefficients. *J. R. Stat. Soc. B* Suppl. 6, 169–173. (p. 287)

Bartlett, M. S. (1947). Multivariate analysis. *J. R. Stat. Soc. B* 9, 176–197. (pp. 100 and 363)

Bartlett, M. S. (1951). The effect of standardization on a χ^2 approximation in factor analysis (with an appendix by Lederman, W.). *Biometrika* 38, 337–344. (p. 229)

Bartlett, M. S. (1954). A note on multiplying factors for various chi-squared approximations. *J. R. Stat. Soc. B* 16, 296–298. (p. 269)

Bartlett, M. S. (1965). Multivariate statistics. In T. H. Waterman and H. J. Morowitz (Eds.), *Theoretical and Mathematical Biology*, New York, pp. 201–224. Blaisdell. (pp. 291 and 345)

Beale, E. M. L. (1970). Selecting an optimum subset. In J. Abadie (Ed.), *Integer and Nonlinear Programming*, Amsterdam, 451–462. North-Holland. (p. 177)

Beale, E. M. L., M. G. Kendall, and D. W. Mann (1967). The discarding of variables in multivariate analysis. *Biometrika* 54, 357–366. (pp. 177 and 241)

Berce, R. and R. Wilbaux (1935). Recherche statistique des relations existant entre le rendment des plantes de grande cultures et les facteurs meteorologiques en Belgique. *Bull. Inst. Agron. Stn. Rech. Gembloux.* 4, 32–81. (p. 249)

Besag, J. (1974). Spatial interaction and the statistical analysis of lattice systems. *J. R. Stat. Soc. B* 36, 192–225. (p. 188)

Bezdek, J. C. (1981). *Pattern Recognition with Fuzzy Objective Function Algorithms*. New York: Plenum Press. (p. 388)

Bhattacharyya, A. (1946). On a measure of divergence between two multinomial populations. *Sankhya* 7, 401–406. (p. 400)

Biernacki, C., G. Celeux, and G. Govaert (2003). Choosing starting values for the EM algorithm for getting the highest likelihood in multivariate Gaussian mixture models. *Comput. Stat. Data Anal.* 41, 561–575. (p. 385)

Bishop, C. M. (1995). *Neural Networks for Pattern Recognition*. Oxford: Clarendon Press. (p. 339)

Bishop, C. M. (2006). *Pattern Recognition and Machine Learning*. New York: Springer. (p. 339)

Bock, R. D. (1975). *Multivariate Statistical Methods in Behaviourial Research*. New York: McGraw-Hill. (p. 375)

Bollen, K. A. (1989). *Structural Equations with Latent Variables*. New York: Wiley. (p. 277)

Bookstein, F. L. (2018). *A Course in Morphometrics for Biologists. Geometry and Statistics for Studies of Organismal Form*. U.K.: Cambridge University Press. (pp. xx and 465)

Bookstein, F. L., A. P. Streissguth, P. D. Sampson, P. D. Connor, and H. M. Barr (2002). Corpus callosum shape and neuropsychological deficits in adult males with heavy fetal alcohol exposure. *NeuroImage* 15, 233–251. (p. 465)

Box, G. E. P. (1949). A general distribution theory for a class of likelihood criteria. *Biometrika* 30, 317–346. (pp. 136, 139, 141, and 229)

Box, G. E. P. and G. C. Tiao (1973). *Bayesian Inference in Statistical Research*. Reading, MA: Addison Wesley. (pp. 117, 122, and 178)

Breiman, L., J. H. Friedman, R. A. Olshen, and C. J. Stone (1984). *Classification and Regression Trees*. Belmont, CA: Wadsworth International Group. (pp. 329–332)

Brownrigg, R. (2018). *mapdata: Extra Map Databases*. Original S code by Richard A. Becker and Allan R. Wilks. R package version 2.3.0. (p. 511)

Butler, R. W. and R. L. Paige (2011). Exact distributional computations for Roy's statistic and the largest eigenvalue of a Wishart distribution. *Stat. Comput.* 21, 147–157. (p. 92)

Cain, M. K., Z. Zhang, and K. Yuan (2017). Univariate and multivariate skewness and kurtosis for measuring nonnormality: prevalence, influence and estimation. *Behav. Res. Methods* 49, 1716–1735. (p. 147)

Caliński, T. and J. Harabasz (1974). A dendrite method for cluster analysis. *Commun. Stat.-Theor. Meth.* 3, 1–27. (p. 383)

Cambanis, S., S. Huang, and G. Simons (1981). On the theory of elliptically contoured distributions. *J. Multivariate Anal.* 11, 368–385. (p. 56)

Cattell, R. B. (1966). The scree test for the number of factors. *Multivariate Behav. Res.* 1, 245–276. (p. 215)

Cavalli-Sforza, L. L. and A. W. F. Edwards (1967). Phylogenetic analysis: models and estimation procedures. *Evolution* 21, 550–570. (p. 401)

Chacón, J. E. and T. Duong (2018). *Multivariate Kernel Smoothing and its Applications.* Boca Raton, FL: CRC Press, Taylor & Francis Group. (p. 322)

Chernoff, H. (1973). Using faces to represent points in k-dimensional space graphically. *J. Am. Stat. Assoc.* 68, 361–368. (p. 18)

Chiani, M. (2016). Distribution of the largest root of a matrix for Roy's test in multivariate analysis of variance. *J. Multivariate Anal.* 143, 467–471. (p. 92)

Chipman, H. A., E. I. George, and R. E. McCulloch (2010). BART: Bayesian additive regression trees. *Ann. Appl. Stat.* 4, 266–298. (p. 325)

Christensen, R. (2006). *Log-Linear Models and Logistic Regression.* Springer Science & Business Media. (p. 200)

Christensen, W. F. and A. C. Rencher (1997). A comparison of type I error rates and power levels for seven solutions to the multivariate Behrens–Fisher problem. *Commun. Stat. Simul. Comput.* 26, 1251–1273. (p. 144)

Clifford, P. and J. M. Hammersley (1971). Markov fields on finite graphs and lattices. https://ora.ox.ac.uk/objects/uuid:4ea849da-1511-4578-bb88-6a8d02f457a6. (p. 188)

Cochran, W. G. (1934). The distribution of quadratic forms in a normal system, with applications to the analysis of variance. *Proc. Camb. Philos. Soc.* 30, 178–191. (p. 78)

Cochran, W. G. (1962). On the performance of the linear discriminant function. *Bull. Inst. Intern. Statist.* 39, 345–447. (p. 346)

Comrey, A. L. and H. B. Lee (1992). *A First Course in Factor Analysis.* New York: Taylor & Francis. (p. 264)

Cook, R. D. (2022). A slice of multivariate dimension reduction. *J. Multivariate Anal.* 188, Article 104812. (pp. 457 and 461)

Cook, R. D. and X. Zhang (2015). Simultaneous envelopes for multivariate linear regression. *Technometrics* 57, 11–25. (p. 464)

Cook, R. D., I. S. Helland, and Z. Su (2013). Envelopes and partial least squares regression. *J. R. Stat. Soc. B* 75, 851–877. (p. 462)

Cormack, R. M. (1971). A review of classification. *J. R. Stat. Soc. A* 134, 321–367. (p. 408)

Cornish, E. A. (1954). The multivariate t-distribution associated with a set of normal sample deviates. *Aust. J. Phys.* 7, 531–542. (pp. 49 and 67)

Cortes, C. and V. Vapnik (1995). Support-vector networks. *Mach. Learn.* 20, 273–297. (p. 342)

Cover, T. M. and P. E. Hart (1967). Nearest neighbor pattern classification. *IEEE Trans. Inf. Theory* 13, 21–27. (p. 320)

Cox, D. R. (1972). The analysis of multivariate binary data. *Appl. Stat.* 21, 113–120. (p. 52)

Cox, D. R. and L. Brandwood (1959). On a discriminatory problem connected with the works of Plato. *J. R. Stat. Soc. B* 21, 95–200. (pp. 316–318)

Cox, D. R. and D. V. Hinkley (1974). *Theoretical Statistics*. London: Chapman and Hall. (pp. 109, 146, and 362)

Craig, A. T. (1943). A note on the independence of certain quadratic forms. *Ann. Math. Stat.* 14, 195–197. (p. 79)

Craven, P. and G. Wahba (1979). Smoothing noisy data with spline functions. *Numer. Math.* 31, 377–403. (p. 469)

Csardi, G. and T. Nepusz (2006). The igraph software package for complex network research. *InterJournal, Complex Systems*, 1695. (p. 511)

Dagnelie, P. (1975). *Analyse Statistique a Plusieurs Variables*. Brussels: Vander. (pp. 230, 249–251, 253, and 307)

Davis, A. W. (1980). On the effects of moderate multivariate nonnormality on Wilks's likelihood ratio criterion. *Biometrika* 63, 661–670. (p. 147)

Davis, A. W. (1982). On the effects of moderate multivariate nonnormality on Roy's largest root test. *J. Am. Stat. Assoc.* 77, 896–900. (p. 147)

Day, N. E. (1969). Estimating the components of a mixture of normal distributions. *Biometrika* 56, 301–312. (p. 413)

De Boor, C. (2001). *A Practical Guide to Splines*. Springer. (p. 468)

Deemer, W. L. and I. Olkin (1951). The Jacobians of certain matrix transformations useful in multivariate analysis. *Biometrika* 38, 345–367. (p. 33)

Delany, M. J. and M. J. Healy (1966). Variation in white-toothed shrews in the British Isles. *Proc. R. Soc. B* 164, 63–74. (pp. 365, 366, 377, 392, and 393)

Deming, W. E. and F. F. Stephan (1940). On a least squares adjustment of a sampled frequency table when the expected marginal totals are known. *Ann. Math. Stat.* 11, 427–444. (pp. 192 and 198)

Dempster, A. P. (1969). *Elements of Continuous Multivariate Analysis*. Reading, MA: Addison-Wesley. (p. 55)

Dempster, A. P. (1971). An overview of multivariate data analysis. *J. Multivariate Anal.* 1, 316–346. (p. 53)

Dempster, A. P. (1972). Covariance selection. *Biometrics* 28, 157–175. (pp. 53 and 194)

Demšar, J. (2006). Statistical comparisons of classifiers over multiple data sets. *J. Mach. Learn. Res.* 7, 1–30. (p. 318)

Denison, D. G. T., B. K. Mallick, and A. F. M. Smith (1998). A Bayesian CART algorithm. *Biometrika* 85, 363–377. (p. 325)

Di Marzio, M. and C. C. Taylor (2005). Kernel density classification and boosting: an L_2 analysis. *Stat. Comput.* 15, 113–123. (p. 322)

Diaconis, P. and B. Efron (1983). Computer-intensive methods in statistics. *Sci. Am.* 248, 116–131. (p. 225)

Diaz Garcia, J. A., R. G. Jaimez, and K. V. Mardia (1997). Wishart and pseudo-Wishart distributions and some applications to shape theory. *J. Multivariate Anal.* 63, 73–87. (p. 83)

Dickey, J. M. (1967). Matrix variate generalisations of the multivariate t distribution and the inverted multivariate t distribution. *Ann. Math. Stat.* 38, 511–518. (p. 92)

Diestel, R. (1987). Simplicial decompositions of graphs — some uniqueness results. *J. Comb. Theory, Ser. B* 42, 133–145. (p. 187)

Dobson, A. J. and A. G. Barnett (2018). *An Introduction to Generalized Linear Models*. Chapman and Hall/CRC. (p. 336)

Donoho, D. L. (1982). Breakdown properties of multivariate location estimators. *PhD qualifying paper. Dept. Statistics*, Harvard University, Boston, MA. (p. 115)

Donoho, D. L. and P. J. Huber (1983). The notion of breakdown point. In P. Bickel, K. Doksum, and J. Hodges (Eds.), *A Festschrift for Erich L. Lehmann*, Belmont, CA, pp. 157–184. Wadsworth. (pp. 113 and 114)

Dryden, I. L. (2021). *shapes: Statistical Shape Analysis*. R package version 1.2.6. (p. 511)

Dryden, I. L. and K. V. Mardia (2016). *Statistical Shape Analysis, with Applications in R* (2nd ed.). Chichester: Wiley. (pp. xx and 442)

Dunnett, C. W. and M. Sobel (1954). A bivariate generalization of student's t distribution with tables for certain cases. *Biometrika* 41, 153–169. (pp. 50 and 67)

Dutilleul, P. (1999). The MLE algorithm for the matrix normal distribution. *J. Stat. Comput. Simul.* 64, 105–123. (p. 112)

Eaton, M. L. (1972). *Multivariate Statistical Analysis*. University of Copenhagen: Institute of Mathematics and Statistics. (p. 98)

Ebner, B. and N. Henze (2020). Tests for multivariate normality — a critical review with emphasis on weighted L^2 -statistics. *Test* 29, 845–892. (p. 147)

Edwards, A. W. F. and L. L. Cavalli-Sforza (1965). A method of cluster analysis. *Biometrics* 21, 362–375. (p. 383)

Efron, B. and T. Hastie (2016). *Computer Age Statistical Inference: Algorithms, Evidence and Data Science*. New York: Cambridge University Press. (p. 454)

Efron, B. and R. J. Tibshirani (1994). *An Introduction to the Bootstrap*. Monographs on Statistics and Applied Probability. New York: Chapman and Hall/CRC. (p. 224)

Efron, B., T. Hastie, I. Johnstone, and R. Tibshirani (2004). Least angle regression. *Ann. Stat.* 32, 407–499. (p. 454)

el Bouhaddani, S., H.-W. Uh, C. Hayward, G. Jongbloed, and J. Houwing-Duistermaat (2018). Probabilistic partial least squares model: identifiability, estimation and application. *J. Multivariate Anal.* 167, 331–346. (p. 465)

Enis, P. (1973). On the relation $E(Y) = E(E(X|Y))$. *Biometrika* 60, 432–433. (p. 45)

Everitt, B. S. and T. Hothorn (2011). *An Introduction to Applied Multivariate Analysis with R*. New York: Springer. (p. 509)

Everitt, B. S., S. Landau, and M. Leese (2001). *Cluster Analysis* (4th ed.). London: Arnold. (pp. 379, 383, 386, and 403)

Evett, I. W. and E. J. Spiehler (1987). Rule induction in forensic science. Technical report, Central Research Establishment, Home Office Forensic Science Service. (p. 405)

Farebrother, R. W. (1976). Further results on the mean square error of ridge regression. *J. R. Stat. Soc. B* 38(3), 248–250. (p. 473)

Ferraty, F. and P. Vieu (2006). *Nonparametric Functional Data Analysis: Theory and Practice*. Springer Science & Business Media. (p. 466)

Fisher, R. A. (1936). The use of multiple measurements in taxonomic problems. *Ann. Eugen.* 7, 179–188. (pp. 7, 314, and 315)

Fisher, R. A. (1947). The analysis of covariance method for the relation between a part and a whole. *Biometrics* 3, 65–68. (pp. 22 and 111)

Fisher, R. A. (1970). *Statistical Methods for Research Workers* (14th ed). Edinburgh: Oliver and Boyd. (p. 104)

Foley, J. D. and A. Van Dam (1982). *Fundamentals of Interactive Computer Graphics.* Addison-Wesley Systems Programming Series. Reading, MA: Addison-Wesley. (p. 367)

Forbes, P. G. M. and S. Lauritzen (2015). Linear estimating equations for exponential families with application to Gaussian linear concentration model. *Linear Algebra Appl.* 473, 261–283. (p. 194)

Fraley, C. and A. E. Raftery (2002). Model-based clustering, discriminant analysis, and density estimation. *J. Am. Stat. Assoc.* 97(458), 611–631. (p. 385)

Fréchet, M. (1951). Sur les tableaux de correlation dont les mages sont donnees. *Ann. Univ. de Lyon, Sect. A, Ser. 3* 14, 53–77. (pp. 45 and 63)

Frets, G. P. (1921). Heredity of head form in man. *Genetica* 3, 193–384. (p. 126)

Friedman, H. P. and J. Rubin (1967). On some invariate criteria for grouping data. *J. Am. Stat. Assoc.* 62, 1159–1178. (p. 383)

Friedman, J. H. and J. W. Tukey (1974). A projection pursuit algorithm for exploratory data analysis. *IEEE Trans. Comput.* 100(9), 881–890. (p. 244)

Gabriel, K. R. (1970). On the relation between union intersection and likelihood ratio tests. In R. C. Bose (Ed.), *Essays in Probability and Statistics*, Rayleigh, pp. 251–266. University of North Carolina. (p. 146)

Gabriel, K. R. (1971). The biplot graphic display of matrices with application to principal component analysis. *Biometrika* 58, 453–467. (pp. 218 and 220)

Gamage, J., T. Mathew, and S. Weerahandi (2004). Generalized *p*-values and generalized confidence regions for the multivariate Behrens-Fisher problem and MANOVA. *J. Multivariate Anal.* 88, 177–189. (p. 144)

Geiger, D. and J. Pearl (1993). Logical and algorithmic properties of conditional independence and graphical models. *Ann. Stat.* 21, 2001–2021. (p. 186)

Gentleman, R., E. Whalen, W. Huber, and S. Falcon (2020). *graph: graph: A Package to Handle Graph Data Structures.* R package version 1.68.0. (p. 511)

Giri, N. (1968). On tests of the equality of two covariance matrices. *Ann. Math. Stat.* 39, 275–277. (p. 146)

Girshick, M. A. (1939). On the sampling theory of roots of determinantal equations. *Ann. Math. Stat.* 10, 203–224. (pp. 138 and 252)

Glass, D. V. (1954). *Social Mobility in Britain.* London: Routledge and Kegan Paul. (p. 289)

Gnanadesikan, R. (1977). *Methods for Statistical Data Analysis of Multivariate Observations.* New York: Wiley. (p. 18)

Gnanadesikan, R. and S. S. Gupta (1970). A selection procedure for multivariate normal distributions in terms of the generalised variances. *Technometrics* 12, 103–117. (p. 11)

Goodman, N. R. (1963). Statistical analysis based on a certain multivariate complex Gausian distribution (an introduction). *Ann. Math. Stat.* 34, 152–177. (p. 92)

Goodman, L. A. (1972). Some multiplicative models for the analysis of cross classified data. In *Proceedings of the 6th Berkeley Symposium on Mathematical Statistics and Probability*, Volume 1, pp. 649–494. (p. 289)

Goodman, L. A. and W. H. Kruskal (1954). Measures of association for cross classifications. *J. Am. Stat. Assoc.* 49, 732–764. (p. 409)

Govindaraju, V. and C. R. Rao (2013). Editors: *Machine Learning: Theory and Applications, Handbook of Statistics*, Vol. 31, Elsevier/Academic Press. (p. 299)

Govindaraju, V., A. S. R. Srinivasa Rao and C. R. Rao (2023), Editors: *Deep Learning, Handbook of Statistics*, Vol. 48, Elsevier/Academic Press. (p. 299)

Gower, J. C. (1966). Some distance properties of latent root and vector methods in multivariate analysis. *Biometrika* 53, 315–328. (p. 424)

Gower, J. C. (1968). Adding a point to vector diagrams in multivariate analysis. *Biometrika* 55, 582–585. (p. 444)

Gower, J. C. (1971a). A general coefficient of similarity and some of its properties. *Biometrics* 27, 857–874. (pp. 404 and 417)

Gower, J. C. (1971b). Statistical motheds of comparing different multivariate analyses of the same data. In F. R. Hodson, D. G. Kendall, and P. Tautu (Eds.), *Mathematics in the Archaeological and Historical Sciences*, Edinburgh, pp. 138–149. Edinburgh University Press. (p. 440)

Gower, J. C. and G. J. S. Ross (1969). Minimum spanning trees and single linkage cluster analysis. *Appl. Stat.* 18, 54–64. (pp. 142, 229 and 238)

Graybill, F. A. (1969). *Introduction to Matrices with Applications in Statistics*. Belmont, CA: Wadsworth. (p. 475)

Green, B. F. (1952). The orthogonal approximation of an oblique structure in factor analysis. *Psychometrika* 17, 429–440. (p. 440)

Green, P. J., N. L. Hjort, and S. Richardson (2003). *Highly Structured Stochastic Systems*. New York: Oxford University Press. (p. 119)

Green, P. J. and A. Thomas (2013). Sampling decomposbale graphs using a Markov chain on junction trees. *Biometrika* 100, 91–110. (p. 188)

Greenacre, M. J. (1984). *Theory and Applications of Correspondence Analysis*. London: Adademic Press. (p. 230)

Greenacre, M. (2010). *Biplots in Practice*. Number 2011113 in Books. Fundacion BBVA / BBVA Foundation. (p. 230)

Greenacre, M. (2016). *Correspondence Analysis in Practice* (3rd ed.). New York: Chapman and Hall/CRC. (p. 230)

Gupta, R. D. and D. S. P. Richard (1985). Hypergeometric functions of scalar matrix argument are expressible in terms of classical hypergeometric functions. *SIAM J. Math. Anal.* 16, 852–858. (p. 92)

Guttman, L. (1968). A general non-metric technique for finding the smallest co-ordinate space for a configuration of points. *Psychometrika* 33, 469–506. (p. 408)

Haberman, S. J. (1974). Log-linear models for frequency tables with ordered classifications. *Biometrics* 30, 589–600. (p. 192)

Hand, D. J. (1982). *Kernel Discriminant Analysis*. Chichester: Wiley. (p. 322)

Hand, D. J. and V. Vinciotti (2003). Choosing *k* for two-class nearest neighbour classifiers with unbalanced classes. *Pattern Recognit. Lett.* 24, 1555–1562. (p. 321)

Hartigan, J. A. (1975). *Clustering Algorithms*. New York: Wiley. (p. 383)
Hartigan, J. A. (1978). Asymptotic distributions for clustering criteria. *Ann. Stat.* 6, 117–131. (p. 383)
Hastie, T. and B. Efron (2013). *lars: Least Angle Regression, Lasso and Forward Stagewise*. R package version 1.2. (pp. 455 and 511)
Hastie, T. and B. Efron (2022). *lars: Least Angle Regression, Lasso and Forward Stagewise*. R package version 1.3. (p. 455)
Hastie, T., R. Tibshirani, and J. H. Friedman (2009). *Elements of Statistical Learning*. New York: Springer. (p. 298)
Hastie, T., R. Tibshirani, and M. Wainwright (2015). *Statistical Learning with Sparsity: The Lasso and Generalizations*. Chapman & Hall/CRC monographs on statistics & applied probability, p. 143. Boca Raton, FL: CRC Press. (p. 454)
Helland, I. S. (1992). Maximum likelihood regression on relevant components. *J. R. Stat. Soc. B* 54, 637–647. (p. 462)
Helwig, N. E. (2018). *ICA: Independent Component Analysis*. R package version 1.0-2. (pp. 244, 247, and 511)
Hill, M. O. (1974). Correspondence analysis: a neglected multivariate method. *Appl. Stat.* 23, 340–354. (p. 230)
Hill, R. C., T. B. Formby, and S. R. Johnson (1977). Component selection norms for principal components regression. *Commun. Stat.-Theor. Meth. A* 6, 309–334. (p. 243)
Hjort, N. (1986). Notes on the theory of statistical symbol recognition. Technical Report 778, Norwegian Computing Centre, Oslo. (p. 343)
Hoerl, A. E. and R. W. Kennard (1970). Ridge regression: biased estimation for nonorthogonal problems. *Technometrics* 12, 55–67 and 69–82. (pp. 243, 256, 295, and 453)
Höffding, W. (1940). Masstabinvariante korrelationstheorie. *Schr. Math. Inst. Inst. Angew. Math. Univ. Berl.* 5, 181–233. (p. 63)
Højsgaard, S. (2013). gRim (version 0.1-17): Graphical interaction models. CRAN.R-project .org/package=gRim. (p. 195)
Højsgaard, S. and S. Lauritzen (2007). *Inference in Graphical Gaussian Models with Edge and Vertex Symmetries with the gRc Package for R*. (p. 194)
Højsgaard, S. and S. L. Lauritzen (2008). Graphical gaussian models with edge and vertex symmetries. *J. R. Stat. Soc. B* 70, 1005–1027. (pp. 194 and 195)
Højsgaard, S., D. Edwards, and S. Lauritzen (2012). *Graphical Models with R*. New York: Springer. (pp. 184, 195, and 511)
Holgate, P. (1964). Estimation for the bivariate Poisson distribution. *Biometrika* 51, 152–177. (p. 66)
Hooper, J. W. (1959). Simultaneous equations and canonical correlation theory. *Econometrica* 27, 245–256. (p. 171)
Hopkins, J. W. (1966). Some considerations in multivariate allometry. *Biometrics* 22, 747–760. (p. 237)
Horst, P. (1965). *Factor Analysis of Data Matrices*. New York: Holt, Rinehart and Winston. (p. 271)
Horváth, L. and P. Kokoszka (2012). *Inference for Functional Data with Applications*, Volume 200. Springer Science & Business Media. (p. 466)
Hotelling, H. (1931). The generalization of student's ratio. *Ann. Math. Stat.* 2, 360–378. (p. 83)

Hotelling, H. (1933). Analysis of a complex of statistical variables into principal components. *J. Educ. Psychol.* 24, 417–441, 498–520. (p. 207)
Hotelling, H. (1936). Relations between two sets of variables. *Biometrika* 28, 321–377. (pp. 138, 171, and 294)
Huber, P. J. (1972). Robust statistics: a review. *Ann. Math. Stat.* 43, 1041–1067. (p. 113)
Huber, P. J. (1985). Projection pursuit. *Ann. Stat.* 13, 435–475. (p. 244)
Hubert, M., P. J. Rousseeuw, and S. Van Aelst (2008). High-breakdown robust multivariate methods. *Stat. Sci.* 23, 92–119. (p. 116)
Husson, F., S. Lê, and J. Pagès (2017). *Exploratory Multivariate Analysis by Example Using R* (2nd ed.). Boca Raton, FL: Chapman & Hall. (p. 509)
Hyvärinen, A. (1999). Fast and robust fixed-point algorithms for independent component analysis. *Neural Netw.* 10, 626–634. (p. 247)
Hyvärinen, A. and E. Oja (2000). Independent component analysis: algorithms and applications. *Neural Netw.* 13, 411–430. (p. 247)
Hyvärinen, A., J. Karhunen, and E. Oja (2001). *Independent Component Analysis*. New York: Wiley. (p. 245)
Jain, A. K. and R. C. Dubes (1988). *Algorithms for Clustering Data*. Prentice-Hall, Inc. (p. 379)
Jain, A., R. Duin, and J. Mao (2000). Statistical pattern recognition: a review. *IEEE Trans. Pattern Anal. Mach. Intell.* 22(1), 4–37. (p. 318)
James, A. T. (1964). Distributions of matrix variates and latent roots derived from normal samples. *Ann. Math. Stat.* 35, 475–501. (p. 92)
Jardine, N. and R. Sibson (1971). *Mathematical Taxonomy*. New York: Wiley. (pp. 396 and 403)
Jeffers, J. N. R. (1967). Two case studies in the application of principal component analysis. *Appl. Stat.* 16, 225–236. (pp. 175, 217, 241, and 243)
Jeffreys, H. (1961). *Theory of Probability*. Oxford: Clarendon Press. (p. 118)
Joe, H. (2014). *Dependence Modeling with Copulas*. CRC Press. (pp. 62 and 63)
Johnson, N. L. and S. Kotz (1972). *Distribution in Statistics: Continuous Multivariate Distributions*. New York: Wiley. (p. 51)
Johnstone, I. (2001). On the distribution of the largest eigenvalue in principal components analysis. *Ann. Stat.* 29, 295–327. (p. 248)
Jolicoeur, P. and Mosimann (1960). Size and shape variation in the painted turtle: a principal component analysis. *Growth* 24, 339–354. (pp. 142 and 238)
Jolliffe, I. T. (1972). Discarding variables in principal component analysis. I: Artificial data. *Appl. Stat.* 21, 160–173. (p. 241)
Jolliffe, I. T. (1973). Discarding variables in principal component analysis. II: Real data. *Appl. Stat.* 22, 21–31. (p. 241)
Jones, M. C. and R. Sibson (1987). What is projection pursuit? *J. R. Stat. Soc. A* 150, 1–37. (p. 244)
Joreskog, K. G. (1967). Some contributions to maximum likelihood factor analysis. *Psychometrika* 32, 443–482. (pp. 266–268 and 279)
Joreskog, K. G. (1970). A general method for analysis of covariance structures. *Biometrika* 57, 239–251. (p. 277)
Joreskog, K. G. and D. Sorbom (1977). Statistical models and methods for analysis of longitudinal data. In D. J. Aigner and A. S. Goldberger (Eds.), *Latent Variables in Socio-economic Modules*, Amsterdam, pp. 235–285. North-Holland. (p. 277)

Jung, S. and J. S. Marron (2009). Pca consistency in high dimension, low sample size context. *Ann. Stat.* 37, 4104–4130. (pp. 247–249)

Kagan, A. M., Y. V. Linnik, and C. R. Rao (1973). *Characterization Problems in Mathematical Statistics*. New York: Wiley. (pp. 52 and 96)

Kaiser, H. F. (1958). The varimax criterion for analytic rotation in factor analysis. *Psychometrika* 23, 187–200. (pp. 215, 271, and 272)

Karlis, D. and E. Xekalaki (2003). Choosing initial values for the EM algorithm for finite mixtures. *Comput. Stat. Data Anal.* 41, 577–590. (p. 385)

Kass, G. V. (1980). An exploratory technique for investigating large quantities of categorical data. *Appl. Stat.* 29, 119–127. (p. 327)

Kaufman, L. and P. J. Rousseeuw (1990). *Finding Groups in Data: An Introduction to Cluster Analysis*. New York: Wiley. (pp. 391 and 409)

Kelker, D. (1970). Distribution theory of spherical distributions and a location-scale parameter generalization. *Sankhyā* 32, 419–430. (p. 56)

Kendall, D. G. (1971). Seriation from abundance matrices. In F. R. Hodson, D. G. Kendall, and P. Tautu (Eds.), *Mathematics in the Archaeological and Historical Sciences*, Edinburgh, pp. 215–251. Edinburgh University Press. (pp. 417, 421, 437, and 440)

Kendall, M. G. (1975). *Multivariate Analysis*. London: Griffin. (pp. xxiii, 254, and 273)

Kendall, M. G. and A. Stuart (1967). *The Advanced Theory of Statistics*, Volume 2. London: Griffin. (p. 214)

Kendall, M. G. and A. Stuart (1973). *The Advanced Theory of Statistics*, Volume 3. London: Griffin. (pp. 128 and 144)

Kent, J. T. and K. V. Mardia (2022). *Spatial Analysis*. Hoboken, NJ: Wiley. (pp. 186, 188 and 466)

Kent, J. T. and D. E. Tyler (1988). Maximum likelihood estimation for the wrapped Cauchy distribution. *J. Appl. Stat.* 15, 247–254. (p. 117)

Kepner, J. L., S. Z. Keith, and J. D. Harper (1989). A note on evaluating a certain orthant probability. *Am. Stat.* 43(1), 48–49. (p. 46)

Khatri, C. G. (1965). Classical statistical analysis based on a certain multivariate complex Gaussian distributions. *Ann. Math. Stat.* 36, 98–114. (p. 92)

Khatri, C. G. and K. C. S. Pillai (1965). Some results on the non-central multivariate beta distribution and moments of traces of two matrices. *Ann. Math. Stat.* 36, 1511–1520. (pp. 92 and 93)

Kiefer, J. and R. Schwartz (1965). Admissible Bayes character of t^2, r^2 and other fully invariant tests for classical multivariate normal problems. *Ann. Math. Stat.* 36, 747–770. (p. 146)

Koller, D. and N. Friedman (2009). *Probabilistic Graphical Models: Principles and Techniques*. Cambridge, MA: MIT Press. (p. 204)

Komsta, L. and F. Novomestky (2015). *moments: Moments, Cumulants, Skewness, Kurtosis and Related Tests*. R package version 0.14. (p. 511)

Korin, B. P. (1968). On the distribution of a statistic used for testing a covariance matrix. *Biometrika* 55, 171–178. (pp. 130 and 136)

Korkmaz, S., D. Goksuluk, and G. Zararsiz (2014). MVN: An R package for assessing multivariate normality. *R Journal* 6(2), 151–162. (p. 147)

Kotz, S. (1975). Multivariate distributions at a cross road. In G. P. Patil, S. Kotz, and J. K. Ord (Eds.), *A Modern Course on Statistical Distributions in Scientific Work*, Volume 1, Dordrecht, pp. 247–270. Springer Netherlands. (p. 57)

Krantz, S. G., A. S. R. Srinivasa Rao, and C. R. Rao (2023). Editors: *Artificial Intelligence, Handbook of Statistics*, Vol. 49, Academic Press. (p. 299)

Krishnamoorthy, A. S. (1951). Multivariate binomial and Poisson distributions. *Sankhyā* 11, 117–124. (p. 66)

Krishnamoorthy, K. and J. Yu (2004). Modified Nel and Van der Merwe test for the multivariate Behrens-Fisher problem. *Stat. Probab. Lett.* 66, 161–169. (pp. 144 and 145)

Kruskal, W. H. (1958). Ordinal measures of association. *J. Am. Stat. Assoc.* 53, 814–861. (p. 69)

Kruskal, J. B. (1964). Non-metric multidimensional scaling. *Psychometrika* 29, 1–27, 115–129. (pp. 438 and 439)

Kshirsagar, A. M. (1960). Some extensions of multivariate *t*-distribution and the multivariate generalization of the distribution of the regression coefficient. *Proc. Camb. Philos. Soc.* 57, 80–86. (pp. 92 and 93)

Kshirsagar, A. M. (1961). The non-central multivariate beta distribution. *Ann. Math. Stat.* 32, 104–111. (p. 92)

Kshirsagar, A. M. (1972). *Multivariate Analysis*. New York: Marcell Dekker. (pp. xxiii, 222, 228, 287, and 375)

Lachenbruch, P. and M. Mickey (1968). Estimation of error rates in discriminant analysis. *Technometrics* 10, 1–11. (p. 303)

Lancaster, H. O. (1969). *The Chi-squared Distribution*. New York: Willey. (p. 79)

Lance, G. N. and W. J. Williams (1967). A general theory of classificatory sorting strategies: 1. Hierarchical systems. *Comput. J.* 9, 373–380. (p. 414)

Lauritzen, S. L. (1996). *Graphical Models*. Oxford University Press. (pp. 184, 187, 188, 190, and 192)

Lawley, D. N. (1959). Test of significance in canonical analysis. *Biometrika* 46, 59–66. (p. 295)

Lawley, D. N. and A. E. Maxwell (1971). *Factor Analysis as a Statistical Method* (2nd ed.). London: Butterworths. (pp. 264 and 271)

Lévy, P. (1937). *Théorie de l'Addition des Variables Aléatoires*. Paris: Gauthier-Villars. (p. 57)

Lewis, T. (1970). *"The Statistician as a Member of Society". Inaugural lecture*. Ph.D. thesis, University of Hull. (pp. 290–292)

Liland, K. H., B.-H. Mevik, and R. Wehrens (2021). *pls: Partial Least Squares and Principal Component Regression*. R package version 2.8-0. (pp. 456, 474, and 511)

Lim, T.-S., W.-Y. Loh, and Y.-S. Shih (2000). A comparison of prediction accuracy, complexity, and training time of thirty-three old and new classification algorithms. *Mach. Learn.* 40, 203–228. (p. 318)

Lingoes, J. C. (1970). Some boundary conditions for a monotone analysis of symmetric matrices. *Psychometrika* 36, 195–203. (p. 446)

Lopuhaä, H. P. and P. J. Rousseeuw (1991). Breakdown points of affine equivariant estimators of multivariate location and covariance matrices. *Ann. Stat.* 19, 229–248. (p. 113)

Lubischew, A. A. (1962). On the use of discriminant functions in taxonomy. *Biometrics* 18, 455–477. (p. 344)

Lucy, D. (2012). *blighty: United Kingdom Coastlines*. R package version 3.1-4. (p. 511)

Lumley, T. (2020). *leaps: Regression Subset Selection*. Based on Fortran code by Alan Miller. R package version 3.1. (pp. 456 and 511)

MacFarlane, A., N. Dattani, R. Gibson, G. Harper, P. Martin, M. Scanlon, M. Newburn, and M. Cortina-Borja (2019). Births and their outcomes by time, day and year: a retrospective birth cohort data linkage study.. *Health Serv. Delivery Res.* 7(18). doi: 10.3310/hsdr07180. (p. 255)

MacLachlan, G. J. (1992). *Discriminant Analysis and Statistical Pattern Recognition*. Wiley Series in Probability and Statistics. New York: Wiley. (pp. 332, 333, and 343)

MacLachlan, G. J. and T. Krishnan (1997). *The EM Algorithm and Extensions*. New York: Wiley. (pp. 384 and 414)

Maddala, G. S. and K. Lahiri (2009). *Introduction to Econometrics* (4th ed.). New Delhi, India: Wiley. (p. xx)

Madsen, M. (1976). Statistical analysis of multiple contingency tables. Two examples. *Scand. J. Stat.* 3, 97–106. (p. 196)

Marchenko, V. A. and L. A. Pastur (1967). Distribution of eigenvalues for some sets of random matrices. *Math. USSR Sb.* 1, 457–483. (p. 248)

Mardia, K. V. (1962). Multivariate Pareto distributions. *Ann. Math. Stat.* 33, 1008–1015. (pp. 44, 50, 66, and 123)

Mardia, K. V. (1964a). Exact distributions of extremes, ranges and midranges in samples from any multivariate population. *J. Indian Stat. Assoc.* 2, 126–130. (p. 46)

Mardia, K. V. (1964b). Some results on the order statistics of the multivariate normal and Pareto type I populations. *Ann. Math. Stat.* 35, 1815–1818. (pp. 50, 66, and 123)

Mardia, K. V. (1967a). A non-parametric test for the bivariate two-sample location problem. *J. R. Stat. Soc. B* 29, 320–342. (p. 151)

Mardia, K. V. (1967b). Some contributions to contingency-type bivariate distributions. *Biometrika* 54, 235–249. (pp. 45, 64, 68, and 69)

Mardia, K. V. (1968). Small sample power of a non-parametric test for the bivariate two-sample location problem. *J. R. Stat. Soc. B* 30, 83–92. (p. 151)

Mardia, K. V. (1969). On the null distribution of a non-parametric test for the two-sample problem. *J. R. Stat. Soc. B* 31, 98–102. (p. 151)

Mardia, K. V. (1970a). *Families of Bivariate Distributions*. London (No. 27 of Griffin's Statistical Monographs and Courses): Griffin. (p. 44)

Mardia, K. V. (1970b). Measures of multivariate skewness and kurtosis with applications. *Biometrika* 57, 519–530. (pp. 20, 24, 29, 30, 146, 147, and 149)

Mardia, K. V. (1970c). A translation family of bivariate distributions and Fréchet bounds. *Sankhyā* A, 32, 119–122. (pp. 45 and 65)

Mardia, K. V. (1971). The effect of non-normality on some multivariate tests and robustness to non-normality in the linear model. *Biometrika* 58, 105–121. (pp. 146, 149, and 358)

Mardia, K. V. (1972). A multisample uniform scores test on a circle and its parametric competitor. *J. R. Stat. Soc. B* 34, 102–113. (p. 151)

Mardia, K. V. (1974). Applications of some measures of multivariate skewness and kurtosis in testing normality and robustness studies. *Sankhyā B* 36, 115–128. (pp. 65, 147, and 149)

Mardia, K. V. (1975). Assessment of multinormality and the robustness of Hotelling's T^2 test. *Appl. Stat.* 24, 163–171. (pp. 52 and 149)

Mardia, K. V. (1978). Some properties of classical multidimensional scaling. *Commun. Stat.-Theor. Meth. A* 7, 1223–1241. (pp. 146, 431, 432, and 446)

Mardia, K. V. (1980). Tests of univariate and multivariate normality. In P. Krishnaiah (Ed.), *Handbook of Statistics, Analysis of Variance*, Volume 1, Amsterdam, pp. 279–320. North-Holland. (p. 147)

Mardia, K. V. (1985). *Mardia's Test of Multinormality*, Volume 5, pp. 217–221. New York: Wiley. (p. 147)

Mardia, K. V. (1987). Discussion to "What is projection pursuit?" By Jones, M. C. and Sibson, R. *J. R. Stat. Soc. A* 150, 22–23. (pp. 147, 148, and 244)

Mardia, K. V. (2023). Fisher's pioneering work on discriminant analysis and its impact on AI. https://arxiv.org/abs/2309.04774. (pp. 5, 18, 298, 315, and 475)

Mardia, K. V. and C. R. Goodall (1993). Spatial-temporal analysis of multivariate environmental monitoring data. *Multivariate Environ. Stat.* 6, 347–386. (pp. 41 and 112)

Mardia, K. V. and D. Holmes (1980). A statistical analysis of megalithic data under elliptic patterns. *J. R. Stat. Soc. A* 143, 293–302. (pp. 56 and 68)

Mardia, K. V. and P. E. Jupp (2000). *Directional Statistics*. Chichester: Wiley. (pp. xx, 151, and 383)

Mardia, K. V. and J. T. Kent (1991). Rao score tests for goodness of fit and independence. *Biometrika* 78, 355–363. (pp. 147 and 148)

Mardia, K. V. and A. D. Riley (2021). The classical multidimensional scaling revisited. (pp. 434 and 446)

Mardia, K. V. and B. D. Spurr (1977). On some tests for the bivariate two-sample location problem. *J. Am. Stat. Assoc.* 72, 994–995. (p. 151)

Mardia, K. V. and J. W. Thompson (1972). Unified treatment of moment formulae. *Sakhyā A* 34, 121–132. (p. 45)

Mardia, K. V. and P. J. Zemroch (1978). *Tables of the F- and Related Distributions with Algorithms*. London: Academic Press. (p. 91)

Mardia, K. V., F. Bookstein, and I. Moreton (2000). Statistical assessment of bilateral symmetry. *Biometrika* 87, 285–300. Correction: *Biometrika* (2005), 92, 249–250. (p. 155)

Mardia, K. V., H. Wiechers, B. Eltzner, and S. F. Huckemann (2022). Principal component analysis and clustering on manifolds. *J. Multivariate Anal.* 188, Article 104862. (p. 390)

Maronna, R. A. (1976). Robust M-estimators of multivariate location and scatter. *Ann. Stat.* 4, 51–67. (p. 113)

Marriott, F. H. (1952). Tests of significance in canonical analysis. *Biometrika* 39, 58–64. (p. 287)

Marriott, F. H. (1971). Practical problems in a method of cluster analysis. *Biometrics* 27, 501–514. (p. 383)

Marshall, A. W. and I. Olkin (1967). A multivariate exponential distribution. *J. Am. Stat. Assoc.* 62, 30–44. (pp. 52 and 66)

Massart, D. L., B. G. M. Vandeginste, L. M. C. Buydens, S. de Jong, P. J. Lewi, and J. Smeyers-Verbeke (1998). *Handbook of Chemometrics and Qualimetrics: Part B*. Elsevier. (p. 462)

Massy, F. W. (1965). Principal component analysis in exploratory data research. *J. Am. Stat. Assoc.* 60, 234–256. (p. 242)

Maxwell, A. E. (1961). Recent trends in factor analysis. *J. R. Stat. Soc. A* 124, 49–59. (p. 279)

McNeil, A. J., R. Frey, and P. Embrechts (2015). *Quantitative Risk Management: Concepts, Techniques and Tools – revised edition*. Princeton University Press. (p. 62)

McRae, D. J. (1971). MICKA, a FORTRAN IV iterative *k*-means cluster analysis program. *Behav. Sci.* 16, 423–424. (p. 383)

Meijer, R. J. and J. J. Goeman (2013). Efficient approximate *k*-fold and leave-one-out cross-validation for ridge regression. *Biom. J.* 55(2), 141–155. (p. 453)

Meilă, M. (2007). Comparing clusterings — an information based distance. *J. Multivariate Anal.* 98, 873–895. (p. 409)

Mercer, J. (1909). Functions of positive and negative type, and their connection the theory of integral equations. *Philos. Trans. R. Soc. A* 209, 415–446. (p. 467)

Mevik, B.-H. and R. Wehrens (2007). The PLS package: principal component and partial least squares regression in R. *J. Stat. Softw.* 18, 1–24. (p. 462)

Meyer, D., E. Dimitriadou, K. Hornik, A. Weingessel, and F. Leisch (2021). *e1071: Misc Functions of the Department of Statistics, Probability Theory Group (Formerly: E1071), TU Wien*. R package version 1.7-9. (p. 511)

Michie, D., D. J. Spiegelhalter, and C. C. Taylor (1994). *Machine Learning, Neural and Statistical Classification*. Ellis Horwood. (pp. 318 and 366)

Miller, A. (2002). *Subset Selection in Regression*. Chapman & Hall. (p. 174)

Miller, M. B. (2018). *Quantitative Financial Risk Management*. Wiley. (p. 62)

Milligan, G. W. (1981). A Monte Carlo study of thirty internal criterion measures for cluster analysis. *Psychometrika* 46, 187–199. (p. 409)

Milligan, G. W. and M. C. Cooper (1985). An examination of procedures for determining the number of clusters in a data set. *Psychometrika* 50, 159–179. (p. 383)

Mitra, S. K. (1969). Some characteristic and non-characteristic properties of the Wishart distribution. *Sakhyā A* 31, 19–22. (pp. 92 and 95)

Morgenstern, D. (1956). Einfache beispiele zweidimesionaler verteilungen. *Mitt. Math. Stat.* 8, 234–235. (pp. 26 and 30)

Morrison, D. F. (1976). *Multivariate Statistical Methods* (2nd ed.). New York: McGraw-Hill. (pp. xxiii, 347, 369, 373, and 375)

Mosimann, J. E. (1970). Size allometry: size and shape variables with characterizations of the lognormal and generalized gamma distributions. *J. Am. Stat. Assoc.* 65, 930–945. (p. 237)

Nadarajah, S. (2003). The Kotz-type distribution with applications. *Statistics* 37, 341–358. (p. 57)

Narain, R. D. (1950). On the completely unbiased character of tests of independence in multivariate normal system. *Ann. Math. Stat.* 21, 293–298. (p. 138)

Nelsen, R. B. (2007). *An Introduction to Copulas*. Springer Science & Business Media. (p. 68)

Nielsen, T. D. and F. V. Jensen (2009). *Bayesian Networks and Decision Graphs*. Springer Science & Business Media. (p. 203)

Nychka, D., R. Furrer, J. Paige, and S. Sain (2017). fields: Tools for Spatial Data. R package version 12.5. (p. 511)

Ogawa, J. (1949). On the independence of linear and quadratic forms of a random sample from a normal population. *Ann. Inst. Math. Stat.* 1, 83–108. (p. 80)

O'Hagan, A. and J. J. Forster (2004). *Kendall's Advanced Theory of Statistics, volume 2B: Bayesian Inference, second edition*. London: Arnold. (p. 119)

Oja, H. (2010). *Multivariate Nonparametric Methods with R: An Approach Based on Spatial Signs and Ranks*. New York: Springer. (p. 149)

Oja, H. and R. H. Randles (2004). Multivariate nonparametric tests. *Stat. Sci.* 19, 598–605. (p. 149)

Olkin, I. and T. L. Tomsky (1975). A new class of multivariate tests based on the union-intersection principle. *Bull. Inst. Stat. Inst.* 46, 202–204. (p. 137)

Ossani, P. C. and M. A. Cirillo (2021). *Projection Pursuit*. R package version 1.0.2. (p. 511)

Park, T. and G. Casella (2008). The Bayesian lasso. *J. Am. Stat. Assoc.* 103(482), 681–686. (p. 454)

Pearce, S. C. (1965). The measurement of a living organism. *Biometrie-Paraximetrie* 6, 143–152. (p. 239)

Pearson, E. S. (1956). Some aspects of the geometry of statistics. *J. R. Stat. Soc. A* 119, 125–146. (p. 16)

Pearson, K. (1901). On lines and planes of closest fit to systems of points in space. *Philos. Mag.* (6)2, 559–572. (p. 207)

Pearson, E. S. and H. O. Hartley (1972). *Biometrika Tables for Statisticians*, Volume 2. Cambridge: Cambridge University Press. (pp. 91, 92, 127, 130, and 141)

Penrose, K. W., A. G. Nelson, and A. G. Fisher (1985). Generalized body composition prediction equation for men using simple measurement techniques. *Med. Sci. Sports Exerc.* 17, 189. (p. 239)

Pillai, K. C. S. (1955). Some new test criteria in multivariate analysis. *Ann. Math. Stat.* 26, 117–121. (p. 146)

Pillai, K. C. S. and K. Jayachandran (1967). Power comparison of tests of two multivariate hypotheses based on four criteria. *Biometrika* 54, 195–210. (p. 146)

Plackett, R. L. (1965). A class of bivariate distributions. *J. Am. Stat. Assoc.* 60(310), 516–522. (pp. 64 and 68)

Presidential Commission on the Space Shuttle Challenger Accident (1986). Report of the presidential commission on the space shuttle challenger accident (vols. 1 & 2). http://history .nasa.gov/rogersrep/genindex.htm. (p. 352)

Press, S. J. (1972). *Applied Multivariate Analysis*. New York: Holt, and Rinehart, and Winston. (pp. xxiii, 57, 118, and 124)

Puri, M. L. and P. K. Sen (1971). *Non-parametric Methods in Multivariate Analysis*. New York: Wiley. (p. 151)

Quinlan, J. R. (1993). *C4.5: Programs for Machine Learning*. San Mateo, CA: Morgan Kaufmann. (pp. 329 and 330)

R Core Team (2020). *R: A Language and Environment for Statistical Computing*. Vienna, Austria: R Foundation for Statistical Computing. (p. xx)

Ramabhadran, V. K. (1951). A multivariate gamma-type distribution. *Sankhyā* 11, 45–46. (p. 50)

Ramsay, J. O. and B. W. Silverman (2002). *Applied Functional Data Analysis*. New York: Springer. (pp. 466 and 467)

Ramsay, J. O., S. Graves, and G. Hooker (2020). *fda: Functional Data Analysis*. R package version 5.1.9. (p. 511)

Rand, W. M. (1971). Objective criteria for the evaluation of clustering methods. *J. Am. Stat. Assoc.* 66, 846–850. (p. 409)

Randles, R. H. (2000). A simpler, affine-invariant multivariate, distribution-free sign test. *J. Am. Stat. Assoc.* 95, 1263–1268. (p. 149)

Rao, C. R. (1948). Tests of significance in multivariate analysis. *Biometrika* 35, 58–79. (pp. 10, 11, and 152)

Rao, C. R. (1951). An asymptotic expansion of the distribution of Wilks' criterion. *Bull. Inst. Intern. Statist.* 33, 177–180. (p. 100)

Rao, C. R. (1964). The use and interpretation of principal component analysis in applied research. *Sankhyā* 26, 329–358. (p. 238)

Rao, C. R. (1966). Covariance adjustment and related problems in multivariate analysis. In P. R. Krishnaiah (Ed.), *Multivariate Analysis*, New York, pp. 87–103. Academic Press. (pp. 153 and 154)

Rao, C. R. (1971). Taxonomy in anthropology. In F. R. Hodson, D. G. Kendall, and P. Tautu (Eds.), *Mathematics in the Archaeological and Historical Sciences*, Edinburgh, pp. 19–29. Edinburgh University Press. (p. 237)

Rao, C. R. (1973). *Linear Statistical Inference and its Applications*, New York, pp. 19–29, Wiley. (pp. xxiii, 43, 52, 100, 106, 301, 312, 360, and 475)

Rao, C. R. and S. K. Mitra (1971). *Generalised Inverse of Matrices and its Applications*. New York: Wiley. (pp. 286 and 304)

Reeve, E. C. R. (1941). A statistical analysis of taxonomic differences within the genus Tamandu Gray. *(Xenorthra). Proc. Zool. Soc. Lond., A* 111, 279–302. (pp. 357 and 376)

Reinsch, C. H. (1967). Smoothing by spline functions. *Numer. Math.* 10, 177–183. (p. 469)

Ripley, B. D. (1996). *Pattern Recognition and Neural Networks*. Cambridge: Cambridge University Press. (p. 339)

Rohe, K. and M. Zeng (2023). Vintage factor analysis with Varimax performs statistical inference (with discussion). *J. R. Stat. Soc. B* 85, 1037–1060. (p. 275)

Roś, B., F. Bijma, J. C. de Munck, and M. C. M. de Gunst (2016). Existence and uniqueness of the maximum likelihood estimator for models with a Kronecker product covariance structure. *J. Multivariate Anal.* 143, 345–361. (p. 112)

Rosenblatt, M. (1952). Remarks on a multivariate transformation. *Ann. Math. Stat.* 23, 470–472. (p. 33)

Rothkopf, E. Z. (1957). A measure of stimulus similarity and errors in some paired-associate learning tasks. *J. Exp. Psychol.* 53, 94–101. (p. 420)

Rousseeuw, P. J. (1984). Least median of squares regression. *J. Am. Stat. Assoc.* 79, 871–880. (pp. 113 and 115)

Rousseeuw, P. J. (1987). Silhouettes: a graphical aid to the interpretation and validation of cluster analysis. *J. Comput. Appl. Math.* 20, 53–65. (p. 409)

Rousseeuw, P. J. and K. van Driessen (1999). A fast algorithm for the minimum covariance determinant estimator. *Technometrics* 41, 212–223. (p. 113)

Rousseeuw, P. J. and B. C. van Zomeren (1990). Unmasking multivariate outliers and leverage points. *J. Am. Stat. Assoc.* 85, 633–639. (p. 113)

Roy, S. N. (1957). *Some Aspects of Multivariate Analysis*. New York: Wiley. (pp. 132 and 166)

Salmon, F. (2012). The formula that killed Wall Street. *Significance* 9(1), 16–20. (p. 62)

Sammon, J. W. (1969). A nonlinear mapping for data structure analysis. *IEEE Trans. Comput.* 18, 401–409. (p. 440)

Sanghvi, L. D. (1957). Comparison of genetical and morphological methods for a study of biological differences. *Am. J. Phys. Anthropol.* 53, 94–101. (p. 400)

Sanghvi, L. D. and Balakrishnan (1972). Comparison of different measures of genetic distance between human populations. In J. A. Weiner and H. Huizinga (Eds.), *The Assessment of Population Affinities in Man*, Oxford, pp. 25–36. Clarendon Press. (p. 400)

Santosa, F. and W. W. Symes (1986). Linear inversion of band-limited reflection seismograms. *SIAM J. Sci. Stat. Comput.* 7, 1307–1330. (p. 454)

Schatzoff, M. (1966). Exact distributions of Wilks's likelihood ratio criterion. *Biometrika* 53, 347–358. (p. 146)

Schoenberg, I. J. (1935). Remarks to maurice Fréchet's article "Sur la definition axiomatique d'une classe d'espace distancieés vectoriellement applicable sur l'espace de Hilbert". *Ann. Math.* 36, 724–732. (p. 424)

Scott, A. J. and M. J. Symonds (1971). Clustering methods based on likelihood ratio criteria. *Biometrics* 27, 387–397. (pp. 383, 384, 386, and 412)

Scrucca, L., M. Fop, T. B. Murphy, and A. E. Raftery (2016). mclust 5: clustering, classification and density estimation using Gaussian finite mixture models. *R Journal* 8, 289–317. (pp. 385 and 511)

Seal, H. L. (1964). *Multivariate Analysis for Biologists.* London: Methuen. (p. 152)

Sell, T., T. B. Berrett, and T. I., Cannings (2023). Nonparametric classification with missing data. https://arxiv.org/abs/2305.11672. (p. 299)

Shaver, R. H. (1960). The Pennsylvanian ostracode *bairdia oklahomaensis* in indiana. *J. Paleontol.* 34, 656–670. (p. 150)

Shepard, R. N. (1962a). The analysis of proximities: multidimensional scaling with an unknown distance function i. *Psychometrika* 27, 125–139. (p. 438)

Shepard, R. N. (1962b). The analysis of proximities: multidimensional scaling with an unknown distance function II. *Psychometrika* 27, 219–246. (p. 438)

Shepard, R. N. (1963). The analysis of proximities as a technique for the study of information processing in man. *Hum. Factors* 5, 19–34. (p. 440)

Sibson, R. (1978). Studies in the robustness of multidimensional scaling: procrustes statistics. *J. R. Stat. Soc. B* 40, 234–238. (p. 442)

Silverman, B. W. (1981). Using kernel density estimates to investigate bimodality. *J. R. Stat. Soc. B* 43, 97–99. (p. 407)

Silverman, B. W. (1986). *Density Estimation for Statistics and Data Analysis.* London: Chapman and Hall. (p. 321)

Silvey, S. D. (1970). *Statistical Inference.* Baltimore, MD: Penguin. (p. 128)

Siskind, V. (1972). Second moments of inverse Wishart-matrix elements. *Biometrika* 59, 691–692. (p. 92)

Sklar, A. (1959). Fonctions de rèpartition á *n* dimensions et leurs marges. *Publ. Inst. Stat. Univ. Paris* 8, 229–231. (p. 60)

Smith, C. A. B. (1947). Some examples of discrimination. *Ann. Eugen.* 13, 272–282. (pp. 309 and 311)

Smith, T. W., M. Davern, J. Freese, and S. Morgan (2018). *General Social Surveys.* University of Chicago. (pp. 233 and 234)

Sokal, R. R. and C. D. Michener (1958). A statistical method for evaluating systematic relationships. *Univ. Kansas Sci. Bull.* 38, 1409–1438. (p. 403)

Späth, H. (1980). *Cluster Analysis Algorithms for Data Reduction and Classification of Objects.* Chichester: Ellis Horwood. (p. 387)

Spearman, C. (1904). "General intelligence," objectively determined and measured. *Am. J. Psychol.* 15, 201–292. (p. 259)

Speed, T. P. and H. T. Kiiveri (1986). Gaussian Markov distributions over finite graphs. *Ann. Stat.* 14, 138–150. (pp. 186 and 192)

Sprent, P. (1969). *Models in Regression and Related Topics.* London: Methuen. (p. 237)

Sprent, P. (1972). The mathematics of size and shape. *Biometrics* 28, 23–37. (p. 237)

Stahel, W. A. (1981). *Robuste Schätzungen: infinitesimale Optimalität u. Schätzungen von Kovarianzmatrizen.* Ph.D. thesis, Eidgenössischen Technischen Hochschule, Zürich. (p. 115)

Stein, C. M. (1956). Inadmissibility of the usual estimator for the mean of a multivariate normal distribution. In *Proceedings of the 3rd Berkeley Symposium on Mathematical Statistics and Probability*, Volume 1, pp. 197–206. (p. 109)

Subrahmaniam, K. and K. Subrahmaniam (1973). *Multivariate Analysis. A Selected and Abstracted Bibliography 1957–1972*. New York: Marcel Dekker. (p. xxii)

Sugiyama, T. and H. Tong (1976). On a statistic useful in dimensionality reduction in multivariate linear stochastic system. *Commun. Stat.-Theor. Meth. A* 5, 711–721. (p. 228)

Swierenga, H., A. P. de Weijer, R. J. van Wijk, and L. M. C. Buydens (1999). Strategy for constructing robust multivariate calibration models. *Chemom. Intell. Lab. Syst.* 49, 1–17. (p. 456)

ter Braak, C. and I. Prentice (1988). A theory of gradient analysis. *Adv. Ecol. Res.* 18, 271–317. (p. 235)

Therneau, T. and B. Atkinson (2019). *rpart: Recursive Partitioning and Regression Trees*. R package version 4.1-15. (pp. 329, 331, and 511)

Thompson, M. L. (1978). Selection of variables in multiple regression I, II. *Int. Stat. Rev.* 46, 1–20, 129–146. (p. 174)

Tibshirani, R. (1996). Regression shrinkage and selection via the lasso. *J. R. Stat. Soc. B* 58, 267–288. (p. 454)

Tibshirani, R. and G. Walther (2005). Cluster validation by prediction strength. *J. Comput. Graph. Stat.* 14, 511–528. (p. 410)

Todorov, V. (2021). *rrcov: Scalable Robust Estimators with High Breakdown Point*. R package version 1.6-0. (p. 511)

Torgerson, W. S. (1958). *Theory and Methods of Scaling*. New York: Wiley. (p. 424)

Tracy, C. A. and H. Widom (1996). On orthogonal and symplectic matrix ensembles. *Commun. Math. Phys.* 177, 727–754. (p. 248)

Tuddenham, R. D. and M. M. Snyder (1954). Physical growth of California boys and girls from birth to age 18. *Univ. Calif. Publ. Child Dev.* 1, 183–364. (p. 466)

Turgeon, M. (2018). *rootWishart: Distribution of Largest Root for Single and Double Wishart Settings*. R package version 0.4.1. (pp. 511 and 520)

Tyler, D. E. (1987). A distribution-free M-estimator of multivariate scatter. *Ann. Stat.* 15, 234–251. (p. 117)

Tyler, D., F., Critchley, L., Dümbgen, and H. Oja (2009). Invariant co-ordinate selection (with discussion). *J. R. Stat. Soc. B* 71, 549–592. (p. 245)

Venables, W. N. and B. D. Ripley (2002). *Modern Applied Statistics with S* (4th ed.). New York: Springer. ISBN 0-387-95457-0. (pp. 339, 340, 440, and 511)

Vinod, H. D. (1976). Canonical ridge and econometrics of joint production. *J. Econom.* 4, 147–166. (p. 295)

Vinod, H. D. (2011). *Hands-On Matrix Algebra Using R*. World Scientific. (p. 509)

von Rosen, D. and T. Kollo (2022). Editorial foreword. *J. Multivariate Anal.* (50th Anniversary Jubilee Edition) 188, 104859. (p. xx)

Wagner, S. and D. Wagner (2007). Comparing clusterings — an overview. Technical Report 2006-04, Fakultät für Informatik, Universität Karlsruhe. (p. 409)

Wand, M. (2020). *KernSmooth: Functions for Kernel Smoothing supporting Wand & Jones (1995)*. R package version 2.23-17. (p. 511)

Wand, M. P. and M. P. Jones (1995). *Kernel Smoothing*. London: Chapman and Hall. (p. 321)

Ward, J. H. Jr. (1963). Hierarchical grouping to optimize an objective function. *J. Am. Stat. Assoc.* 68, 236–244. (p. 392)

Wegman, E. J. (1990). Hyperdimensional data analysis using parallel coordinates. *J. Am. Stat. Assoc.* 85, 664–675. (p. 20)

Weihs, C., U. Ligges, K. Luebke, and N. Raabe (2005). klaR analyzing german business cycles. In D. Baier, R. Decker, and L. Schmidt-Thieme (Eds.), *Data Analysis and Decision Support*, Berlin, pp. 335–343. Springer-Verlag. (pp. 314 and 511)

Weinman, D. G. (1966). *A multivariate extension of the exponential distribution*. Ph.D. thesis, Arizona State University. (p. 51)

Welch, B. L. (1939). Note on discriminant functions. *Biometrika* 31, 218–220. (p. 302)

West, D. B. (2001). *Introduction to Graph Theory* (2nd ed.). Prentice Hall. (p. 187)

Whittaker, J. (1990). *Graphical Models in Applied Multivariate Statistics*. Wiley Publishing. (pp. 183, 184, and 192)

Wickham, H., D. Cook, and H. Hofmann (2015). Visualizing statistical models: removing the blindfold. *Statistical Analysis and Data Mining* 8, 203–225. (p. 15)

Wilk, M. B., R. Gnanadesikan, and M. J. Huyett (1962). Probability plots for the gamma distribution. *Technometrics* 4, 1–20. (p. 153)

Wilkinson, J. H. (1965). *The Algebraic Eigenvalue Problem*. Oxford: Clarendon Press. (p. 475)

Wishart, D. (1969). Mode analysis: a generalization of nearest neighbor which reduces chaining effects. In E. J. Cole (Ed.), *Numerical Taxonomy*, Volume 76, New York, pp. 282–311. Academic Press. (p. 406)

Wisnowski, J. W., J. R. Simpson, and D. C. Montgomery (2002). A performance study for multivariate location and shape estimators. *Qual. Reliab. Eng. Int.* 18, 117–129. (p. 113)

Wold, H. (1966). Estimation of principal components and related models by iterative least squares. In P. Krishnaiah (Ed.), *Multivariate Analysis*, pp. 391–420. Academic Press. (p. 457)

Wolfe, J. H. (1971). A Monte Carlo study of the sampling distribution of the likelihood ratio for mixtures of multinormal distributions. Technical report, Navel Personnel and Training Res. Lab. Tech. Bull. (p. 385)

Wood, S. N. (2017). *Generalized Additive Models: An Introduction with R* (2nd ed.). New York: Chapman and Hall/CRC. (p. 466)

Wooding, R. A. (1956). The multivariate distribution of complex normal variables. *Biometrika* 43, 329–350. (p. 92)

Wooldridge, J. M. (2019). *Introductory Econometrics: A Modern Approach* (7th ed.). Mason, OH: South-Western. (p. xx)

World Bank Group (2014). The World Bank: working for a world free of poverty. databank .worldbank.org/data/views/reports/tableview.aspx. (p. 410)

Wright, S. (1954). The interpretation of multivariate systems. In O. Kempthorne (Ed.), *Statistics and Mathematics in Biology*, pp. 11–33. Iowa State University Press. (p. 256)

Xu, P.-F., J. Guo, and X. He (2011). An improved iterative proportional scaling procedure for gaussian graphical models. *J. Comput. Graph. Stat.* 20, 417–431. (p. 192)

Young, G. and A. S. Householder (1938). Discussion of a set of points in terms of their mutual distance. *Psychometrika* 3, 19–22. (p. 424)

Zuo, Y. (2006). Multidimensional trimming based on projection depth. *Ann. Stat.* 34, 2211–2251. (p. 114)

Index

k

l

m

t

u